Ecology of the
Shortgrass Steppe

LONG-TERM ECOLOGICAL RESEARCH NETWORK SERIES
LTER Publications Committee

Grassland Dynamics: Long-Term Ecological Research in Tallgrass Prairie
Editors: Alan K. Knapp, John M. Briggs, David C. Hartnett, and Scott L. Collins

Standard Soil Methods for Long-Term Ecological Research
Editors: G. Philip Robertson, David C. Coleman, Caroline S. Bledsoe, and Phillip Sollins

Structure and Function of an Alpine Ecosystem: Niwot Ridge, Colorado
Editors: William D. Bowman and Timothy R. Seastedt

Climate Variability and Ecosystem Response at Long-Term Ecological Sites
Editors: David Greenland, Douglas G. Goodin, and Raymond C. Smith

Biodiversity in Drylands: Toward a Unified Framework
Editors: Moshe Shachak, James R. Gosz, Steward T. A. Pickett, and Avi Perevolotsky

Long-Term Dynamics of Lakes in the Landscape: Long-Term Ecological Research on North Temperate Lakes
Editors: John J. Magnuson, Timothy K. Kratz, and Barbara J. Benson

Alaska's Changing Boreal Forest
Editors: F. Stuart Chapin, III, Mark W. Oswood, Keith Van Cleve, Leslie A. Viereck, and David L. Verbyla

Structure and Function of a Chihuahuan Desert Ecosystem: The Jornada Basin Long-Term Ecological Research Site
Editors: Kris M. Havstad, Laura F. Huenneke, and William H. Schlesinger

Principles and Standards for Measuring Primary Production
Editors: Timothy J. Fahey and Alan K. Knapp

Agrarian Landscapes in Transition: Comparisons of Long-Term Ecological and Cultural Change
Editors: Charles L. Redman and David R. Foster

Ecology of the Shortgrass Steppe: A Long-Term Perspective
Editors: William K. Lauenroth and Ingrid C. Burke

Ecology of the Shortgrass Steppe

A Long-Term Perspective

Edited by

WILLIAM K. LAUENROTH
INGRID C. BURKE

LTER

OXFORD
UNIVERSITY PRESS

2008

OXFORD
UNIVERSITY PRESS

Oxford University Press, Inc., publishes works that further
Oxford University's objective of excellence
in research, scholarship, and education.

Oxford New York
Auckland Cape Town Dar es Salaam Hong Kong Karachi
Kuala Lumpur Madrid Melbourne Mexico City Nairobi
New Delhi Shanghai Taipei Toronto

With offices in
Argentina Austria Brazil Chile Czech Republic France Greece
Guatemala Hungary Italy Japan Poland Portugal Singapore
South Korea Switzerland Thailand Turkey Ukraine Vietnam

Copyright © 2008 by Oxford University Press, Inc.

Published by Oxford University Press, Inc.
198 Madison Avenue, New York, New York 10016

www.oup.com

Oxford is a registered trademark of Oxford University Press.

All rights reserved. No part of this publication may be reproduced,
stored in a retrieval system, or transmitted, in any form or by any means,
electronic, mechanical, photocopying, recording, or otherwise,
without the prior permission of Oxford University Press.

Library of Congress Cataloging-in-Publication Data
Ecology of the shortgrass steppe : a long-term perspective / edited by
William K. Lauenroth and Ingrid C. Burke.
   p. cm.
Includes bibliographical references and index.
ISBN 978-0-19-513582-4
1. Grassland ecology—North America. 2. Steppe ecology—North America.
I. Lauenroth, William K. II. Burke, Ingrid C.
QH102.S56 2009
577.4'4097—dc22   2008006014

9 8 7 6 5 4 3 2 1

Printed in the United States of America
on acid-free paper

Acknowledgments

This book is the result of more than four decades of research and represents the efforts of many dedicated individuals, some authors and some not. In many significant ways, Dr. George Van Dyne began the program that continues today as the Shortgrass Steppe Long-Term Ecological project; his vision permeates our presentation. Countless field, laboratory, and program assistants and graduate students have contributed their hard work and insight to our understanding of the shortgrass steppe, and this book is richer as a result.

Special thanks to Bob Flynn, Judy Hendryx, Nicole Kaplan, Mark Lindquist, Kim Melville-Smith, Jeri Morgan, Sallie Sprague, Petra Lowe, Sonia Hall, Mark Gathany, and especially Becky Riggle, who assisted in the final preparation of the manuscript. To all of you and more—many thanks.

Contents

Contributors xi

1 The Shortgrass Steppe
   The Region and Research Sites 3
   William K. Lauenroth, Ingrid C. Burke, and Jack A. Morgan

2 Climate of the Shortgrass Steppe 14
   Roger A. Pielke, Sr., and Nolan J. Doesken

3 Soil Development and Distribution in the Shortgrass Steppe Ecosystem 30
   Eugene F. Kelly, Caroline M. Yonker, Steve W. Blecker, and Carolyn G. Olson

4 Land-Use History on the Shortgrass Steppe 55
   Richard H. Hart

5 Vegetation of the Shortgrass Steppe 70
   William K. Lauenroth

6 The Role of Disturbances in Shortgrass Steppe Community and Ecosystem Dynamics 84
   Debra P. C. Peters, William K. Lauenroth, and Ingrid C. Burke

7 Simulation of Disturbances and Recovery in Shortgrass Steppe Plant Communities 119
   Debra P. C. Peters, and William K. Lauenroth

viii   Contents

8 Ecology of Mammals of the Shortgrass Steppe   132
   Paul Stapp, Beatrice Van Horne, and Mark D. Lindquist

9 Birds of the Shortgrass Steppe   181
   John A. Wiens, and Nancy E. McIntyre

10 Insect Populations, Community Interactions,
   and Ecosystem Processes in the Shortgrass Steppe   215
   Thomas O. Crist

11 Trophic Structure and Nutrient Dynamics of the Belowground
   Food Web within the Rhizosphere of the Shortgrass Steppe   248
   John C. Moore, Jill Sipes, Amanda A. Whittemore-Olson,
   H. William Hunt, Diana H. Wall, Peter C. de Ruiter,
   and David C. Coleman

12 Net Primary Production in the Shortgrass Steppe   270
   William K. Lauenroth, Daniel G. Milchunas,
   Osvaldo E. Sala, Ingrid C. Burke, and Jack A. Morgan

13 Soil Organic Matter and Nutrient Dynamics of Shortgrass
   Steppe Ecosystems   306
   Ingrid C. Burke, Arvin R. Mosier, Paul B. Hook, Daniel G. Milchunas,
   John E. Barrett, Mary Ann Vinton, Rebecca L. McCulley,
   Jason P. Kaye, Richard A. Gill, Howard E. Epstein, Robin H. Kelly,
   William J. Parton, Caroline M. Yonker, Petra Lowe,
   and William K. Lauenroth

14 Soil–Atmosphere Exchange of Trace Gases in the Colorado
   Shortgrass Steppe   342
   Arvin R. Mosier, William J. Parton, Roberta E. Martin,
   David W. Valentine, Dennis S. Ojima, David S. Schimel,
   Ingrid C. Burke, E. Carol Adair, and Stephen. J. Del Grosso

15 The Shortgrass Steppe and Ecosystem Modeling   373
   William J. Parton, Stephen J. Del Grosso,
   Ingrid C. Burke, and Dennis S. Ojima

16 Effects of Grazing on Vegetation   389
   Daniel G. Milchunas, William K. Lauenroth,
   Ingrid C. Burke, and James K. Detling

17 Cattle Grazing on the Shortgrass Steppe   447

Richard H. Hart and Justin D. Derner

18 Effects of Grazing on Abundance and Distribution of Shortgrass Steppe Consumers   459

Daniel G. Milchunas and William K. Lauenroth

19 The Future of the Shortgrass Steppe   484

Ingrid C. Burke, William K. Lauenroth, Michael F. Antolin, Justin D. Derner, Daniel G. Milchunas, Jack A. Morgan, and Paul Stapp

Index   511

Contributors

E. Carol Adair, Department of Forest Sciences, Graduate Degree Program in Ecology Colorado State University, Fort Collins, CO 80523

Michael F. Antolin, Department of Biology, Colorado State University, Fort Collins, CO 80523

John E. Barrett, Department of Biological Sciences, Virginia Tech, Blacksburg, VA 24061

Steve W. Blecker, U.S. Geological Survey, Mackay School of Earth Science and Engineering, University of Nevada, Reno, NV 89557

Ingrid C. Burke, Graduate Degree Program in Ecology, Colorado State University, Fort Collins, CO 80523

David C. Coleman, Institute of Ecology, University of Georgia, Athens, GA 30602

Thomas O. Crist, Department of Zoology, Miami University, Oxford, OH 45056

Stephen J. Del Grosso, U.S. Department of Agriculture–Agricultural Research Service, Soil Plant Nutrient Research, Fort Collins, CO 80526

Peter C. de Ruiter, Alterra, Wageningen University and Research Centre, Wageningen, The Netherlands

Justin D. Derner, U.S. Department of Agriculture–Agricultural Research Service, High Plains Grasslands Research Station, Cheyenne, WY 82009

James K. Detling, Department of Biology, Colorado State University, Fort Collins, CO 80523

Nolan J. Doesken, Atmospheric Science Department and Colorado Climate Center, Colorado State University, Fort Collins, CO 80523

Howard E. Epstein, Environmental Science Department, University of Virginia, Charlottesville, VA 22904

Richard A. Gill, School of Earth and Environmental Sciences, Washington State University, Pullman, WA 99164

Richard H. Hart, U.S. Department of Agriculture–Agricultural Research Service, High Plains Grasslands Research Station, Cheyenne, WY 82009

Paul B. Hook, Intermountain Aquatics, Inc., Driggs, ID 83422

H. William Hunt, Natural Resource Ecology Laboratory, Colorado State University, Fort Collins, CO 80524

Jason P. Kaye, Department of Crop and Soil Sciences, Penn State University, University Park, PA 16802

Eugene F. Kelly, Department of Soil and Crop Sciences, Colorado State University, Fort Collins, CO 80523

Robin H. Kelly, Colorado State University, Fort Collins, CO 80523

William K. Lauenroth, Graduate Degree Program in Ecology, Warner College of Natural Resources, Colorado State University, Fort Collins, CO 80523

Mark D. Lindquist, Shortgrass Steppe Long-Term Ecological Research, Colorado State University, Fort Collins, CO 80523

Petra Lowe, Shortgrass Steppe Long-Term Ecological Research, Colorado State University, Fort Collins, CO 80523

Roberta E. Martin, Department of Global Ecology, Carnegie Institution, Stanford, CA 94305

Rebecca L. McCulley, Plant and Soil Sciences, University of Kentucky, Lexington, KY 40546

Nancy E. McIntyre, Department of Biological Sciences, Texas Tech University, Lubbock, TX 79409

Daniel G. Milchunas, Department of Forest, Rangeland, and Watershed Stewardship, Colorado State University, Fort Collins, CO 80523

John C. Moore, Natural Resource Ecology Laboratory, Colorado State University, Fort Collins, CO 80524

Jack A. Morgan, U.S. Department of Agriculture–Agricultural Research Service, Fort Collins, CO 80526

Arvin R. Mosier, U.S. Department of Agriculture–Agricultural Research Service, Soil–Plant–Nutrient Research Unit, Fort Collins, CO 80521

Dennis S. Ojima, Colorado State University, Natural Resource Ecology Laboratory, Fort Collins, CO 80523

Carolyn G. Olson, U.S. Department of Agriculture–Natural Resource Conservation Service, Lincoln, NE 68508

William J. Parton, Natural Resource Ecology Laboratory, Colorado State University, Fort Collins, CO 80523

Debra P. C. Peters, U.S. Department of Agriculture–Agricultural Research Service, Jornada Exp. Range, Las Cruces, NM 88003

Roger A. Pielke, Sr., Cooperative Institute for Research in Environmental Sciences and Department of Atmospheric and Ocean Sciences, Boulder, CO 80309

Osvaldo E. Sala, Center for Environmental Studies, Brown University, Providence, RI 02912

David S. Schimel, Terrestrial Sciences/Climate and Global Dynamics Division, National Center for Atmospheric Research, Boulder, CO 80307

Jill Sipes, School of Biological Sciences, University of Northern Colorado, Greeley, CO 80639

Paul Stapp, Department of Biological Science, California State University Fullerton, Fullerton, CA 92834

David W. Valentine, Department of Forest Sciences, University of Alaska, Fairbanks, AK 99775

Beatrice Van Horne, U.S. Department of Agriculture Forest Service, Wildlife, Fish, Watershed, and Air Research, Arlington, VA 22209

Mary Ann Vinton, Department of Biology, Creighton University, Omaha, NE 68178

Diana H. Wall, Natural Resource Ecology Laboratory, Colorado State University, Fort Collins, CO 80524

Amanda A. Whittemore-Olson, School of Biological Sciences, University of Northern Colorado, Greeley, CO 80639

John A. Wiens, The Nature Conservancy, Arlington, VA 22203

Caroline M. Yonker, Department of Soil and Crop Science, Colorado State University, Fort Collins, CO 80523

Ecology of the
Shortgrass Steppe

1

The Shortgrass Steppe
The Region and Research Sites

William K. Lauenroth
Ingrid C. Burke
Jack A. Morgan

> On the shortgrass prairie, the green hills roll away toward the distant horizon, uncluttered by buildings, signs, or paved roads. Pronghorn herds graze peacefully while prairie dogs chatter from their burrows. At night coyotes howl, horned owls call, and poorwills sing, just as they have for thousands of years. The land seems eternal.
>
> —Cushman and Jones (1988, p. 9)

The central grassland region of North America (Fig. 1.1) is the largest contiguous grassland environment on earth. Prior to European settlement, it was a vast, treeless area characterized by dense head-high grasses in the wet eastern portion, and very short sparse grasses in the dry west. As settlers swept across the area, they replaced native grasslands with croplands, most intensively in the east, and less so in the west (Fig. 1.2). The most drought-prone and least productive areas have survived as native grasslands, and the shortgrass steppe occupies the warmest, driest, least productive locations. James Michener (1974) provided an apt description of the harshness of the shortgrass region in his book *Centennial*:

> It is not a hospitable land, like that farther east in Kansas or back near the Appalachians. It is mean and gravelly and hard to work. It lacks an adequate topsoil for plowing. It is devoid of trees or easy shelter. A family could wander for weeks and never find enough wood to build a house. (p. 64)

The objective of this chapter is to introduce the shortgrass steppe (Fig. 1.3) and its record of ecological research. First we present an ecological history of the shortgrass steppe since the Tertiary, and provide the geographic and climatic context for the region. Second we describe the major research sites, and the history of

4 Ecology of the Shortgrass Steppe

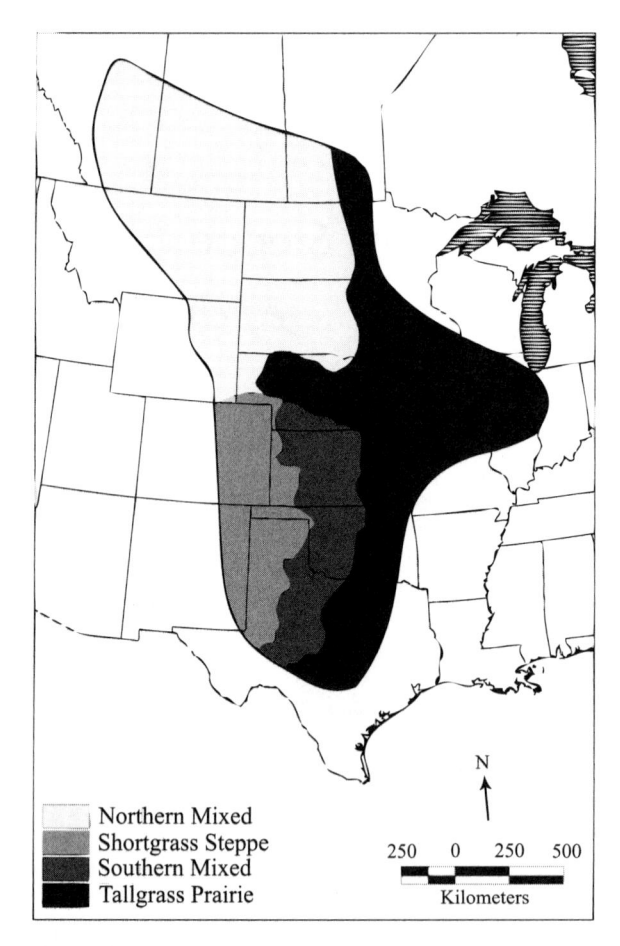

Figure 1.1 Map of the central grassland region of North America (Lauenroth et al., 1999).

the three major entities or programs that have shaped much of the science done in the shortgrass steppe: the U.S. Department of Agriculture (USDA)–Agricultural Research Service (ARS), the International Biological Programme (IBP), and the Long-Term Ecological Research (LTER) Program.

The Shortgrass Steppe

Grasses have been an important component of the shortgrass steppe of North America since the Miocene (5–24 million years ago) (Axelrod, 1985; Stebbins, 1981). Before that, during the Paleocene and Eocene (34–65 million years ago), the vegetation was a mixture of temperate and tropical mesophytic forests. Two causes have been proposed as explanations for this ancient change from forest

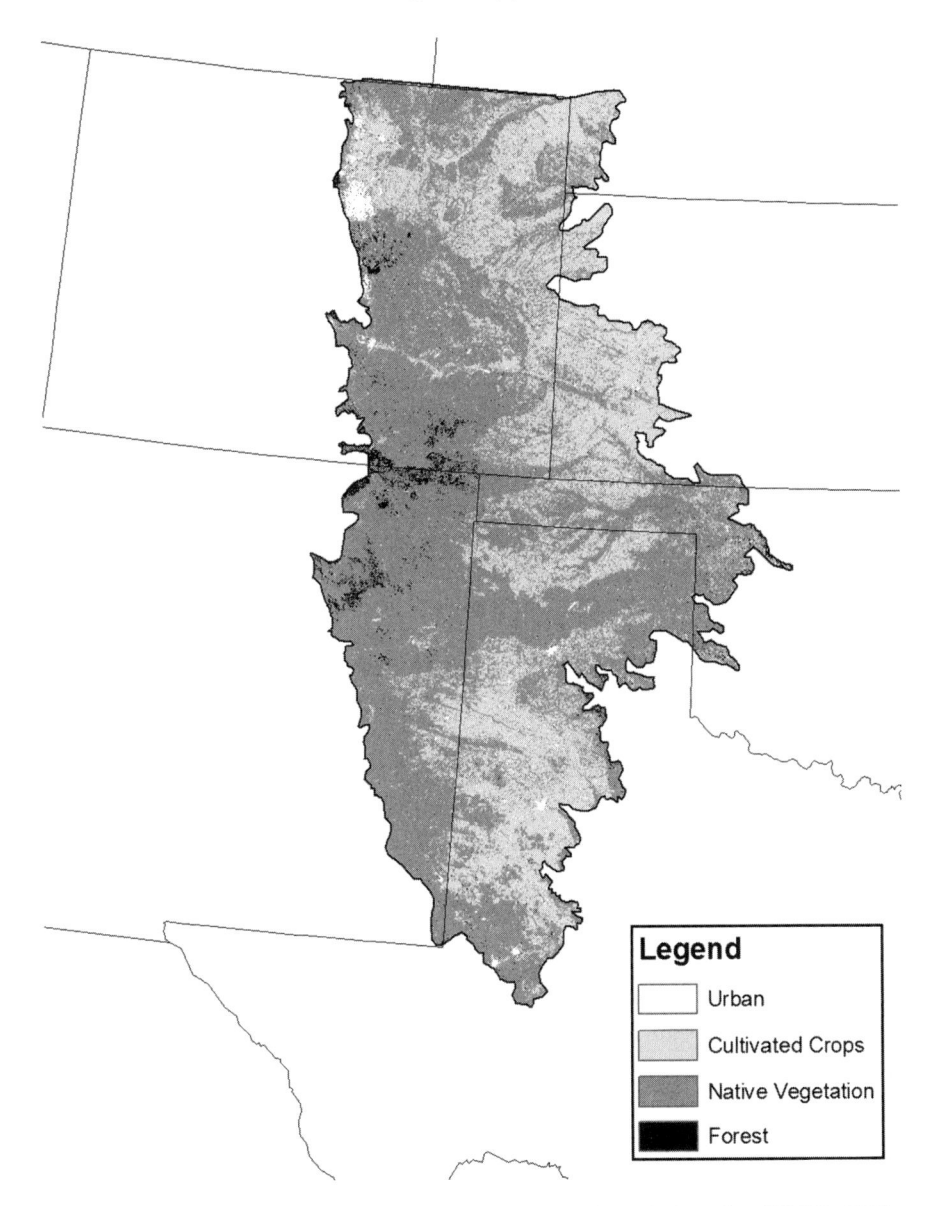

Figure 1.2 Current land use in the shortgrass steppe of North America (NLCD, 2001).

to grassland. First, global temperatures decreased rapidly during the Oligocene (24–34 million years ago), creating conditions for a drier climate. These drier conditions, combined with a renewal of the uplift of the Rocky Mountains that had begun during the Paleocene, left the Great Plains in a rain shadow. Thus, the Plains became dry and cool, causing a shift from forest to grassland vegetation. Currently, the vegetation is dominated by C_4 grasses, a characteristic of many

6 Ecology of the Shortgrass Steppe

Figure 1.3 Shortgrass steppe landscape with the Ogallala escarpment in the background. (Photo by Sallie Sprague.)

world grasslands today that is attributed to the relatively low atmospheric CO_2 concentrations that have prevailed from the late Miocene up until the early 20th century and that tend to favor C_4 over C_3 metabolism (Ehleringer et al., 1997).

After the rise to dominance by grasses in the Great Plains, there were four major episodes of continental glaciation during the Pleistocene, beginning with the Nebraskan glacial period and terminating with the Wisconsin, which ended 10,000 years ago. Separating these glacial periods were warmer interglacial periods. This glacial/interglacial alternation lasted for a million years and had a substantial impact on the region. The shortgrass steppe was never under ice, but landscapes in the region currently demonstrate effects of the glacial/interglacial cycles; soils are commonly composed of mixtures of alluvium deposited as outwash from the melting of mountain glaciers. These cycles also had an enormous effect on the fauna. During the past 20,000 years, with the termination of the most recent glacial period, 32 genera of mammals have disappeared. These extinctions are thought by some researchers to have been the result of human hunters who entered North America from Siberia across the Bering Land Bridge during the Wisconsin period (Brown and Lomolino, 1998; Owen-Smith, 1987; Stuart, 1991).

The Holocene (the past 11,000 years) has been characterized by significant climatic fluctuation in the northern shortgrass steppe. Muhs (1985) used data from dune fields in northeastern Colorado to reconstruct climate during the mid to late Holocene to suggest that the eolian sand was deposited during the Altithermal (from 8000 to 5000 years before present [BP]), a warm and dry period throughout much of western and central North America. The Altithermal was followed by

2000 years of cooler and wetter conditions, during which rates of soil formation exceeded rates of erosion. From 3000 to 1000 years BP, the climate again became drier, and erosion apparently exceeded soil formation. This second period of dune formation is when the modern dune fields in northeastern Colorado were created. Since 1500 years BP, conditions have been cooler and wetter, again allowing soil formation to exceed erosion, resulting in the development of modern soils and vegetation (Kelly, 1989; Kelly et al., chapter 3, this volume).

Geographic Features

The shortgrass steppe covers approximately 3.4×10^5 km² in the central and southern Great Plains, representing 11% of the central grassland region (Fig. 1.1) (Küchler, 1964; Lauenroth et al., 1999). The northern boundary between the shortgrass steppe and the northern mixed prairie is approximately the Colorado–Wyoming border, at 41°N latitude, and coincides with the boundary between the Colorado Piedmont and the High Plains sections of the Great Plains. At this location, the High Plains is, on average, 200 m higher than the Colorado Piedmont. This elevation change affects the seasonality of both temperature and precipitation (Lauenroth and Milchunas, 1992). The northern boundary is slightly different from the northern boundary of Küchler's *Bouteloua–Buchloë* type (Küchler, 1964; Lauenroth, chapter 5, this volume). The steppe extends southward to latitude 32°N in western Texas, where it grades into southern mixed prairie to the east and Chihuahuan desert shrub savanna to the west (Küchler, 1964; Lauenroth et al., 1999). The western edge of the shortgrass steppe is the foothills of the Rocky Mountains, where it abuts coniferous woodland and forest. The key conifer species are *Pinus ponderosa, Juniperus scopulorum,* and *Pseudotsuga menziesii*. The eastern boundary of the shortgrass steppe with the southern mixed prairie occurs in western Kansas, western Oklahoma, and western Texas, reaching a maximum eastward extension of 100°W longitude in Oklahoma.

Climatic Features

The shortgrass steppe occurs in the driest and warmest portion of the Great Plains (Lauenroth and Burke, 1995; Pielke and Doesken, chapter 2, this volume). Mean annual precipitation ranges from 300 to 600 mm, with the lowest amounts occurring in the northwestern and southwestern portions, and the largest amounts along the eastern edge. Late spring to early summer is the wettest time of year; winter is the driest. Mean annual temperatures range from less than 9 °C in the north to more than 16 °C in the south. The northern portion of the region has an average of more than 170 days per year with daily temperatures reaching 0 °C or less, whereas the southern portion has fewer than 70 such days.

The shortgrass steppe lies entirely within the semiarid zone of the central grassland region as defined by Bailey (1979) (Lauenroth et al., 1999). The boundary between the shortgrass steppe and the southern mixed prairie is roughly the boundary between the semiarid and the dry subhumid zones in Kansas, Oklahoma, and Texas. The climate is discussed in detail by Pielke and Doesken in chapter 2 (this volume).

Vegetation

Shortgrass steppe vegetation is characterized by the dominance of two C_4 grasses: *Bouteloua gracilis* and *Buchloë dactyloides*. Both are perennial caespitose grasses, but *B. dactyloides* has the capability to produce stolons, giving it the potential to form a matlike stand and spread rapidly after disturbances (Aguiar and Lauenroth, 2001). *Bouteloua gracilis* is an archetypical caespitose grass that spreads by tillering.

Approximately 70% of the shortgrass steppe remains in natural vegetation, and the majority of that is used for livestock grazing (Fig. 1.4). The rest of the area is used for row crops, divided between irrigated and dryland crops (Lauenroth et al., 1999). The major irrigated crops are *Medicago* spp. (alfalfa), *Zea mays* (corn), *Beta vulgaris* (sugar beet), and *Gossypium* spp. (cotton). Much of the irrigated land is adjacent to the major river channels or lies over the Ogallala aquifer. The dominant dryland crop is *Triticum aestivum* (wheat), which is grown using a summer fallow rotation system in which a crop is grown every other year. The land lies fallow during the alternate year, storing water for the crop year.

The dominant grasses of the shortgrass steppe (*B. gracilis* and *B. dactyloides*) occur throughout the central grassland region, although *B. dactyloides* is less

Figure 1.4 Cattle on the shortgrass steppe. (Photo by Sallie Sprague.)

common in the northernmost part (Epstein et al., 1998). In many locations in the mixed prairies, heavy grazing by domestic livestock can result in a shift in dominance from the normally dominant midheight grasses to one or both of the short grasses. This has led to widespread use of the term *shortgrass prairie* (*prairie* is defined as having continuous plant cover whereas *steppe* is characterized by discontinuous cover) to describe grasslands from Canada to the Gulf of Mexico. Many of these areas are fundamentally different from the shortgrass steppe in that they have the potential to support midheight grasses with an understory of shortgrasses. The true shortgrass steppe lacks such potential on upland sites. The vegetation is described in detail by Lauenroth in chapter 5 (this volume).

Research Sites and Research Approaches

The information contained in this book draws heavily on the results of long-term interdisciplinary studies conducted within one area encompassing the USDA ARS Central Plains Experimental Range (CPER) and the adjacent USDA Forest Service Pawnee National Grasslands (PNG); this area is designated as the Shortgrass Steppe Long-Term Ecological Research (SGS LTER) site (http://sgslter.colostate.edu) (Fig. 1.5). Two other locations have been sites of short-duration, concentrated research activity. The Texas Tech University Research Farm, 24 km east of Amarillo, Texas, was the focus of considerable research activity during the late 1960s and early 1970s, associated with the Grassland Biome project of the IBP. The other site, located near Springfield, Colorado, was a Colorado State University (CSU) Agricultural Experiment Station from 1957 to1998.

The CPER encompasses 6280 ha of shortgrass steppe located approximately 8 km north of Nunn, Colorado (40°49′N latitude, 107°47′W longitude), and is

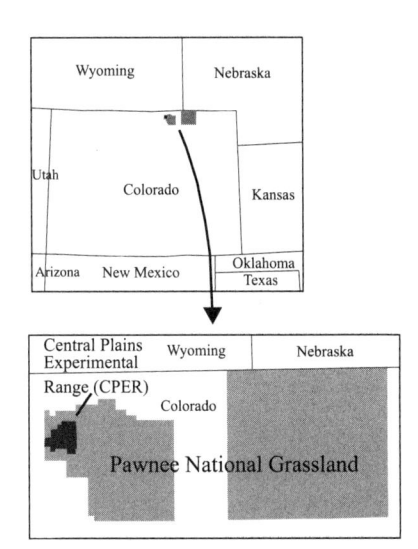

Figure 1.5 Location of the Shortgrass Steppe Long-Term Ecological Research site and the two administrative units within it: the Central Plains Experimental Range (CPER) and the Pawnee National Grasslands (PNG). Image courtesy of U.S. National Forest Service, Pawnee National Grasslands.

operated by the Rangeland Resources Research Unit of the USDA–ARS. Average elevation is 1650 m, mean annual precipitation is 321 mm, and mean annual temperature is 8.6 °C. The majority of annual precipitation falls as rain during the May to September growing season. Mean monthly temperatures range from–5 °C in January to 22 °C in July.

The CPER was established in 1939 on land taken over by the USDA Forest Service after drought, overgrazing, and dust storms had taken their toll on ranches and farms in eastern Colorado, forcing their abandonment (Shoop et al., 1989). Forest Service scientists began a research program directed toward developing sustainable management practices, with the particular objective of enhancing livestock production from shortgrass rangelands. From the beginning, all livestock for research were contributed by the Crow Valley Livestock Cooperative, Inc., composed today of approximately 40 individual ranches. As a result, much of the station's research has been conducted in close collaboration and with the strong support of the local ranching community. The first formal research began in May 1939, with a grazing intensity study that is still underway today. In 1953, the Agricultural Act transferred the CPER to the USDA–ARS. Early research at the CPER determined critical relationships among stocking rate, forage production, and animal gain that provided foundational information for establishing recommended grazing practices for the region (Bement, 1969; Hyder et al., 1975; Klipple and Costello, 1960). Research by the ARS in the 1960s and 1970s focused on reseeding, legume genetics, range improvements, and plant–grazing interactions (Shoop et al., 1989; Townsend et al., 1995). Many of the results used in this book come either directly from publications by ARS scientists or from experiments that they initiated.

Researchers from CSU and from other nonfederal organizations had a relatively small presence at the CPER until the late 1960s, when the National Science Foundation (NSF) initiated funding of the U.S. IBP Grassland Biome project. The Grassland Biome project was led by the visionary scientist Dr. George Van Dyne, and for a period of 9 years it received approximately $16 million, which funded the work of some 200 scientists (Golley, 1993; Van Dyne and Anway, 1976). A large proportion of that funding was spent on research at the CPER. Van Dyne's vision was that the Grassland Biome project be the first research project to take a systems approach to grasslands, including a major role for simulation modeling (Van Dyne and Anway, 1976). Although the Grassland Biome project was not successful in achieving all of its objectives, in retrospect it was one of the most important building blocks of modern ecosystem science. Despite the fact that he was naive about many of the details, Van Dyne's belief in the importance of a systems approach to studying ecology has been upheld since he first articulated it. Much of the data included in this volume were collected during the Grassland Biome project.

Funding for the IBP formally ended in 1974, and from then until 1982 research on the CPER was continued by ARS and CSU scientists who were funded by individual research grants. In 1982, the CPER was funded as an NSF LTER site (Franklin et al., 1990). The overall objective of the SGS LTER project has been to understand the processes that account for the origin and maintenance of the structure and function of shortgrass steppe ecosystems. During the first 15 years of the project (1982–1996), effectively all the research was conducted on the CPER. After 1996, the research area was expanded to include portions of the PNG to incorporate more of the region's variability in soils, vegetation, and land management (Burke and Lauenroth, 1993).

Initially, the SGS LTER was under the direction of Drs. Robert Woodmansee and William Lauenroth, both CSU scientists. Both Woodmansee and Lauenroth were trained under Van Dyne's influence during the IBP project, Lauenroth as a PhD student and Woodmansee as a postdoctoral fellow. Many of the participants of the newly funded LTER project had been members of the IBP team. The original leadership also included Dr. William Laycock, Research Leader of the Forage and Range Research Unit of the ARS. Thus, the SGS LTER integrated the long-term experience and experiments of the ARS with those of the IBP.

The ARS range program began shifting in the 1980s more toward sustainability, conservation, and environmental concerns, and continues in that direction today. However, current research still addresses the early interests in beef cattle production (Derner and Hart, 2005; Hart and Ashby, 1998), shrub ecology (Cibilis et al., 2003a, b), and range improvements (Shoop et al., 1985, 1989). Recent initiatives in trace gases (Mosier et al., 1991, 1996), climate change (Morgan et al., 2004, 2007), and plant–animal interactions (Derner et al., 2006) have helped forge a closer vision between ARS and CSU scientists, and have enhanced their ability to conceptualize and implement relevant ecological rangeland research for a diverse set of clients.

The SGS LTER project has made a number of major contributions in the area of grazing ecology. The CPER, with its long-term grazing experiments initiated in the 1930s and maintained since then by the ARS, has been a key ingredient to the group's success. Perhaps more important than the actual field experiments and facilities has been the gathering of agriculturalists, ecologists, modelers, and many others with an interest in native grasslands at a single site where critical and honest discourse are encouraged. This combination of resources has been instrumental in advancing a systems approach to grassland science. This systems approach, which has guided the SGS LTER project since its inception in 1982, contributed substantially to the training of the first principal investigators of the SGS LTER project and is a direct result of the influence of George Van Dyne and the Grassland Biome project.

In this book, shortgrass steppe researchers describe the history, environment, ecology, and vulnerabilities of the shortgrass steppe as we understand them today. Our hope is that future generations will find our analyses to be useful stepping-stones from which to continue research into the mysteries of the shortgrass steppe.

References

Axelrod, D. I. 1985. Rise of the grassland biome. *Botanical Review* **51**:163–201.
Aguiar, M. R., and W. K. Lauenroth. 2001. Local and regional differences in abundance of co-dominant grasses in the shortgrass steppe: a modeling analysis of potential causes. *Plant Ecology* **156**:161–171.

Bailey, H. P. 1979. Semi-arid climates: Their definition and distribution, pp. 73–97. In: A. E. Hall, G. H. Cannell, and H. W. Lawton (eds.), *Agriculture in semi-arid environments*. Springer-Verlag, Berlin.

Bement, R. R. 1969. A stocking-rate guide for beef production on blue-grama range. *Journal of Range Management* **22**:83–86.

Burke, I. C., and W. K. Lauenroth. 1993. What do LTER results mean? Extrapolating from site to region and decade to century. *Ecological Modelling* **67**:19–35.

Brown, J. H., and M. V. Lomolino. 1998. *Biogeography*. Sinauer Associates, Inc., Sunderland.

Cibils, A. F., D. M. Swift, and R. H. Hart. 2003a. Changes in shrub fecundity in fourwing saltbush browsed by cattle. *Journal of Range Management* **58**:39–46.

Cibils, A. F., D. M. Swift, and R. H. Hart. 2003b. Female-biased herbivory in fourwing saltbush browsed by cattle. *Journal of Range Management* **56**:47–51.

Cushman, R. C., and S. R. Jones. 1988. *The shortgrass prairie*. Pruett Publishing, Boulder, Colo.

Derner, J. D., J. K. Detling, and M. F. Antolin. 2006. Are livestock gains affected by blacktailed prairie dogs? *Frontiers in Ecology and Environment* **4**:459–464.

Derner, J. D., and R. H. Hart. 2005. Heifer performance under two stocking rates on fourwing saltbush-dominated rangeland. *Rangeland Ecology and Management* **58**:489–494.

Ehleringer, J. R., T. E. Cerling, and B. R. Helliker. 1997. C_4 photosynthesis, atmospheric CO_2, and climate. *Oecologia* **112**:285–299.

Epstein, H. E., W. K. Lauenroth, I. C. Burke, and D. P. Coffin. 1998. Regional productivities of plant species in the Great Plains of the United States. *Plant Ecology* **134**:173–195.

Franklin, J. F., C. S. Bledsoe, and J. T. Callahan. 1990. Contributions of the Long-Term Ecological Research program. *BioScience* **40**:509–523.

Golley, F. B. 1993. *A history of the ecosystem concept in ecology*. Yale University Press, New Haven, Conn.

Hart, R. H., and M. M. Ashby. 1998. Grazing intensities, vegetation, and heifer gains: 55 years on shortgrass. *Journal of Range Management* **51**:392–398.

Hyder, D. N., R. E. Bement, E. E. Remmenga, and D. F. Hervey. 1975. Ecological responses of native plants and guidelines for management of shortgrass range. USDA technical bulletin no. 1503. USDA. Washington, D.C.

Kelly, E. F. 1989. *A study of the influence of climate and vegetation on the stable isotope chemistry of soils in grassland ecosystems of the Great Plains*. PhD diss., University of California, Berkeley, Calif.

Klipple, G. E., and D. F. Costello. 1960. Vegetation and cattle responses to different intensities of grazing on shortgrass ranges on the Central Great Plains. USDA technical bulletin no. 1216. USDA. Washington, D.C.

Küchler, A. W. 1964. *Potential natural vegetation of the conterminous United States*. American Geographical Society, New York.

Lauenroth, W. K., and I. C. Burke. 1995. Great Plains: Climate variability, pp. 237–249. In: W. A. Nierenberg (ed.), *Encyclopedia of environmental biology*. Academic Press, New York.

Lauenroth, W. K., I. C. Burke, and M. P. Gutmann. 1999. The structure and function of ecosystems in the central North American grassland region. *Great Plains Research* **9**:223–259.

Lauenroth, W. K., and D. G. Milchunas. 1992. Short-Grass Steppe, pp. 183–226. In: R. T. Coupland (ed.), *Natural grasslands: Introduction and western hemisphere*. Ecosystems of the world, vol. 8A. Elsevier, Amsterdam.

Michener, J. A. 1974. *Centennial*. Fawcett Publishing, Greenwich, Conn.
Morgan, J. A., D. G. Milchunas, D. R. LeCain, M. West, and A. Mosier. 2007. Carbon dioxide enrichment alters plant community structure and accelerates shrub growth in the shortgrass steppe. *Proceedings of the National Academy of Sciences* **104**:14724–14729.
Morgan, J. A., A. R. Mosier, D. G. Milchunas, D. R. LeCain, J. A. Nelson, and W. J. Parton. 2004. CO_2 enhances productivity, alters species composition, and reduces forage digestibility of shortgrass steppe vegetation. *Ecological Applications* **14**:208–219.
Mosier, A. R., W. J. Parton, D. W. Valentine, D. S. Ojima, D. S. Schimel, D. J. Delgado. 1996. CH_4 and N_2O fluxes in the Colorado Shortgrass Steppe: I. Impact of landscape and nitrogen addition. *Global Biogeochemical Cycles* **10**:387–399.
Mosier, A. R., D. S. Schimel, D. Valentine, K. Bronson, and W. J. Parton. 1991. Methane and nitrous oxide fluxes in native, fertilized and cultivated grasslands. *Nature* **350**:330–332.
Muhs, D. R. 1985. Age and paleoclimatic significance of Holocene sand dunes in northeastern Colorado. *Annals of the Association of American Geographers* **75**:566–582.
NLCD. 2001. National Land Cover Database. Online. Available at http://www.mrlc.gov/mrlc2k_nlcd.asp.
Owen-Smith, N. 1987. Pleistocene extinctions: The pivotal role of megaherbivores. *Paleobiology* **13**:351–362.
Shoop, M. C., R. C. Clark, W. A. Laycock, and R. J. Hansen. 1985. Cattle diets on shortgrass ranges with different amount of fourwing saltbush. *Journal of Range Management* **38**:443–449.
Shoop, M., S. Kanode, and M. Calvert. 1989. Central Plains Experimental Range: 50 years of research. *Rangelands* **11**:112–117.
Stebbins, G. L. 1981. Coevolution of grasses and herbivores. *Annals of the Missouri Botanical Garden* **68**:75–86.
Stuart, A. J. 1991. Mammalian extinctions in the late Pleistocene of northern Eurasia and North America. *Biological Review of the Cambridge Philosophical Society* **66**:453–562.
Townsend, C. E., S. Wand, and T. Tsuchiya. 1995. Registration C-25, C-26, and C-27 alfalfa germplasms. *Crop Science* **35**:289.
Van Dyne, G. M., and J. C. Anway. 1976. A program for and the process of building and testing grassland ecosystem models. *Journal of Range Management* **29**:114–122.

2

Climate of the Shortgrass Steppe

Roger A. Pielke, Sr.
Nolan J. Doesken

The climate of a region involves the short- and long-term interaction among the atmospheric, hydrologic, ecologic, oceanographic, and cryospheric components of the earth's environmental system (Hayden, 1998; Pielke, 1998, 2001a,b). These interactions occur across all spatial and temporal scales, from turbulence generated by diurnal cycles at a landscape scale, to global-scale circulation. The establishment of particular ecosystem types is associated with a nonlinear feedback between the atmosphere and the underlying vegetation (Pielke and Vidale, 1995). Wang and Eltahir (2000) and Claussen (1998) have demonstrated that vegetation patterning cannot be accurately simulated in a model unless vegetation–atmosphere feedbacks are included.

In this chapter we summarize the climate system of the shortgrass steppe. This is a region of large seasonal contrasts, and of interannual and longer term variability. It is also a region that has undergone major human impacts during the past 150 years. We present both average conditions and examples of extreme events in the shortgrass steppe to illustrate the variable climate of this interesting ecosystem.

Geographic Factors Controlling the Climate

Geographic factors play a large role in determining the climatic characteristics of the shortgrass steppe (Lauenroth and Burke, 1995; Lauenroth and Milchunas, 1992; Lauenroth et al., 1999). Key factors for this region include its mid-latitude position, its relatively high elevations, its interior continental location, and its proximity to the Rocky Mountains, a substantial north–south-oriented mountain barrier immediately to the west. Air masses affecting the region consist of continental polar air from the north, humid continental air masses from the east, humid subtropical air masses from the southeast and south, and Pacific maritime air masses from the west. The latter can be significantly modified as they cross a series of mountain ranges and interior dry regions before reaching the shortgrass steppe region. Each of these geographic and atmospheric features contributes to the climate of the region. Latitude determines day length and sun angle, and, hence, solar insolation. This, in turn, greatly affects air temperature. Upper level westerly winds increase over the mid-latitudes in the fall and winter in response to strengthening north–south temperature gradients in the atmosphere. Pacific air masses are carried eastward over the Rocky Mountains, depositing considerable cool-season precipitation in the mountains, but rarely on the shortgrass steppe.

The high elevation of the region contributes to its low humidity and intense solar insolation. This also means that infrared radiational heat losses from the earth to space at night lead to rapid cooling of the air near the ground, resulting in large diurnal temperature variations. The interior continental location of the region contributes further both to large day-to-day and large seasonal temperature changes. Atmospheric moisture must travel long distances to reach the area, unlike the coastal regions of the continent.

The mountain ranges to the west play a huge role in the region's climate by serving as an effective block or *rain shadow* from storms carrying Pacific moisture from the west. The mountains impose both upward motion on the windward side and downward vertical motion on the leeward side. This is critical to cloud formation and precipitation. Throughout the winter season, westerly winds are dominant aloft over the mid-latitudes, bringing predominant *downslope* winds to the shortgrass steppe as the air descends the eastern slopes of the Rocky Mountains. This translates into abundant sunshine, low humidity, low precipitation, and periods of strong winds in the shortgrass steppe. But occasionally, for brief periods of time ranging from a few hours to a few days, easterly upslope winds ascend the High Plains, bringing low clouds and widespread precipitation to the region. This occurs in the fall and winter, but most frequently in March, April, and May, when slow-moving storms are able to tap into moisture from the Gulf of Mexico and carry it into the region from the southeast.

During summer, winds aloft become weak. Air masses advect more slowly, and local factors become more important. Convection becomes the primary mechanism for producing upward motion, which leads to cloud formation and precipitation. Dry air from the intermountain area of the southwestern United States is the most common warm-season air mass in the region, but much more humid air lies just east and southeast of the shortgrass steppe. Under certain weather patterns, this moist air is pushed westward, fueling occasional heavy, local storms. During late summer, the North American monsoon, which is strongly influenced by the Mexican Plateau and elevated terrain in the southwest United States (Castro et al., 2001), provides an additional source of moisture, particularly over the southern portion of the area.

Overview of the Weather Patterns in the Shortgrass Steppe

The geographic factors described earlier work together to produce a climate characterized by a strong annual cycle, large daily variations, frequent and persisting dry weather, and occasional very vigorous storms.

Winter

The winter weather over the shortgrass steppe is dominated by frequent migratory high- and low-pressure systems that are associated with the polar jet stream and its associated polar front. When the jet stream flow is zonal, air masses associated with the front are Pacific in origin, producing substantial snows in the mountains and relatively mild air in the shortgrass steppe just to the lee of these barriers. Further east, the weather in the Great Plains is cooler, with some air from Canada entrained south into the region west of the low-pressure systems. Occasionally, Arctic high-pressure systems travel southward over the region, producing the area's coldest weather. These intrusions of Arctic air occur when the polar jet stream travels across Alaska and northwest Canada, before plunging south over the Great Plains. Upslope snows frequently occur in the western High Plains during these cold outbreaks. The shortgrass steppe coincides with a preferred region of strong cool-season cyclogenesis (formation zone for low-pressure areas) (Davis et al., 1999). Very strong winds (Weaver, 1999) and rapidly changing weather conditions occur as these storm systems develop and sweep across the Plains and Midwest.

Spring

Spring is a transition season when high sun angles can produce warm days, yet cold air masses still occasionally travel southward bringing heavy snows to the Plains. Extreme weather changes are common in spring as cold and warm air masses alternate. In the western High Plains, March and April are the snowiest months of the year. Ferocious blizzards are not uncommon during the spring—decreasing from south to north as the sun climbs higher in the sky. Increased solar heating of the elevated land surfaces rapidly heats the ground while temperatures aloft remain cool. This thermal instability effectively mixes the atmosphere, helping bring strong winds down to the surface. The result is many very windy days during spring.

At this time in the eastern High Plains, thunderstorms become common. Thunderstorms can often be quite intense during spring as a result of strong solar surface insolation, relatively cold air aloft, and sharp contrasts in air masses. The tornado season in the shortgrass steppe region peaks during April through June as a still-vigorous polar jet stream provides large changes in wind speed and direction with altitude. This large wind shear provides the initial horizontal wind circulation for tornadic thunderstorms, particularly when the wind shear is tilted on its side by intense thunderstorm updrafts and downdrafts. A tornado can subsequently be produced when this horizontal wind circulation is concentrated into a small area by intense updrafts.

However, spring is also characterized by slower migration of large, horizontal-length waves in the atmosphere that control the motion of surface lows and highs. This results in slow-moving storms that have time to tap into abundant humidity east of the region. Spring storms bring occasional episodes of precipitation (either rain or snow) that may last for 1 to 3 days and cover broad areas. This moisture, falling just as the area is greening up in the spring, is important to local vegetation.

Summer

By summer, the polar jet has typically migrated far to the north. Rainfall becomes dominated by topographically heated upslope flows, weak migratory low-pressure systems, local air mass boundaries, and storm outflows. During this period, weather pattern changes are relatively slow. Precipitation is characterized by thunderstorms that bring high-intensity rainfall, but over relatively small areas for short periods of time. As storms progress eastward and tap into more humid air, they produce more widespread rainfall. Organized clusters of thunderstorms, called mesoscale convective systems (MCSs), can develop and are associated with weak frontal boundaries or higher terrain (Cotton, 1999). These MCSs usually move eastward in response to the weak westerly winds in the middle and upper troposphere during this time of the year.

A dryline boundary usually forms in the western Plains, separating humid air coming from the Gulf of Mexico from dry air originating in the desert Southwest and northern Mexico (Ziegler, 1999). During summer, thunderstorms frequently form along this dryline boundary. In late summer, the North American monsoon starts to affect the western portion of this region. Substantial rains often occur as moisture originating in the tropical Pacific Ocean or the southern Gulf of Mexico is advected northward into the southern Plains and southern Rocky Mountains.

Autumn

The North American monsoon flow regime weakens in late August, and relatively dry weather, dominated by persistent and often nearly stationary high-pressure systems, begins to dominate fall months. Long stretches of fair weather are interrupted occasionally by cool, cloudy, and wet periods as mid-latitude storm systems develop in response to the changing seasons. Sometimes, widespread heavy rains occur as moisture from tropical storms is entrained. Snowstorms can occur as early as September in the northern and central Great Plains and Rocky Mountains, when the polar jet stream migrates southward in early fall.

Summary

The climate of the shortgrass steppe displays a strong seasonal cycle. Each season is characterized by significantly different meteorological conditions that are described by weather events such as the passage of frontal systems, development of leeside low-pressure troughs and low-pressure centers, and the production of

18 Ecology of the Shortgrass Steppe

convective storms. These weather events operate on timescales ranging from several hours to a few days, resulting in exciting and changeable weather conditions.

Temperature

The shortgrass steppe is an area of large and rapid temperature changes. A single year's daily temperature data for the SGS LTER site provides a clear illustration of the large diurnal and seasonal temperature changes that are characteristic of the region (Fig. 2.1).

The shortgrass steppe region is also characterized by temperature gradients from west to east and from north to south (Fig. 2.2). Hot temperatures are common across the area during summer, especially at low elevations and over the easternmost portions of the region. Record maxima in this region have exceeded 42 °C (Fig. 2.3). There is a negative correlation between precipitation and temperature. When summer precipitation has been significantly lower than average, summer temperatures are especially hot. Under these circumstances, more of the solar insolation is used in heating the ground and adjacent air whereas less is utilized in evaporating water (Pielke, 2001b).

The shortgrass steppe is also characterized by periods of cold temperatures in winter. The coldest weather occurs in the northern and highest elevation portion of the steppe. During an average year, along the northern edge of the shortgrass steppe, temperatures occasionally fall to around –34 °C. In extreme years,

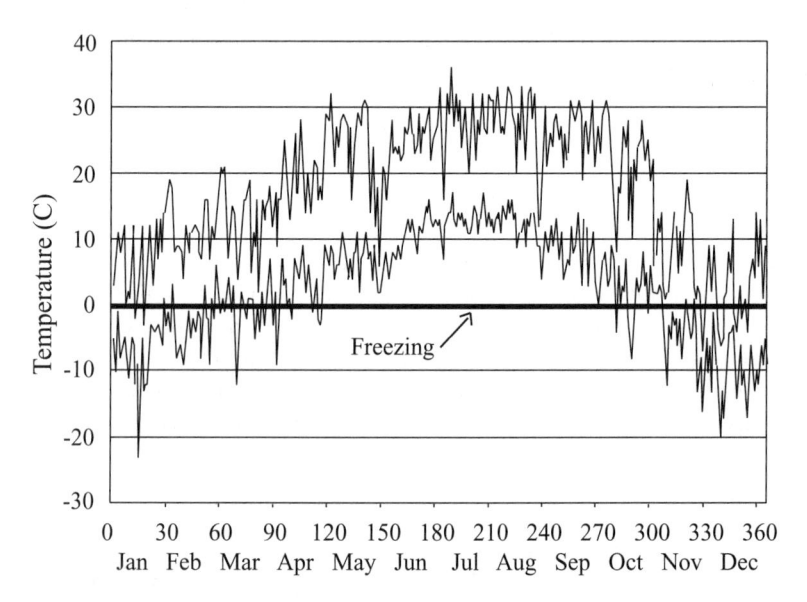

Figure 2.1 Daily maximum and minimum temperatures for an entire year from the Shortgrass Steppe LTER site.

Climate of the Shortgrass Steppe   19

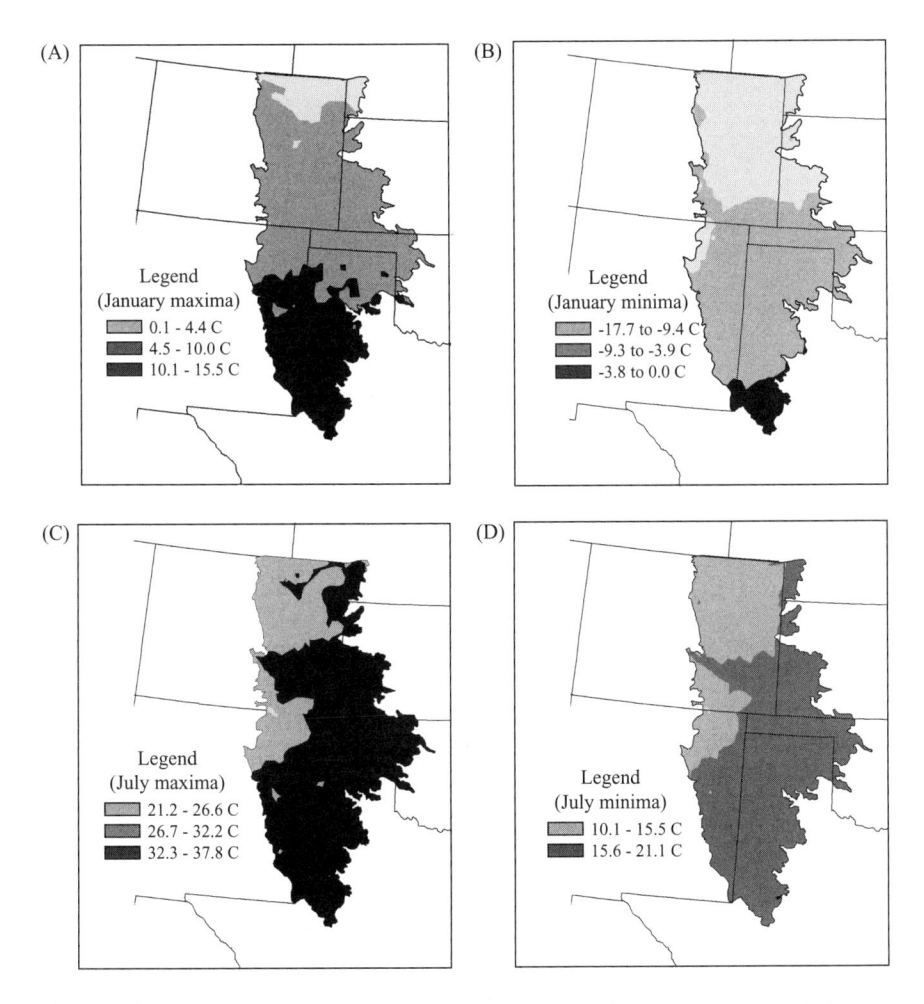

Figure 2.2 (A) January mean maximum temperatures. (B) January mean minimum temperatures. (C) July mean maximum temperatures. (D) July mean minimum temperatures for the shortgrass steppe. (Adapted from NCDC [2002].)

temperatures less than −40 °C can occur (Fig. 2.3). Even in the southern portion of the shortgrass steppe, temperatures can fall to less than −12 °C in extreme years. The large range between summer and winter temperatures, which can exceed 49 °C in the northern portion of the steppe (Fig. 2.2), influence which animals and plants can survive in this region.

Growing season characteristics are another indicator of a region's climate. Vegetation, especially many agricultural crops, can be sensitive to the first and last occurrence of 0 °C during the year (Fig. 2.4). The entire shortgrass steppe experiences periods of the year with subfreezing temperatures ranging from just

20 Ecology of the Shortgrass Steppe

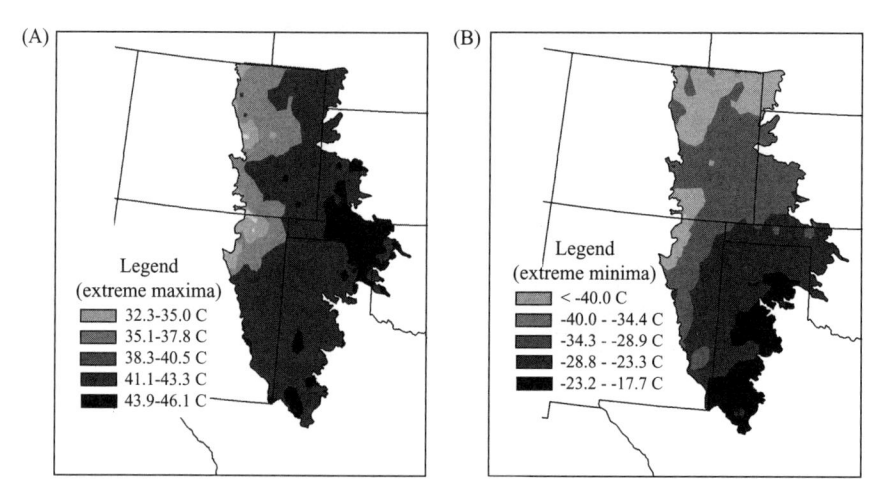

Figure 2.3 (A) Annual record extreme maximum temperatures. (B) Annual record extreme minimum temperatures for the shortgrass steppe. (Adapted from NCDC [2002].)

more than 200 days in southeast Wyoming to fewer than 90 days in western Texas and southeastern New Mexico (NCDC, 2002). In addition, the region experiences many freeze–thaw cycles through winter as temperatures warm during the day but cool quickly at night. The frequent freezing and thawing of the surface topsoil plays an interesting and important role in soil conditioning, potentially making bare soil very vulnerable to wind erosion during late winter and spring (Doesken, 1988).

Humidity

Humidity over the shortgrass steppe is characteristically low much of the year, particularly when the region is dominated by air masses crossing the mountains from the west or moving down the continent from the north. However, high-humidity air often lies just east and southeast of the shortgrass steppe and does move westward under appropriate meteorological conditions (NCDC, 2002). Relative humidity is a common measure of atmospheric moisture; however, it can be misleading because it is very much an inverse function of air temperature. The dew point temperature is a more appropriate measure of humidity, because it is the temperature at which condensation would begin to occur if the air is cooled without changing its pressure or water content. Thus, because it is an absolute measure of the amount of water vapor in the air, it is the index of atmospheric moisture most often used by meteorologists for tracking and comparing humidity. The annual average maximum dew point temperatures decrease greatly from southeast to northwest across the shortgrass steppe (Fig. 2.5) as elevation, latitude, and distance from the Gulf of Mexico moisture increases.

Climate of the Shortgrass Steppe 21

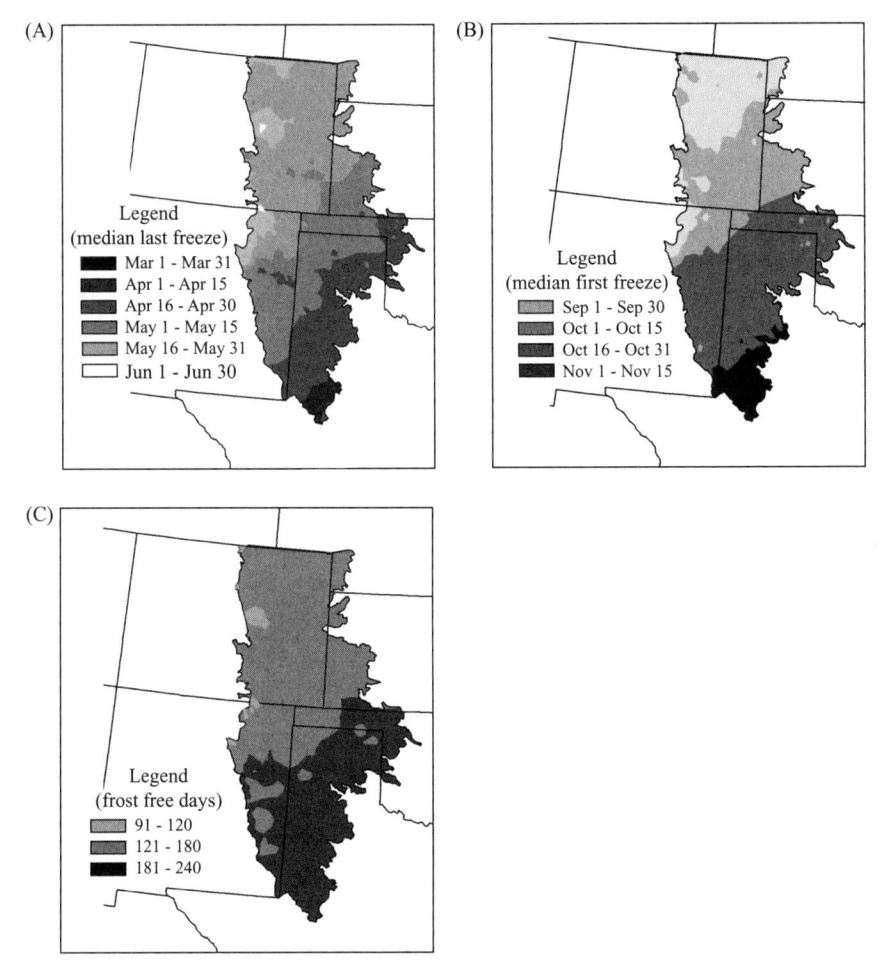

Figure 2.4 (A) Median date of last freeze in spring. (B) First freeze in fall. (C) Length of frost-free period for the shortgrass steppe. (Adapted from NCDC [2002].)

Precipitation

Precipitation is arguably the single most important climatic variable controlling the ecology of the shortgrass steppe (Lauenroth and Sala, 1992). Lu et al. (2001) showed that the response of vegetation to precipitation in the central U.S. grasslands is the area's largest weather sensitivity. This followed up on work by Sala et al. (1988) and Lauenroth and Sala (1992), showing that precipitation is the single largest predictor of shortgrass steppe productivity both across regions and among years. Pielke et al. (1999) have shown that soil moisture at the beginning of the growing season, resulting from winter and fall precipitation, influences subsequent spring and summer rainfall in the shortgrass steppe, with moister soils

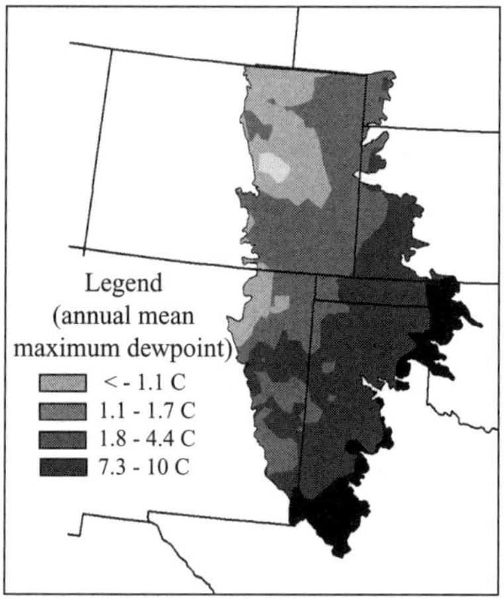

Figure 2.5 Annual mean maximum dew point temperatures for the shortgrass steppe. (Adapted from NCDC [2002].)

favoring enhanced rainfall. This positive feedback occurs as enhanced evaporation and transpiration from the wetter soils provide additional energy to fuel more rainfall-producing thunderstorms. Eastman et al. (2001a,b) have shown how settlement and changes in land use, such as the removal of bison and the switch from livestock to crop production, has altered precipitation in the region.

Mean annual precipitation of the shortgrass steppe varies from 508 to 635 mm over the easternmost portions to less than 305 mm in some locations farther west (Fig. 2.6). Precipitation is highly seasonal. Growing season (April–September) precipitation contributes from 70% to 82% of the average annual moisture across the region (NCDC, 2002).

Unlike temperature, humidity, and solar radiation, which are continuous weather variables, precipitation is episodic. Precipitation only falls in 2% to 4% of the hours of the year in the shortgrass steppe (Doesken and Eckrich, 1987). As in many semiarid regions, a few storms during the year produce a large portion of the annual precipitation (Sala et al., 1992). Cowie and McKee (1986) showed that in eastern Colorado, 20% of the days with measurable precipitation account for at least 50% of the accumulated precipitation. At the SGS LTER site, less than 10% of the precipitation events produce more than 20 mm of water, but these large events contribute more than 30% of annual precipitation (Sala and Lauenroth, 1982). This pattern extends throughout most of the shortgrass steppe region (Fig. 2.7).

Very intense, potentially erosive, but fairly localized rainfall is a common trait of the shortgrass steppe (Hjelmfelt, 1999). Storm rainfall that occurs for a 6-hour duration only has a 1% probability of occurrence at a specific point in any given year. The size of these events ranges from about 76 mm over the northern portion of the region up to more than 127 mm over the southeastern portion of the region. Storms equal to or greater than this occur several times almost every year somewhere in the area (NCDC, 2002). Vegetation becomes very important in minimizing the water erosion associated with these heavy downpours. Doesken

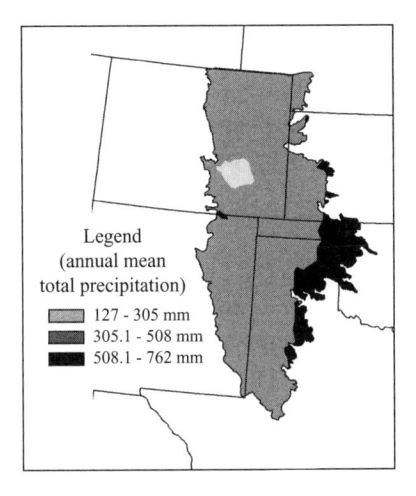

Figure 2.6 Annual mean total precipitation for the shortgrass steppe. (Adapted from NCDC [2002].)

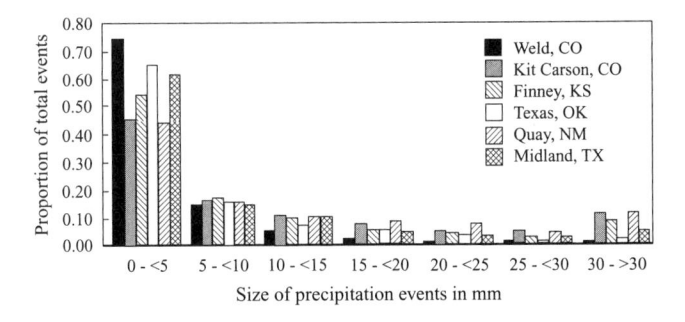

Figure 2.7 Frequency of precipitation event sizes for stations from 6 counties in the shortgrass steppe.

and McKee (1999) showed that mean annual precipitation in this area can vary within a number of years from less than half the long-term average to double the average. Precipitation within seasons is also highly variable.

Evaporation

With hot summer temperatures, low humidity, moderate to strong winds, and plenty of solar insolation, the shortgrass steppe is a region of high potential evaporation (Lauenroth and Milchunas, 1992). Potential evaporation rates greatly exceed precipitation across the shortgrass steppe (Lauenroth and Burke, 1995). At the SGS LTER site, Sala et al. (1992) evaluated the frequency of periods when the ratio of precipitation to potential evapotranspiration exceeded one and found that even at the daily scale the ratio was greater than one for only 10% of the days for 30 years of records. Lauenroth and Bradford (2006), also at the SGS

LTER site, reported that average daily potential evapotranspiration exceeded average daily precipitation by a factor of three during the growing season and by a factor of 10 for the rest of the year. They found coefficients of variation of daily precipitation to be 333% whereas CV for potential evapotranspiration were 17%, which emphasizes the lack of temporal variability in atmospheric demand for water.

A very high fraction of the annual precipitation is quickly evaporated and transpired back into the atmosphere (Lauenroth and Bradford, 2006; Sala et al., 1992). Estimates of the proportions of total evapotranspiration accounted for by bare soil evaporation and transpiration suggest that water loss is equally divided between the two processes (Lauenroth and Bradford, 2006). The key controls of this partitioning in order of importance are amount of aboveground biomass, its seasonality, and soil texture (Lauenroth and Bradford, 2006).

Shortgrass steppe annual potential evapotranspiration ranges from 1200 mm in the extreme northwestern portion to more than 1800 mm in the southeastern portion. The major controls on this variability are latitude and elevation. Annual potential evapotranspiration over the shortgrass steppe is closely related to air temperature. Lauenroth and Burke (1995) evaluated the relationship between mean annual potential evapotranspiration and mean annual temperature for the Great Plains and found that mean annual temperature accounted for 72% of the variability. Throughout the region, mean annual temperature is closely related to both elevation and latitude (Lauenroth and Burke, 1995).

Snowfall

As a result of the colder winter temperatures across the northwestern portion of the shortgrass steppe, the mean total snowfall is higher there than farther south (Fig. 2.8). Only the northernmost portion of the shortgrass steppe typically has snow cover in January (Fig. 2.8). January snow depth is more common in the northern mixed prairie immediately to the north of the shortgrass steppe. This snow cover provides insulation to the plants and soils. In the southernmost portion of the shortgrass steppe, the average January snow cover is less than 13 mm (Fig. 2.8), although all regions of the shortgrass steppe have had large occasional January snowstorms of 300 mm or more (NCDC, 2002).

The first snow of the season can come as early as September in northern portions of the shortgrass steppe, and accumulating snows can occur as late as May (NCDC, 2002). The amount and duration of snowfall is directly related to the high elevations of the region. Although the snow season is long, the actual snowfall totals are low to moderate. Average snowfall ranges from less than 300 mm in western Texas and southwestern Kansas to more than 1200 mm at the north edge of the region and at the western edge, where the mountains begin to rise above the Plains (Fig. 2.8). On average, the water content of snowfall contributes a small amount to annual precipitation. It ranges from less than 5% over southeastern portions of the region to more than 25% in portions of northern and central Colorado.

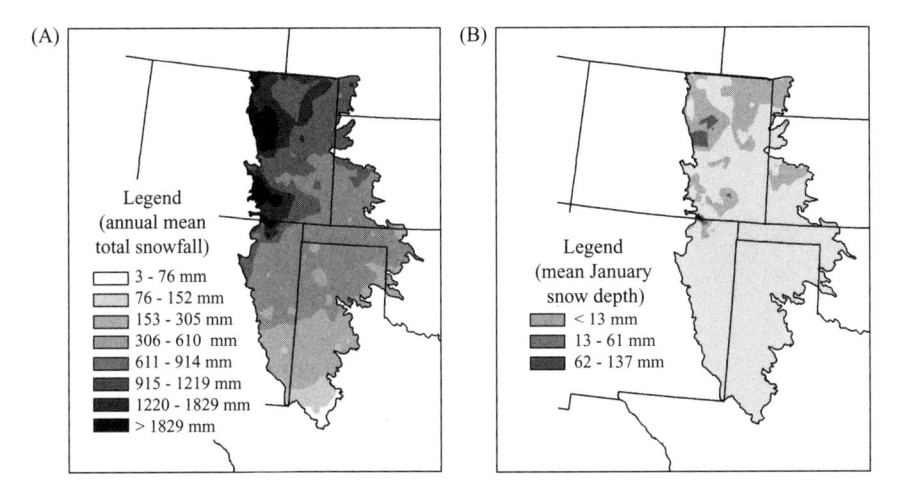

Figure 2.8 (A) Annual mean total snowfall. (B) Mean January snow depth for the shortgrass steppe. (Adapted from NCDC [2002].)

Wind, Thunderstorms, Hail, and Tornadoes

The Great Plains is a windy region and the shortgrass steppe at its western edge shares much of that windiness (Fig. 2.9). These high winds occur because of the relatively high elevation and the vast open spaces without tall vegetation to slow the winds at ground level. Contributing to the overall windiness of the region are the frequent occurrences of intense thunderstorms with associated strong winds during the warm season in the southern and central portion of the shortgrass steppe, and strong winter storms (leeside cyclogenesis) in the central and northern part of the steppe, including blizzards, in the cold season.

Winds are moderate to strong year round, with average speeds exceeding $4 \text{ m} \cdot \text{s}^{-1}$ in most areas (Fig. 2.9). Spring is the windiest season. Throughout the shortgrass steppe, dust storms are common during dry periods in winter and spring. South to southeast winds are widespread over the region during summer, whereas strong northwesterly winds become more common during winter months.

The number of days with thunder is between 30 and 50 throughout the shortgrass steppe (NCDC, 2002). Even though the region is semiarid, there is often enough moisture for high-based thunderstorms, which can produce strong surface winds (Serafin et al., 1999). Dangerous cloud-to-ground lightning is a common occurrence with these storms (Lyons, 1999).

Thunderstorms over the shortgrass steppe often produce hail, and the central western High Plains is one of a few sites worldwide with a high frequency of large hailstones (NCDC, 2002). Such hail often causes a large amount of crop damage and can even be hazardous to livestock and humans. Changnon (1999) provided an excellent overview of hail characteristics in the region. Tornadoes can also result from these thunderstorms and are most intense in the eastern and southern

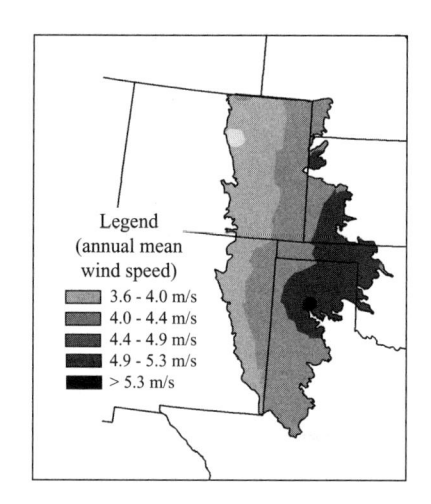

Figure 2.9 Annual mean wind speed for the shortgrass steppe. (Adapted from NCDC [2002].)

Legend (annual mean wind speed)
- 3.6 - 4.0 m/s
- 4.0 - 4.4 m/s
- 4.4 - 4.9 m/s
- 4.9 - 5.3 m/s
- > 5.3 m/s

portion of the shortgrass steppe, where the moisture to fuel severe thunderstorms is generally the greatest (Golden, 1999).

Drought

Drought is the major climatic extreme on the shortgrass steppe. McKee et al. (1999) used a standardized precipitation index (SPI) to investigate short- and long-term drought in Colorado. The SPI evaluates precipitation anomalies with respect to the historical probability distributions of precipitation at each measurement site. Their analysis of the fraction of Colorado that had a drought of at least 48 months documented the drought decades of the 1930s and 1950s. Between the early 1980s and late 1990s, the state was in an unusually long period without long-term, statewide drought (McKee et al., 1999). Drought returned to Colorado in 2000 and continues to the present (Henz et al., 2004).

Multidecadal droughts, fortunately, have not occurred since colonization of the shortgrass steppe in the 1800s. However, as reported by Stahle et al. (2000) using tree ring data, a widespread, multidecadal severe drought occurred in the shortgrass steppe in the 16th century. Fires were also more frequent during this time period (Clark, 1990). Such a climatic extreme would have enormous social and economic consequences if it reoccurred.

Summary

The shortgrass steppe is a region of extreme climatic conditions. The semiaridity of the region and its location inland and to the east of a large mountain barrier results in large inter- and intraseasonal ranges of temperature. Its proximity to a rich source of moisture in the Gulf of Mexico results in frequent, severe thunderstorms in the warm season with accompanying hail, strong winds, and tornadoes, whereas the absence of a terrain barrier to the north permits occasional, severely cold arctic air masses to intrude into the region. This potpourri of weather is the environment in which the organisms of the shortgrass steppe have evolved, and into which humans have introduced grazing, row–crop agriculture, and urbanization. The biophysical, biogeochemical, and biogeographic processes within the steppe, along with human disturbance, interact with the atmosphere to produce the unique climate of the region. The shortgrass steppe climate of the future will depend on these regional earth system interactions, as well as changes in the larger scale climate system.

Acknowledgments We acknowledge the Colorado Agricultural Experiment Station, the Colorado Climate Center, and LTER grant no. DEB 9632852.

References

Calef, W. 1950. The winter of 1948–49 in the Great Plains. *Annals of the Association of American Geographers XL* **4**:267–292.

Castro, C. L., T. B. McKee, and R. A. Pielke, Sr. 2001. The relationship of the North American monsoon to tropical and North Pacific sea surface temperatures as revealed by observational analyses. *Journal of Climate* **14**:4449–4473.

Changnon, S. A. 1999. Impacts of hail in the United States, pp. 163–191. In: R. A. Pielke and R. A. Pielke (eds.), *Storms*. Vol. II. Routledge, New York.

Chase, T. N., R. A. Pielke, Sr., T. G. F. Kittel, J. S. Baron, and T. J. Stohlgren. 1999. Potential impacts on Colorado Rocky Mountain weather due to land use changes on the adjacent Great Plains. *Journal of Geophysical Research* **104**:16673–16690.

Clark, J. S. 1990. Fire and climate change during the last 750 yr in northwestern Minnesota. *Ecological Monographs* **60**(2):135–159.

Claussen, M. 1998. On multiple solutions of the atmosphere–vegetation system in present-day climate. *Global Change Biology* **4**:549–560.

Cotton, W. R. 1999. An overview of mesoscale convective systems, pp. 3–25. In R. A. Pielke and R. A. Pielke (eds.), *Storms*. Vol. II. Routledge, New York.

Cowie, J. R., and T. B. McKee. 1986. *Colorado precipitation event and variability analysis*. Climatology report 86-3, atmospheric science paper 400. Department of Atmospheric Science, Colorado State University, Fort Collins, Colo.

Davis, R. E., P. J. Michaels, and B. P. Hayden. 1999. Overview of extratropical cyclones, pp. 401–426. In: R. A. Pielke and R. A. Pielke (eds.), *Storms*. Vol. I. Routledge, New York.

Doesken, N. J. 1989. The Colorado freeze-thaw see-saw. *Colorado Climate Summary Water-Year Series* (Oct. 1987–Sep. 1988), Climatology Report No. 89–1, Atmospheric Science Department, Colorado State University, Fort Collins, CO, pp. 50–59.

Doesken, N. J. 1995. How hard can it rain? *Colorado Climate* **June:**107–109.

Doesken, N. J., and W. P. Eckrich. 1987. How often does it rain where you live? *Weatherwise* **40**:200–203.

Doesken, N. J., and T. B. McKee. 1999. Drought in Colorado. *Colorado Climate* **1**:13–20.

Eastman, J. L., M .B. Coughenour, and R. A. Pielke. 2001a. Does grazing affect regional climate? *Journal of Hydrometeorology* **2**:243–253.

Eastman, J. L., M. B. Coughenour, and R. A. Pielke. 2001b. The effects of CO2 and landscape change using a coupled plant and meteorological model. *Global Change Biology* **7**:797–815.

Golden, J. H. 1999. Tornadoes, pp. 103–132. In: R. A. Pielke and R. A. Pielke (eds.), *Storms*. Vol. II. Routledge, New York.

Gruntfest, E. 1999. Flash floods in the United States, pp. 192–206. In: R. A. Pielke and R. A. Pielke (eds.), *Storms*. Vol. II. Routledge, New York.

Hayden, B. P. 1998. Ecosystem feedbacks on climate at the landscape scale. *Philosophical Transactions of the Royal Society of London, Series B, Biological Sciences* **353**:5–18.

Henz, J., S. Turner, W. Badini, and J. Kenny. 2004. Historical perspectives on Colorado drought. In: *Colorado drought and water supply assessment*. Colorado Water Conservation Board, Colo.

Hjelmfelt, M. 1999. Flash floods associated with convective storms, pp. 80–102. In: R. A. Pielke and R. A. Pielke (eds.), *Storms*. Vol. II. Routledge, New York.

Lauenroth, W. K., and J. B. Bradford. 2006. Ecohydrology and the partitioning AET between transpiration and evaporation in a semiarid steppe. *Ecosystems* **9**:756–767.

Lauenroth, W. K., and I. C. Burke. 1995. Great Plains, climate variability, pp. 237–249. In: W. A. Nierenberg (ed.), *Encyclopedia of environmental biology*. Vol. 2. Academic Press, San Diego, Calif.

Lauenroth, W. K., I. C. Burke, and M. P. Gutmann. 1999. The structure and function of ecosystems in the central North American grassland region. *Great Plains Research* **9**:223–259.

Lauenroth W. K., and D. G. Milchunas. 1992. Short-Grass Steppe, pp. 183–226. In: R. T. Coupland (ed.), *Natural grasslands: Introduction and western hemisphere*. Ecosystems of the world, vol. 8A. Elsevier, Amsterdam.

Lauenroth, W. K., and O. E. Sala. 1992. Long-term forage production of North American shortgrass steppe. *Ecological Applications* **2**:397–403.

Lu, L., R. A. Pielke, Sr., G. E. Liston, W. J. Parton, D. Ojima, and M. Hartman. 2001. Implementation of a two-way interactive atmospheric and ecological model and its application to the central United States. *Journal of Climate* **14**:900–919.

Lyons, W. A. 1999. Lightning, pp. 60–79. In: R. A. Pielke and R. A. Pielke (eds.), *Storms*. Vol. II. Routledge, New York.

McKee, T. B., N. J. Doesken, and J. Kleist. 1999. *Historical dry and wet periods in Colorado*. Part A: Technical report. Climatology report no. 99–1 A. Colorado Climate Center.

NCDC. 2002. *Climate atlas of the United States*. CD. U.S. Department of Commerce, National Oceanic and Atmospheric Administration, National Climatic Data Center, Asheville, N.C.

Pielke, Sr., R. A. 1998. Climate prediction as an initial value problem. *Bulletin of the American Meteorological Society* **79**:2743–2746.

Pielke, Sr., R. A. 2001a. Earth system modeling: An integrated assessment tool for environmental studies, pp. 311–338. In: T. Matsuno and H. Kida (eds.), *Present and future of modeling global environmental change: Toward integrated modeling*. TERRAPUB, Tokyo, Japan.

Pielke, Sr., R. A. 2001b. Influence of the spatial distribution of vegetation and soils on the prediction of cumulus convective rainfall. *Reviews of Geophysics* **39**:151–177.

Pielke, R. A., G. E. Liston, J. L. Eastman, L. Lu, and M. Coughenour. 1999. Seasonal weather prediction as an initial value problem. *Journal of Geophysical Research* **104**:19463–19479.

Pielke, R. A., and P. L. Vidale. 1995. The boreal forest and the polar front. *Journal of Geophysical Research* **100**:25755–25758.

Sala, O. E., and W. K. Lauenroth. 1982. Small rainfall events: An ecological role in semi-arid regions. *Oecologia* **55**:301–304.

Sala, O. E., W. K. Lauenroth, and W. J. Parton. 1992. Long-term soil water dynamics in the shortgrass steppe. *Ecology* **73**:1175–1181.

Serafin, R. J., J. W. Wilson, J. McCarthy, and T. T. Fujita. 1999. Progress in understanding windshear animplications on aviation, pp. 237–252. In: R. A. Pielke and R. A. Pielke (eds.), *Storms*. Vol. II. Routledge, New York.

Stahle, D. W., E. R. Cook, M. K. Cleaveland, M. D. Therrell, D. M. Meko, H. D. Grissino-Maer, E. Watson, and B. H. Luckman. 2000. Tree-ring data document 16th century megadrought over North America. *EOS* **81**:121, 125.

Wang, G., and E. A. B. Eltahir. 2000. Role of vegetation in enhancing the low-frequency variability of the Sahel rainfall. *Water Resources Research* **36**:1013–1021.

Weaver, J. F. 1999. Windstorms associated with extratropical cyclones, pp. 449–460. In R. A. Pielke and R. A. Pielke (eds.), *Storms*. Vol. I. Routledge, New York.

Ziegler, C. L. 1999. Issues in forecasting mesoscale convective systems: An observational and modeling perspective, pp. 26–42. In: R. A. Pielke and R. A. Pielke (eds.), *Storms*. Vol. II. Routledge, New York.

3

Soil Development and Distribution in the Shortgrass Steppe Ecosystem

Eugene F. Kelly
Caroline M. Yonker
Steve W. Blecker
Carolyn G. Olson

Beneath the gently rolling, seemingly mundane topography that characterizes the shortgrass steppe is a complex mosaic of soils. Many of these soils are superimposed upon older, buried soils that formed in other millennia under different climatic regimes. The nature of this soil mosaic reveals much about the past and dictates much about the future of the shortgrass steppe.

There is considerable heterogeneity among soils of the shortgrass steppe, yet they maintain a high degree of homogeneity when contrasted with soils of other ecosystems. The driving forces that make these soils alike are a semiarid climate and a resilient plant community (Pielke and Doesken, chapter 2, this volume; and Lauenroth, chapter 5, this volume). The combined effects of vegetation and climate on soil development yield generally predictable results. Shortgrass steppe soils are characterized by the accumulation of organic matter in the surface (0–20 cm). Approximately 60% of the graminoid root mass resides in the first 10 cm of mineral soil (Schimel et al., 1986); 90% is contained in the surface 20 cm (Schimel et al., 1985). Surface horizons typically are darker hued than underlying horizons and have organic carbon contents that average 1% to 3% (Yonker et al., 1988). Shortgrass steppe soils maintain a high-percent base saturation (and high pH) because of low leaching and weathering potentials that result from semiarid conditions. Zones of secondary calcium carbonate accumulation are common in subsurface horizons and may appear as threads, seams, or nodules (Blecker et al., 1997). In addition, these soils are characterized by zones of secondary clay accumulation in subsurface horizons; clay accumulations are a result of either the in situ weathering of primary minerals or the translocation of clay minerals leached from the surface horizon. In either case, the maximum depth of accumulation gives some indication of the time-averaged depth of the wetting front in the soil profile (Blecker et al., 1997).

The factors that produce considerable heterogeneity among the soils of the shortgrass steppe are related to parent material, the age of the soil, and the subtleties of topography. These factors vary at a finer scale than either vegetation or climate. This is illustrated through the comparison of small-scale (1:7,000,000 and 1:7,500,000) maps of natural vegetation (Küchler, 1964), soils (USDA–Soil Conservation Service, 1967), geology (Kinney, 1966), and physiography (Fenneman, 1946). These maps indicate that the boundaries of the abiotic variables do not coincide with those of the natural vegetation. In fact, there are multiple map units of each of the abiotic variables contained within the single map unit depicting the shortgrass steppe (Fig. 3.1).

Jenny (1941) formalized the concept that soil formation is a function of five factors: parent material, time, topography, biota, and climate. In this chapter we define these factors for that portion of the shortgrass steppe known as the Pawnee National Grasslands (PNG). More specifically, emphasis is placed on the relationships observed from our intensive characterization of soils at the Central Plains Experimental Range (CPER), located on the western fringe of the PNG.

Parent Material

The shortgrass steppe lies entirely within the boundary of the Great Plains physiographic province (Fenneman, 1946), a region characterized by low topographic relief yet containing areas of structural highs and lows. Much of northeastern Colorado lies within the Colorado Piedmont, which can be distinguished from the remainder of the shortgrass steppe because of its unique geological history. An understanding of the geological history of the Colorado Piedmont (Osterkamp et al., 1987; Wayne et al., 1991) is requisite to understanding the distribution of sediments that serve as parent materials for the soils of northeastern Colorado.

The floor of the Colorado Piedmont consists of thousands of meters of interbedded sandstones and shales deposited after the uplift of the Rocky Mountains. Prior to the uplift, the mid continent was covered by shallow seas. Uplift, which commenced during the late Cretaceous (Table 3.1), forced the retreat of the sea. Streams flowing eastward from the mountains deposited sediment across a widening coastal plain. As uplift continued into the Tertiary, eroded sediments continued to accumulate, forming a depositional surface that extended across eastern Colorado, southern Wyoming, and western Nebraska and Kansas.

During the late Tertiary, the existing river system began to cut into part of this vast surface. The prehistoric Arkansas and South Platte rivers and their tributaries slowly excavated the Tertiary sedimentary deposits down to the older Cretaceous surface. The excavated area, known as the Colorado Piedmont, extends from the Colorado–Wyoming border south 700 km and from the eastern edge of the Rocky Mountains east 500 km. Consequently, the Colorado Piedmont is topographically lower than the High Plains to the north and east.

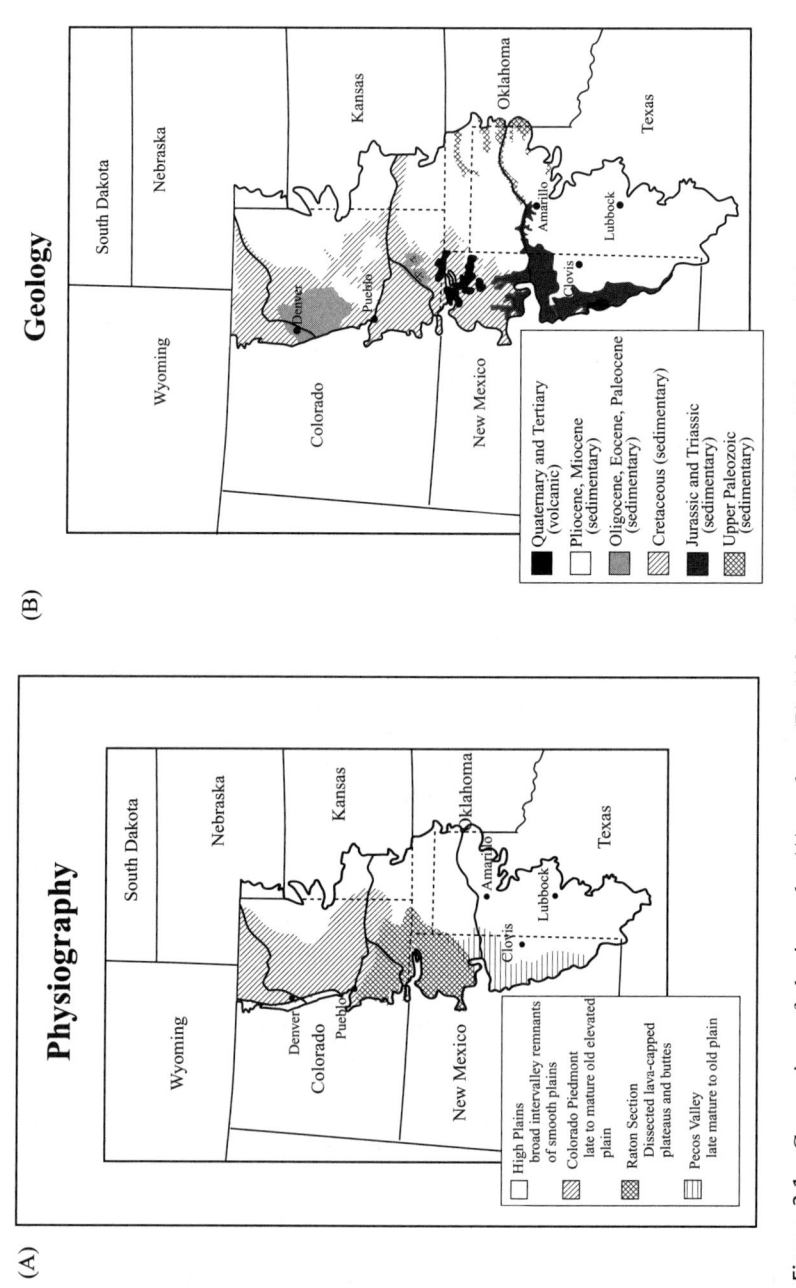

Figure 3.1 Comparison of physiography (A), geology (B). (After Fenneman [1946] and Kinney [1966]).

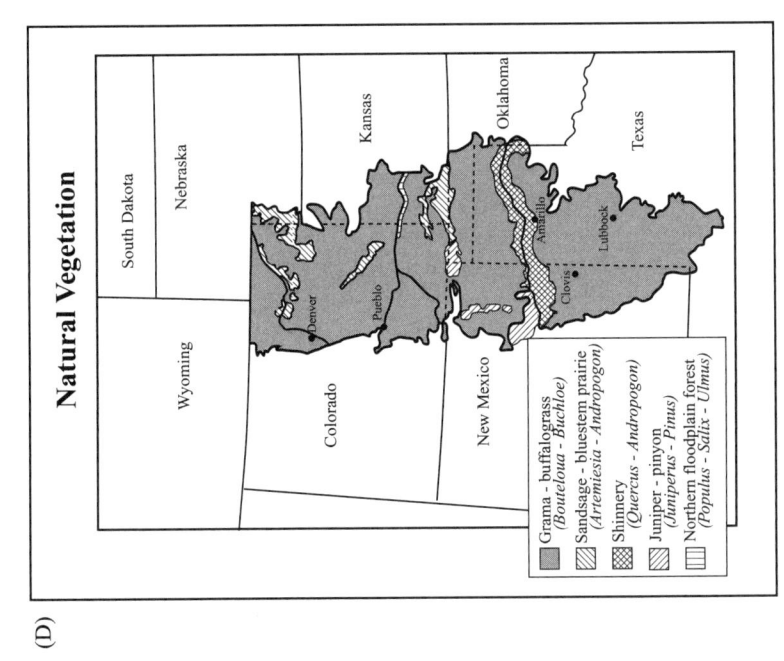

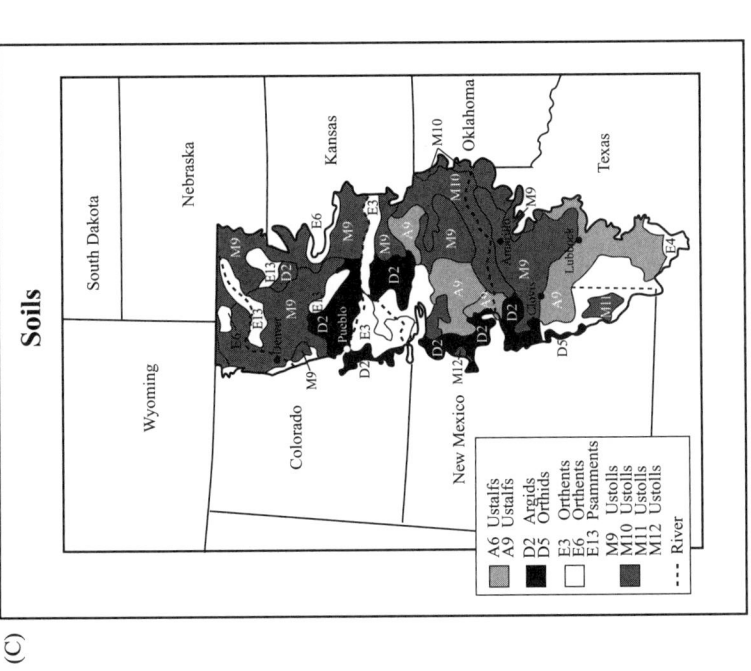

Figure 3.1 (*cont.*) Comparison of soils (C), natural vegetation (D) for the shortgrass steppe. (After USDA–Soil Conservation Service [1967] and Küchler [1965]).

Table 3.1 Geological Timescale for the Colorado Piedmont

Periods	Epochs	Local Formations
Quaternary	Recent (Holocene)	Stream terraces
	Pleistocene	Glacial terraces
Tertiary	Pliocene	Ogallala groups
	Miocene	Arikaree group
	Oligocene	White River group
	Eocene	Denver group
Cretaceous	Upper Cretaceous	Laramie
		Foxhills
		Pierre
		Niobrara
		Benton
		Dakota

The South Platte River system continued its influence throughout the Pleistocene, eroding the Cretaceous deposits of the Piedmont to a low and gently rolling topography. Although the Piedmont itself was never glaciated, glacial outwash and alluvium from the Rocky Mountains were deposited throughout the Piedmont in varying thicknesses along stream channels. Alluvial deposition generally occurred during pluvial climatic intervals and, in particular, as climate shifted toward more arid conditions. Episodes of stream downcutting, combined with uplift of the mountain front, resulted in a series of well-established terraces along the Front Range (Holliday, 1987; Osterkamp et al., 1987; Wayne et al., 1991; Zier et al., 1994). These terraces, which are quite distinct close to the mountain front, lose their topographic expression as they extend farther east into the Piedmont, but do occur as mappable remnants (Scott, 1982). The youngest stream terraces, which are Holocene in age, generally consist of reworked alluvium derived from local sources.

Eolian deposition generally occurred during more arid intervals of the Pleistocene and Holocene. The most abundant loess of the Piedmont was deposited during the late Pleistocene and occurs in deposits as thick as 4.5 m within 60 km of the Front Range; Holocene loess deposits, derived from local drainages, occur in northeastern Colorado but are sparse (Madole, 1995).

Since the end of the Pleistocene, considerable amounts of sand have been mobilized and deposited as well (Forman and Maat, 1990; Forman et al., 1992, 1995; Muhs, 1985). Eolian deposits cover approximately 60% of eastern Colorado; one third of those deposits are Holocene eolian sand (Madole, 1995). Prominent deposits of eolian sand, derived primarily from the South Platte River and its tributaries, are located in dune fields south of the river. The source of sediment north of the river is from small drainage basins that carry smaller sediment loads. Consequently, the northernmost part of the Colorado Piedmont lacks significant areas of contiguous, thick, eolian sand deposits.

Case Study

Central Plains Experimental Range parent materials consist almost exclusively of Holocene alluvium and eolian materials derived from local sources. Very few of the soils have developed in residuum of the underlying Cretaceous deposits. These deposits, which consist locally of interbedded sandstones and shales, are exposed where contemporary drainages have eroded into the bedrock. Such outcrops occur over less than 1% of the area. Davidson (1988) identified and described the prominent alluvial and eolian deposits of the CPER, and depicted the chronology of depositional events, beginning with the oldest identifiable Pleistocene deposit (Fig. 3.2). It is notable that the alluvium of the Q4 and Q3 episodes are poorly exposed across the site, because both are veneered by younger eolian material. Q4, which occupies less than 2% of the area, is preserved only above 1650 m as terrace remnants. Q3 occupies approximately 79% of the site area. Q2 alluvium is perhaps the best exposed of all the alluvial deposits, but occupies only 15% of

A) 160,000 B.P. Formation of Q4 erosion surface and deposition of Q4 alluvium.

B) 90,000 B.P. Downcutting and depostion of Q3 alluvium.

C) Less than 90,000 B.P. Continued downcutting and Q3 deposition.

D) 8,000 to 5,000 B.P. Deposition of older eolian material and Q2 valley fill.

E) 5,000 to 3,000 B.P. Continued depostion of Q2 and soil formation

F) Approximately 3,000 B.P. Partial removal of eolian material and soil.

G) 3,000 to 1,500 B.P. Deposition of younger eolian material. After 1,500 B.P. continued downcutting and deposition of Q1 alluvium.

H) 1,500 B.P. Soil development on younger eolian. Incision of Q1 surface.

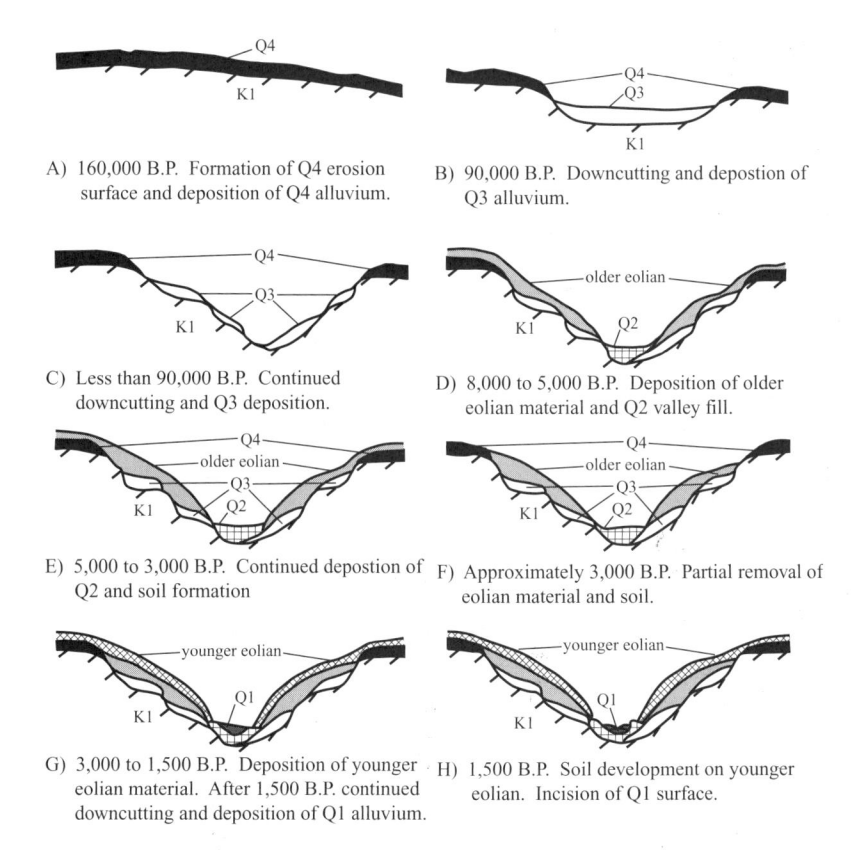

Figure 3.2 Chronological schematic of surficial deposits, CPER. Deposits are labeled as Q or K, signifying Quaternary and Cretaceous, respectively. Deposits are also identified numerically, with the highest number indicating the oldest deposit of that period. (After Davidson [1988].)

36   Ecology of the Shortgrass Steppe

the area. All these deposits are discontinuous and, because of extensive reworking by wind, are not easily distinguished from eolian deposits. Central Plains Experimental Range eolian deposits are identified on the basis of their stratigraphy as opposed to the occurrence of prominent features (e.g., dunes) or grain size frequency distributions. Central Plains Experimental Range surficial sediments have grain size frequency distributions similar to both alluvial and eolian deposits (Blecker, 1994), and are easily confused with one another.

Selected soil profile characteristics for two CPER soils—the Edgar and the Valent—illustrate differences attributed to parent material (Fig. 3.3). Both soils

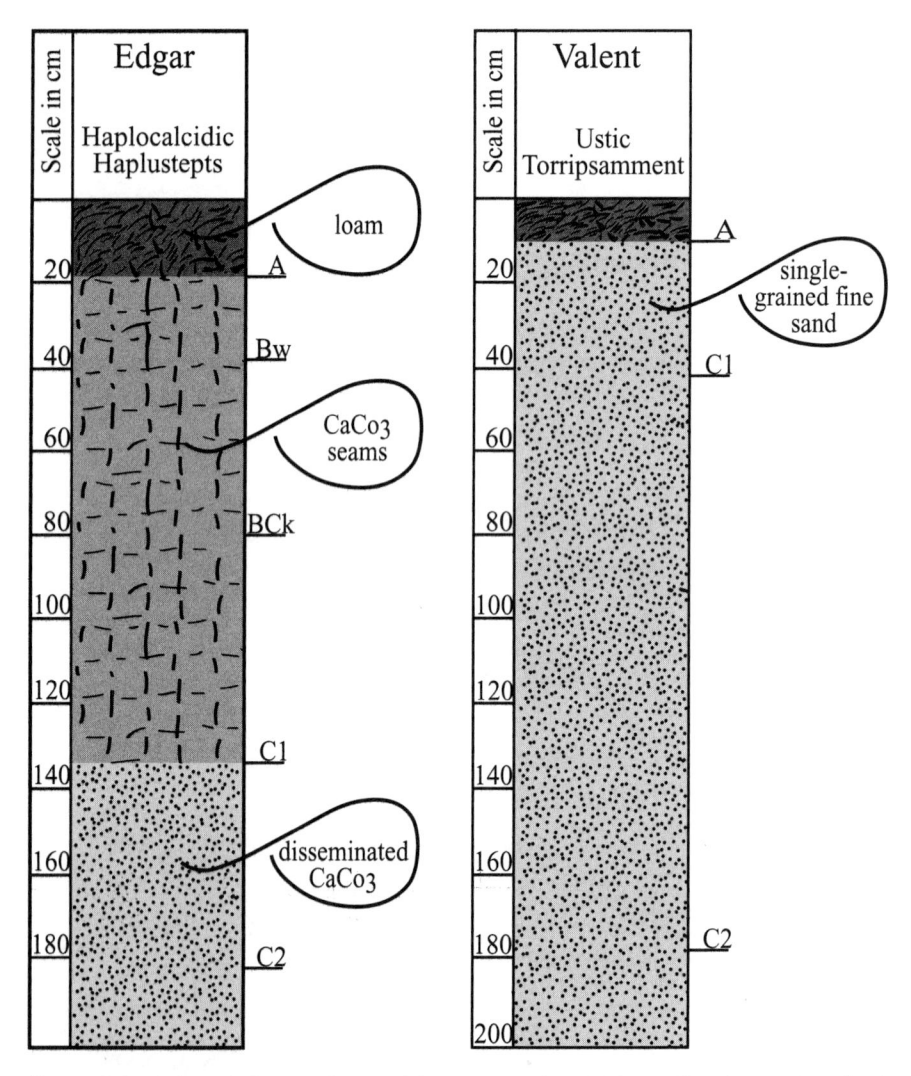

Figure 3.3   Selected characteristics of the Edgar and the Valent soil series, as described at the CPER. The soils formed in loamy mixed alluvium with some loess and eolian sand, respectively.

are late Holocene in age and occur on level landscape positions. The Edgar soil formed in mixed alluvium with some loess and is loamy in texture throughout, whereas the Valent soil formed in eolian sand and is sandy in texture throughout. Edgar is an Aridisol with a relatively thick surface horizon containing significant organic matter and subsoil B horizons. In contrast, the Valent soil is an Entisol, with minimal soil development. Available water-holding capacity for the finer textured Edgar soil is 26 cm $H_2O \cdot 150$ cm^{-1} compared with 5.6 cm $H_2O \cdot 150$ cm^{-1} soil for the coarser Valent (CPER Soil Survey, USDA- Natural Resource Conservation Service (NRCS), unpublished). This difference is reflected in the plant community composition. Species associated with the Valent soil include *Sporobolus cryptandrus* (sand dropseed), *Bouteloua gracilis* (blue grama), *Muhlenbergia torreyi* (ring muhly), *Muhlenbergia cuspidate* (plains muhly), *Opuntia polyacantha* (prickly pear), *Yucca glauca*, and *Calamovilfa longifolia* (prairie sandreed). Species associated with the Edgar soil include *Agropyron smithii* (western wheatgrass), *Distichlis spicata* (saltgrass), *Atriplex canescens* (fourwing saltbush), *B. gracilis, M. torryei, Buchloë dactyloides* (buffalograss), and *O. polyacantha* (CPER Soil Survey, USDA-NRCS, unpublished).

Time

Time is the most intangible of the factors that influence soil formation. Although parent material has a direct effect on soil development, time does not. Time, rather, controls the duration over which soil development processes operate, and thus we can differentiate between young and old soils on the basis of degrees of chemical and physical weathering. From an ecological perspective, there may be an optimal soil age for a particular environment. For the shortgrass steppe, the optimal soil age might be where sufficient time has elapsed to allow for the weathering of primary minerals to increase availability of plant nutrients, lead to accumulation of organic matter, and cause the development of "horizonation" that enhances water-holding capacity. From a paleoclimatic perspective, characteristics of both young and old soils serve as indicators of changing climate through time.

The age of a soil is constrained by the age of the geological deposit in which it formed. Recently published stratigraphic studies indicate that eolian deposits of northeastern Colorado are exclusively late Pleistocene and Holocene in age (Forman et al., 1995; Madole, 1995). Archaeological data for the Front Range of Colorado suggest that some of the eolian sand was deposited as early as 22,500 years BP; relative age criteria indicate the most abundant loess was deposited as early as 21,000 years BP (Madole, 1995). Alluvium of Pleistocene age is preserved as patchy terrace remnants in northeastern Colorado and is of limited geographic extent (Scott, 1982). Where this alluvium has not been blanketed by a more contemporary eolian or alluvial deposit, we find the soils have been developing for perhaps as long as 600,000 years (Loadholt, 2001). Few of the soils of northeastern Colorado, however, have remained undisturbed for the past 20 millennia.

Indeed, Madole (1994) stated that more than half the eolian sand present on the land surface has been mobilized during the past 1000 years.

Radiocarbon ages of paleosol organic matter are used to determine maximum limiting dates of soil burial, thereby establishing the age of the superjacent deposit and providing some chronological control on the duration of soil formation. Radiocarbon dates of paleosols throughout the region suggest that much of the land surface has been buried by wind- or water-deposited materials during the Holocene (Forman et al., 1995; Gould et al., 1979; Kelly et al., 1993; Madole, 1995; Yonker et al., 1988; Zier et al., 1994). Muhs (1985) summarized the existing archaeological and geomorphological evidence for an early Holocene drought spanning approximately 8000 to 5000 years BP, and further proposed that a late Holocene drought occurred from approximately 3000 to 1500 years BP. Presumably, soil burial, as opposed to soil development, dominated during these intervals as a result of enhanced eolian activity. The aforementioned paleosol ^{14}C ages tend to cluster around those intervals before and after the proposed droughts. Paleosols are localized and discontinuous. Their patchy preservation suggests that many of the superjacent deposits may be localized as well. Regardless, each of these buried soils represents an interval of relative landscape stability during which environmental conditions were conducive to soil development.

In many locations, the soil profile consists of multiple soils stacked one upon the other, each of which represents only a few thousand years of soil development before burial and the subsequent development of a superimposed soil. This phenomenon of soil burial has implications for the ecology of the shortgrass steppe. Buried organic carbon is an unaccounted component of the global carbon pool. The pool may be significant to ecosystem recovery, should significant drought again occur, reducing plant cover and inducing the mobilization of eolian sand sheets in the region (Yonker et al., 1988).

Case Study

Within the CPER, there is high variability in soil age (Fig. 3.4), even for soils both formed from alluvium and occurring in similar topographic settings, such as the Bankard and Remmit soils. The Bankard soil is located on the modern floodplain and is very young (<1500 years). The Remmit series is located on higher terraces and upland plains, and, based on its landscape position and association with soils of known age, is assumed to be at least 3000 years old, or older. The Bankard soil is relatively undeveloped compared with the Remmit. The latter has more complex horizonation, reflecting the subsoil accumulation of both carbonates and clay. Available water-holding capacity for the Bankard soil is 3.3 cm $H_2O \cdot$ 150 cm^{-1} soil compared with approximately 17.5 cm $H_2O \cdot$150 cm^{-1} for Remmit (CPER Soil Survey, USDA-NRCS, unpublished). Organic matter content also differs significantly between these soils; surface horizon accumulations range from 0.25% to 0.75% and 0.8% to 1.5% for the Bankard and Remmit soils, respectively.

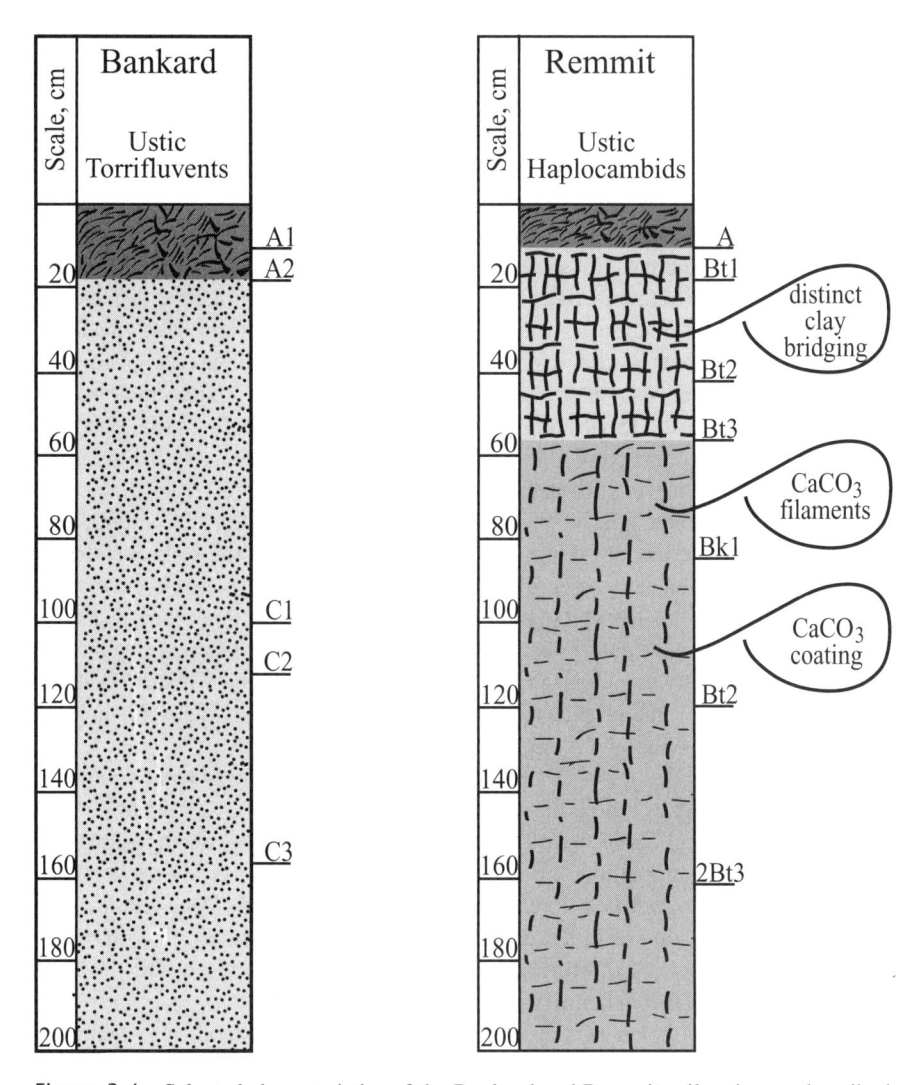

Figure 3.4 Selected characteristics of the Bankard and Remmit soil series, as described at the CPER. The ages of the soils are less than 1500 years and more than 3000 years, respectively.

Topography

Topography encompasses the qualitative description of a surface (flat, rolling, mountainous, and so forth) as well as the quantitative measure of inclination at a given point in space. Topography influences soil formation by moderating the movement of both wind and water, and the materials they carry, across the land surface.

In relatively humid environments water moves both vertically through the soil profile as well as laterally across topographic gradients. The lateral movement of water from high to low points of the land surface results, over time, in the differentiation of soils across that gradient. Quite typically, soil formation on hillslopes in humid environments results in a general increase in organic matter, soil depth, and the accumulation of various chemical constituents with position down the slope. That is, the shallowest, least productive soils are typically located at the convex portion of hillslopes, where erosion of soil and runoff of water is greatest; the deepest, most productive soils are typically located at the concave or linear base of the hillslope, where deposition of eroded soil and run-on water is maximized.

In semiarid environments like the shortgrass steppe, such hillslope differentiation does occur but is not as pronounced. The reasons for this are twofold. First, runoff-producing precipitation events are a very small percentage of total rainfall events. Microwatersheds constructed at the CPER collected no measurable runoff over a multiyear period (Striffler, 1971, 1972). Likewise, rainfall simulations have failed to generate runoff from CPER hillslopes despite the relatively heavy application of water (Gary Frazier, personal communication, August 1997). Because rainfall events produce relatively small amounts of precipitation, the topography is nearly level to moderately sloping (seldom exceeding 10%), the soils are usually at a soil moisture deficit (Sala et al., 1992), and storm intensities rarely exceed the infiltration capacity of the soils (Lapitan and Parton, 1996); thus, long-term soil water data show no evidence that soil water differs across landscape positions (Singh et al., 1998). Second, wind counters any redistribution of materials down hillslopes with the force of gravity by moving materials up and across hillslopes based on wind direction and local topography.

Common to the Colorado Piedmont are topographic features locally referred to as playas. A playa is defined as the usually dry, nearly level, lowest portion of a closed depression. These areas may be temporarily flooded during unusually wet periods, creating shallow, intermittent lakes. Soils in these lowest positions often are stratified and reflect intermittent saturation (Fig. 3.5). As a result of the long-term accumulation of fine sediments (clay) from upland areas, they may also exhibit vertic properties of shrink and swell wherein the surface soil develops cracks as deep as 0.6 m when dry.

Although the evolution of playas is not completely understood, it is generally attributed to wind action. Stose (1912) recognized these depressions and attributed them to wind erosion, but also suggested that in some cases they may have formed when surficial geological deposits obstructed drainageways. Muhs (1985), working in dunal terrain, also cites eolian processes as a means of formation, reasoning that their orientation parallels that of the presumed Quaternary prevailing wind direction.

In an effort to understand better the evolution of the playas on the Piedmont, we drilled and sampled five sites on the CPER to a depth of up to 16 m. Holes adjacent to playas were drilled to shale bedrock when possible, and were found to contain significant alluvial material and buried soils. Similar holes drilled in adjacent drainageways contained thick sections of sands and alluvium. Cores from sites adjacent to playas contained shrink–swell clayey material to depth of 2.4 to 2.7 m. The clayey material also exhibited evidence of restricted water movement. In one core at the edge of a playa, a thick sequence of buried soils was present, beginning at a depth of 2.9 m. Charcoal and roots were present at 3.8 m and large fragments

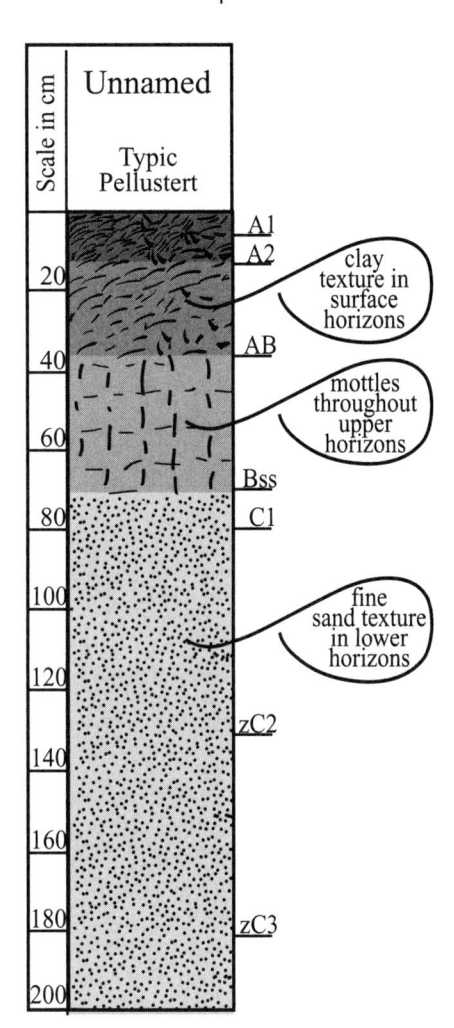

Figure 3.5 Selected characteristics of a playa soil, unnamed soil series, as described at the CPER. The surface 0 to 70 cm contains 45% to 50% clay, resulting in very poor drainage, as evidenced by mottling.

of wood from 4.1 to 4.6 m. At 4.27 m, large wood fragments were radiocarbon dated at approximately 38,850 years BP (Kelly and Olson, unpublished data).

From this evidence it is clear that the area currently occupied by the playas provided an environment for vegetative growth at various times in the Quaternary. Whether these were former valleys or only depressions such as we see today is yet to be determined. Additional field and laboratory measurements will be necessary to begin to assess the processes surrounding the development of these features and their relation to the hydrological system of the Colorado Piedmont.

Case Study

Durar (1978) described soil profiles at 5-m intervals down an east-facing hillslope at the CPER (Fig. 3.6). The slope varied from approximately 2% at the convex

42 Ecology of the Shortgrass Steppe

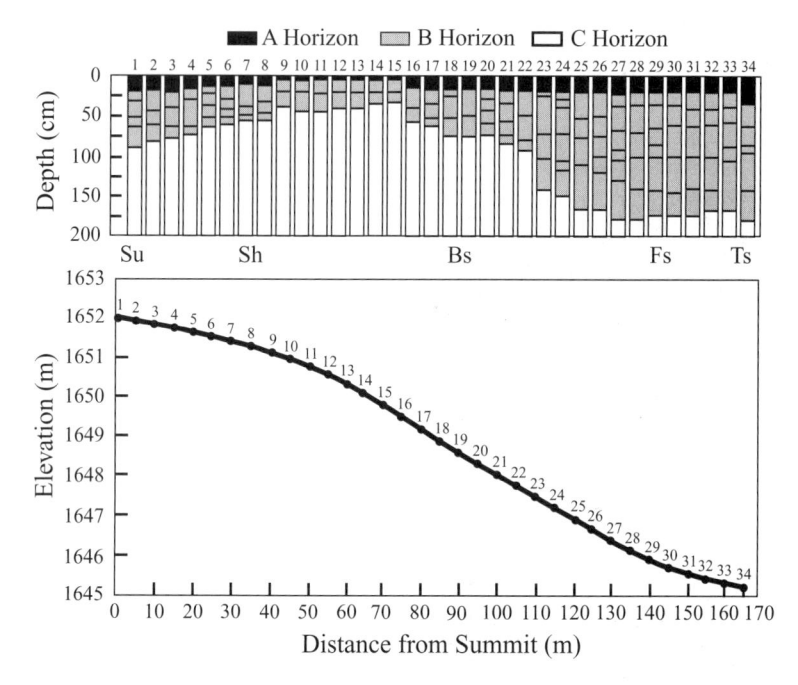

Figure 3.6 Depth and horizonation of soils forming in a CPER hillslope. The parent material is alluvium and is uniform across the hillslope. (After Durar [1978].)

portion of the hill slope to a maximum of 6% on the backslope. The soil parent material is sandy alluvium throughout; for that reason, one can conclude that the soil differences observed in Figure 3.6 are primarily the result of topography moderating the movement of water, soil material, and other constituents down the hillslope and are not the result of overthickening from localized alluvial or eolian deposits. Detailed observation of the land surface was conducted to assess the degree of water erosion (overland transport). These observations confirmed that the linear back portion of the hillslope was moderately erosional, and the concave base of the hillslope was depositional. The data indicate that both mineral and organic matter have been redistributed through time as a function of topography (Burke et al., chapter 13, this volume). Although productivity measurements were not made, organic matter content suggests the redistribution of organic constituents and increased plant production on depositional hillslope segments. Organic matter ranged from 1.4% to 2.0% on the nearly level summit (pedons, 1–4), 1.0 to 1.2% on the linear backslope (pedons, 9–16), and 1.3% to 2.9% on the base of the hillslope (pedons 31–34) (Durar, 1978).

Biota

Although animals affect soil formation, primarily through disturbance, this section focuses primarily on long-term effects of plants on pedogenesis. The short-term (life span of an individual plant) influence of plants on soils is addressed in Milchunas et al. (chapter 16, this volume). The influence of biota is primarily related to controls on inputs and outputs of organic and inorganic materials. The quantity and chemical nature of soil organic matter is primarily related to rates of plant material input (plant death) and output (decomposition of dead plant matter). The balance between these opposing processes governs soil organic matter content. The importance of organic matter to the biogeochemical functioning of the ecosystem has been well established. The importance of biota in regulating inorganic components of the soil system can be complex and more difficult to quantify. For example, biological controls on textural differentiation arise from processes that include the translocation of colloidal material, erosion and deposition by both wind and water, and bioturbation. In most environments, including the shortgrass steppe, these processes are conditioned by the presence or absence of plants (Hook et al. 1991; Burke et al., chapter 13, this volume; Lauenroth, chapter 5, this volume).

Case Study

The effect of vegetation type on the formation of soils in the Front Range of the Rockies can be observed by contrasting a soil forming under grassland vegetation with that of a soil forming under forest (Fig. 3.7). Vegetation at the grassland site consists of those species common to the CPER; vegetation at the forest site consists of lodgepole pine (*Pinus contorta latifolia*) and aspen (*Populus tremuloides*). The grassland soil formed in alluvium and has a sandy loam texture throughout the profile. The forest soil formed in colluvium and has a sandy loam texture in all but the Bt1 horizon, which is sandy clay loam. Mean annual precipitation and mean annual temperature for the grassland and forest soils are 30 cm and 9 °C (Kittel, 1988) and approximately 58 cm and 8 °C, respectively.

Clearly, the differences in soil morphology are a function of several factors in addition to vegetation; however, several vegetation-induced differences are notable. Inputs to the soil organic matter pool are disproportionately aboveground in the forest soil, consisting of needle and leaf litter; consequently, the zone of maximum organic matter accumulation is a layer of partially decomposed organic material on top of the first mineral soil layer. The grassland soil receives organic matter inputs that are largely belowground; consequently, the zone of maximum organic matter accumulation is in the first mineral horizon. Comparing the surface mineral horizon of grassland and forest, respectively, the A horizon has more than twice the organic matter content of the E horizon. The forest soil has a uniformly more acidic pH throughout its profile, and a significantly lower cation exchange capacity (CEC). Differences in CEC are attributed to the differences in clay mineralogy between the sites, which results from differences in the weathering environment.

Biota and Paleoclimate

It is often difficult to separate the influence of biota on soil formation from that of climate, because these factors covary. This covariation has been used in

44 Ecology of the Shortgrass Steppe

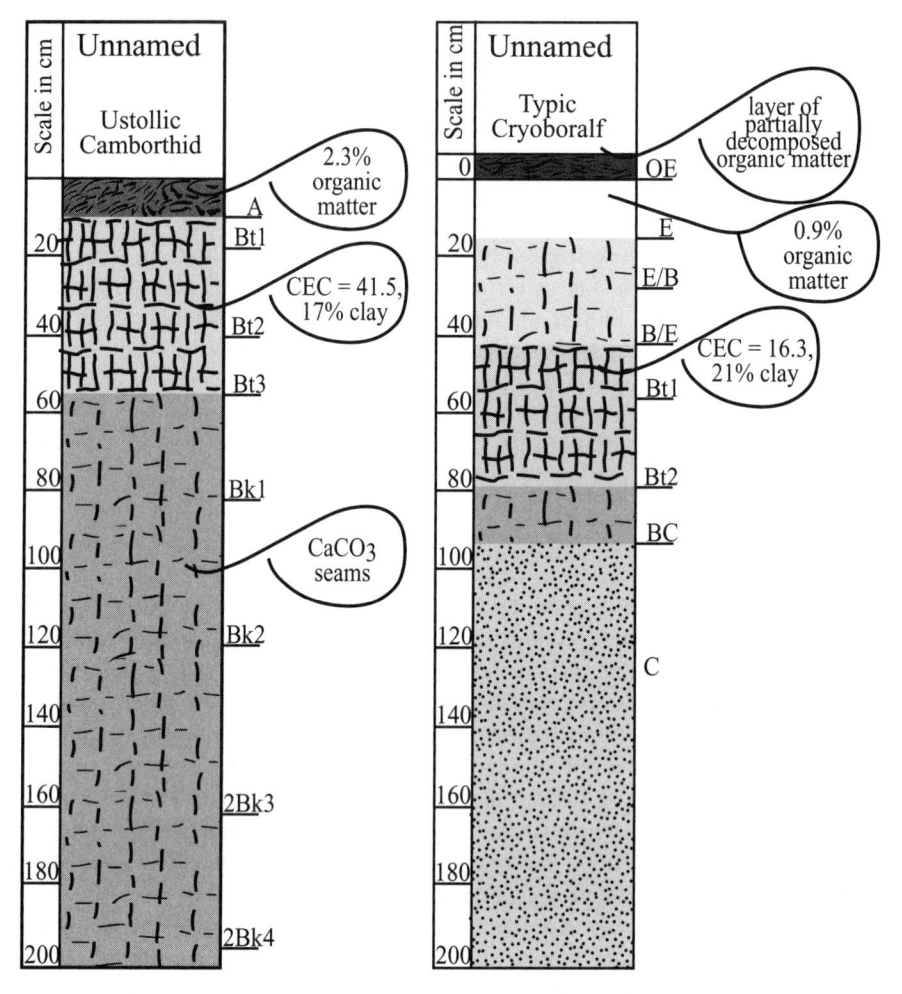

Figure 3.7 Selected soil characteristics of soils, unnamed soil series, forming under grassland (left) and forest (right). CEC is cation exchange capacity.

paleoenvironmental research, because determination of paleovegetative conditions can lead to an understanding of paleoclimate. Indeed, this technique has fostered a considerable amount of research directed at paleoenvironmental reconstruction at many locations throughout the world. For instance, a fundamental assumption, easily documented in modern ecosystems, is that the environmental conditions necessary to sustain trees differ from those under which grasses are dominant. Numerous studies have utilized this relationship to assess the paleoenvironment by reconstructing forest–grassland transitions in temperate and tropical regions (Chadwick et al., 2007; Emanuel et al., 1985; MacDonald et al., 1993; Newnham, 1992; Pentico and Tieszen, 1991).

Boundaries of the shortgrass steppe are associated with the dominance of C_4 plants (Lauenroth, chapter 5, this volume). The relative proportion of C_3 and C_4

Soil Development and Distribution in the Shortgrass Steppe Ecosystem   45

plants growing at a site reflects both local and regional climatic conditions, with higher minimum summer temperature and lower soil moisture favoring higher proportions of C_4 plants (Ehleringer, 1978; Epstein et al., 1997; Paruelo and Lauenroth, 1996; Terri and Stowe, 1976; Tieszen et al., 1979). Studies of contemporary soils indicate that the stable C isotope composition of soil organic matter, pedogenic carbonates, and biogenic minerals (phytoliths) are closely related to the relative proportions of C_3 and C_4 plants (Balesdent et al., 1987; Cerling et al., 1989; Kelly et al., 1991a, b). Application of these relationships to buried soils has proved to be useful in assessing changes in plant composition (C_3 and C_4 proportions) during the Holocene and in making inferences to the paleoclimate of the Piedmont region of the shortgrass steppe (Fig. 3.8) (Kelly et al., 1993, 1998).

Case Study

We have applied a Holocene chronology presented by Muhs (1985), coupled with radiocarbon ages of buried soils in the Colorado Piedmont (Table 3.2), to our understanding of soil formation during the past 10,000 years. The data compiled indicate distinct periods of stability and soil formation from approximately 10,000 to 8000 years BP, 5000 to 3500 years BP, and 1500 years BP to the present, with intervening periods of instability (drought) resulting in dune formation and soil truncation or burial.

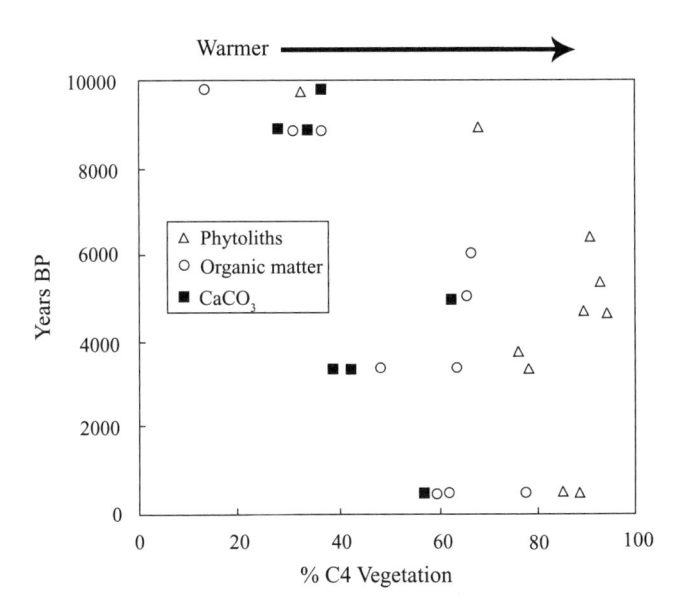

Figure 3.8 The percentage of C_4 vegetation versus time as calculated from stable C isotope ratios and ^{14}C dates, CPER and central Great Plains sites. (After Kelly et al. [1991b].)

Table 3.2 Some Soil Properties Affected by Plant Presence or Absence

Property	Imparted by	Significance
Organic matter	Accumulation and decomposition of plant material	Storage and cycling of nutrients
Root abundance	Species abundance and, composition	Soil CO_2, soil erodibility, channels for flow of water and air
Soil depth	Plant structure	Dust and sediment traps, increased surface horizon thickness under plant canopy
Soil structure	Root pressure, root exudates	Flow of water and air, soil erodibility

Data from the CPER (Blecker et al., 1996; Gould et al., 1979, Kelly et al., 1993; Yonker et al., 1988) indicate that early Holocene paleosols (those dating between 10,000 and 8000 years BP) generally contain more organic carbon than contemporary surface soils. In contrast, mid-Holocene paleosols dated between 5000 and 3500 years BP all contain less organic carbon than contemporary surface soils. It is difficult to relate organic carbon content alone to changes in productivity, because it is unlikely that all microbial decomposition was arrested upon burial of the soil. Kelly (1989) suggested that phytolith content can be used, however, as a proxy for plant production. If this model is indeed applicable, the data indicate higher plant production during both the early and mid Holocene than currently, because the buried soils contain more than a twofold increase in phytoliths over their contemporary counterparts.

The stable isotope composition of organic matter, carbonates, and phytoliths indicates a variable plant community composition through the Holocene. When comparisons are made between contemporary and buried soils, the $\delta^{13}C$ values of organic carbon during the early Holocene are consistently more negative (a decrease in ^{13}C) than both the contemporary surface and mid-Holocene soils. These data indicate that greater amounts of C_3 vegetation persisted during this period of time. The substantially more negative $\delta^{13}C$ values of phytoliths during the early Holocene also indicate greater proportions of C_3 vegetation relative to the mid-Holocene and contemporary soils (Fig. 3.9).

Holocene variability in plant production and plant community composition has led to inferences about Holocene climate change. Organic carbon and phytolith mass suggest a wetter climate during the early Holocene than the present (Blecker et al., 1996). The dominance of C_3 vegetation during that period suggests a cooler climate than the present. Additional isotopic data, derived from a broader geographic context, are required to assess Holocene climate change confidently in the shortgrass steppe. Stable isotopic analyses must be coupled with radiocarbon dating and stratigraphic analyses to provide both chronological control and realm of inference.

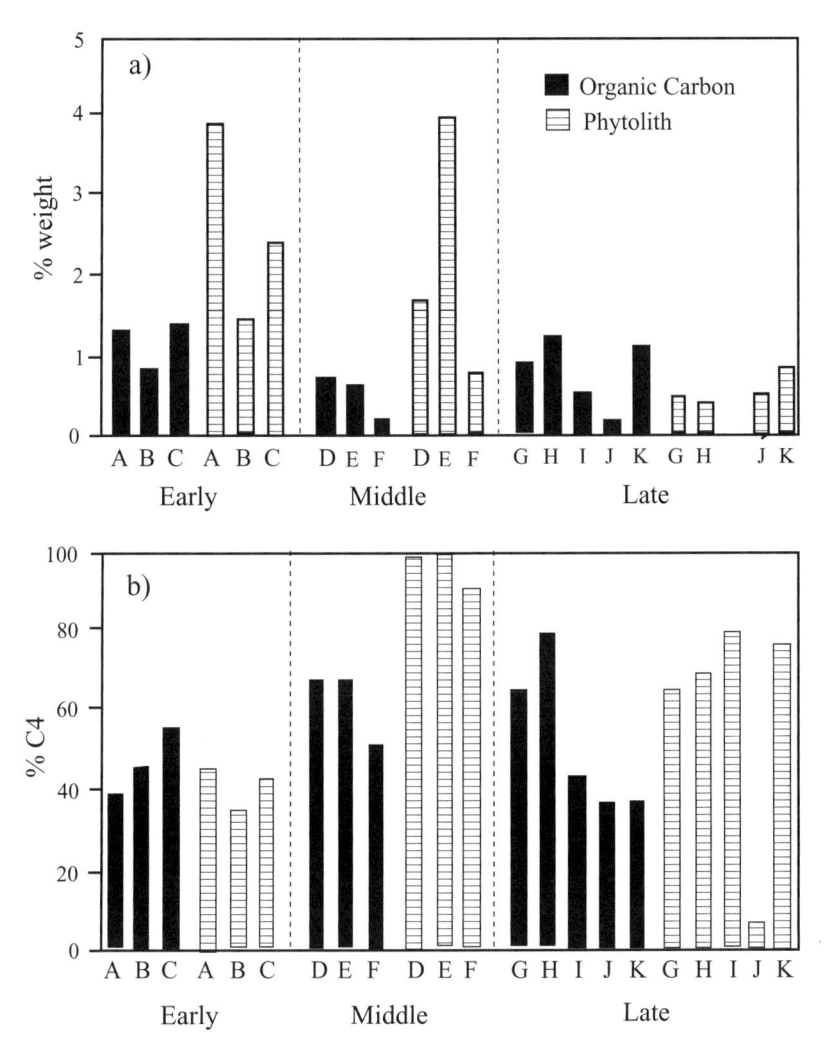

Figure 3.9 (a, b) Quantities of organic C and phytoliths, and percentages of C_4 vegetation estimated from isotopic composition of soil organic C and phytoliths in 11 Holocene paleosols (n=2), CPER. (After Kelly et al. [1998].)

Climate

Of the five factors of soil formation, the seasonal, annual, and even millennial fluctuation and interaction of temperature and precipitation tend to have the greatest overall impact on soil development. Soil characteristics derived from topographic and geological variation are ultimately dictated by the prevailing climate. Parent material influences often dominate development in relatively young soils. Climate, however, has controlled soil formation and the type of vegetation present and, in a broad, regional context, is responsible for the topography and often the type of parent material deposited.

Paleosols are valuable tools that provide clues regarding climate variation within the Holocene (Kelly et al., 1993). The differing morphologies of soils that developed within periods of temperature and precipitation patterns different from those of the present lend insight into pedogenic and ecological responses to climate. Understanding the relationship between soil development and past climate change provides a useful tool for modeling the potential impact that future changes in climate may have on soil development and, ultimately, the entire shortgrass steppe ecosystem.

Pielke and Doesken (chapter 2, this volume) describe the semiarid climate of the shortgrass steppe. Deviations from the mean climate have significant short-term impacts on vegetation, but less immediate effects on pedological parameters. These short-term climatic extremes, potentially episodic in magnitude (e.g., floods and dust storms), could play a significant role in the pedological additions and translocations over longer time frames.

Given the semiarid climate, only limited soil weathering and development occurs. The majority of soils formed in this region are Aridic Argiustolls (Mollisols) and Ustic Haplargids (Aridisols). Slight variations in organic matter content of the surface horizons is the primary difference between these two soil types. Microclimatic variations due, in part, to the topographic position and aspect as well as soil age and stability may play a more significant role in soil organic matter content than the prevailing climate. Soils with northern and eastern aspects in these same positions are also more likely to have higher organic matter contents than soils on drier southern and western aspects.

Concerning subsurface soil development, the current climatic regime currently supports cambic and calcic horizons in this region, as these diagnostic horizons require minimal weathering. Cambic horizons show minimal parent material alteration, typically in the form of a color or texture change. Calcic horizons are zones of secondary carbonate accumulation typical of arid and semiarid climates. Soil moisture data for the current climate (Sala et al., 1992) suggest that there is insufficient infiltrating soil moisture to translocate clays from the surface horizons to form an argillic horizon. However, soil surveys indicate that argillic horizons are fairly common to the shortgrass steppe. The occurrence of Bk, Bt, and even Btk horizons within the same soil profile suggest previous climatic fluctuations, particularly involving periods during which precipitation exceeded evapotranspiration. Specifically, the presence of Btk horizons, where carbonate has moved into previously developed argillic horizons (Blecker, 1994), suggests a shift toward a drier climate during the Holocene.

The morphological complexity and presence of paleosols in the soils of the shortgrass steppe have sparked great interest in paleoclimate research (Blecker et al., 1996; Forman et al., 1992; Kelly et al., 1993; Madole, 1995; Muhs, 1985). Various field (soil classification, geomorphic classification) and laboratory (radiocarbon dates, mineralogy, stable carbon isotopes, granulometric analyses, remote sensing) techniques have been used to identify probable climatic shifts in the recent geological history of the Colorado Piedmont. Such climatic variation in the past is thought to be necessary to permit burial of an existing soil by an eolian or alluvial deposit, thereby creating a paleosol, and to start a new episode of soil formation in the new surficial deposit, as the current climatic conditions are not likely to have driven the large-scale erosion and deposition of material. Given the abundance of argillic horizons and the variable depth to the argillic horizon (8–90 cm), more significant soil moisture than currently exists was necessary to permit clay translocation (Blecker, 1994). Stable carbon isotopes of organic matter and calcium carbonate (Kelly et al., 1993) suggest a general warming trend since the onset of the Holocene. Thus, the cooler and moister periods of the early and middle Holocene, interspersed with warmer and drier periods, would have led to the various combinations of Bt, Bk, and Btk horizonations found throughout the shortgrass steppe.

In addition to driving pedological processes, climate shifts also affected Holocene geomorphic processes. Dune deposits and alluvial terraces of variable age have been identified throughout the region (Blecker, 1994; Davidson, 1988; Forman et al., 1992; Madole, 1995; Maes, 1990; Muhs, 1985). The typically complex stratigraphy of these intimately associated eolian and alluvial deposits lend evidence to the proposed climatic shifts.

Landscape stability provides the connection between geomorphic activity and soil development. Numerous lithologic discontinuities, ubiquitous to the soils of the Colorado Piedmont, have developed in deposits of both eolian and fluvial origin (Blecker, 1994). Difficulties in distinguishing one definite mode of deposition could be explained by the probable fluvial reworking of eolian deposits and/or the eolian reworking of fluvial deposits. Truncation and/or burial of soil horizons during these climate shifts could account for the lack of significant horizon development in some of the shortgrass steppe soils, particularly those soils in close proximity to contemporary and paleodrainageways. Further geomorphic examination could provide clarification of the extent and significance of past climate change.

Case Study

Because climate is relatively uniform within the shortgrass steppe of northeastern Colorado, here we contrast the influence of climate on soils formed under shortgrass steppe (Olney series) and tallgrass prairie (Hastings series of Nebraska; Fig. 3.10). The Olney soil formed in loamy alluvium and loess on nearly level to gentle slopes and is Holocene in age. The Hastings soil formed in loess on nearly level to gentle slopes and is Holocene to late Pleistocene in age. Although differences exist among the factors of soil formation, general trends in soil formation attributable to climate, such as clay translocation and carbonate precipitation, can be seen.

The northeastern Colorado site has a mean annual precipitation of 30 cm and a mean annual temperature of 9 °C (Kittel, 1988) compared with a mean annual precipitation of approximately 70 cm and a mean annual temperature of 11 °C (USDA–Soil Conservation Service, 1974, 1982) for the southeastern Nebraska site. Although typically cultivated, the Hastings soil exhibits greater depth and amount

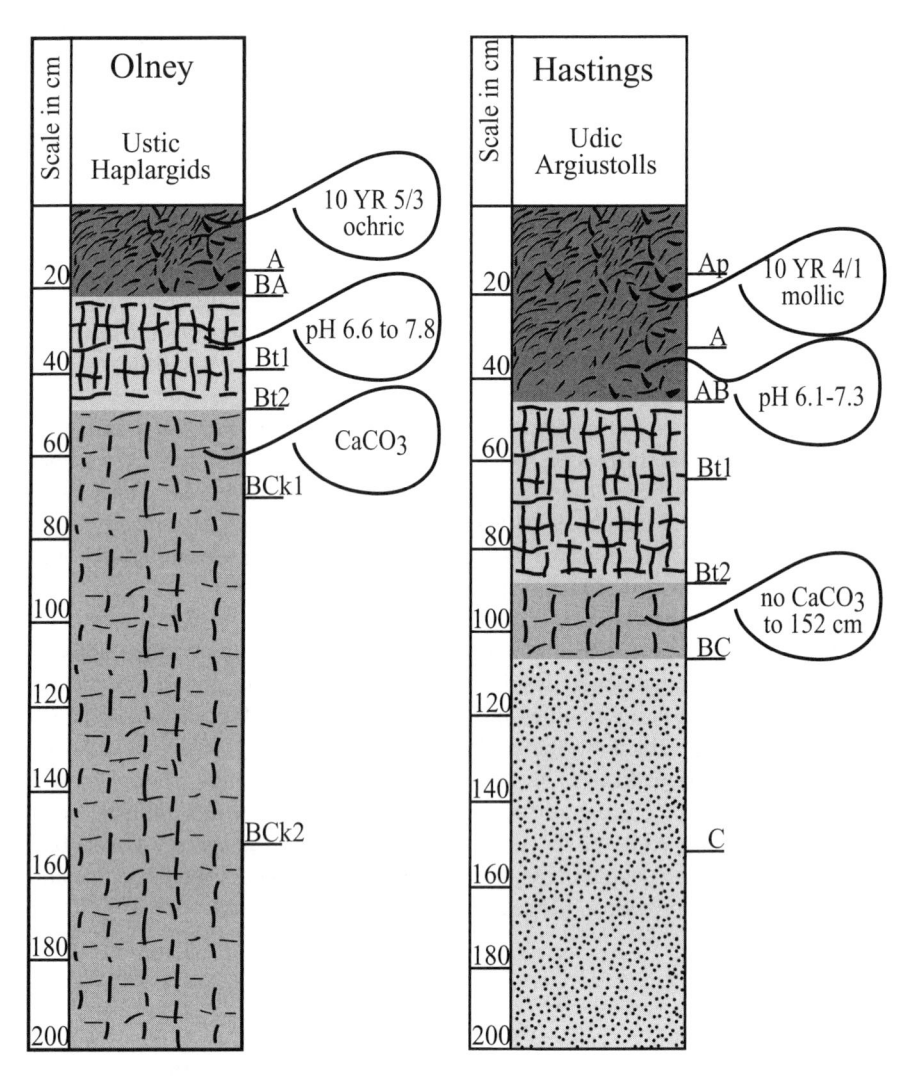

Figure 3.10 Selected characteristics of the Olney and Hastings soil series, as described at the CPER and York County, Nebraska, respectively. The soils formed under 30.9 cm precipitation (Olney) and 69.8 cm precipitation (Hastings).

of soil organic matter. Even though both soils contain an argillic horizon, the argillic horizon of the Hastings soil is both thicker and deeper than that in the Olney soil. Finally, the depth to carbonates in the Olney soil is 46 cm, whereas no carbonates are present in the Hastings soil (at least to 152 cm). Because the difference in mean annual precipitation between the two regions is much greater than that in mean annual temperature, it appears as though precipitation is the driving force behind the major pedological differences. Without prior knowledge of the vegetation present at each site, these differences alone would suggest that the Hastings soil, with a greater available water-holding capacity and nutrient reservoir, would most likely support a more robust grassland community than the Olney soil.

Potential Future Impacts

A reason for studying past climate change is the desire to predict responses to potential future climate shifts. A number of scenarios have been proposed for the shortgrass steppe in light of possible global warming. General circulation models have indicated that this region is susceptible to an increase in temperature coupled with either a slight decrease or even an increase in precipitation (Lauenroth and Burke, 1995). Such climatic changes would likely affect vegetation first through changes in plant composition and density. Potential changes in geomorphic activity could assume the form of reactivated dune fields (drier scenario) or alteration of ephemeral streams to a more intermittent nature (wetter scenario). Burial or truncation of soils, or alterations in the accumulation of carbonate and clays in soils, could potentially alter the pedogenic pathway of the shortgrass steppe soils. Either situation would most likely result in overall landscape changes. The extent and form of ecosystem-level changes would depend largely on the existing climate–vegetation–soil relationship (as was briefly illustrated through the Olney soil–Hastings soil comparison), because certain communities would be potentially more adaptable to any future climate change. Ideally, as knowledge of ecosystem-level responses to climate change continues to grow, a greater understanding of the future of the shortgrass steppe will emerge.

References

Balesdent, J., A. Mariotti, and B. Guillet. 1987. Natural ^{13}C abundance as a tracer for studies of soil organic matter dynamics. *Soil Biology and Biochemistry* **19**:25–30.

Blecker, S. W. 1994. Pedologic and geomorphic indicators of Holocene environments in the Colorado Piedmont. Masters thesis, Colorado State University, Fort Collins, Colo.

Blecker, S. W., C. M Yonker, C. G. Olson, and E. F. Kelly. 1997. Paleopedologic and geomorphic evidence for Holocene climate variation, Shortgrass Steppe, Colorado, USA. *Geoderma* **76**:113–130.

Cerling, T. E., J. Quade, Y. Wang, and J. R. Bowman. 1989. Carbon isotopes in soils and paleosols as ecology and paleoecology indicators. *Nature* **341**:138–139.

Chadwick, O. A., E. F. Kelly, S. C. Hotchkiss, and P. Vitousek. 2007. Pre-contact vegetation and soil nutrient status in the shadow of Kohala Vocano, Hawaii. *Geomorphology,* **80**:70–83.

Davidson, J. M. 1988. Surficial geology and Quaternary history of the Central Plains Experimental Range, Colorado. Masters thesis, Colorado State University, Fort Collins, Colo.

Durar, A. 1978. Using soil continuum landscape relationships to evaluate soil erosion. Masters thesis, Colorado State University, Fort Collins, Colo.

Ehleringer, J. R. 1978. Implications of quantum yield differences on the distribution of C_3 and C_4 grasses. *Oecologia* **31**:255–267.

Emanuel, W. R., H. A. Shugart, and M. P. Stevenson. 1985. Climate change and the broad scale distribution of terrestrial ecosystem complexes. *Climate Change* **7**:29–43.

Epstein, H. E., W. K. Lauenroth, I. C. Burke, and D. P. Coffin. 1997. Productivity patterns of C_3 and C_4 functional types in the Great Plains of the U.S. *Ecology* **78**:722–731.

Fenneman, N. M. 1946. Physical divisions of the United States. US Department of Interior, Geological Survey, Reston, VA.

Forman, S. L., A. F. H. Geotz, and R. H. Yuhas. 1992. Large-scale stabilized dunes on the High Plains of Colorado: Understanding the landscape response to Holocene climates with the aid of images from space. *Geology* **20**:145–148.

Forman, S. L., and P. Maat. 1990. Stratigraphic evidence for late Quaternary dune activity near Hudson on the Piedmont of northern Colorado. *Geology* **18**:745–748.

Forman, S. L., R. Oglesby, V. Markgraf, and T. Stafford. 1995. Paleoclimatic significance of late Quaternary eolian deposition on the Piedmont and High Plains, Central United States. *Global and Planetary Change* **11**:35–55.

Gould, W. D., R. V. Anderson, J. F. McClellan, D. C. Coleman, and J. L. Gurnsey. 1979. Characterization of a paleosol: Its biological properties and effect on overlying soil horizons. *Soil Science* **128**:201–210.

Holliday, V. T. 1987. Geoarchaeology and late Quaternary geomorphology of the middle South Platte River, northeastern Colorado. *Geoarchaeology* **2**:317–329.

Hook, P. B., I. C. Burke, and W. K. Lauenroth. 1991. Heterogeneity of soils and plant N and C associated with plants and openings in North American shortgrass steppe. *Plant and Soil* **138**:247–256.

Jenny, H. 1941. *Factors of soil formation*. McGraw-Hill, New York.

Kelly, R. H. 1995. Soil organic matter responses to variation in plant inputs on shortgrass steppe. Masters thesis, Colorado State University, Fort Collins, Colo.

Kelly, E. F. 1989. A study of the influence of climate and vegetation on the stable isotope chemistry of soils in grassland ecosystems of the Great Plains. Ph.D. dissertation, University of California, Berkeley, Calif.

Kelly, E. F., R. G. Amundson, B. D. Marino, and M. J. DeNiro. 1991a. Environmental and geological influences on the stable isotope composition of carbonate in Holocene grassland soils. *Soil Science Society of America Journal* **55**:1651–1658.

Kelly, E. F., R. G. Amundson, B. D. Marino, and M. J. DeNiro. 1991b. The stable isotope ratios of carbon in phytoliths as a quantitative method of monitoring vegetation and climatic change. *Quaternary Research* **35**:222–223.

Kelly, E. F., B. D. Marino, and C. M Yonker. 1993. The stable carbon isotope composition of paleosols: an application to the Holocene. *Geophysical Monographs* **78**:233–240.

Kelly, E. F., S. W. Blecker, C. M. Yonker, C. G. Olson, E. E. Wohl, and L. C. Todd. 1998. Stable isotope composition of soil organic matter and phytoliths as paleoenvironmental indicators. *Geoderma* **82**:59–81.

Kinney, D. M. 1966. *Geology*. U.S. Geological Survey, Washington, D.C.

Kittel, G. F. 1988. Climate variability in the shortgrass steppe. In: D. Greenland and L. W. Swift, Jr. (eds.), *Climate variability and ecosystem response: Proceedings of a Long Term Ecological Research workshop*. Boulder, Colo. pp 67–75. General Technical Report SE-65. USFS Southeastern Forest Experiment Station.

Küchler, A. W. 1964. *Potential natural vegetation of the conterminous United States*. American Geographical Society, New York.

Lapitan, R. L., and W. J. Parton. 1996. Seasonal variabilities in the distribution of the microclimatic factors and evapotranspiration in a shortgrass steppe. *Agricultural and Forest Meteorology* **79**:113–130.

Lauenroth, W. K., and I. C. Burke. 1995. Great Plains, climate variability. *Encyclopedia of Environmental Biology* **2**:237–249.

Loadholt, S. W. 2001. Soil physiography of a terrace chronosequence in the Pawnee National Grasslands, Colorado. Masters thesis, Colorado State University, Fort Collins, Colo.

MacDonald, G. M., T. W. D. Edwards, K. A. Moser, R. Pienitz, and J. P. Smol. 1993. Rapid response of treeline vegetation and lakes to past climate warming. *Nature* **361**:243–246.

Madole, R. F. 1994. Stratigraphic evidence of desertification in the west–central Great Plains within the past 1000 yr. *Geology* **22**:483–486.

Madole, R. F. 1995. Spatial and temporal patterns of late Quaternary eolian deposition, eastern Colorado, U.S.A. *Quaternary Science Reviews* **14**:155–177.

Maes, T. J. 1990. An investigation of Holocene and recent eolian features on the Central Plains Experimental Range, Colorado. Masters thesis, Colorado State University, Fort Collins, Colo.

Muhs, D. R. 1985. Age and paleoclimatic significance of Holocene sand dunes in northeastern Colorado. *Annals of the American Association of Geographers* **75**:566–582.

Newnham, R. M. 1992. A 30,000 year pollen, vegetation and climate record from Otakairangi (Hikurangi), Northland, New Zealand. *Journal of Biogeography* **19**:541–544.

Osterkamp, W. R., M. M. Fenton, T. C. Gustavson, R. F. Hadley, V. T. Holliday, R. B. Morrison, and T. J. Toy. 1987. Great Plains, pp. 163–210. In: W. Graaf (ed.), *Geomorphic systems of North America*. Centennial special, vol. 2. Geological Society of America, Boulder, Colo.

Paruelo, J. M., and W. K. Lauenroth. 1996. Relative abundance of plant functional types in grasslands and shrublands of North America. *Ecological Applications* **6**:1212–1224.

Pentico, E. D., and L. L. Tieszen. 1991. Community and soil organic matter stable isotope ratios: Forest–grassland transitions at Wind Cave National Park. *Proceedings of South Dakota Academy of Science* **70**:69.

Sala, O. E., W. K. Lauenroth, and W. J. Parton. 1992. Long-term soil water dynamics in the shortgrass steppe. *Ecology* **73**:1175–1181.

Schimel, D. S., W. J. Parton, F. J. Adamsen, R. G. Woodmansee, R. L. Senft, and M. A. Stillwell. 1986. The role of cattle in the volatile loss of nitrogen from a shortgrass steppe. *Biogeochemistry* **2**:39–52.

Schimel, D. S., M. A. Stillwell, and R. G. Woodmansee. 1985. Biogeochemistry of C, N, and P in a soil catena of the shortgrass steppe. *Ecology* **66**:276–282.

Scott, G. R. 1982. *Paleovalley and geologic map of northeastern Colorado*. U.S. Geological Survey miscellaneous investigation series. Map I-1378. U.S. Geologic Survey, Washington, D.C.

Singh, J. S., D. G. Milchunas, and W. K. Lauenroth. 1998. Soil water dynamics and vegetation patterns in a semiarid grassland. *Plant Ecology* **134**:77–89.

Stose, G. W. 1912. *Apishapa folio*, p. 11. U.S. Geological Survey, Washington, D.C.

Striffler, W. D. 1971. *Hydrologic data 1970, Pawnee Grasslands*. U.S. IBP Grassland Biome technical report no. 75. Colorado State University, Fort Collins, Colo.

Striffler, W. D. 1972. *Hydrologic data 1971, Pawnee Grasslands*. U.S. IBP Grassland Biome technical report no. 196. Colorado State University, Fort Collins, Colo.

Terri, M. A., and L. G. Stowe. 1976. Climatic patterns and the distribution of C_4 grasses in North America. *Oecologia* **23**:1–2.

Tieszen, L. L., M. M. Senyimba, S. K. Imbamba, and J. H. Troughton. 1979. The distribution of C_3 and C_4 grass species along an altitudinal and moisture gradient in Kenya. *Oecologia* **37**:337–350.

USDA–Soil Conservation Service. 1967. *Distribution of principal kinds of soils: Orders, suborders and great groups*. National Cartographic Center, Fort Worth, Texas.

USDA–Soil Conservation Service. 1974. *Soil survey of Seward County, Nebraska.* U.S. Government Printing Office, Washington, D.C.

USDA–Soil Conservation Service. 1982. *Soil survey of Weld County, Colorado, northern part.* U.S. Government Printing Office, Washington, D.C.

Wayne, W. J., J. S. Aber, S. S. Agard, R. N. Bergantino, J. P. Bluemle, D. A. Coates, M. E. Cooley, R. F. Madole, J. E. Martin, B. Mears, R. B. Morrison, Jr., and W. M. Sutherland. 1991. Quaternary geology of the northern Great Plains. In: R. B. Morrison (ed.), *Quaternary nonglacial geology; conterminous U.S.; Boulder, Colorado.* Geological Society of America, The Geology of North America, v. K-2.

Yonker, C. M., D. S. Schimel, E. Paroussis, and R. D. Heil. 1988. Patterns of organic carbon accumulation in a semiarid shortgrass steppe, Colorado. *Soil Science Society of America Journal* **52**:478–483.

Zier, C. J., M. McFaul, L .K. Traugh, G. D. Smith, and W. Doering. 1994. Geoarchaeologic analysis of South Platte River terraces: Kersey, Colorado. *Geoarchaeology* **9**:345–374.

4

Land-Use History on the Shortgrass Steppe

Richard H. Hart

As described in chapter 1 of this volume, the grasslands of central North America began to expand at the end of the Wisconsin period (about 10,000 years BP), and continued their expansion through the warming trend that persisted until about 3000 years BP, occupying their maximum territory at that time (Dix, 1964). Currently, the region still supports trees on escarpments, along streams, and at other sites protected from fire, but centuries ago, fires caused by lightning or kindled by Native Americans may have eliminated relict stands of forest and savanna on the open plains. Large browsers and grazers also may have played a part in eliminating trees as well as grasses sensitive to grazing pressure (Axelrod, 1985). Throughout millennia, bison in particular were likely to have shaped the plant communities of the shortgrass steppe, and thus were an essential component of the system (Larson, 1940).

Bison appeared as early as 300,000 years BP; bison, mammoths, mastodons, camels, horses, and other grazers were numerous by 20,000 years BP. Humans arrived in North America perhaps as early as 60,000 years BP, but certainly by 15,000 years ago. Fires and bison may have achieved maximum impact as recently as the past 500 years (Axelrod, 1985; Looman, 1977). The roles of climate, fire, and grazing in the development of North American grasslands have been examined by Ellison (1960), Coupland (1979), Dyer et al. (1982), Anderson (1982), and Tetlyanova et al. (1990).

The Shortgrass Steppe before European Contact

The earliest known human sites on the shortgrass steppe date to about 13,000 years BP (Wedel, 1979) and are found in the vicinity of fossil glacial lakes.

Figure 4.1 Excavating at the Kaplan-Hoover Bison Bone Bed in northern Colorado. (Photo by Lawrence Todd.)

The population of these mammoth hunters was apparently sparse and scattered. Soon after 11,000 years BP, many of the large mammalian species such as the mammoth, native horse, camel, and ground sloth vanished, and the hunters turned to bison. Bone beds representing mass kills of bison have been found below *buffalo jumps* (Fig. 4.1) and even in the remains of wood or stone corrals, but single kills must have been much more common.

Few archeological sites are known from about 7000 to 4000 years BP (Wedel, 1979). This corresponds to the Alti-thermal, when the shortgrass steppe may have been even hotter and drier than it is today. About 2000 years BP, the first evidences of pottery and of agriculture began to appear on the eastern margins of the Great Plains.

By the eighth or ninth century CE (common era, formerly AD) the Plains Village tradition, distinguished by small communities of no more than 50 to 60 people, was well established. *Zea mays*, *Phaseolus* spp., *Cucurbita* spp., and *Helianthus* spp. were grown on perhaps 1 ha/family. Bison were regularly utilized, as were dozens of other species of mammals, birds, reptiles, amphibians, and fish (Wedel, 1979).

Native Americans began to acquire horses from the Spanish settlements of the Southwest in the late 17th to early 18th centuries. By the mid 18th century, horses had spread from the Rio Grande to the Saskatchewan. With the coming of the horse, many Plains Indian tribes abandoned their permanent villages for a nomadic life (Fig. 4.2), and relied more and more on the bison for subsistence.

Figure 4.2 Pawnee Indians on the shortgrass steppe. The introduction of the horse led many Plains Indians to adopt a nomadic way of life based on bison hunting. (Photo courtesy of the Denver Public Library [X-32649], Denver, Colorado.)

Modern estimates assume between 15 and 20 million bison on the Great Plains at the time of first European contact (Cushman and Jones, 1988).

European Exploration

Some of the earliest European explorers, such as De Vaca (Bandelier, 1905) and La Salle (Hudson, 1993, Joutel, 1962), only skirted the southern fringes of the shortgrass steppe. Coronado, in 1540 to 1542, led the first party of Europeans to penetrate the Great Plains (Castañeda, 1966). They may have traveled as far as Kansas from the Spanish settlements in New Mexico, although Donoghue (1929) maintains that Coronado never got out of Texas. However, it is certain that they reached the shortgrass steppe: A recent translation of Coronado's writings states: "The country we passed through is spacious and level.... The grass grows tall near the lakes; away from them it is very short, a span or less...." (Castañeda et al. 1990, p. 59).

Early explorers of the southern Plains had little time or inclination to botanize, but later and better informed visitors agreed that the shortgrass steppe, then as now, was dominated by *Bouteloua gracilis* and *Buchloë dactyloides*, with *Hilaria* spp. in the south. Pattie (1831) grumbled about the hard surface of the shortgrass steppe, writing: "The plains are covered with a short, fine grass, about four inches high, of such a kind, as to be very injurious to the hoofs of animals...." In 1843, Fremont (1843), on the Republican River, about 8 days' travel above the mouth of the Smoky Hill Fork, reported: "Among a variety of grasses which today made their first appearance, I noticed...buffalo grass...," and later mentions "the short sward of the buffalo grass...." Cooke (1857) commented "...[dormant] buffalo grass...looks like curled gray horsehair..." and echoes Pattie (1831): "...its sod is a near approach to wooden pavements."

Such historical observations are also useful for documenting the presence of certain plant species in the time prior to livestock grazing. For instance, cacti of the *Opuntia* genus are today often regarded as an indicator of mismanagement, but they were present before cattle or sheep grazing occurred on the shortgrass steppe. Long-term data show, in fact, that cacti do not increase with grazing (Milchunas et al., chapter 16, this volume). On the South Platte just above the forks, the Long expedition found: "prickly pears [*Opuntia*] profusely covering extensive tracks" (pp. 225–226)" (James, 1823).

Prosopis spp. have also been regarded as an indicator of overgrazing by livestock, but Hurtado (Bolton, 1916) reported "many mesquite bushes" along the Canadian River in New Mexico in 1715. In 1806, Pike reported "musqueet" between present-day San Antonio and Austin, Texas (Pike, 1966). Marcy found "mezquite" trees all across the Texas Panhandle in 1849 and on the North Fork of the Red River in southwest Oklahoma in 1842 (Marcy, 1850, 1938).

Early European Settlement

During the early 19th century, several forts and trading posts were established where the shortgrass steppe of what is now Colorado rises into the foothills of the Rocky Mountains. Crops and livestock were produced in the immediate vicinity of these posts, for the use of their residents and the occasional hunter, trapper, or exploring party (Steinel and Working, 1926). Farnham (1841) found several farmers growing grain and vegetables on the Arkansas River about 8 km above Bent's Fort in 1839. Fremont (1843) reported "a comfortable farm" with hogs, cattle, poultry, and a garden at Fort Lupton and "a fine stock of cattle" and "an abundance of milk" in a settlement where Fountain Creek joins the Arkansas. In 1846, Parkman (1901) reported "wide cornfields and green meadows, where cattle were grazing" (p. 330) in the vicinity of Pueblo, Colorado.

Crop and domestic livestock agriculture were difficult while millions of bison roamed freely over the shortgrass steppe. Occasional herds of tens of thousands of bison were still encountered as late as the 1870s, but by 1882 the Plains Indians were starving because bison, their nutritional base, had become so rare (Hodgson, 1994). In 1895, only an estimated 800 bison remained in North America (Cushman and Jones, 1988).

Crop agriculture did not take hold on the shortgrass steppe in Colorado, except for its western fringe, until after the passage of the Homestead Act in 1862. The first irrigated claim on the shortgrass of southern Colorado was filed in 1866, east of Trinidad. The first major irrigation project was the Union Colony (near what is now Greeley), founded in 1870. By 1875, 74 km of main canal had been built, and something near 40,000 ha were in cultivation (Fig. 4.3) (Steinel and Working, 1926).

Figure 4.3 The building of irrigation systems that brought mountain waters to the shortgrass steppe led to increasing crop agriculture. (Photo courtesy of the Denver Public Library [MCC-2403], Denver, Colorado.)

In the meantime, the range livestock industry was developing on the shortgrass steppe outside the limited areas of irrigation (Fig. 4.4). As noted earlier, a few cattle were in Colorado in the early 1800s, and in the late 1850s several entrepreneurs bought up work oxen and fattened them for sale as beef. After the Civil War, longhorn "Spanish" cattle were brought in from Texas, and "American" cattle were brought in from the East. The first trainload of finished cattle from Colorado was loaded out of Kit Carson in Cheyenne County in 1869 (Steinel and Working, 1926). The cattle population of Colorado was still concentrated on the Plains in 1877, when 74,440 head were shipped east, principally to Kansas City, Chicago, and Omaha.

The first attempts at nonirrigated crop agriculture on the shortgrass of eastern Colorado were promoted by the Kansas Pacific and the Atchison, Topeka & Santa Fe railroads in the early 1870s. *Boomers* employed by the railroads or local newspapers inflated estimates of annual precipitation, proclaiming that "Rain follows the plow," and spread tales of the wonderful crops to be produced without irrigation (Emmons, 1971; Lyman, 1906; Union Pacific Railroad, 1894). Campbell (1907) contended that, with his methods, wheat and other crops could be grown profitably on the shortgrass steppe. Stockmen were ordered to take down illegal

Figure 4.4 Cattle drive. (Photo courtesy of the Colorado Historical Society [CHS-J1467].)

fences on government land to clear the way for 65-ha homesteads. The government-mandated 65-ha homestead was adequate for irrigated agriculture, but not for dryland farming. The 1036-ha recommended by John Wesley Powell (Stegner, 1954) would have been more realistic, allowing a combination of rangeland grazing and dryland grain farming. Nevertheless, "[a]lmost the whole of eastern Colorado was settled quite thickly during the years from 1886 to 1889," according to J. E. Payne, superintendent of a research station at Cheyenne Wells (Steinel and Working, 1926).

The shortgrass area of extreme western Kansas was settled about the same time as that of eastern Colorado. Malin (1961) gives no figures for numbers of farms west of the 100th meridian in Kansas until 1895, but at this time there were 723 farms in the area. Of these, 360 were in Wallace County, on the Kansas–Colorado border. Most of these farms were too small for raising stock, but there was still plenty of public land on which to graze livestock.

Settlement of the shortgrass steppe of the Texas Panhandle, and adjacent areas of the Oklahoma Panhandle and northeastern New Mexico, lagged behind that of Colorado. The Texas Panhandle was organized into numerous counties in 1876, but because there was "no politically significant population" (Rathjen, 1973, p. 71), all were administered from Jack and Clay counties. The first documented permanent settlers moved into the western Panhandle along the Canadian River in November 1876 with about 4500 sheep and a few horses and cattle. About the same time, Charles Goodnight founded the JA ranch in Palo Duro Canyon (Haley,

1936). During the next few years, settlements grew up around Mobeetie, Tascos, and Clarendon (Rathjen, 1973).

In the 1880 census, the 26 Texas Panhandle counties recorded 1600 people, 97,000 cattle, and 108,000 sheep. Cattle increased to 250,000 head by 1900, whereas sheep dropped to 18,000 head by 1890 and nearly disappeared by 1900. "What damage was done the land through overgrazing and poor land management is anybody's guess...the damage probably was as horrendous as might be expected from any immature, ill-disciplined, unregulated, and exploitive industry" (Rathjen, 1973, p. 190).

The Early Twentieth Century

The panhandles of both Texas and Oklahoma were initially dominated by cattlemen. However, the first *Triticum aestivum* (dryland wheat) was grown in the Oklahoma Panhandle in 1903 and, by 1927 Texas County, one of three Oklahoma Panhandle counties, was the leading *T. aestivum* producer in the state. In 1921, 1926, 1928, and 1929 it produced more *T. aestivum* than any other county in the nation (Green, 1977). Within the shortgrass steppe, wheat was grown in rotation with a year-long fallow period, such that 1 year of wheat was grown on a plot of land and, the next year, that land was kept fallow to store soil water for the following year's crop. During these early years, large fields were managed as either wheat or fallow, leaving them vulnerable to soil erosion.

A similar expansion in land planted to *T. aestivum* occurred in the Texas Panhandle, from 33,000 ha in 1909 to 236,000 ha in 1919 and 793,000 in 1929 (Nall, 1973). Area of other crops also increased; approximately 230,000 ha of grain varieties and 160,000 ha of forage varieties of *Sorghum vulgare* (sorghum), more than 40,000 ha each in *Z. mays* (corn) and *Avena sativa* (oats), and more than 80,000 ha of *Gossypium hirsutum* (cotton) were planted in the Texas Panhandle in 1919 (Nall, 1973). However, most of the 6.5 million ha of the Texas Panhandle remained in rangeland. The increase in crop area was accompanied by an increase in livestock numbers. Cattle numbers in the Texas Panhandle increased to 763,000 in 1920, then dropped to 640,000 in 1925, paralleling a drop in prices. Low cattle prices and a high demand for potential cropland led to the breakup of some ranches. During the 1920s, the number of farms in the Texas Panhandle increased by 71%, whereas the average farm size dwindled from 443 ha in 1920 to 284 ha in 1929 (Nall, 1974). Farms/ranches of 400 ha or more fell from 18% of the total in 1920 to 10% in 1929.

Similar increases in cropland at the expense of rangeland were occurring in the shortgrass counties of Colorado and Kansas. For example, Colorado harvested 103,000 Mg (metric tonnes) of spring wheat varieties and 293,000 Mg of winter wheat varieties of *T. aestivum* in 1925 (Fig. 4.5). From the average yields given by Steinel and Working (1926); it can be calculated that more than 2.1 million ha were planted with *T. aestivum* in 1925, nearly all on former shortgrass steppe. The area planted with *Z. mays* rose from 131,000 ha in 1909 to 188,000 in 1915 and 598,000 in 1925, of which only 55,000 ha were irrigated (Steinel and Working, 1926).

Figure 4.5 As dryland agriculture increased on the shortgrass steppe, harvesting wheat became a yearly sight. (Photo courtesy of the Denver Public Library [MCC-2926], Denver, Colorado.)

In southwest Kansas, from 1910 to 1935, the *T. aestivum* area increased from 15,890 to 58,610 ha in Seward County, from 610 to 56,930 ha in Grant County, and from 200 to 61,680 ha in Stanton County (Hurt, 1981). Ford County had 113,000 ha of grazing land in 1915, but only 59,000 ha of grass remained by 1930.

Despite the increase in crop area, during the 1920s, livestock grazing was still the major operation on the 4.2 million ha of shortgrass steppe in Colorado (Gutmann et al., 2005). Steinel and Working (1926) estimate that about 700,000 cattle and 100,000 sheep were in the shortgrass counties of Colorado in 1925.

The Dust Bowl

However, the conversion of shortgrass steppe rangeland to cropland set the stage for the destructive wind erosion of the Dust Bowl of the 1930s. Other contributing factors were the lack of soil conservation practices, improper crop selection, drought, and wind.

Drought, wind, and dust storms, like blizzards and grasshopper plagues, were no novelty on the shortgrass steppe, and had occurred whenever summers were markedly hotter or drier than normal (Gutmann and Cunfer, 1999). The dust storm of March 26–27, 1880, extended from Las Cruces, New Mexico, to Iowa. In April 1895, several storms in eastern Colorado deposited so much sand and dust on railroad tracks that trains were stalled until the tracks could be cleared with snow plows and shovels. These same storms killed as many as 5000 horses and cows, smothered by dust between Lamar, Colorado, and Larned, Kansas (Hurt, 1981).

Localized dust storms occurred on the Plains in 1932 and 1933, but storms of great intensity and extent began in the spring of 1934. Storms in May stripped soil to the depth of the plow furrows and deposited it as far away as Washington, DC, and New York City (Hurt, 1981). Storms turned day into night, frosted window glass, sandblasted the paint from automobiles, sifted into the tightest houses, and piled drifts of soil over the tops of fences.

Drought on the southern Plains from 1934 through 1937 was the most severe on record, and storms grew worse in 1935, 1936, and 1937 (Fig. 4.6). The boundaries of the Dust Bowl during 1935 to 1936 (Hurt, 1981) were very similar to the boundaries of the shortgrass steppe (Lauenroth and Milchunas, 1992), omitting only the northeastern counties of Colorado and the eastern counties of the Texas Panhandle. Gutmann and Cunfer (1999) concluded that drought and high temperatures, as well as cropping practices, contributed to the severity of the dust storms of the 1930s. In the 1950s, weather conditions were similar, but wind erosion was greatly reduced by tillage practices that increased litter cover and surface roughness.

Crop agriculture was devastated by the drought and dust storms. *Triticum aestivum* harvests across the Dust Bowl counties averaged only 125 to 325 kg·ha^{-1} compared with the usual 1700 to 2400 kg·ha^{-1}. But still, crop area expanded (Gutmann et al., 2005). In western Kansas, 73% of the land was in crops in 1930, increasing to 78% in 1935 (Hurt, 1981).

Figure 4.6 Dust storm in Colorado, 1935. (Photo courtesy of the Denver Public Library [X-17607], Denver, Colorado.)

The livestock industry also suffered. By the summer of 1934, cattle and sheep were dying of starvation, heat, and suffocation on the southern Plains. The Forest Service reported range forage production at 50% to 90% below normal in 1936, and warned that rangelands with a capacity of 10.8 million animal units were carrying 17.3 million units (Baker, 1936). The federal government bought more than eight million cattle (Worster, 1979), nearly a million of them from the drought-stricken counties of the shortgrass steppe (Hurt, 1981), to provide financial aid to cattleman and salvage some value from the cattle. In Cimarron County, Oklahoma, cattle numbers fell from 37,590 in 1930 to 14,876 in 1940 (Worster,1979). However, in the 30 shortgrass counties of Colorado, cattle numbers increased every year during the 1930s, from 565,000 in 1930 to 885,000 in 1940 (Colorado State Planning Commission, 1935, 1941).

Under the Land Utilization Project, submarginal land too abused or unproductive to support agriculture was purchased by the federal government and taken out of production. When the program ended in 1940, nearly 400,000 ha had been purchased, and grass had been restored on 140,000 ha of former cropland, which was then leased back to cattlemen (Worster, 1979). Most of the purchased land is now part of the National Grasslands system. Aside from these National Grasslands, very little land on the shortgrass steppe remains in federal ownership (Hart, 1994).

During the drought of the 1930s, the shortgrass steppe spread 160 to 240 km eastward into what was previously tallgrass prairie (Weaver and Bruner, 1954). On the original shortgrass, plant communities were badly damaged; 50% to 70% of the grass cover was killed on ungrazed areas, and even more on grazed areas (Hurt, 1981). Photographs in McGinnies et al. (1991) show shortgrass rangeland in eastern Colorado, vigorous and productive in 1907, nearly denuded of all vegetation in 1937. Much of the land purchased by the government was planted into nonnative bunchgrasses, which have gradually given way to *B. gracilis* and other native species (Coffin et al., 1996).

After the Dust Bowl

The drought of the 1930s, coupled with advanced pumping technology and the availability of rural electric power, encouraged the use of underground water for crop irrigation on the shortgrass steppe. The Texas Panhandle led the way; by 1940, approximately 100,000 ha were being irrigated from 2180 wells (Fite, 1979). The most important crops included *Beta vulgaris* (sugar beets), *Z. mays* (corn), and *T. aestivum* (winter wheat) (Fite, 1979). By 1976, 38 counties in the High Plains of West Texas had about 71,000 wells pumping water to irrigate more than 2.5 million ha.

With the increase in irrigation came a corresponding increase in the production of feed grains. In 1944, Texas High Plains farmers harvested 1.53 million Mg of *S. vulgare* for grain; in 1964, they harvested 3.15 million Mg (Nall, 1982). The industry continued to grow for several decades, and is now important for both land and livestock of the shortgrass steppe.

Today on the Shortgrass Steppe

Falling water tables and increasing water and energy costs may eventually preclude irrigation on the shortgrass steppe (Parton et al., 2007). Water tables are falling at 0.3 to 1.0 m·y^{-1} in the Texas Panhandle, and 1.5 to 3.0 m·y^{-1} in southwestern Kansas (National Agriculture Statistics Service, USDA, 1997c). Competition for water with urban areas is also increasing, particularly in areas where the source of irrigation water is snowfall in the mountains (Parton et al., 2007). Energy costs, also increasing rapidly, will influence irrigation in the region. In 1997, only about 1.3 million ha were under irrigation in 38 counties of the Texas Panhandle (National Agriculture Statistics Service, USDA, 1997c) versus 2.5 million ha in 1976 (Fite, 1979). Less than 4% of the land is irrigated in most of the counties on the shortgrass steppe; up to 10% is irrigated in a few counties on the fringes of the shortgrass where water is more readily available. Sorghum, corn, wheat, and alfalfa are the main irrigated crops today, with a considerable area of cotton in the lower Texas Panhandle. Total cropland, including both dryland and irrigated, is less than 30% of the area of nearly every shortgrass county (National Agriculture Statistics Service, USDA, 1997a, b, c). In the majority of counties, cropland is less than 15% and, in south–central Colorado, less than 5% of the area is cropped.

Livestock production is still the dominant land use on most of the shortgrass steppe, but total numbers of cattle can be misleading. The 21 shortgrass counties report more than 100,000 cattle, but most of these same counties report that less than 20% of the cattle are cows and heifers that have calved (National Agriculture Statistics Service, USDA, 1997a, b, c). Most of the cattle are in feedlots, and thus are dependent upon feed produced on irrigated cropland. At any given time, only a very small fraction, probably less than 1%, of all the beef cows in the entire United States are found grazing on the shortgrass steppe. Sheep have almost disappeared from the shortgrass. Only a handful of counties report as many as 1000 sheep, and what remains of fat lamb production is concentrated in Weld County in north–central Colorado.

The Future

Since European settlement, the shortgrass steppe has been characterized by a *boom-and-bust* economy. Agriculture prospered or suffered with fluctuations in precipitation, prices of agricultural products, and costs of energy, machinery, and other necessities of production. Farmers and ranchers sought to combat these influences by increasing efficiency, but greater efficiency was not always enough. Thousands were thrown into bankruptcy by the drought of the 1890s, the depression of the 1920s, the dust storms of the 1930s, the drought of the 1950s, the skyrocketing energy prices of the 1970s, and the collapse of land prices in the 1980s. Government subsidies enabled many to hang on or even thrive. In 1967, 1110 of the 1310 farms in Hale County, Texas, in the south–central Panhandle, received stabilization and conservation payments totaling $15.6 million (Fite, 1979). Nevertheless, the economy of many shortgrass counties continued to decline.

Popper and Popper (1987) identified six signs of "land-use distress" in Great Plains counties: (1) current population less than half that in 1930, (2) current population less than 90% of that in 1980, (3) population of four or fewer people per 2.5 km², (4) median age 35 years or older, (5) poverty rate at least 20%, and (6) per-capita investment in construction in 1986 less than $50. As many as 110 counties on the Plains, including a dozen or so in the shortgrass steppe of eastern Colorado and the Texas Panhandle, showed at least three of the six signs as of 1987. The rural Plains population seemed to be concentrated in fewer places, and rural counties fell short of the national average in income and employment growth. However, a recent in-depth analysis of U.S. Census and land-use data over the past century (Parton et al., 2007) shows that rural populations in the Great Plains are stable, with agricultural production maintaining economic incomes through increased crop yields, government payments, and animal feed production. Thus, although farmland in rural areas has decreased to some extent during the past several decades as a result of urbanization, total rural income and population have been relatively stable (Parton et al., 2003), leading to the conclusion that population and agriculture are not declining in the shortgrass steppe. Near the urban fringe, recreation and tourism are becoming important land uses (see chapter 19, this volume).

Predicted global warming and doubling of the CO_2 concentration of the atmosphere may produce major shifts in the functioning and productivity of the shortgrass ecosystem. The magnitude of these effects is still under study and debate. It is not debatable that the shortgrass steppe will continue to change in the future, and land use will inevitably change with it.

References

Anderson, R. C. 1982. An evolutionary model summarizing the roles of fire, climate, and grazing animals in the origin and maintenance of grasslands: An end paper, pp. 297–308. In: J. R. Estes, R. J. Tyrl, and J .N. Brunken (eds.), *Grasses and grasslands: Systematics and ecology.* University of Oklahoma Press, Norman, Okla.

Axelrod, D. I. 1985. Rise of the grassland biome, central North America. *Botanical Review* **51**:163–201.

Baker, W. 1936. *The western range.* Senate Document 199, 74th Congress, 2d session. U.S. Government Printing Office, Washington, D.C.

Bandelier, F. R. 1905. *The journey of Alvar Nunez Cabeza de Vaca and his companions from Florida to the Pacific, 1528–1536.* A. S. Barnes, New York.

Bolton, H. E. 1916. *Spanish exploration in the Southwest.* Charles Scribner's Sons, New York.

Campbell, Hardy W. 1907. *Campbell's 1907 soil culture manual; a complete guide to scientific agriculture as adapted to the semiarid regions. The proper fitting of the soil for the conservation and control of moisture and the development of soil fertility; how moisture moves in the soil by capillary attraction, percolation and evaporation; the relation of water and air to plant growth, and how this may be regulated by cultivation.* H. W. Campbell, Lincoln, Neb.

Castañeda, P., P. de Castañeda de Nájera, P., G. P. Winship, and F. W. Hodge. 1990. *The Journey of Coronado.* Dover Publications, New York.

Coffin, D. P., W. K. Lauenroth, and I. C. Burke. 1996. Recovery of vegetation in a semiarid grassland 53 years after disturbance. *Ecological Applications* **6**(2):538–555.
Colorado State Planning Commission. 1935. *Colorado agricultural statistics 1935*. U.S. Department of Agriculture, Bureau of Agricultural Economics, & Colorado State Planning Commission, Denver, Colo.
Colorado State Planning Commission. 1941. *Colorado agricultural statistics 1941*. U.S. Department of Agriculture, Bureau of Agricultural Economics, & Colorado State Planning Commission, Denver, Colo.
Cooke, P. St. G. 1857. *Scenes and adventures in the Army or romance of military life*. Lindsay & Blakiston, Philadelphia, Pa.
Coupland, R. T. 1979. Distribution of grasses and grasslands of North America, pp. 77–83. In: M. Numata (ed.), *Ecology of grasslands and bamboolands in the world*. Dr. W. Junk Publisher, The Hague, the Netherlands.
Cushman, R. C., and S. R. Jones. 1988. *The shortgrass prairie*. Pruett Publishing, Boulder, Colo.
de Castañeda, P. 1966. *The journey of Coronado*. G. P. Winship (trans.). University Microfilms, Ann Arbor, Mich.
Dix, R. L. 1964. A history of biotic and climatic changes within the North American grassland, pp. 71–89. In: D. J. Crisp (ed.), *Grazing in terrestrial and marine environments*. Blackwell Scientific, Oxford, UK.
Donoghue, D. 1929. The route of the Coronado expedition in Texas. *Southwest Historical Quarterly* **32**(3):181–192.
Dyer, M. I., J. K. Detling, D. C. Coleman, and D. W. Hilbert. 1982. The role of herbivores in grasslands, pp. 255–295. In: J. R. Estes, R. J. Tyrl, and J. N. Brunken (eds.), *Grasses and grasslands: Systematics and ecology*. University of Oklahoma Press, Norman, Okla.
Ellison, L. 1960. The influence of grazing on plant succession of rangelands. *Botanical Review* **16**:1–78.
Emmons, D. H. 1971. *Garden in the grasslands: Boomer literature of the central Great Plains*. University of Nebraska Press, Lincoln, Neb.
Farnham, T. J. 1841. *Travels in the great western prairies*. Poughkeepsie, N.Y.
Fite, G. C. 1979. The Great Plains: Promises, problems, and prospects, pp. 187–203. In: B. W. Blouet and F. C. Luebke (eds.), *The Great Plains: Environment and culture*. University of Nebraska Press, Lincoln, Neb.
Fremont, J. C. 1843. *A report on an exploration of the country lying between the Missouri River and the Rocky Mountains, on the line of the Kansas and Great Platte Rivers*. Senate document 243, 27th Congress, Washington, D.C.
Green, D. E. 1977. Beginnings of wheat culture in Oklahoma, pp. 56–73. In: D. E. Green (ed.), *Rural Oklahoma*. Oklahoma Historical Society, Oklahoma City, Okla.
Gutmann, M. P., and G. Cunfer. 1999. A new look at the causes of the Dust Bowl. Publication 99–1. International Center for Arid and Semiarid Land Studies, Texas Tech University, Lubbock, Texas.
Gutmann, M. P., W. J. Parton, G. Cunfer, and I. C. Burke. 2005. Population and environment in the U.S. Great Plains, pp. 84–105. In: B. Entwisle and P. G. Stern (eds.), *Population, land use, and environment: Research directions*. National Academy of Sciences, Washington, D.C.
Haley, J. E. 1936. *Charles Goodnight*. Houghton Mifflin, New York.
Hart, R. H. 1994. Rangeland, pp. 491–501. In: J. Arntzen and E. M. Ritter (eds.), *Encyclopedia of agricultural science*. Vol. 3. Academic Press, New York.
Hodgson, B. 1994. Buffalo: Back home on the range. *National Geographic* **186**(5):64–89.

Hudson, C. 1993. Reconstructing the de Soto expedition route west of the Mississippi River: Summary and contents, pp. 143–154. In: G. A. Young and M. P. Hoffman (eds.), *The expedition of Hernando de Soto west of the Mississippi, 1541–1543*. University of Arkansas Press, Fayetteville, Ark.

Hurt, R. D. 1981. *The Dust Bowl: An agricultural and social history*. Nelson-Hall, Chicago, Ill.

James, E. 1823. Account of an expedition from Pittsburgh to the Rocky Mountains, performed in the years 1819, 1820, vol. XIV–XVII. In: R. G. Thwaites (ed.), *Early western travels, 1748–1846*. Arthur H. Clark, Cleveland, Ohio.

Joutel, H. 1962. *A journal of La Salle's last voyage*. New York.

Larson, F. 1940. The role of bison in maintaining the short grass plains. *Ecology* **21**:113–121.

Lauenroth, W. K., and D. G. Milchunas. 1992. Short-grass steppe, pp. 183–226. In: R. T. Coupland (ed.), *Natural grasslands: Introduction and western hemisphere*. Elsevier Science, Amsterdam.

Looman, J. 1977. Applied phytosociology in the Canadian prairies and parklands, pp. 317–356. In: W. Krause (ed.), *Application of vegetation science to grassland husbandry*. Dr. W. Junk Publishers, The Hague, the Netherlands.

Lyman, C. A. 1906. *The fertile lands of Colorado and northern New Mexico. A concise description of the vast areas of agricultural, horticultural and grazing lands located on the line of the Denver and Rio Grande railroad in the state of Colorado and the territory of New Mexico. Full information for intending settlers*. Denver & Rio Grande RR, Denver, Colo.

Malin, J. C. 1961. *The grassland of North America: Prolegomena to its history with addenda*. James C. Malin, Lawrence, Kans.

Marcy, R. B. 1850. *Route from Fort Smith to Santa Fe*. U.S. 31st Congress, 1st Session, House executive document 45:26–83.U.S. Government Printing Office, Washington, D.C.

Marcy, R. B. 1938. *Adventure on Red River: Report on the exploration of the Red River*. University of Oklahoma Press, Norman, Okla.

McGinnies, W. J., H. L. Shantz, and W. G. McGinnies. 1991. *Changes in vegetation and land use in eastern Colorado*. USDA-ARS 85. U.S. Government Printing Office, Washington, D.C.

Nall, G. L. 1973. Panhandle farming in the "Golden Era" of American agriculture. *Panhandle-Plains Historical Review* **46**:68–93.

Nall, G. L. 1974. Specialization and expansion: Panhandle farming in the 1920's. *Panhandle-Plains Historical Review* **47**:46–67.

Nall, G. L. 1982. The cattle-feeding industry on the Texas High Plains, pp. 106–115. In: H. C. Dethloff and I. M. May, Jr. (eds.), *Southwestern agriculture: Pre-Columbian to modern*. Texas A & M University Press, College Station, Texas.

National Agriculture Statistics Service, USDA. 1997a. *1997 Census of agriculture*. Vol. 1, part 6, Colorado. State & county data. U.S. Government Printing Office, Washington, D.C.

National Agriculture Statistics Service, USDA. 1997b. *1997 Census of agriculture*. Vol. 1, part 36, Oklahoma. State & county data. U.S. Government Printing Office, Washington, D.C.

National Agriculture Statistics Service, USDA. 1997c. *1997 Census of agriculture*. Vol. 1, part 43, Texas. State & county data. U.S. Government Printing Office, Washington, D.C.

Parkman, F. 1901. *The California and Oregon trail*. T. Y. Crowell, New York.

Parton, W. J., M. P. Gutmann, and D. Ojima. 2007. Long-term trends in population, farm income, and crop production in the Great Plains. *BioScience* **57**:738–747.

Parton, W. J., M. P. Gutmann, and W. R. Travis. 2003. Sustainability and historical land use change in the Great Plains: The case of eastern Colorado. *Great Plains Research* **13**:97–125.

Pattie, J. O., T. Flint, and C. Malte-Brun. 1831. *The personal narrative of James O. Pattie, of Kentucky.* John H. Wood, Publisher, Cincinnati, Ohio.

Pike, Z. M. 1966. *The journals of Zebulon Montgomery Pike.* University of Oklahoma Press, Norman, Okla.

Popper, J. E., and F. J. Popper. 1987. The Great Plains: From dust to dust. *Planning* **53**(12):12–18.

Rathjen, F. W. 1973. *The Texas Panhandle frontier.* University of Texas Press, Austin, Texas.

Stegner, W. 1954. *Beyond the hundredth meridian.* Houghton Mifflin, Boston, Mass.

Steinel, A. T., and D. W. Working. 1926. *History of agriculture in Colorado, 1858–1926.* Colorado State Agricultural College, Fort Collins, Colo.

Tetlyanova, J. F., R. I. Zlotin, and N. R. French. 1990. Changes in structure and function of temperate-zone grasslands under the influence of man, pp. 301–334. In: A. Breymeyer (ed.), *Managed grasslands.* Elsevier Science, Amsterdam.

Union Pacific Railroad. 1894. *The resources and attractions of Colorado for the home seeker, capitalist and tourist. Facts on climate, soil, farming, stock raising, dairying, fruit growing, lumbering, mining, scenery, game and fish.* Woodward & Tiernan, St. Louis, Mo.

Weaver, J. E., and W. E. Bruner. 1954. Nature and place of transition from true prairie to mixed prairie. *Ecology* **35**:117–126.

Wedel, W. R. 1979. Holocene cultural adaptations in the Republican River basin, pp. 1–25. In: B. W. Blouet and F. C. Luebke (eds.), *The Great Plains: Environment and culture.* University of Nebraska Press, Lincoln, Neb.

Worster, D. 1979. Dust Bowl: *The southern Plains in the 1930s.* Oxford University Press, New York.

5

Vegetation of the Shortgrass Steppe

William K. Lauenroth

Two species are most characteristic of the vegetation of the shortgrass steppe: *Bouteloua gracilis* and *Buchloë dactyloides*. Both are perennial C_4 grasses and are informally called shortgrasses. Technically, this means that they are both culmless grasses in which, for the majority of the tillers, the apical meristem remains at or near the soil surface and is protected by a succession of enveloping leaf sheaths for the entire growing season (Dahl, 1995; Dahl and Hyder, 1977). This morphological characteristic makes these two grasses well adapted to withstand turnover of aboveground organs as a result of intraseasonal drought or grazing by large generalist herbivores such as cattle.

The current dominance of *B. gracilis* and *B. dactyloides* in the region is clear, but their ecological status has been controversial. In the earliest assessment of the vegetation of the Great Plains, Shantz and Zon (1924) mapped the western portion of the grasslands from the Canadian border to the panhandle of Texas as shortgrass or plains grassland, with *B. gracilis* as the characteristic species. However, Weaver and Clements (1938) argued that it had been proved that Shantz and Zon's shortgrass plains was not a climax plant community, but rather a disclimax as a result of overgrazing by livestock. Although these two views persist today, a 1964 map of the potential natural vegetation of the United States presented an alternative interpretation and a compromise between the views of Shantz and Zon (1924) and Weaver and Clements (1938). Küchler (1964) divided the large shortgrass area into a northern portion that is consistent with Weaver and Clements's mixed prairie and a southern portion in which the potential vegetation is dominated by shortgrasses. Our definition of the shortgrass steppe is identical to Küchler's (1964) *Bouteloua–Buchloë* vegetation type, except along the northern border.

His *Bouteloua–Buchloë* type extends into the southeastern corner of Wyoming and the panhandle of Nebraska, but we draw the boundary between the shortgrass steppe and the northern mixed prairie at the Colorado–Wyoming border because of the sharp elevation gradient between the Colorado Piedmont and the High Plains that occurs at approximately the Colorado–Wyoming border, and the associated changes in both the environment and the vegetation (Lauenroth and Milchunas, 1992).

Approximately 70% of the shortgrass steppe remains in natural vegetation, and the majority of that is used for cattle grazing (Hart, chapter 4, this volume). The remainder of the area is divided between dryland crops on uplands and irrigated crops where surface or groundwater is available (Dornbusch et al., 1995; Huntzinger, 1995; Lauenroth et al., 1999). The major dryland crop is winter wheat (*Triticum aestivum*) and the irrigated crops are alfalfa (*Medicago* spp.), corn (*Zea mays*), sugar beet (*Beta vulgaris*), and cotton (*Gossypium* spp.).

Major Community Types across the Region[1]

Küchler (1964) identified four potential natural vegetation types within the boundaries of the shortgrass steppe region. The most extensive of these is the one that defines the region: the *Bouteloua–Buchloë* type. *Bouteloua gracilis* and *B. dactyloides* are the characteristic and dominant species over this area. Within the larger matrix of the *Bouteloua–Buchloë* type, on sandy soils along the Canadian River in New Mexico, Oklahoma, and Texas, Küchler (1964) recognized a second community dominated by a shrubby oak (*Quercus havardii*) and the grass *Schizachyrium scoparium* (previously *Andropogon scoparium*). He identified a third community on the eolian sands associated with the North Canadian River in Oklahoma and Colorado, and the South Platte River in Colorado. In addition, other areas of deep sand in Colorado and Kansas are occupied by an *Artemisia–Schizachyrium* community. The characteristic species are the shrub *Artemisia filifolia* and the grass *S. scoparium*. The fourth type is a piñon–juniper woodland dominated by *Juniperus osteosperma* and/or *Pinus edulis*.

A detailed map of the natural vegetation of Colorado prepared by the USDA Soil Conservation Service (Natural Resource Conservation Service) (McGinnies et al., 1983) is based upon Küchler (1964) but identifies six community types for the shortgrass steppe portion of Colorado. Five are grasslands or shrub–grassland mixtures, and one is woodland. The area that Küchler identified as *Bouteloua–Buchloë* type is divided into three types. The most extensive community type is dominated by *B. gracilis*. Two less common types are interspersed within the *Bouteloua* type: a *B. gracilis–Agropyron smithii* community and a *B. gracilis–Sporobolus airoides–Hilaria jamesii—B. dactyloides* type. In the Rocky Mountain foothills from Colorado Springs northward, a type dominated by cool-season (C_3) grasses occurs, called the *A. smithii–Stipa* spp.*–Calamovilfa longifolia–Bouteloua curtipendula* type. This community is distinguished from the shortgrass steppe because it has very little *B. gracilis* and an occasional occurrence of an overstory of *P. ponderosa*

---

[1] This section relies heavily on Lauenroth and Milchunas (1992).

or shrubs. Deep eolian sands are dominated by a community of tall and midheight grasses including *Andropogon hallii, S. scoparium, B. curtipendula, Sorghastrum nutans, C. longifolia,* and *Stipa comata,* and the shrub *A. filifolia.* The final type is a woodland community of *J. osteosperma* and *P. edulis,* either singly or together. This type occupies an area of lithic soils mostly near the foothills in the southern portion of the state. The understory species in this community are characteristic of the surrounding plains vegetation.

In southeastern Colorado, northeastern New Mexico, and western Oklahoma, and in the grasslands of the Mesa de Mayo, the most common dominant grasses are *B. gracilis* and *B. hirsuta,* (Rogers, 1953). *Buchloë dactyloides* is a frequent associate on dry and heavily grazed sites in these areas, whereas *A. smithii* is common in swales. Zonal soils in the panhandle of western Oklahoma and in southeastern Colorado were described by Heerwagen and Aandahl (1961) as supporting communities dominated by *B. gracilis* and *B. dactyloides.*

Bouteloua gracilis was reported by Lotspeich and Everhart (1962) to occur on all sites in the Llano Estacado south of the Canadian River in Texas and eastern New Mexico during their study of soil–plant community relationships. Sites with fine-texture soils were dominated by *B. gracilis* and *B. dactyloides.* Sites with silt loam to sandy loam soils were dominated by *B. curtipendula, B. gracilis,* and *Sporobolus cryptandrus.* Sandy soils in arroyos supported a diverse community of tall grasses, including *Panicum virgatum, S. scoparium,* and *S. nutans,* mixed with shorter grasses. Limestone breaks were dominated by *B. curtipendula, B. hirsuta,* and *S. scoparium.*

Beavis et al. (1982) used cluster analyses to describe two community types in east–central New Mexico based on data from 51 stands at 13 sites. Their *B. gracilis* type had a composition (based on canopy cover) of *B. gracilis,* 40%; *Gutierrezia sarothrae,* 8%; *H. jamesii,* 13%; forbs, 7%; *S. cryptandrus,* 8%; and 14 other grasses, 24%. Their *Stipa neomexicana* community was composed of: *S. neomexicana,* 15%; *B. hirsuta,* 8%; *G. sarothrae,* 26%; forbs, 12%; *B. gracilis,* 9%; and 15 other grasses, 29%. The key environmental difference between the *Bouteloua* and the *Stipa* types was soil texture. The soils under the *Bouteloua* stands were deeper and finer in texture than those under the *Stipa* stands.

At the Texas Tech University Research Farm, 24 km east of Amarillo, Texas, upland sites, regardless of grazing regime, are dominated by a *B. gracilis–Opuntia polyacantha* community type (Pettit, 1974). Subordinate species include *Aristida purpurea, B. dactyloides, Sphaeralcea coccinea, S. cryptandrus,* and several species of *Amaranthus.* Warm-season (C_4) grasses account for 75% of average growing-season aboveground biomass on both grazed and ungrazed sites (Sims et al., 1978). Cool-season (C_3) grasses contribute 5%; forbs, 10%; and cacti, 10%.

Vegetation of the SGS LTER Site

Diversity and Dominance

Five hundred twenty-one species of vascular plants occur on the SGS LTER site (85,933 ha), most of which are native perennials (Table 5.1 [Hazlett, 1998]). Native

Table 5.1 Distribution of Plant Species over the SGS LTER Site with Respect to Six Life Forms and Three Life History Categories

	Graminoids	Herbs	Trees	Shrubs	Aquatics/ Succulents	Vines/ Parasites	Total
Native							
Annual	6	70	—	—	1/4	0/0	81
Biennial	0	23	—	—	—	—	23
Perennial	65	190	6	21	6/6	4/4	299
subtotal	71	283	6	21	7/10	4/4	406
Exotic							
Annual	13	53	—	—	0/2	1/0	68
Biennial	0	14	—	—	—	—	14
Perennial	9	18	3	1	0/0	1/0	32
subtotal	22	85	3	1	0/2	2/0	115
Total	93	368	9	22	7/12	6/4	521
Percent	18	71	2	4	3	2	100

Adapted from Hazlett (1998).

perennial graminoids (grasses, sedges, and rushes) make up 12% of the species; native perennial forbs (herbs), 36%; native shrubs, 4%; and native trees, 1%. Most of the native annuals and biennials are forbs. Twenty-two percent or 115 vascular plant species on the site are exotics, meaning that they were introduced to the area by human activity (Table 5.1 [Hazlett, 1998]). The majority of the exotic species are either forbs (74%) or grasses (19%).

The vast majority of the site is upland steppe (Hazlett [1998] estimated more than 80%), but only 20% of the vascular plant species occur here (Table 5.2 [Hazlett, 1998]). The remaining 80% of the species occur on less than 20% of the area, and most of these occur in riparian areas. Riparian areas constitute a small fraction of the total area, very likely less than 5%, but contain 145 species of native vascular plants, or 28% of the total native species on the site. This unequal distribution of species across habitats for native species is even more dramatic for exotic species (Table 5.2). None of the exotics have the upland steppe as their primary habitat. They are almost evenly split between riparian areas (51 species) and roadsides/disturbances (62 species). Betz (2001) sampled roadsides and adjacent upland steppe throughout the western section of the PNG and found abundant exotic plants and their seeds in the soil seedbank on roadsides, but few of either on uplands. These results were similar with and without summer grazing by cattle.

A similar study to the one reported by Hazlett (1998) provided complementary data for the 6280-ha Central Plains Experimental Range (CPER). The CPER is adjacent to the western edge of the Pawnee National Grassland (PNG), and together the CPER and PNG make up the SGS LTER site (Lauenroth et al., chapter 1, this volume, Fig. 1.6). Three hundred thirty-seven plant species occur on the CPER, 85% of which are natives (Table 5.3 [Kotanen et al., 1998]). Grasses account for 36 species and graminoids account for 49 species, which represents 14% of the total number of species. Similar to the data for the entire site, the

Table 5.2 Distribution of Plant Species across Six Habitat Categories at the SGS LTER Site

	Upland Steppe	Sandy Soils	Breaks/ Barrens	Cliffs/ Ravines	Riparian	Roadsides/ Disturbances	Total
Native							
Annual	22	12	0	4	23	20	81
Biennial	5	1	1	4	10	2	23
Perennial	80	19	28	50	112	13	299
subtotal	107	32	29	58	145	35	406
Exotic							
Annual	0	2	0	0	26	41	68
Biennial	0	0	0	0	5	9	14
Perennial	0	0	0	0	20	12	32
subtotal	0	2	0	0	51	62	115
Total	107	34	29	58	196	97	521
Persent	21	6	6	11	38	18	100

Adapted from Hazlett (1998).
Species occur in only their primary habitat category.

Table 5.3 Distribution of Vascular Plant Species across Habitat Categories and Similarity Coefficients Comparing Species Lists for Pairs of Habitats

	Riparian	Roadsides	Ravine	Disturbed	Steppe
No. of species					
Natives	88	37	30	44	150
Exotics	35	37	1	15	0
Total	123	74	31	59	150
Similarity*					
Riparian	—	0.216	0.000	0.028	0.000
Roadsides	—	—	0.010	0.243	0.023
Ravine	—	—	—	0.000	0.000
Disturbed	—	—	—	—	0.222

*Similarity values were calculated using Jaccard's index (Legendre and Legendre 1983).
Species can occur in more than one habitat category.
Adapted from Kotanen et al. (1998).

portions of the landscape that account for the smallest proportion of the area contain the largest number of species. Forty-four percent of the species are found on the upland steppe. This number is not directly comparable with the analogous number from Hazlett (1998), because the habitat categories were not exclusive in the CPER data as they were in the total site data. The lack of high similarity between pairs of habitats in the CPER data emphasizes the important differences that occur in the distribution of species across the shortgrass steppe landscape (Table 5.3 [Kotanen et al., 1998]). None of the pairs of habitats shared more than 25% of their species, and several habitats (ravines and riparian, disturbed and ravine, steppe and riparian, and steppe and ravine) had zero overlap in their species lists.

The distribution of species among families is heavily skewed with a relatively small number of families having the largest number of species (Fig. 5.1) (Kotanen et al., 1998). Three families—Asteraceae, Poaceae, and Fabaceae—account for 46% of the vascular plant species at the CPER. They are also among the largest families in the region, with the Asteraceae having 378 species; the Poaceae, 265 species; and the Fabaceae, 143 species (Fig. 5.1 [Weber and Wittmann, 1996]). These three families represent 34% of the species in the region.

At the CPER, native grasses, sedges, and rushes, which are often lumped into the category graminoids, account for 49 of the total 286 native species. The majority of the remaining 237 species are herbaceous dicots. This means that plant species richness in the steppe, at any spatial scale, is controlled almost entirely by herbaceous dicots, not by graminoids. A very different situation occurs for standing crop and net primary production (NPP). For standing crop, the key contributors are graminoids, a single species of succulent (*O. polyacantha*), and four species of dwarf shrubs (Fig. 5.2). The 232 species of herbaceous dicots contribute substantially less than 10% of the aboveground standing crop. Net primary production is even more biased away from herbaceous dicots and toward graminoids (Lauenroth et al., chapter 12, this volume). Liang et al. (1989) sampled five sites in a single year at the CPER and reported that graminoids contributed an average of 65% of aboveground net primary production (ANPP; Fig. 5.3). Succulents contributed 13%, dwarf shrubs, 16%; and forbs (herbaceous dicots), 6%.

In the presence of the current dominant species, the group of plants that contributes the most to plant species richness is unrelated to NPP or any other quantitative indicator of importance. At small spatial scales (<3 m^2), species richness is most heavily influenced by the presence or absence of the dominant species: *B. gracilis* and *B. dactyloides* (Singh et al., 1996). At scales larger than 3 m^2, species richness is a function of the total number of species in the local area (Singh et al. (1996). Species richness increases as a nonlinear function of area sampled, but because species are not randomly distributed over the site, species area relationships will likely be different among habitats. The species area curves developed for a series of upland locations by Singh et al. (1996) suggested that one would have to sample approximately 4000 m^2 to encounter the 107 species Hazlett (1998) reported for the upland habitat.

Major Community Types

Similar to the rest of the shortgrass steppe region, the vegetation of the SGS LTER site is largely dominated by *B. gracilis* and *B. dactyloides,* although there is variability in their relative importance depending upon landscape position, soil texture, and land-use history. Upland sites are most frequently dominated by a *B. gracilis–B. dactyloides–O. polyacantha*–dwarf shrub community. The common dwarf shrubs are *Artemisia frigida, Chrysothamnus nauseosus, Eriogonum effusum, and G. sarothrae*, occurring singly or in a mixture. The average growing-season aboveground plant biomass at the CPER is composed of 48% C$_4$ grasses, 8% C$_3$ grasses, 9% forbs, 11% dwarf shrubs, and 24% *O. polyacantha* (Sims et al., 1978).

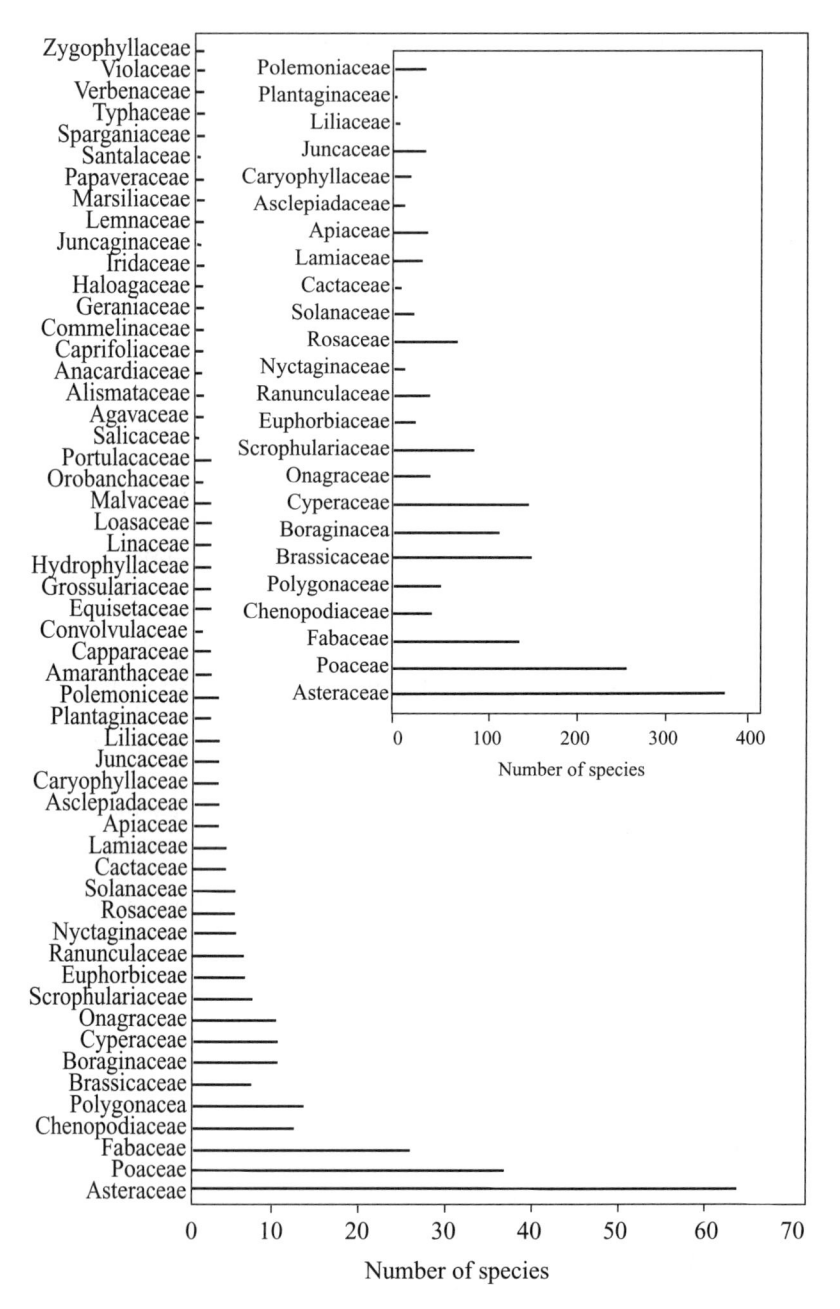

Figure 5.1 Distribution of plant species among families at the CPER and for the region (inset). (Regional data from Weber and Whittman [1996].)

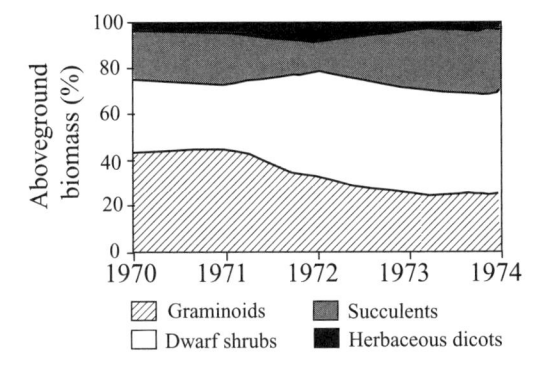

Figure 5.2 Partitioning of aboveground biomass among plant functional groups in the shortgrass steppe. (Data from Dodd and Lauenroth [1979].)

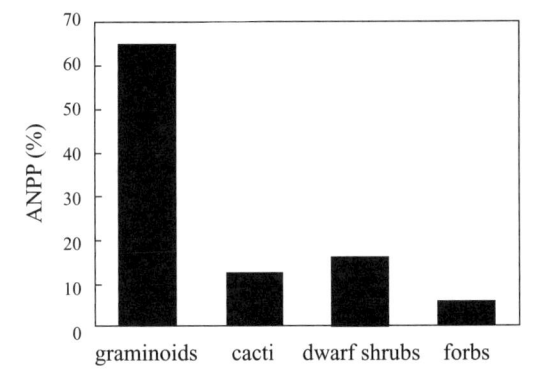

Figure 5.3 Distribution of aboveground net primary production (ANPP) among plant functional groups in the shortgrass steppe. (Data from Liang et al. [1989].)

Coffin et al. (1996) sampled 15 sites covering much of the variability present in the SGS LTER site. The relative basal cover of *B. gracilis* ranged from 25% to more than 80% (Fig. 5.4). Even on sites with low relative cover of *B. gracilis,* it was usually the dominant species, except for a single site in which *B. dactyloides* was dominant. Sites with low basal cover of *B. gracilis* tended to have high cover of *B. dactyloides,* so that the relative cover of total shortgrasses ranged from 46% to 87%. To represent the other species on the sites, Coffin et al. (1996) also sampled the density of individuals (Fig. 5.5). Basal cover tends to emphasize the importance of those species that spread over the soil surface, such as bunchgrasses, and underestimate the importance of single-stemmed grass, sedge, and herbaceous dicot species. Annuals were the dominant group in terms of relative density on seven of the sites, and C_3 graminoids, principally *A. smithii* and *Carex* spp., were the dominant group on the remaining eight sites. Perennial forbs accounted for as little as 3% to as much as 37% of the relative density on the sites.

Hyder et al. (1966) described the vegetation associated with different soil groups and topographic positions at the CPER (Table 5.4). Uplands, regardless of specific soil type, were dominated by *B. gracilis* and *O. polyacantha* with *Aristida*

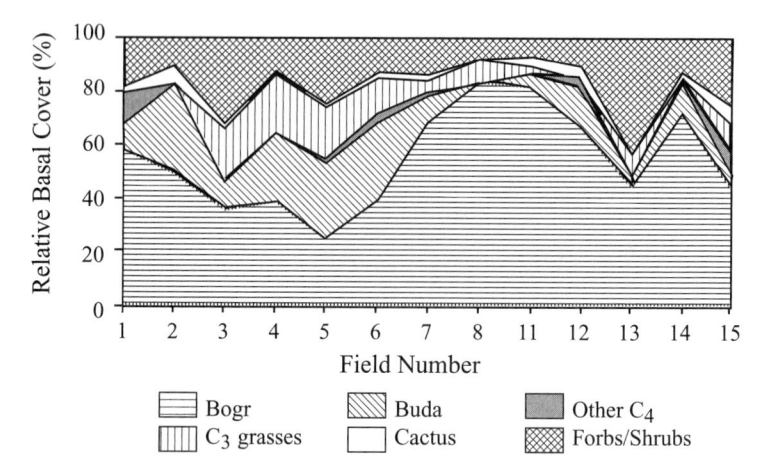

Figure 5.4 Relative basal cover for major species and functional groups for 15 locations in the shortgrass steppe. Bogr = *Bouteloua gracilis*; Buda = *Buchloe dactyloides*. (Data from Coffin et al. [1996].)

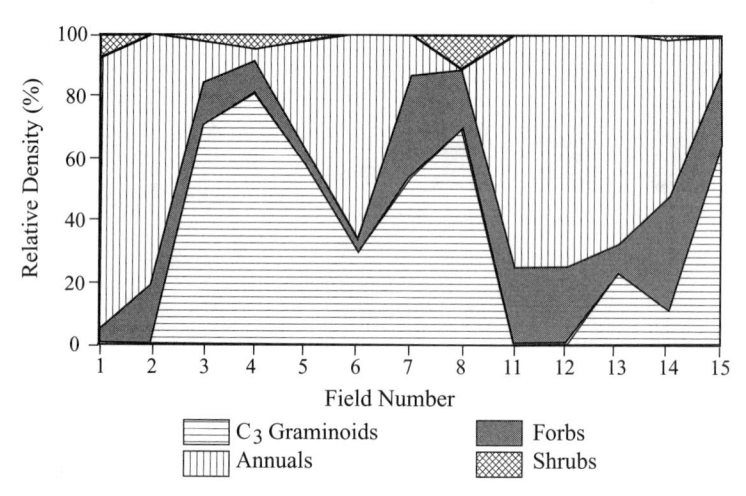

Figure 5.5 Relative density for major species and functional groups for 15 locations in the shortgrass steppe. (Data from Coffin et al. [1996].)

longiseta, A. smithii, C. heliophila, B. dactyloides, E. effusum, S. coccinea, and *Plantago patagonica* as important associated species. Nonsaline lowlands were dominated by either *B. gracilis, B. dactyloides*, or *A. smithii*, with *C. heliophila*, and *Iva laxiflora* as important associated species. *Opuntia polyacantha* was substantially less abundant in lowlands than uplands. Lowlands with saline–alkali soils were dominated by *A. smithii* and *Distichlis spicata*, with lesser amounts of *B. gracilis* and *S. airoides*.

Moir and Trlica (1976) identified six community types on the CPER based on a cluster analysis of canopy cover. The distinctive feature of all the communities

Table 5.4 Mean Frequency Percentages of 30 Plant Species by Soil Group and Landscape Position

Species	\multicolumn{7}{c}{Soil Groups}						
	1	2	3	4	5	6	7
Perennial graminoides							
Bouteloua gracilis	71	75	75	72	76	18	33
Aristida longiseta	27	15	16	6	1	1	3
Agropyron smithii	26	8	6	10	33	52	70
Carex heliophila	13	16	40	29	22	63	4
Stipa comata	9	6	1	1	1	1	3
Sporobolus cryptandrus	7	4	4	2	1	0	3
Buchloë dactyloides	3	7	7	36	10	10	3
Distichilis spicata	0	0	0	0	5	0	81
Sporobolus airoides	0	0	0	0	0	0	36
Shrubs and cacti							
Opuntia polyacantha	26	32	35	43	22	2	4
Eriogonum effusum	11	23	4	3	1	0	3
Atriplex canescens	4	1	1	0	7	1	6
Artemisia frigida	3	1	7	1	1	2	1
Ceratoides lanata	1	1	0	0	9	0	0
Perennial forbs							
Sphaeralcea coccinea	29	32	41	43	39	9	12
Bahia oppositifolia	12	1	1	4	4	3	2
Gaura coccinea	2	6	2	4	2	1	3
Lygodesmia juncea	1	5	1	1	3	0	4
Sophora sericea	1	1	1	1	7	6	0
Astragalus bisulcatus	1	1	1	0	0	7	0
Iva axillaris	0	0	0	1	1	38	11
Talinum parviflorum	0	0	0	0	0	1	13
Annuals							
Vulpia octoflora	9	31	65	53	2	11	40
Aster tanacetifolius	6	9	3	7	1	0	2
Lepidium densiflorum	4	7	4	3	1	1	6
Plantago patagonica	2	10	22	26	1	3	12
Gilia laxiflora	2	4	6	2	0	1	1
Chenopodium leptophyllum	1	1	1	1	2	1	5
Cryptantha minima	1	3	4	1	1	0	5
Ambrosia tomentosa	0	0	0	0	0	12	0

Soil groups 1 through 4 are sandy loams, 5 and 6 are clay loams, and 7 is a saline–alkali soil.
After Hyder et al. (1966).

except one was the high cover and constancy of *B. gracilis*. They sampled 102 transects, and for 84 of them *B. gracilis* was the most important species. On only two transects, *B. gracilis* was not an important species in the community. The most extensive community was the *B. gracilis–C. eleocharis* type. Next in extent was the *B. gracilis–A. frigida* community, followed by the *B. gracilis–A. longiseta* type. Three minor communities were identified: a *C. eleocharis–B. gracilis* type, a *B. dactyloides–B. gracilis* type, and one named for *A. smithii*, but actually codominated by *A. smithii, B. dactyloides,* and *C. eleocharis*. Thirty-year

80 Ecology of the Shortgrass Steppe

exclosures were dominated by either the *C. eleocharis–B. gracilis* type or the *B. gracilis–C. eleocharis* type. Lightly and moderately grazed pastures consisted of a mixture of the *B. gracilis–A. longiseta* type and the *B. gracilis–C. eleocharis* type. Community types reported for a heavily grazed pasture included *B. gracilis–C. eleocharis*, *C. eleocharis–B. gracilis*, and *B. dactyloides–B. gracilis*.

Cover Types

A generalization that holds for the SGS LTER site, as well as for the entire shortgrass steppe region, is that *B. gracilis* is the dominant species across all soil types, except those with extreme characteristics such as high sodium content or high clay content. These extreme conditions account for only a very small fraction of the area of the SGS LTER site or of the shortgrass steppe. Although *B. gracilis* is the dominant plant species in shortgrass ecosystems, as much as 30% to 70% of the soil surface characteristically remains unvegetated, either as areas of bare ground or litter. The relationship between the vegetated and unvegetated portions of the soil surface varies over both space and time. In the unvegetated portion, the balance between litter and bare ground is heavily influenced by livestock grazing (Milchunas et al., 1988; Milchunas et al., chapter 16, this volume).

Coffin et al. (1996) reported bare ground and litter cover in addition to plant cover on 15 grazed locations across the SGS LTER site (Fig. 5.6). Bare ground (including rocks) accounted for a minimum of 63% and a maximum of 81% of the surface cover of these sites. Litter on these grazed locations contributed 3% to 7% of total cover. Not surprisingly, the covers of *B. gracilis* and bare ground

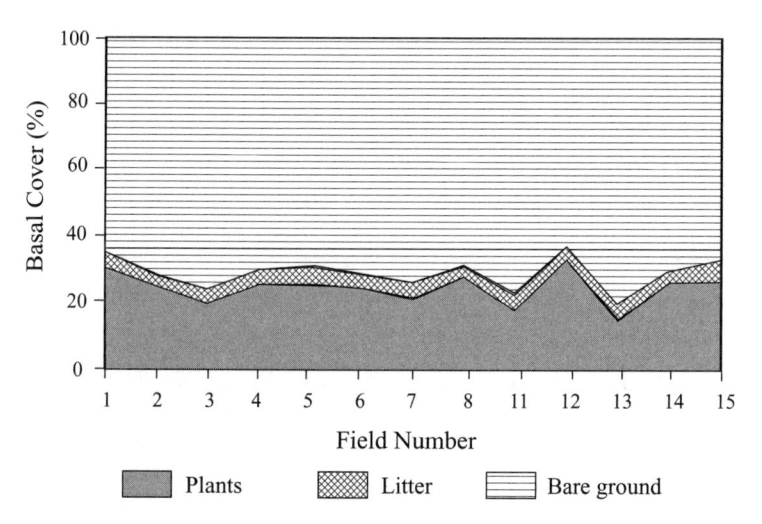

Figure 5.6 Cover of plants, litter, and bare ground for 15 locations in the shortgrass steppe. (Data from Coffin et al. [1996].)

were negatively correlated ($r = -0.64$, $P < .05$). Sites with the highest cover of *B. gracilis* had the lowest cover of bare ground. Spatial variability in plant, litter, and bare ground cover is controlled by plant demographics, location characteristics, and management. In most cases, appropriate data are not available to explore these relationships. Soil texture, which influences water and nutrient balance, was available for the locations sampled by Coffin et al. (1996). Sand and clay content explained only 17% of the variability in bare ground cover among sites. However, within a single catena site at the CPER, sand and clay (averaged over the depth of the soil) explained 36% of the variability in bare ground plus rock cover.

Temporal variability in surface cover, in the absence of changes in management, is largely caused by fluctuation in weather. Seven years of data from an exclosure at the CPER indicated fluctuations of up to 16% in plant cover, 18% in litter, and 7% in bare ground (Fig. 5.7). Comparison of these data with those in Figure 5.6 point out how important livestock grazing can be in influencing the cover of litter. At the locations studied by Coffin et al. (1996), litter cover averaged 4%, whereas in the CPER exclosure it averaged 42%. Regression of cover categories against precipitation suggests a relationship with the previous year's precipitation (Fig. 5.8). Litter cover was negatively related to the previous year's precipitation, and plant cover was positively related. The simple explanation of these relationships is that dry years produce litter by accelerating the turnover of plant parts, and wet years increase the basal area of plants by promoting tillering.

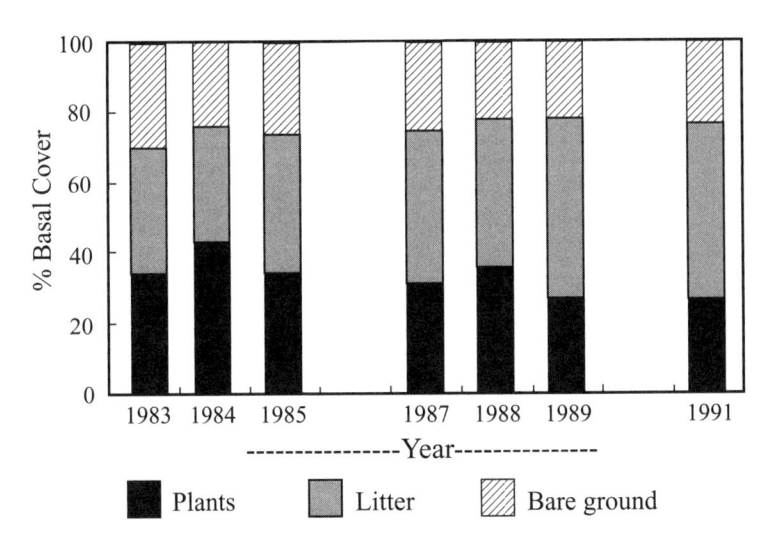

Figure 5.7 Seven years of cover data for plants, litter, and bare ground from an ungrazed location at the Central Plains Experimental Range.

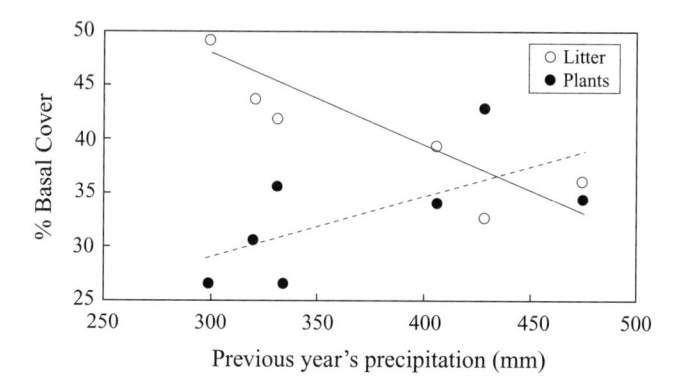

Figure 5.8 Relationship between cover of litter and plants, and previous year's precipitation for an ungrazed location at the Central Plains Experimental Range. Plant cover = 11.9 + 0.06 × PrevPrecip; $r^2 = 0.42$. Litter cover = 73.5 − 0.09 × PrevPrecip; $r^2 = 0.69$.

References

Beavis, W. D., J. C. Owens, J. A. Ludwig, and E. W. Huddleston. 1982. Grassland communities of east–central New Mexico and density of range caterpillar *Hemileuca olivea*. *Southwestern Naturalist* **27**:335–343.

Betz, D. 2001. Dynamics of exotic species in the Pawnee National Grasslands, CO, USA. Masters thesis, Colorado State University, Fort Collins.

Coffin, D. P., W. K. Lauenroth, and I. C. Burke. 1996. Recovery of vegetation in a semi-arid grassland 53 years after disturbance. *Ecological Applications* **6**:538–555.

Dahl, B. E. 1995. Developmental morphology of plants, pp. 22–58. In: D. J. Bedunah and R. E. Sosebee (eds.), *Wildland plants: Physiological ecology and developmental morphology*. Society for Range Management, Denver, Colo.

Dahl, B. E., and D. N. Hyder. 1977. Developmental morphology and management implications, pp. 257–290. In: R. E. Sosebee (ed.), *Rangeland plant physiology*. Society for Range Management, Denver, colo.

Dodd, J. L., and W. K. Lauenroth. 1979. Analysis of the response of a grassland ecosystem to stress, pp.43–59. In N. R. French (ed.), *Perspectives in grassland ecology*. Ecology Studies, vol. 32. Springer-Verlag, New York.

Dornbusch, A. J., Jr., B. M. Vining, and J. L. Kearney. 1995. Total resource management plan for addressing groundwater concerns, pp. 231–251. In: S. R. Johnson and A. Bouzaher (eds.), *Conservation of Great Plains ecosystems*. Kluwer Academic Publishers, Amsterdam.

Hazlett, D. L. 1998. *Vascular plant species of the Pawnee National Grassland*. USDA general technical report RMRS-GTR-17. Rocky Mountain Research Station, Fort Collins, Colo.

Heerwagen, A., and A. R. Aandahl. 1961. Utility of soil classification units in characterizing native grassland plant communities in the southern Great Plains. *Journal of Range Management* **14**:207–213.

Huntzinger, T. L. 1995. Surface water: A critical resource of the Great Plains, pp. 253–273. In: S. R. Johnson and A. Bouzaher (eds.), *Conservation of Great Plains ecosystems*. Kluwer Academic Publishers, Amsterdam.

Hyder, D. N., R. E. Bement, E. E. Remenga, and C. Termilliger. 1966. Vegetation, soils and vegetation grazing relations from frequency data. *Journal of Range Management* **19**:11–17.

Kotanen, P. M., J. Bergelson, and D. L. Hazlett. 1998. Habitats of native and exotic plants in Colorado shortgrass steppe: A comparative approach. *Canadian Journal of Botany* **76**:664–672.

Küchler, A. W. 1964. *Potential natural vegetation of the conterminous United States.* American Geographical Society, New York.

Lauenroth, W. K., I. C. Burke, and M. P. Gutmann. 1999. The structure and function of ecosystems in the central North American grassland region. *Great Plains Research* **9**:223–259.

Lauenroth, W. K., and D. G. Milchunas. 1992. Shortgrass steppe, pp. 183–226. In: R. T. Coupland (ed.), *Natural grasslands: Introduction and western hemisphere.* Elsevier Scientific, Amsterdam.

Liang, Y., D. L. Hazlett, and W. K. Lauenroth. 1989. Water use efficiency of five plant communities in the shortgrass steppe. *Oecologia (Berl.)* **80**:148–153.

Lotspeich, F. B., and M. E. Everhart. 1962. Climate and vegetation as soil forming factors on the Llano Estacado. *Journal of Range Management* **15**:134–141.

McGinnies, W. J., W. G. Hassel, and C. H. Wasser. 1983. *A summary of range seeding trials in Colorado.* Colorado State University Experiment Station special series 21. Cooperative Extension Service, Fort Collins.

Milchunas, D. G., O. E. Sala, and W. K. Lauenroth. 1988. A generalized model of the effects of grazing by large herbivores on grassland community structure. *American Naturalist* **132**:87–106.

Moir, W. H, and M. J. Trlica. 1976. Plant communities and vegetation pattern as affected by various treatments in shortgrass prairies of northeastern Colorado. *Southwestern Naturalist* **21**:359–371.

Pettit, R. D. 1974. *Herbage dynamics studies at the Pantex site, 1972.* Technical paper 250. U.S. IBP Grassland Biome, Colorado State University, Fort Collins.

Rogers, C. M. 1953. The vegetation of the Mesa de Mayo region of Colorado, New Mexico and Oklahoma. *Lloydia* **16**:257–290.

Shantz, H. L., and R. Zon. 1924. Natural vegetation, pp. 3–28. In: O. E. Baker (ed.), *Atlas of American agriculture, section E.* U.S. Government Printing Office, Washington, D.C.

Sims, P. L., J. S. Singh, and W. K. Lauenroth. 1978. The structure and function of ten western North American grasslands. I. Abiotic and vegetational characteristics. *Journal of Ecology* **66**:251–285.

Singh, J. S., P. Bourgeron, and W. K. Lauenroth. 1996. Plant species richness and species–area relations in a shortgrass steppe. *Journal of Vegetation Science* **7**:645–650.

Weaver, J. E., and F. E. Clements. 1938. *Plant ecology.* McGraw-Hill, New York.

Weber, W. A., and R. C. Whittmann. 1996. *Colorado flora: Eastern slope.* University Press of Colorado, Niwot.

6

The Role of Disturbances in Shortgrass Steppe Community and Ecosystem Dynamics

Debra. P. C. Peters
William K. Lauenroth
Ingrid C. Burke

The disturbance regime of an ecosystem consists of a number of different types of disturbance agents operating over a range of spatial and temporal scales (Pickett and White, 1985). Each type of disturbance has its own set of characteristics, including size, frequency of occurrence, intensity, and attributes associated with location, including soil texture, topographic position, and grazing intensity by cattle. These characteristics result in different short-term localized effects on ecosystems as well as long-term broad-scale effects as the disturbances accumulate through time. Disturbance effects occur at multiple levels of organization, from individuals to populations, communities, and the ecosystem. Effects of disturbance can also vary for different types of organisms or processes associated with plants, animals, and soils. Understanding interactions among the characteristics of a disturbance and the properties associated with the response variable is key to understanding and predicting recovery patterns through time and space.

Although successional studies have been conducted in grasslands for more than a century, our understanding of the roles of different kinds of disturbances in generating and maintaining patterns in vegetation and in determining species dominance in shortgrass ecosystems has developed only since the 1980s. Referred to as gap dynamics, our current view of the role of disturbance is a dynamic one, in which the recovery of vegetation depends upon interactions among disturbance characteristics and the life history traits of plants. This gap dynamics conceptualization provides an alternative view of vegetation dynamics compared with traditional successional models based on Clements (1916, 1928). Much of the recent work on disturbances in the shortgrass steppe focuses on the relationships between disturbance characteristics and plant life history traits to test the different Clementsian-based models.

Old Field Models of Succession

Most successional studies of shortgrass communities prior to the 1980s focused on recovery after large-scale disturbances and, in particular, cultivation and subsequent abandonment of agricultural fields (Costello, 1944; Judd, 1974; Judd and Jackson, 1939; Savage and Runyon, 1937). The earliest studies were based upon a Clementsian model (Clements, 1916, 1928) that formed the traditional view of succession in these communities (Fig. 6.1). This model predicted that shortgrasses would dominate cover within 25 to 50 years after abandonment. More recent studies conducted during the mid 1900s observed a delayed recovery by shortgrass

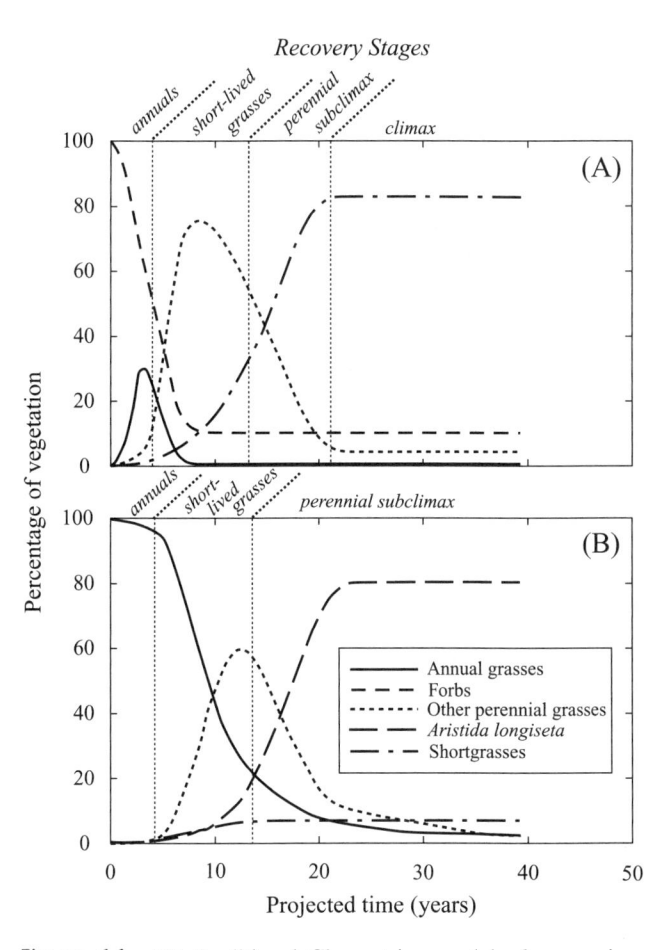

Figure 6.1 (A) Traditional Clementsian model of succession. (Adapted from Judd and Jackson [1939].) (B) Conventional model modified for eastern Colorado. (Adapted from Costello [1944], Hyder and Everson [1968], Hyder et al. [1971], and Laycock [1989, 1991].)

species. An important conclusion from these studies was that the dominant species in shortgrass communities, *Bouteloua gracilis*, either recovers very slowly or is unable to recover after disturbance (Hyder et al., 1971). Characteristics of *B. gracilis* that contribute to its slow recovery include climatic constraints on seed germination and seedling establishment (Briske and Wilson, 1977, 1978; Hyder et al., 1971; Riegel, 1941), low and variable seed production (Coffin and Lauenroth, 1992), lack of seed storage in the soil (Coffin and Lauenroth, 1989c), and very slow tillering rates (Samuel, 1985). These constraints were thought to be especially important in areas such as northeastern Colorado, which receives less than 380 mm annual precipitation (Laycock, 1989, 1991). Based upon observations of plant recovery in northeastern Colorado, a modified old field model was proposed, in which communities dominated by the perennial bunchgrass *Aristida purpurea* are maintained indefinitely (Costello, 1944; Hyder and Everson, 1968; Hyder et al., 1971) and form an alternative state of the system (Laycock, 1989, 1991). These studies provide the basis for the conventional view of succession in these communities (Fig. 6.1).

Gap Dynamics Model of Succession

In the 1980s we proposed an alternative view of the role of disturbance in shortgrass communities that focuses on the dynamics of individual plants and interactions between disturbance characteristics and plant life history traits to explain successional patterns (Coffin and Lauenroth, 1988, 1990a; Lauenroth and Coffin, 1992). Our gap dynamics approach is a Gleasonian view of communities that emphasizes plant population processes (Gleason, 1926), and the inherent variability associated with different types of disturbances. Watt (1947) conceptualized plant communities as a mosaic of patches, each in one of four stages of development: pioneer (gap), building, mature, and degenerate. Most species become established during the gap phase, which is created by the death of a full-size individual of the dominant species in the mature phase. Plant death results in an increase in availability of above- and belowground resources that can support the establishment and growth of new plants. The long-term result of gap dynamics processes is a landscape composed of a shifting mosaic of patches, each undergoing its own successional dynamics (Bormann and Likens, 1979).

Our gap dynamics approach has been expanded from Watt's (1947) description to incorporate a patch dynamics conceptualization that emphasizes patches scaled to the size of a disturbance (Pickett and White, 1985). Disturbances that kill one individual of the dominant species produce a gap in resource space as well as a patch in community structure. As disturbance size increases beyond the size of one plant, the effect of killing one individual is overwhelmed by changes resulting from disturbance size, such as microclimatic variation across the patch, seed availability and dispersal constraints, and incomplete mortality of individuals. Furthermore, recovery through time is affected by spatial variation across the disturbed plot as well as by interactions among plants. Although disturbances that only kill parts of a plant (i.e., tillers) can also produce patches, they do not produce gaps. These small areas release insufficient resources to support a full-size individual of the dominant species. Thus, recovery occurs primarily through vegetative spread by remaining tillers, rather than through successional dynamics and the replacement of one species by another through time.

A gap dynamics conceptualization of shortgrass communities focuses on gaps in resource space produced by the death of individual *B. gracilis* plants. A plant is defined for this purpose as all tillers currently connected by a crown (Coffin and Lauenroth, 1988). Because soil water is the most frequently limiting resource for plant growth and community structure in semiarid grasslands (Lauenroth et al., 1978; Noy-Meir, 1973), we hypothesized that death of a full-size *B. gracilis* plant results in a gap in belowground resource space that initiates successional dynamics (Coffin and Lauenroth, 1990a). This hypothesis has been supported by both field experiments and simulation modeling analyses. Mortality of established *B. gracilis* plants locally increases soil water content and is required for seedling regeneration to occur (Aguilera and Lauenroth, 1995). Intensity of interference by *B. gracilis* roots is high in openings less than 30 cm across, and results in reduced recruitment of *B. gracilis* seedlings as a result of competition from neighbors (Aguilera and Lauenroth, 1993b). This effect declines dramatically as opening size increases to more than 50 cm (Hook et al., 1994). Furthermore, both seedling establishment and vegetative colonization are enhanced in 50-cm-diameter areas (Aguilera, 1992; Coffin and Lauenroth, 1989b); the status of 30- to 50-cm openings is uncertain. The presence of a large number of small *B. gracilis* plants in neighborhoods with a high proportion of bare soil supports the gap dynamics view of the shortgrass community being composed of patches, each undergoing its own successional dynamics (Aguilera and Lauenroth, 1993a).

Many openings caused by disturbance are similar in size to full-size *B. gracilis* plants (5–20 cm in diameter [Hook et al., 1994]), a size range in which interference from neighboring plants changes from strong to weak (Aguilera and Lauenroth, 1993b). Most of the area in shortgrass ecosystems (>99.5%) is within the range of the root system of *B. gracilis* (Hook and Lauenroth, 1994). Because most openings large enough for regeneration are explored by roots of *B. gracilis*, gap dynamics in the shortgrass steppe involve constraints on water use by *B. gracilis* root systems. Large openings (>50 cm across), which are known to enhance regeneration by *B. gracilis,* have low root density (Hook et al., 1994). Such openings occupy 2% of the area, and nearly all are caused by disturbance. Because large openings are rare, variation in belowground competition in the abundant, small openings may be most important for regeneration (Hook et al., 1994).

We developed a simulation model based on a gap dynamics approach to represent the dynamics of shortgrass steppe communities (Coffin and Lauenroth, 1990a). We used this model, STEPPE, to evaluate the effects of different environmental conditions (climate, soil texture) on vegetation dynamics for a range of disturbance sizes (Coffin and Lauenroth, 1989a, 1994; Coffin et al., 1993). Our model results have also been used to identify key processes limiting recovery after disturbance and to generate testable hypotheses concerning these key processes (Coffin and Lauenroth, 1992; Coffin et al., 1996). Our use of this model to simulate the effects of different disturbance types on shortgrass steppe dynamics is described more fully in Peters and Lauenroth (chapter 7, this volume).

Disturbance Characteristics and Plant Recovery Dynamics

Our conceptual model focuses on the size (small, large) and intensity of the disturbance (Fig. 6.2). Small disturbances (<1 m²) occur naturally as a result of the activities of mammals and arthropods. These disturbances may or may not kill individual plants. Large disturbances (>1 m²) include both natural and human causes, such as cultivation. Intensity has two dimensions: whether *B. gracilis* plants are killed and whether the soil structure is altered. Because of the importance of *B. gracilis* plants to shortgrass communities, and the relevance of plant size to processes associated with gap dynamics, we focus on disturbances sufficiently large to kill *B. gracilis* plants (Fig. 6.3A). In general, the relationship between size and frequency of occurrence is inverse; thus, the smallest disturbances can potentially affect the largest amounts of area as their effects accumulate through time (Fig. 6.3B). Each disturbance may be grouped into one of three sizes: small disturbances most frequently kill individual *B. gracilis* plants, intermediate-size disturbances always kill a number of *B. gracilis* plants and produce patches within a topographic position or pasture, and large disturbances always kill a large number of *B. gracilis* plants and can affect entire landscapes within a topographic unit or pasture. Each disturbance type may be described in terms of its most important characteristics, including size, frequency of occurrence, landscape location, its effects on vegetation, and recovery patterns through time (with a focus on vegetation).

Small-Scale Disturbances

There are four types of small-scale disturbances (<25 cm^{-2} m in diameter) that affect shortgrass ecosystems: fecal pats deposited by cattle, harvester ant nest

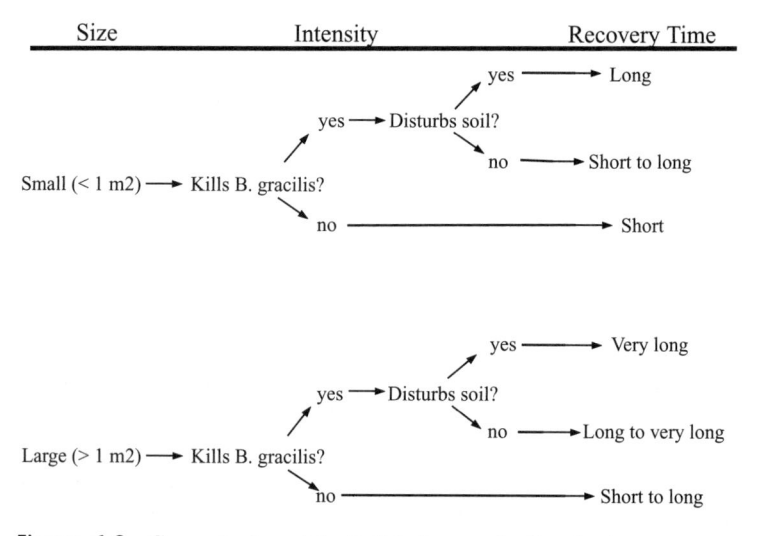

Figure 6.2 Conceptual model of disturbances in the shortgrass steppe that includes categories for size and intensity of a disturbance.

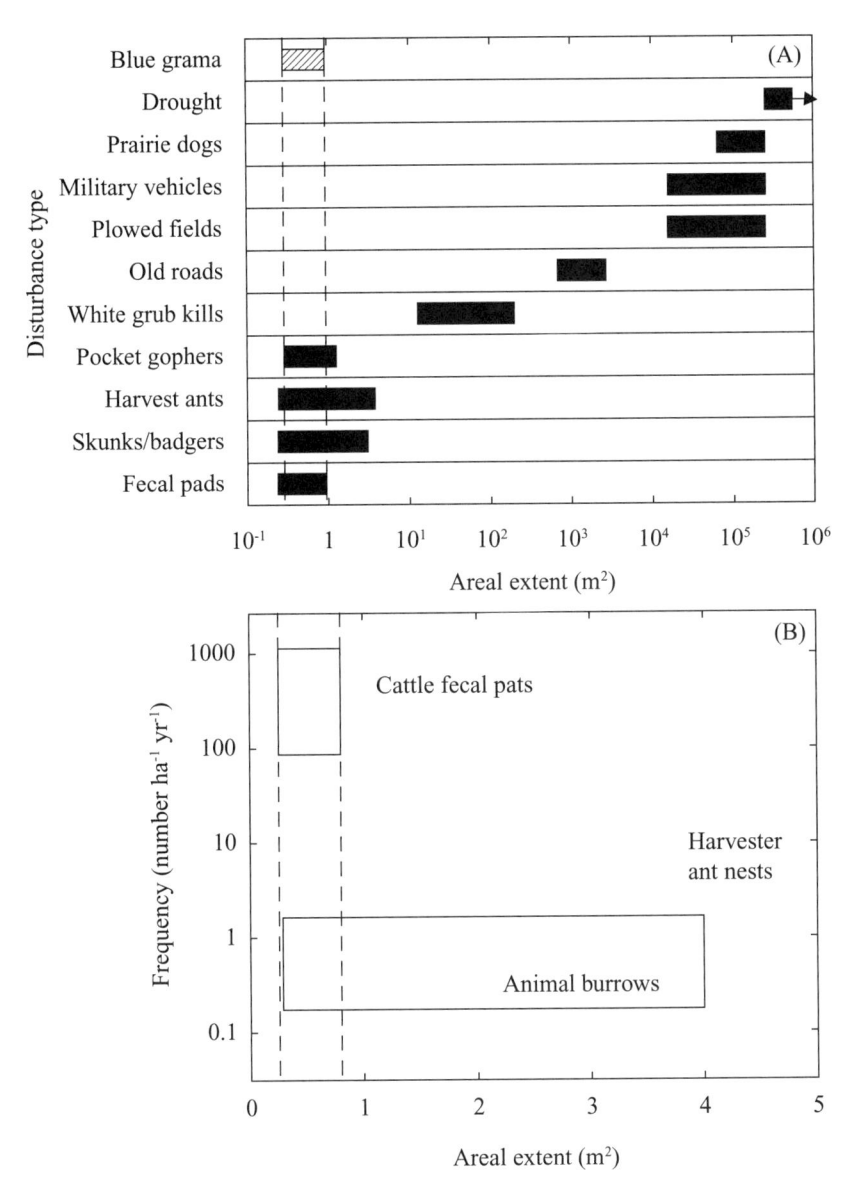

Figure 6.3 Disturbance regime of shortgrass steppe ecosystems showing size range of each disturbance type (A), and size and frequency of occurrence for the three smallest disturbance types (B).

sites, burrows produced by skunks and badgers, and burrows produced by pocket gophers (Fig. 6.3A). Each has its characteristic size and frequency of occurrence as well as different initial effects on plants, with resulting differences in plant recovery. Each of these disturbance types overlaps the size distribution of *B. gracilis* plants and has the potential to kill an entire plant of this species. Although animals, such as thirteen-lined ground squirrels, can also reduce plant biomass through their digging activities; their holes are not sufficiently large to affect entire *B. gracilis* plants and result in the initiation of gap dynamics.

Cattle Fecal Pats

Fecal pats from cattle are an important disturbance to shortgrass plant communities in large part because of their effects on *B. gracilis* plants. The low stature of *B. gracilis* plants ensures that the portion of a plant beneath a pat is unlikely to survive (tiller mortality), yet the entire plant must be covered by a pat before mortality occurs (Fair et al., 2001). The average size of fecal pats overlaps the size of *B. gracilis* plants (Fig. 6.4 [Coffin and Lauenroth, 1988]). The average sizes of

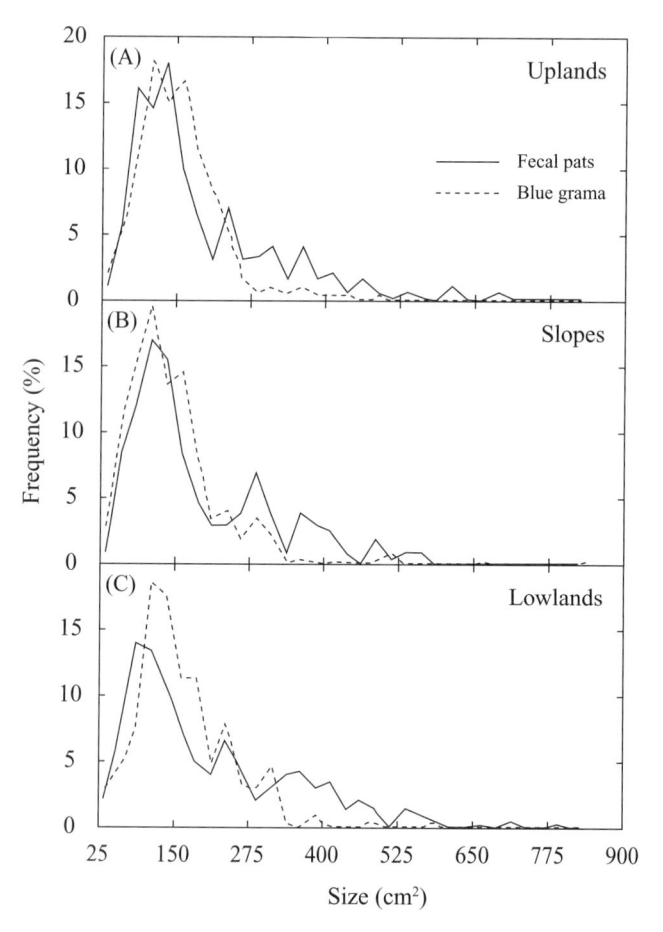

Figure 6.4 Frequency distribution of sizes (in square centimeters) of *B. gracilis* and fecal pats from cattle for three topographic positions: uplands (A), slopes (B), and lowlands (C). (Adapted from Coffin and Lauenroth [1988].)

pats are also related to topographic position: 134 cm², uplands; 167 cm², slopes; and 190 cm², lowlands. Estimated frequencies of occurrence of fecal pats increase with grazing intensity within each topographic position, and increase from uplands to slopes to lowlands within each grazing treatment (Fig. 6.5A [Coffin and Lauenroth, 1988]), because cattle tend to spend more time in lowlands than in other topographic positions (Schwartz, 1977; Senft, 1983). We used the frequencies of occurrence of pats to simulate their effects on *B. gracilis* relative to other small disturbances. Based on these simulation model analyses, we found that pats affect a larger amount of plant cover than do larger but less frequent disturbances

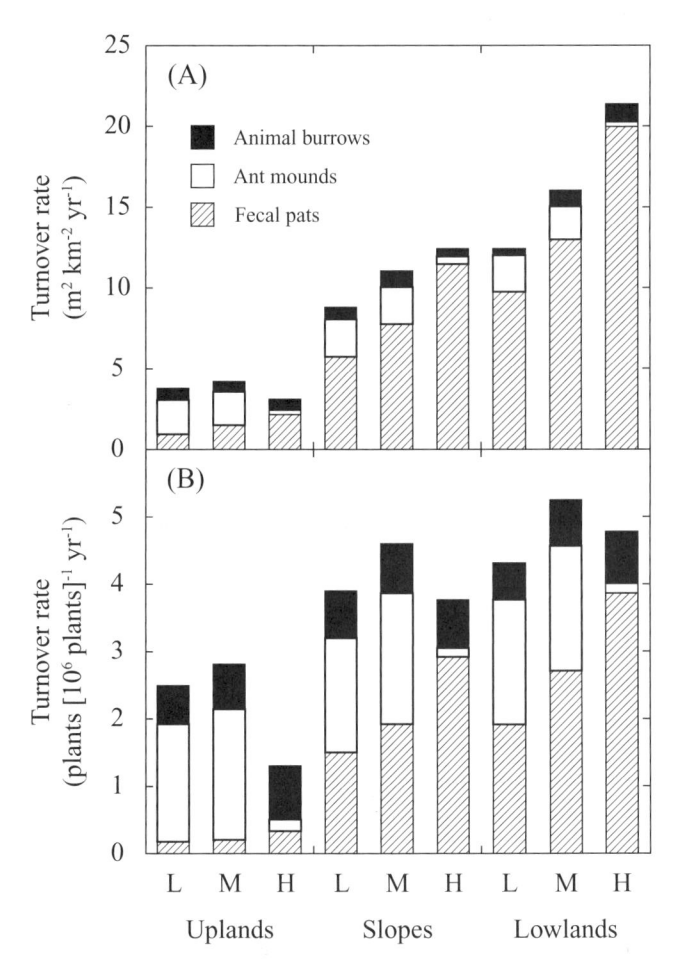

Figure 6.5 Turnover rates of plant cover (A) and plants (B) as a result of three types of disturbances for nine locations based on topographic position and cattle grazing intensity (L, light grazing; M, moderate grazing; H, heavy grazing). (Redrawn from Coffin and Lauenroth [1988].)

such as ant nest sites or animal burrows (Coffin and Lauenroth, 1988). However, the turnover rate of plants resulting from pats is lower than that because of ant nests or animal burrows for most grazing intensity–topographic positions, with the exception of heavily grazed pastures (Fig. 6.5B [Coffin and Lauenroth, 1988]). Because pats are similar in size to *B. gracilis* plants but must cover the entire plant before death occurs, each deposition event always reduces cover by killing at least some tillers, but does not always kill the plant (Fair et al., 2001).

After pats are deposited, there is a time delay before plant recovery can begin; the length of this period is associated with the decomposition of pats. Although decomposition begins immediately, most pats (90%) require 2 years to decompose fully (Fig. 6.6A). We found that the recovery pattern of plants after pat decomposition depends on the species. Within 2 years after deposition of pats, *B. gracilis*

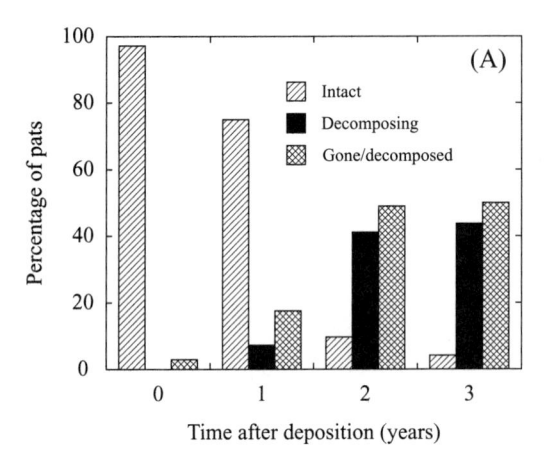

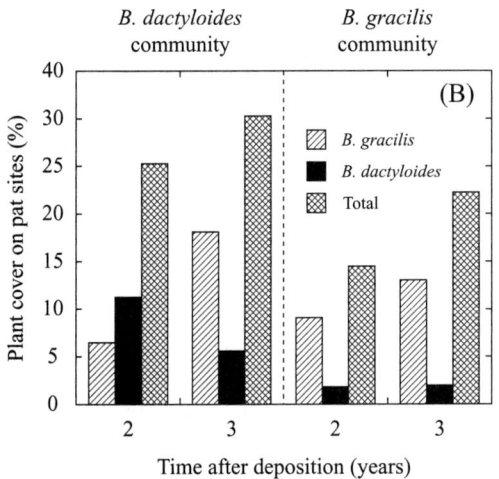

Figure 6.6 Percentage of cattle fecal pats in one of three states of decomposition for time (years) after deposition (A), and plant communities dominated by either *B. gracilis* or *B. dactyloides* (B).

can dominate cover on pats (Fig. 6.6B). In other situations, *Buchloë dactyloides* recovers rapidly on areas under pats by year 2, but then decreases as the cover of *B. gracilis* increases. Patterns in plant recovery are due both to vegetative spread by surviving tillers and adjacent plants, and to seedling establishment. Because cattle often consume seed heads of grasses, and many seeds survive passage through their digestive tracts, we found viable seeds of many grass species in fecal pats (Fraleigh, 1999). If the opening created by a pat is sufficiently large to create a resource gap, these seeds can germinate and become established (Aguilera and Lauenroth, 1993b).

Ant Nest Sites

Western harvester ants (*Pogonomyrex occidentalis*) produce disturbances by clearing vegetation from areas around their nests as well as by foraging for seeds and collecting them in underground chambers. Nest sites consist of a cone-shaped mound (0–0.2 m high, 0–0.7 m wide at the base) over the nest, with the mound being surrounded by a circular bare area called the disk (0.2–1.9 m in diameter; Fig. 6.7) (Lavigne, 1969; Rogers, 1972). Of 77 nest sites sampled in a lightly grazed pasture, we found that most covered about 2 m^2 in surface area, but ranged in size from 0.25 to 4 m^2 (Fig. 6.8 [Coffin and Lauenroth, 1990b]).

Nest density is related to grazing intensity. Based on counts of colonies in areas of 8 ha, Rogers and Lavigne (1974) reported 28 colonies/ha (light grazing), 31 colonies/ha (moderate), 3 colonies/ha (heavy), and 23 colonies/ha (ungrazed) whereas the proportion of soil surface area affected by ant nests ranged from 0.02% to 0.3%. In a more recent study, a similar pattern in density values with grazing intensity was found. Based on counts in areas of 65 to 146 ha, densities were 15.5 colonies/ha (light grazing), 9 colonies/ha (moderate), and 5.8 colonies/ha (heavy) (Crist and Wiens, 1996). These densities were based on ranges of values that also varied with grazing intensity: 1 to 36 mounds/ha, light; 0 to 50 mounds/ha, moderate; and 0 to 13 mounds/ha, heavy. Variation in the density of mounds

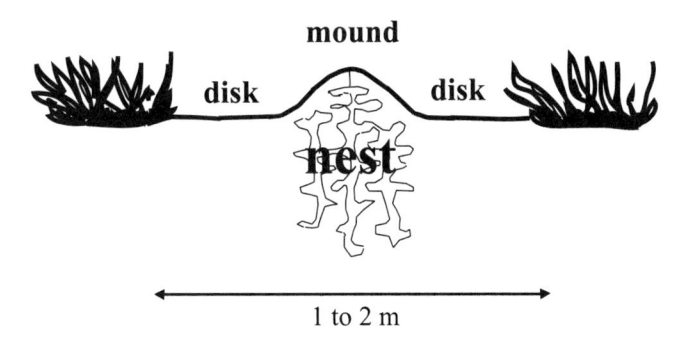

Figure 6.7 Diagram of Western harvester ant nest site showing the disk, mound, and location of nest chambers. See Figure 10.2, this volume, for a photo.

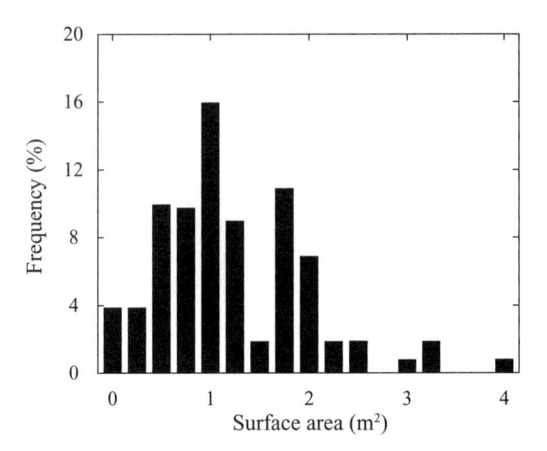

Figure 6.8 Frequency distribution of sizes of nest sites of Western harvester ants. New, immature, full-size, and old nest sites were combined. (From Coffin and Lauenroth [1990b].)

in this study was related to topography and soils as well as to grazing intensity. Higher densities of mounds were found on level uplands with deep, well-drained sandy loam soils. Lower densities occurred on ridges and slopes with soils that were shallow and/or had high clay content. However, in heavily grazed areas, densities were found to be uniform regardless of variations in soils.

Our results (Coffin and Lauenroth, 1989b) showed that plant species composition on ant nest sites 1 year after recovery began included high density and cover of perennial graminoids and forbs. This suggests that rhizomes (e.g., *Carex heliophila*) and tap roots (e.g., *Sphaeralcea coccinea*) are not killed by ants (Fig. 6.9). Although no *B. gracilis* seeds were found in the mound or disk, *B. gracilis* plants were found in small amounts. We also found that abandoned ant nest sites had a high density of annuals (Fig. 6.9B [Coffin and Lauenroth, 1990b]), perhaps germinated from seeds previously collected and stored the by ants, and increased growth of plants surrounding nest sites (Coffin and Lauenroth, 1990b; Rogers, 1974; Rogers and Lavigne, 1974). The average number of seeds capable of being germinated found near the soil surface of nest sites (6756 seeds/m^2 [Coffin and Lauenroth. 1990b]) was significantly larger than the number of seeds at a coarse-textured control site (2748 seeds/m^2) sampled at the same time (Fig. 6.10 [Coffin and Lauenroth, 1989c]). Within a nest site, densities of annuals and seeds of annuals were significantly greater on mounds than disks, whereas densities of perennial plants and seeds were similar for the two nest site locations. Differences in soil moisture and nutrient concentrations between the mound and disk areas of nest sites may affect these plant dynamics (Rogers and Lavigne, 1974). Because ant colony densities are spatially variable, their effects on the seed pool through selective granivory are also spatially heterogeneous (Crist and Wiens, 1994).

We also found that the aboveground biomass of perennial graminoids and forbs as well as *B. gracilis* was significantly greater in the 15-cm-wide ring surrounding

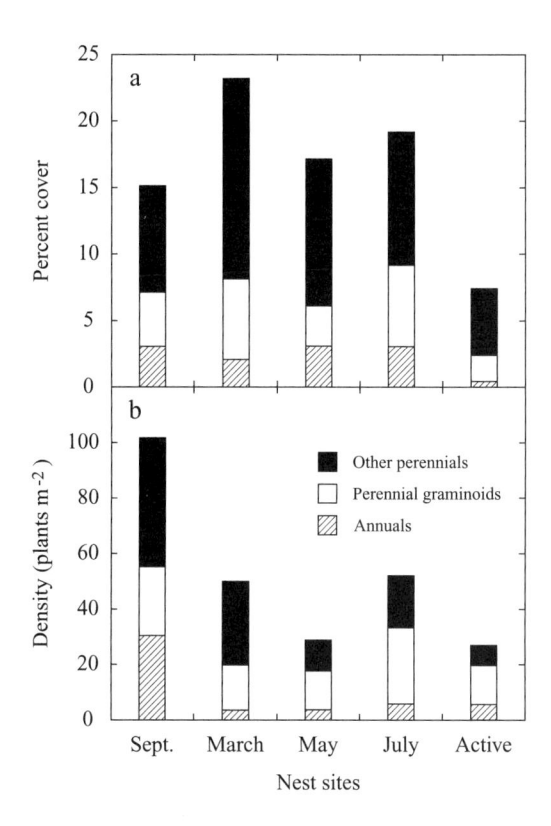

Figure 6.9 Cover (A) and density (B) of annuals, perennial graminoids, and other perennials for ant nest sites abandoned on one of four dates, and active nest sites. (Adapted from Coffin and Lauenroth [1990b].)

the disk than that found in undisturbed vegetation (Coffin and Lauenroth, 1990b). Because ants clip vegetation on disk areas, this increase in plant growth is likely the result of higher water availability in the areas immediately surrounding disks. Kelly et al. (1996) found a 91% reduction in total root biomass 30 cm from the edge of a mound. Because a *P. occidentalis* nest site may exist for 30 to 60 years (Coffin and Lauenroth, 1990b; Keeler, 1993), and because successional dynamics are not initiated until after a nest site is abandoned (Clark and Comanor, 1975), these nest sites have important long-term effects on shortgrass plant communities. Age of nest site at time of abandonment also affects successional dynamics because nutrient availability may vary among nest sites of different ages. Carbon and nitrogen mineralization, microbial biomass nitrogen, and root biomass are significantly less on mature and old nest sites compared with immature nest sites (Kelly et al., 1996). Most loss of soil organic matter is the result of plant removal by ants during the first 5 to 10 years after nest establishment (Kelly et al., 1996). Furthermore, we observed that new nest sites occur within areas dominated by *B. gracilis* plants. The small average size of these plants (131 cm^2 [Coffin and Lauenroth, 1988]) suggests that at least one plant is killed before the nest site is

Figure 6.10 Density of seeds found in the soil for two microsites associated with Western harvester ant nest sites and the reference vegetation. (Adapted from Coffin and Lauenroth [1989c, 1990b].)

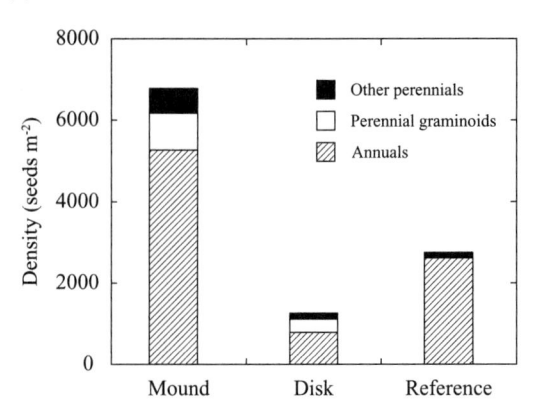

abandoned. Thus, estimates of the amount of cover killed each year by harvester ants (0.03–0.07 m^2·ha^{-1}; Fig. 6.5 [Coffin and Lauenroth, 1988]) are small.

Burrows of Badgers and Skunks

Burrows from badgers and skunks have not been as well studied as the effects of other small animals. Badgers and skunks produce small mounds of soil (1–2 m in diameter, 50 cm tall) that can cover and kill plants or parts of plants. Using simulation modeling, we estimated that, on average, badger and skunk burrows reduce cover of *B. gracilis* by 0.20 m^2·ha^{-1}·y^{-1} and kill 14 *B. gracilis* plants·ha^{-1}·y^{-1} (Coffin and Lauenroth, 1988). Recovery of vegetation occurs after the mound production activity ceases. We found that cover on recovering animal burrows is dominated by perennial graminoids, forbs, shrubs, and succulents after 1 year of recovery (Fig. 6.11 [Coffin and Lauenroth, 1989b]). High cover (6%) by the dominant plant species, *Bouteloua gracilis*, after 1 year is primarily the result of vegetative regrowth of this species even when it is partially covered by soil. In contrast, *B. gracilis* is not found in significant amounts on ant nest sites or on artificially disturbed plots. Our results suggest that recovery time for *B. gracilis* is less on badger and skunk burrows than for ant nest sites or artificial plots, as well as less than on large disturbed areas, such as old fields (Coffin et al., 1996).

Northern Pocket Gophers

Pocket gophers (*Thomomys talpoides attenuatus* Hall and Montague) create disturbance through their burrowing activities that result in mounds of soil that can cover and kill plants and/or plant parts. As part of their foraging behavior, pocket gophers consume stems and leaves of plants they have buried, and they forage on the surface near the entrance of burrows (Vaughan, 1967). Thus, gophers cause plant mortality both directly through consumption and indirectly through burial.

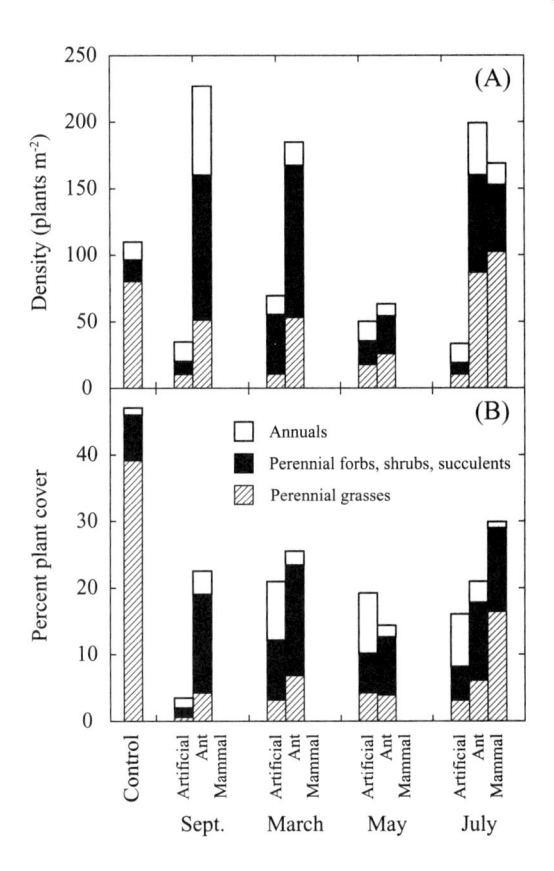

Figure 6.11 Density (A) and cover (B) of three groups of species on three types of disturbed areas by recolonization date and the control vegetation. (Adapted from Coffin and Lauenroth [1989b].)

In a study by Carlson and Crist (1999), individual mounds ranged in size from 0.3 to 1.3 m^2; size varied among years, grazing treatments, and topographic position (Fig. 6.12). Because they found mounds were produced in clusters of 3 to 26 m^2 with frequencies ranging from 4 to 57 clusters/ha, they were able to estimate the percentage of soil surface area disturbed by gophers as being between 0.1% and 5.6% (Fig. 6.12 [Carlson and Crist, 1999]). They also found that pocket gophers seem to prefer south-facing slopes and uplands to north-facing slopes and lowlands, where less than 1% of the area disturbed was disturbed by gopher burrowing. Individual mounds were found to be larger in heavily grazed pasture but their number was lower, hence the total amount of area disturbed was similar to that found in lightly grazed pasture where mounds were smaller but more frequent. An earlier study by Grant et al. (1980) produced higher estimates of surface area disturbance by gophers: 8% in heavily grazed pasture and 2.5% in lightly grazed pasture. This suggests that other factors, such as soil characteristics and plant species composition, may interact with grazing intensity and topographic position to influence patterns of pocket gopher disturbance. Furthermore, the amount of

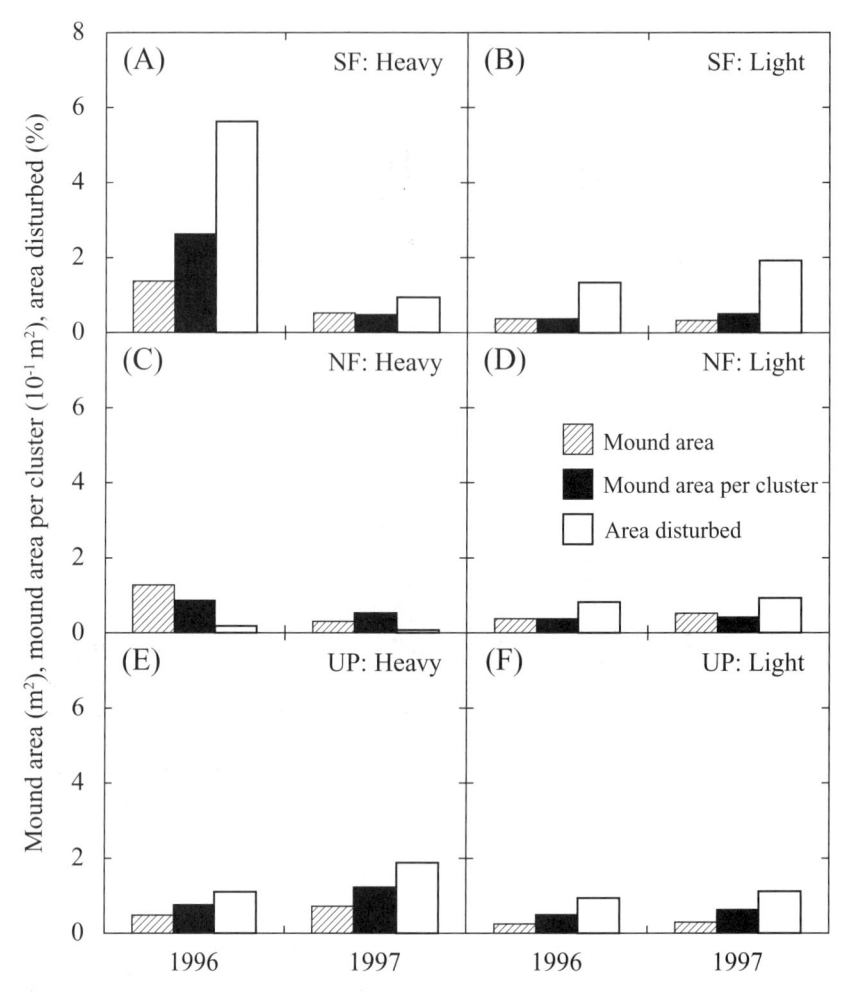

Figure 6.12 Mound area, mound area per cluster, and percentage area disturbed by pocket gophers for three topographic positions (NF, north-facing slopes; SF, south-facing slopes; UP, uplands) and two grazing intensities (heavy, light). (Adapted from Carlson and Crist [1999].)

plant cover affected and the number of plants killed are determined by interactions among the size and frequency of occurrence of burrows, and the size and cover of plants.

Patterns of vegetation on pocket gopher mounds primarily reflect the recovery potential of plant species, rather than the negative effects of gophers resulting from their food preferences. During a 2-year sampling period of active mounds, annuals and perennial forbs were more abundant, although perennial grasses were significantly lower, in gopher-disturbed areas than in undisturbed vegetation (Foster and Stubbendieck, 1980). Although it is expected that plant cover on mounds would be similar to diets, perennial forbs and succulents (in particular, *Opuntia polyacantha*) are a major part of gopher diets; significant amounts of perennial grasses are also consumed (Vaughan, 1967). Frequency of perennial grasses increase, and frequency of annuals and perennial forbs decrease as the age of an abandoned mound increases from 1 to 4 years (Foster and Stubbendieck, 1980). Although total plant cover is similar across topographic position (slopes and uplands), cover of *B. gracilis* is significantly less on upland gopher mounds (Carlson and Crist, 1999). This decrease in dominance by *B. gracilis* corresponds with an increase in cover of other perennial grasses (*A. longiseta*, *Stipa comata*) and subshrubs (*Artemisa frigida*). Mounds in a heavily grazed pasture have a greater cover of *B. gracilis* and higher yearly variation resulting from increased cover of annuals and reduced cover of shrubs compared with mounds in a lightly grazed pasture. Plant species richness is similar for mounds across topographic position and grazing intensity (Carlson and Crist, 1999). Similar to harvester ants, pocket gophers may act to redistribute plant productivity rather than to reduce it. Aboveground biomass of vegetation is significantly greater directly adjacent to mounds compared with vegetation more than 40 cm from the mound edge (Grant et al., 1980). This increase is likely the result of reduced competition for water and nutrients resulting from low plant density and cover on mounds. Mounds have been reported to reduce bare soil evaporation and increase soil water permeability compared with undisturbed areas (Grant et al. 1980). This increase in biomass may also affect seed production and availability of seeds to colonize mounds through time, although these processes have not yet been studied on or around mounds.

Artificial Plots

Because naturally occurring disturbances occur over a range of sizes, intensities, time of occurrence, and location by soil texture and topography, it is difficult to determine the effects of each disturbance characteristic. One approach to dealing with this natural variation is to create artificial disturbances with known characteristics. We studied disturbance size by producing artificial plots comparable in size and shape to the range of sizes observed for ant mounds and animal burrows (0.2–1.8 m^2). Seasonality was evaluated by four colonization dates: September 1, 1984; and March 1, May 1, and July 1, 1985 (dates when the artificial disturbances were produced). We selected two locations with similar topography and climate yet different soil texture: coarse-textured soil (sandy loam) and fine-textured soil (sandy clay loam). Regardless of colonization date, we found that cover was dominated by annuals after 1 year of recovery (Fig. 6.11 [Coffin and Lauenroth, 1989b]). This result is similar to plant communities on old roads and abandoned fields within 5 years of the beginning of plant recovery (Costello, 1944; Judd, 1974; Reichhardt, 1982; Shantz, 1917). Perennials also colonized the plots, as indicated by comparable densities of annuals and perennial grasses and forbs. Annuals and perennials responded similarly to initiation date, but differently to disturbance size. Annual cover decreased as plot size decreased as a result of increased competition from surrounding plants. Perennials recovered either through seedling establishment or vegetative growth. Because we found

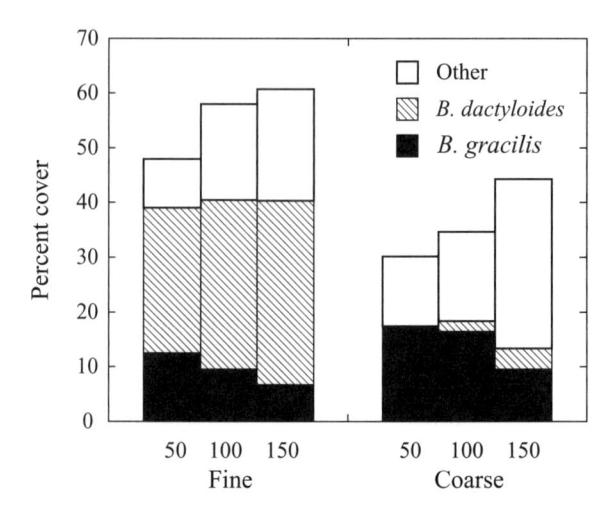

Figure 6.13 Percent cover of *B. gracilis*, *B. dactyloides*, and the total on artificially disturbed plots after 8 years of recovery for two sites based on soil texture and three disturbance sizes (50, 100, and 150 cm in diameter). (Adapted from Martinez-Turanzas et al. [1997].)

few perennial seeds stored in the soil compared with annual seeds (Coffin and Lauenroth, 1989c), most recovery was assumed to be through vegetative spread as ingrowth from the edge of plots. The greater perimeter-to-area ratio of small compared with large plots resulted in significantly greater cover and density of perennials on the smallest plots.

In 1993, after 8 years of recovery, plant cover on these artificial plots was significantly greater on fine- compared with coarse-textured plots, and was comparable with the range of cover found for undisturbed communities (30% to 60%; Fig. 6.13 [Martinez-Turanzas et al., 1997]). Cover on plots at the fine-textured site was dominated by *B. dactyloides*, but this species had very low cover on plots at the coarse-textured site. Cover of *B. gracilis* was significantly greater on plots at the coarse-textured site within each disturbance size, and was significantly greater on small plots compared with large plots. Cover of *B. gracilis* after 8 years of recovery (7% to 17%) was also higher than predicted based upon the old field models of succession (Fig. 6.1), and was more similar to that expected based upon the gap dynamics model (Coffin et al., 1996).

Microtopography associated with plants and interspaces, a key characteristic of the shortgrass steppe, developed within the first 9 years of recovery (Martinez-Turanzes et al., 1997). We found that differences in height between plant crowns and bare soil openings were similar for undisturbed and disturbed areas, although complete recovery had not yet occurred, because the absolute heights of the microtopography on disturbed plots were lower than on undisturbed microsites. Microtopographic differences between plants and interspaces increased as disturbance size increased, suggesting an increasing importance of soil erosion/ deposition and/or plant accumulation of soil with increasing disturbance size. Our results lead to the conclusion that rapid soil erosion occurs during the early stages of recovery after disturbance. Microtopographic relief can be attributed to the redistribution of soil resulting from erosion from bare soil openings and the deposition and accumulation of soil beneath plants as they grow through time. In the shortgrass steppe, both active and total soil organic matter are concentrated in the upper 1 to 3 cm of soil beneath *B. gracilis* plants (Burke et al., 1999; Hook et al., 1991). After the death of a plant, we found that as erosion occurs, microsites are created with low, depleted total soil organic matter (Kelly and Burke, 1997).

Intermediate-Size Disturbances

Two types of disturbances are intermediate in size between small and large disturbances: areas affected by larvae of June beetles (white grubs) and abandoned roads. Both of these are sufficiently large always to kill at least one *B. gracilis* plant, yet are sufficiently small to produce patches across the landscape.

White Grubs

Larvae of June beetles (*Phyllophaga fimbripes* LeConte) feed on roots of perennial grasses to create disturbances through plant mortality. Disturbed areas are of variable size and shape, and consist of complete and incomplete mortality of affected plants (Coffin et al., 1998; Ueckert, 1979; Weiner and Capinera, 1980). Our results show that patches killed by grubs range in size from 2 to 8 m in diameter in grazed pastures (average, 4.5 m) compared with adjacent ungrazed exclosures (average, 3.9 m; Fig. 6.14 [Coffin et al., 1998]). Larger areas affected by white grubs (>1 ha) have also been observed (Ueckert, 1979). Frequencies of occurrence of patches are unknown, although outbreaks of grubs and subsequent

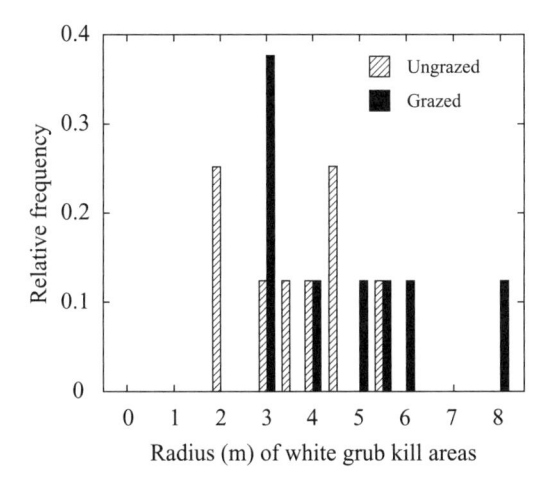

Figure 6.14 Size distribution of areas killed by white grubs for grazed and ungrazed enclosures. (From Coffin et al. [1998]. Reprinted with kind permission of Springer Science and Business Media.)

patches of plant mortality occur periodically with different intensities and densities (Anonymous, 1959). Most information about effects of grubs on shortgrass communities was obtained from one outbreak that occurred in fall 1977 (Weiner and Capinera, 1980). High grub densities and low densities of nematodes, both in the dead patches and near their margins with live vegetation, indicated that grubs were the most likely factor responsible for plant death (Stanton et al., 1984; Weiner and Capinera, 1980). Grub densities decreased markedly from fall 1977 to fall 1978, likely as a result of a summer drought (Weiner and Capinera, 1980).

In a 14-year study of successional dynamics after the outbreak in 1977, we found that vegetation patterns through time on patches killed by white grubs were similar to patterns on other types of disturbances (Coffin et al., 1998). Annuals initially dominate the vegetation, followed by short-lived perennials and long-lived perennial grasses. However, the rate of recovery on grub-killed patches is faster than higher intensity disturbances where all plants are killed. Cover of *B. gracilis* ranged from 20% to 30% and was more than 40% of the total cover on both grazed and ungrazed patches after only 5 years of recovery (Fig. 6.15 [Coffin et al., 1998]). *Bouteloua gracilis* dominated plant cover starting in 1980

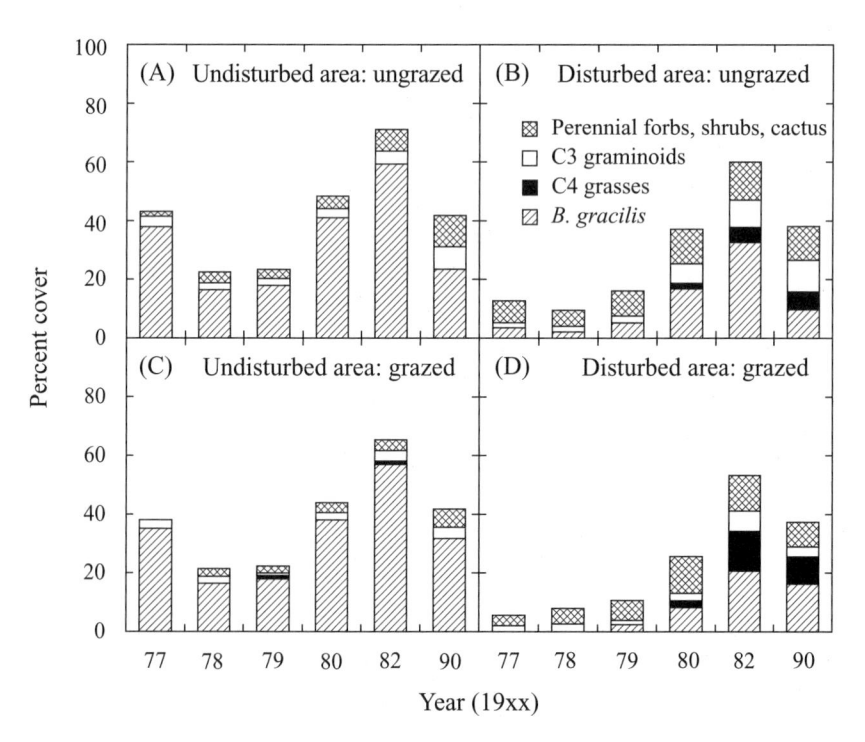

Figure 6.15 Basal cover of *B. gracilis* and three species groups for grazed and ungrazed undisturbed vegetation and for patches affected by white grubs for 6 sample years. (From Coffin et al. [1998]. Reprinted with kind permission of Springer Science and Business Media.)

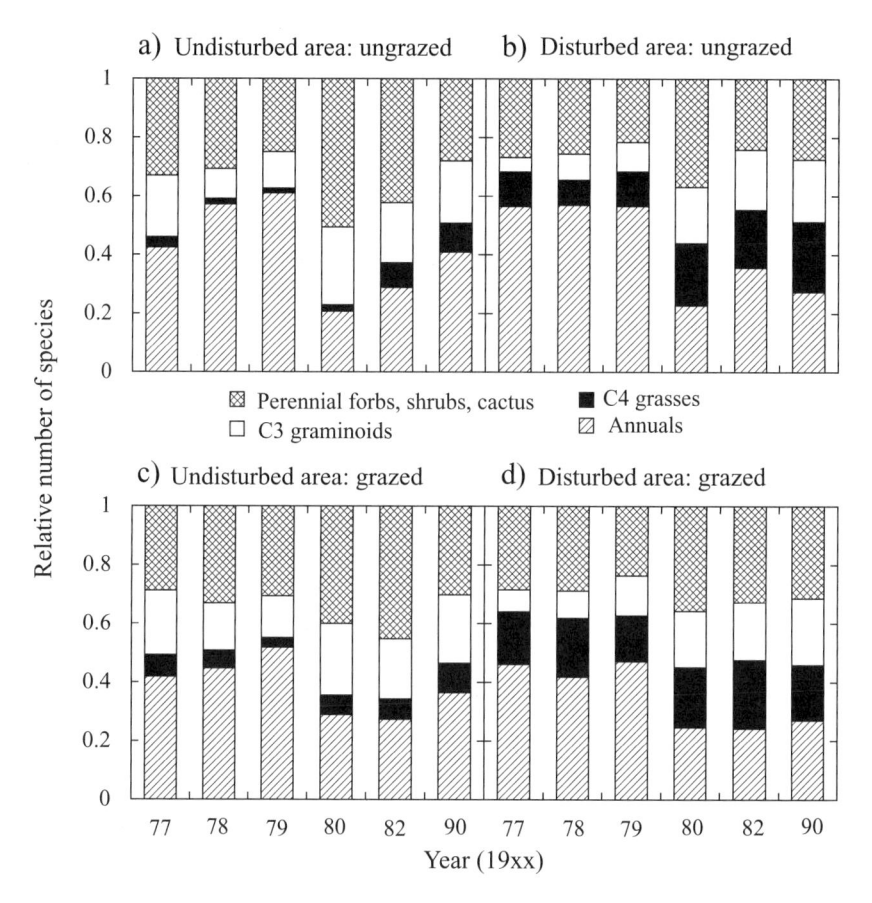

Figure 6.16 Number of species of four species groups relative to the total for grazed and ungrazed undisturbed vegetation and for patches affected by white grubs for 6 sample years.

on ungrazed patches and by 1982 on grazed patches. Cover or density by species or species group was infrequently related to grazing. In general, cover and density were similar to or higher for ungrazed compared with grazed patches. Species richness also was not affected by grazing. Most species on undisturbed and disturbed patches were annuals and perennial forbs, shrubs, and succulents (Fig. 6.16). However, disturbed patches had greater numbers of C_4 grass species throughout the time of the study compared with undisturbed patches.

Our results showed that patches located in ungrazed enclosures had higher initial survival of *B. gracilis* tillers (65%) compared with patches in grazed pastures (41%) (Coffin et al., 1998). These differences in tiller survival have important effects on recovery of this species through time. Degree of survival explained 70% of the variability in *B. gracilis* cover through time on ungrazed patches, which indicates that initial conditions are important for as many as 14 years after recovery begins. For grazed patches, tiller survival explained more than 60% of variability in cover for the first 3 years, but less than 41% of the variability for the remaining years. Thus, grazing becomes more important through time compared with the continued importance of initial conditions for ungrazed patches.

Old Roads

During the settlement of the shortgrass steppe, roads were produced that were later abandoned as fences were erected along section lines. Although much of the area now contains a network of roads with varying use, sections of two-track roads continue to be generated and abandoned to avoid low-lying areas and other patches of soil that become impassable during periods of heavy rainfall. Little is known about these abandoned roads relative to their high frequency of occurrence. In one study conducted in eastern Colorado, succession on old roads was found to follow the typical sequences attributed to other disturbance types (Shantz, 1917). Roads abandoned from 1 to 3 years were dominated by annual forbs such as *Salsola kali* and *Plantago patagonica*, and the annual grass *Vulpia octoflora*. Roads abandoned from 2 to 5 years were also dominated by annual forbs, but in addition had short-lived perennial grasses such as *Schedonnardus paniculatus* and the subshrub *Gutierrezia sarothrae*. After 4 to 8 years, *S. paniculatus* dominated and *G. sarothrae* became increasingly important. *Gutierrezia sarothrae* dominated during years 7 to 14, after which it was replaced by *B. dactyloides*, which dominated until year 23. Meanwhile, *B. gracilis* was established and gradually increased in importance until, by year 50, the community was codominated by *B. gracilis* and *B. dactyloides*. These trends were obtained using observations of roads abandoned for varying amounts of time, which also varied in other characteristics, such as how long they were in use before abandonment, soil type, and initial vegetation. Age since abandonment was estimated based on fence construction. Thus, this study suffered from many of the same limitations as studies conducted on abandoned agricultural fields. Plant species associated with the successional stages were found to correspond to other types of disturbances; however, the number of years in each stage and the recovery time to a *B. gracilis*-dominated community are likely to be overestimates because of the use of chronological sequences of roads.

Large-Scale Disturbances

Several types of natural and anthropogenic disturbances affect shortgrass steppe ecosystems at broader scales. These disturbances always kill hundreds to thousands of plants and affect entire landscape units.

Fertilization

Effects of nutrient enrichments on shortgrass communities have been assessed both through nitrogen fertilization and application of sewage sludge. Although these applications may be more appropriately termed *stresses*, they are included under disturbances because of their potential short- and long-term effects on community structure and species composition. A study initiated in 1970 was conducted to evaluate the effects of water, nitrogen, and water plus nitrogen on properties of shortgrass steppe communities. These plots have been followed since the cessation of the applications in 1975. At the end of the treatments in 1975, the nitrogen treatment showed little effect on plant community composition whereas the water treatment increased total biomass and reduced succulent populations (Lauenroth et al., 1978). The water-plus-nitrogen treatment had the largest effect on biomass, particularly that of C_4 perennial grasses and subshrubs, although the communities were still dominated by shortgrasses. However, these initial trends did not continue through time (Milchunas and Lauenroth, 1995), even though our results show that all communities were significantly different from controls 7 years after the end of treatments. At that time, exotics were abundant on both water-plus-nitrogen and nitrogen treatments, whereas native grasses other than *B. gracilis* had increased on water treatments. During the next 5 years, all communities approached the structure of the control communities. However, during the subsequent 3 years (1988–1991), native species decreased in the water-plus-nitrogen treatments and exotic species increased in the water treatments, but similar trends were not found in the nitrogen treatments during the same time period. There may be important feedbacks between exotic species establishment and nitrogen availability that leads to their persistent dominance after nitrogen enrichment (Vinton and Burke, 1995). Thus, the short-term effects of stressors may not be the same as long-term effects that are not predictable from climatic conditions. Based on these long-term observations, we proposed biotic feedbacks associated with the accumulation and decomposition of litter (and subsequent effects on weed recruitment) as an alternative mechanism for time lags in community responses (Milchunas and Lauenroth, 1995).

Application of municipal sewage sludge (biosolids) onto the soil surface is another stress that is a recent disturbance to shortgrass communities. A one-time application of 22.5 to 45.0 Mg·ha^{-1} of anaerobically digested sewage sludge resulted in an increase in both cover and biomass of *B. gracilis*, and did not lead to heavy metal contamination of soil and plant tissue (Fresquez et al., 1990a,b, 1991). Thus, the application of sludge to grasslands provides a recycling option for biodegradable organic matter that also contains large amounts of inorganic material.

Tracking by Military Vehicles

The use of the shortgrass steppe by the U.S. military for training exercises has resulted in a recent disturbance type: tracking by heavy vehicles. Tracking has immediate effects from killing plants, and long-term effects from soil compaction and erosion (Lathrop, 1982). The amount of area affected and the frequency of occurrence of this type of disturbance are difficult to determine, because tracking is limited to military training areas, such as the Piñon Canyon Maneuver site in southern Colorado. From 1985 to 1987, tracking at this site decreased plant basal and litter cover, and increased the proportion of bare ground (Shaw and Diersing, 1990). Tracking also reduced cover of *B. gracilis,* and increased annual grasses (*V. octoflora, Hordeum pusillum*) and forbs (*Helianthus annuus, S. kali, Kochia scoparia*). After 10 years of military maneuvers, trends in vegetation resulting from vehicle tracking were compensated for, at least in part, by the removal of cattle (Milchunas et al., 1999). Decreases in long-lived perennials resulted in increases in short-lived perennials, and decreases in C_4 grasses were followed by increases in C_3 grasses.

Prairie Dogs

Black-tailed prairie dogs (*Cynomys ludovicianus*) affect vegetation and soils through their burrowing and grazing activities. Colonies range in size from tens to hundreds of hectares (Dahlsted et al., 1981; Knowles, 1986) at average densities of 10 to 55 animals/ha (Archer et al., 1987; Knowles, 1986; O'Meilia et al., 1982), although densities vary considerably depending on environmental conditions and status of the colony. Burrows can extend to depths of 1 to 3 m; prairie dogs move subsoil to the surface in mounds near the burrow entrance. Mounds are 1 to 2 m in diameter in towns that range from 50 to 300 per ha (Archer et al., 1987; White and Carlson, 1984). In addition to killing plants beneath the mounds of soil, prairie dogs also denude the area surrounding the burrow system by repeatedly grazing and clipping herbaceous vegetation (Archer et al., 1987). Much of our understanding of the effects of prairie dogs on semiarid grasslands comes from studies conducted in the northern mixed-grass prairie of South Dakota, where the direct effects from these animals led to indirect effects of their grazing (Whicker and Detling, 1988). Current studies on the shortgrass steppe are underway. Grazing by prairie dogs can lead to dominance by shorter, more prostrate plant populations compared with off-colony populations, as well as changes in species composition (Archer et al., 1987; Coppock et al., 1983; Detling and Painter, 1983). Grazing can also affect microclimate, nutrient cycling, and belowground production (Archer and Detling, 1986; Carlson and White, 1987; Coppock et al., 1983; Hartley, 2006; Jaramillo and Detling, 1988; Whicker and Detling, 1988).

As in mixed-grass prairie communities, prairie dog colonies have large effects on plant species composition in shortgrass steppe communities. Higher species richness of perennials and annuals are found in the areas surrounding prairie dog mounds compared with off-mound locations (Bonham and Lerwick, 1976). Cover of *B. gracilis* is significantly greater outside than inside colonies, whereas cover of *B. dactyloides* and annual forbs is consistently greater inside colonies than outside. Two subshrubs—*Eriogonum microthecum* and *G. sarothrae*—have greater cover outside than inside the colony. Several species, including *C. eleocharis* and *Sphaeralcea coccinea*, are not affected by the presence of prairie dogs (Bonham and Lerwick, 1976). Differences in plant species composition between on- and off-colony locations have been attributed to diet selection by prairie dogs. The most important plants in their annual diets are *Carex* spp. (36%), *B. gracilis* (20%), *Sporobolus cryptandrus* (13%), *A. frigida* (8%), and *S. coccinea* (7%) (Hansen and Gold, 1977). However, clipping and feeding activities of desert cottontails in prairie dog colonies can affect these vegetation patterns (Hansen and Gold, 1977). Little is known about recovery of vegetation after abandonment of areas by prairie dogs. In one study, cover of several species of perennial grasses,

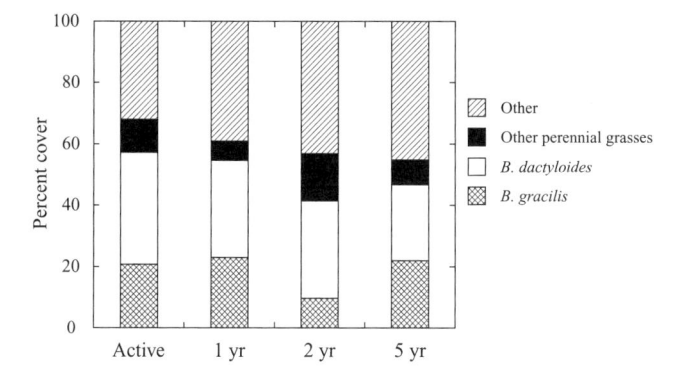

Figure 6.17 Percent cover of *B. gracilis*, *B. dactyloides*, other perennial grasses, and the total for one active prairie dog colony and three colonies abandoned for different lengths of time from 1 to 5 years. (From Klatt and Hein [1978].)

annual grasses, forbs, and shrubs was similar among an active colony and three colonies abandoned for from 1 to 5 years (Fig. 6.17 [Klatt and Hein, 1978]).

Abandoned Agricultural Fields

Cultivation and subsequent abandonment have occurred in the shortgrass steppe since the time of European settlement. In the 1860s, before cultivation, many of the early settlers used the land for grazing by cattle (Klipple and Costello, 1960). Gradually, by 1920, much of the area was settled by homesteaders who shifted the land-use emphasis from grazing to farming (Ubbelohde et al., 1988). Although this region is characterized by low, variable precipitation and high temperatures during the growing season, homesteaders planted crops that were typical of wetter areas. During periods of drought, repeated crop failures led to widespread abandonment of fields. An estimated 20% to 30% of what is now the Pawnee National Grassland of northeastern Colorado was cultivated and abandoned before or during the drought of the 1930s, and subsequently acquired by the federal government by 1941. Abandoned fields varied in size from 16 to 260 ha, and occurred over a range of climatic conditions and soil textures characteristic of the region. After abandonment, these fields were susceptible to wind and water erosion resulting from low vegetative cover (Albertson and Weaver, 1944b). Thus, the effects of cultivation and abandonment included large-scale plant mortality, losses of seeds from the soil, increased soil erosion coupled with localized areas of deposition, and reduction in soil organic matter (Burke et al., 1989; Costello, 1944; Weaver and Mueller, 1942).

Several studies were conducted during the early and mid 1900s to evaluate recovery patterns and to estimate recovery rates on old fields abandoned prior to 1941 (Costello, 1944; Judd, 1974; Judd and Jackson, 1939; Savage and Runyon, 1937). The results of these studies can be summarized into four general stages of succession: (1) an annual grasses and forbs stage; (2) a short-lived perennial grass stage typically dominated by *S. paniculatus*, *Munroa squarrosa*, and *Sitanion hystrix;* (3) a perennial subclimax stage dominated by the grasses *A. purpurea*, *S. comata*, and *Muhlenbergia torreyi*, and the dwarf shrub *A. frigida;* and (4) the climax stage dominated by *B. gracilis*, *B. dactyloides*, and the perennial sedges *C. heliophila* and *C. eleocharis*. This final stage was predicted to occur 25 to more than 50 years after abandonment, based on inferences from chronological sequences of fields of varying ages, but most of which had been abandoned for less than 10 years. This view of shortgrass succession represents the traditional Clementsian model (Clements, 1916, 1928). In studies in eastern Colorado, a modified old-field model was suggested in which the stage dominated by *A. purpurea* lasts indefinitely (Hyder and Everson, 1968; Hyder et al., 1971), and by doing so represents an alternative state of the system (Laycock, 1989, 1991). This model provides the basis for the conventional view of succession in shortgrass communities. Observations from two fields that had been abandoned for more than 40 years supported this model, suggesting that far greater than 50 years are required for the recovery of *B. gracilis* (Reichhardt, 1982). Attempts to redirect secondary succession past an *Aristida*-dominated stage using nitrogen applications had limited success (Horn and Redente, 1998; Hyder and Bement, 1972).

Our more recent analysis using an objective selection of 13 fields with similar lengths of time after abandonment (53 years) found very different results (Coffin et al., 1996). In these 13 fields, cover of shortgrasses and sedges ranged from 12% to 88% (Fig. 6.18). Only two fields (fields 11 and 12) had high shortgrass cover similar to predictions from the Clementsian model and only two (fields 4 and 8) had low shortgrass cover similar to that predicted by the conventional model. Most fields had intermediate values and did not fit either model. Furthermore, *B. gracilis* was found on all fields sampled and even dominated cover on two fields (Fig. 6.19C). *Buchloë dactyloides* was an important component of vegetation on all fields, and dominated cover on nine fields (Fig. 6.19A, B). Other vegetation in fields dominated by *B. dactyloides* showed similar trends in cover, density, and species richness to those found in fields dominated by *B. gracilis*, indicating that these species respond similarly to dominance by either shortgrass species. The two remaining fields were dominated by a variety of other C_3 and C_4 perennial graminoids, including *A. purpurea*, *S. cryptandrus*, *S. comata*, *S. hystrix*, *A. smithii*, and *Carex* spp. (Fig. 6.19D). Although *A. purpurea* was found on 12 of the 13 fields, a steady state of vegetation dominated by this species, as predicted by the modified old-field model, was not found for any field. This species had less than 4% cover and represented less than 20% of total plant cover on all fields.

Thus, neither the Clementsian nor the conventional model of succession is an accurate representation of vegetation recovery on these fields. The ability of *B. gracilis* to recover after large disturbances, such as cultivation and abandonment, led us to an alternative view of the role of disturbance in these systems that focuses on interactions between individual plants and their environment, and in particular on the importance of disturbance characteristics in determining recovery rate and pattern. Our results show that previous studies of succession using a chronological sequence of fields or sampling only a small number of fields that

The Role of Disturbances in Community and Ecosystem Dynamics 109

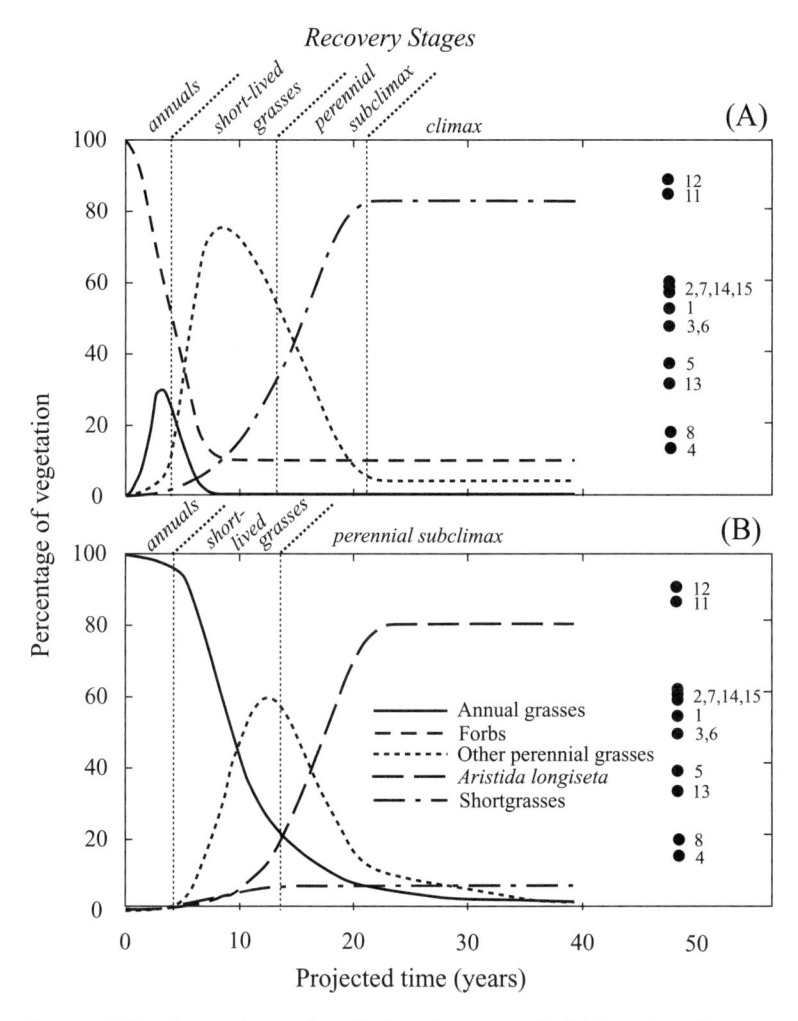

Figure 6.18 Comparisons of predictions from two old-field models of succession and actual shortgrass cover found on fields sampled 53 years after abandonment using the traditional, Clementsian model (A) and the conventional model (B). Predictions are shown as percentages of the vegetation by species or group for the two models. Numbers on the right identify specific fields. (From Coffin et al. [1996].)

obviously had been plowed have underestimated recovery rates of these communities, and in particular of *B. gracilis*.

Furthermore, the ability of *B. gracilis* to recover at large distances (>75 m) on some old fields within 50 years is in sharp contrast to predicted recovery rates based only on its life history traits. Slow tillering rates limit vegetative spread of *B. gracilis* to those areas of abandoned fields that abut areas of undisturbed vegetation, or to clonal growth after seedling establishment within the fields

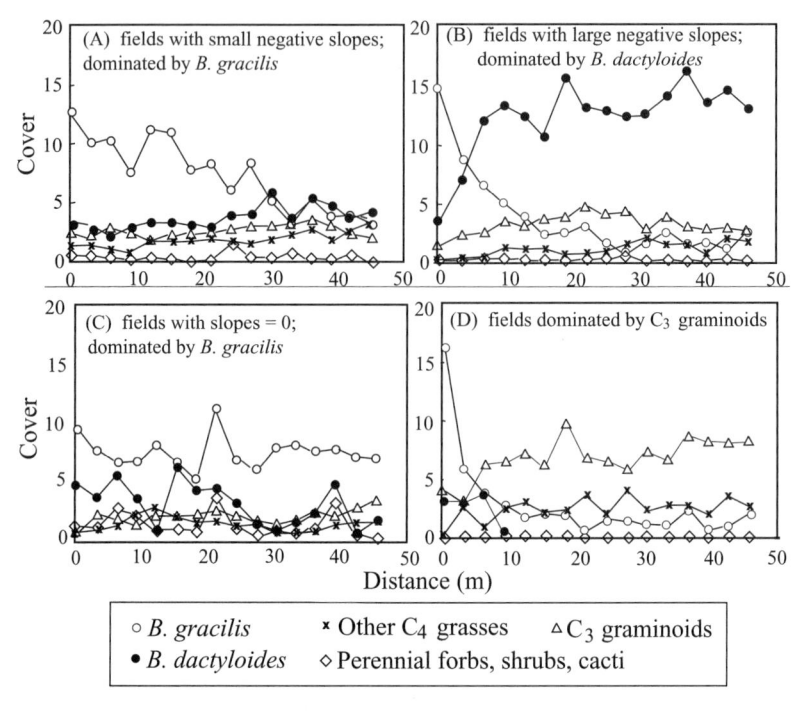

Figure 6.19 (A–D) Cover of *B. gracilis*, *B. dactyloides*, and three functional types with distance from the edge for four groups of fields based on dominance by *B. gracilis*. Cover is the average across fields within each group or distance. (From Coffin et al. [1996].)

(Samuel, 1985). Recovery at large distances from field edges likely results from the dispersal of seed from surrounding undisturbed areas with subsequent germination and establishment. Similar to many grasses, *B. gracilis* seeds are usually dispersed by wind. However, the short stature of *B. gracilis* plants (<30 cm) limits the distance that seeds can travel through wind alone. High wind speeds (>5.4 m·s^{-1}) can disperse *B. gracilis* seeds for distances up to 12 m, but low wind speeds result in dispersal distances of less than 1 m (Fraleigh, 1999). Dispersal of *B. gracilis* seeds on hair and in feces of cattle provide additional mechanisms for long-distance dispersal (>100 m) that could account for recovery patterns on old fields (Fraleigh, 1999). Similar studies of dispersal vectors for *B. dactyloides* found that dispersal by wind is limited to the immediate area around a plant, but that dispersal by cattle feces and hair may provide long-distance movement of seeds (Fraleigh, 1999).

Another poorly understood aspect of old-field dynamics in the shortgrass steppe is the long-term recovery of soil organic matter and soil fertility after abandonment. Our companion studies of carbon and nitrogen dynamics conducted on the same 13 fields have provided important information on soil processes and properties. Our analyses compared native (never cultivated), abandoned (cultivated until 1937), and currently cultivated fallow fields. Our results show that 50 years after abandonment, surface (0–10 cm) and subsurface soils (0–30 cm) are 30% to 40% lower in carbon and nitrogen compared with nearby native fields (Burke et al., 1995; Ihori et al., 1995b). Microbial biomass and nitrogen mineralization, as indexed by potential net carbon and nitrogen mineralization, are also significantly reduced (Burke et al., 1995). By contrast, rates of nitrogen mineralization and turnover are highest in cultivated fields, likely as a result of higher soil water content (Ihori et al., 1995a). Microbial biomass, potentially mineralizable nitrogen, and respirable carbon are not significantly different between abandoned and native fields (Burke et al., 1995). Small-scale heterogeneity in soil carbon and nitrogen associated with individual *B. gracilis* plants also recovered on abandoned fields within the 53-year recovery period. Measured variation in soil properties is less on abandoned fields than variation in vegetation, because sampling was only conducted under *B. gracilis* plants. High spatial variation within and among fields would likely have been found if sampling sites had not been limited to this one species. For example, we found that annuals do not accumulate significant levels of nutrients in soils beneath their canopies (Vinton and Burke, 1995).

Our results also show that recovery of vegetation and soil properties on old fields is highly variable, and depends upon a number of historical factors as well as more recent climatic conditions. The number of years and intensity of cultivation, subsequent grazing intensity, soil texture, proximity to native seed sources, slope and topography, kind and degree of erosion, amount of soil deposition, amount and distribution of rainfall, and distance, direction, and speed of wind have been identified as potentially important to recovery rates (Costello, 1944; Judd, 1974). Most abandoned fields in the region were reseeded at some time between 1940 and 1955 with a mixture of native and introduced species, including *Agropyron cristatum*. Differential success of these reseeding efforts on different fields may also have influenced the recovery rate of native grasses. We have very little understanding of how these various factors affect recovery rates, in large part because of the lack of historical information about each field.

Drought

Periodic drought occurs in the central Great Plains, including eastern Colorado (Pielke and Doesken, chapter 2, this volume). Low precipitation, high temperatures, high winds, and dust storms combine to kill plants as a result of severe water limitation and burial of plants by loose soil (Albertson and Weaver, 1942, 1944a). Two of the most important droughts during the past century (1933–1939 and 1952–1955) have been studied in terms of plant death and vegetation recovery. In eastern Colorado, effects of drought were more severe in the 1950s than in the 1930s. In the 1950s, east–central Colorado (average losses in cover of 89%) was more severely affected than northeastern Colorado (losses of 49% to 82%) (Albertson et al., 1957). During both droughts, *B. gracilis* was more tolerant of extreme conditions than *B. dactyloides* throughout eastern Colorado and western Kansas (Albertson and Weaver, 1944b; Albertson et al., 1957). *Sporobolus cryptandrus,* through abundant seedling establishment, also survived the droughts.

Rate of recovery of dominant grasses after drought is highly variable and is related to a number of factors, including plant composition of predrought communities, intensity of drought, degree of plant survival, grazing intensity, and amount and distribution of precipitation after the cessation of drought (Albertson and Weaver, 1944b). In general, rate of recovery is much faster for *B. dactyloides* than for *B. gracilis*. In one pasture, *B. dactyloides* increased from 6.1% to 78.2% whereas *B. gracilis* only doubled its basal cover (7.3% to 15.1%) during the same time period (1940–1943). Our more recent studies have found that *B. gracilis* can respond rapidly to small rainfall events, and the ability of this species to respond after drought is related to the intensity and length of the drought (Sala and Lauenroth, 1982; Sala et al., 1981).

The stages of succession after drought are similar to those for other disturbance types and consist of (1) annual forbs and grasses, including *S. kali*, *Chenopodium* spp., *Plantago* spp., *H. annuus*, and *V. octoflora;* (2) short-lived perennial grasses dominated by *S. cryptandrus*, *A. smithii*, *S. paniculatus*, and *M. squarrosa;* and (3) long-lived perennial grasses that included *A. longiseta*, *S. hystrix*, *B. dactyloides*, and *B. gracilis* (Albertson and Weaver, 1944b). Although actual rates of recovery were not determined, it was estimated that cover of dominant grasses could be restored within 5 years under wet periods with reasonable grazing management, but that more than 20 years would be required for pastures that were overgrazed prior to the drought (Albertson et al., 1957). Faster recovery compared with abandoned agricultural fields likely results from the survival of some small plants and tillers that could respond after the cessation of drought.

Summary and Conclusions

The disturbance regime of shortgrass ecosystems consists of a rich array of disturbance types, each with its unique set of characteristics having important effects on individual plants and communities. Recovery patterns and rates after disturbance are dependent upon interactions among disturbance characteristics and the life history traits of plants. One of the most important characteristics is the size of the disturbed area. In general, patterns of recovery are similar among different sizes of disturbances, in that annuals and short-lived perennials colonize initially, with long-lived perennials eventually dominating the community. However, our results show that the rate of recovery is affected by disturbance size. Recovery rates increase in a nonlinear way as size increases. Disturbance intensity also modifies recovery patterns and rates. Low-intensity disturbances, such as areas killed by grubs or covered with soils by burrowing animals, do not completely kill perennial plants or their belowground organs, and thus have faster rates of recovery compared with high-intensity disturbances, such as old fields, where all plants are killed. Although grazing by cattle, soil texture, and topographic position have important effects on disturbance frequencies and sizes, very little is known about how these factors affect plant recovery.

Our gap dynamics conceptualization of plant communities provides important insights into the role of disturbance in shortgrass ecosystems and is a better representation of shortgrass steppe dynamics than either the traditional Clementsian model or the conventional model modified for eastern Colorado. Actual recovery rates of *B. gracilis* on disturbances of different sizes can only be explained through our gap dynamics approach that explicitly includes scale-dependent processes interacting with disturbance characteristics. High variation in *B. gracilis* recovery between old fields with similar abandonment date and soil texture indicates the importance of historical factors, such as grazing management, reseeding practices, and local variations in precipitation and temperature, to recovery. Very fast rates of recovery by *B. gracilis* on some fields (>90 m within 52 years) suggest that biotic factors, such as dissemination of seed by cattle and ants, as well as abiotic events, such as extreme wind gusts, may also be important in the recovery of this species. Because of the importance of *B. gracilis* to shortgrass steppe ecosystems, and the controversy concerning its recovery after disturbance, much of the process-level research has focused on this species. Little is known about key processes limiting the response of other species to disturbance. Further research is needed to elucidate this information, which could be important in the management of these ecosystems, especially in terms of preserving biodiversity at multiple spatial and temporal scales.

Acknowledgments   This research was supported by an NSF grant (BSR 9011659) to CSU as part of the SGS LTER program. We thank Brandon Bestelmeyer for assistance with the figures, and Tom Crist and other anonymous reviewers for helpful comments on the manuscript.

References

Aguilera, M. O. 1992. *Intraspecific interactions in blue grama*. PhD diss., Colorado State University, Fort Collins, Colo.

Aguilera, M. O., and W. K. Lauenroth. 1993a. Neighborhood interactions in a natural population of the perennial bunchgrass *Bouteloua gracilis*. *Oecologia* **94**:595–602.

Aguilera, M. O., and W. K. Lauenroth. 1993b. Seedling establishment constraints in adult neighborhoods: Intraspecific constraints in the regeneration of the bunchgrass *Bouteloua gracilis*. *Journal of Ecology* **81**:253–261.

Aguilera, M. O., and W. K. Lauenroth. 1995. Influence of gap disturbances and type of microsites on seedling establishment in *Bouteloua gracilis*. *Journal of Ecology* **83**:87–97.

Albertson, F. W., G. W. Tomanek, and A. Riegel. 1957. Ecology of drought cycles and grazing intensity in grasslands of central Great Plains. *Ecological Monographs* **27**:27–44.

Albertson, F. W., and J. E. Weaver. 1942. History of the native vegetation of western Kansas during seven years of continuous drought. *Ecological Monographs* **12**:23–51.

Albertson, F. W., and J. E. Weaver. 1944a. Effects of drought, dust, and intensity of grazing on cover and yield of short-grass pastures. *Ecological Monographs* **14**:1–29.

Albertson, F. W., and J. E. Weaver. 1944b. Nature and degree of recovery of grassland from the Great Drought of 1933 to 1940. *Ecological Monographs* **14**:394–479.

Anonymous. 1959. Control of common white grubs. USDA Farmers bulletin no. 1798. U. S. Department of Agriculture, Washington, D.C.

Archer, S., and J. K. Detling. 1986. Evaluation of potential herbivore mediation of plant water status in a North American mixed-grass prairie. *Oikos* **47**:287–291.

Archer, S., M. G. Garrett, and J. K. Detling. 1987. Rates of vegetation change associated with prairie dog (*Cynomys ludovicianus*) grazing in North American mixed-grass prairie. *Vegetatio* **72**:159–166.

Bonham, C. D., and A. Lerwick. 1976. Vegetation changes induced by prairie dogs on shortgrass range. *Journal of Range Management* **29**:221–225.

Bormann, F. H., and G. E. Likens. 1979. *Pattern and process in a forested ecosystem.* Springer-Verlag, New York.

Briske, D. D., and A. M. Wilson. 1977. Temperature effects on adventitious root development in blue grama seedlings. *Journal of Range Management* **30**:276–280.

Briske, D. D., and A. M. Wilson. 1978. Moisture and temperature requirements for adventitious root development in blue grama seedlings. *Journal of Range Management* **31**:174–178.

Burke, I. C., W. K. Lauenroth, and D. P. Coffin. 1995. Soil organic matter recovery in semiarid grasslands: Implications for the Conservation Reserve Program. *Ecological Applications* **5**:793–801.

Burke, I. C., W. K. Lauenroth, R. Riggle, P. Brannen, B. Madigan, and S. Beard. 1999. Spatial variability of soil properties in the shortgrass steppe: The relative importance of topography, grazing, microsite, and plant species in controlling spatial patterns. *Ecosystems* **2**:422–438.

Burke, I. C., C. M. Yonker, W. J. Parton, C. V. Cole, K. Flach, and D. S. Schimel. 1989. Texture, climate, and cultivation effects on soil organic matter content in U.S. grassland soils. *Soil Science Society of America Journal* **53**:800–805.

Carlson, J. M., and T. O. Crist. 1999. Plant responses to pocket-gopher disturbances across pastures and topography. *Journal of Range Management* **52**:637–645.

Carlson, D. C., and E. M. White. 1987. Effects of prairie dogs on mound soils. *Soil Science Society of America Journal* **51**:389–393.

Clark, W. H., and P. L. Comanor. 1975. Removal of annual plants from the desert ecosystem by Western harvester ants, *Pogonomyrex occidentalis*. *Environmental Entomology* **4**:52–56.

Clements, F. E. 1916. Plant succession: An analysis of the development of vegetation. Publication no. 242. Carnegie Institute of Washington, Washington, D.C.

Clements, F. E. 1928. *Plant succession and indicators: A definitive edition of plant succession and plant indicators.* H. W. Wilson, New York.

Coffin, D. P., and W. K. Lauenroth. 1988. The effects of disturbance size and frequency on a shortgrass plant community. *Ecology* **69**:1609–1617.

Coffin, D. P., and W. K. Lauenroth. 1989a. Disturbances and gap dynamics in a semiarid grassland: A landscape-level approach. *Landscape Ecology* **3**(1):19–27.

Coffin, D. P., and W. K. Lauenroth. 1989b. Small scale disturbances and successional dynamics in a shortgrass community: Interactions of disturbance characteristics. *Phytologia* **67**(3):258–286.

Coffin, D. P., and W. K. Lauenroth. 1989c. The spatial and temporal variability in the seed bank of a semiarid grassland. *American Journal of Botany* **76**(1):53–58.

Coffin, D. P., and W. K. Lauenroth. 1990a. A gap dynamics simulation model of succession in the shortgrass steppe. *Ecological Modelling* **49**:229–266.

Coffin, D. P., and W. K. Lauenroth. 1990b. Vegetation associated with nest sites of Western harvester ants (*Pogonomyrex occidentalis*) in a semiarid grassland. *American Midland Naturalist* **123**:226–235.

Coffin, D. P., and W. K. Lauenroth. 1992. Spatial variability in seed production of the perennial bunchgrass *Bouteloua gracilis* (H.B.K.) Lag. ex Griffiths. *American Journal of Botany* **79**:347–353.

Coffin, D. P., and W. K. Lauenroth. 1994. Successional dynamics of a semiarid grassland: Effects of soil texture and disturbance size. *Vegetatio* **110**:67–82.

Coffin, D. P., W. K. Lauenroth, and I. C. Burke. 1993. Spatial dynamics in recovery of shortgrass steppe ecosystems. *Lectures on Mathematics in the Life Sciences* **23**:75–108.

Coffin, D. P., W. K. Lauenroth, and I. C. Burke. 1996. Recovery of vegetation in a semiarid grassland 53 years after disturbance. *Ecological Applications* **6**:538–555.

Coffin, D. P., W. A. Laycock, and W. K. Lauenroth. 1998. Disturbance intensity and above- and belowground herbivory effects on long-term (14y) recovery of a semiarid grassland. *Plant Ecology* **139**:221–233.

Coppock, D. L., J. K. Detling, J. E. Ellis, and M. I. Dyer. 1983. Plant–herbivore interactions in a North American mixed-grass prairie. I. Effects of black-tailed prairie dogs on intraseasonal aboveground plant biomass and nutrient dynamics and plant species diversity. *Oecologia* **56**:1–9.

Costello, D. F. 1944. Natural revegetation of abandoned plowed land in the mixed prairie association of northeastern Colorado. *Ecology* **25**:312–326.

Crist, T. O., and J. A. Wiens. 1994. Scale effects of vegetation on forager movement and seed harvesting by ants. *Oikos* **69**:37–46.

Crist, T. O., and J. A. Wiens. 1996. The distribution of ant colonies in a semiarid landscape: Implications for community and ecosystem processes. *Oikos* **76**:301–311.

Dahlsted, K. J., S. Sather-Blair, B. K. Worcester, and R. Klukas. 1981. Application of remote sensing to prairie dog management. *Journal of Range Management* **34**:218–223.

Detling, J. K., and E. L. Painter. 1983. Defoliation responses of western wheatgrass populations with diverse histories of prairie dog grazing. *Oecologia* **57**:65–71.

Fair, J. L., D. P. C. Peters, and W. K. Lauenroth. 2001. Response of *Bouteloua gracilis* (Gramineae) plants and tillers to small disturbances. *American Midland Naturalist* **145**:147–158.

Foster, M. A., and J. Stubbendieck. 1980. Effects of the plains pocket gopher (*Geomys bursarius*) on rangeland. *Journal of Range Management* **33**:74–78.

Fraleigh, H. D., Jr. 1999. *Seed dispersal of two important perennial grasses in the shortgrass steppe*. Masters thesis, Colorado State University, Fort Collins, Colo.

Fresquez, P. R., R. Aguilar, R. E. Francis, and E. F. Aldon. 1991. Heavy metal uptake by blue grama growing in a degraded semiarid soil amended with sewage sludge. *Journal of Water, Air, and Soil Pollution* **57–58**:903–912.

Fresquez, P. R., R. E. Francis, and G. L. Dennis. 1990a. Effects of sewage sludge on soil and plant quality in a degraded semiarid grassland. *Journal of Environmental Quality* **19**:324–329.

Fresquez, P. R., R. E. Francis, and G. L. Dennis. 1990b. Soil and vegetation responses to sewage sludge on a degraded semiarid broom snakeweed/blue grama plant community. *Journal of Range Management* **43**:325–331.

Gleason, H. A. 1926. The individualistic concept of the plant association. *Bulletin of the Torrey Botanical Club* **53**:7–26.

Grant, W. E., N. R. French, and L. J. Folse, Jr. 1980. Effects of pocket gopher mounds on plant production in shortgrass prairie ecosystems. *Southwestern Naturalist* **25**:215–224.

Hansen, R. M., and I. K. Gold. 1977. Blacktail prairie dogs, desert cottontails and cattle trophic relations on shortgrass range. *Journal of Range Management* **30**:210–214.

Hartley, L. M. 2006. *Plague and the black-tailed prairie dog: An introduced disease mediates the effects of an herbivore on ecosystem structure and function*. PhD diss, Colorado State University, Fort Collins, Colo.

Hook, P. B., I. C. Burke, and W. K. Lauenroth. 1991. Heterogeneity of soil and plant N and C associated with individual plants and openings in North American shortgrass steppe. *Plant and Soil* **138**:247–256.

Hook, P. B., and W. K. Lauenroth. 1994. Root system response of a perennial bunchgrass to neighborhood-scale soil water heterogeneity. *Journal of Ecology* **8**:738–745.

Hook, P. B., W. K. Lauenroth, and I. C. Burke. 1994. Spatial patterns of roots in a semiarid grassland: Abundance of canopy openings and regeneration gaps. *Journal of Ecology* **82**:485–494.

Horn, B. E., and E. F. Redente. 1998. Soil nitrogen and plant cover of an old-field on the shortgrass steppe in southeastern Colorado. *Arid Soil Research and Rehabilitation* **12**:193–206.

Hyder, D. N., and R. E. Bement. 1972. Controlling red threeawn on abandoned cropland with ammonium nitrate. *Journal of Range Management* **25**:443–446.

Hyder, D. N., and A. C. Everson. 1968. Chemical fallow of abandoned croplands on the shortgrass prairie. *Weed Science* **16**:531–533.

Hyder, D. N., A. C. Everson, and R. E. Bement. 1971. Seedling morphology and seeding failures with blue grama. *Journal of Range Management* **24**:287–292.

Ihori, T., I. C. Burke, and P. B. Hook. 1995a. Nitrogen fertilization in native cultivated and abandoned fields in shortgrass steppe. *Plant and Soil* **171**:203–208.

Ihori, T., I. C. Burke, W. K. Lauenroth, and D. P. Coffin. 1995b. Effects of cultivation and abandonment on soil organic matter in northeastern Colorado. *Soil Science Society of America Journal* **59**:1112–1119.

Jaramillo, V. J., and J. K. Detling. 1988. Grazing history, defoliation, and competition: Effects on shortgrass production and nitrogen accumulation. *Ecology* **69**:1599–1608.

Judd, I. B. 1974. Plant succession on old fields in the Dust Bowl. *Southwestern Naturalist* **19**:227–239.

Judd, I. B., and M. L. Jackson. 1939. Natural succession of vegetation on abandoned farmland in the Rosebud soil area of western Nebraska. *Journal American Society of Agronomy* **39**:541–547.

Keeler, K. H. 1993. Fifteen years of colony dynamics in *Pogonomyrex occidentalis*, the Western harvester ants, in western Nebraska. *Southwestern Naturalist* **38**:286–289.

Kelly, R. H., and I. C. Burke. 1997. Heterogeneity of soil organic matter following death of individual plants in shortgrass steppe. *Ecology* **78**:1256–1261.

Kelly, R. H., I. C. Burke, and W. K. Lauenroth. 1996. Soil organic matter and nutrient availability responses to reduced plant inputs in shortgrass steppe. *Ecology* **77**:2516–2527.

Klatt, L. E., and D. Hein. 1978. Vegetative differences among active and abandoned towns of black-tailed prairie dogs (*Cynomys ludovicianus*). *Journal of Range Management* **31**:315–317.

Klipple, G. E., and D. F. Costello. 1960. *Vegetation and cattle responses to different intensities of grazing on shortgrass ranges of the Central Great Plains*. USDA–ARS technical bulletin no. 1216. U.S. Department of Agriculture, Washington, D.C.

Knowles, C. J. 1986. Some relationships of black-tailed prairie dogs to livestock grazing. *Great Basin Naturalist* **46**:198–203.

Lathrop, E. 1982. Recovery of perennial vegetation in military maneuver areas, pp. 269–276. In: R. H. Webb and H. G. Wilshire (eds.), *Environmental effects of off-road vehicles: Impacts and management in arid regions*. Springer-Verlag, New York.

Lauenroth, W. K., and D. P. Coffin. 1992. Belowground processes and the recovery of semiarid grasslands from disturbance, pp. 131–150. In: M. K. Wali (ed.), *Ecosystem rehabilitation: Preamble to sustainable development*. Vol. 2: *Ecosystem analysis and synthesis*. SPB Academic Publishing, The Hague, Netherlands.

Lauenroth, W. K., J. L. Dodd, and P. L. Sims. 1978. The effects of water- and nitrogen-induced stresses on plant community structure in a semiarid grassland. *Oecologia* **36**:211–222.

Lavigne, R. J. 1969. Bionomics and nest structure of *Pogonomyrex occidentalis* (Hymenoptera: Formicidae). *Annals Entomological Society of America* **62**:1166–1175.

Laycock, W. A. 1989. Secondary succession and range condition criteria: Introduction to the problem, pp. 1–15. In: W. K. Lauenroth and W. A. Laycock (eds.), *Secondary succession and the evaluation of rangeland condition*. Westview Special Studies in Agriculture Science and Policy. Westview Press, Boulder, Colo.

Laycock, W. A. 1991. Stable steady states and thresholds of range condition on North American rangelands: A viewpoint. *Journal of Range Management* **44**:427–433.

Martinez-Turanzas, G., D. P. Coffin, and I. C. Burke. 1997. Development of microtopographic relief in a semiarid grassland: Effects of disturbance size and soil texture. *Plant and Soil* **191**:163–171.

Milchunas, D. G., and W. K. Lauenroth. 1995. Inertia in plant community structure: State changes after cessation of nutrient-enrichment stress. *Ecological Applications* **5**:452–458.

Milchunas, D. G., K. A. Schultz, and R. B. Shaw. 1999. Plant community responses to disturbance by mechanized military maneuvers. *Journal of Environmental Quality* **28**:1533–1547.

Milchunas, D. G, W. K. Lauenroth, P. L. Chapman, and M. K. Kazempour. 1990. Community attributes along a perturbation gradient in a shortgrass steppe. *Journal of Vegetation Science* **1**:375–384.

Noy-Meir, I. 1973. Desert ecosystems: Environment and producers. *Annual Review of Ecology and Systematics* **4**:25–51.

O'Meilia, M. E., F. L. Knopf, and J. C. Lewis. 1982. Some consequences of competition between prairie dogs and beef cattle. *Journal of Range Management* **35**:580–585.

Pickett, S. T. A., and P. S. White. 1985. *The ecology of natural disturbance and patch dynamics*. Academic Press, New York.

Reichhardt. K. L. 1982. Succession of abandoned fields on the shortgrass prairie, northeastern Colorado. *Southwestern Naturalist* **27**:299–304.

Riegel, A. 1941. Life history habits of blue grama. *Kansas Academy of Sciences Transactions* **44**:76–83.

Rogers, L. E. 1972. *The ecological effects of the Western harvester ant* (Pogonomyrex occidentalis) *in the shortgrass prairie ecosystem*. PhD diss., University of Wyoming, Laramie, Wyo.

Rogers, L. E. 1974. Foraging activity of the Western harvester ant in the shortgrass plains ecosystem. *Environmental Entomology* **3**:420–424.

Rogers, L. E., and R. J. Lavigne. 1974. Environmental effects of Western harvester ants on the shortgrass plains ecosystem. *Environmental Entomology* **3**:994–997.

Sala, O. E., and W. K. Lauenroth. 1982. Small rainfall events: An ecological role in semi-arid regions. *Oecologia* **53**:301–304.

Sala, O. E., W. K. Lauenroth, and W. J. Parton. 1981. Plant recovery following prolonged drought in a shortgrass steppe. *Agricultural Meteorology* **27**:49–58.

Samuel, M. J. 1985. Growth parameter differences between populations of blue grama. *Journal of Range Management* **38**:339–342.

Savage, D. A., and H. E. Runyon. 1937. Natural revegetation of abandoned farmland in the central and southern Great Plains, pp. 178–182. In: Grassland ecology. Section 1, Fourth International Grassland Congress, Aberystwyth, UK.

Schwartz, C. C. 1977. *Pronghorn grazing strategies on the shortgrass prairie, Colorado*. PhD diss., Colorado State University, Fort Collins, Colo.

Senft, R. L. 1983. *The redistribution of nitrogen by cattle*. PhD diss., Colorado State University, Fort Collins, Colo.

Shantz, H. L. 1917. Plant succession on abandoned roads in eastern Colorado. *Journal of Ecology* **5**:19–42.

Shaw, R. B., and V. E. Diersing. 1990. Tracked vehicle impacts on vegetation at the Pinon Canyon Maneuver site, Colorado. *Journal of Environmental Quality* **19**:234–243.

Stanton, N. L., D. Morrison, and W. A. Laycock. 1984. The effect of phytophagous nematode grazing on blue grama die-off. *Journal of Range Management* **37**:447–450.

Ubbelohde, C., M. Benson, and D. A. Smith. 1988. *A Colorado history*. Pruett, Boulder, Colo.

Ueckert, D. N. 1979. Impact of a white grub (*Phyllophaga crinita*) on a shortgrass community and evaluation of selected rehabilitation practices. *Journal of Range Management* **32**:445–448.

Vaughan, T. A. 1967. Food habits of the northern pocket gopher on shortgrass prairie. *American Midland Naturalist* **77**:176–189.

Vinton, M. A., and I. C. Burke. 1995. Interactions between individual plant species and soil nutrient status in shortgrass steppe. *Ecology* **76**:1116–1133.

Watt, A. S. 1947. Pattern and process in the plant community. *Journal of Ecology* **35**:1–22.

Weaver, J. E., and I. M. Mueller. 1942. Role of seedlings in recovery of midwestern ranges from drought. *Ecology* **23**:275–294.

Weiner, L. F., and J. L. Capinera. 1980. Preliminary study of the biology of the white grub *Phyllophaga fimbripes* (LeConte) (Coleoptera: Scarabaeidae). *Journal of the Kansas Entomological Society* **53**:701–710.

Whicker, A. D., and J. K. Detling. 1988. Ecological consequences of prairie dog disturbances. *BioScience* **38**:778–785.

White, E. M., and D. C. Carlson. 1984. Estimating soil mixing by rodents. *Proceedings South Dakota Academy Science* **63**:34–37.

7

Simulation of Disturbances and Recovery in Shortgrass Steppe Plant Communities

Debra P. C. Peters
William K. Lauenroth

Simulation modeling is a complementary tool to field observation and experimentation in understanding ecological systems (Lauenroth et al., 1998). The overall objective of our plant community modeling is to allow us to evaluate the importance of gap dynamics concepts of succession for understanding shortgrass plant community recovery after disturbances. A gap dynamics approach focuses on individual plants, and the interactions between disturbance characteristics and plant life history traits in explaining successional patterns (Watt, 1947). Simulation models have been used extensively to evaluate the importance of gap dynamics processes to short- and long-term vegetation dynamics in temperate and tropical forests (e.g., Botkin et al., 1972; Shugart, 1984).

We developed a gap dynamics model for shortgrass steppe plant communities (STEPPE [Coffin and Lauenroth, 1990]) based upon the conceptual and modeling framework provided by forest models, modifying it to represent Great Plains grasslands (Coffin and Lauenroth, 1996; Coffin and Urban, 1993). We used STEPPE in several capacities: (1) to synthesize and integrate existing knowledge to improve our understanding of recovery processes after disturbance, (2) to identify key processes limiting recovery, and (3) to predict long-term recovery dynamics for different climate and disturbance characteristics—in particular, soil texture and disturbance size. Our approach to modeling shortgrass community dynamics was to incorporate only the most important processes needed to address specific research questions. We added processes through time either because the model did not sufficiently represent ecosystem dynamics or because we posed more complicated research questions.

STEPPE Model Description

STEPPE simulates the recruitment, growth, and mortality of individual plants on a small plot through time at an annual time step (Fig. 7.1) (Coffin and Lauenroth, 1990). Recruitment and mortality both have stochastic elements. Growth is deterministic and is based upon competition for resources among plants. A key difference between STEPPE and the forest models from which it was derived is that belowground resources are the most frequently limiting resources in semiarid grasslands compared with aboveground resources (light) in forests (Lauenroth and Coffin, 1992). Because of the importance of *Bouteloua gracilis* to shortgrass communities, we hypothesized that the death of a full-size *B. gracilis* plant results in a gap in belowground resource space that initiates the successional processes of gap dynamics. Thus, the simulated plot size (0.12 m^2) is based upon the belowground resource space associated with the roots of a full-size *B. gracilis* plant (Coffin and Lauenroth, 1991).

The model simulates 15 groups of species chosen to represent the life history traits of the more than 300 species found in shortgrass communities at the SGS LTER site. These groups are further aggregated into five resource groups based on spatial distributions of root biomass. Size and age of each plant on a plot are simulated at an annual time step. Driving variables include precipitation, temperature, and soil texture. Model parameters and driving variables are obtained primarily from studies conducted at the SGS LTER site. The model can simulate

STEPPE (individual plant model)

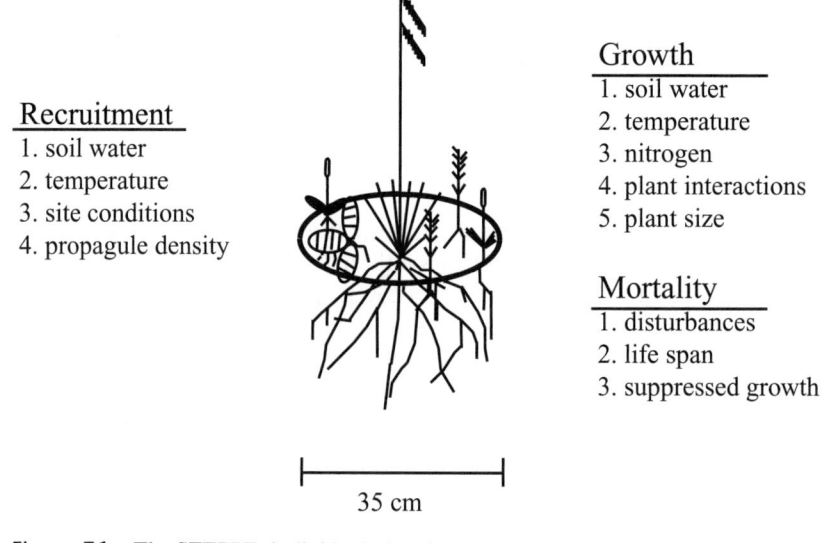

Recruitment
1. soil water
2. temperature
3. site conditions
4. propagule density

Growth
1. soil water
2. temperature
3. nitrogen
4. plant interactions
5. plant size

Mortality
1. disturbances
2. life span
3. suppressed growth

35 cm

Figure 7.1 The STEPPE, individual plant-based gap dynamics model for shortgrass plant communities.

either a single plot or a grid of plots that interact through seed dispersal. Because the model is stochastic, a large number of replicate plots or grids (25–50) are typically simulated, and the results are averaged for each time step.

Recruitment

Each year there is a probability of establishment of either a seedling or a vegetative propagule for each species. In single-plot simulations, we assume that seeds of all species are present on a plot and resources are available for establishment every year. In grid simulations, seed availability is dependent upon proximity to plots on which seeds are produced. The probability that a seedling from a particular species will become established is based either on the occurrence of suitable microenvironmental conditions or on the relative abundance of seeds on a plot. Establishment of the dominant species, *B. gracilis*, is based on the probability ($0.125 \cdot y^{-1}$) that microenvironmental conditions for establishment would occur each year (Lauenroth et al., 1994). For all other species, the probability of establishment is based on the relative abundance of seeds produced, which is estimated from seed production data collected at the SGS LTER site. We assume that one to five species establish each year, with one to three seedlings added for each species. The exact number of species and seedlings is determined by drawing random numbers. All seedlings are added at the estimated size of a 1-year-old plant.

Vegetative propagation occurs in clonal species or those with deep tap roots. We assume that plants have a 75% to 90% chance of regrowth after death of aboveground parts, depending on the source of mortality. If vegetative propagation occurs, then one to three plants of that species are added to the plot.

Mortality

The probability of mortality for each current individual is determined by the disturbance rate, the longevity of the species, and a risk of death associated with slow-growing plants. Effects of cattle fecal pats, western harvester ant mounds, and burrows from small animals are incorporated into STEPPE using their frequencies of occurrence, as reported in Coffin and Lauenroth (1988). Similar to forest gap models, we assume that each species or group has an age-independent intrinsic likelihood of mortality, such that only 1% of a cohort growing under optimal conditions will reach maximum age. We also assume that slow-growing plants have a greater risk of death from disease, insects, and severe environmental conditions than do plants having average growth rates. This probability of mortality ($0.368 \cdot y^{-1}$) results in slow-growing plants having a 1% chance of surviving 10 years. Clonal plants, such as *B. gracilis* and *Opuntia polyacantha*, are excluded from this source of mortality. Mortality of *B. gracilis* clumps occurs as a result of insufficient resources or of disturbances. *Opuntia polyacantha* plants experience these same sources of mortality with high growing season precipitation being an additional source (Dodd and Lauenroth, 1975).

Growth

Growth of plants occurs annually as a function of optimum growth rates, effects of precipitation and temperature, and interactions with other plants for belowground resources. An optimum growth rate estimated from the literature for each species is used to calculate the amount of resources required for each plant in a resource group to grow optimally. Soil water and interactions with other plants determine the actual amount of resources available to each group. STEPPE associates a particular proportion of belowground resource space with individual species or groups of species based on root distribution with depth, distribution of resources with depth, and the temporal variation in both distributions. Root distributions by depth are estimated from the literature and field data. Because soil water is the most frequent control on plant growth and community structure in semiarid grasslands (Lauenroth et al., 1978; Noy-Meir, 1973), the distribution of resources is based on soil water availability with depth in the soil profile. We have two different methods of determining soil water availability, depending upon whether STEPPE is run alone or run in conjunction with SOILWAT, a daily time step soil water simulation model (Parton, 1978; Sala et al., 1992). In the standalone version, temporal variation in water availability to each species depends on both the amount of annual precipitation and the size of plants of a particular species relative to the size of other plants. In the SOILWAT–STEPPE simulations, we integrate daily soil water values from SOILWAT to determine the amount of water available to each species on an annual basis. This is elaborated in the section on STEPPE–SOILWAT simulations. Regardless of the source of information about soil water availability, the actual growth rate for a plant is based on the relationship between resources required for optimum growth and resources available to the plant.

Single-Plot Simulations

We conducted a set of single-plot simulations to test whether processes that result in plot–plot interactions are important for determining the structure of shortgrass steppe plant communities. Single-plot simulations assume that each plot is independent of all plots that surround it. This means that propagules of each species are always available from either the seedbank or from propagule production. Our assumption is appropriate for an isolated single gap disturbance in which the plot is surrounded by undisturbed vegetation. STEPPE can be used in this mode to evaluate successional dynamics and the time required for *B. gracilis* to dominate total plant cover after disturbance (Coffin and Lauenroth, 1990). Disturbance rates for the SGS LTER sites are based on frequencies of occurrence of cattle fecal pats, nest sites of Western harvester ants, and burrows from small animals (Coffin and Lauenroth, 1988; Peters et al., chapter 6, this volume). In our initial approach to conducting these simulations, we further assumed that the occurrence of a disturbance killed all plants on the plot, and that all species had an equal probability of establishing from seed.

In single-plot mode, we found that although basal cover on the simulated plot is eventually dominated by *B. gracilis*, during the initial 20 years after a

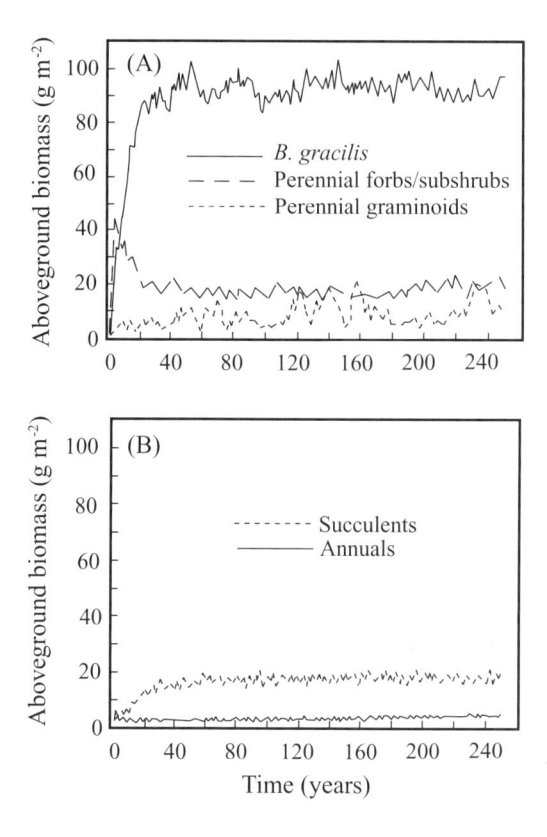

Figure 7.2 STEPPE results for the average of 50 plots for 250 years assuming seeds always available for all species: aboveground biomass of *B. gracilis*, perennial graminoids, and perennial forbs and shrubs (A); and aboveground biomass of annuals and succulents (B). (Coffin and Lauenroth [1990a].)

disturbance other perennial graminoids are the dominants (Fig. 7.2). Although the relative proportions of aboveground biomass and average biomass predicted for each species group by these simulations are comparable with the composition of shortgrass communities in northeastern Colorado (Coffin and Lauenroth, 1989b), the time required for *B. gracilis* to recover is much faster in the simulations than has been reported from field studies. Field and laboratory experiments indicated that a restrictive set of temperature and soil water conditions are required for *B. gracilis* seed germination and establishment (Briske and Wilson, 1977, 1978), suggesting that our assumption about equal probability of seedling establishment is incorrect. Using historical weather and Monte Carlo simulations with SOILWAT, we estimated the probability of appropriate conditions for seedling establishment of *B. gracilis* (Lauenroth et al., 1994). Using this probability of establishment, simulation prediction of average recovery time for *B. gracilis* after a disturbance increased from 20 to 65 years, which is approximately what has been found in field studies (Fig. 7.3). The problem with these results is that the model predicts the exact same recovery time and dynamics regardless of the size of the disturbance. Observations from abandoned agricultural fields make it clear that this prediction of recovery time is too fast, and suggests that other processes have important influences on recovery at scales larger than a single gap.

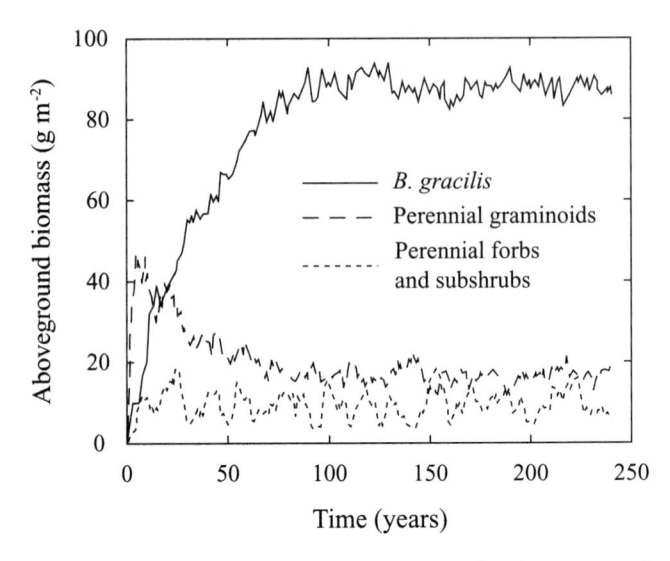

Figure 7.3 Simulated aboveground biomass for the average of 50 plots for 250 years under the condition that *B. gracilis* seeds have a probability less than 1.0 of being present on the plot. (Coffin and Lauenroth [1990a].)

Spatially Explicit Simulations

Seed Dispersal

The discrepancy between our simulated recovery time of *B. gracilis* on an individual plot and the predicted time based upon observations of old fields led us to hypothesize that spatially explicit processes are important as disturbance size increases beyond a single gap. We incorporated spatial structure into STEPPE by considering that each disturbed area consists of a grid of plots in which processes on one plot could affect processes on all neighboring plots (Coffin and Lauenroth, 1989a). The key spatially explicit process we added to the model was seed dispersal. In addition to constraints on germination and establishment, in these simulations the availability of *B. gracilis* seeds to each plot is based on probabilities associated with the production and dispersal of seeds. Seed production is assumed to be a function of *B. gracilis* biomass on each plot and the amount of annual precipitation received the previous year (Coffin and Lauenroth, 1992). Because storage of *B. gracilis* seeds in the soil is low and variable through time and space (Coffin and Lauenroth, 1989c), we assumed that seed availability is primarily a function of seeds produced during the previous year. The probability of seeds dispersing to each plot is assumed to be a function of the distance from the nearest source of seeds, the release height of seeds from the inflorescence, average wind speed, and the aerodynamic properties of *B. gracilis* seeds.

Simulations of disturbances using the spatially explicit version for sizes ranging from 2 to 16 m² show that recovery time of *B. gracilis* increases as disturbance

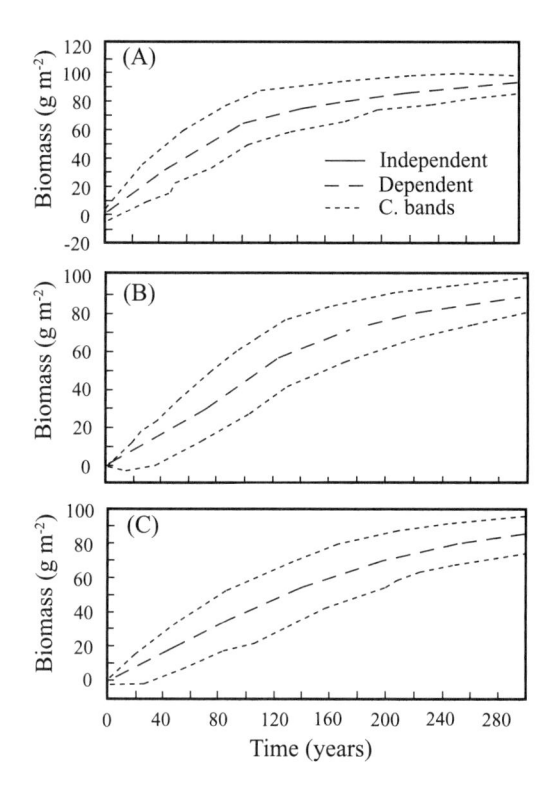

Figure 7.4 Simulated aboveground biomass of *B. gracilis* (average and 95% confidence intervals) for 300 years for two types of landscapes and three disturbance sizes: 2 m² (A), 18 m² (B), and 49 m² (C). (From Coffin and Lauenroth [1989a]. Reprinted with kind permission of Springer Science and Business Media.)

size increases, and that estimates of recovery time are much greater from the spatially explicit than from the spatially independent version (Fig. 7.4 [Coffin and Lauenroth, 1989a]). Recovery in these spatially explicit simulations was defined as the time at which the 95% confidence interval of the simulated biomass includes the steady-state value predicted by the independent simulations (Fig. 7.4). The inclusion of spatial processes associated with seed dispersal increased the recovery time of *B. gracilis* because plots that are not within the maximum dispersal distance after the disturbance cannot receive *B. gracilis* seeds until nearby plots have been colonized. The time required for a particular plot to be colonized by *B. gracilis* is proportional to its distance from the closest plot that has sufficient *B. gracilis* to produce seeds. The most relevant unit for measuring this is D, the maximum dispersal distance. These spatially explicit recovery times are in line with results from studies of abandoned agricultural fields.

Soil Texture

Because disturbances occur over a range of soil textures, and soil properties have important effects on plant processes, we hypothesized that soil texture would influence the recovery of *B. gracilis* for different disturbance sizes (Coffin and

Lauenroth, 1994). We used results from Monte Carlo simulations based on historical weather data and SOILWAT to determine the average annual probability of seedling establishment for five soil texture classes (Lauenroth et al., 1994). We found that the probability of establishment of *B. gracilis* increases as silt content of the soil increases. We simulated effects of soil texture in the same manner as effects of resource availability on rate of growth using a statistical relationship between annual precipitation, aboveground net primary production, and water-holding capacity of the soil (Sala et al., 1988). Our model results show that soil texture is more important than disturbance size to simulated recovery of *B. gracilis*, and constraints on recruitment are more important than constraints on growth (Coffin and Lauenroth, 1994). Fastest recovery occurs on soils with the highest silt content (silt loam); slowest recovery occurs on soils with low silt content and either high (clay) or low (loamy sand) water-holding capacity (Fig. 7.5). Biomass and recovery rate of *B. gracilis* decrease as disturbance size (2–16 m^2) increases and as distance from the disturbed plot to the edge of undisturbed vegetation increases.

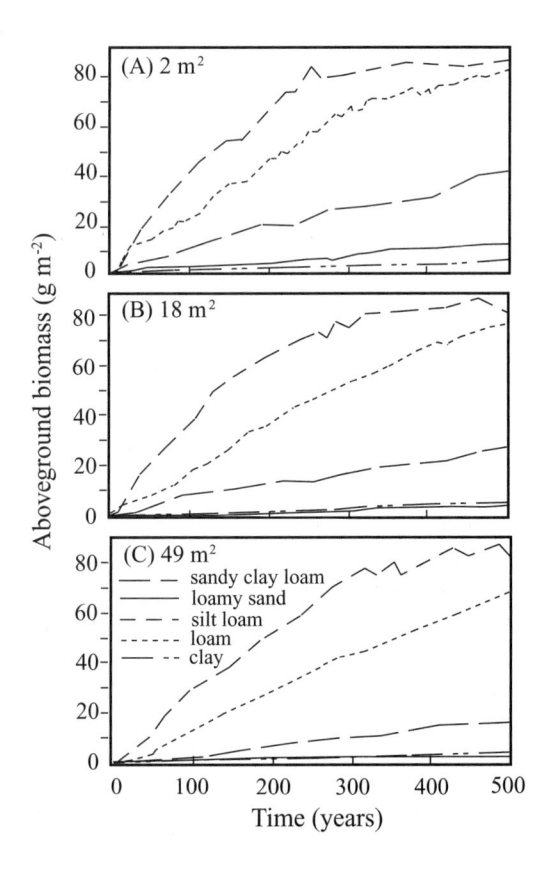

Figure 7.5 Simulated average aboveground biomass of *B. gracilis* for 500 years for three disturbance sizes and five soil texture classes. (From Coffin and Lauenroth [1994]. Reprinted with kind permission of Springer Science and Business Media.)

STEPPE–SOILWAT Simulations

We used a spatially explicit modeling approach to link STEPPE with a model of soil water dynamics to simulate vegetation dynamics and recovery rates of *B. gracilis* on abandoned agricultural fields (Coffin et al., 1993). We evaluated the effects of seed dispersal, weather, soil texture, and nitrogen availability on recovery. Availability of soil water or nitrogen to each plant was simulated as a function of its root distribution with depth, the distribution of resources by depth, and temporal variation in the distributions. We represented plant growth and soil water interactions dynamically using a one-to-one correspondence between a plot simulated by STEPPE and one simulated by SOILWAT. Nitrogen availability was calculated within STEPPE based on a relationship between soil texture and time after abandonment, which was obtained by running simulations with the CENTURY soil process model (Parton et al., 1987; Parton et al., chapter 15, this volume). Plots were arrayed into a grid that allowed them to be interconnected to represent old fields. We simulated two sites in northeastern Colorado with different long-term precipitation and temperature. Within each site, we simulated fields with different soil textures.

Our results show that simulated recovery patterns vary both between and within fields (Coffin et al., 1993). Variability in patterns between fields is the result of differences in soil texture, with the fastest recovery occurring on fields with silt loam soils (Fig. 7.6). Precipitation is less important than soil texture, even though the fastest recovery occurred on fields with the highest precipitation. Distance from the source of seeds and soil texture both have important effects on spatial patterns of recovery within each field. At any point in time, biomass of *B. gracilis* decreases as distance from the edge of a field increases. Biomass also decreases with decreasing silt content.

Although the general pattern of decrease in *B. gracilis* cover with distance from the edge of old fields is represented by the model, our simulated rate of recovery is slow compared with observed patterns (Coffin et al., 1996). Our model was developed assuming that recovery is dependent solely upon wind dispersal of seeds from the undisturbed edge of the field. Thus, these old-field simulations indicate that a mode of dispersal that can operate at distances much greater than wind may be important to the recovery process. Fraleigh (1999) evaluated dispersal distances for *B. gracilis* seeds over a range of wind speeds up to 19 m·s^{-1} and found that 98% of the seeds are found with 8 m of the release point with only a small fraction (0.3%) found beyond 11 m. The maximum dispersal distance by wind in the model is 8 m. Increasing the maximum dispersal distance to 11 m did not speed up recovery enough to match the field data. Biotic factors, such as the dissemination of seeds on the fur of cattle or through consumption and subsequent deposition in fecal pats, are other possible candidates for long-distance seed dispersal into abandoned fields. Fraleigh (1999) found that *B. gracilis* seeds can be transported on the hair of cattle for distances of at least 100 m and probably much further. Fraleigh (1999) also found substantial numbers of *B gracilis* seeds, that could be germinated, in cattle fecal pats. Thus, we have concluded that cattle may be the important link to explain the speed of recovery on abandoned

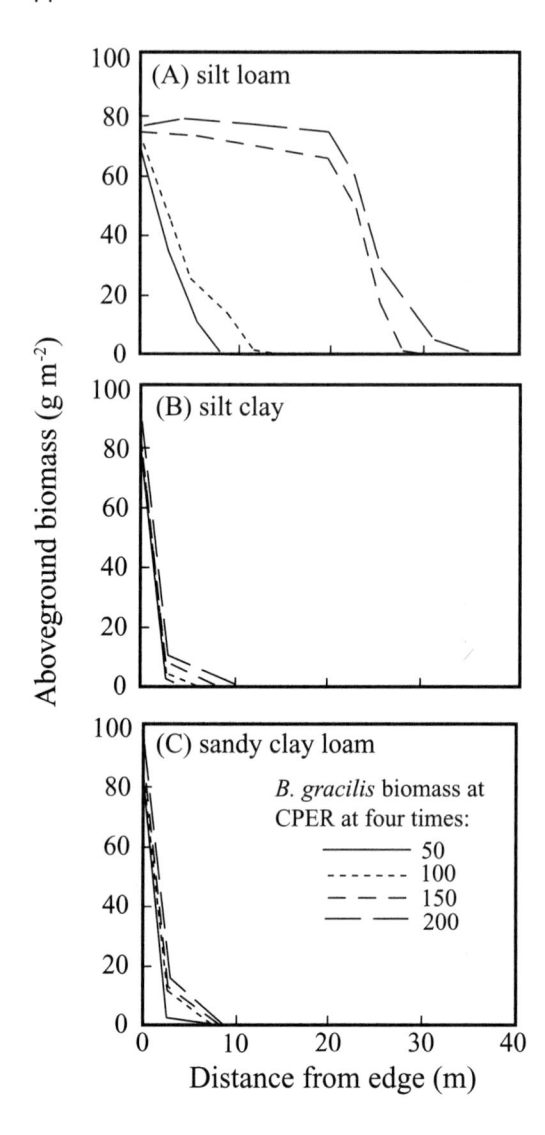

Figure 7.6 Simulated aboveground biomass of *B. gracilis* at the CPER by distance from the edge of field at four times after abandonment (50, 100, 150, 200 years) for three fields with different soil textures. (From Coffin et al. [1993].)

agricultural fields in the shortgrass region. Incorporation of the effects of cattle as dispersal agents is the next logical step in simulating the recovery of *B. gracilis* on large disturbances.

Summary and Conclusions

Simulation modeling provides a powerful approach to synthesizing information about ecological properties and processes to identify key controls on community dynamics and to predict long-term recovery patterns after disturbance.

Our analyses in the shortgrass steppe show that seed dispersal is a key process limiting recovery of *B. gracilis* across a range of disturbance sizes. Our simulation results also indicate the importance of soil properties, especially silt content, to recovery rates of *B. gracilis*, and the greater importance of seedling establishment compared with seedling growth; however, these hypotheses have yet to be tested in the field. For large disturbances, the decrease in *B. gracilis* cover with increasing distance from the source of seeds at the field edge agrees with field and simulation results from small- and intermediate-size disturbances (Coffin and Lauenroth, 1989a, b; 1994). However, our simulated recovery times for large disturbances are longer than for small disturbances, and recovery is not a simple linear function of disturbance size. As disturbance size increases, surface area increases faster than the effective radius over which seeds are dispersed, with a corresponding exponential decrease in probability of seed dispersal into the disturbed area. High between-field variation observed by sampling 13 fields in northeastern Colorado (Coffin et al., 1996) was not captured by this modeling exercise, indicating the importance of other factors not currently included in the model.

Simulation modeling will remain a powerful tool for understanding shortgrass steppe plant communities in the future as the global environment changes and temperatures continue to increase with increases in atmospheric CO_2. In many cases, the ecological consequences of a changing climate are unknown, and the direction and amount of change in precipitation has a high degree of uncertainty for semiarid regions. Models can be used to explore future dynamics under different climate scenarios that include changes in both the amount and timing of precipitation as well as interactions with changes in temperature for a range of soil textures and disturbance sizes and types. These multifactorial, long-term simulations are easy to conduct using models, yet are very challenging and in some cases impossible to examine using experiments. Combining model results with focused short- and long-term experiments and observations will provide new insights into complex ecosystem dynamics.

Acknowledgments This research was supported by an NSF grant (BSR 9011659) to CSU as part of the SGS LTER program. We thank Brandon Bestelmeyer for assistance with the figures, and Tom Crist and one anonymous reviewer for helpful comments on the manuscript.

References

Botkin, D. B., J. F. Janak, and J. R. Wallis. 1972. Some ecological consequences of a computer model of forest growth. *Journal of Ecology* **60**:849–873.

Briske, D. D., and A. M. Wilson. 1977. Temperature effects on adventitious root development in blue grama seedlings. *Journal of Range Management* **30**:276–280.

Briske, D. D., and A. M. Wilson. 1978. Moisture and temperature requirements for adventitious root development in blue grama seedlings. *Journal of Range Management* **31**:174–178.

Coffin, D. P., and W. K. Lauenroth. 1988. The effects of disturbance size and frequency on a shortgrass plant community. *Ecology* **69**:1609–1617.

Coffin, D. P., and W. K. Lauenroth. 1989a. Disturbances and gap dynamics in a semiarid grassland: A landscape-level approach. *Landscape Ecology* **3**(1):19–27.

Coffin, D. P., and W. K. Lauenroth. 1989b. Small scale disturbances and successional dynamics in a shortgrass community: Interactions of disturbance characteristics. *Phytologia* **67**(3):258–286.

Coffin, D. P., and W. K. Lauenroth. 1989c. The spatial and temporal variability in the seed bank of a semiarid grassland. *American Journal of Botany* **76**(1):53–58.

Coffin, D. P., and W. K. Lauenroth. 1990. A gap dynamics simulation model of succession in the shortgrass steppe. *Ecological Modelling* **49**:229–266.

Coffin, D. P., and W. K. Lauenroth. 1991. Effects of competition on spatial distribution of roots of blue grama. *Journal of Range Management* **44**:67–70.

Coffin, D. P., and W. K. Lauenroth. 1992. Spatial variability in seed production of the perennial bunchgrass *Bouteloua gracilis* (H.B.K.) Lag. ex Griffiths. *American Journal of Botany* **79**:347–353.

Coffin, D. P., and W. K. Lauenroth. 1994. Successional dynamics of a semiarid grassland: Effects of soil texture and disturbance size. *Vegetatio* **110**:67–82.

Coffin, D. P., and W. K. Lauenroth. 1996. Regional analysis of transient responses of grasslands to climate change. *Climate Change* **34**:269–278.

Coffin, D. P., W. K. Lauenroth, and I. C. Burke. 1996. Recovery of vegetation in a semiarid grassland 53 years after disturbance. *Ecological Applications* **6**:538–555.

Coffin, D. P., W. K. Lauenroth, and I. C. Burke. 1993. Spatial dynamics in recovery of shortgrass steppe ecosystems. *Lectures on Mathematics in the Life Sciences* **23**:75–108.

Coffin, D. P., and D. L. Urban. 1993. Implications of natural history traits to system-level dynamics: Comparisons of a grassland and a forest. *Ecological Modelling* **67**:147–178.

Dodd, J. L., and W. K. Lauenroth. 1975. Responses of *Opuntia polyacantha* to water and nitrogen perturbations in the shortgrass prairie, pp. 229–240. In: M. K. Wali (ed.), *Prairie: A multiple view*. University of North Dakota, Grand Forks, N. Dak.

Fraleigh, H. D., Jr. 1999. *Seed dispersal of two important perennial grasses in the shortgrass steppe*. Masters thesis, Colorado State University, Fort Collins, Colo.

Lauenroth, W. K., C. D. Canham, A. P. Kinzig, K. A. Poiani, W. M. Kemp, and S. W. Running. 1998. Simulation modeling in ecosystem science, pp. 404–415. In: M. L. Pace and P. M. Groffman (eds.), *Successes, limitations and frontiers in ecosystem science*. Springer-Verlag, New York.

Lauenroth, W. K., and D. P. Coffin. 1992. Belowground processes and the recovery of semiarid grasslands from disturbance, pp. 131–150. In: M. K. Wali (ed.), *Ecosystem rehabilitation*. Vol 2. Ecosystem analysis and synthesis. SPB Academic Publishing, The Hague, the Netherlands.

Lauenroth, W. K., J. L. Dodd, and P. L. Sims. 1978. The effects of water- and nitrogen-induced stresses on plant community structure in a semiarid grassland. *Oecologia* **36**:211–222.

Lauenroth, W. K., O. E. Sala, D. P. Coffin, and T. B. Kirchner. 1994. The importance of soil water in the recruitment of *Bouteloua gracilis* in the shortgrass steppe. *Ecological Applications* **4**:741–749.

Noy-Meir, I. 1973. Desert ecosystems: Environment and producers. *Annual Review of Ecology and Systematics* **4**:25–51.

Parton, W. J. 1978. Abiotic section of ELM, pp. 31–53. In: G. S. Innis (ed.), *Grassland simulation model*. Springer-Verlag, New York.

Parton, W. J., D. S. Schimel, C. V. Cole, and D. S. Ojima. 1987. Analysis of factors controlling soil organic matter levels in Great Plains grasslands. *Soil Science Society of America Journal* **51**:1173–1179.

Sala, O. E., W. K. Lauenroth, and W. J. Parton. 1992. Long term soil water dynamics in the shortgrass steppe. *Ecology* **73**:1175–1181.

Sala, O. E., W. J. Parton, L. A. Joyce, and W. K. Lauenroth. 1988. Primary production of the central grassland region of the United States. *Ecology* **69**:40–45.

Shugart, H. H. 1984. *A theory of forest dynamics*. Springer-Verlag, New York.

Watt, A. S. 1947. Pattern and process in the plant community. *Journal of Ecology* **35**:1–22.

8

Ecology of Mammals of the Shortgrass Steppe

Paul Stapp
Beatrice Van Horne
Mark D. Lindquist

At first glance, the shortgrass steppe seems to offer little in the way of habitat for mammals. The expansive rolling plains, with little topographic relief or vegetative cover, provide minimal protection from predators or the harsh weather typical of the region. The short stature of the dominant native grasses prevents the development of any significant litter layer, and although snowfall can often be significant, too little accumulates to form the subnivean habitats that support small mammal populations in forests and more productive grasslands in winter. As a consequence, ecologists have typically considered the vertebrate fauna of the shortgrass steppe to be depauperate compared with other Great Plains grasslands, a hardy collection of generalists living in sparse populations. Although this characterization may generally be accurate, it has led mammalian ecologists to overlook the fauna of the shortgrass steppe in favor of that of other grasslands. It is precisely these circumstances, however, that suggest that a long-term approach may be necessary to understand the dynamics of mammal populations here. Relatively few such studies have been completed to date, but we can use the comparative and experimental results that are available to begin to determine what factors might be important.

Here we review research on mammals in the shortgrass steppe, with the goal of identifying the general patterns and processes that contribute to them. Our review is roughly divided into four parts. We begin by describing the mammal communities and their broad habitat associations in shortgrass steppe environments. We then review the history of mammal research in the region to synthesize what these studies (many unpublished) have taught us about the most important determinants of the distribution and abundance of native species. Studies of mammal populations in the northern shortgrass steppe have spanned nearly 40 years, and we next describe some major patterns that have emerged from studies during this period. Much of this past research focused on the role of mammals in the structure and function of shortgrass steppe ecosystems, and we revisit this issue in some detail, with special emphasis on the important and sometimes controversial role of prairie dogs and other burrowing rodents. Finally, we end by considering how humans, and especially agriculture and its related activities, affect the diversity, abundance, and persistence of resident mammal populations.

We make three caveats at the outset. First, most of the long-term research on shortgrass steppe mammals, as well as our own work, was conducted as part of Grassland Biome studies of the U.S. IBP and the NSF SGS LTER project. Our review thus emphasizes results from research at a single site in north–central Colorado, which currently encompasses only about 23% of the environmental variation in the shortgrass steppe region (Burke and Lauenroth, 1993), and thus may not be representative of all shortgrass steppe ecosystems. Second, because of their relative abundance and ease of capture and/or detection, many of the studies conducted during these programs focused on small mammals, principally rodents and rabbits. Although some information is available on species of economic importance (i.e., carnivores and ungulates), our research and this review reflects a bias toward small mammals. Last, studies of the shortgrass steppe have focused explicitly on areas with grassland cover. We advocate a broader view of the shortgrass steppe as a mosaic of habitat types, which includes shrublands, riparian zones, escarpments associated with permanent streams, and row–crop agricultural lands. Several mammals, including most bats, shrews, and species associated with eastern grasslands (e.g., eastern cottontails [*Sylvilagus floridanus*], eastern woodrats [*Neotoma floridana*], Virginia opossum [*Didelphis virginiana*]) are restricted largely to riparian areas and are not considered in detail here. However, most native species are found in multiple habitat types, all of which may be important for understanding population and community dynamics at landscape and regional scales.

The Shortgrass Steppe as a Habitat for Mammals

As defined by Lauenroth and Milchunas (1991), the shortgrass steppe is located in the central Great Plains, bounded on the west by the Front Range of the Rocky Mountains and on all other sides by mixed-grass prairie or by desert. Geopolitically, the shortgrass steppe encompasses approximately the eastern third of both Colorado and New Mexico, the western Texas and Oklahoma panhandles, and southwestern Kansas, as well as small areas of southern Wyoming and Nebraska. The dominant plant communities have been described elsewhere (Lauenroth, chapter 5, this volume; Lauenroth and Milchunas, 1991), but a few points merit mention here because of their likely influence on mammal populations. First, with the exception of breaks and riparian strips near large waterways such as the South Platte, Arkansas, and Canadian rivers, most of the shortgrass steppe is characterized by rolling hills of shortgrass vegetation, dominated by two perennial grasses: blue grama (*Bouteloua gracilis*) and buffalograss (*Buchloë dactyloides*). Succulents (e.g., prickly pear [*Opuntia polyacantha*]), midgrasses, dwarf shrubs, and forbs are interspersed among grass plants, but most of the vegetation is short (<25 cm) and provides almost no overhead cover. For most mammals, the spreading growth form of the dominant grasses and relatively compact soil between individual plants makes burrowing difficult, so that many species may depend on burrows constructed by other animals. Moreover, because of this growth form and the relatively small seeds of the two dominant grasses, the forb and grass species that are important seed sources for granivores in other ecosystems are relatively scarce in the shortgrass steppe.

Because of the scarcity of suitable cover and seeds in most upland areas, the patchy areas of shrub vegetation associated with low-lying areas, ridgetops, and sandy soils are often centers of activity and abundance for many mammals. In north–central Colorado, fourwing saltbush (*Atriplex canescens*) is the most abundant large shrub, and is usually found in wide bands near ephemeral streams and other bottomlands. Farther east and throughout the remainder of the region, sand–sage (*Artemesia filifolia*) and soapweed (*Yucca glauca*) are the dominant large shrubs. Additionally, areas where soils are more coarsely textured and where animal disturbances are more frequent support a greater variety and higher densities of grasses and forbs that produce palatable seeds.

The predominant human land use of the shortgrass steppe is cattle grazing, but direct and indirect consequences of other agricultural activities also influence availability and quality of habitats for mammals. Areas of high primary productivity associated with cultivation and rural land development, in particular, contrast strongly with the surrounding grasslands. These include roadside and fence-line right-of-ways, abandoned and fallow croplands, and small areas of native prairie left by irregular crop or irrigation patterns. Such areas are often occupied by small mammals and serve as important foraging areas for carnivores. Windbreaks and abandoned homesteads provide roosting and nesting sites for bats and for the raptors that prey on mammals.

Mammals of the Shortgrass Steppe

A total of 68 species of mammals inhabit the shortgrass steppe region (Table 8.1); 25 of these tend to be restricted to well-developed riparian zones. This list does not include species such as bison (*Bison bison*), black-footed ferrets (*Mustela nigripes*), or gray wolves (*Canis lupus*), which were extirpated during the past two centuries, or introduced commensal species such as the house mouse (*Mus musculus*) that occasionally are found near agricultural or populated areas. Because of the juxtaposition of the shortgrass steppe with the Rocky Mountains, mixed- and tallgrass prairies, and southwestern deserts, the native fauna are associated with several biogeographic provinces. Armstrong (1972) categorized most shortgrass steppe species in Campestrian or Chihuahuan faunal elements, reflecting their affinities with other grassland and arid environments. Some 30%, however, are widespread species associated with multiple biogeographic regions and with distributions covering much of North America.

Table 8.1 Mammals of the Shortgrass Steppe

Order	Families, No. of Species	Representative Species
Marsupialia	Didelphidae (1)	Virginia opossum (*Didelphis virginiana*)
Insectivora	Soricidae (5)	Least shrew (*Cryptotis parva*)
	Talpidae (1)	Eastern mole (*Scalopus aquaticus*)
Chiroptera	Vespertilionidae (11)	Western small-footed myotis (*Myotis ciliolabrum*)
		Big brown bat (*Eptesicus fuscus*)
	Molossidae (2)	Brazilian free-tailed bat (*Tadarida brasiliensis*)
Xenarthra	Dasypodidae (1)	Nine-banded armadillo (*Dasypus novemcinctus*)
Lagomorpha	Leporidae (4)	Desert cottontail (*Sylvilagus audubonii*)
		Black-tailed jackrabbit (*Lepus californicus*)
		White-tailed jackrabbit (*L. townsendii*)
Rodentia	Sciuridae (3)	Spotted ground squirrel (*Spermophilus spilosoma*)
		Thirteen-lined ground squirrel (*S. tridecemlineatus*)
		Black-tailed prairie dog (*Cynomys ludovicianus*)
	Geomyidae (4)	Plains pocket gopher (*Geomys bursarius*)
	Heteromyidae (7)	Ord's kangaroo rat (*Dipodomys ordii*)
		Hispid pocket mouse (*Chaetodipus hispidus*)
		Silky pocket mouse (*Perognathus flavus*)
		Plains pocket mouse (*P. flavescens*)
	Castoridae (1)	American beaver (*Castor canadensis*)
	Muridae (13)	Northern grasshopper mouse (*Onychomys leucogaster*)
		Deer mouse (*Peromyscus maniculatus*)
		Western harvest mouse (*Reithrodontomys megalotis*)
		Plains harvest mouse (*R. montanus*)
		Prairie vole (*Microtus ochrogaster*)
		Southern Plains woodrat (*Neotoma micropus*)
		Cotton rat (*Sigmodon hispidus*)
	Erethizontidae (1)	Common porcupine (*Erethizon dorsatum*)
Carnivora	Canidae (3)	Coyote (*Canis latrans*)
		Swift fox (*Vulpes velox*)
	Procyonidae (1)	Raccoon (*Procyon lotor*)
	Mustelidae (6)	American badger (*Taxidea taxus*)
		Long-tailed weasel (*Mustela frenata*)
		Striped skunk (*Mephitis mephitis*)
		Eastern spotted skunk (*Spilogale putorius*)
	Felidae (1)	Bobcat (*Felis rufus*)
Artiodactyla	Cervidae (2)	Mule deer (*Odocoileus hemionus*)
	Antilocapridae (1)	Pronghorn (*Antilocapra americana*)
Total no. of species	68	
Excluding strictly riparian species	43	

This list includes species found in grasslands, shrublands, escarpments, and riparian cottonwood and willow stands, but omits exotic species and those that have been extirpated. (Compiled from Armstrong [1972], Blair [1954], Caire et al. [1989], Findley et al. [1975], and Fitzgerald et al. [1994]. Taxonomy and vernacular names follow Jones et al. [1997].)

Most species occur across the entire shortgrass steppe region. Desert cottontails (*Sylvilagus audubonii*) and black-tailed jackrabbits (*Lepus californicus*) are the most widespread rabbits, reaching highest population densities in areas with extant burrows or the cover provided by shrubs or taller grasses (Dano, 1952; Flinders and Hansen, 1975). Rodents have the highest diversity (29 species), with sigmodontines (Muridae) and sciurids comprising most of the individuals and most of the small-mammal biomass. Excluding gophers, six to eight species of rodents may be present on a given location (Hall and Willig, 1994; McCulloch, 1959; Mohamed, 1989; Moulton et al., 1981b), although most species are relatively rare. Northern grasshopper mice (*Onychomys leucogaster*), deer mice (*Peromyscus maniculatus*), Western harvest mice (*Reithrodontomys megalotis*), and Ord's kangaroo rats (*Dipodomys ordii*) are the most frequently captured nocturnal rodents, but thirteen-lined ground squirrels (*Spermophilus tridecemlineatus*) and, farther east and south, spotted ground squirrels (*S. spilosoma*) comprise most of the rodent biomass (Fig. 8.1) (Green, 1969; McCulloch, 1959; Mohamed, 1989). Throughout the shortgrass steppe, black-tailed prairie dogs (*Cynomys ludovicianus*) form densely populated colonies where soil and topographic conditions permit establishment of preferred plant species. Swift foxes (*Vulpes velox*), coyotes (*Canis latrans*), badgers (*Taxidea taxus*), and pronghorn (*Antilocapra americana*) are the most common and conspicuous larger mammals (Fig. 8.2).

Many of the remaining fauna show particular affinities to northern or southern areas of the shortgrass steppe, divided roughly by the elevated mesas and broken country of the Raton section of southeastern Colorado and western Oklahoma. Several genera have representatives unique to northern or southern regions, including shrews (*Sorex merriami, Notiosorex crawfordi*), pocket mice (*Perognathus fasciatus, P. merriami*), pocket gophers (*Thomomys talpoides, Cratogeomys castanops*), and woodrats (*N. cinerea, N. micropus*). Some species are restricted to the north, such as white-tailed jackrabbits (*L. townsendii*) and meadow voles (*Microtus pennsylvanicus*), whereas others are primarily southern in distribution but may be expanding their ranges north and west, such as armadillos (*Dasypus novemcinctus*) and cotton rats (*Sigmodon hispidus*).

Biogeographic patterns reflect the operation of ecological interactions in the context of historical changes in climate and vegetation. French et al. (1976) developed a conceptual model that described the ecological and evolutionary relationships among small-mammal communities in North American grasslands. They noted that the small-mammal fauna of the shortgrass steppe is dominated by omnivorous rodents, notably sciurids, most of which occur at relatively low population densities. With increasing aridity after the Pleistocene glaciation, folivores such as voles followed retreating grasslands northward, and were in turn replaced by granivores in the southern prairies and deserts. They suggested that grass eaters had a competitive advantage by nature of their high reproductive rates, but their abandonment of the shortgrass steppe led to invasion by the omnivores and the few granivores that characterize the current fauna. Grant and Birney (1979) reached similar conclusions in a comparative analysis of small-mammal community structure. They asserted that the small-mammal fauna of North American grasslands could be categorized on the basis of the amount of vegetative cover

Figure 8.1 Thirteen-lined ground squirrels (A) and northern grasshopper mice (B) are the most common and widespread small mammals in northern areas of shortgrass steppe. (Photos A and B by Paul Stapp.)

(Table 8.2). They concluded that the shortgrass steppe is a marginal habitat for herbivores because of the lack of herbage supply and quantity, and the absence of a significant litter layer. Most specialized granivores are limited by the low seed-to-vegetation production ratios. As a result, omnivores predominate, but both density and diversity are relatively low. Although recent studies confirm that

Figure 8.2 (A–C) The swift fox (A), American badger (B), and pronghorn (C) are the most conspicuous large mammals. (Photos A and B by Paul Stapp.)

Table 8.2 Patterns of Small-Mammal Community Structure in North American Grasslands

	Vegetative Cover	Density and Biomass	Species Diversity	Representative Groups
Tallgrass prairie	High	High	Low	Herbivores with high reproductive potential
Shortgrass steppe	Intermediate	Low	Low	Generalist omnivores, long-lived and seasonally active
Desert grasslands	Low	High	High	Granivores, long-lived pulse breeders

Superimposed upon the gradient from high to low cover is a north–south gradient that reflects variation in the harshness of weather conditions and hence seasonal activity of small mammals.
(From Grant and Birney [1979].)

population densities are relatively low (discussed later), we suggest that a broader landscape view of the shortgrass steppe offers a more complete view of mammalian diversity. Ignoring species that are largely restricted to riparian areas, some 43 mammalian species are found in the region, including 22 species of rodents (Table 8.1). These totals compare favorably with species richness in the tallgrass prairie (46 species, 19 rodents [Finck et al., 1986]) and in desert grasslands (56 species, 33 rodents [Parmenter and Van Devender, 1995]).

Mammal Research in the Shortgrass Steppe

Although mammals of the shortgrass steppe have been the focus of distributional and natural history studies for more than a century, concerted efforts to understand the patterns and dynamics of mammal populations and their role in the shortgrass steppe began with the U.S. IBP Grassland Biome studies during the late 1960s. Two sites represented shortgrass ecosystems: the Pawnee site, in north–central Colorado; and the Pantex site, in the western Texas Panhandle, representing the southern shortgrass steppe. The Pawnee site, located on the USDA–ARS CPER, was the most intensively studied of the 14 Grassland Biome sites, principally by researchers from CSU. Pantex was considered a second-order research site, and was studied primarily by scientists at Texas Tech University.

A primary goal of the Grassland Biome project was to develop and parameterize simulation models to capture the structure and dynamics of grassland ecosystems. The scope of the mammal research reflected this goal. Because of the comparative nature of the IBP studies, early research focused on describing the natural history and distribution of abundance and biomass of representative mammals. Excluding grazing studies, the vast majority of the research effort was focused on small mammals. Packard (1972, 1975) compared temporal changes in abundance and species composition of small-mammal communities on the

Pantex and Jornada Desert Grassland sites. On the Pawnee, Flake (1971, 1973, 1974) compiled information on the diet, reproductive cycles, and abundance of rodents, whereas Gross (1969), Donoho (1971), and Flinders and Hansen (1972, 1973, 1975) analyzed diet and habitat associations of rabbits. Marti (1974) and Olendorff (1973) studied ecology of raptors and owls, respectively, which are major predators of small mammals. Others (Ellis and Travis, 1975; Peden et al., 1974; Schwartz and Ellis, 1981; Schwartz and Nagy, 1976; Sparks, 1972) documented and compared diet and foraging behavior of native ungulates with cattle.

In 1982, the Pawnee site was designated an LTER area (SGS LTER) by the NSF. It was expanded in 1997 to include the adjacent USDA Forest Service PNG. Unlike the Grassland Biome studies, there was little mammal-related research early during the SGS LTER project. Between 1987 and 1989, L. McEwen of CSU trapped small mammals on grassland and saltbush-dominated sites that had previously been sprayed with pesticides. McEwen's population studies continued from 1990 to 1995 on saltbush sites, with the objective of tracking changes in population density and species composition over time.

Recognizing that there was little current information on the status of mammals on the site, in 1994 we began long-term population monitoring programs to track changes in abundance and species composition of mammals in representative cover types (Stapp, 1996). The goals of these long-term studies were to facilitate comparative studies among other grassland and desert sites in the LTER network using the same research design, and to provide baseline information for other comparative and experimental research on the SGS LTER site. These studies tracked populations of three major groups: nocturnal rodents, rabbits, and mammalian carnivores. Rodents were live-trapped twice per year on 3.14-ha trapping webs at three upland shortgrass prairie sites and three sites in saltbush-dominated areas. Rabbits were surveyed once each season by spotlight counts along a 32-km route of pasture and gravel county roads. At the same time, we counted scats of coyotes and swift foxes along the same route as an index of temporal changes in the activity and abundance of these carnivores. These studies were ongoing at the time of writing, and over time will provide a clearer picture of the temporal dynamics of mammal populations in the shortgrass steppe and the ecological factors that contribute to them.

Researchers from outside the immediate scope of the Grassland Biome and SGS LTER projects have also contributed much to our knowledge of shortgrass steppe mammals. For example, Fitzgerald and his students from the University of Northern Colorado provided crucial information on population ecology and management of swift foxes in eastern Colorado (Cameron, 1984; Eussen, 1999; Finley, 1999; Loy, 1981; Roell, 1999). The Piñon Canyon Maneuver Site in southeastern Colorado, which contains some areas of shortgrass steppe grasslands, was the focus of recent studies of carnivore and ungulate populations (e.g., Covell, 1992; Firchow, 1986; Gerlach, 1987; Gese et al., 1988, 1989; Rongstad et al., 1989). Pojar et al., in the Colorado Division of Wildlife (Pojar, 1988; Pojar et al., 1995) updated estimates of pronghorn population densities in the eastern Plains. Choate and Fleharty from Fort Hays State University, and Moulton from Texas Tech University, contributed greatly to our knowledge of the natural history of small mammals in native and agricultural areas of the southern shortgrass steppe (Choate and Pinkham, 1988; Choate and Reed, 1988; Fleharty and Navo, 1983; Lovell et al., 1985; Mellott and Fleharty, 1986; Moulton et al., 1981a, b, 1983). Lastly, much of what we know about the ecology of prairie dogs and associated plant and animal communities from the south–central shortgrass steppe is the result of work by Cully and his students at Kansas State University (e.g., Cully and Williams, 2001; Kretzer and Cully, 2001a, b; Winter et al., 2002, 2003).

What Factors Determine the Distribution and Abundance of Shortgrass Steppe Mammals?

Based on the ecological studies that have been conducted to date, we can begin to evaluate the relative importance of various factors as determinants of local distribution and abundance of resident mammals. These can be conveniently divided into physical and biotic factors, although many of these are obviously interrelated.

Physical Factors: Weather and Soils

The climate of the shortgrass steppe is semiarid, with most of the 311 mm of average annual precipitation falling in convective rainstorms during the April-to-September growing season (Pielke and Doesken, chapter 2, this volume). Daytime temperatures exceeding 40 °C are common in July and August, winter minimum temperatures in the north are well below freezing for several months, and wind is omnipresent. To reduce exposure to inclement weather and predators, many year-round residents seek shelter belowground or restrict activity to localized areas of dense vegetation.

To date there is little consensus on the influence of weather on mammal populations, largely because we lack the long-term data necessary to evaluate these effects. Moreover, responses to weather are, of course, confounded with the indirect changes in food availability and habitat quality that usually accompany seasonal and annual changes in weather patterns. Some herbivorous mammals (e.g., pocket gophers [Vaughan, 1967], and cotton rats and harvest mice [Packard, 1975]) increase in number after wet years, whereas abundance of other species (thirteen-lined ground squirrels [Packard, 1972] and white-tailed jackrabbits [Lim, 1987]) may be inversely related to the amount of precipitation. On both grassland and salt-bush sites in the northern shortgrass steppe, the amount of area covered by mounds of northern pocket gophers during early summer is highly correlated ($r=.91$) to winter–spring precipitation the previous year, suggesting that weather-related increases in plant productivity and/or overwinter survival result in an increase in gopher populations (n=7 years, 1999–2005; P. Stapp, unpublished data).

The size and growth of black-tailed prairie dog colonies also appears to be related to weather, although the mechanisms involved are still not clear. In an analysis of variation in an active area of colonies for a 20-year period on the Pawnee Grasslands, Stapp et al. (2004) found that extinctions of colonies were associated with periods of higher than average winter temperature and precipitation that often accompany El Niño Southern Oscillation (ENSO) events. Most extinctions were attributed to plague, a flea-vectored disease caused by the introduced bacterium *Yersinia pestis*, which causes more than 90% mortality in prairie dogs (Cully and Williams, 2001). Large colonies were as likely to go extinct as small ones, presumably because transmission of plague is density dependent. The change in colony active area between years was inversely related to both fall–winter (October–April) minimum temperature ($r = -.60$, $P = .003$, $n = 22$ years) and precipitation ($r = -.55$, $P = .003$, $n = 24$ years [1981–2003]), suggesting that colonies increased in size after colder, drier periods. On the surface, this result is surprising, because one would expect that warm, wet conditions would result in higher plant productivity and, therefore, higher food availability for prairie dogs, and hence higher population size. This argument assumes, however, that there is a positive relationship between colony active area and population density, which has not been established. Alternatively, increases in active area may actually reflect a *reduction* in population density if, during dry years, prairie dogs consume all the forage in the colony center and are forced to move to the periphery, expanding the colony perimeter and, therefore, the estimated active area. If large colonies do not necessarily support higher population densities than smaller colonies, then the apparent increased risk of plague extinction for large colonies may, in part, be the result of some factor other than the density of prairie dog hosts, such as increased activity of carnivores or other small mammals that may be involved in the spread of plague.

By comparison, the effects of soil properties on grassland mammal populations are better understood. Soil texture and depth determine the ease with which mammals can construct and maintain burrows, which may have a significant influence on the diversity and abundance of flora and fauna of the shortgrass steppe. Prairie dogs tolerate a range of soil types, but colonies tend to be located in deep, well-drained alluvial soils with medium to fine texture, and less often in sandy or coarse rocky soils (Koford, 1958). In northern Colorado, most colonies are located in loam or fine sandy loam soils associated with swales and ephemeral drainages (Table 8.3). Soil texture and depth are particularly important to pocket gophers, which create small mounds or mound networks as a result of their tunneling and foraging belowground. Miller (1964) indicated that all four species of pocket gopher in the shortgrass steppe prefer light-textured soils, and suggested that differences in competitive ability explained local distribution of species in areas of sympatry. Moulton et al. (1983) similarly reported competition between Botta's (*T. bottae*) and yellow-faced (*Pappogeomys = C. castanops*) pocket gophers, but concluded that the disjunct biogeographic distribution of *C. castanops* and plains pocket gopher (*Geomys bursarius*) resulted from differences in soil preference and anthropogenic changes in land-use practices rather than competition. *Thomomys talpoides* is the most common species of pocket gopher in the northwestern shortgrass steppe and in the SGS LTER study area, although the larger *G. bursarius* inhabits deeper alluvial soils associated with riparian and disturbed areas (Fitzgerald et al., 1994). Both the size and density of mounds increases with increasing soil particle size (Table 8.3), although *T. talpoides* tend

Table 8.3 Mean Density, Area, and Percent Cover of Disturbances of Pocket Gophers and Prairie Dogs in Different Soil Types on the Northern Shortgrass Steppe

Species and Soil Type	No. of Sites	Burrow/Mound Density, no./ha^{-1}	Mound Area, m^2	Cover of Mounds and Burrows, %
Prairie dogs (within colony)				
Loam/fine sandy loam	10	95	1.49	1.41
Pocket gophers				
Loam	10	119	0.64	0.77
Fine sandy loam	10	239	0.84	2.01
Sandy loam	8	388	1.73	6.73
Loamy sand	9	994	1.62	16.12

Percent cover was calculated by multiplying mound density by mound area. Only gopher mounds >0.2 m^2 in area were included. Soil are types based on Soil Conservation Survey map units. (P. Stapp, unpublished data.)

to be rare or absent in most sandy soils. Edaphic factors are also important for species that live aboveground, but these effects are mediated by the effects of soils on plant species and functional type diversity, and hence on habitat and food.

Biotic Factors: Vegetation Structure

As mentioned earlier, the availability of vegetative cover is probably the most significant single determinant of mammalian abundance, biomass, and diversity in the shortgrass steppe (Fig. 8.3). The presence of taller grasses, forbs, and shrubs alters small-mammal species composition and productivity by providing cover from predators and inclement weather, as well as nest materials for species such as harvest mice. At the SGS LTER site, rodent species diversity and density are approximately two times higher in areas with saltbush than without (Table 8.4), although total rodent biomass is roughly equivalent. Soils in the saltbush areas tend to be coarsely textured, and intermediate-height grasses such as *Stipa comata*, *Agropyron smithii*, and *Oryzopsis hymenoides* are also common. In addition to ground squirrels, four species of nocturnal rodents are captured regularly in saltbush areas, and prairie voles (*M. ochrogaster*) and hispid pocket mice (*Chaetodipus hispidus*) sometimes invade our trapping sites after wet years (P. Stapp, unpublished data). Stapp and Van Horne (1997) found that the population density of deer mice was correlated with the density and spatial pattern of shrubs in the shortgrass steppe. Mice tended to orient their movements toward shrubs in areas with few shrubs, but on sites with higher shrub canopy cover (>10%), mice showed no detectable preference for shrubs, possibly because they were able to achieve the protective benefits of shrub cover without actually moving beneath them.

In contrast to saltbush sites, grasshopper mice and thirteen-lined ground squirrels are the only species captured consistently on upland shortgrass sites (Table 8.4). Grasshopper mice are more numerous in saltbush areas and their abundance increases in both shrub and grassland sites with the abundance of gopher mounds (Stapp, 1997b). Ground squirrels are found in both grassland and saltbush habitats,

Figure 8.3 Areas with greater vegetative cover as such large shrubs (*Atriplex canescens*) and roadside verges are centers of activity and diversity. (Photo by Mark Vandever.)

but tend to be more abundant and have higher productivity in grasslands (Table 8.4). Black-tailed jackrabbits and desert cottontails favor areas with greater vegetative cover (Flinders and Hansen, 1975). More than 60% of black-tailed jackrabbits sighted during our roadside surveys were recorded in saltbush-dominated areas, even though saltbush represents only 27% of the vegetative cover along the survey route (P. Stapp, unpublished data). Desert cottontails tend to be most common near roadside ditches, in active and abandoned prairie dog colonies, and near livestock corrals.

Table 8.4 Rodent Population Densities (Measured in Number per Hectare) in Upland Prairie and Saltbush-Dominated Grasslands

Species	Weight, g	Upland Prairie	Saltbush Grasslands
Northern grasshopper mouse (O/I)	33	0.92	1.69
Deer mouse (O)	21	0.17	1.89
Western harvest mouse (O/G)	12	0	0.58
Ord's kangaroo rat (G)	69	0	2.98
Thirteen-lined ground squirrel (O)	124	3.69	1.84
Nocturnal rodent species richness	—	1.84	4.04
Mean rodent density, no./ha^{-1}	—	4.81	10.20
Mean rodent biomass, g·ha^{-1}	—	492	575

For nocturnal species, values are means based on captures from three 3.14-ha trapping webs in each habitat type sampled in May and September trapping sessions from 1994 to 2006, beginning in September 1994 (n = 25 trapping sessions). Each web consisted of 124 traps (12 lines of 10 traps at a 10-m spacing, with four traps at the center), which were set for four consecutive nights in each session. For ground squirrels, values are means of numbers of unique individuals captured in late June, prior to the emergence of young-of-year, from 1999 to 2006. Traps were set at 20-m intervals (62 traps/perweb) and checked for four consecutive mornings. For both nocturnal species and squirrels, density was calculated by dividing the number of unique individuals captured by web area (3.14 ha). Plains harvest mice, silky pocket mice (upland prairie), hispid pocket mice, and prairie voles (saltbush grasslands) are also occasionally captured but are rare. Letters in parentheses indicate trophic level: G, granivore; I, insectivore; O, omnivore.

Roadside vegetation is an important habitat for many rodents as well. These narrow habitats are only grazed or graded irregularly, and the overgrown vegetation, friable soils, and high seed densities support populations of rodents that are otherwise missing from shortgrass areas (Fig. 8.4 [Abramsky, 1978]). Roadside areas may also function as dispersal corridors and provide refugial habitats during periods of environmental stress. For example, deer mice were absent on all long-term trapping plots in late summer 1997, but a few individuals remained in roadside vegetation nearby. These individuals likely contributed to the recovery of deer mouse populations on plots the following year. In response to the high abundance of prey, carnivores such as swift foxes travel and hunt in roadside and other disturbed habitats (c.f., Cameron, 1984; Roell, 1999), and the availability of perches and roosts associated with roads and fence lines attract raptors such as owls (Zimmerman et al., 1996). For small mammals, life in roadside habitats therefore may reflect a tradeoff between higher cover and food availability and increased risk of encountering predators (Stapp and Lindquist, 2007).

These comparative studies show the striking differences in small-mammal populations between areas that differ in vegetative cover. Experimental manipulations of nutrient and water stress demonstrated that changes in primary productivity of shortgrass vegetation also markedly affect small-mammal productivity and community structure. Between 1971 and 1974, Lauenroth et al. (1978) added nitrogen, water, and the two in combination to 1-ha plots, with two replicates of each treatment and two plots serving as controls, to determine effects of nutrient and water limitation on primary productivity and plant communities. At the same time, Grant et al. (1977) examined changes in small-mammal productivity and community structure in response to treatment-related changes in vegetation

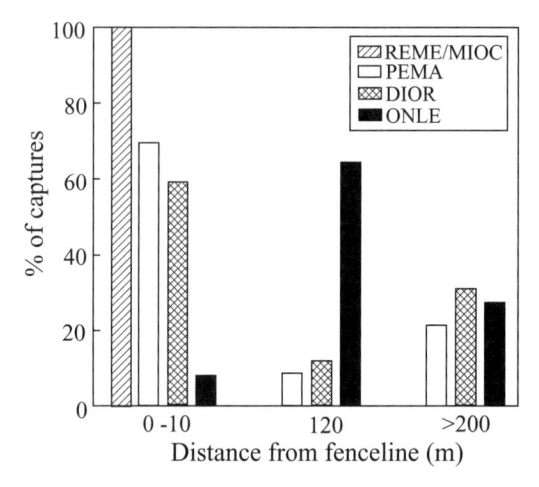

Figure 8.4 Frequency of captures of small mammals at different distances from weedy vegetation along fence lines adjacent to gravel roads in shrub-dominated areas of the shortgrass steppe. Small mammals were live-trapped on line transects placed at the fence line and 120 m into the pasture in 1997 and 1998. Abundance at more than 200 m was estimated concurrently on trapping webs as part of the SGS LTER population monitoring programs (P. Stapp, unpublished data). Values are percentage of captures per 100 trap nights on all areas. Species abbreviations: DIOR, Ord's kangaroo rat; MIOC, prairie vole; ONLE, northern grasshopper mouse; PEMA, deer mouse; REME, western harvest mouse.

and arthropod productivity. Prairie voles and western harvest mice colonized the water and nitrogen-plus-water plots in response to increased plant biomass and invasion of these plots by weedy plants (Grant et al., 1977). Northern grasshopper mice, which prefer areas of bare soil and sparse vegetation (Egoscue, 1960; Kaufman and Fleharty, 1974), were mostly absent. Two dietary and habitat generalists, the deer mouse and thirteen-lined ground squirrel, were present on all treatments but responded differently; deer mice increased dramatically in number on the "wet" treatments (water, nitrogen plus water), whereas ground squirrels were significantly more abundant on the "dry" nitrogen and control plots. Grant et al. (1977) concluded that small-mammal community structure was closely tied to vegetative structure.

Biotic Factors: Food Availability

The results of the nutrient and water addition studies emphasized the importance of vegetative structure for small mammals, but the change in productivity might also have directly affected the abundance and quality of food. Vegetation and food availability are obviously linked for herbivores and granivores, but areas with greater structural complexity also support more arthropods and, hence, more prey for insectivores and omnivores. Grant et al. (1977) estimated that small mammals consumed less than 4% of plant material and 34% of available arthropods, and concluded that neither herbivorous nor omnivorous rodent populations were limited by food availability. On a plot adjacent to the nitrogen and water experiment described earlier, Abramsky (1978) added alfalfa pellets and whole oats to examine the effects of food limitation directly. None of the resident species increased in number or biomass in response to supplemental food, which supports the conclusions of Grant et al. (1977). However, kangaroo rats colonized the food addition, apparently attracted by the novel food resource. Kangaroo rats are relatively uncommon in most areas of the shortgrass steppe, although a few individuals were actually captured earlier on the ecosystem stress plots (French and Grant, 1974). Like prairie voles and harvest mice, kangaroo rats likely dispersed onto plots from weedy vegetation along the adjacent roads, which underscores the importance of considering landscape context in interpreting results of these small-scale experiments.

Abramsky's (1978) experiment suggests that, of the resident small mammals, granivores such as kangaroo rats and pocket mice are one rodent group whose numbers may be closely tied to food availability. These species are uncommon in most shortgrass steppe areas except in sandy or disturbed soils, which support a greater diversity of forbs and midgrass species, and presumably a richer seedbank. Pocket gophers (*T. talpoides*) are another group whose distribution in the shortgrass steppe seems to be closely linked to its food source. Prickly pear cactus (*O. polyacantha*) makes up 50% of the annual diet of *T. talpoides* and may be particularly important in winter, when it was the dominant food eaten (79% of diet) and also an important source of water (Vaughan, 1967). A survey of the spatial distribution of gopher mounds and prickly pear cactus showed that gopher mounds are significantly associated with prickly pear—more than would be expected by chance (L. Dempsey and P. Stapp, unpublished data).

Most studies of mammalian carnivores have focused largely on describing habitat relationships and diet composition, but have not evaluated possible factors that may limit their abundance in the shortgrass steppe. The density of mammalian predators (Table 8.5) is relatively low compared with other grasslands and shrub steppe ecosystems (e.g., Bekoff, 1982; Goodrich and Buskirk, 1998; Linzey, 1982; Messick and Hornocker, 1981), presumably as a result of the relatively low availability of vertebrate prey. Except for small mustelids such as skunks and

Table 8.5 Mean Population Density of Common Medium-Size and Large Mammals in the Shortgrass Steppe

Species	Density, no./km²	References
Coyote	0.13–0.17	Flinders and Hansen (1975), Gese et al. (1989), Stapp (unpublished data)
Swift fox	0.07–0.23	Covell (1992), Roell (1999), Rongstad et al. (1989), Stapp (unpublished data)
American badger	0.21	Flinders and Hansen (1975)
Pronghorn	0.25–0.67	Pojar (1988), Pojar et al. (1995)

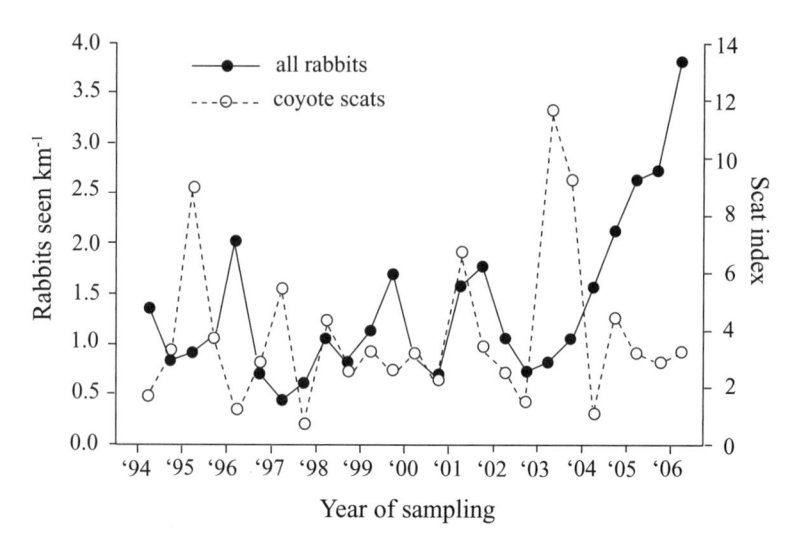

Figure 8.5 Changes in relative abundance of rabbits and coyotes on the SGS LTER from spotlight and scat counts, respectively, along the same 32-km transect in north–central Colorado (P. Stapp, unpublished data). The scat index is the number of scats deposited per day per kilometer of transect × 100. Values are combined from surveys in spring (April/May) and autumn (October).

weasels, most of the resident carnivores are able to construct burrows (Fitzgerald et al., 1994) and thus are not obviously limited by habitat. Our monitoring studies show no close tracking of coyote populations with those of rabbits (Fig. 8.5), although these data are limited by the reliability of scat counts alone as an index of carnivore density (Schauster et al., 2002). Rabbit numbers typically peak in summer or autumn, whereas highest numbers of coyote scats are usually found in winter (Fig. 8.5), perhaps because of a slower rate of scat decomposition during winter months. Comprehensive studies of population dynamics of coyotes, swift foxes, and mustelids in the shortgrass steppe are badly needed.

Biotic Factors: Species Interactions

In a system with such low primary and secondary productivity, interactions among species arguably may be less significant in determining population densities than abiotic factors or the abundance of habitat or food. Still, in areas where habitat conditions permit coexistence, interactions can affect local population densities if individuals limit access to key resources such as food or cover. We know of no studies that have tested whether predation regulates or limits mammal populations in the shortgrass steppe. Predation by coyotes has been shown to be a major source of mortality for swift foxes (Kamler et al., 2003; Roell, 1999; Sovada et al., 1998). Peak rabbit numbers are often associated with periods of low coyote activity (Fig. 8.5), suggesting that rabbit numbers may be influenced by coyote abundance. In contrast, interspecific competition has been invoked regularly to explain patterns of apparent niche segregation in grassland mammals, including the shortgrass steppe. Miller (1964), for example, interpreted the disjunct distributions of pocket gophers as the result of competition mediated by species-specific differences in preference for soil type. Moulton et al. (1983) disagreed with some of Miller's (1964) conclusions and emphasized that changes in land use since the Dust Bowl may have led to competitive displacement of previously widespread species. Similarly, Burnett (1925 [cited in Fitzgerald et al., 1994]) suggested that conversion of native grasslands to agricultural and shrub-dominated vegetation favored black-tailed jackrabbits, which subsequently outcompeted native white-tailed jackrabbits (Flinders and Hansen, 1972).

Throughout the 1970s and 1980s, small-mammal communities were widely used as experimental systems for studies of competition (Dueser et al., 1989), and one of the most widely cited pieces of evidence for the importance of competition arose from experiments conducted during the Grassland Biome studies. One unexpected outcome of the nitrogen and water addition "stress" experiment described earlier was the relatively small response of deer mice to changes in vegetative structure (Grant et al., 1977). Research from other systems (e.g., Grant, 1972; Redfield et al., 1977) has demonstrated that deer or white-footed mice often compete with voles for food and/or space, and Abramsky et al. (1979) speculated that deer mice and other "native" species avoided highly disturbed plots because of the competitive dominance of the invading species, particularly prairie voles. They subsequently removed both voles and harvest mice from one nitrogen-plus-water replicate, leaving the other replicate as a control. Deer mice increased in number after the removals, whereas grasshopper mice still avoided the nitrogen-plus-water plots. Abramsky et al. (1979) concluded that voles excluded deer mice via exploitative or interference competition, whereas the avoidance of dense vegetation by grasshopper mice reflected habitat specialization that resulted from past competition.

The studies by Abramsky et al. (1979) were conducted in an experimentally manipulated area of the shortgrass steppe (Abramsky, 1976, 1978; Abramsky and Tracy, 1979, 1980; Abramsky and Van Dyne, 1980), but interactions among rodent species may also be important in areas of native vegetation. Stapp (1997a) studied the role of competitive and predatory interactions between deer mice and grasshopper mice as determinants of local abundance of each species (Stapp, 1996). Unlike deer mice, which prefer shrub cover (Stapp and Van Horne, 1997), grasshopper mice show no affinity for shrubs and instead prefer open microhabitats (Stapp, 1997a). Grasshopper mice are known to prey opportunistically on other rodents, including deer mice (Bailey and Sperry, 1929; Flake, 1973; Rebar and Conley, 1983), and Stapp (1997a) speculated that one explanation for the preference of deer mice for shrubs, and their rarity in most grassland areas (Table 8.4), was avoidance of grasshopper mice. Alternatively, because insects are important prey for both deer mice and grasshopper mice during the spring, competition might explain habitat use of both species. In response to experimental removal of grasshopper mice, deer mice remained higher on removal plots than on controls (Stapp 1997a); declines in deer mouse abundance were negatively correlated with abundance of grasshopper mice and with the amount of shrub cover on study plots. Deer mice actually increased their use of shrubs when grasshopper mice were most abundant, suggesting that shrubs may provide some refuge from the more stocky-bodied grasshopper mice. Because consumption of insects by deer mice was unaffected by the removal experiment, and because granivorous rodents (kangaroo rats and harvest mice) also increased in abundance on removal plots, aggressive or predatory interference, rather than exploitative competition, was the most likely explanation (Stapp, 1997a). Predation by grasshopper mice is probably not an important source of mortality for adult deer mice and other rodents; however, high densities of grasshopper mice may affect local abundance of other rodents directly (through opportunistic predation on juveniles or litters in burrows) or indirectly (by affecting activity and habitat use).

Overall, patterns of the local distribution and abundance of small mammals emerge from the interactions among species and their resources in the context of the range of environmental conditions in a given area (Fig. 8.6). For species such as deer mice, western harvest mice, and probably prairie voles, local population density is determined largely by the amount of vegetative cover, which provides both food and protection from predators. Granivorous rodents such as kangaroo rats and the less common pocket mice (*Chaetodipus, Perognathus*) respond primarily to soil type through its effect on the production and availability of palatable seeds. The distribution of grasshopper mice also reflects edaphic factors, but indirectly via the effects of soil friability on the density and activity of burrowing rodents, and as a consequence, the availability of arthropod prey. In the shrub-dominated areas of the shortgrass steppe, grasshopper mice may modify the behavior and population dynamics of deer mice and other rodents, although

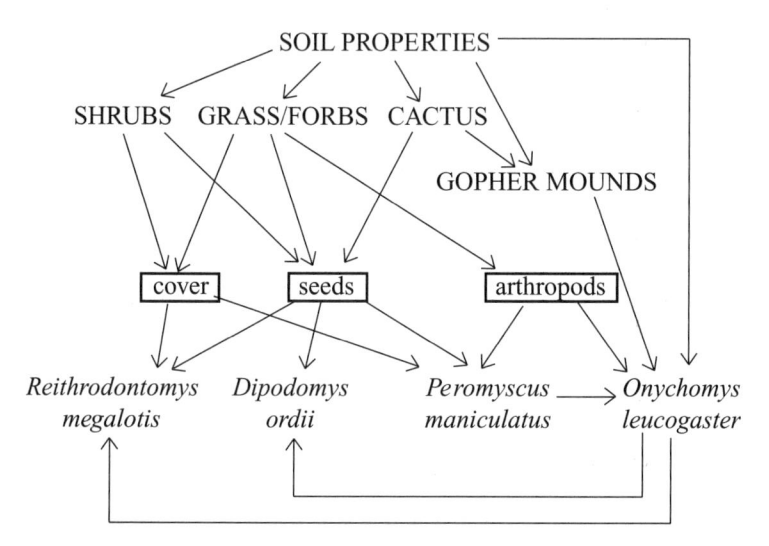

Figure 8.6 Ecological relationships among habitat characteristics (uppercase), resources (lowercase), and interactions among the four most common nocturnal rodents (italics) in the northern shortgrass steppe. (Modified from Stapp [1996].)

these effects will likely vary seasonally with the availability of insect prey. Only grasshopper mice and diurnal ground squirrels are typically present in most shortgrass habitats, however, and only at very low densities (Table 8.4), which makes it logistically difficult to conduct intensive population studies. These circumstances suggest that a long-term, comparative approach is necessary to understand what factors limit the dynamics and distribution of mammal populations.

Long-Term Dynamics: Integrating IBP Grassland Biome and SGS LTER Studies

Although more than a decade passed between the Grassland Biome and SGS LTER mammal studies, the fact that small mammals were the focus of research on the same site during both projects provides an opportunity to make some long-term comparisons. We compared temporal variation in the relative abundance of the two most common nocturnal rodents—deer mice and grasshopper mice (Fig. 8.7)—for a span of 35 years. As in other systems (e.g., Brown and Heske, 1990; Kesner and Linzey, 1997), *Peromyscus* populations tended to fluctuate from year to year, whereas grasshopper mouse numbers remained relatively constant over most of the study period. Deer mice are more generalized in their diet than grasshopper mice, which are larger, longer lived, and have a lower reproductive rate (McCarty, 1978). Fluctuations in abundance of generalist species such as deer

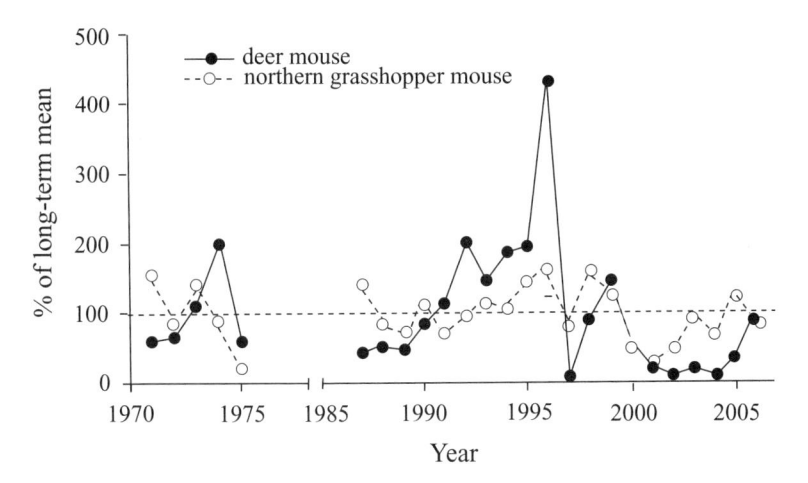

Figure 8.7 Temporal changes in the relative abundance of deer mice and grasshopper mice on the Pawnee/SGS LTER site in north–central Colorado. Values are expressed as a percentage of the mean abundance from Grassland Biome studies (1971–1975 [Abramsky, 1976, French and Grant, 1974]), saltbush trapping grids (1987–1993 [L. McEwen, unpublished data]), and SGS LTER long-term monitoring studies (September trapping sessions in saltbush sites, 1994–2006 [P. Stapp, unpublished data]).

152 Ecology of the Shortgrass Steppe

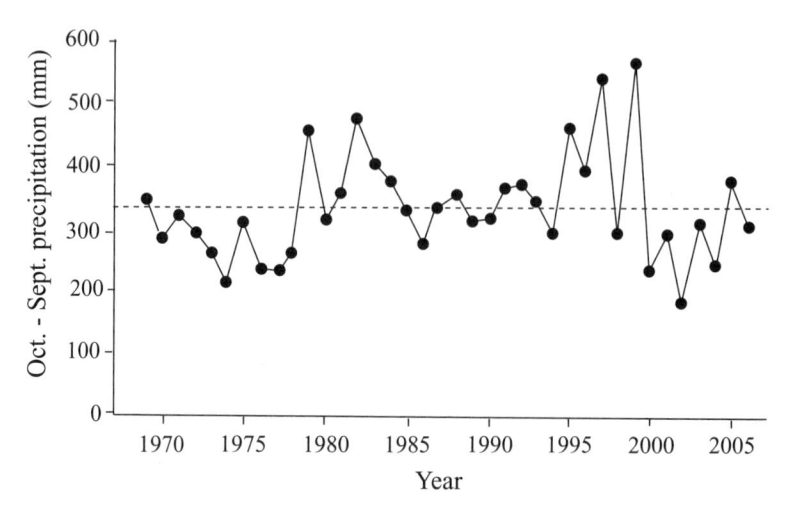

Figure 8.8 Annual precipitation (measured in millimeters) expressed as totals from October through September, from the Grassland Biome and SGS LTER meteorological station (station 11). The dashed line represents the annual mean between October 1968 to September 2006.

mice may be associated with weather-related variation in food or cover; the 1997 population crash followed the wettest summer in 35 years of records for our study area (Fig. 8.8). The 1975 decline in grasshopper mouse numbers (Fig. 8.7) may have reflected higher plant cover on control plots of the nitrogen and water addition experiment, which had been removed from cattle grazing for 6 years and may have been affected by seed production of exotic species in the neighboring treatments. Deer mice and, to a lesser extent, grasshopper mice experienced marked declines in 2000 in association with the onset of drought conditions that persisted through at least 2004.

We also compared our more recent estimates of rabbit density with those conducted during the Grassland Biome studies (Table 8.6). These results suggest that desert cottontails may be more abundant now than during the IBP studies, but that, until recently, jackrabbit densities were similar to those 30 years ago. Our recent surveys also suggest that white-tailed jackrabbits may be more abundant now than during the Grassland Biome studies (P. Stapp, unpublished data). Since approximately 2004, abundance of cottontails and black-tailed jackrabbits has increased dramatically (Fig. 8.5), which may reflect the combined effects of the exponential increase in density of prairie dogs on the CPER, and the period of drought from 2000 to 2004 (Fig. 8.8). Interestingly, J. Fitzgerald of UNC (December 1997) indicated that both cottontails and jackrabbits were much more abundant during the late 1970s, the last time when annual precipitation was less than 300 mm for multiple years.

The nitrogen and water addition experiment conducted during the Grassland Biome studies showed that the addition of these resources to native shortgrass

Table 8.6 Population Densities of Rabbits in Northern Shortgrass Steppe

Site	Density, no./km^2	Total, %
Grassland Biome (1970–1971)		
Black-tailed jackrabbit	5.86	57.73
White-tailed jackrabbit	1.63	16.06
Desert cottontail	2.66	26.21
SGS LTER (1994–1996)		
Black-tailed jackrabbit	6.04	44.12
White-tailed jackrabbit	1.23	8.98
Desert cottontail	6.42	46.90

Population densities of rabbits in the late 1990s in the northern shortgrass steppe were similar to those during the Grassland Biome studies of the early 1970s. Densities were calculated from Flinders and Hansen (1973, 1975) and from SGS LTER population monitoring studies (P. Stapp, 1996, unpublished data). Numbers of both cottontails and, especially, black-tailed jackrabbits, increased dramatically starting in 2004 (see Fig. 8.5).

plots had dramatic short-term effects on both plant and rodent communities (Grant et al., 1977; Lauenroth et al., 1978). Milchunas and Lauenroth (1995) investigated the long-term consequences of these treatments by continuing to sample vegetation on the experimental plots after the termination of the experiment in 1975. They found that the initial disturbance had far-reaching and unpredictable effects on plant community structure, driven primarily by time lags in the effects of litter accumulation in water and nitrogen-plus-water plots. Given these dynamic shifts in the plant community, we asked: How have resident rodent populations responded to changes in vegetation and habitat structure during the past two decades? In September 1995 and 1996, we trapped nocturnal small mammals on the original eight plots and compared abundance and species composition with results from the last years of the Grassland Biome experiment (1974–1975) (French and Grant, 1974; Grant et al., 1977). Recognizing that changes in abundance might reflect differences in habitat characteristics among treatments, we also collected data on vegetation (percentage cover, species richness, densities of shrubs, cacti, and exotics) and substrate (percentage bare soil, density and area of animal disturbances) on each plot.

Recall that, during the U.S. IBP studies, changes in rodent populations paralleled shifts in plant biomass between "wet" (water, nitrogen-plus-water) and "dry" (nitrogen, control) treatments, with invasion by voles and harvest mice onto highly productive, but disturbed, water and nitrogen-plus-water plots (Grant et al., 1977). A cluster analysis on habitat variables measured in 1996 indicated that clear differences in vegetation and substrate characteristics between wet and dry treatments still persisted some 20 years later (Fig. 8.9). A cluster analysis using rodent densities from 1974 to 1975 produced a similar grouping of sites (P. Stapp, unpublished data). By 1995 to 1996, the rodent communities of these sites had changed dramatically (Fig. 8.10). In response to the long-term disturbance created by the treatments, both native and exotic species had dispersed among plots, some

Figure 8.9 Cluster analysis showing continuing similarity in vegetation and substrate in 1996 among replicate experimental plots where nitrogen (N), water (W), or both (NW) were added in 1971 as part of the Grassland Biome studies. C, controls. Wet (W, NW) and dry (N, C) treatments were still recognizably different 25 years after the experimental treatments began.

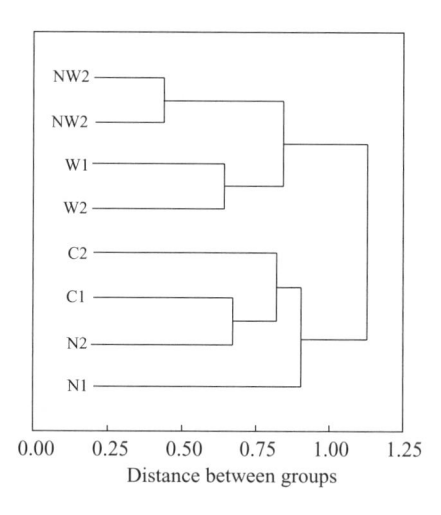

had gone locally extinct, and plots had been invaded by an additional three species of rodents (plains harvest mice, *R. montanus*; hispid pocket mice; and house mice) that were never captured during any of the Grassland Biome studies. These effects are in part confounded by the removal of grazing, because the pasture containing the plots had been fenced off from cattle in 1969 (Lauenroth et al., 1978). The extent of disturbance caused by long-term grazing exclusion is clear when one compares plant species composition and abundance on the control plots (C in Fig. 8.10) with those from outside the exclosure on native prairie (SG in Fig. 8.10) and with other long-term grazing exclosures (EX in Fig. 8.10). Curiously, kangaroo rats, which were captured previously on treatment plots (Abramsky, 1978; French and Grant, 1974), and which were frequently seen foraging along the adjacent gravel road during our trapping, were never captured by us on any of the experimental plots in 1995 to 1996.

We conclude from these results that long-term responses to disturbance of both plants and small mammals of the shortgrass steppe are highly dynamic, persistent, and unpredictable. Even though the effects of treatments were still detectable 25 years later, in the absence of grazing, both manipulated and control plots clearly were disturbed compared with the adjacent grasslands. Given the small size and close proximity of the study plots, and in light of increasing recognition of the movement abilities of small mammals (e.g., Stapp, 1997b), we speculate that by 1995, rodents may have perceived habitat features of all plots in the pasture as similar, regardless of the initial treatment. Although some patterns from the earlier studies remain (e.g., more exotic plant species on nitrogen-plus-water plots), the apparent differences in rodent abundance and species composition among plots may best reflect random variation in the location and capture rates of individuals interacting with habitat at a larger scale. Future studies should take into account the relative homogeneity of shortgrass steppe vegetation for most species and the overwhelming role that less common habitats such as shrublands, road margins, and fence lines might play in dynamics at the landscape scale.

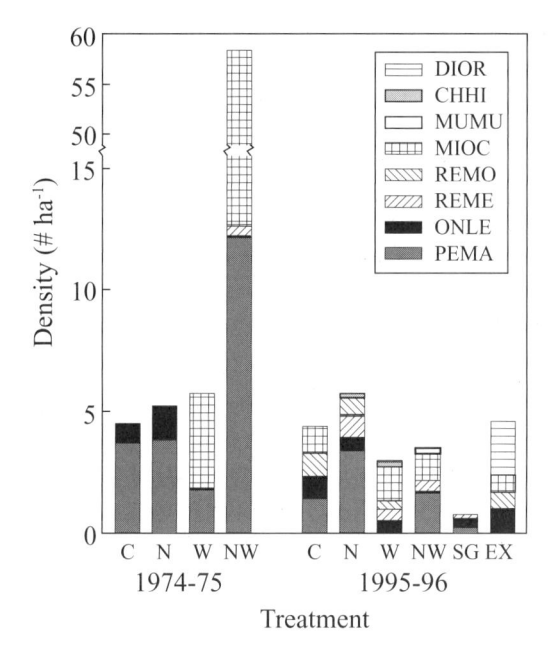

Figure 8.10 Comparisons between population density and species composition of nocturnal small mammals during and 20 years after the end of nutrient stress experiments to examine the effects of nitrogen (N), water (W), and nitrogen and water combined (NW), on plant and small-mammal communities of the shortgrass steppe. C, controls. Values are means from two replicate plots of each treatment. Density estimates from two nearby upland shortgrass prairie plots (SG) and from two long-term grazing exclosures trapped in 1996 (EX) are provided for comparison. Species abbreviations: CHHI, hispid pocket mouse; DIOR, Ord's kangaroo rat; MIOC, prairie vole; MUMU, house mouse; ONLE, northern grasshopper mouse; PEMA, deer mouse; REME, Western harvest mouse; REMO, plains harvest mouse.

Role of Mammals in the Shortgrass Steppe Ecosystem

Much of research during the Grassland Biome and SGS LTER projects was focused on understanding the role of mammals in grassland ecosystems. These studies have demonstrated that mammals perform two central functions that have influenced the evolution and ecology of the shortgrass steppe: removal of primary productivity by herbivore grazing, and burrow and mound construction by small mammals.

Herbivory

Larson (1940) was one of the first to argue that grazing by native herbivores such as bison was integral in the evolution and maintenance of the shortgrass steppe, citing historical records of expansive shortgrass plains populations prior to widespread introduction of livestock and the relative tolerance of the dominant perennial grasses to overgrazing. Although wild bison were extirpated in the shortgrass steppe during the early 20th century (Fitzgerald et al., 1994), low to moderate grazing by cattle may perform a similar function in maintaining plant communities in the shortgrass steppe (Milchunas et al., 1989).

Hart and Derner (chapter 17, this volume) and Milchunas et al. (chapters 16 and 18, this volume) provide excellent reviews of the history and effects of herbivory by mammals, especially native ungulates and livestock, so our discussion here will be brief and limited to the general effects of small mammals on ecosystems. Unlike more productive grasslands, small rodent herbivores such as voles and cotton rats are relatively uncommon in the shortgrass steppe, and therefore have minimal impacts on vegetation. Prairie dogs typically occupy low-lying areas with relatively fine-textured soils (Table 8.3), where sedges and buffalograss predominate, but they can affect vegetation structure and plant species composition in these areas by selectively consuming certain plant species and by providing sites of establishment for forbs and exotic species (Barko et al., 1999; Bonham and Lerwick, 1976; Hartley, 2006; Klatt and Hein, 1978; Severe, 1977; Stapp, 2007; Winter et al., 2002). Comparatively little is known of the effects of browsing by rabbits, which are the dominant small herbivores in terms of biomass, on grasses and shrubs (Lauenroth and Milchunas, 1991). Both jackrabbit species prefer western wheatgrass (*A. smithii*), and Flinders and Hansen (1972) suggested that consumption by jackrabbits may significantly affect the vegetation in their habitats. Similarly, in coarsely textured soils and in disturbed areas, selective storage and consumption of seeds by granivores can influence the distribution of plants. Working in abandoned agricultural plots on the CPER, Hoffman (1992; Hoffman et al., 1995) showed that seed predation by kangaroo rats and pocket mice significantly affected seedling establishment of large-seeded grass species.

Grant and French (1980) used simulation models to evaluate the role of small mammals in grassland ecosystems. They concluded that, even under high population densities, direct consumption by small mammals has an insignificant effect on primary production. This is particularly true for the shortgrass steppe, where herbivorous rodents (*Microtus, Sigmodon*) are absent from most habitats. Their simulations suggested that small mammals can affect other ecosystem components in two ways. First, consumption may have a significant effect on aboveground arthropod biomass dynamics, which may be particularly significant in the shortgrass steppe, with fauna and biomass that are dominated by omnivores and insectivores (Table 8.4). Second, the upward translocation of soil to the ground surface caused by the burrowing of small mammals significantly alters the percentage of bare surface soil and, especially, the amount and rate of processing of soil organic matter to make nutrients available to plants. The effects of fossorial and semifossorial rodents on the flora, and especially the fauna of the shortgrass steppe are described in the next section.

Burrowing and Mound Construction

The most successful mammals in the shortgrass steppe construct burrows to minimize exposure to inclement weather and predators or as a by-product of their foraging activities. In the process, they transport shallow soil organic matter and mineral soils to the surface, increase soil erosion and filtration rates, and alter surface microclimates (Koford, 1958). At normal densities, small mammals introduce an estimated 5 to 6 kg nitrogen (N)·ha^{-1}·y^{-1} to the top soil layer, which is similar to or greater than that contributed by precipitation and other sources (3–5 kg N·ha^{-1}·y^{-1}) (Grant and French, 1980). These benefits are balanced by the detrimental effects of digging and mound construction on the mortality of individual plants (Peters et al., chapter 6, this volume). In some cases, mounds also serve as seedbank and establishment sites for less common herbaceous and grass species that otherwise could not compete with *B. gracilis* and *B. dactyloides*. More important, burrows and mounds provide critical refuge habitats for a variety of grassland animals.

The effects of burrowing mammals on grassland ecosystem function and diversity differ among species according to their tolerances for different soil and topographic conditions, to the distribution of preferred food plants, and to their population densities. Swift foxes, badgers, and coyotes all construct and maintain holes (e.g., Egoscue, 1979; Linzey, 1982) that can have significant effects (Platt, 1975), but in the shortgrass steppe, small mammals are the most significant source of burrows. Of these, prairie dogs (*Cynomys*) and pocket gophers (*Thomomys, Geomys, Cratogeomys*) are probably the most important because of the intensity and persistence of disturbances they create.

Prairie Dogs

It is widely recognized that prairie dogs (Fig. 8.11), through their grazing and burrowing activities, significantly alter structure and function of grassland ecosystems

Figure 8.11 Black-tailed prairie dogs play an important and often controversial role in the ecology of many areas of the shortgrass steppe. (Photo by Stephen J. Dinsmore.)

in areas where soil and topographic conditions favor the establishment of colonies (Whicker and Detling, 1988). Prairie dog burrows also provide shelter and nesting locations for a variety of other grassland animals (Stapp, 1998), including birds such as Burrowing Owls (e.g., Butts and Lewis, 1982; Desmond et al., 2000; Orth and Kennedy, 2001; Sidle et al., 2001a) and Mountain Plovers (*Charadrius montanus* [Dinsmore et al., 2005; Dreitz et al., 2005]), both species of conservation concern in the western Great Plains. Prairie dogs are the primary prey of the endangered black-footed ferret, which faces extinction largely because of the widespread extirpation of prairie dogs (Anderson et al., 1986; Calahane, 1954). Prairie dogs are also an important food source for many grassland raptors (Cully, 1991; Plumpton and Andersen, 1997, 1998; Schmutz and Fyfe, 1987; Seery and Matiatos, 2000; Weber, 2001). The recent increase in shortgrass steppe populations of Ferruginous and Swainson's Hawks (Wiens and McIntyre, chapter 9, this volume) may, in part, reflect the recovery of prairie dog colonies in eastern Colorado during the past two decades.

Despite their ecological importance, prairie dogs are often considered rangeland pests because of their effects on production and availability of forage for livestock (Derner et al., 2006; Hansen and Gold, 1977; O'Meilia et al., 1982). Prairie dogs still occupy most of their historical geographic range, but local populations have been eliminated or reduced by extensive poisoning and shooting campaigns associated with ranching and farming, and by sylvatic plague (Miller et al., 1990). Existing colonies are thought to cover less than 10% of the 41 million ha occupied in 1900 (Anderson et al., 1986), although there is some debate about the historical coverage of prairie dog colonies (Forrest, 2005; Knowles et al., 2002; Vermeire et al., 2004; Virchow and Hygnstrom, 2002). Although black-tailed prairie dogs occur throughout the shortgrass steppe, populations appear to be relatively small and disjunct compared with historical accounts. On the PNG in northeastern Colorado, most (67%) colonies are less than 20 ha in size, and the total area occupied there by prairie dogs in 2006 (1149 ha) represented only about 1.5% of the total area (78,000 ha) (USDA Forest Service, unpublished data). This coverage seems typical for colonies of black-tailed prairie dogs elsewhere (Butts and Lewis, 1982; Clark et al., 1982; Sidle et al., 2001b), although larger colonies occurred historically. Because populations are naturally fragmented by variation in topography, soils, and abundance of preferred plants (Koford, 1958), it seems likely that colonies never covered more than 10% to 20% of the landscape. However, through their burrowing and grazing activities and by serving as prey, prairie dogs may have a disproportionately large impact on both shortgrass steppe vegetation and native fauna, at both local and landscape scales.

Between 1997 and 1999, we conducted comparative studies to determine the ecological role of prairie dogs in the northern shortgrass steppe. Plots (1.2 ha) were established on five different prairie dog colonies and similar control plots on neighboring areas with the same soil type, topography, vegetation, and land-use history, but without prairie dogs. We hypothesized that the effects of prairie dogs on the shortgrass steppe differed from those on mixed-grass prairie in South Dakota and Montana, where much of the earlier research had been conducted. In these more productive grasslands, the activities of prairie dogs create patches of low grassland in a sea of taller vegetation, creating greater landscape-scale heterogeneity. By comparison, in most areas of the shortgrass steppe, the contrast between the vegetation of colonies and that of native prairie would be less obvious to us and to the resident fauna. However, because of the scarcity of belowground refuges and low consumer densities on the shortgrass steppe, prairie dog burrows provide critical habitat for other animals, and prairie dogs themselves are important prey for top vertebrate predators (Stapp, 1998).

Our comparative studies suggest that plant communities and fauna of prairie dog colonies differ in several ways from grasslands where prairie dogs are absent (Table 8.7). As found in other studies in the shortgrass steppe (Bonham and Lerwick, 1976; Severe, 1977), burrowing by prairie dogs seems to alter plant species composition in the area of the mounds by providing germination sites for exotic and other species associated with disturbance. Plant species associated exclusively with mounds included *Cleome serrulata*, *Solanum triflorum*, *Portulaca oleracea*, *Chenopodium album*, *Euphorbia glyptosperma,* and *Salsola iberica*. Unlike Bonham and Lerwick's (1976) results, however, there were no consistently significant differences in plant species richness between colonies and uncolonized grasslands. Grazing and burrowing by prairie dogs produced slight but significant changes in vegetation height, and an increase in the percentage cover and patchiness of bare soil (Bonham and Hannan, 1978), but did not significantly alter the microtopography relative to the controls (Table 8.7). These results are similar to those reported by Stapp (2007) in a comparison of 18 active colonies with seven uncolonized control sites on the PNG, and are generally consistent with other studies of vegetation responses to prairie dogs in the southern shortgrass steppe (Barko et al., 2001; Winter et al., 2002).

Except for grasshoppers, which tended to be more abundant outside of prairie dog colonies than inside colonies (J. R. Junell and B. Van Horne, unpublished data), changes in habitat structure apparently had little effect on abundance of arthropods or most of their vertebrate predators (Table 8.7) (Kretzer and Cully, 2001b). There was one exception, in that lesser earless lizards (*Holbrookia maculata*) were more abundant on prairie dog colonies than off. However, there were no striking differences in abundance of other amphibian or reptile species (Davis and Theimer, 2003; Kretzer and Cully, 2001a). Rodent burrows, including those of prairie dogs, are known to serve as primary overwintering sites for western rattlesnakes (*Crotalus viridis*) and other reptiles (Hammerson, 1999), but we were not able to confirm that, because our field surveys were limited to the warmer months (May–September).

Horned Larks (*Eremophila alpestris*), a bird species of open grasslands, were more common on prairie dog colonies than on adjacent uncolonized grassland plots, whereas Lark Buntings (*Calamospiza melanocorys*) were significantly less abundant on prairie dog colonies than off (Table 8.7). Avian species richness was significantly higher on prairie dog colonies, presumably as a result of raptors preying on prairie dogs (M. Andre and P. Stapp, unpublished data). Similarly, Barko et al. (1999) recently reported higher avifaunal abundance and diversity in five Oklahoma prairie dog colonies than in grasslands without prairie dogs, but only during the plant growing season (Winter et al., 2003). Smith and Lomolino (2004)

Table 8.7 Ecological Effects of Prairie Dogs in the Northern Shortgrass Steppe Based on SGS LTER Comparative Studies from Four to Five Paired Colony and Grassland Control Plots in 1997 and 1998

Group	Colonies (vs. Grassland Controls)
Plants and cover	
Mound scale	More exotic and disturbance species on mounds; higher density of *Sphaeralcea coccinea*; lower density of shortgrass species, *Plantago patagonica*, and lichens; no difference in plant species richness or the number of unique species compared with off mounds
Colony scale	No difference in plant species richness; shorter grasses and shrubs; greater percentage cover and higher spatial variation of bare soil; no difference in microtopographic variation
Arthropods	No difference in relative abundance of beetles, crickets, or spiders; fewer grasshoppers[a]; no difference in density of harvester ant mounds
Vertebrates	
Amphibians and lizards[b]	No difference in species richness; no differences in abundance of western chorus frogs, tiger salamanders, or short-horned lizards; higher numbers of lesser earless lizards; fewer northern many-lined skinks
Birds[c]	Higher species richness (resulting from the presence of raptors); horned larks, burrowing owls, raptors more abundant; fewer lark buntings; no differences in abundance of western meadowlarks, McCown's longspurs, or all birds combined; lower densities of passerine nests and lower survival rates of artificial nests
Mammals	Higher small-mammal species richness; higher numbers of northern grasshopper mice and Ord's kangaroo rats; fewer thirteen-lined ground squirrels; higher density of cottontails and coyote scats; no difference in density of fox and badger burrows; no difference in cattle activity (density of fecal pats)

[a]Junell, 2002.
[b]$\alpha \leq 0.20$ because of low capture rates and, therefore, low power.
[c]M. Andre and P. Stapp, (unpublished data).
Unless noted, all results were statistically significant at $P \leq .10$ using paired *t*-tests. Species not mentioned were captured or sighted too infrequently to compare abundance statistically (P. Stapp, unpublished data).

also found strong seasonal differences in bird communities between colonies and other vegetation types in Oklahoma, with Horned Larks showing the most consistent differences in abundance on and off colonies. In their study, species richness and community structure in summer were highest on prairie dog colonies and fallow crop fields, but most differences among vegetation types were reduced during fall. In addition to comparing avian population densities and diversity on and off prairie dog colonies, we conducted artificial nest experiments on paired plots in 1998. Densities of passerine nests were higher, and rates of nest predation were lower, on grassland control plots than on prairie dog colonies (M. Andre and P. Stapp, unpublished data). Baker et al. (1999, 2000) similarly found higher rates of predation of artificial nests on colonies of white-tailed and black-tailed prairie dogs, respectively, than in adjacent grasslands lacking prairie dogs.

Prairie dogs may also have significant effects on other small mammals. Northern grasshopper mice were more abundant in colonies and, on average, more species of small mammals were captured in colonies than in nearby grasslands (Table 8.7). In contrast, thirteen-lined ground squirrels were less abundant in prairie dog colonies. A more comprehensive study of small-mammal communities in active and inactive colonies (Stapp, 2007) similarly found higher numbers of grasshopper mice in colonies than in uncolonized grasslands, and lower numbers of ground squirrels in active colonies than in inactive ones, but no difference in overall species richness or community patterns. The lack of strong differences in rodent communities on and off colonies is consistent with other studies throughout the region, with local vegetation and habitat characteristics likely playing more important roles (Stapp, 2007). The potential for interactions between prairie dogs and ground squirrels, however, warrants additional attention. There was no evidence from microhabitat characteristics at trapping locations that ground squirrels avoided areas of high prairie dog activity (P. Stapp, unpublished data); however, vigilance behavior of juvenile ground squirrels was significantly higher on active prairie dog colonies than on colonies that had been recently extirpated by plague (J. Aldana and P. Stapp, unpublished data). There may be subtle agonistic interactions between prairie dogs and ground squirrels, or prairie dogs may affect food availability or habitat quality for ground squirrels, or they may attract predators to colonies that subsequently prey on squirrels.

We spotted significantly higher numbers of desert cottontails and scats of coyotes on prairie dog colonies than on grassland control plots (Table 8.7). Dano (1952) and Hansen and Gold (1977) also found higher densities of rabbits in prairie dog colonies, which may provide an additional source of prey for both avian and mammalian predators, including coyotes. Shaughnessy and Cifelli (2004) reported few differences in carnivore abundance on and off prairie dog colonies in Oklahoma, with badgers (*T. taxus*) and spotted skunks (*Spilogale putorius*) more common on colonies in Cimarron County but not in the Oklahoma Panhandle (Lomolino and Smith, 2003). Finally, other researchers have suggested that cattle and other ungulates concentrate activity on prairie dog colonies (Coppock et al., 1983; Knowles, 1986; Koford, 1958; Krueger, 1986; Lomolino and Smith 2003), but in our studies, density of fecal pats on prairie dog colonies were similar to those on our control plots (Table 8.7) and to plots in low and moderately grazed pastures (P. Stapp, unpublished data). Guenther and Detling (2003) also reported that cattle neither preferred nor avoided prairie dog colonies at the SGS LTER site.

Collectively, our results to date suggest that prairie dogs may have a significant influence on the flora and fauna of the shortgrass steppe, although the effects of prairie dogs are less striking than reported in studies in more productive grasslands, where the contrast between colonies and surrounding grasslands is more pronounced. Animals that benefit most from the presence of prairie dogs include avian and mammalian predators, species associated with heavy grazing and/or low stature grasslands (e.g., Mountain Plovers, Horned Larks, lesser earless lizards), and species that depend on abandoned burrows for shelter (e.g., Burrowing Owls, northern grasshopper mice, desert cottontails, skunks). Except for a few notable species of conservation concern, all the species recorded in prairie dog colonies are regularly observed in areas without prairie dogs, suggesting that effects of prairie dogs are primarily reflected in changes in relative abundance of the most common consumers. These results are not surprising, given the relatively small amount of our study area occupied by prairie dogs and the relatively low population densities of many animals in the shortgrass steppe. However, combined with the effects of high densities of prairie dogs and their effects on ecosystem functioning, they highlight the importance of prairie dog colonies for increasing grassland biodiversity at larger, landscape scales.

Because prairie dog colonies in the shortgrass steppe tend to be small and somewhat isolated, it is critical to understand the biology and dynamics of prairie dogs to evaluate the ecological consequences of their historical declines. Roach et al. (2001) used microsatellite markers to describe the genetic structure of SGS LTER prairie dog populations. The 13 colonies they studied ranged greatly in active area (1–52 km^2) and age (1–10 years), in part as a result of recent extinctions caused by plague and human eradication efforts. Roach et al. (2001) reported a moderate amount of genetic relatedness among colonies, suggesting some degree of isolation and differentiation, but they found little evidence of inbreeding within colonies. They argued that, despite the distance separating colonies, the genetic evidence indicated a considerable movement of individuals among populations, especially among large, persistent colonies. Colony area and age were highly correlated, and colony size/age was the best predictor of the degree of genetic relatedness among colonies. Small, recently established colonies tended to differ more from one another than larger, more permanent ones, which suggests a *mainland–island* metapopulation structure, in which small satellite sites with high extinction rates are recolonized by dispersers from large colonies. Dispersal was facilitated by a network of swales and ephemeral drainages, areas where preferred food plants (*B. gracilis*, *Carex eleocharis*, *B. dactyloides*, *A. smithii*) (Bonham and Lerwick, 1976; Hansen and Gold, 1977; Koford, 1958) are abundant and where most colonies are located. The physical distance between colonies via drainages explained a small (<5%) but significant amount of variation in genetic distance among colonies. Roach et al. (2001) predicted that increasing isolation caused by habitat loss and eradication programs would lead to a loss of within-colony allelic diversity through inbreeding depression, and recommended regionwide protection for all colonies, as well as for the drainage habitats that may serve as dispersal corridors.

These results have important implications for understanding the ecology and conservation of prairie dogs in the shortgrass steppe. First, despite the large differences in colony area and age, the distances separating colonies, and several recent local extinctions, the level of genetic differentiation among SGS LTER colonies was similar to that of populations of prairie dogs from other grasslands and that of other ground squirrels. The effectiveness of dispersal among seemingly distinct colonies supports the view that prairie dog populations are adapted to the high degree of natural isolation that results from spatial heterogeneity in the availability of suitable soil, vegetation, and topographic conditions. The fact that physical distance measures, including shortest routes along drainages, roads, or in a straight line, explained relatively little or no variation in genetic relatedness among colonies also suggests that prairie dogs perceive matrix habitats outside colonies to be relatively homogeneous and presenting few natural barriers to dispersal. However, this may not be the case in landscapes with greater structural heterogeneity in topography or vegetation (i.e., more productive grasslands or those with more lands under cultivation).

A second key point is that we know little about the factors that determine initial site selection by prairie dogs because most dispersers tend to settle in established or extinct colonies. Because not all habitats are equally suitable, prairie dogs thus seem to match the theoretical model of metapopulations, in which population patches are fixed in space and are classified as either occupied or extinct (Levins, 1970). Thus, if regional persistence depends on some degree of spatial asynchrony in local dynamics, it may be as or more important to preserve potential habitat represented by abandoned colonies than to focus conservation efforts simply on protecting the small populations in these areas from recreational shooting or poisoning.

Last, the relative importance of these small, isolated populations may depend largely on the dynamics of plague, which first appeared in Colorado in the late 1940s (Ecke and Johnson, 1952). The results of Roach et al. (2001) underscore the importance of large colonies as sources of dispersers, and because these are the areas with highest density of burrows and prairie dogs, they may arguably be the most valuable for other species closely associated with prairie dogs (paradoxically, predators attracted to prairie dogs may prey opportunistically on other species, so that large colonies may act as population sinks for some species compared with small or abandoned colonies).

On the surface, these observations suggest that to maintain the structure of prairie dog metapopulations, protection of large colonies should receive higher priority than protecting small, extinction-prone satellite populations. In the presence of plague, however, large, established colonies may be *more*, not less, susceptible to extinction because plague may spread more rapidly through large than small populations. As discussed earlier, an analysis of patterns of extinction and recolonization of colonies on PNG during the past 20 years suggests that this is the case (Stapp et al., 2004): Both small (<1 ha) and large (>15 ha) colonies were more likely to go extinct than midsize ones. Moreover, the probability of extinction was higher when large, neighboring colonies also went extinct, regardless of intercolony distance, suggesting that plague may be spread by prairie dogs dispersing to the nearest active colony, or by wide-ranging carnivores (Salkeld and Stapp, 2006; Salkeld et al., 2007). Thus, although most dispersal currently seems to be in the direction of large to small colonies, in the future these small, isolated colonies may be critical to regional persistence if their dispersers can repopulate large colonies that have been decimated by plague. More research on the ecology of plague and the degree of connection and temporal asynchrony in population dynamics among colonies is needed.

Pocket Gophers

The mounds constructed by prairie dogs are conspicuous, but they actually cover a relatively small proportion (1.4%) of the area of a colony. By comparison, mounds created by pocket gophers cover a much larger area, are more aggregated, and occur across a broader range of soil and vegetation types than those of prairie dogs (Table 8.3). In one study (Grant et al., 1980), northern pocket gophers (*T. talpoides*) transported an estimated 11+ metric tons of soil·ha^{-1}·y^{-1} aboveground, with mounds covering 3% to 8% of the surface. The latter value is similar to our estimates of mound coverage in the most widespread soil types on the SGS LTER site (sandy loam and fine sandy loam; Table 8.3) during the mid to late 1990s. More recently, coverage by gopher mounds in upland prairie sites has declined by 98%, presumably reflecting the period of extended drought that began in 2000 (P. Stapp, unpublished data).

Like prairie dogs, burrowing and mound building by pocket gophers significantly alter nutrient dynamics, and plant and animal communities of North American grasslands (Grant and French, 1980; Huntly and Inouye, 1988). Compared with undisturbed areas, gopher mounds have higher rates of water infiltration and of nitrogen mineralization (L. Dempsey, I. Burke, and P. Stapp, unpublished data), which contribute to more dense plant growth adjacent to mounds (Grant et al., 1980). Gopher mounds also support a higher diversity of forb species, but fewer individual plants, than in the area immediately off mounds. The close association between *T. talpoides* and prickly pear cactus (Vaughan, 1967) may contribute to higher plant productivity or diversity. In the shortgrass steppe, prickly pear tends to collect wind-blown litter and may reduce grazing pressure or create favorable microclimates for both grasses and forbs (Bayless et al., 1996; Rebollo et al., 2002). It seems likely that the disturbed soil on gopher mounds increases rates of germination and seedling establishment for these plants. Therefore, in combination, gopher mounds and prickly pear may facilitate less common plants that are palatable to grazers, and represent islands of locally high diversity in some grazed areas of the shortgrass steppe. Mounds and cacti affect forb and grass species diversity differently (L. Dempsey and P. Stapp, unpublished data), but this potential interaction remains largely unstudied.

Like prairie dog burrows, mounds and burrow networks created by pocket gophers provide refuges and nest sites for a variety of grassland animals (Vaughan, 1961). Stapp (1997b) found that population density of grasshopper mice was positively related to the abundance and number of *T. talpoides* mounds, which provide these mice with dust-bathing sites and easy access to burrows. More important, mounds are also areas of high concentrations of the arthropod prey of grasshopper mice, especially large beetles and crickets, which burrow each night into the tilled soil or hide in reexcavated holes. The presence of gopher mounds tends to be more important for mice in areas where soils are more finely textured and therefore less friable (Stapp, 1997b, 1999). Belowground temperatures at the typical depths of gopher and ground squirrel burrows typically remain at or above freezing throughout the winter (Stapp, 1996), which make mounds ideal locations for wintering arthropods, amphibians, and reptiles (Hammerson, 1999), as well as other small mammals. Finally, although pocket gophers are mostly nocturnal and mostly active belowground, dispersing individuals may be important prey for some raptors (e.g., Great Horned Owls, *Bubo virginianus* [Zimmerman et al., 1996]).

Although pocket gophers have received comparatively less attention than prairie dogs, they may have a similar yet underappreciated role in the dynamics of the shortgrass steppe and other grassland ecosystems. However, because pocket gophers often occur at lower population densities than prairie dogs, they therefore are less likely to attract and sustain large predators and charismatic specialist species such as black-footed ferrets. Gopher burrows are much smaller than those of prairie dogs and are usually plugged by their builder, which may make them less suitable for species such as Burrowing Owls. However, because disturbances created by pocket gophers are more widespread across shortgrass steppe soil types (Table 8.3) than those of prairie dogs, the total amount of area affected by them may be comparable with or even exceed that of prairie dogs. Gopher burrows are usually shallow (<30–50 cm) (Miller, 1964; Moulton et al., 1983) and thus their mounds have the effect of concentrating and mixing mostly organic soil material. The soil nutrient dynamics, and hence, nutrients available to plants on gopher mounds, therefore probably differ from those on prairie dog mounds, which consist of a mixture of deep mineral and organic soils (Koford, 1958). Moreover, the high level of surface activity by prairie dogs compacts the soil on mounds and nearby, which further alters the hydrology and nutrient dynamics in ways that differ from the less intensive surface activity of pocket gophers. Comparative studies of the ecology of prairie dogs and pocket gophers in areas where they co-occur would clarify the unique contributions of each to structure and function of prairie ecosystems.

Effects of Human Activities

Exploitation

For as long as humans have inhabited North American prairies, they have taken mammals as sources of food, clothing, and tools. During the 19th and early 20th centuries, many of the larger species that humans prized as food, feared as predators, or loathed as livestock competitors were directly or indirectly extirpated from the southern Great Plains (elk, bison, gray wolves, grizzly bears, black-footed ferrets) or substantially reduced in number (deer, pronghorn, prairie dogs). Populations of remaining large wildlife species are currently managed by state and federal agencies, which regulate hunting, shooting, and trapping. In some cases, implementation of strict harvest regulations has allowed modest recovery of native populations of pronghorn (Fig. 8.12). Managing populations of species such as prairie dogs and coyotes, which are often considered nuisances by agricultural interests, has been controversial because of significant indirect consequences of control programs on other native mammals. For example, swift foxes historically have declined in number throughout their range as an indirect result of poisoning, trapping, and shooting campaigns directed at coyotes (Egoscue, 1979; Samuel and Nelson, 1982). The extinction of black-footed ferrets from most of their range, including the shortgrass steppe, has been directly linked to the effectiveness of prairie dog eradication programs (Miller et al., 1990). Prairie dog populations

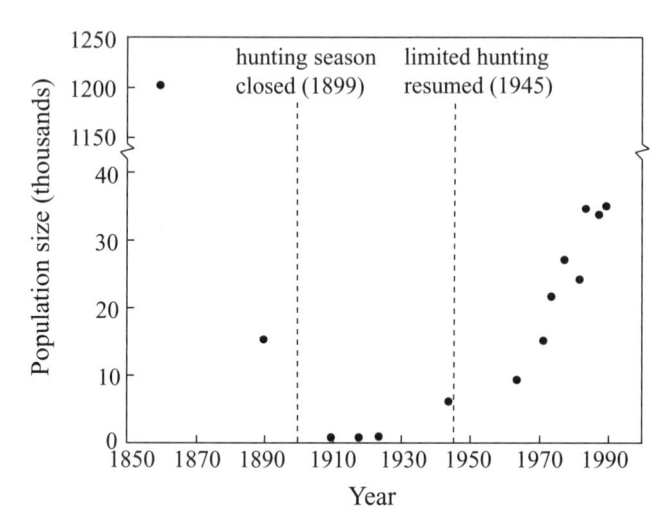

Figure 8.12 Historical changes in estimated population size of pronghorn in eastern Colorado in response to changes in hunting pressure. (Values calculated from Yoakum [1968], Olson [1972], Myers [1974], Pojar [1984, 1988], and Duvall [1990].)

continue to be extirpated on some private lands, despite increasing recognition of the contributions of prairie dogs to biotic and landscape diversity of grassland ecosystems (Kotliar et al., 1999; Stapp, 1998). Recently, some attention has been given to the potential effects of recreational shooting on prairie dog populations. As a control measure, the effectiveness of shooting prairie dogs is still in question (Knowles, 1988), but it does alter aboveground behavior (Vosburgh and Irby, 1998), which, at a minimum, makes demographic and ecological studies of prairie dogs more difficult (Roach, 1999). These concerns, in part, led to a brief moratorium on shooting of prairie dogs on public lands in Colorado, but it was subsequently lifted in 2006 because of the marked increase in acreage of active colonies.

Livestock Grazing

Livestock grazing is the primary agricultural use of the shortgrass steppe. The degree to which grazing by cattle mimics historical grazing patterns of bison and other native herbivores is a topic of some debate, but plant communities of the shortgrass steppe are generally recognized as being more resistant to destabilizing effects of moderate grazing than other grasslands (Lauenroth et al., 1994). Hart (chapters 4 and 17, this volume) reviews the history of livestock production in the shortgrass steppe and Milchunas et al. (chapter 16, this volume) describe the ecological consequences of grazing on ecosystem processes and native plants and animals. Here we focus primarily on the effects of livestock grazing on other native mammals.

The presence of cattle influences the shortgrass steppe, and hence mammal populations, in at least three ways. The first and most obvious is a reduction in height and change in species composition of plant communities. Throughout this chapter we have underscored the important influence of vegetation structure on grassland mammals. Grant et al. (1982) used data from the Grassland Biome project to compare small-mammal communities under different grazing regimes in grasslands across North America. Not surprisingly, communities in grasslands with more standing biomass (tallgrass and montane grasslands) were affected more by grazing than those with little vegetative cover. In the shortgrass steppe, grazing had no significant effect on species or functional group composition of small-mammal communities. Grant et al. (1980) also found little difference in the surface coverage of pocket gopher mounds among pastures with low, moderate, and high intensity of grazing.

It is important to recognize that, in contrast to work at other IBP grassland locations, rodent populations on the Pawnee site were censused in lightly grazed rather than ungrazed pastures (Grant et al., 1982). Thus, although these conclusions may accurately depict the response of small mammals to grazing in some shortgrass-dominated uplands, our research and observations suggest that in areas of the shortgrass steppe that potentially support taller vegetation, the effects of grazing can be substantial. Long-term grazing exclosures on the CPER tend to support higher densities and diversity of small mammals in these ungrazed areas than on adjacent grazed pastures. Differences in abundance and species composition of small mammals on controls of the nitrogen and water addition experiment compared with moderately grazed pastures can be explained in part by a 30-year absence of grazing on these treatments (Fig. 8.10). In another area, western harvest mice were the dominant species in an area of dense saltbush and western wheatgrass (*A. smithii*), but after several months of intensive grazing by both horses and yearling heifers, no harvest mice were captured and the area had been colonized by grasshopper mice (P. Stapp, unpublished data). Working on a mixture of active and inactive prairie dog colonies and grassland areas without prairie dogs, all of which were moderately grazed by cattle, Stapp (2007) found that rodent diversity was positively related to mean grass height. McCulloch (1959) reported higher densities of grasshopper mice and kangaroo rats in grazed sand–sage grasslands than in ungrazed areas in western Oklahoma.

Second, removal of forage by cattle may reduce food availability for native herbivores. Hansen and Gold (1977) found that cattle reduced by approximately 10% the potential forage available to prairie dogs and desert cottontails. Flinders and Hansen (1975) suggested that the ratio of abundance of the three rabbit species in the shortgrass steppe is, in part, a function of grazing intensity, and that population densities could be affected by changes in grazing management. Black-tailed jackrabbits preferred moderate and light summer grazing treatments, and lowland pastures with heavy cover, whereas white-tailed jackrabbits showed no preference for particular grazing regimes on upland areas. Desert cottontails were most abundant in moderately grazed pastures. For species such as rabbits that depend on vegetation for protection from predators as well as for food, it may be difficult to separate the effects of forage reduction versus elimination of cover.

A third major consequence of livestock production is the disturbance and compaction of the soil in areas where cattle congregate. Highly disturbed areas include watering and feeding locations, along fence lines, and in loafing sites such as riparian strips. The degree of disturbance in these areas often leads to invasion by exotics and weedy plants. However, these conditions are fairly localized except at high stocking densities, and there are few studies of their effects on resident animal populations. Knowles (1986, 1993) found that prairie dogs are often attracted to areas of high cattle activity, especially watering tanks and homesteads. In areas without much standing water, stock tanks provide a source of water for native mammals, serve as foraging areas for insectivorous bats, and provide habitat for grassland amphibians such as tiger salamanders.

Fire is used as a management tool to increase forage value for livestock in more productive grassland systems but has received little attention in the shortgrass steppe. Ford (1999) reviewed the effects of fire on *B. gracilis* and *B. dactyloides* and concluded that, in most cases, the effects of fire were neutral or positive, depending on season and past precipitation. Fires that remove most live vegetative cover as well as litter can reduce small-mammal communities in the short term, but some small mammals respond positively to the increased availability of forb seeds and insects after a fire. Kaufman et al. (1990) described the effects of fire on rodent communities in tallgrass prairie and categorized species as fire positive and fire negative. Rodents that eat plant foliage or litter-dwelling invertebrates, or depend on dense vegetation or litter for cover (i.e., voles, harvest mice) respond negatively to fire, whereas those that live in burrows prefer open microhabitats and feed on seeds or surface-dwelling arthropods (grasshopper mice, ground squirrels), recover quickly from the immediate effects of fire.

Differences in vegetation and rodent communities between the shortgrass steppe and other grasslands suggest that fire may affect the shortgrass steppe differently, but relatively few studies have been conducted to date. P. Ford and her associates at the USDA Forest Service recently began long-term studies of the effects of fire on fauna and plant communities of the Kiowa National Grasslands. Her early results suggest that effects of fire on small-mammal populations and diversity are mostly temporary (2 months or less), which is not surprising given the low amount of standing aboveground biomass in even unburned areas of shortgrass vegetation (Ford and McPherson, 1999). As in tallgrass prairie, grasshopper mice and ground squirrels responded positively or were largely unaffected by fire, whereas harvest mice *(R. montanus)* tended to be less abundant on burned plots (Ford 2007). Fire is currently used in the northern shortgrass steppe to create nesting habitat for Mountain Plovers.

Crop Production

Winter wheat (*T. aestivum*) and alfalfa (*Medicago sativa*) are the major agricultural crops in arable regions of the shortgrass steppe. Compared with grazing, however, cultivation substantially alters hydrology, soil nutrient dynamics, and belowground food webs, causing dramatic changes to plant communities and ecosystem function that persist for decades (Burke et al., 1997). These changes have significant effects on resident mammal populations as well. Lovell et al. (1985), for example, speculated that agricultural development that followed the construction of Barr Lake dam resulted in the local extinction of 50% of the native mammal species. Besides the obvious replacement of native grasslands with monoculture, crop production likely affects native mammals in at least two ways. First, the presence of palatable forage may attract and concentrate in space populations of wild herbivores, increasing local population density. Pronghorn commonly congregate in winter on wheat and alfalfa fields in Colorado, although they apparently do relatively little damage to plantings (Liewer, 1988 [cited in Fitzgerald et al., 1994]). Species such as voles, rabbits, and harvest mice also take advantage of available forage and dense cover of crops during the growing season, as well as the weedy vegetation commonly found along adjacent fences and roadsides. The north and westward range expansion of cotton rats and least shrews, for example, typically followed agricultural right-of-ways and riparian vegetation (Mellott and Fleharty, 1986; Owen and Hamilton, 1986). Second, the increase in herbivore numbers and biomass may attract and support high populations of avian and mammalian predators. The increase in regional predator abundance may influence native prey populations if predators are mobile and spill over into neighboring uncultivated areas.

Compared with agricultural regions of the Midwest, there have been few studies of mammal communities in cultivated areas of the shortgrass steppe. Alfalfa and winter wheat were the second most important food items in the annual diets of black-tailed and white-tailed jackrabbits, respectively, in northern Colorado (Flinders, 1971). Fleharty and Navo (1983) described small-mammal diversity and abundance in Kansas cornfields and in the disturbed but unplowed areas associated with center pivot irrigation. The latter habitat had significantly higher species diversity and community evenness than cornfields and native sand–sage prairie. Although only a few species were common in native prairie (northern grasshopper mice, Ord's kangaroo rats, spotted ground squirrels), none of the 14 species of small mammals present locally were absent from cultivated areas. Fleharty and Navo (1983) argued that remnant habitats such as fallow croplands, roadside and roadside right-of-ways, fence rows, and abandoned farmsteads may act as refugia and breeding habitat for small mammals in agricultural landscapes. Moulton et al. (1983) suggested that, early during the 20th century in southeastern Colorado, cultivation and the resulting changes in vegetation led to the replacement of the yellow-faced pocket gopher by the plains pocket gopher, which in the past had been limited to sandy river bottoms and disturbed areas.

Pesticide application often accompanies crop production and may have immediate, direct effects on survival of native species, including mammals. McEwen and Ellis (1975) applied malathion and toxaphene separately to two live-trapping grids and compared abundance of rodents prior to and after application, and with changes during the same period on two unmanipulated control plots. Although postspray abundance of all species combined was similar on pesticide and control plots, none of the individuals captured prespray on the toxaphene treatment plots were recaptured during the postspray sampling period. However, other species and new individuals colonized the toxaphene plots, suggesting that effects of toxaphene were primarily short term. Toxaphene also killed Western Meadowlarks and nestlings of Horned Larks and McCown's Longspurs, reducing density of native songbirds by 30%.

Much of the shortgrass steppe under cultivation at the turn of the century was abandoned during the Dust Bowl era, leaving behind highly degraded crop and rangelands (Hart, chapter 4, this volume). In 1985, the USDA established the Conservation Reserve Program (CRP) to stabilize and conserve soil and water resources in agricultural lands. Through the CRP, highly erodible lands are removed from crop production and are planted with exotic or native grasses. Hall and Willig (1994) compared the diversity and composition of small mammals between CRP grasslands of different successional stages with native, ungrazed shortgrass prairie in the southern High Plains of Texas. Conservation Reserve Program grasslands were similar to native grasslands in species diversity but not species composition, which changed from year to year with successional changes in vegetation. Grasshopper mice were most abundant immediately after plantings, but were replaced by harvest mice and cotton rats on second-year sites, when cover of exotic grasses was thick. Kangaroo rats were only captured regularly during the third year, when grass plants had become more dispersed. Hall and Willig (1994) suggested that permitting fire or grazing might restore native species composition, but these management actions would be counter to the primary objectives of the CRP. Moulton et al. (1981a) also examined species composition of rodent communities on three formerly cultivated sites 40 to 50 years after they were reseeded with nonnative grasses. Two of the three sites had as many small-mammal species as native prairie, which led them to conclude that vegetative structure, not floristic composition, had a significant influence on species occurrence.

Summary and Conclusions

Mammals have attracted the attention of researchers in the shortgrass steppe for nearly four decades. Their studies suggest that the distribution and abundance of most species are strongly influenced by edaphic factors and the availability of cover, although the presence of plant species with large, palatable seeds increases diversity by improving conditions for granivores. Because primary productivity—and hence, vegetative structure—is low and the weather can be quite harsh, the most successful species are omnivorous and live belowground in burrows constructed by them or other species, or they are large enough in body size to minimize energetic costs. Compared with other ecosystems, most small-mammal populations are at low (less than four individuals/ha) densities and local communities are composed of a few common species; however, on a regional scale, the diversity of mammals is comparable with that in other grasslands. The scarcity of prey is translated into low densities of most carnivorous mammals, many of which suffered greatly from historical efforts at predator control. Large mammals such as bison and pronghorn historically had a major influence on native vegetation, but their numbers are still greatly diminished. Small mammals consume little of the standing plant biomass. Burrowing mammals, however, significantly alter the structure and function of shortgrass steppe ecosystems by translocating soil nutrients, altering soil microclimate and hydrology, and creating patchiness in vegetation. Moreover, the mounds and burrows created by these mammals modify plant communities by providing establishment sites for rarer plants, and refuge for other animals from predators as well as from weather conditions.

In our view, the outlook for the future of the shortgrass steppe is more favorable than for other grasslands in the Great Plains because the system is relatively resistant to livestock grazing, and because many areas can only marginally support consistent crop production. Our review suggests that many of the native mammals are relatively resilient to the types of disturbance represented by the current types of human land use, although the conversion of native prairie to large-scale cultivation would be disastrous for most resident populations. For these reasons, and because populations of some of the native large mammals have recovered, albeit modestly, from the past decades of exploitation, we are cautiously optimistic about the continued persistence of the native mammalian fauna in most areas of the shortgrass steppe. However, our review has ignored probably the most significant threat to remaining large tracts of the shortgrass steppe: the development and conversion of native prairie to urban and exurban development. Despite their resilience and ability to tolerate the harsh climatic and resource conditions characteristic of the shortgrass steppe, we cannot reasonably expect the native fauna to persist intact in the altered landscapes that accompany development of the scale currently envisioned, for example, along the densely populated corridor adjacent to the Rocky Mountain Front Range. Our review reveals that ecologists have so far anticipated well the questions necessary to understand the factors that affect distribution and abundance of shortgrass steppe mammals and their responses to the most frequent human activities. Discovering ways to integrate and protect broad expanses of shortgrass steppe and the native fauna successfully from the rapidly expanding human population in the West represents the most significant challenge for the future.

Acknowledgments This research has been funded by the American Society of Mammalogists, the American Museum of Natural History and, especially, the support of W. Lauenroth, I. Burke, and E. Kelly through the NSF SGS LTER project. A large number of undergraduate and graduate students have helped us with their research during the past decade, and although we are grateful for all their efforts, we particularly thank A. Anderson, D. Smith, L. Clippard, L. Hartley, S. Hauser, D. Lett, N. Kaplan, S. Ratering, J. Stevens, A. Suazo, J. Weaver, M. Andre, B. Frank, D. Hanni, B. Hendrixson, L. Higgins, and J. Roach for their assistance in the field. Our knowledge and appreciation of the shortgrass steppe has been enriched by thoughtful conversations with D. Milchunas, J. Fitzgerald, R. Ryder, J. Detling, and J. Moore.

References

Abramsky, Z. 1976. *Small mammal studies in natural and manipulated shortgrass prairie*. PhD diss., Colorado State University, Fort Collins, Colo.

Abramsky, Z. 1978. Small mammal community ecology: Changes in species diversity in response to manipulated productivity. *Oecologia* **34**:113–123.

Abramsky, Z., M. I. Dyer, and P. D. Harrison. 1979. Competition among small mammals in experimentally perturbed areas of the shortgrass prairie. *Ecology* **60**:530–536.

Abramsky, Z., and C. R. Tracy. 1979. Population biology of a "noncycling" population of prairie voles and a hypothesis on the role of migration in regulating microtine cycles. *Ecology* **60**:349–361.

Abramsky, Z., and C. R. Tracy. 1980. Relation between home range size and regulation of population size in *Microtus ochrogaster*. *Oikos* **34**:347–355.

Abramsky, Z., and G. M. Van Dyne. 1980. Field studies and a simulation model of small mammals inhabiting a patchy environment. *Oikos* **35**:80–92.

Anderson, E., S. Forrest, T. W. Clark, and L. Richardson. 1986. Paleobiology, biogeography, and systematics of the black-footed ferret, *Mustela nigripes* (Audubon and Bachman), 1851. *Great Basin Naturalist Memoirs* **8**:11–62.

Armstrong, D. M. 1972. Distribution of mammals in Colorado. *Monographs of the Kansas Museum of Natural History* **3**:1–415.

Bailey, V., and C. C. Sperry. 1929. Life history and habits of the grasshopper mice, genus *Onychomys*. *United States Department of Agriculture Technical Bulletin* **145**:1–19.

Baker, B. W., T. R. Stanley, and G. E. Plumb. 2000. Nest predation on black-tailed prairie dog colonies. *Journal of Wildlife Management* **64**:776–784.

Baker, B. W., T. R. Stanley, and J. A. Sedgwick. 1999. Predation on artificial ground nests on white-tailed prairie dog colonies. *Journal of Wildlife Management* **63**:270–277.

Barko, V. A., M. W. Palmer, J. G. Stewart, and D. M. Engle. 2001. Vascular plant communities associated with black-tailed prairie dog colonies in southern shortgrass prairie. *Proceedings of the Oklahoma Academy of Science* **81**:11–19.

Barko, V. A., J. H. Shaw, and D. M. Leslie, Jr. 1999. Birds associated with black-tailed prairie dog colonies in southern shortgrass prairie. *Southwestern Naturalist* **44**:484–489.

Bayless, M., W. K. Lauenroth, and I. C. Burke. 1996. Refuge effect of the *Opuntia polyacantha* (plains prickly pear) on grazed areas of the shortgrass steppe. *Bulletin of the Ecological Society of America* **77**:28.

Bekoff, M. 1982. Coyote, pp. 447–459. In: J. A. Chapman and G. A. Feldhammer (eds.), *Wild mammals of North America*. Johns Hopkins University Press, Baltimore, Md.

Blair, W. F. 1954. Mammals of the Mesquite Plains Biotic District in Texas and Oklahoma, and speciation in central grasslands. *Texas Journal of Science* **3**:235–264.

Bonham, C. D., and J. S. Hannan. 1978. Blue grama and buffalo grass patterns in and near a prairie dog town. *Journal of Range Management* **31**:63–65.

Bonham, C. D., and A. C. Lerwick. 1976. Vegetation changes induced by prairie dogs on a shortgrass prairie. *Journal of Range Management* **29**:221–225.

Brown, J. H., and E. J. Heske. 1990. Temporal changes in a Chihuahuan desert rodent community. *Oikos* **59**:290–302.

Burke, I. C., and W. K. Lauenroth. 1993. What do LTER results mean? Extrapolating from site to region and decade to century. *Ecological Modeling* **67**:19–35.

Burke, I. C., W. K. Lauenroth, and D. G. Milchunas. 1997. Biogeochemistry of managed grasslands in central North America, pp. 85–102. In: E. A. Paul, K. Paustian, E. T. Elliot, and C. V. Cole (eds.), *Soil organic matter in temperate agroecosystems*. CRC Press, Boca Raton, Fla.

Burnett, W. L. 1925. Jack rabbits of Colorado with suggestions for their control. *Circular of the Colorado State Entomologist* **48**:1–11.

Butts, K. O., and J. C. Lewis. 1982. The importance of prairie dog towns to Burrowing Owls in Oklahoma. *Proceedings of the Oklahoma Academy of Science* **62**:46–52.

Caire, W., J. D. Tyler, B. P. Glass, and M. A. Mares. 1989. *Mammals of Oklahoma.* University of Oklahoma, Norman, Okla.

Calahane, V. H. 1954. Status of the black-footed ferret. *Journal of Mammalogy* **35**:418–424.

Cameron, M. W. 1984. *The swift fox* (Vulpes velox) *on the Pawnee National Grasslands: Its food habits, population dynamics, and ecology.* Masters thesis, University of Northern Colorado, Greeley, Colo.

Choate, J. R., and J. B. Pinkham. 1988. Armadillos in north-eastern Colorado. *Prairie Naturalist* **20**:174.

Choate, J. R., and M. P. Reed. 1988. Least shrew, *Cryptotis parva*, in southwestern Kansas and southeastern Colorado. *Southwestern Naturalist* **33**:361–362.

Clark, T. W., T. M. Campbell, III, D. G. Socha, and D. E. Casey. 1982. Prairie dog colony attributes and associated vertebrate species. *Great Basin Naturalist* **42**:572–582.

Coppock, D. L., J. E. Ellis, J. K. Detling, and M. I. Dyer. 1983. Plant–herbivore interactions in a North American mixed-grass prairie. II. Responses of bison to modification of vegetation by prairie dogs. *Oecologia* **56**:10–15.

Covell, D. F. 1992. *Ecology of the swift fox* (Vulpes velox) *in southeastern Colorado.* PhD diss., University of Wisconsin, Madison, Wisc.

Cully, J. F., Jr. 1991. Response of raptors to reduction of a Gunnison's prairie dog population by plague. *American Midland Naturalist* **125**:140–149.

Cully, J. F., Jr. and E. S. Williams. 2001. Interspecific comparisons of sylvatic plague in prairie dogs. *Journal of Mammalogy* **82**:894–905.

Dano, L. E. 1952. *Cottontail rabbit* (Sylvilagus auduboni baileyi) *populations in relation to prairie dog* (Cynomys ludovicianus ludovicianus) *towns.* Masters thesis, Colorado State University, Fort Collins, Colo.

Davis, J. R., and T. C. Theimer. 2003. Increased lesser earless lizard (*Holbrookia maculata*) abundance on Gunnison's prairie dog colonies and short term responses to artificial prairie dog burrows. *American Midland Naturalist* **150**:282–290.

Derner, J. D., J. K. Detling, and M. F. Antolin. 2006. Are livestock weight gains affected by black-tailed prairie dogs? *Frontiers in Ecology and the Environment* **4**:459–464.

Desmond, M. J., J. K. Savidge, and K. M. Eskridge. 2000. Correlations between Burrowing Owl and black-tailed prairie dog declines: A 7-year analysis. *Journal of Wildlife Management* **64**:1067–1075.

Dinsmore, S. J., G. C. White, and F. L. Knopf. 2005. Mountain Plover population responses to black-tailed prairie dogs in Montana. *Journal of Wildlife Management* **69**:1546–1553.

Donoho, H. S. 1971. *Dispersion and dispersal of white-tailed and black-tailed jackrabbits, Pawnee National Grasslands.* U.S. IBP Grassland Biome technical report no. 96. Colorado State University, Fort Collins, Colo.

Dreitz, V. J., M. B. Wunder, and F. L. Knopf. 2005. Movements and home ranges of Mountain Plovers raising broods in three Colorado landscapes. *Wilson Bulletin* **117**:128–132.

Dueser, R. D., J. H. Porter, and J. L. Dooley. 1989. Direct tests for competition in North American rodent communities: Synthesis and progress, pp. 105–125. In: D. W. Morris, Z. Abramsky, B. J. Fox, and M. R. Willig (eds.), *Patterns in the structure of small mammal communities.* Texas Tech University Press, Lubbock, Texas.

Duvall, A. C. 1990. Colorado pronghorn antelope status report, pp. 6–7. In: *Proceedings of the 14th Biennial Pronghorn Antelope Workshop.* Silver Creek, Colo.

Ecke, D. H., and C. W. Johnson. 1952. Plague in Colorado, pp. 1–37. In: *Plague in Colorado and Texas.* Public Health monograph 6, U.S. Public Health Service, Washington, D.C.

Egoscue, H. J. 1960. Laboratory and field studies of the northern grasshopper mouse. *Journal of Mammalogy* **41**:91–110.

Egoscue, H. J. 1979. *Vulpes velox. Mammalian Species* **122**:1–5.

Ellis, J. E., and M. Travis. 1975. Comparative aspects of foraging behaviour of pronghorn antelope and cattle. *Journal of Applied Ecology* **12**:411–420.

Eussen, J. T. 1999. *Food habits of the kit fox* (Vulpes macrotis) *and swift fox* (Vulpes velox) *in Colorado*. Masters thesis, University of Northern Colorado, Greeley, Colo.

Finck, E. J., D. W. Kaufman, G. A. Kaufman, S. K. Gurtz, B. K. Clark, L. J. McLellan, and B. S. Clark. 1986. Mammals of the Konza Prairie Research Natural Area, Kansas. *Prairie Naturalist* **18**:153–166.

Findley, J. S., A. H. Harris, D. E. Wilson, and C. Jones. 1975. Mammals of New Mexico. University of New Mexico Press, Albuquerque, N.Mex.

Finley, D. J. 1999. *Distribution of the swift fox* (Vulpes velox) *on the eastern plains of Colorado*. Masters thesis, University of Northern Colorado, Greeley, Colo.

Firchow, K. M. 1986. *Ecology of pronghorns on the Piñon Canyon Maneuver Site, Colorado*. Masters thesis, Virginia Polytechnic Institute and State University, Blacksburg, Va.

Fitzgerald, J. P., C. A. Meaney, and D. M. Armstrong. 1994. *Mammals of Colorado*. Denver Museum of Natural History, Denver, Colo.

Flake, L. D. 1971. *An ecological study of rodents in a shortgrass prairie in northeastern Colorado*. U.S. IBP Grassland Biome technical report no. 100. Colorado State University, Fort Collins, Colo.

Flake, L. D. 1973. Food habits of four species of rodents on a shortgrass prairie in Colorado. *Journal of Mammalogy* **54**:636–647.

Flake, L. D. 1974. Reproduction of four rodent species in a shortgrass prairie of Colorado. *Journal of Mammalogy* **55**:213–216.

Fleharty, E. D., and K. W. Navo. 1983. Irrigated cornfields as habitat for small mammals in the sandsage prairie region of western Kansas. *Journal of Mammalogy* **64**:367–379.

Flinders, J. T. 1971. *Diets and feeding habits of jackrabbits within a shortgrass ecosystem*. PhD diss., Colorado State University, Fort Collins, Colo.

Flinders, J. T., and R. M. Hansen. 1972. *Diets and habits of jackrabbits in northeastern Colorado*. Science Series 1, Range Science Department, Colorado State University, Fort Collins, Colo.

Flinders, J. T., and R. M. Hansen. 1973. Abundance and dispersion of leporids within a shortgrass ecosystem. *Journal of Mammalogy* **54**:287–291.

Flinders, J. T., and R. M. Hansen. 1975. Spring population responses of cottontails and jackrabbits to cattle grazing shortgrass prairie. *Journal of Range Management* **28**:290–293.

Ford, P. L. 1999. Response of buffalograss (*Buchloë dactyloides*) and blue grama (*Bouteloua gracilis*) to fire. *Great Plains Research* **9**:1–16.

Ford, P. L. 2007. Shared community patterns following experimental fire in a semiarid grassland. Abstracts of the Fourth International Wildland Fire Conference, 13–17 May 2007, Sevilla, Spain.

Ford, P. L., and G. R. McPherson. 1999. Ecology of fire in shortgrass prairie communities of the Kiowa National Grassland, pp. 71–76. In: C. Warwick (ed.), *Proceedings of the 15th North American Prairie Conference*. Natural Areas Association, Bend, Ore.

Forrest, S. 2005. Getting the story right: A response to Vermeire and colleagues. *BioScience* **55**:526–530.

French, N. R., and W. E. Grant. 1974. *Summary report of small mammal project grid livetrapping data*. U.S. IBP Grassland Biome technical report no. 258. Colorado State University, Fort Collins, Colo.

French, N. R., W. E. Grant, W. Grodzinski, and D. M. Swift. 1976. Small mammal energetics in grassland ecosystems. *Ecological Monographs* **46**:201–220.

Gerlach, T. P. 1987. *Ecology of mule deer on the Piñon Canyon Maneuver Site, Colorado.* Masters thesis, Virginia Polytechnic Institute and State University, Blacksburg, Va.

Gese, E. M., O. J. Rongstad, and W. R. Mytton. 1988. Home range and habitat use of coyotes in southeastern Colorado. *Journal of Wildlife Management* **52**:640–646.

Gese, E. M., O. J. Rongstad, and W. R. Mytton. 1989. Population dynamics of coyotes in southeastern Colorado. *Journal of Wildlife Management* **53**:174–181.

Goodrich, J. M., and S. W. Buskirk. 1998. Spacing and ecology of North American badgers (*Taxidea taxus*) in a prairie-dog (*Cynomys leucurus*) complex. *Journal of Mammalogy* **79**:171–179.

Grant, P. R. 1972. Interspecific competition among rodents. *Annual Review of Ecology and Systematics* **3**:79–105.

Grant, W. E., and E. C. Birney. 1979. Small mammal community structure in North American grasslands. *Journal of Mammalogy* **60**:23–36.

Grant, W. E., E. C. Birney, N. R. French, and D. M. Swift. 1982. Structure and productivity of grassland small mammal communities related to grazing-induced changes in vegetative cover. *Journal of Mammalogy* **63**:248–260.

Grant, W. E., and N .R. French. 1980. Evaluation of the role of small mammals in grassland ecosystems: A modelling approach. *Ecological Modelling* **8**:15–37.

Grant, W. E., N. R. French, and L. J. Folse, Jr. 1980. Effects of pocket gophers on plant production in shortgrass prairie ecosystems. *Southwestern Naturalist* **25**:215–224.

Grant, W. E., N. R. French, and D. M. Swift. 1977. Response of a small mammal community to water and nitrogen treatments in a shortgrass prairie ecosystem. *Journal of Mammalogy* **58**:637–652.

Green, N. E. 1969. *Occurrence of small mammals on the sandhill rangelands in eastern Colorado.* Masters thesis, Colorado State University, Fort Collins, Colo.

Gross, J. E. 1969. *Jackrabbit demographic and life history studies, Pawnee site.* U.S. IBP Grassland Biome technical report no. 16. Colorado State University, Fort Collins, Colo.

Guenther, D. A., and J. K. Detling. 2003. Observations of cattle use of prairie dog towns. *Journal of Range Management* **56**:410–417.

Hall, D. L., and M. R. Willig. 1994. Mammalian species composition, diversity, and succession in Conservation Reserve Program grasslands. *Southwestern Naturalist* **39**:1–10.

Hammerson, G. A. 1999. *A guide to amphibians and reptiles of Colorado.* University Press of Colorado, Niwot, Colo.

Hansen, R. M., and I. K. Gold. 1977. Blacktail prairie dogs, desert cottontails, and cattle trophic relations on shortgrass range. *Journal of Range Management* **30**:210–214.

Hartley, L. 2006. *Plague and the black-tailed prairie dog: An introduced disease mediates the effects of an herbivore on ecosystem structure and function.* PhD diss., Colorado State University, Fort Collins, Colo.

Hoffman, L. A. 1992. *Small mammal granivory and herbivory: Trophic constraints on the establishment of native grasses.* PhD diss., Colorado State University, Fort Collins, Colo.

Hoffman, L. A., E. F. Redente, and L. C. McEwen. 1995. Effects of selective seed predation by rodents on shortgrass establishment. *Ecological Applications* **5**:200–208.

Huntly, N., and R. Inouye. 1988. Pocket gophers in ecosystems: Patterns and mechanisms. *BioScience* **38**:786–793.

Jones, C., R. S. Hoffman, D. W. Rice, M. D. Engstrom, R. D. Bradley, D. J. Schmidly, C. A. Jones, and R. J. Baker. 1997. *Revised checklist of North American mammals north of Mexico, 1997.* Occasional Papers of the Museum of Texas Tech University, Lubbock, Texas.

Kamler, J. F., W. B. Ballard, R. L. Gilliland, and K. Mote. 2003. Spatial relationships between swift foxes and coyotes in northwestern Texas. *Canadian Journal of Zoology* **81**:168–172.

Kaufman, D. W., E. J. Finck, and G. A. Kaufman. 1990. Small mammals and grassland fires, pp. 46–80. In: S. L. Collins and L. L. Wallace (eds.), *Fire in North American tallgrass prairies*. University of Oklahoma Press, Norman, Okla.

Kaufman, D. W., and E. D. Fleharty. 1974. Habitat selection by nine species of rodents in north–central Kansas. *Southwestern Naturalist* **18**:443–452.

Kesner, M. H., and A. V. Linzey. 1997. Modeling population variation in *Peromyscus leucopus*: An exploratory analysis. *Journal of Mammalogy* **78**:643–654.

Klatt, L. E., and D. Hein. 1978. Vegetative differences among active and abandoned towns of black-tailed prairie dogs (*Cynomys ludovicianus*). *Journal of Range Management* **31**:315–317.

Knowles, C. J. 1986. Some relationships of black-tailed prairie dogs to livestock grazing. *Great Basin Naturalist* **46**:198–203.

Knowles, C. J. 1988. An evaluation of shooting and habitat alteration for control of black-tailed prairie dogs. *Proceedings of the Great Plains Wildlife Damage Control Conference* **8**:53–56.

Knowles, C. J. 1993. Association of black-tailed prairie dog colonies with cattle point attractants in the northern Great Plains. *Great Basin Naturalist* **53**:385–389.

Knowles, C. J., J. D. Proctor, and S. C. Forrest. 2002. Black-tailed prairie dog abundance and distribution in the Great Plains based on historic and contemporary information. *Great Plains Research* **12**:219–254.

Koford, C. 1958. Prairie dogs, whitefaces and blue grama. *Wildlife Monographs* **3**:1–78.

Kotliar, N. B., B. W. Baker, A. D. Whicker, and G. Plumb. 1999. A critical review of assumptions about the prairie dog as a keystone species. *Environmental Management* **24**:177–192.

Kretzer, J. E., and J. F. Cully, Jr. 2001a. Effects of black-tailed prairie dogs on reptiles and amphibians in Kansas shortgrass prairie. *Southwestern Naturalist* **46**:171–177.

Kretzer, J. E., and J. F. Cully, Jr. 2001b. Prairie dog effects on harvester ant species diversity and density. *Journal of Range Management* **54**:11–14.

Krueger, K. 1986. Feeding relationships among bison, pronghorn, and prairie dogs: An experimental analysis. *Ecology* **67**:760–770.

Larson, F. 1940. The role of bison in maintaining the shortgrass plains. *Ecology* **21**:113–121.

Lauenroth, W. K., J. L. Dodd, and P. L. Sims. 1978. The effects of water- and nitrogen-induced stresses on plant community structure in a semiarid grassland. *Oecologia* **36**:211–222.

Lauenroth, W. K., and D. G. Milchunas. 1991. Shortgrass steppe, pp. 183–226. In: R. T. Coupland (ed.), *Ecosystems of the world 8A: Natural grasslands*. Elsevier, Amsterdam.

Lauenroth, W. K., D. G. Milchunas, J. L. Dodd, R. H. Hart, R. K. Heitschmidt, and L. R. Rittenhouse. 1994. Effects of grazing on ecosystems in the Great Plains, pp. 69–100. In: M. Vavra, W. A. Laycock, and R. D. Pieper (eds.), *Ecological implications of livestock herbivory in the West*. Society for Range Management, Denver, Colo.

Levins, R. 1970. Extinction. *Lecture in Mathematics in the Life Sciences* **2**:75–107.

Liewer, J. A. 1988. *Pronghorn grazing impacts on winter wheat*. Masters thesis, Colorado State University, Fort Collins, Colo.

Lim, B. K. 1987. *Lepus townsendii. Mammalian Species* **288**:1–6.
Linzey, F. G. 1982. Badger, pp. 653–663. In: J. A. Chapman and G. A. Feldhammer (eds.), *Wild mammals of North America*. Johns Hopkins University Press, Baltimore, Md.
Lomolino, M. V., and G. A. Smith. 2003. Terrestrial vertebrate communities at black-tailed prairie dog towns. *Biological Conservation* **115**:89–100.
Lovell, D. C., J. R. Choate, and S. J. Bissell. 1985. Succession of mammals in a disturbed area of the Great Plains. *Southwestern Naturalist* **30**:335–342.
Loy, R. R. 1981. *An ecological investigation of the swift fox* (Vulpes velox) *on the Pawnee National Grasslands, Greeley, Colorado*. Masters thesis, University of Northern Colorado, Greeley, Colo.
Marti, C. D., Jr. 1974. Feeding ecology of four sympatric owls in Colorado. *Condor* **76**:45–61.
McCarty, R. 1978. *Onychomys leucogaster. Mammalian Species* **87**:1–6.
McCulloch, C. Y. 1959. *Populations and range effects of rodents on the sand sagebrush grasslands of western Oklahoma*. PhD diss., Oklahoma State University, Stillwater, Okla.
McEwen, L. C., and J. O. Ellis. 1975. Field ecology investigations of the effects of selected pesticides on wildlife populations. Technical report no. 289. U.S. IBP Grassland Biome, Fort Collins, Colo.
Mellott, R. S., and E. D. Fleharty. 1986. Distribution status of the cotton rat (*Sigmodon hispidus*) in Colorado. *Transactions of the Kansas Academy of Science* **89**:75–77.
Messick, J. P., and M. G. Hornocker. 1981. Ecology of the badger in southwestern Idaho. *Wildlife Monographs* **76**:1–53.
Milchunas, D. G., and W. K. Lauenroth. 1995. Inertia in plant community structure: State changes after cessation of nutrient-enrichment stress. *Ecological Applications* **5**:452–458.
Milchunas, D. G., W. K. Lauenroth, P. L. Chapman, and M. K. Kazempour. 1989. Effects of grazing, topography, and precipitation on the structure of a semiarid grassland. *Vegetatio* **80**:11–23.
Miller, R. S. 1964. Ecology and distribution of pocket gophers (Geomyidae) in Colorado. *Ecology* **45**:256–271.
Miller, B., C. Wemmer, D. Biggins, and R. Reading. 1990. A proposal to conserve black-footed ferrets and the prairie dog ecosystem. *Environmental Management* **14**:763–769.
Mohamed, R. M. 1989. Ecolo*gy and biology of* Perognathus *species in a grassland ecosystem of northeastern Colorado*. PhD diss., University of Northern Colorado, Greeley, Colo.
Moulton, M. P., J. R. Choate, and S. J. Bissell. 1983. Biogeographic relationships of pocket gophers in southeastern Colorado. *Southwestern Naturalist* **28**:53–60.
Moulton, M. P., J. R. Choate, and S. J. Bissell. 1981a. Small mammals on revegetated agricultural land in eastern Colorado. *Prairie Naturalist* **13**:99–104.
Moulton, M. P., J. R. Choate, S. J. Bissell, and R. A. Nicholson. 1981b. Associations of small mammals on the central High Plains of eastern Colorado. *Southwestern Naturalist* **26**:553–557.
Myers, G. T. 1974. Antelope in Colorado, pp. 5–6. In: *Proceedings of the 6th Biennial Antelope States Workshop*. Salt Lake City, Utah.
Olendorff, R. R. 1973. *The ecology of nesting birds of prey of northeastern Colorado*. Technical report no. 211. U.S. IBP Grassland Biome, Fort Collins, Colo.
Olson, P. 1972. Antelope management in Colorado, pp. 51–52. In: *Proceedings of the 5th Biennial Antelope States Workshop*, Billings, Mont.

O'Meilia, M. E., F. L. Knopf, and J. C. Lewis. 1982. Some consequences of competition between prairie dogs and beef cattle. *Journal of Range Management* **35**:580–585.

Orth, P. B., and P. L. Kennedy. 2001. Do land use patterns influence nest-site selection by Burrowing Owls (*Athene cunicularia hypugaea*) in northeastern Colorado? *Canadian Journal of Zoology* **79**:1038–1045.

Owen, R. D., and M. J. Hamilton. 1986. Second record of *Cryptotis parva* (Soricidae: Insectivora) in New Mexico, with review of its status on the Llano Estacado. *Southwestern Naturalist* **31**:403–405.

Packard, R. L. 1972. *Small mammal studies on Jornada and Pantex sites, 1970–71.* Technical report no. 188. U.S. IBP Grassland Biome, Fort Collins, Colo.

Packard, R. L. 1975. *Small mammal studies on Jornada and Pantex sites, 1972.* Technical report no. 277. U.S. IBP Grassland Biome, Fort Collins, Colo.

Parmenter, R. R., and T. R. Van Devender. 1995. Diversity, spatial variability, and functional roles of vertebrates in the desert grassland, pp. 196–229. In: M. P. McClaran and T. R. Van Devender (eds.), *The desert grassland.* University of Arizona Press, Tucson, Ariz.

Peden, D. G., G. M. Van Dyne, R. W. Rice, and R. M. Hansen. 1974. The trophic ecology of *Bison bison* L. on shortgrass plains. *Journal of Applied Ecology* **11**:489–498.

Platt, W. J. 1975. The colonization and formation of equilibrium plant species associations on badger disturbances in a tall-grass prairie. *Ecological Monographs* **45**:285–305.

Plumpton, D. L., and D. E. Andersen. 1997. Habitat use and time budgeting by wintering Ferruginous Hawks. *Condor* **99**:888–893.

Plumpton, D. L., and D. E. Andersen. 1998. Anthropogenic effects on winter behavior of Ferruginous Hawks. *Journal of Wildlife Management* **62**:340–346.

Pojar, T. M. 1984. Colorado pronghorn status report, pp. 12–13. In: *Proceedings of the 11th Pronghorn Antelope Workshop.* Corpus Christi, Texas.

Pojar, T. M. 1988. Colorado state report, pp. 15–17. In: *Proceedings of the 13th Biennial Pronghorn Antelope Workshop.* Hart Mountain, Ore.

Pojar, T. M., D. C. Bowden, and R. B. Gill. 1995. Aerial counting experiments to estimate pronghorn density and herd structure. *Journal of Wildlife Management* **59**:117–128.

Rebar, C., and W. Conley. 1983. Interactions in microhabitat use between *Dipodomys ordii* and *Onychomys leucogaster. Ecology* **64**:984–988.

Rebollo, S., D. G. Milchunas, I. Noy-Meir, and P. L. Chapman. 2002. The role of a spiny plant refuge in structuring grazed shortgrass steppe plant communities. *Oikos* **98**:53–64.

Redfield, J. A., C. J. Krebs, and M. J. Taitt. 1977. Competition between *Peromyscus maniculatus* and *Microtus townsendii* in grass-lands of coastal British Columbia. *Journal of Animal Ecology* **46**:607–616.

Roach, J. L. 1999. *Genetic structure of a black-tailed prairie dog* (Cynomys ludovicianus) *metapopulation in shortgrass steppe.* Masters thesis, Colorado State University, Fort Collins, Colo.

Roach, J. L., B. Van Horne, P. Stapp, and M. F. Antolin. 2001. Genetic structure of a black-tailed prairie dog metapopulation. *Journal of Mammalogy* **82**:946–959.

Roell, B. J. 1999. *Demography and spatial use of swift fox* (Vulpes velox) *in northeastern Colorado.* Masters thesis, University of Northern Colorado, Greeley, Colo.

Rongstad, O. J., T. R. Larion, and D. E. Anderson. 1989. *Ecology of swift fox on the Piñon Canyon Maneuver Site, Colorado.* Unpublished final report to the Directorate on Engineering and Housing. Fort Carson, Colo.

Salkeld, D. J., R. J. Eisen, P. Stapp, A. P. Wilder, J. Lowell, D. W. Tripp, D. Albertson, and M. F. Antolin. 2007. The role of swift foxes (*Vulpes velox*) and their fleas in prairie-dog plague. *Journal of Wildlife Diseases* **43**:425–431.

Salkeld, D. J., and P. Stapp. 2006. Seroprevalence rates and transmission of plague (*Yersinia pestis*) in mammalian carnivores. *Vector-Borne and Zoonotic Diseases* **6**:231–239.

Samuel, D. E., and B. B. Nelson. 1982. Foxes, pp. 475–490. In: J. A. Chapman and G. A. Feldhammer (eds.), *Wild mammals of North America*. Johns Hopkins University Press, Baltimore, Md.

Schauster, E. R., E. M. Gese, and A. M. Kitchen. 2002. An evaluation of survey methods for monitoring swift fox abundance. *Wildlife Society Bulletin* **30**:464–477.

Schmutz, J. K., and R. W. Fyfe. 1987. Migration and mortality of Alberta Ferruginous Hawks. *Condor* **89**:169–174.

Schwartz, C. C., and J. E. Ellis. 1981. Feeding ecology and niche separation in some native and domestic ungulates on the shortgrass prairie. *Journal of Applied Ecology* **18**:343–353.

Schwartz, C. C., and J. Nagy. 1976. Pronghorn diets relative to forage availability in northeastern Colorado. *Journal of Wildlife Management* **40**:469–478.

Seery, D. B., and D. J. Matiatos. 2000. Response of wintering buteos to plague epizootics in prairie dogs. *Western North American Naturalist* **60**:420–425.

Severe, D. S. 1977. *Revegetation of blacktail prairie dog mounds in shortgrass prairie*. Masters thesis, Colorado State University, Fort Collins, Colo.

Shaugnessy, M. J., Jr., and R. E. Cifelli. 2004. Influence of black-tailed prairie dog towns (*Cynomys ludovicianus*) on carnivore distributions in the Oklahoma Panhandle. *Western North American Naturalist* **64**:184–192.

Sidle, J. G., M. Ball, T. Byer, J. E. Chynoweth, G. Foli, R. Hodorff, G. Moravek, R. Peterson, and D. N. Svingen. 2001a. Occurrence of Burrowing Owls in black-tailed prairie dog colonies on Great Plains National Grasslands. *Journal of Raptor Research* **35**:316–321.

Sidle, J. G., D. H. Johnson, and B. R. Euliss. 2001b. Estimate areal extent of colonies of black-tailed prairie dogs in the northern Great Plains. *Journal of Mammalogy* **82**:928–936.

Smith, G. A., and M. V. Lomolino. 2004. Black-tailed prairie dogs and the structure of avian communities on the shortgrass plains. *Oecologia* **138**:592–602.

Sovada, M. A., C. C. Roy, J. B. Bright, and J. R. Gillis. 1998. Causes and rates of mortality of swift foxes in western Kansas. *Journal of Wildlife Management* **62**:1300–1306.

Sparks, K. L. 1972. *Grazing behavior of bison and cattle on a shortgrass prairie*. U.S. IBP Grassland Biome technical report no. 149. ,Colorado State University, Fort Collins, Colo.

Stapp, P. 1996. *Determinants of habitat use and community structure of rodents in northern shortgrass steppe*. PhD diss., Colorado State University, Fort Collins, Colo.

Stapp, P. 1997a. Community structure of shortgrass-prairie rodents: Competition or risk of intraguild predation? *Ecology* **78**:1519–1530.

Stapp, P. 1997b. Habitat selection by an insectivorous rodent: Patterns and mechanisms across multiple scales. *Journal of Mammalogy* **78**:1128–1143.

Stapp, P. 1998. A reevaluation of the role of prairie dogs in Great Plains grasslands. *Conservation Biology* **12**:1253–1259.

Stapp, P. 1999. Size and habitat characteristics of home ranges of northern grasshopper mice (*Onychomys leucogaster*). *Southwestern Naturalist* **44**:101–105.

Stapp, P. 2007. Rodent communities in active and inactive colonies of black-tailed prairie dogs in shortgrass steppe. *Journal of Mammalogy* **88**:241–249.

Stapp, P., M. F. Antolin, and M. Ball. 2004. Patterns of extinction in prairie-dog metapopulations: Plague outbreaks follow El Niño events. *Frontiers in Ecology and the Environment* **2**:235–240.

Stapp, P., and M. D. Lindquist. 2007. Roadside foraging by kangaroo rats in a grazed shortgrass prairie landscape. *Western North American Naturalist* **67**:368–377.

Stapp, P., and B. Van Horne. 1997. Response of deer mice (*Peromyscus maniculatus*) to shrubs in shortgrass prairie: Linking small-scale movements and the spatial distribution of individuals. *Functional Ecology* **11**:644–651.

Vaughan, T. A. 1961. Vertebrates inhabiting pocket gopher burrows in Colorado. *Journal of Mammalogy* **42**:171–174.

Vaughan, T. A. 1967. Food habits of the northern pocket gopher on shortgrass prairie. *American Midland Naturalist* **77**:176–189.

Vermeire, L. T., R. K. Heitschmidt, P. S. Johnson, and B. F. Sowell. 2004. The prairie dog story: Do we have it right? *BioScience* **54**:689–695.

Virchow, D. R., and S. E. Hygnstrom. 2002. Distribution and abundance of black-tailed prairie dogs in the Great Plains: A historical perspective. *Great Plains Research* **12**:197–218.

Vosburgh, T. C., and L. R. Irby. 1998. Effects of recreational shooting on prairie dog colonies. *Journal of Wildlife Management* **62**:363–372.

Weber, D. 2001. Winter raptor use of prairie dog towns in the Denver, Colorado vicinity. *Journal of the Colorado Field Ornithologists* **35**:86–91.

Whicker, A., and J. K. Detling. 1988. Ecological consequences of prairie dog disturbances. *BioScience* **38**:778–785.

Winter, S. L., J. F. Cully, Jr., and J. S. Pontius. 2002. Vegetation of prairie dog colonies and non-colonized shortgrass prairie. *Journal of Range Management* **55**:502–508.

Winter, S. L., J. F. Cully, Jr., and J. S. Pontius. 2003. Breeding season avifauna of prairie dog colonies and non-colonized areas in shortgrass prairie. *Transactions of the Kansas Academy of Science* **106**:129–138.

Yoakum, J. 1968. Antelope management in Colorado, pp. 4–14. In: *Proceedings of the 3rd Biennial Antelope States Workshop.* Casper, Wyo.

Zimmerman, G., P. Stapp, and B. Van Horne. 1996. Seasonal variation in the diet of Great Horned Owls (*Bubo virginianus*) in shortgrass prairie. *American Midland Naturalist* **136**:149–156.

9

Birds of the Shortgrass Steppe

John A. Wiens
Nancy E. McIntyre

Birds are part of the special magic of grasslands. Birds such as McCown's Longspurs (scientific names are given in the Appendix) or Horned Larks, which seem to disappear against the background of grass, soil, and stones when they are on the ground, launch breathtaking courtship flights punctuated by tinkling songs and mothlike flutterings. Male Lark Buntings, incongruously black and white (Fig. 9.1A) against the subdued tones of the grassland, may break into their morning territorial displays or gather together spontaneously in melodious group choruses. Mountain Plovers may burst from underfoot into utterly convincing broken-wing distraction displays. Ferruginous and other hawks (Fig. 9.1B) may suddenly plummet from the blue skies above. Sightings of relatively rare species such as Chestnut-collared Longspurs (Fig. 9.1C) may bring joy to dedicated bird-watchers. Birds give the shortgrass steppe an aura that *Bouteloua* alone cannot.

Yet birds have not figured importantly in most discussions of grassland ecology. They are generally drab and brownish, so they have not attracted much attention from the general public, and their contributions to ecosystem production and energy flow are small, so they have not been of much interest to ecologists studying ecosystem processes. However, grassland birds are showing the most widespread and consistent population declines of any group of North American birds (Herkert, 1995; Knopf, 1994; Peterjohn and Sauer, 1999). As a consequence, they have become a focus of conservation concern (Brennen and Kulvesky, 2005; Vickery and Herkert, 2001).

The history of ornithological research in the shortgrass steppe is closely intertwined with the broadly interdisciplinary work conducted during the IBP in the

Figure 9.1 Avian visitors to the shortgrass steppe. (A) Lark Bunting. (Photo by William Paff. Courtesy of Cornell Lab of Ornithology.) (B) Ferruginous Hawk. (Photo by Mark Lindquist.) (C) Chestnut-collared Longspur. (Photo by M. and B. Schwarzchild. Courtesy of Cornell Lab of Ornithology.)

late 1960s and early 1970s, and more recently (since 1982) as part of the NSF LTER program. In this chapter we describe the birds of the shortgrass steppe and summarize pertinent research that has been conducted on them during the past 40+ years. Our aim is to synthesize this information to provide a perspective on how environmental factors may relate to population fluctuations, on spatiotemporal shifts in community composition, and on patterns of habitat occupancy among the birds of the shortgrass steppe. We conclude by noting some continuing research priorities that have become more critical as conservation concerns about these birds have heightened.

Community Attributes

The regional species pool of the shortgrass steppe includes approximately 280 species from 17 orders that have been documented as breeding on, overwintering on, or migrating through the region (see the Appendix at the end of the chapter). Most of these species are uncommon or rare; only slightly more than 100 species can be regarded as common or abundant on the shortgrass steppe (30 year-round residents, 40 summering species, eight wintering species, and 28 migrants).

Although most of the birds that occur in the shortgrass steppe are passerines, the region also supports a large number of raptors. Most of these feed primarily on lagomorphs and rodents, with seasonal shifts in diet composition corresponding to changes in the abundance of prey species (Marti, 1969; Ryder, 1972; Zimmerman et al., 1996). Raptors (particularly Rough-legged Hawks, Golden Eagles, Prairie Falcons, Ferruginous Hawks, and Northern Harriers) comprise most of the avian biomass on the shortgrass steppe in winter (Ryder, 1970). During the breeding season, raptor numbers are apparently limited by the availability of nest sites and perches, and by nest failure (Olendorff, 1972, 1973). More generally, limited nest site and perch availability may account for the patchy distribution on the steppe of many bird species, such as Brewer's and Cassin's Sparrows (Marti, 1969; Olendorff, 1972).

Shortgrass steppe landscapes include a variety of distinct habitat types. In addition to the grasslands that dominate the landscape, the steppe contains riparian woodlands, wetlands, and permanent and ephemeral ponds (playas). This habitat variety enhances the avifaunal diversity of the shortgrass steppe. Of the common and abundant species, for example, 40% are riparian species, whereas an additional 27% are associated with wetland habitats. Some 16% of the common and abundant species are wide-ranging habitat generalists (chiefly raptors and corvids), and only 17% can be considered to be true *grassland* species. Clearly, the diversity of habitats over the landscape contributes significantly to the overall avian diversity of the shortgrass steppe, and riparian habitats may be especially important (Laubhan and Fredrickson, 1997; Skagen, 1997).

In common with several other arid North American ecosystems, the shortgrass steppe is generally depauperate with respect to birds (Cody, 1966, 1985; Ryder, 1970). Grasslands in general possess relatively low densities of birds relative to other habitats in North America (Wiens, 1971). For example, grasslands generally

Table 9.1 General Bird–Community Attributes in Grassland and Shrub Steppe Ecosystems of North America

Grassland	n	No. of Species $\bar{x}$	SD	Density, Individuals/km² $\bar{x}$	SD	Biomass, g·ha⁻¹ $\bar{x}$	SD
Tallgrass	9	3.9	1.6	319.0	151.4	150.1	82.3
Northern mixed	10	4.7	1.4	237.2	130.6	92.4	41.6
Shortgrass	15	4.2	1.3	294.0	79.1	108.4	25.4
Palouse	4	3.0	0.0	205.2	51.1	91.3	15.6
Shrub steppe	6	4.1	0.8	266.3	139.7	64.4	17.2

SD, standard deviation.
From Wiens (1971, 1978; Wiens and Dyer, 1975).

support less than one quarter of the breeding densities found in coniferous forests and perhaps 40% of the biomass, with coniferous forests containing more small but abundant species (Wiens, 1978). Average densities of breeding birds on the shortgrass steppe are somewhat less than those in tallgrass prairie, but are somewhat greater than those in other grassland or shrub steppe habitats (Table 9.1 [Wiens, 1971, 1978; Wiens and Dyer, 1975]). However, there is substantial variation in both density and biomass among regions within the major grassland types, among locations within a given site, and even between study plots located a short distance apart within a single pasture (Wiens and Dyer, 1975).

Characteristic Species

Just as the term *grassland* encompasses a wide range of vegetation types (Joern and Keeler, 1995), so *grassland birds* includes a variety of species, many of which occur only infrequently in the shortgrass steppe. In fact, the distinctiveness of the shortgrass steppe in terms of vegetation and climate (Lauenroth, chapter 5, this volume; Lauenroth and Milchunas, 1991; Pielke and Doesken, chapter 2, this volume) does not extend to its avifauna. No species has a breeding or wintering distribution that closely matches the biogeographic limits of the *Bouteloua–Buchloë* shortgrass steppe (sensu Küchler, 1964), and many of the species that do occur there are widespread over the Great Plains or throughout North America. For example, two of the most abundant breeding species on the shortgrass steppe (Horned Lark and Western Meadowlark) are also common over much of the Great Plains and western shrublands and deserts, and even species such as Brewer's Sparrows, which have a locally patchy distribution in the shortgrass steppe, are distributed broadly throughout the shrub steppe of the intermountain West (Fig. 9.2).

Nonetheless, we can distinguish a suite of species that are more characteristic of the shortgrass steppe than other biomes. For example, Knopf (1996) recognized nine species as primary or endemic species of the Great Plains grasslands as a whole. Six of these species (Ferruginous Hawk, Mountain Plover, Cassin's Sparrow, Lark Bunting, McCown's Longspur, and Chestnut-collared Longspur)

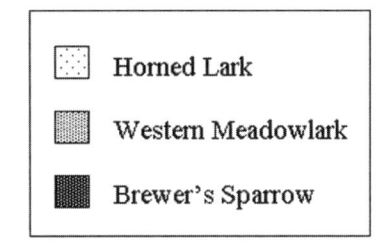

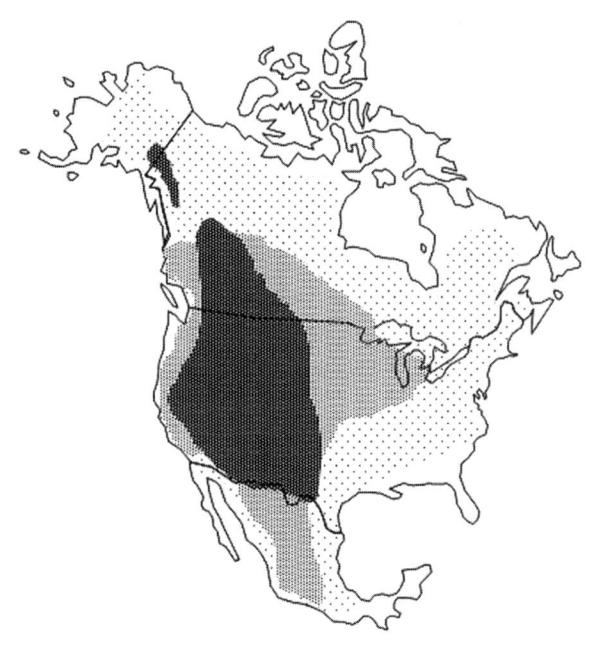

Figure 9.2 Range maps (breeding and year-round occurrences combined) of Horned Lark, Western Meadowlark, and Brewer's Sparrow. (Data are from the North American Breeding Bird Survey.)

occur with some regularity in at least some portions of the shortgrass steppe. Cassin's Sparrows, for example, are found during the breeding season in an area that is largely concordant with the shortgrass steppe (Fig. 9.3), although they are year-round residents in a much larger area of the southwestern United States and north–central Mexico. The breeding ranges of other species, such as Mountain Plover and McCown's Longspur, include the northern part of the shortgrass steppe but often extend northward into the more arid, western portion of the northern mixed-grass prairie (Fig. 9.3). The breeding range of Chestnut-collared Longspurs includes some of the northern mixed-grass prairie, whereas that of the Lark Bunting incorporates some of the southern mixed-grass prairie.

This lack of distributional concordance between birds and floristic regions of the Great Plains is not particularly surprising. The distinctions among the shortgrass steppe, northern mixed-grass prairie, and southern mixed-grass prairie are

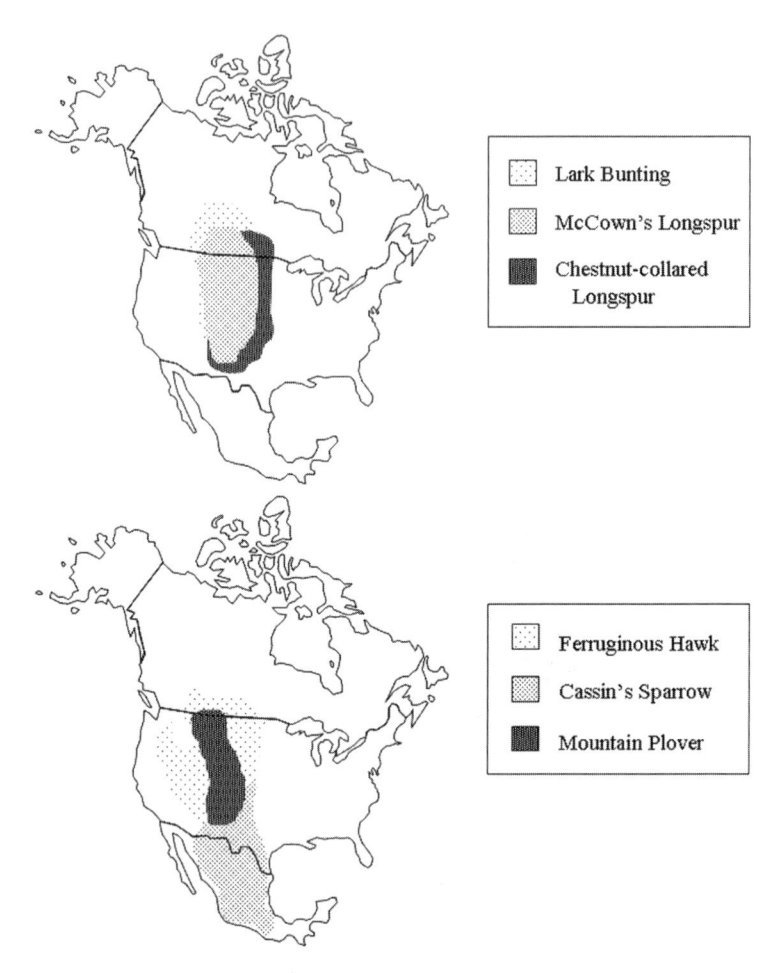

Figure 9.3 Range maps (breeding and year-round occurrences combined) of Lark Bunting, McCown's Longspur, and Chestnut-collared Longspur (A); and Ferruginous Hawk, Cassin's Sparrow, and Mountain Plover (B). (Data are from the North American Breeding Bird Survey.)

based on which plant species comprise the potential natural vegetation in upland sites (Lauenroth and Milchunas, 1991). In fact, in many of these areas (especially the western part of the northern mixed-grass prairie), heavy grazing by domestic livestock results in dominance of short grasses of a variety of species, and the actual physiognomy of these grasslands is in many places very similar to that of the shortgrass steppe. Birds generally respond much more strongly to vegetation structure or physiognomy than to floristic composition (Rotenberry, 1985), and because they also respond to actual rather than potential vegetation, bird distributions may not correspond closely to the region that has traditionally been called the shortgrass steppe (see also Vickery et al., 1999).

One can also define characteristic species on the basis of what one might reasonably expect to see on a visit to the shortgrass steppe. There may be considerable differences in breeding birds from place to place at both local and regional scales within the shortgrass steppe. At the CPER site in north–central Colorado, for example, Giezentanner (1970) recorded more Western Meadowlarks, Lark Buntings, and Chestnut-collared Longspurs as breeding species in plots subjected to moderate summer grazing, whereas nearby pastures that received heavy summer grazing supported more Mountain Plovers and McCown's Longspurs (Table 9.2). Over this site as a whole, five passerine species (Lark Bunting, Horned Lark, Western Meadowlark, McCown's Longspur, and Brewer's Sparrow) account for nearly 95% of the total nesting bird population (Gietzentanner, 1970), with the first three comprising the bulk (84%) of this total (R. Ryder, unpublished data). Lark Buntings and Horned Larks are the most abundant birds at this site in the summer, accounting for more than 75% of individuals. Horned Larks are among the few year-round residents of the shortgrass steppe, accounting for more than 85% of individuals observed in winter (Ryder, 1970). In winter, the other species most often observed are (in decreasing order of abundance): Lapland Longspur, Rough-legged Hawk, Golden Eagle, Prairie Falcon, Ferruginous Hawk, and Northern Harrier (Ryder, 1970).

Many of these species share common features of ecology and life history. For example, most shortgrass steppe birds are migratory, which allows them to exploit favorable conditions while avoiding harsh seasons. Nearly three quarters of the species that occur in the shortgrass steppe occupy the area only seasonally or as transients, with some species (e.g., Mountain Bluebirds) being locally abundant during migration (Andrews and Righter, 1992). Many species are omnivorous, being primarily granivorous and but seasonally insectivorous, with Gramineae seeds accounting for most of their winter, spring, and autumn diets; and beetles, spiders, lepidopterans (mainly caterpillars), and grasshoppers comprising most of the summer diet (McIntyre and Thompson, 2003). There is high dietary overlap among species with respect to the types of foods consumed, but there are differences in the relative proportions of each type of food in the diets of different species (Baldwin, 1970, 1971; Baldwin et al., 1971; Wiens and Rotenberry, 1979; Wiens et al., 1974b,

Table 9.2 Average Number of Individuals for Replicate Plots Subjected to Different Grazing Regimes

Species	Heavy summer	Moderate summer	Light summer	Heavy winter	Moderate winter	Light winter
Mountain Plover	37.1	0	0	0	0	0
Horned Lark	111.1	59.3	56.8	66.7	79.1	61.8
Western Meadowlark	0	37.1	29.7	27.2	49.4	49.4
Lark Bunting	0	76.6	118.6	69.2	71.7	74.1
McCown's Longspur	81.5	0	81.5	0	9.9	0
Chestnut-collared Longspur	0	24.7	12.3	0	0	0
Brewer's Sparrow	0	0	0	24.7	71.7	56.8

Data from Giezentanner (1970).

1975). Diet composition varies substantially among seasons and from year to year (Wiens and Rotenberry, 1979), and many species appear to respond opportunistically to short-term variations in prey availability (Rotenberry, 1985).

Within the grassland, some species exhibit clear habitat preferences. Brewer's Sparrows, for example, are generally associated with areas that have at least some shrub cover. Mountain Plovers appear to favor areas of short grass in playas or intensively grazed areas (Knopf, 1996; Knopf and Miller, 1994). Indeed, Knopf and Miller (1994) characterized Mountain Plovers as "disturbed prairie or semidesert species rather than a specific associate of grasslands" (p. 505) on the basis of their predilection to nest in areas with considerable coverage of bare ground. Most species, however, are habitat generalists, occupying a wide array of sites differing in exposure, grazing intensity, floristic composition, or vegetation structure (Cody, 1968). Habitat partitioning on the basis of vertical structure (Wiens, 1970a), nest site microhabitat requirements, and timing of nesting (Creighton and Baldwin, 1974) also produces some ecological differentiation among species with generally similar habitat affinities. These mechanisms, coupled with direct and indirect environmental limiting factors, result in significant spatial and temporal heterogeneity in avian occurrence and abundance on the shortgrass steppe (Giezentanner, 1970; Giezentanner and Ryder, 1969; Ryder, 1972; Wiens, 1970a, 1971, 1974a; Wiens and Dyer, 1975; Wiens et al., 1974a).

Courtship displays often incorporate ritualistic flight patterns, which can be seen for some distance on the relatively flat and treeless landscape. Owing to the scarcity of woody vegetation in which to build nests, most birds of the shortgrass steppe are, of necessity, ground-nesters. Most species raise one to two broods a year (Greer, 1988; Strong, 1971). They often build nests in grass tufts, in the lee of cactus clumps, or wherever they can find some cover (With and Webb, 1993). Most young are altricial and require 2 to 3 weeks of parental care. Reproductive failure is chiefly a result of weather (e.g., hail) and predation, particularly by thirteen-lined ground squirrels (*Spermophilus tridecemlineatus*) (Graul, 1972; Strong, 1971; With, 1994) and raptors (Yackel Adams et al., 2001). Nest failure is thought to be one of the primary factors limiting passerine numbers on the shortgrass prairie (Wiens, 1974a; With, 1994).

Relationships between Birds and Environmental Features

The physical characteristics of the shortgrass steppe affect the organisms that live there. Vegetation (including modifications to vegetation induced by grazing) and climate are the two most important environmental features that influence bird diversity and abundance. Relative to other grasslands, fire has historically been of little importance in shaping biological patterns on the shortgrass steppe.

Vegetation

Most of the vegetation on the shortgrass steppe is, appropriately enough, short in stature, effectively making the steppe a one-layered macrohabitat for its larger inhabitants, including birds. Therefore, we might expect variation on a finer spatial scale to affect the distribution and abundance of birds on the shortgrass steppe. An indication of the nature of this fine-scale spatial heterogeneity in distribution and abundance is evident from surveys conducted along 2-km belt transects during the breeding season of 2000 (J. A. Wiens, unpublished data). Occurrences of several species were tallied in 50-m segments of the transect (here aggregated into 100-m segments; Fig. 9.4) at two sites that differed in slope, exposure, and vegetation cover. There were differences in the evenness of distribution of individuals over the transect length among species and for a given species between the two sites (Fig. 9.4). Lark Buntings, for example, were considerably more abundant and more evenly distributed at site 1 than at site 2, whereas Horned Larks were more abundant at site 2 than at site 1, but occurred over only part of the transect. For all species there was substantial spatial variation in distribution and abundance.

It is likely that the patterns of species distributions in Figure 9.4 are related to underlying variations in habitat features. For example, the horizontal patterning of vegetation is a good predictor of avian biomass (Wiens, 1970b, 1974a,b). Most relationships between avian community attributes and habitat heterogeneity at a fine (square meter) scale, however, seem to be weak or inconclusive (Porter, 1974; Rotenberry and Wiens, 1980a,b; Wiens and Rotenberry, 1981; Wiens et al., 1972).

Heterogeneity on a broader spatial scale does appear to affect richness and density, however. With increasing fragmentation of North American grasslands by human activities has come the recognition of the importance of the size of grassland patches

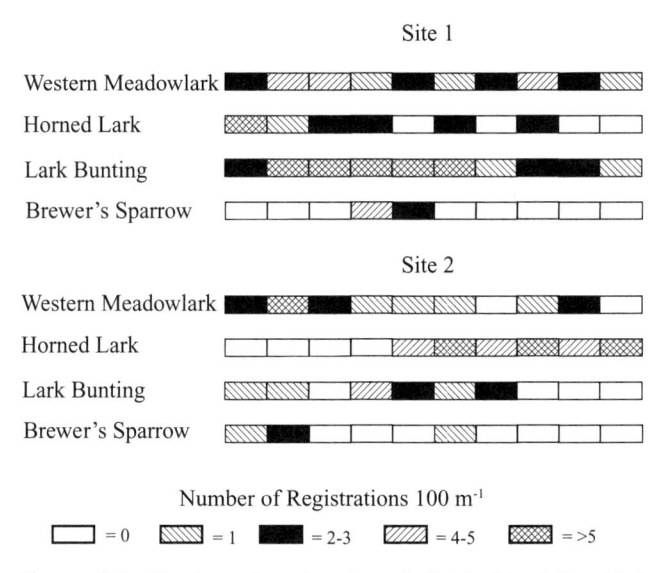

Figure 9.4 Number of registrations (individuals) of four bird species in 100-m segments of 2-km-long transects at two study sites in the shortgrass steppe of northeastern Colorado. (From Wiens et al. [unpublished data].)

in preserving bird numbers and diversity (Helzer, 1996; Herkert, 1994; Winter and Faaborg, 1999). Most studies have agreed that grassland species occur at their highest densities in large, contiguous patches, probably from the maintenance of a consistent microclimate and microhabitat, or perhaps as a result of lower deleterious edge effects such as the increased predation and brood parasitism rates found in forest fragments and forest–grassland ecotones (Andrén and Angelstam, 1988; Andrén et al., 1985; Gates and Gysel, 1978; Wilcove 1985; but see Skagen et al., 2005). Certain species appear to be highly sensitive to patch area. For example, even though Grasshopper Sparrows maintain individual territories less than1 ha in size, they tend to occur only in patches more than 10 ha in area (Herkert, 1994; Wiens, 1969).

Fragmentation as a result of agriculture in particular has been implicated in the decline of numerous grassland bird populations (Reynolds et al., 1994), although cropland does provide useful foraging habitat for some species (particularly Mountain Plover [Knopf and Rupert, 1999]). Prairie reclamation through projects like the USDA's CRP has successfully halted the declines of some species in some areas, especially Grasshopper Sparrows, Western Meadowlarks, Field Sparrows, and Chestnut-collared Longspurs (Best et al., 1998; Coppedge et al., 2001; Johnson and Schwartz, 1993; McCoy et al., 1999; Reynolds et al., 1994). Other species, however, do not appear to be recovering from population declines incurred by conversion of grassland to cropland even several years after the land was enrolled in the CRP (e.g., Dickcissels, Red-winged Blackbirds [McCoy et al., 1999; Reynolds et al., 1994]). Because there have been relatively few examinations of the CRP in shortgrass areas, however, our understanding of how grassland reclamation affects bird populations is incomplete (Berthelsen and Smith, 1995; Thompson, 2003).

One of the key differences between CRP and native prairie is in vegetation stature. In the shortgrass steppe, CRP lands are often seeded with nonnative grass species, resulting in vegetation that is taller than in native shortgrass. Howard et al. (2001) examined the effects of fragmentation of both native prairie and CRP lands in northeastern Colorado on predation rates on both natural and artificial nests. They were unable to document statistically significant differences in nest losses related to indices of habitat fragmentation, although daily mortality of artificial (but not natural) nests did decline with increasing vegetation height. In this area, fragmentation of the grasslands is not extensive (62% of the study area was occupied by native grasslands), and habitat loss may not have reached the point at which nest predation rates are affected.

Grazing

Some of the spatial variation in the local distribution of shortgrass steppe birds may be related to grazing by domestic livestock and native herbivores. Livestock may trample eggs or fledglings, and their dung attracts insect prey and creates nesting windbreaks (Ryder, 1980). Native herbivores such as black-tailed prairie dogs (*Cynomys ludovicianus*) dig burrows that serve as nesting cavities for Burrowing Owls (Desmond and Savidge, 1999; Hughes, 1993; Plumpton, 1992).

The primary impact of grazing on birds of the shortgrass steppe, however, has been indirect, through changes in vegetation structure. Because vegetation

Table 9.3 Breeding Bird Community Attributes in Grazed and Ungrazed Plots in Three North American Grassland Types

Grassland	Grazing	n	No. of Species $\bar{x}$	SD	Density, Individuals/km² $\bar{x}$	SD	Biomass, g·ha⁻¹ $\bar{x}$	SD
Shortgrass	Ungrazed	4	4.1	1.9	245.3	52.3	97.3	17.6
	Grazed	11	4.3	1.1	311.7	81.6	112.5	27.1
Northern, mixed	Ungrazed	5	5.0	1.3	220.0	91.6	93.1	42.3
	Grazed	5	4.5	1.6	254.4	171.0	91.5	45.9
Tallgrass	Ungrazed	3	3.0	1.7	193.1	94.1	60.8	21.6
	Grazed	6	4.3	1.5	381.9	137.3	179.8	71.7

"Ungrazed" includes lightly grazed areas; "grazed" includes moderately to heavily grazed sites. SD, standard deviation.
(From Wiens and Dyer [1975]).

lushness may be greater in ungrazed areas in tallgrass or mixed-grass prairies than in the shortgrass steppe, we might expect differences between grazed and ungrazed areas to be less in the shortgrass steppe than elsewhere. Indeed, although breeding bird densities appear to be somewhat greater in heavily grazed areas than in ungrazed or lightly grazed areas throughout the grasslands, differences with grazing are greater in tallgrass prairie than in shortgrass or mixed-grass prairies (Table 9.3 [Wiens and Dyer, 1975]). Bird biomass also tends to be greater in heavily grazed areas, at least in the shortgrass steppe and tallgrass prairie. These patterns are based on small sample sizes and there is substantial variance, however, so the relationships between grazing and avian community structure in grasslands remain to be defined rigorously. Additionally, changes in habitat physiognomy may have more profound effects if they occur at the start of the growing season, so the *timing* of grazing may have a greater impact on species occurrence than the *intensity* of grazing (Milchunas et al., chapter 16, this volume; Wiens, 1970a, 1971).

Several of the characteristic bird species of the shortgrass steppe, however, exhibit clear associations with grazing regimes. In north–central Colorado, Mountain Plovers, Horned Larks, and Ferruginous Hawks reach their greatest abundances in sites subjected to heavy summer grazing (Table 9.2 and Fig. 9.5 [Giezentanner, 1970; Wiens, 1973]), where the vegetation is reduced to an almost lawnlike stature (Giezentanner, 1970; Knopf, 1996; Knopf and Miller, 1994; Knopf and Samson, 1997; Ryder, 1980; Ryder and Cobb, 1969; Wiens, 1971). Other species, such as Lark Buntings, Western Meadowlarks, Northern Harriers, and (especially in mixed-grass prairie) Chestnut-collared Longspurs, appear to favor the taller vegetation characteristic of ungrazed or only lightly or moderately grazed areas (Fig. 9.5). Brewer's Sparrows and, in more southern areas, Cassin's Sparrows occur chiefly in areas with moderate shrub cover; in these rangelands, such areas are usually only lightly to moderately grazed. For example, notice from Figure 9.5 that Brewer's Sparrows at the CPER occur only in winter-grazed pastures, regardless of grazing intensity. This reflects the behavior of ranchers, who

Summer grazing

4% 6% 0% (Brewer's Sparrow)
23%
44%
16%
7%

▨ Mountain Plover
▧ Horned Lark
☐ Western Meadowlark
☰ Lark Bunting
■ McCown's Longspur
▨ Chestnut-collared Longspur
▦ Brewer's Sparrow

Winter grazing

0% (Mountain Plover)
(Chestnut-collared Longspur) 0%
18%
33%
1%
35%
13%

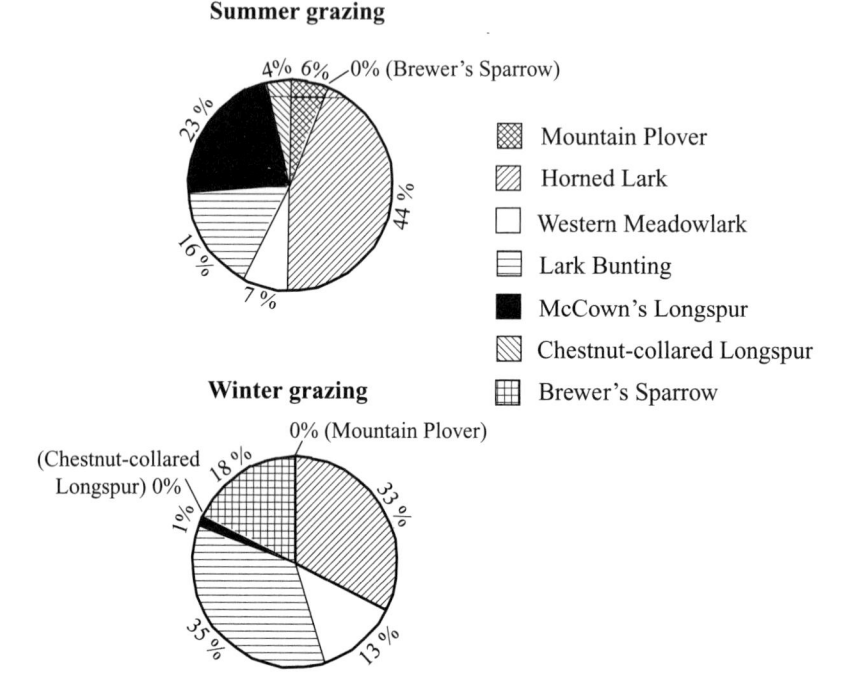

Figure 9.5 Pie diagrams depicting proportion of species by grazing regime. (Data from Giezentanner [1970] and Wiens [1973].)

typically use shrub-dominated areas only for winter grazing. Differences among species in response to grazing may reflect species-specific responses to reduced visibility from ungrazed vegetation (Bock et al., 1993; Wiens, 1970a). Such differences may explain why some species (e.g., Chestnut-collared Longspur, Lark Bunting, and Western Meadowlark) respond positively to grazing in tallgrass and mixed-grass prairie, where grazing would create favorable close-cropped conditions, but negatively to grazing in shortgrass prairie, where grazing would eliminate an already low level of cover (Bock et al., 1993).

These patterns run counter to the popular perception that grazing has detrimental effects on grassland fauna (Fleischner, 1994). Historically, the shortgrass steppe and northern mixed-grass prairie contained vast numbers of bison (*Bison bison*), and grazing pressures are thought to have been intense but spatially and temporally patchy (Hartnett et al., 1997). Although the details of grazing patterns of domestic livestock differ from those of bison, the effects of creating a landscape mosaic of areas differing in vegetation stature may be similar. This heterogeneity may enhance the suitability of grassland habitats to the bird species that are most characteristic of the semiarid grasslands, which, after all, evolved in the context of grazing by large mammalian herbivores (Knopf, 1996; see also Temple et al., 1999).

Climate and Microclimate

At a broad scale, the distribution of North American grasslands (and of the plant species that define the shortgrass steppe) is determined by regional climatic patterns (Laurenroth and Milchunas, 1991). Within the region there are also well-defined gradients in the amount and periodicity of precipitation and in temperature. The broad, seasonal changes in climate determine the timing and duration of the breeding and wintering seasons. In the shortgrass steppe, breeding species usually begin to arrive in late April and early May, and most summer residents have departed by late September or October. There are usually two peaks in abundance each year, corresponding to the arrival of migrants and breeders in late spring/early summer and the production of young of the year in late summer (Giezentanner, 1970). These factors produce phenological changes in the composition of the avian community that are dramatic between years but may also be substantial within a year (Giezentanner, 1970; Giezentanner and Ryder, 1969) (Fig. 9.6). Although a large number of factors may contribute to these variations, weather may play the greatest role in interannual variation. Cold, wet springs may delay the onset of breeding. This, in turn, may affect reproductive success, and these effects may be expressed in population levels in subsequent years (Giezentanner, 1970; R. Ryder

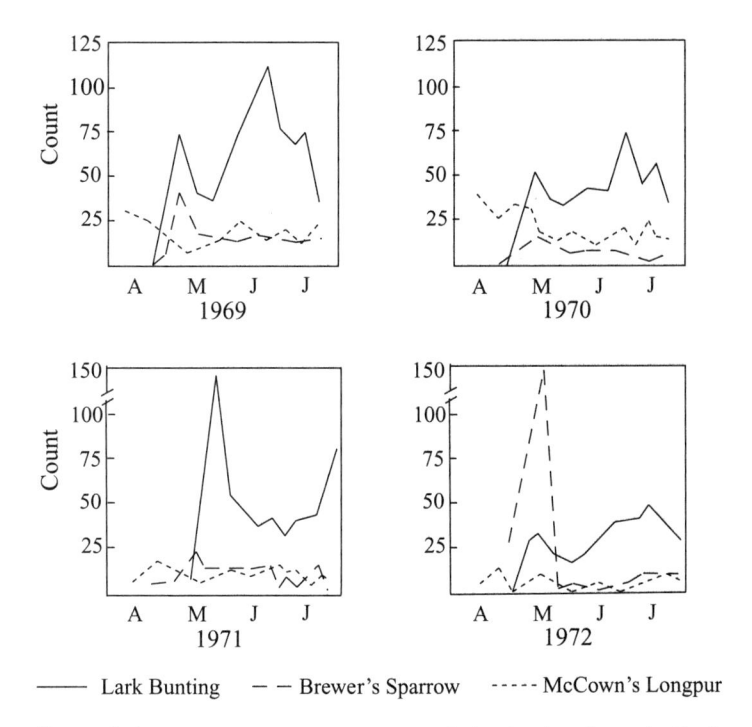

Figure 9.6 Within- and between-year variation in abundance for Lark Bunting, Brewer's Sparrow, and McCown's Longspur on the CPER from April through July, 1969 to 1972. (Modified from Wiens [1975, Fig. 4].)

unpublished). For example, Wiens (1974a) found that Horned Larks increased in density in the year after a drought whereas Grasshopper Sparrows, Lark Buntings, and Western Meadowlarks declined in number during the same period. On the other hand, George et al. (1992) reported that a severe drought in western North Dakota grasslands significantly affected the breeding densities and nesting success of several bird species during the year of the drought, but many of these species recovered to predrought population levels a year later.

Such variations in responses to weather (or to habitat changes or other factors) may be generally characteristic of the dynamics of arid and semiarid systems (Wiens, 1974a) and may complicate attempts to define clear associations between climate and biological variables (O'Connor et al., 1999). Nonetheless, to explore whether there might be a general relationship between climate and avian diversity in the shortgrass steppe, we conducted a Pearson correlation analysis of the number of breeding bird species and mean annual temperature, precipitation, and net primary productivity from 1969 to 1995 (bird data are from the North American Breeding Bird Survey [BBS] route at the CPER [Rockport route]; weather data were collected daily at the CPER). At the scale of this analysis (the 40-km length of the BBS survey route), no significant relationships emerged (temperature: $r^2 = 0.123$, $P = 0.690$; precipitation: $r^2 = 0.105$, $P = 0.733$; NPP: $r^2 = 0.358$, $P = 0.230$). Even incorporating time lags and examining relationships between diversity and conditions of the previous year revealed no significant correlations (temperature: $r^2 = 0.012$, $P = 0.971$; precipitation: $r^2 = 0.356$, $P = 0.255$; NPP: $r^2 = 0.194$, $P = 0.546$).

Although there is little evidence that climatic patterns affect the birds of the shortgrass steppe at such a scale, climatic *variability* may be extremely important. The shortgrass steppe experiences a relatively short growing season, the duration of which is directly linked to the timing and amount of precipitation (Lauenroth and Burke, 1995; Pielke and Doesken, chapter 2, this volume; Rosenzweig, 1968). Both are highly unpredictable and variable from year to year, at least at a local scale (Lauenroth and Milchunas, 1991). Variability in precipitation translates to variability in primary productivity (Lauenroth and Sala, 1992) and thus in resource availability (Wiens, 1974a). At a regional scale, the lower and more inconsistent levels of productivity in the shortgrass steppe compared with tallgrass prairie are expressed in an avian community that is composed of relatively few, small-bodied, migratory species at low densities (Wiens, 1974a). Thus, although birds may be directly affected by these major geographic patterns of variation in temperature and precipitation (Giezentanner, 1970; Graul, 1972), at a local scale abiotic factors influence the birds of the shortgrass steppe primarily in an indirect manner through effects on vegetation composition and structure.

Features of the physical environment, such as temperature and precipitation, may have a more direct influence on birds at the microclimate scale (i.e., a few square centimeters to square meters). Because of the openness of the shortgrass steppe, organisms are exposed to the full force of environmental extremes. Birds can use their mobility to escape unfavorable conditions, either through seasonal migration or by more localized movements within seasons. When they are breeding, however, birds are tied to a nest site, and escape from extremes is not so easy. For several species, characteristics of the microclimate at nest sites may be important determinants of habitat suitability and reproductive success (Pleszczynska, 1977; With and Webb, 1993).

Long-Term Trends

Bird populations and communities in the shortgrass steppe are not stable over time. Population trends have a great deal of interannual variance (Ryder, 1972, 1980), more so than in mixed- or tallgrass prairies (Wiens, 1974a). Although these dynamics are at odds with the equilibrium paradigm that once dominated ecological thinking (Pickett et al., 1994; Wiens, 1977, 1984), they are what we have come to expect of environments that experience the sorts of wide environmental variations that characterize the shortgrass steppe (Lauenroth and Milchunas, 1991; Wiens, 1974a). We can probe the matter more deeply, however, and ask whether there is evidence of long-term, multiyear trends that are superimposed on these short-term yearly variations. This issue is particularly germane to conservation, where declines in the abundances of some species have raised immediate concerns.

There has been considerable debate about exactly how one should go about measuring population trends, and there is no consensus regarding the best ways to standardize records across multiple observers or to distinguish between population variability and a true population trend (Fewster et al., 2000; James et al., 1996; Link and Sauer, 1997, 1998; Peterjohn et al., 1995; Thomas, 1996; Thomas and Martin, 1996). Despite this debate, it is clear that many species of grassland birds *have* declined in abundance at a continental scale during recent decades, in concert with the loss of more than 80% of native grasslands in most parts of North America during the past century (Knopf and Samson, 1997; Peterjohn and Sauer, 1999). This loss is mirrored globally. Temperate grasslands and savannas throughout the world are at greater risk, in terms of the magnitude of habitat conversion relative to the area under protection, than any other biome or major habitat (Hoekstra et al., 2005).

Grassland birds in general (Askins, 1993, 2000; Herkert, 1995; Knopf, 1994; Vickery et al., 1999), and what Knopf (1996) categorizes as endemic grassland birds in particular, "have shown steeper, more consistent, and more geographically widespread declines than any other behavioral or ecological guild" of North American birds (p. 142). Overall, Knopf's (1996) analysis indicates that one third of the North American grassland bird species are currently declining at statistically significant rates. Such trends led Samson and Knopf (1994) to conclude that the prairies of the Great Plains are the most endangered ecosystems in North America.

How are these trends expressed in the subset of bird species that are characteristic of the shortgrass steppe? According to analyses conducted by Sauer et al. (2005), most of these species declined significantly between 1966 and 2005, although Ferruginous Hawks exhibited increases across North America (Fig. 9.7).

These trends were assessed at a continental scale and not merely in the shortgrass steppe, however. Within the shortgrass steppe, bird diversity and abundance

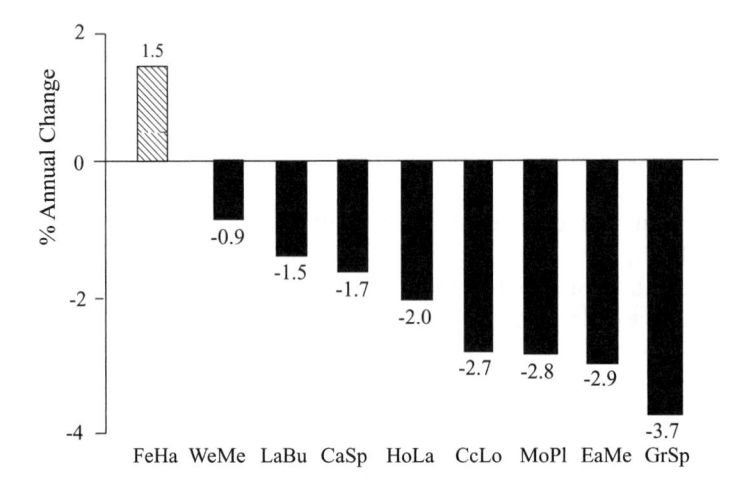

Figure 9.7 Significant percent annual population change of grassland bird species from 1966–2005. Species with hatched bars showed significant increases in abundance, whereas species with black bars showed significant declines. FeHa, Ferruginous Hawk; WeMe, Western Meadowlark; LaBu, Lark Bunting; CaSp, Cassin's Sparrow; HoLa, Horned Lark; CcLo, Chestnut-collared Longspur; MoPl, Mountain Plover; EaMe, Eastern Meadowlark; GrSp, Grasshopper Sparrow. (Data are from the Breeding Bird Survey [Sauer et al., 2005].)

have been systematically monitored at the CPER site since the 1960s. Counts of wintering birds conducted as part of the National Audubon Society's annual Christmas Bird Count were continuous from 1969 through 2005. Taken at face value, these counts suggest an apparent linear increase in the number of species observed ($r^2 = 0.463$, $P \leq 0.0001$). Because the number of observers and time spent counting also increased during this period, however ($r^2 = 0.103$, $P = 0.043$), we standardized the analysis to the number of species observed per 10 party-hours (Fig. 9.8). After this standardization, there were no systematic trends in the number of wintering species observed ($r^2 = 0.003$, $P = 0.742$).

Richness of breeding birds at the CPER during this period varied considerably among years but showed no long-term trends ($r^2 = 0.058$, $P = 0.225$) (Fig. 9.9). However, four of the five species most closely associated with the shortgrass steppe (Ferruginous Hawk, Lark Bunting, McCown's Longspur, and Mountain Plover) showed significant declines in breeding population numbers from 1969 to 2005 (Fig. 9.10). Although Mountain Plovers exhibited an overall decline during this period at the CPER, marked differences in trends can be seen prior to 1980 compared with the subsequent 25 years (Fig. 9.10). (A similar apparent difference in Chestnut-collared Longspur trends was not significant.)

Interpretation of such long-term trends (or their absence) is difficult. Changes in land use and fragmentation of native grassland habitat in breeding areas may be important determinants of population trends. For example, conversion of Illinois

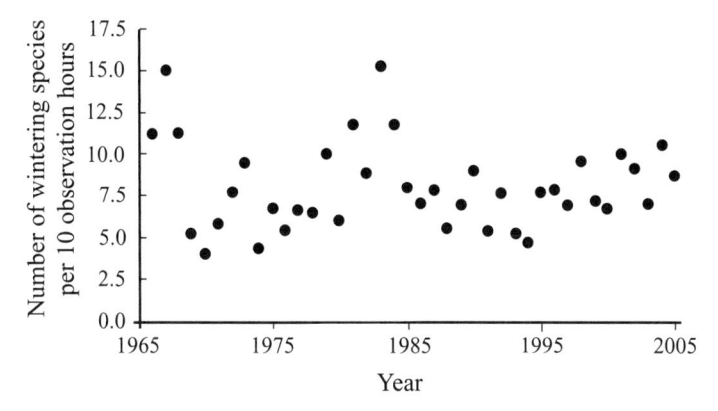

Figure 9.8 Number of wintering bird species (including winter migrants) per 10 observation hours on the CPER, 1966 to 2005. (Data are from the Nunn [Colorado] count circle of the annual Audubon Christmas Bird Count.)

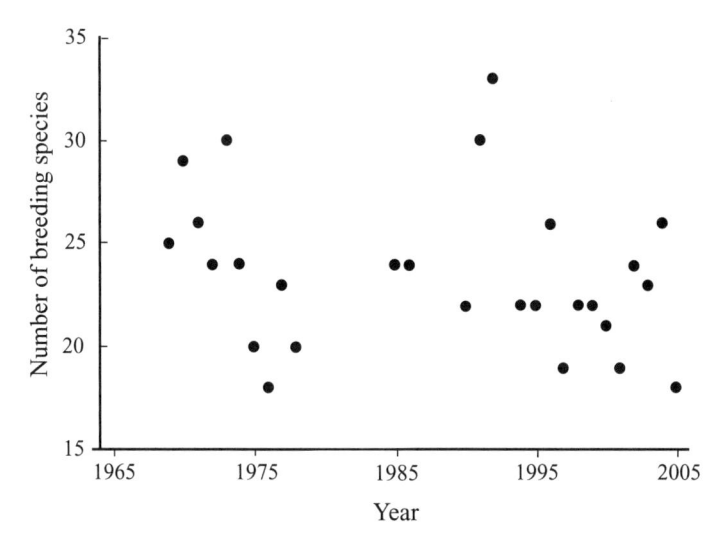

Figure 9.9 Number of breeding bird species present on the CPER, 1966 to 2005 (no data were collected in 1968, 1979–1984, 1987–1989, 1993). (Data are from the North American BBS, taken during the first 2 weeks in June [standardized by one observer, 50 stops, Rockport route].)

grasslands to agricultural land during a 3-year period was implicated in population declines for 11 of 15 local breeding species (Herkert, 1994). Many of these species showed pronounced area sensitivity, being found only in large grassland remnants, and other species showed a strong aversion to habitat edges, nesting well away from grassland–cropland boundaries. In their analysis of Mountain Plover population changes (a 63% decrease in abundance between 1966 and 1991),

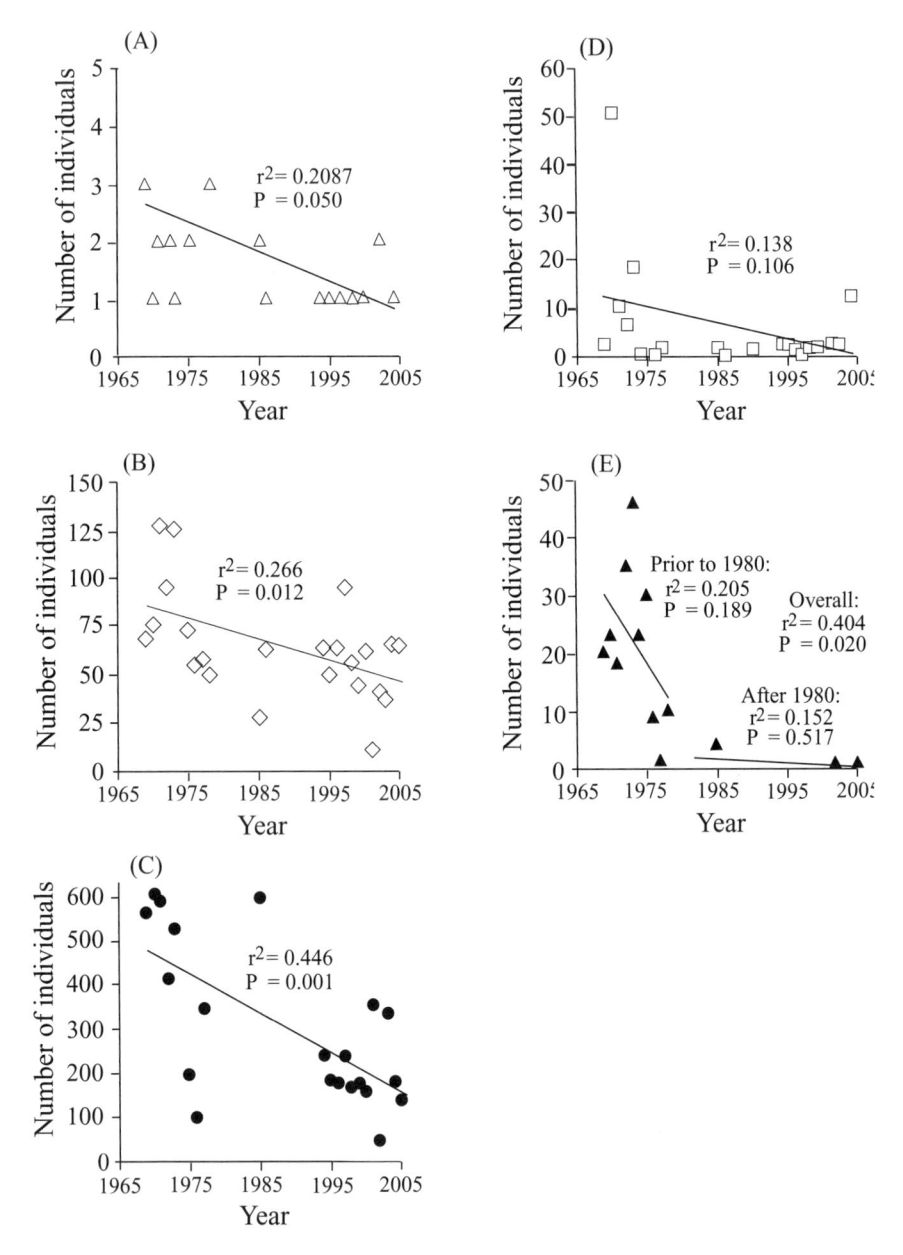

Figure 9.10 Number of individuals present on the CPER for Ferruginous Hawk (A), McCown's Longspur (B), Lark Bunting (C), Chestnut-collared Longspur (D), and Mountain Plover (E) from 1969 to 2005 (with no data collected in 1979–1984, 1987–1989, 1993). Lines are linear regressions. (Data are from the North American BBS, taken during the first 2 weeks in June [standardized by one observer, 50 stops, Rockport route].)

however, Knopf and Rupert (1996) concluded that population declines appeared to be independent of any recent landscape fragmentation in the breeding stronghold in northeastern Colorado.

In some cases, the causes of population declines may not be linked to proximate changes in breeding areas. It is increasingly acknowledged that ecological systems are open to external influences, especially at local scales (DeAngelis and Waterhouse, 1987). Thus, attributing changes in the abundance of species or the composition of communities at particular localities to events or changes occurring at those localities is likely to be difficult or misleading (Wiens, 1981). For migratory species, changes in habitat in the wintering areas or along the migratory route may influence the abundances that we tally in breeding surveys (Basili and Temple, 1999; Fretwell, 1972; Sherry and Holmes, 1996). For instance, Knopf and Rupert (1995) evaluated habitat conditions and availability to Mountain Plovers on the wintering grounds in California. They found that some core areas of wintering habitat (e.g., alkali flats) had all but disappeared and that birds wintering in the San Joaquin Valley were forced to use cultivated lands. Nonetheless, Knopf et al. (Knopf, 1996; Knopf and Rupert, 1995, 1996; Miller and Knopf, 1993) concluded that the recent dramatic declines in Mountain Plover populations could not be linked to either fragmentation of the breeding habitat or the loss of native habitat in the California wintering area, and they surmised that "continental declines appear attributable to longer-term processes on the breeding grounds" (Knopf and Rupert, 1995, p. 750). Knopf and Rupert did not identify what those processes might be, but global climate changes are projected to have especially profound effects in the semiarid grasslands of North America (Parton et al., 1994), and the plovers may be telling us that such changes are already underway. It is apparent that any effort aimed at conserving shortgrass steppe birds must integrate events that occur outside the shortgrass steppe itself.

The Role of Birds in the Shortgrass Steppe Ecosystem

Much of the impetus for the scientific study of the shortgrass steppe came from the Grassland Biome project of the IBP during the late 1960s and early 1970s. The focus of this program was on developing computer simulation models of grassland ecosystems and, through the models, an understanding of how the structure of these systems relates to their functioning (Golley, 1993). The currencies that tied such models together were energy flow and nutrient dynamics. Much of what we now know about the ecology of birds in the shortgrass steppe grew out of studies that were initiated as part of the IBP.

Understandably, a large part of the initial research effort was devoted to documenting the role of birds in the trophic dynamics and energetics of grassland ecosystems. Diets of the most abundant breeding species were documented to determine the pathways of energy flow, and a computer simulation model (BIRD [Wiens and Innis, 1974]) was developed to estimate energetics at the population and community levels. The broader ecological insights that we have described earlier emerged largely as by-products of the ecosystem-centered research.

Because the biomass and productivity of the shortgrass steppe avifauna are low in comparison with those of other consumer groups (Wiens, 1971), birds may affect ecosystem structure and functioning more indirectly than directly (sensu McCullough, 1970). For example, birds may regulate levels of pest insects such as grasshoppers that might otherwise experience population outbreaks and reduce gross primary production (Baldwin, 1970; Fowler et al., 1991; Wiens, 1971). Birds are themselves consumed by mammalian, avian, and reptilian predators. In general, however, flows of energy through birds are minuscule compared with flows through other ecosystem components (Wiens, 1971). Thus, although birds may be greatly affected by the properties of the shortgrass steppe ecosystem, their presence apparently does little to alter ecosystem structure or functioning.

Future Needs

Maintenance of bird populations and communities (and of the magic they provide) on the shortgrass steppe depends on maintaining the physical characteristics of the prairie itself (Knopf and Samson, 1997). Spatial heterogeneity at multiple scales must be preserved. This requires recreating the factors (such as grazing) that generated and maintained the shortgrass steppe's avian community. This simplistic prescription is a difficult one to fill, however, for we know very little of historical patterns of heterogeneity and bird occurrence. We can, however, recommend the following:

- Preserve the remaining large (>60 ha) parcels of native shortgrass steppe against conversion to cropland or urban areas (Boren et al., 1997; Walk and Warner, 1999).
- Reclaim former grassland areas (e.g., through the CRP).
- Preserve a variety of macrohabitat types (e.g., riparian areas, shortgrass uplands) within these landscapes.
- Focus conservation and management actions on functioning landscapes, not just isolated parcels of grassland.
- Recognize the value of grazing as a *bona fide*, ecologically sound management practice, and maintain both grazed and ungrazed areas.
- Develop a broader understanding of the factors and areas that limit grassland bird populations.
- Develop networks of linked conservation and management areas to enhance connectivity and promote conservation of wintering and breeding grounds outside the shortgrass steppe (e.g., The Nature Conservancy's Prairie Wings Program [www.nature.org/initiatives/programs/birds/explore/]).
- Support legislative protection of rare, threatened, and endangered bird species of the shortgrass steppe.
- Promote continued monitoring of bird diversity and density on the shortgrass steppe.

Clearly, we need to know more about the mechanisms that determine population fluctuations (particularly for declining species), differences in habitat occupation, and the sensitivity to changes in habitat structure and heterogeneity that may be produced by changing land use and climate change. To place the findings of such studies on a mechanistic foundation, studies are also needed of long-term reproductive success and survivorship under various grazing regimes.

The research tradition of the NSF's LTER program should foster continued ornithological studies at the SGS LTER site and elsewhere (see Collins, 2001). Our limited understanding of avian community and population patterns and the mechanisms accounting for these patterns has come from years of effort by numerous biologists and birders. Further advances in our ornithological knowledge will undoubtedly take a similarly strenuous effort, as the birds of the shortgrass steppe hold their secrets tightly.

Concluding Remarks

In the end, it may miss the point to ask what birds *do* in ecosystems. Perhaps the notion that all species must have some function in the ecosystem reflects an overly enthusiastic application of a holistic philosophy of nature, the conviction that everything in nature has a purpose. But the challenge may go more deeply. Tight structuring in variable systems such as grasslands may simply not be possible, and highly vagile consumers such as birds may live, in a sense, off the variance in production, without offering a large regulatory or functional role in return. Perhaps, after all, the role of birds is to embody the special magic of grasslands, not as systems to be dissected and modeled, but as landscapes to be experienced and cherished.

Acknowledgments We are indebted to Ron Ryder for providing us with species lists and population data. Chris Helzer graciously provided information on avian responses to grassland fragmentation from his Masters of Science research. We thank Jim Miller, Susan Skagen, Paul Stapp, and Peter Vickery for critiquing manuscript drafts; Rich Strauss for statistical advice; and Fritz Knopf for discussions on Mountain Plovers.

Appendix: Checklist of Bird Species Occurring at the CPER Site, 1962–2005

Key: *A*, abundant; *C*, common; *M*, species present only as migrants; *R*, rare species; *S*, species that breed or are present in summer; *U*, uncommon; *W*, species that overwinter; *Y*, year-round residents. Terminology follows National Audubon Society usage and standards.

202 Ecology of the Shortgrass Steppe

Gaviiformes

Common Loon (*Gavia immer*)	W	R

Podicipediformes

Pied-billed Grebe (*Podilymbus podiceps*)	S	U
Horned Grebe (*Podiceps auritus*)	W	R
Eared Grebe (*Podiceps nigricollis*)	M	U
Western Grebe (*Aechmophorus occidentalis*)	M	U
Clark's Grebe (*Aechmophorus clarkii*)	M	R

Pelecaniformes

American White Pelican (*Pelecanus erythrorhynchos*)	S	U
Double-crested Cormorant (*Phalacrocorax auritus*)	S	U

Ciconiiformes
American Bittern (*Botaurus lentiginosus*)	M	R
Great Blue Heron (*Ardea herodias*)	S	U
Snowy Egret (*Egretta thula*)	S	R
Green Heron (*Butorides virescens*)	S	R
Black-crowned Night Heron (*Nycticorax nycticorax*)	S	U
Yellow-crowned Night Heron (*Nycticorax violacea*)	S	R
White-faced Ibis (*Plegadis chihi*)	M	U

Anseriformes

Greater White-fronted Goose (*Anser albifrons*)	M	R
Snow Goose (*Chen caerulescens*)	M	U
Ross' Goose (*Chen rossii*)	M	R
Brant (*Branta bernicla*)	M	R
Cackling Goose (*Branta hutchinsii*)	W	U
Canada Goose (*Branta canadensis*)	Y	U
Wood Duck (*Aix sponsa*)	Y	R
Green-winged Teal (*Anas crecca*)	M	C
Mallard (*Anas platyrhynchos*)	Y	C
Northern Pintail (*Anas acuta*)	M	C
Blue-winged Teal (*Anas discors*)	S	C
Cinnamon Teal (*Anas cyanoptera*)	S	C
Northern Shoveler (*Anas clypeata*)	S	U
Gadwall (*Anas strepera*)	Y	C
American Wigeon (*Anas americana*)	M	U
Canvasback (*Aythya valisineria*)	M	R
Redhead (*Aythya americana*)	S	U
Ring-necked Duck (*Aythya collaris*)	M	U
Greater Scaup (*Aythya marila*)	M	R
Lesser Scaup (*Aythya affinis*)	M	C
Common Goldeneye (*Bucephala clangula*)	M	C
Bufflehead (*Bucephala albeola*)	M	C
Common Merganser (*Mergus merganser*)	M	C
Red-breasted Merganser (*Mergus serrator*)	M	R
Ruddy Duck (*Oxyura jamaicensis*)	S	U

Falconiformes

Turkey Vulture (*Cathartes aura*)	S	U
Osprey (*Pandion haliaetus*)	M	R

Mississippi Kite (*Ictinia mississippiensis*)	M	R
Bald Eagle (*Haliaeetus leucocephalus*)	W	R
Northern Harrier (*Circus cyaneus*)	Y	C
Sharp-shinned Hawk (*Accipiter striatus*)	Y	U
Cooper's Hawk (*Accipiter cooperii*)	M	U
Northern Goshawk (*Accipiter gentilis*)	M	R
Broad-winged Hawk (*Buteo platypterus*)	M	R
Red-tailed Hawk (*Buteo jamaicensis*)		
Krider's race	Y	C
Harlan's race	W	R
Swainson's Hawk (*Buteo swainsoni*)	S	C
Rough-legged Hawk (*Buteo lagopus*)	W	C
Ferruginous Hawk (*Buteo regalis*)	Y	C
Golden Eagle (*Aquila chrysaetos*)	Y	C
American Kestrel (*Falco sparverius*)	Y	C
Merlin (*Falco columbarius*)	W	C
Peregrine Falcon (*Falco peregrinus*)	M	R
Gyrfalcon (*Falco rusticolus*)	W	R
Prairie Falcon (*Falco mexicanus*)	Y	C

Galliformes

Ring-necked Pheasant (*Phasianus colchicus*)	Y	U
Sharp-tailed Grouse (*Tympahuchus phasianellus*)	Y	R
Northern Bobwhite (*Colinus virginianus*)	Y	U

Gruiformes

Virginia Rail (*Rallus limicola*)	M	U
Sora (*Porzana carolina*)	M	U
American Coot (*Fulica americana*)	Y	C
Sandhill Crane (*Grus canadensis*)	M	C

Charadriiformes

Snowy Plover (*Charadrius alexandrinus*)	M	R
Semipalmated Plover (*Charadrius semipalmatus*)	M	R
Killdeer (*Charadrius vociferus*)	Y	C
Mountain Plover (*Charadrius montanus*)	S	C
American Avocet (*Recurvirostra americana*)	S	C
Greater Yellowlegs (*Tringa melanoleuca*)	M	C
Lesser Yellowlegs (*Tringa flavipes*)	M	C
Solitary Sandpiper (*Tringa solitaria*)	M	U
Willet (*Tringa semipalmata*)	M	C
Spotted Sandpiper (*Actitis macularius*)	S	C
Upland Sandpiper (*Bartramia longicauda*)	S	R
Whimbrel (*Numenius phaeopus*)	M	R
Long-billed Curlew (*Numenius americanus*)	S	U
Hudsonian Godwit (*Limosa haemastica*)	M	R
Marbled Godwit (*Limosa fedoa*)	M	C
Sanderling (*Calidris alba*)	M	R
Semipalmated Sandpiper (*Calidris pusilla*)	M	U
Western Sandpiper (*Calidris mauri*)	M	U
Least Sandpiper (*Calidris minutilla*)	M	C
White-rumped Sandpiper (*Calidris fuscicollis*)	M	R

204 Ecology of the Shortgrass Steppe

Baird's Sandpiper (*Calidris bairdii*)	M	C
Pectoral Sandpiper (*Calidris melanotos*)	M	U
Stilt Sandpiper (*Calidris himantopus*)	M	R
Long-billed Dowitcher (*Limnodromus scolopaceus*)	M	C
Wilson's Snipe (*Gallinago delicata*)	Y	U
Wilson's Phalarope (*Phalaropus tricolor*)	S	C
Red-necked Phalarope (*Phalaropus lobatus*)	M	R
Franklin's Gull (*Larus pipixcan*)	M	A
Ring-billed Gull (*Larus delawarensis*)	Y	C
California Gull (*Larus californicus*)	S	C
Forster's Tern (*Sterna forsteri*)	S	U
Least Tern (*Sterna antillarum*)	M	R
Black Tern (*Chlidonias niger*)	S	C

Columbiformes

Rock Dove (*Columba livia*)	Y	C
White-winged Dove (*Zenaida asiatica*)	M	R
Mourning Dove (*Zenaida macroura*)	Y	C

Cuculiformes

Black-billed Cuckoo (*Coccyzus erythropthalmus*)	S	R
Yellow-billed Cuckoo (*Coccyzus americanus*)	S	U

Strigiformes

Common Barn-owl (*Tyto alba*)	S	U
Eastern Screech-owl (*Otus asio*)	Y	U
Great Horned Owl (*Bubo virginianus*)	Y	C
Snowy Owl (*Nyctea scandiacus*)	W	R
Burrowing Owl (*Athene cunicularia*)	S	C
Long-eared Owl (*Asio otus*)	Y	U
Short-eared Owl (*Asio flammeus*)	W	R
Northern Saw-whet Owl (*Aegolius acadicus*)	W	R

Caprimulgiformes

Common Nighthawk (*Chordeiles minor*)	S	C
Common Poorwill (*Phalaenoptilus nuttallii*)	S	R

Apodiformes

Chimney Swift (*Chaetura peliagica*)	S	R
White-throated Swift (*Aeronautes saxatalis*)	S	U
Calliope Hummingbird (*Stellula calliope*)	M	R
Broad-tailed Hummingbird (*Selasphorus platycercus*)	S	U
Rufous Hummingbird (*Selasphorus rufus*)	M	R

Coraciiformes

Belted Kingfisher (*Ceryle alcyon*)	Y	U

Piciformes

Lewis' Woodpecker (*Melanerpes lewis*)	M	R

Red-headed Woodpecker (*Melanerpes erythrocephalus*)	S	U
Yellow-bellied Sapsucker (*Sphyrapicus varius*)	W	R
Red-naped Sapsucker (*Sphyrapicus nuchalis*)	M	U
Williamson's Sapsucker (*Sphyrapicus thyroideus*)	M	R
Ladder-backed Woodpecker (*Picoides scalaris*)	M	R
Downy Woodpecker (*Picoides pubescens*)	Y	C
Hairy Woodpecker (*Picoides villosus*)	W	U
Northern Flicker (*Colaptes auratus*)		
Red-shafted race	Y	C
Yellow-shafted race	Y	U

Passeriformes

Cassin's Kingbird (*Tyrannus vociferans*)	S	R
Western Kingbird (*Tyrannus verticalis*)	S	C
Eastern Kingbird (*Tyrannus tyrannus*)	S	C
Scissor-tailed Flycatcher (*Tyrannus forficatus*)	M	R
Western Wood-pewee (*Contopus sordidulus*)	S	C
Olive-sided Flycatcher (*Contopus cooperi*)	M	R
Willow Flycatcher (*Empidonax traillii*)	S	R
Least Flycatcher (*Empidonax minimus*)	S	R
Hammond's Flycatcher (*Empidonax hammondii*)	M	U
Dusky Flycatcher (*Empidonax oberholseri*)	M	U
Cordilleran Flycatcher (*Empidonax occidentalis*)	M	U
Say's Phoebe (*Sayornis saya*)	S	C
Ash-throated Flycatcher (*Myiarchus cinerascens*)	S	R
Horned Lark (*Eremophilia alpestris*)	Y	A
Purple Martin (*Progne subis*)	M	R
Tree Swallow (*Tachycineta bicolor*)	M	U
Violet-green Swallow (*Tachycineta thalassina*)	M	U
Northern Rough-winged Swallow (*Stelgidopteryx serripennis*)	S	C
Bank Swallow (*Riparia riparia*)	S	C
Barn Swallow (*Hirundo rustica*)	S	C
Cliff Swallow (*Hirundo pyrrhonota*)	S	C
Blue Jay (*Cyanocitta cristata*)	Y	C
Pinyon Jay (*Gymnorhinus cyanocephalus*)	M	R
Clark's Nutcracker (*Nucifraga columbiana*)	W	R
Black-billed Magpie (*Pica hudsonia*)	Y	C
American Crow (*Corvus brachyrhynchos*)	Y	C
Common Raven (*Corvus corax*)	W	R
Black-capped Chickadee (*Parus atricapillus*)	Y	C
Mountain Chickadee (*Parus gambeli*)	W	R
Red-breasted Nuthatch (*Sitta canadensis*)	W	C
White-breasted Nuthatch (*Sitta carolinensis*)	Y	U
Brown Creeper (*Certhia americana*)	W	C
House Wren (*Troglodytes aedon*)	S	C
Rock Wren (*Salpinctes obsoletus*)	S	C
Bewick's Wren (*Thryomanes bewickii*)	M	R
Golden-crowned Kinglet (*Regulus satrapa*)	W	U
Ruby-crowned Kinglet (*Regulus calendula*)	M	C
Blue-gray Gnatcatcher (*Polioptila caerulea*)	M	C
Eastern Bluebird (*Sialia sialis*)	S	U
Western Bluebird (*Sialia mexicana*)	M	R
Mountain Bluebird (*Sialia currucoides*)	M	C
Townsend's Solitaire (*Myadestes townsendi*)	M	U

206 Ecology of the Shortgrass Steppe

Veery (*Catharus fuscescens*)	M	U
Swainson's Thrush (*Catharus ustulatus*)	M	C
Hermit Thrush (*Catharus guttatus*)	M	U
American Robin (*Turdus migratorius*)	Y	C
American Pipit (*Anthus rubescens*)	W	R
Sprague's Pipit (*Anthus spragueii*)	M	R
Gray Catbird (*Dumetella carolinensis*)	S	U
Northern Mockingbird (*Mimus polyglottos*)	S	C
Sage Thrasher (*Oreoscoptes montanus*)	S	C
Brown Thrasher (*Toxostoma rufum*)	S	C
Bohemian Waxwing (*Bombycilla garrulus*)	W	R
Cedar Waxwing (*Bombycilla cedrorum*)	Y	U
Northern Shrike (*Lanius excubitor*)	W	C
Loggerhead Shrike (*Lanius ludovicianus*)	S	C
European Starling (*Sturnus vulgaris*)	Y	C
White-eyed Vireo (*Vireo griseus*)	S	R
Plumbeous Vireo (*Vireo plumbeus*)	M	R
Yellow-throated Vireo (*Vireo flavifrons*)	M	R
Warbling Vireo (*Vireo gilvus*)	M	C
Red-eyed Vireo (*Vireo olivaceus*)	S	R
Golden-winged Warbler (*Vermivora chrysoptera*)	M	R
Tennessee Warbler (*Vermivora peregrina*)	M	R
Orange-crowned Warbler (*Vermivora celata*)	M	U
Nashville Warbler (*Vermivora ruficapilla*)	M	R
Virginia's Warbler (*Vermivora virginiae*)	M	U
Yellow Warbler (*Dendroica petechia*)	S	C
Chestnut-sided Warbler (*Dendroica pensylvanica*)	M	R
Magnolia Warbler (*Dendroica magnolia*)	M	R
Cape May Warbler (*Dendroica tigrina*)	M	R
Black-throated Blue Warbler (*Dendroica caerulescens*)	M	R
Yellow-rumped Warbler (*Dendroica coronata*)		
Myrtle's race	M	C
Audubon's race	M	C
Townsend's Warbler (*Dendroica townsendi*)	M	U
Black-throated Green Warbler (*Dendroica virens*)	M	R
Blackburnian Warbler (*Dendroica fusca*)	M	R
Prairie Warbler (*Dendroica discolor*)	M	R
Palm Warbler (*Dendroica palmarum*)	M	R
Bay-breasted Warbler (*Dendroica castanea*)	M	R
Blackpoll Warbler (*Dendroica striata*)	M	R
Black-and-white Warbler (*Mniotilta varia*)	M	R
American Redstart (*Setophaga ruticilla*)	M	U
Prothonotary Warbler (*Protonotaria citrea*)	M	R
Worm-eating Warbler (*Helmintheros vermivorum*)	M	R
Ovenbird (*Seiurus aurocapillus*)	M	R
Northern Waterthrush (*Seiurus noveboracensis*)	M	R
Kentucky Warbler (*Oporornis formosus*)	M	R
MacGillivray's Warbler (*Oporornis tolmiei*)	M	C
Common Yellowthroat (*Geothlypis trichas*)	S	C
Hooded Warbler (*Wilsonia citrina*)	M	R
Wilson's Warbler (*Wilsonia pusilla*)	M	A
Canada Warbler (*Wilsonia canadensis*)	M	R
Yellow-breasted Chat (*Icteria virens*)	S	U
Summer Tanager (*Piranga rubra*)	M	R
Scarlet Tanager (*Piranga olivacea*)	M	R

Western Tanager (*Piranga ludoviciana*)	M	R
Rose-breasted Grosbeak (*Pheucticus ludovicianus*)	M	R
Black-headed Grosbeak (*Pheucticus melanocephalus*)	S	C
Blue Grosbeak (*Passerina caerulea*)	S	C
Lazuli Bunting (*Passerina amoena*)	S	U
Dickcissel (*Spiza americana*)	S	R
Green-tailed Towhee (*Pipilo chlorurus*)	S	R
Spotted Towhee (*Pipilo maculatus*)	S	U
Cassin's Sparrow (*Aimophila cassinii*)	S	C
American Tree Sparrow (*Spizella arborea*)	W	C
Chipping Sparrow (*Spizella passerina*)	M	A
Clay-colored Sparrow (*Spizella pallida*)	M	C
Brewer's Sparrow (*Spizella breweri*)	S	C
Vesper Sparrow (*Pooecestes gramineus*)	M	C
Lark Sparrow (*Chondestes grammacus*)	S	C
Black-throated Sparrow (*Amphispiza bilineata*)	M	R
Sage Sparrow (*Amphispiza belli*)	M	R
Lark Bunting (*Calamospiza melanocorys*)	S	A
Savannah Sparrow (*Passerculus sandwichensis*)	S	C
Baird's Sparrow (*Ammodramus bairdii*)	M	R
Grasshopper Sparrow (*Ammodramus savannarum*)	S	C
Fox Sparrow (*Passerella iliaca*)	M	R
Song Sparrow (*Melospiza melodia*)	Y	C
Lincoln's Sparrow (*Melospiza lincolnii*)	M	C
Swamp Sparrow (*Melospiza georgiana*)	M	R
White-crowned Sparrow (*Zonotrichia leucophrys*)	W	A
White-throated Sparrow (*Zonotrichia albicollis*)	M	U
Harris' Sparrow (*Zonotrichia querula*)	W	R
Dark-eyed Junco (*Junco hyemalis*)		
White-winged race	W	R
Slate-colored race	W	U
Oregon race	W	C
Gray-headed race	W	U
McCown's Longspur (*Calcarius mccownii*)	S	C
Lapland Longspur (*Calcarius lapponicus*)	W	C
Chestnut-collared Longspur (*Calcarius ornatus*)	S	U
Snow Bunting (*Plectrophenax nivalis*)	W	U
Bobolink (*Dolichonyx oryzivorus*)	S	R
Western Meadowlark (*Sturnella neglecta*)	Y	C
Yellow-headed Blackbird (*Xanthocephalus xanthocephalus*)	S	C
Red-winged Blackbird (*Agelaius phoeniceus*)	Y	C
Brewer's Blackbird (*Euphagus cyanocephalus*)	S	C
Common Grackle (*Quiscalus quiscula*)	S	C
Brown-headed Cowbird (*Molothrus ater*)	S	C
Orchard Oriole (*Icterus spurius*)	S	C
Baltimore Oriole (*Icterus galbula*)	S	R
Bullock's Oriole (*Icterus bullocki*)	S	C
Gray-crowned Rosy-finch (*Leucosticte tephrocotis*)	W	U
Brown-capped Rosy-finch (*Leucosticte australis*)	W	R
Black Rosy-finch (*Leucosticte atrata*)	W	R
Pine Grosbeak (*Pinicola enucleator*)	W	R
Cassin's Finch (*Carpodacus cassinii*)	W	R
House Finch (*Carpodacus mexicanus*)	Y	C
Red Crossbill (*Loxia curvirostra*)	W	U
White-winged Crossbill (*Loxia leucoptera*)	W	R

Common Redpoll (*Carduelis flammea*)	W	U
Pine Siskin (*Carduelis pinus*)	Y	U
American Goldfinch (*Carduelis tristis*)	Y	C
Evening Grosbeak (*Coccothraustes vespertinus*)	W	U
House Sparrow (*Passer domesticus*)	Y	A

References

Andrén, H., and P. Angelstam. 1988. Elevated predation rates as an edge effect in habitat islands: Experimental evidence. *Ecology* **69**:544–547.

Andrén, H., P. Angelstam, E. Lindstrom, and P. Widen. 1985. Differences in predation pressure in relation to habitat fragmentation: an experiment. *Oikos* **45**:273–277.

Andrews, R., and R. Righter. 1992. *Colorado birds*. Denver Museum of Natural History, Denver, Colo.

Askins, R. A. 1993. Population trends in grassland, shrubland and forest birds in eastern North America. *Current Ornithology* **11**:1–34.

Askins, R. A. 2000. *Restoring North American birds: Lessons from landscape ecology*. Yale University Press, New Haven, Conn.

Baldwin, P. H. 1970. Avian food studies at the Pawnee site, pp. 84–85. In: R. T. Coupland and G. M. Van Dyne (eds.), *Grassland ecosystems: Reviews of research*. Range Science Department science series no. 7. Colorado State University, Fort Collins, Colo.

Baldwin, P. H. 1971. *Diet of the killdeer at the Pawnee National Grassland and a comparison with the Mountain Plover, 1970–1971*. U.S. IBP Grassland Biome technical report no. 135. Colorado State University, Fort Collins, Colo.

Baldwin, P. H., P. D. Creighton, and D. S. Kisiel. 1971. *Diet of the mourning dove at the Pawnee National Grassland, 1970–1971*. U.S. IBP Grassland Biome technical report no. 136. Colorado State University, Fort Collins, Colo.

Basili, G. D., and S. A. Temple. 1999. Demographic characteristics of Dickcissels in winter. *Studies in Avian Biology* **19**:281–288.

Berthelsen, P. S., and L. M. Smith. 1995. Nongame bird nesting on CRP lands in the southern High Plains. *Journal of Soil and Water Conservation* **44**:504–507.

Best, L. B., H. Campa, K. E. Kemp, R. J. Robel, M. R. Ryan, J. A. Savidge, H. P. Weeks, and S. R. Winterstein. 1998. Avian abundance in CRP and crop fields during winter in the Midwest. *American Midland Naturalist* **139**:311–324.

Bock, C. E., V. A. Saab, T .D. Rich, and D. S. Dobkin. 1993. Effects of livestock grazing on neotropical migratory landbirds in western North America, pp. 296–309. In: D. M. Finch and P. W. Stangel (eds.), *Status and management of neotropical migratory birds*. USDA Forest Service general technical report RM-229. Rocky Mountain Forest and Range Experiment Station, Fort Collins, Colo.

Boren, J. C., D. M. Engle, and R. E. Masters. 1997. Vegetation cover type and avian species changes on landscapes within a wildland–urban interface. *Ecological Modelling* **103**:251–266.

Brennen, L. A., and W. P. Kulvesky. 2005. North American grassland birds: An unfolding conservation crisis? *Journal of Wildlife Management* **69**:1–13.

Cody, M. L. 1966. The consistency of intra- and inter-continental grassland bird species counts. *American Naturalist* **100**:371–376.

Cody, M. L. 1968. On the methods of resource division in grassland bird communities. *American Naturalist* **102**:107–147.

Cody, M. L. 1985. Habitat selection in grassland and open-country birds, pp. 191–226. In: M. L. Cody (ed.), *Habitat selection in birds*. Academic Press, San Diego, Calif.

Collins, S. L. 2001. Long-term research and the dynamics of bird populations and communities. *Auk* **118**:583–588.

Coppedge, B. R., D. M. Engle, R. E. Masters, and M. S. Gregory. 2001. Avian response to landscape change in fragmented southern Great Plains grasslands. *Ecological Applications* **11**:47–59.

Creighton, P. D., and P. H. Baldwin. 1974. Habitat exploitation by an avian ground-foraging guild. U.S. IBP Grassland Biome technical report no. 263. Colorado State University, Fort Collins, Colo.

DeAngelis, D. L., and J. C. Waterhouse. 1987. Equilibrium and nonequilibrium concepts in ecological models. *Ecological Monographs* **57**:1–21.

Desmond, M. J., and J. A. Savidge. 1999. Satellite burrow use by Burrowing Owl chicks and its influence on nest fate. *Studies in Avian Biology* **19**:128–130.

Fewster, R. M., S. T. Buckland, G. M. Siriwardena, S. R. Baillie, and J. D. Wilson. 2000. Analysis of population trends for farmland birds using generalized additive models. *Ecology* **81**:1970–1984.

Fleischner, T. L. 1994. Ecological costs of livestock grazing in western North America. *Conservation Biology* **8**:629–644.

Fowler, A. C., R. L. Knight, T. L. George, and L. C. McEwen. 1991. Effects of avian predation on grasshopper populations in North Dakota grasslands. *Ecology* **72**:1775–1781.

Fretwell, S. D. 1972. *Populations in a seasonal environment.* Princeton University Press, Princeton, N.J.

Gates, J. E., and L. W. Geysel. 1978. Avian nest dispersion and fledgling success in field–forest ecotones. *Ecology* **59**:871–883.

George, T. L., A. C. Fowler, R. L. Knight, and L. C. McEwen. 1992. Impacts of a severe drought on grassland birds in western North Dakota. *Ecological Applications* **2**:275–284.

Giezentanner, J. B. 1970. *Avian distribution and population fluctuations on the shortgrass prairie of north–central Colorado.* U.S. IBP Grassland Biome technical report no. 62. Colorado State University, Fort Collins, Colo.

Giezentanner, J. B., and R. A. Ryder. 1969. *Avian distribution and population fluctuations, Pawnee site.* U.S. IBP Grassland Biome technical report no. 28. Colorado State University, Fort Collins, Colo.

Golley, F. B. 1993. *A history of the ecosystem concept in ecology: More than the sum of the parts.* Yale University Press, New Haven, Conn.

Graul, W. D. 1972. *Breeding biology of the Mountain Plover* (Charadrius montanus), *1969–1972.* U.S. IBP Grassland Biome technical report no. 199. Colorado State University, Fort Collins, Colo.

Greer, R. D. 1988. *Effects of habitat structure and productivity on grassland birds.* PhD diss., University of Wyoming, Laramie, Wyo.

Hartnett, D. C., A. A. Steuter, and K. R. Hickman. 1997. Comparative ecology of native and introduced ungulates, pp. 72–101. In: F. L. Knopf and F. B. Samson (eds.), *Ecology and conservation of Great Plains vertebrates.* Springer, New York.

Helzer, C. J. 1996. *The effects of wet meadow fragmentation on grassland birds.* Masters thesis, University of Nebraska, Lincoln, Neb.

Herkert, J. R. 1994. The effects of habitat fragmentation on midwestern grassland bird communities. *Ecological Applications* **4**:461–471.

Herkert, J. R. 1995. An analysis of midwestern breeding bird population trends: 1966–1993. *American Midland Naturalist* **134**:41–50.

Hoekstra, J. M., T. M. Boucher, T. H. Ricketts, and C. Roberts. 2005. Confronting a biome crisis: Global disparities of habitat loss and protection. *Ecology Letters* **8**:23–29.

Howard, M. N., S. K. Skagen, and P. L. Kennedy. 2001. Does habitat fragmentation influence nest predation in the shortgrass prairie? *Condor* **103**:530–536.

Hughes, A. J. 1993. *Breeding density and habitat preferences of the Burrowing Owl in northeastern Colorado.* Masters thesis, Colorado State University, Fort Collins, Colo.

James, F. C., C. E. McCulloch, and D. A. Wiedenfeld. 1996. New approaches to the analysis of population trends in land birds. *Ecology* **77**:13–27.

Joern, A., and K. H. Keeler. 1995. Getting the lay of the land: Introducing North American native grasslands, pp. 11–24. In: A. Joern and K. H. Keeler (eds.), *The changing prairie: North American grasslands.* Oxford University Press, New York.

Johnson, D. H., and M. D. Schwartz. 1993. The Conservation Reserve Program and grassland birds. *Conservation Biology* **7**:934–937.

Knopf, F. L. 1994. Avian assemblages on altered grasslands. *Studies in Avian Biology* **15**:232–246.

Knopf, F. L. 1996. Prairie legacies: Birds, pp. 135–148. In: F. B. Samson and F. L. Knopf (eds.), *Prairie conservation: Preserving North America's most endangered ecosystem.* Island Press, Covelo, Calif.

Knopf, F. L., and B. J. Miller. 1994. *Charadrius montanus*: Montane, grassland, or bareground plover? *Auk* **111**:505–506.

Knopf, F. L., and J. R. Rupert. 1995. Habits and habitats of Mountain Plovers in winter. *Condor* **97**:743–751.

Knopf, F. L., and J. R. Rupert. 1996. Reproduction and movements of Mountain Plovers breeding in Colorado. *Wilson Bulletin* **108**:28–35.

Knopf, F. L., and J. R. Rupert. 1999. Use of cultivated fields by breeding Mountain Plovers in Colorado. *Studies in Avian Biology* **19**:81–86.

Knopf, F. L., and F. B. Samson. 1997. Conservation of grassland vertebrates, pp. 273–289. In: F. L. Knopf and F. B. Samson (eds.), *Ecology and conservation of Great Plains vertebrates.* Springer, New York.

Küchler, A. W. 1964. *Potential natural vegetation of the coterminous United States.* Special publication 36. American Geographical Society, New York.

Laubhan, M. K., and L. H. Fredrickson. 1997. Wetlands of the Great Plains: Habitat characteristics and vertebrate aggregations, pp. 20–48. In: F. L. Knopf and F. B. Samson (eds.), *Ecology and conservation of Great Plains vertebrates.* Springer, New York.

Lauenroth, W. K., and I. C. Burke. 1995. Great Plains, climate variability, pp. 237–249. In: W. A. Nierenberg (ed.), *Encyclopedia of environmental biology.* Vol. 2. Academic Press, New York.

Lauenroth, W. K., and D. G. Milchunas. 1991. Shortgrass steppe, pp. 183–226. In: R. T. Coupland (ed.), *Ecosystems of the world 8A: Natural grasslands.* Elsevier, Amsterdam.

Lauenroth, W. K., and O. E. Sala. 1992. Long term forage production of North American shortgrass steppe. *Ecological Applications* **2**:397–403.

Link, W. A., and J. R. Sauer. 1997. New approaches to the analysis of population trends in land birds: Comment. *Ecology* **78**:2632–2634.

Link, W. A., and J. R. Sauer. 1998. Estimating population change from count data: Applications to the North American Breeding Bird Survey. *Ecological Applications* **8**:258–268.

Marti, C. D. 1969. Some comparisons of the feeding ecology of four owls in north–central Colorado. *Southwestern Naturalist* **14**:163–170.

McCoy, T. D., M. R. Ryan, E. W. Kurzejeski, and L. W. Burger. 1999. Conservation Reserve Program: Source or sink habitat for grassland birds in Missouri? *Journal of Wildlife Management* **63**:530–538.

McCullough, D. R. 1970. Secondary production of birds and mammals, pp. 107–130. In: D. Reichle (ed.), *Analysis of temperate forest ecosystems.* Springer-Verlag, New York.

McIntyre, N. E., and T. R. Thompson. 2003. A comparison of Conservation Reserve Program habitat plantings with respect to arthropod prey for grassland birds. *American Midland Naturalist* **150**:291–301.

Miller, B. J., and F. L. Knopf. 1993. Growth and survival of Mountain Plovers. *Journal of Field Ornithology* **64**:500–506 and **65**:193.

O'Connor, R .J., M. T. Jones, R. B. Boone, and T. B. Lauber. 1999. Linking continental climate, land use, and land patterns with grassland bird distribution across the conterminous United States. *Studies in Avian Biology* **19**:45–59.

Olendorff, R. R. 1972. *The large birds of prey of the Pawnee National Grassland: Nesting habits and productivity, 1969–1971.* U.S. IBP Grassland Biome technical report no. 151. Colorado State University, Fort Collins, Colo.

Olendorff, R. R. 1973. *The ecology of the nesting birds of prey of northeastern Colorado.* U.S. IBP Grassland Biome technical report no. 211. Colorado State University, Fort Collins, Colo.

Parton, W. J., D. S. Ojima, and D. S. Schimel. 1994. Environmental change in grasslands: Assessment using models. *Climatic Change* **28**:111–141.

Peterjohn, B. G., and J. R. Sauer. 1999. Population status of North American grassland birds from the North American Breeding Bird Survey, 1966–1996. *Studies in Avian Biology* **19**:27–44.

Peterjohn, B. G., J. R. Sauer, and C. S. Robbins. 1995. Population trends from the North American Breeding Bird Survey, pp. 3–39. In: T. E. Martin and D. M. Finch (eds.), *Ecology and management of neotropical migratory birds: A synthesis and review of critical issues.* Oxford University Press, New York.

Pickett, S. T. A., J. Kolasa, and C. G. Jones. 1994. *Ecological understanding: The nature of theory and the theory of nature.* Academic Press, Orlando, Fla.

Pleszczynska, W. K. 1977. *Polygyny in the Lark Bunting.* PhD diss., University of Toronto, Toronto, Ont.

Plumpton, D. L. 1992. *Aspects of nest site selection and habitat use by Burrowing Owls at the Rocky Mountain Arsenal, Colorado.* Masters thesis, Texas Tech University, Lubbock, Texas.

Porter, D. K. 1974. *Accuracy in censusing breeding passerines on the shortgrass prairie.* U.S. IBP Grassland Biome technical report no. 254. Colorado State University, Fort Collins, Colo.

Reynolds, R. E., T. L. Shaffer, J. R. Sauer, and B. G. Peterjohn. 1994. Conservation Reserve Program: Benefit for grassland birds in the northern Plains. *Transactions of the North American Wildlife Natural Resources Conference* **59**:328–336.

Rosenzweig, M. L. 1968. Net primary productivity of terrestrial communities: Predictions from climatological data. *American Naturalist* **102**:67–74.

Rotenberry, J. T. 1985. The role of habitat in avian community composition: Physiognomy or floristics? *Oecologia* **67**:213–217.

Rotenberry, J. T., and J. A. Wiens. 1980a. Habitat structure, patchiness, and avian communities in North American steppe vegetation: A multivariate approach. *Ecology* **61**:1228–1250.

Rotenberry, J. T., and J. A. Wiens. 1980b. Temporal variation in habitat structure and shrubsteppe bird dynamics. *Oecologia* **47**:1–9.

Ryder, R. A. 1970. Avian populations at the Pawnee site, pp. 84–85. In: R. T. Coupland and G. M. Van Dyne (eds.), *Grassland ecosystems: Review of research.* Range Science Department science series no. 7. Colorado State University, Fort Collins, Colo.

Ryder, R. A. 1972. *Avian population studies on the Pawnee site, 1968–1971.* U.S. IBP Grassland Biome technical report no. 171. Colorado State University, Fort Collins, Colo.

Ryder, R. A. 1980. Effects of grazing on bird habitats, pp. 51–66. In: R. M. Degraff and N. G. Tilghman (eds.), *Management of western forests and grasslands for nongame birds.* USDA Forest Service general technical report INT-86. Intermountain Forest and Range Experiment Station, Ogden, Utah.

Ryder, R. A., and D. A. Cobb. 1969. Birds of the Pawnee National Grassland in northern Colorado. *Journal of the Colorado–Wyoming Academy of Science* **6**:5. (Abst.).

Samson, F. B., and F. L. Knopf. 1994. Prairie conservation in North America. *BioScience* **44**:418–421.

Sauer, J. R., J. E. Hines, and J. Fallon. 2005. *The North American Breeding Bird Survey, results and analysis 1966–2005, version 6.2.2006.* USGS Patuxent Wildlife Research Center, Laurel, Md. Also available at http://www.mbr-pwrc.usgs.gov/bbs/bbs.htm.

Sherry, T. W., and R. T. Holmes. 1996. Winter habitat quality, population limitation, and conservation of neotropical–neoarctic migrant birds. *Ecology* **77**:36–48.

Skagen, S. K. 1997. Stopover ecology of transitory populations: The case of migrant shorebirds, pp. 244–269. In: F. L. Knopf and F. B. Samson (eds.), *Ecology and conservation of Great Plains vertebrates.* Springer, New York.

Skagen, S. K., A. A. Yackel Adams, and R. D. Adams. 2005. Nest survival relative to patch size in a highly fragmented shortgrass prairie landscape. *Wilson Bulletin* **117**:23–34.

Strong, M. A. 1971. *Avian productivity on the shortgrass prairie of north–central Colorado.* Masters thesis, Colorado State University, Fort Collins, Colo.

Temple, S. A., B. M. Fevold, L. K. Paine, D. J. Undersander, and D. W. Sample. 1999. Nesting birds and grazing cattle: Accommodating both on midwestern pastures. *Studies in Avian Biology* **19**:196–202.

Thomas, L. 1996. Monitoring long-term population change: Why are there so many analysis methods? *Ecology* **77**:49–58.

Thomas, L., and K. Martin. 1996. The importance of analysis method for Breeding Bird Survey population trend estimates. *Conservation Biology* **10**:479–490.

Thompson, T. R. 2003. *The effectiveness of the Conservation Reserve Program's native seeding requirement in providing breeding and wintering habitat for grassland birds in the southern High Plains of Texas.* Masters thesis, Texas Tech University, Lubbock, Texas.

Vickery, P. D., and J. R. Herkert. 2001. Recent advances in grassland bird research: Where do we go from here? *Auk* **188**:11–15.

Vickery, P. D., P. L. Tubaro, J. M. Cardoso da Silva, B. G. Peterjohn, J. R. Herkert, and R. B. Cavalcanti. 1999. Conservation of grassland birds in the western hemisphere. *Studies in Avian Biology* **19**:2–26.

Walk, J. W., and R. E. Warner. 1999. Effects of habitat area on the occurrence of grassland birds in Illinois. *American Midland Naturalist* **141**:339–344.

Wiens, J. A. 1969. An approach to the study of ecological relationships among grassland birds. *Ornithological Monographs* **8**:1–93.

Wiens, J. A. 1970a. *Avian populations and patterns of habitat occupancy at the Pawnee site, 1968–1969.* U.S. IBP Grassland Biome technical report no. 63. Colorado State University, Fort Collins, Colo.

Wiens, J. A. 1970b. Habitat heterogeneity and avian consumer populations in grasslands, pp. 77–83. In: R. T. Coupland and G. M. Van Dyne (eds.), *Grassland ecosystems:*

Reviews of research. Range Science Department science series no. 7. Colorado State University, Fort Collins, Colo.

Wiens, J. A. 1971. *Avian ecology and distribution in the Comprehensive Network, 1970.* U.S. IBP Grassland Biome technical report no. 77. Colorado State University, Fort Collins, Colo.

Wiens, J. A. 1973. Pattern and process in grassland bird communities. *Ecological Monographs* **43**:237–270.

Wiens, J. A. 1974a. Climatic instability and the "ecological saturation" of bird communities in North American grasslands. *Condor* **76**:385–400.

Wiens, J. A. 1974b. Habitat heterogeneity and avian community structure in North American grasslands. *American Midland Naturalist* **91**:195–213.

Wiens, J. A. 1975. Rangeland avifaunas: Their composition, energetics, and role in the ecosystem, pp. 146–182. In: D. R. Smith (tech. coord.), *Proceedings of the Symposium on Management of Forest and Range Habitats for Nongame Birds*. USDA Forest Service general technical report WO-1. Washington, D.C.

Wiens, J. A. 1977. On competition and variable environments. *American Scientist* **65**:590–597.

Wiens, J. A. 1978. Nongame bird communities in northwestern coniferous forests, pp. 19–20. In: R. M. DeGraaf (ed.), *Proceedings of the Workshop on Nongame Bird Habitat Management in the Coniferous Forests of the Western United States*. USDA Forest Service general technical report PNWS-64. Portland, Ore.

Wiens, J. A. 1981. Single-sample surveys of communities: Are the revealed patterns real? *American Naturalist* **117**:90–98.

Wiens, J. A. 1984. The place of long-term studies in ornithology. *Auk* **101**:202–203.

Wiens, J. A., and M. I. Dyer. 1975. Rangeland avifaunas: Their composition, energetics and role in the ecosystem, pp. 146–182. In: D. R. Smith (tech. coord.), *Proceedings of the Symposium on Management of Forest and Range Habitats for Nongame Birds*. USDA Forest Service general technical report WO-1. Washington, D.C.

Wiens, J. A., and G. S. Innis. 1974. Estimation of energy flow in bird communities: Population bioenergetics model. *Ecology* **55**:730–746.

Wiens, J. A., and J. T. Rotenberry. 1979. Diet niche relationships among North American grassland and shrubsteppe birds. *Oecologia* **42**:253–292.

Wiens, J. A., and J. T. Rotenberry. 1981. Habitat associations and community structure of birds in shrubsteppe environments. *Ecological Monographs* **51**:21–41.

Wiens, J. A., J. T. Rotenberry, and J. F. Ward. 1972. *Avian populations at IBP Grassland Biome sites: 1971.* U.S. IBP Grassland Biome technical report no. 205. Colorado State University, Fort Collins, Colo.

Wiens, J. A., J. T. Rotenberry, and J. F. Ward. 1974a. *Bird populations at ALE, Pantex, Osage, and Cottonwood, 1972.* U.S. IBP Grassland Biome technical report no. 267. Colorado State University, Fort Collins, Colo.

Wiens, J. A., J. F. Ward, and J. T. Rotenberry. 1974b. *Dietary composition and relationships among breeding bird populations as US/IBP Grassland Biome sites, 1970.* U.S. IBP Grassland Biome technical report no. 262. Colorado State University, Fort Collins, Colo.

Wilcove, D. S. 1985. Nest predation in forest tracts and the decline of migratory songbirds. *Ecology* **66**:1211–1214.

Winter, M., and J. Faaborg. 1999. Patterns of area sensitivity in grassland-nesting birds. *Conservation Biology* **13**:1424–1436.

With, K. A. 1994. The hazards of nesting near shrubs for a grassland bird, the McCown's Longspur. *Condor* **96**:1009–1019.

With, K. A., and D. R. Webb. 1993. Microclimate of ground nests: The relative importance of radiative cover and wind breaks for three grassland species. *Condor* **95**:401–413.

Yackel Adams, A. A., S. K. Skagen, and R. D. Adams. 2001. Movements and survival of Lark Bunting fledglings. *Condor* **103**:643–647.

Zimmerman, G., P. Stapp, and B. Van Horne. 1996. Seasonal variation in the diet of Great Horned Owls (*Bubo virginianus*) on shortgrass prairie. *American Midland Naturalist* **136**:149–156.

10

Insect Populations, Community Interactions, and Ecosystem Processes in the Shortgrass Steppe

Thomas O. Crist

Insects are diverse, abundant, and have numerous roles in rangeland ecosystems. More than 1600 species representing 238 families of insects have been recorded in the shortgrass steppe of northeastern Colorado (Kumar et al., 1976). Of this large assemblage, a much smaller subset—perhaps fewer than 50 species—is highly abundant with a large influence on community and ecosystem processes (Lauenroth and Milchunas, 1992). Even within abundant insect groups, such as grasshoppers, some species have far greater effects than others as herbivores (Capinera, 1987).

In this chapter I consider a small number of insect groups that have various influences in shortgrass steppe ecosystems (Table 10.1). I focus on three insect taxa—grasshoppers, beetles, and ants—that are widespread, abundant, and ecologically important in semiarid environments. I also draw attention to neglected groups, such as termites and spiders, for their potentially important roles in the shortgrass steppe. My primary objective is to emphasize the linkages among insect populations, community interactions, and ecosystem function. From this approach stems several related issues: how population distributions affect community interactions, how population abundance affects the processing and redistribution of energy and nutrients in ecosystems, and how abundance and species diversity are important to the functional roles of species in ecosystems. I skirt issues of population regulation in insects, which are reviewed elsewhere (Cappuccino and Price, 1995), and instead consider how temporal and spatial patterns in insect populations relate to community and ecosystem processes.

Understanding relationships among populations, communities, and ecosystems requires approaches that link patterns and processes across scales. Much of what

Table 10.1 Some Abundant Insects and Other Arthropods with Important Roles in the Shortgrass Steppe

Order and Family	Genera	Feeding Mode	Potential Roles in Ecosystem
Araneae			
Gnaphosidae	*Gnaphosa*	Predator	Effects on prey, role in food webs
Lycosidae	*Schizocosa*	Predator	Effects on prey, role in food webs
Coleoptera			
Carabidae	*Harpalus*	Omnivore	Plants/prey, role in food webs
	Pasimachus	Predator	Effects on prey, role in food webs
Scarabaeidae	*Phyllophaga*	Herbivore	Disturbance, primary production
Tenebrionidae	*Eleodes*	Detritivore	Litter processing and redistribution
Diptera			
Asilidae	*Efferia*	Predator	Effects on prey, role in food webs
Hymenoptera			
Formicidae	*Formica*	Predator	Role in food webs, soils
	Pogonomyrmex	Granivore	Disturbance, seed pool, soils
Isoptera			
Rhinotermitidae	*Reticulitermes*	Decomposer	Nutrient cycling, soils
Orthoptera			
Acrididae	*Opeia*	Herbivore	Effects on primary production, litter production, food webs
	Melanoplus	Herbivore	

is known about the roles of insects in the shortgrass steppe is based on studies conducted at relatively fine scales. To link insect population studies to community and ecosystem processes, however, I suggest that insect populations should also be studied across broader scales that encompass topographic variation.

The rolling topography in the shortgrass steppe produces a gradient in soil texture, water availability, and nutrient retention from uplands to lowlands (Clark and Woodmansee, 1992; Schimel et al., 1985). Plant community structure also varies with topography in spatially repeating patterns across the landscape (Milchunas et al., 1989). We know considerably less about how insect distribution and abundance changes along topographic gradients, and how these changes might influence community and ecosystem processes. I discuss ways in which insect distributions might be examined across a range of scales in the shortgrass steppe. Much of this review is based on studies conducted at the CPER in northeastern Colorado. I also draw upon findings from other shortgrass steppe, shrub steppe or mixed-grass ecosystems, to place the CPER studies in a broader regional context.

Population Dynamics and Distribution

Casual observation of insects inhabiting the shortgrass steppe suggests that ground-dwelling beetles (Carabidae, Scarabaeidae, Tenebrionidae; Fig. 10.1), grasshoppers (Acrididae), and ants (Formicidae) are abundant and widely distributed insect groups. In species richness, biomass, and abundance, these taxa may represent a significant fraction of the insect fauna in the shortgrass steppe (Kumar et al., 1976; Lauenroth and Milchunas, 1992; Lavigne and Kumar, 1974). Two taxa, grasshoppers and harvester ants (*Pogonomyrmex* spp.; Fig. 10.2), have long been considered rangeland pests in shortgrass ecosystems because of their abundance and impacts on vegetation and soils (Capinera, 1987; Rogers, 1987). Spiders (Araneae) are also abundant in the shortgrass steppe, but few studies have been conducted to examine their roles as predators in food webs (Lavigne and Kumar, 1974; Weeks and Holtzer, 2000). Similarly, the arid-land subterranean termite (*Reticulitermes tibialis*) is surprisingly common in lowland areas, where they may have an important role as decomposers of woody plant litter (Crist, 1998).

These arthropod groups vary substantially in patterns of distribution and abundance. They also have different feeding modes and functional roles in the shortgrass steppe (Table 10.1). The combined effects of insects on ecosystem functioning may be considerable. The major challenges in assessing these roles are difficulties in extrapolating variable patterns of abundance across time and space, and inferring processes from a complex set of patterns.

Figure 10.1 Several species of darkling beetles (Coleoptera: Tenebrionidae) occur in the shortgrass steppe. Pictured here is the large-bodied *Eleodes hispilabris*. (Photo by T. O. Crist.)

Figure 10.2 Nest mound and clearing of the Western harvester ant (*Pogonomyrmex occidentalis*). Ant colony densities may exceed 30/ha in the shortgrass steppe. (Photo by T. O. Crist.)

Temporal Dynamics

Grasshopper assemblages from various shortgrass sites typically have 25 to 40 species (Onsager, 1987; Przybyszewski and Capinera, 1990; Welch et al., 1991), with three to five species representing most of the variation in overall abundance among years (Capinera and Thompson, 1987; Onsager, 1987; Przybyszewski and Capinera, 1990; Van Horn, 1970; Welch et al., 1991). Many of the dominant grasshopper species in shortgrass are in the subfamily Gomphocerinae, which tend to be more abundant in warmer, drier regions dominated by C_4 grasses. Species of Melanoplinae are more abundant in mixed-grass prairie where C_3 grasses predominate (Belovsky and Joern, 1995; Capinera, 1987). Oedipodinae are locally abundant in more open areas with bare ground (Przybyszewski and Capinera, 1990; Uvarov, 1977). At the CPER, for example, Welch et al. (1991) found that 10 of 34 species comprised 95% of the total grasshopper abundance, four of which were gomphocerines; five, oedipodines; and one, melanopline. Thus, grasshopper species that have large population fluctuations or outbreaks vary among regions according to subfamilies.

A long history of grasshopper outbreaks is documented in western rangelands (Pfadt and Hardy, 1987). During outbreaks, population densities of some species often exceed $20 \cdot m^{-2}$ and can remove 20% to 35% of the standing crop of vegetation (Hewitt and Onsager, 1983; Mitchell and Pfadt, 1974; Onsager, 1987). Severe outbreaks may last from 2 to 6 years, during which much of the green plant biomass may be consumed by grasshoppers (Pfadt and Hardy, 1987). The timing and spatial patterning of outbreaks are variable and some areas have a tendency to outbreak more than others. Lockwood and Lockwood (1991), for instance, report that most outbreaks in Wyoming occur primarily in the northern mixed-grass prairie. Studies at the CPER show that grasshopper densities are usually less than $10 \cdot m^{-2}$ (a value often considered outbreak density) (Lockwood and Lockwood, 1991; Torell and Huddleston, 1987) but often exceed $3.6 \cdot m^{-2}$ (the threshold infestation level) (Capinera and Horton, 1989). From 1980 to 1985, Capinera and Thompson (1987) found that mean densities of grasshoppers in eight pastures changed 10-fold from 0.48 to $4.83 \cdot m^{-2}$. Densities within some pastures were more variable among years, ranging from 0.14 to $15.4 \cdot m^{-2}$. Two species—*Opeia obscura* (Gomphocerinae) and *Melanoplus gladstoni* (Melanoplinae)—were primarily responsible for changes in overall grasshopper densities, and these species had their peak abundances during consecutive years (Capinera and Thompson, 1987).

Darkling beetles (Tenebrionidae) and ground beetles (Carabidae) also exhibit considerable yearly variation in abundance in the shortgrass steppe. Crist and Wiens (unpublished data) recorded beetle abundances from 1990 to 1994 along a 900-m pitfall transect (trap spacing, 10 m) spanning a topographic gradient. As a group, tenebrionids had low abundances in 1990 and 1991, but showed an order of magnitude increase in 1992 and maintained high levels in 1993 and 1994 (Fig. 10.3A). Changes in overall tenebrionid abundances were primarily the result of *Eleodes extricata* and *E. obsoleta* (Fig. 10.3A). These two species, along with *E. hispilabris*, comprised most of the total abundance of the 15 tenebrionid species recorded during the study. Between 1990 and 1992, *E. obsoleta* was most abundant, and seasonal increases in *E. obsoleta* (a late-season species) closely tracked overall tenebrionid abundance. In 1993 and 1994, however, *E. extricata* became more common, and seasonal decreases in tenebrionid abundances were primarily the result of *E. extricata* (an early-season species). This 5-year trend in tenebrionid abundance can be extended with McIntyre's (2000) 4-year study, which showed that densities of *E. extricata* and *E. hispilabris* were high in 1994 (consistent with Fig. 10.3A) and declined steadily until 1997. Decreases in beetle abundances corresponded to a shift from warmer, drier conditions in 1994 and 1995 to cooler, wetter weather in 1996 and 1997. Beetle densities were influenced by temperature and precipitation within a year as well as from the previous year (McIntyre, 2000).

Carabid beetles showed similar yearly patterns of abundance to tenebrionids along the 900-m transect during 1990 to 1994. The overall carabid beetle abundance also increased during 1992 and 1993, but decreased in 1994 (Fig. 10.3B). Carabids consistently increased in abundance from early to late summer. Ground-dwelling spiders showed abundances in pitfall traps that were comparable with both groups of beetles (Fig. 10.3B). More than 80% of the spiders captured in pitfall traps are wolf spiders (Lycosidae: *Schizocosa* spp.) or hunting spiders (Gnaphosidae) (Weeks, 1996; Weeks and Holtzer, 2000). As a group, ground-dwelling spiders appeared to have lower interannual variability in abundance than tenebrionid or carabid beetles.

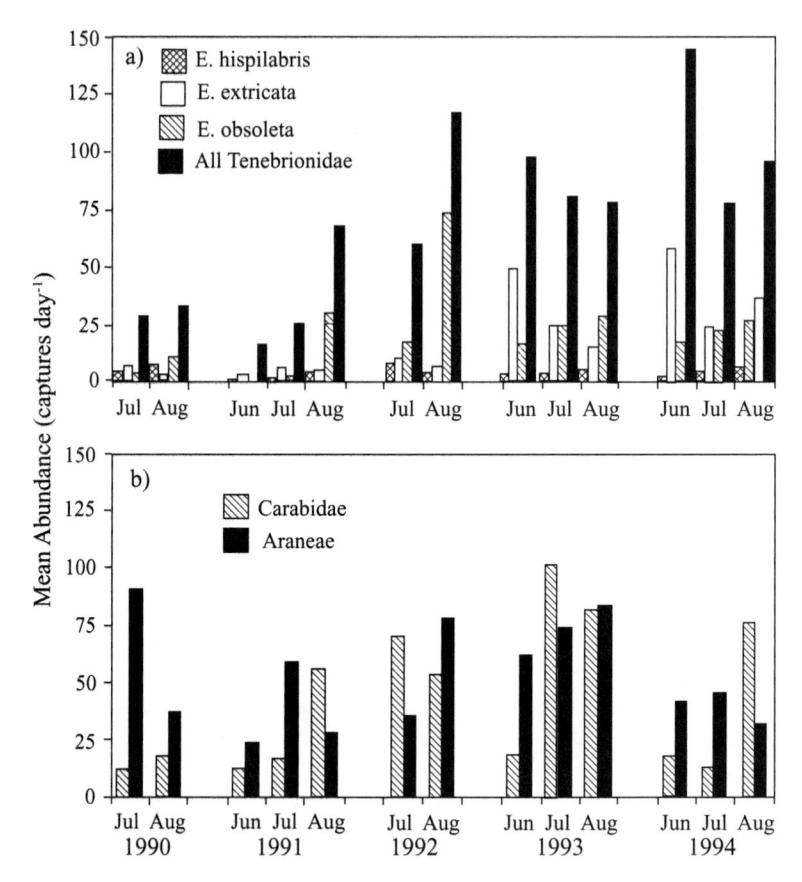

Figure 10.3 Patterns of insect abundance from 1990 to 1994 along a 900-m transect that spans a topographic gradient. Data are expressed as total number of individuals captured per day (based on 3–6 days of trapping) for each sample month. (A) Overall abundance of darkling beetles (Tenebrionidae) and dynamics of three *Eleodes* species. (B) Overall abundance of ground beetles (Carabidae) and ground-dwelling spiders (Araneae).

Densities of harvester ant colonies are reported in several studies (Crist and Wiens, 1996; Rogers and Lavigne, 1974), but few data are available on population dynamics. Porter and Jorgensen (1988) found colony densities of *P. owyheei* remained fairly constant at 40 colonies·ha^{-1} from 1977 to 1986, and 80% of the nests sampled in 1977 were still active in 1986. Keeler (1993) studied 15-year colony dynamics of *P. occidentalis* in western Nebraska. Between 1977 and 1991, colony density varied from 56 to 80 colonies·ha^{-1}. The recruitment of new colonies was 0 to 8 colonies·year^{-1} and the annual mortality was 0 to 5 colonies·year^{-1}. Despite yearly turnover, the population was highly stable from 1982 to 1991, fluctuating narrowly from 74 to 80 colonies·ha^{-1}.

From these data there emerge clear differences among insect taxa in their annual variation in abundance. Grasshoppers are most variable, and their dynamics are more directly tied to annual variation in weather and aboveground primary production (e.g., Belovsky and Slade, 1995) than those of other groups. Tenebrionids are also highly variable in population size, but because of their largely detritivorous food habits, they appear to be less influenced by annual variation in primary production than grasshoppers. The relatively slow development and long life span of tenebrionids (2–3 years) (Allsopp, 1980), however, may delay their responses to food availability or weather. The corresponding trends in yearly abundances of tenebrionids and carabids suggest that similar factors such as annual variation in temperature and precipitation affect beetle abundances (McIntyre, 2000). The most stable population dynamics are shown by harvester ant colonies, which are long-lived (10–35 years) after they become established (Keeler, 1993; Porter and Jorgensen, 1988). Among the insect groups, therefore, greater yearly variability in abundance appears to be associated with shorter life span and generation time (grasshoppers), whereas more stable dynamics are found in longer lived species (carabids, tenebrionids, and especially ants).

Spatial Distribution

Microhabitat Patterns

Patterns of insect movement and microhabitat use can provide a mechanistic basis for understanding population distribution (Crist and Wiens, 1995; Johnson et al., 1992a, b; Wiens et al., 1993b; With and Crist, 1995). Studies of insect movement and microhabitat use, however, are often limited by time-intensive and specialized methodology (Turchin, 1998; Turchin et al., 1991; Wiens et al., 1993a).

Joern (1983b) measured the daily movements of four grasshopper species in the shortgrass steppe of west Texas. Grasshopper movement rates and net displacements were quite similar among species. The average distances moved by individuals were 4.9 to 8.1 m·d^{-1} and 40 to 70 m after 14 days. Species differed in their patterns of movement, depending on their vagility and microhabitat specificity (Joern, 1982, 1983b). With and Crist (1995) examined fine-scale movements of two grasshopper species—*Psolessa delicatula* (Gomphocerinae) and *Xanthoppus corallipes* (Oedopodinae)—at the CPER. Movement rates of each species were measured in vegetation patches that were classified as homogenous, moderately heterogeneous, or very heterogeneous, based on the degree to which *Bouteloua gracilis* was interspersed with other plant cover types such as cacti, shrubs, and forbs. *Psolessa delicatula* is small bodied, has limited vagility, and prefers homogenous patches of *B. gracilis*, whereas *X. corallipes* is larger, more vagile, and prefers more heterogeneous areas (With and Crist, 1995). A simulation model was then used to predict transition probabilities among patches using movement rates measured in the field. Model predictions were consistent with observed grasshopper distributions in the field. *Psolessa delicatula* showed a random distribution and was most abundant in homogenous patches, whereas *X. corallipes* was highly aggregated in heterogeneous patches of vegetation (With and Crist, 1995).

Studies have also examined tenebrionid beetle movements in relation to habitat structure in the shortgrass steppe. Beetle movements were measured in areas that differed in vegetation structure, based on the proportions of grass, bare ground, cacti, and shrubs (Crist and Wiens, 1995; Crist et al., 1992). Rates of movement of *Eleodes* beetles differed among species but were generally greater in the more homogenous grass and bare ground than in heterogeneous cactus and shrub areas. Movement patterns among *Eleodes* species and habitats were nonetheless quite similar when displacements were appropriately rescaled to the overall length of the movement pathway (Crist et al., 1992). An important consequence of differential movement rates is that the population density of some *Eleodes* beetles may be three to four times greater in areas of cactus and shrub than in more homogeneous vegetation because of the greater residence times in cactus–shrub microhabitats (Crist and Wiens, 1995; McIntyre, 1997, 2000; Stapp, 1997).

Broad-Scale Patterns

Broad-scale patterns of insect distribution in the shortgrass steppe can be related to variation in soils and vegetation as they are influenced by topography and grazing. In two respects, the broad-scale spatial patterns are the most critical bits of information in linking insect populations to rangeland ecosystem processes. First, there is considerable broad-scale turnover of species composition and relative abundance among grassland insects and plants as a result of topography or land use (Tscharntke and Greiler, 1995). Second, many ecosystem processes, such as primary production and soil nutrient dynamics, are driven by topography or grazing (Lauenroth and Milchunas, 1992). For these reasons, a major gap in our understanding of the roles of insects in shortgrass steppe ecosystems stems from our lack of information on the broad-scale spatial distributions of insects.

Grasshoppers are strongly influenced by broad-scale changes in vegetation (Kemp, 1992; Kemp et al., 1990; Miller and Onsager, 1991). Przybyszewski and Capinera (1990) found that several grasshopper species reached their highest abundances in different pastures, reflecting variation in vegetation among pastures (resulting from topography, soils, and/or grazing). In an earlier study, Capinera and Sechrist (1982) analyzed grasshopper abundance and plant biomass in six pastures with different grazing intensities. Overall grasshopper densities were $1.35 \cdot m^{-2}$ in high-biomass pastures (which were ungrazed or lightly grazed) and $0.75 \cdot m^{-2}$ in low-biomass pastures (moderately or heavily grazed). Population responses to grazing differed among subfamilies. As grazing intensity increased, gomphocerines decreased by half, melanoplines were similar in abundance, and oedopodines increased. Abundances of Gomphocerinae and Melanoplinae were positively correlated with the biomass of grasses, forbs, and shrubs, whereas Oedipodinae was negatively correlated with all three plant life-forms. Capinera and Sechrist (1982) noted that the observed differences in plant biomass resulting from grazing might be commonly encountered within smaller areas and can result in significant fine-scale changes in species abundances of grasshoppers—a point that deserves further study. Grasshopper responses to grazing can also differ over time. Welch et al. (1991) conducted grasshopper sampling in lightly and heavily grazed pastures and compared their results with those of Van Horn (1970). In the heavily grazed pasture, densities were similar between time periods: $0.86 \cdot m^{-2}$ and $0.78 \cdot m^{-2}$ in 1970 and 1989, respectively. In the lightly grazed pasture, however, densities were

0.99·m^{-2} in 1970 and 1.44·m^{-2} in 1989. Heavily grazed pastures had significantly lower grasshopper densities in 1989 but not in 1970; grazing effects were therefore more apparent when viewed over longer time intervals (Welch et al., 1991).

Broad-scale variation in rangeland is often related to topography, especially in areas such as the shortgrass steppe or mixed-grass prairie, where rolling topography gives rise to spatially repeating patterns of vegetation (Barnes and Harrison, 1982; Milchunas et al., 1989). Landscape patterns of vegetation and soils resulting from topographic variation are central to community and ecosystem studies. Soils generally change from coarse textured in uplands to fine textured in lowlands (Clark and Woodmansee, 1992). Corresponding increases in soil nutrient levels in lowlands (Schimel et al., 1985) produce increased plant biomass and changes in plant species composition (Milchunas et al., 1989). Insect species abundances should also change across a toposequence depending on species responses to changes in soils, vegetation, and/or slope exposure.

A sampling of several insect species across a topographic gradient at the CPER (Fig. 10.4A) reveals distinct patterns of species distributions. The spatial

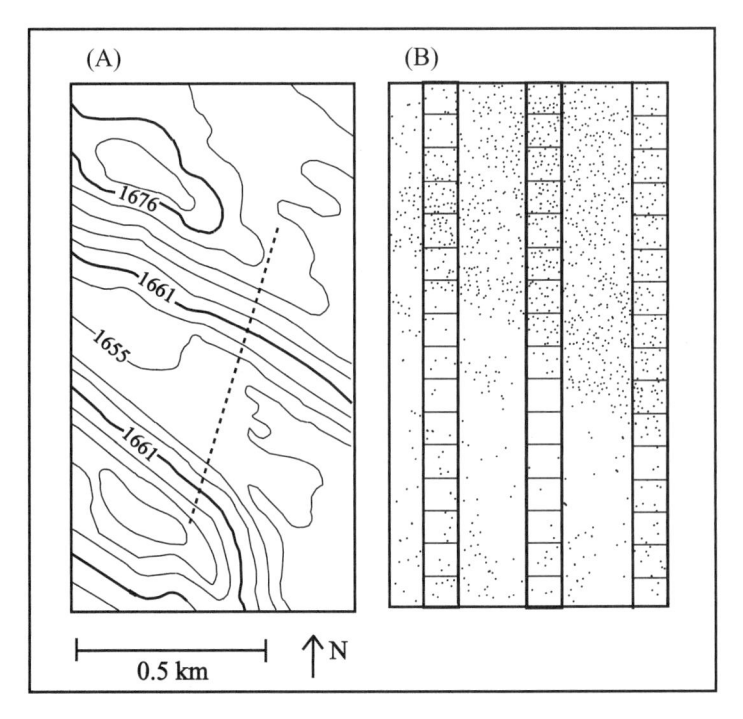

Figure 10.4 (A) Topography within a pasture at the CPER (3-m contour lines) where a 900-m pitfall transect (dotted line) was placed to conduct broad-scale measures of insect distribution and abundance. (B) Map of colonies of harvester ants (*Pogonomyrmex occidentalis*) obtained from digitizing aerial photographs (1:2000 scale) in the same pasture. The three belt transects sectioned into 1-ha segments illustrate how patterns of colony density across the toposequence can differ horizontally (see Fig. 10.6 for analysis).

distribution of harvester ant colonies also varies along the same topographic gradient (Fig. 10.4B). Crist and Wiens (1996) conducted a more extensive mapping of harvester ant colonies (*P. occidentalis*) in five pastures at the CPER using low-level aerial photographs (1:2000 scale). Topography, soils, and grazing intensity influenced ant colony density and spacing patterns. Ants were most abundant on upland plains with coarse-textured soils and in pastures with light or moderate grazing intensity; colony density often exceeded 30 colonies·ha^{-1} on the upland plain, but was generally less than 10 colonies·ha^{-1} in lowland areas (Crist and Wiens, 1996).

Across the same toposequence, *E. extricata* showed higher capture rates in pitfall traps on upper slopes, where a relatively greater proportion of bare ground interrupts the grass matrix; fewer pitfall captures were recorded in the intervening swale (Fig. 10.5A, B). Variation in pitfall captures was also clearly expressed at finer scales, especially in 1993 to 1994, when abundance was higher than in 1991. Although overall abundance of beetles varied considerably among years, the broad-scale spatial patterning was relatively consistent (Fig. 10.3B). Wood ants (*Formica obscuripes*) had a more patchy distribution (Fig. 10.5D). Captures of ants were greatest in the swale where shrubs are more abundant (Fig. 10.5C), partly because *F. obscuripes* uses woody debris for nest construction. In another study, Bestelmeyer and Wiens (2001) recorded a positive spatial association between *F. obscuripes* and saltbush (*Atriplex canescens*) habitats. The arid-land subterranean termite (*R. tibialis*) also had higher abundance in saltbush swales (Fig. 10.5E). A detailed geostatistical analysis showed that termite spatial distribution was strongly associated with saltbush (Crist, 1998). The peak in the probability of termite occurrence, however, was shifted slightly toward the south-facing slope relative to that of shrub proximity; therefore, soils or microclimate also likely affect termite distributions (Fig. 10.5C, E [see also Crist, 1998]).

The spatial scales of variability in abundance of insect distributions can be compared by autocorrelation. *Eleodes extricata* exhibited a high degree of autocorrelation at 10-m and 80-m lags, indicating patchiness in pitfall samples within a topographic position (Fig. 10.6A). A stronger spatial patterning occurred in years with greater abundance (1993 and 1994 compared with 1991). Changes in beetle abundance across slope positions are reflected in the switch from positive to negative autocorrelation (ca., 200 m). *Formica obscuripes* also had a high correlation of ant numbers in adjacent traps with another distinct peak at 120 m in 1993 to 1994 and at 70 m in 1991 (Fig. 10.6B). There is a greater similarity in spatial pattern among years for *F. obscuripes* than for *E. extricata*. This might be expected because *F. obscuripes* is a social insect with long-lived colonies.

Cross-correlation can be used to measure shifts in the spatial pattern of abundance among years. The degree of correlation in *E. extricata* abundance between 1991 and 1993 was higher among traps spaced 50 to 100 m apart (lags, –50 to –100 m) than in the same traps (zero lag) (Fig. 10.7A), indicating shifts in spatial distribution between years. A similar pattern occurred between 1991 and 1994. The cross-correlation between 1993 and 1994, however, was greatest near zero, indicating a similar spatial pattern of captures between these 2 years (Fig. 10.7A). Other studies have shown that landscape patterns of tenebrionid beetles are related to topographic variation in soils and vegetation (Bossenbroek et al., 2004; Crist

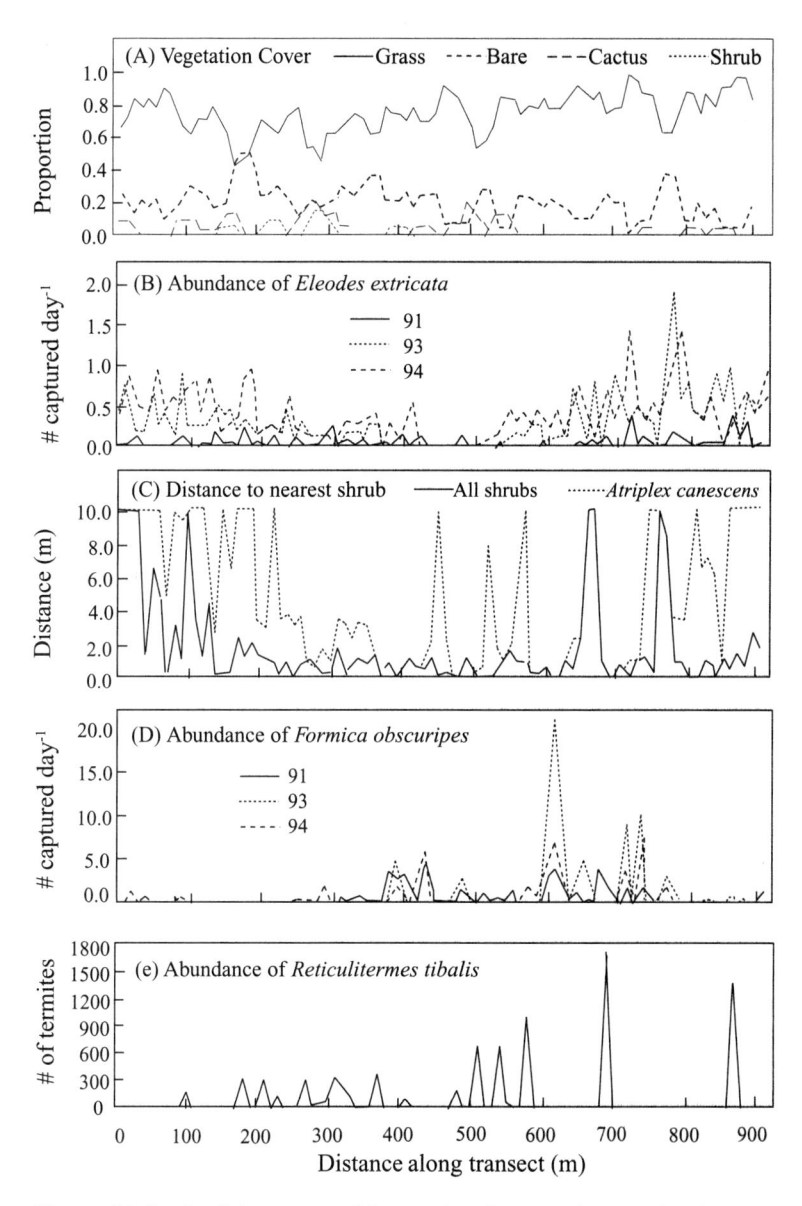

Figure 10.5 Spatial patterns of insect abundance and vegetation features along the 900-m pitfall transect with a 10-m trap spacing (91 total). (A) Vegetative cover determined from eight point samples around each pitfall station. (B) Number of captures per day of the darkling beetle *Eleodes extricata* in 1991, 1993, and 1994 (years that had three sample months; Fig. 10.1). (C) Distance to nearest shrub from each pitfall station (maximum, 10 m) with all shrubs included, and distance to nearest four-wing saltbush, *Atriplex canescens*. (D) Number of workers captured per day of the wood ant, *Formica obscuripes*, during 3 years. (E) Number of arid-land subterranean termites (*Reticulitermes tibialis*) recovered from toilet paper baits positioned on the soil surface at each pitfall–trap station. Values are the combined total from samples in June, July, and August 1993.

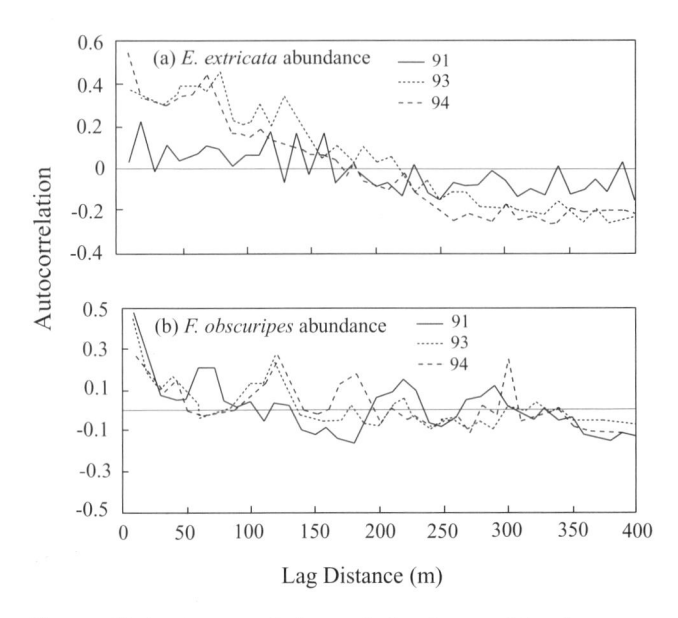

Figure 10.6 Autocorrelations of abundance of beetles, ants, termites across increasing distances between sample pairs (lag length). (A) *Eleodes extricata*. (B) *Formica obscuripes*.

and Wiens, 1995; McIntyre, 2000), but that the strength of the beetle–environment relationship differs across spatial scales (Bossenbroek et al., 2004; Hoffman and Wiens, 2004). Cross-correlation in *F. obscuripes* ants in pitfall traps showed strong positive correlations among years within the same traps (zero lag) and, to a lesser degree, at 30-m lags on each side of zero (Fig. 10.7C), likely a reflection of relative constancy in colony spacing during the study.

Ant colony densities along three transects that span the same topographic gradient (Fig. 10.4B) reveal a shifting boundary in ant colony distribution as one moves horizontally across the pasture (Fig. 10.8A). This variation in ant colony distribution of the entire pasture was analyzed with anisotropic autocorrelation (Fig. 10.8B). The autocorrelation in colony density decreases along the north–south direction (0°), persists across greater lag lengths in the northeast–southwest (45°) and east–west directions (90°), and declines sharply in the southeast–northwest direction (135°), which most closely corresponds to the direction of the topographic gradient (Fig. 10.4B). Directions that run along the topographic contours (45° and 90°) exhibit positive autocorrelation with lags less than 700 m, whereas those that span topographic changes (0° and 135°) become negative at lags more than 400 m (Fig. 10.8B). The degree of change across lag lengths shown by anisotropic autocorrelations is therefore consistent with the position of the topographic gradient. Thus, the slope and aspect are closely tied to the density and distribution of harvester ants. This spatial coupling of ant density with topography, in turn, has important consequences for community interactions and ecosystem processes, as described next (see also Crist and Wiens, 1996; MacMahon et al., 2000).

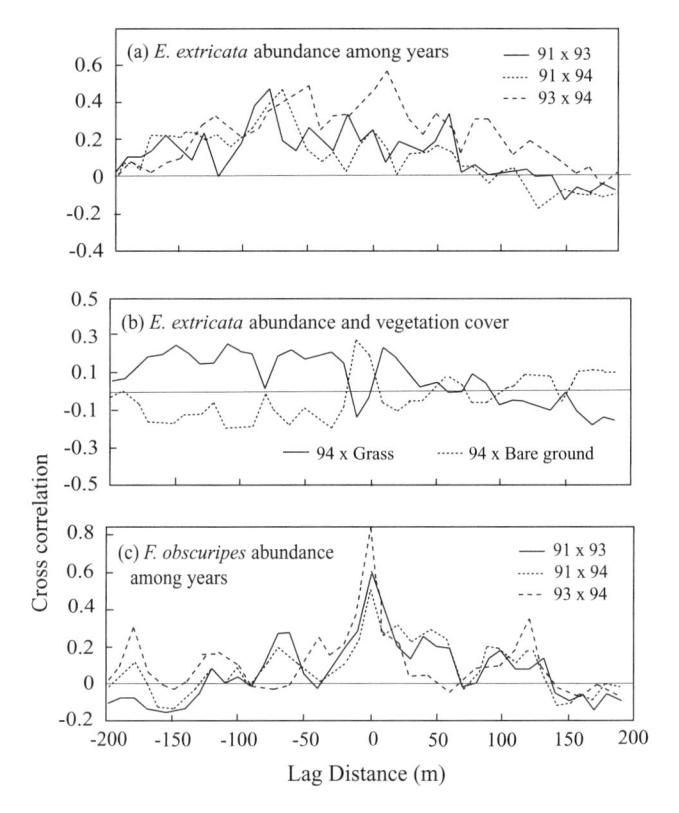

Figure 10.7 Cross-correlations between insect abundances across years. (A) *Eleodes extricata* between different pairs of years. (B) *Formica obscuripes* between different pairs of years. (C) *Formica obscuripes* between different pairs of years.

Community Interactions

Patterns of distribution and abundance impinge on the strength and variability of species interactions within heterogeneous landscapes. For example, population densities and the activities of individual consumers affect the quantity of plant biomass removed by grasshoppers (Joern, 1987; Mitchell and Pfadt, 1974) or the rate at which seeds are harvested by ants (Crist and Wiens, 1994). The patterns of insect distributions in time and space therefore have important implications for community interactions in the shortgrass steppe.

Plant–Herbivore Interactions

Grasshoppers are the most important aboveground insect herbivores in rangeland ecosystems (Watts et al., 1982), often removing 20% to 25% of aboveground

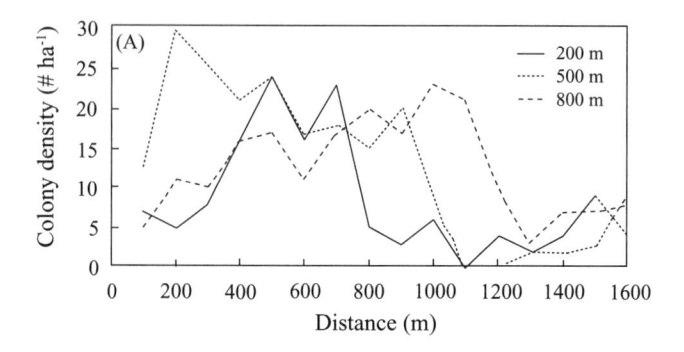

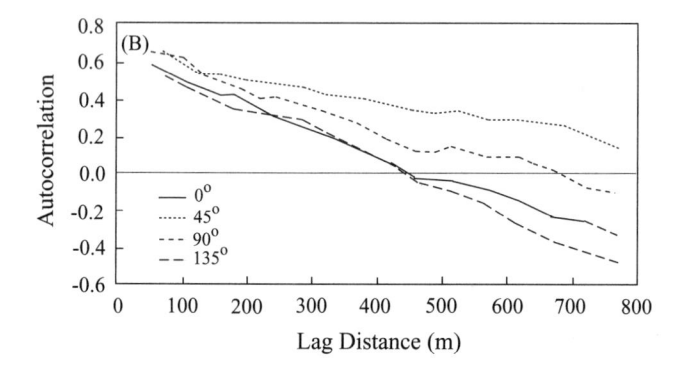

Figure 10.8 (A) Colony density of *Pogonomyrmex occidentalis* across a topographic gradient from three different horizontal positions (see Fig. 10.4B). (B) Anisotropic autocorrelation of colony density from a grid of quadrat counts from the map of the entire pasture. The 0° and 135° directions run across the gradient, whereas the 45° and 90° directions run more parallel to contours.

primary productivity (Hewitt and Onsager, 1983; Mitchell and Pfadt, 1974). Plant consumption by grasshoppers is highly variable, however, and this variability in consumption may be expressed at spatial scales ranging from microhabitat to region and at temporal scales from days to decades (Joern, 1987). Differential effects of grasshopper herbivory on the plant community result from variability in feeding preferences of grasshopper assemblages. For acridids, diet preference and breadth are related to subfamily membership. In mixed-grass prairie, Joern (1983a) found that gomphocerines, which are principally grass feeders, used an average of 8.0 plant taxa in their diets; melanoplines, which feed mostly on forbs, used an average of 17.1 plant taxa. Diet breadth may vary in predictable ways among regions as well. Grasshoppers in the shortgrass steppe feed on fewer plants than in mixed-grass sites (Joern, 1987), possibly because of a decreased plant community dominance by *B. gracilis* as one moves from the shortgrass steppe to mixed-grass communities, which include codominants such as western wheatgrass (*Agropyron smithii*) and little bluestem (*Schizachyrium scoparium*). Despite the breadth of grasshopper diets, some plant species are consistently consumed in greater quantities than predicted by their relative abundance in the plant community (Mitchell, 1975; Mulkern, 1967).

The effects of selective feeding by grasshoppers on plant community structure and ecosystem functioning have received considerably less attention (e.g., Rodell, 1977). Grasshoppers can clearly reduce their food supply, as shown by studies that demonstrate density-dependent regulation of population size (Belovsky and Joern, 1995; Belovsky and Slade, 1995; Kemp and Dennis, 1993). Because many grasshopper species that reach outbreak densities are oligophagous or polyphagous (Joern, 1987), grasshopper feeding may have widespread rather than selective effects within a plant community. Such predictions based on diet preferences could be used to predict the effects of grasshopper herbivory on plant community structure.

In the shortgrass steppe, a significant alteration of plant community structure in response to grasshopper herbivory would likely involve changes in dominance by *B. gracilis*. Dyer and Bokhari (1976) found that laboratory-grown *B. gracilis* plants responded to grasshopper grazing by reallocating energy to belowground processes such as increased tiller production, respiration, and root exudation. This suggests that the effects of grasshopper grazing might be similar to those of livestock grazing in which there is an increased dominance by grazing-tolerant *B. gracilis* (Lauenroth and Milchunas, 1992; Milchunas et al., 1988, 1989); however, this hypothesis has not been tested in the field.

The white grub larvae of June beetles (*Phyllophaga fimbripes* and *P. crinata*) are belowground herbivores that cause considerable mortality of perennial grasses. White grubs may exceed densities of 50/m^2 and have substantial impacts on plant communities in rangelands (Ueckert, 1979; Watts et al., 1982; Wiener and Capinera, 1980). Root feeding by larvae produce killed patches of blue grama that are generally less than 0.1 ha in size but can reach 1 ha or more (Milchunas et al., 1990; Ueckert, 1979). Grub kill areas have altered plant and arthropod communities from those observed in surrounding areas. Rottman and Capinera (1983) found a higher plant diversity in grub kill compared with undamaged areas primarily as a result of an increased number of forb species immediately after grub disturbances (cf. Peters et al., chapter 6, this volume; Ueckert, 1979). Similarly, Milchunas et al. (1990) recorded decreased dominance of *B. gracilis* and *Opuntia polyacantha* in grub disturbances, resulting in higher plant diversity compared with surrounding areas. Overall arthropod diversity was relatively unchanged in grub kill areas, although some taxa showed differential effects to grub disturbance (Rottman and Capinera, 1983). For example, carabid (*Harpalus* spp.) and chrysomelid (*Phyllotreta* spp.) beetles were more abundant in grub disturbances than in surrounding areas; spiders, mites, and collembolans had reduced abundances in disturbances compared with undisturbed areas (Rottman and Capinera, 1983).

Seed–Granivore Interactions

Harvester ants (*Pogonomyrmex* spp.) are widespread and abundant insect granivores in rangelands (MacMahon et al., 2000). Studies on *P. occidentalis* have shown that foraging activity may significantly alter the density and distribution of seeds in the soil (Coffin and Lauenroth, 1990; Crist and MacMahon, 1992; MacMahon et al., 2000; Mull and MacMahon, 1996). Rogers (1974) estimated that 1% to 5% of the total seed pool in soil was harvested by ants at the CPER, which is lower than that found for *P. occidentalis* in other semiarid ecosystems (9% to 26% [Crist and MacMahon, 1992]). Ants selectively harvest seeds so that preferred species show greater losses from soil (Crist and MacMahon, 1992) and greater accumulation in ant nests (Coffin and Lauenroth, 1990) compared with surrounding areas.

Measurements of native seed preferences correlate with ant effects on soil seeds (Crist and MacMahon, 1992), which suggests that seed-choice experiments provide a relative measure of the interaction strength between granivorous ants and various seed species. Crist and Wiens (1994) conducted seed-dish experiments to examine how seed removal by ants was influenced by seed species, plant patch structure, and cattle grazing. Broad-scale (pasture-level) differences in vegetation resulting from grazing and the presence of predators such as horned lizards affected both ant foraging patterns and colony activity, and comprised the largest component of variation in seed harvest. Within pastures, ants removed seeds more rapidly from patches of bare ground or grass than from cacti or shrubs. Within patches, seeds of annual forbs (*Lepidium densiflorum*) were removed far more rapidly than seeds of dominant grasses *B. gracilis* and *Buchloë dactyloides* (Crist and Wiens, 1994). Interestingly, the preference for *L. densiflorum* in seed-dish trials is corroborated by a large number of *L. densiflorum* seeds recovered from ant nests (Coffin and Lauenroth, 1990).

As with insect herbivory, the selective influences of insect granivory on the plant community are largely on subdominant plant species. Changes in the abundance of subdominant plants, however, can have substantial effects on the overall richness and diversity of shortgrass plant communities (Milchunas et al., 1990). In comparison with their effects on dominant grasses, the effects of insect herbivores or granivores on plant species diversity have received far less attention. An increasing emphasis on the management of biodiversity in range ecosystems (Hart, 2001; West, 1994) may renew interest in the effects of insect consumers on plant and arthropod diversity (e.g., Milchunas et al., 1990; Rottman and Capinera, 1983).

Predator–Prey Interactions

I consider two aspects of the roles of insects in aboveground predator–prey interactions: insects as predators and insects as prey. Together, these comprise the complex suite of feeding interactions that structure food webs. Insects and other arthropods are a major part of food webs in the shortgrass steppe (Lauenroth and Milchunas, 1992), but there are few studies of aboveground food web structure.

Insects as Predators

From studies of the IBP Grassland Biome project, the most abundant arthropod predators in the shortgrass steppe include wolf spiders (Lycosidae), hunting spiders

(Gnaphosidae), robber flies (Asilidae), and ground beetles (Carabidae) (Lavigne and Campion, 1978; Lavigne and Kumar, 1974; Lavigne et al., 1971). Searches along 1.6-km transects during 1969 and 1970 showed that "sightable groups" such as lycosids and asilids had abundances roughly 10 times those of mantids (Mantidae), grasshopper wasps (Sphecidae), jumping spiders (Salticidae), or tiger beetles (Cicindelidae). Asilids numbers varied slightly (67–99) along transects, but did not differ among pastures (Lavigne et al., 1971). A total of 120 lycosid spiders were sighted in a lowland area, and 38 to 73 were found in three upland areas with different grazing intensities (Lavigne et al., 1971). In contrast, Weeks and Holtzer (2000) found that lycosids were significantly more abundant in uplands than in lowlands as measured by several pitfall trap arrays in each area. Gnaphosids had similar abundances in the two topographic positions (Weeks and Holtzer, 2000).

Grasshoppers are a major part of the diets of wolf spiders and robber flies (Lavigne et al., 1971), but it is unclear whether spiders or robber flies can significantly reduce grasshopper densities in the shortgrass steppe (Capinera, 1987). Studies in similar systems suggest that arthropod predation on grasshoppers is primarily on early nymphal stages and that vertebrates (especially birds) may have greater effects on population sizes (Belovsky and Slade, 1993; Joern, 1986; Schmitz, 1993). However, spiders and robber flies also feed on other insects, such as leafhoppers and ants, and therefore may be important generalist predators in the shortgrass steppe.

One of the most abundant carabid beetles at the CPER is *H. desertus*, which is an omnivorous species (Lavigne and Campion, 1978). In an ecosystem stress experiment, conducted during 1971 to 1974, *H. desertus* increased dramatically in response to water and nitrogen amendments. The response may have been the result of increased plant biomass, litter, or available prey (Lavigne and Campion, 1978), although microclimate changes also occurred in watered plots. Another carabid, *Pasimachus elongatus*, is a voracious predator capable of slicing large arthropods in half with its powerful jaws. McIntyre (1995) estimated densities of *P. elongatus* to be 1000 to 30,000·ha^{-1} in lowland areas with shrub cover and 30 to 200·ha^{-1} in upland grass sites. The high densities of *P. elongatus* and its ability to feed on a wide variety of insects suggest that it is also an important insect predator, especially in lowland swales.

Insects as Prey

The large abundances and body sizes of grasshoppers make them favored prey for numerous birds and mammals, including some of the most common vertebrates in the shortgrass steppe. Tenebrionid and carabid beetles are also large-bodied insects that form a major part of vertebrate diets. For example, Stapp (1996) found that Coleoptera (mostly scarabs, carabids, and tenebrionids) and Orthoptera (mostly grasshoppers and crickets) comprised 43% to 55% and 10% to 33%, respectively, of the seasonal diet of the grasshopper mouse (*Onychomys leucogaster*); the more omnivorous deer mouse (*Peromyscus maniculatus*) consumed 8% to 31% beetles and 0% to 18% grasshoppers and crickets. Likewise, grasshoppers and beetles comprise a significant fraction of the diets of insectivorous birds (Wiens and

Dyer, 1975). These insects are also important in the diets of rare vertebrates as well. For example, the threatened Mountain Plover (*Charadrius montanus*) consumes tenebrionid beetles as a large part of its diet (F. Knopf, May 2005).

Food Webs

From the feeding relationships described earlier, one can sketch the positions of several insects in aboveground food webs. The herbivores are dominated by grasshoppers and leafhoppers (Andrews, 1979; Lauenroth and Milchunas, 1992; Lavigne and Kumar, 1974). The important arthropods in the predator trophic levels are robber flies and spiders (especially lycosids [Lavigne and Kumar, 1974]). Carabid beetles, such as *P. elongatus* and *H. desertus*, are generalist predators and omnivores, and likely feed on different trophic levels; their roles have been largely ignored in grassland food webs despite their high abundances (Lavigne and Campion, 1978; McIntyre, 1995). Experimental evidence from mixed-grass rangelands indicate that significant top-down effects on insect herbivores are more likely to be from vertebrate rather than arthropod predators (Belovsky and Slade, 1993; Joern, 1986; but see Schmitz, 1993; Schmitz et al., 2000); the critical experiments have yet to be conducted in the shortgrass steppe. Similarly, rodent removal experiments in other semiarid ecosystems suggest that carabid and tenebrionid beetle populations are strongly influenced by rodent predation (Parmenter and MacMahon, 1988a, b). Some evidence suggests that predation by rodents can also influence the distribution and abundance of tenebrionid beetles in the shortgrass steppe (Stapp, 1997).

Links to belowground food webs are also evident among the herbivores, granivores, and detritivores. Harvester ants retrieve large quantities of litter and feces (Rogers, 1974) that decompose in nests and enter the belowground food web—a process that may be facilitated by late-season termite activity in ant nests (Crist and Friese, 1994). Tenebrionid beetles likewise form a connection to belowground food webs because they feed mostly on aboveground detritus that is processed and decomposed in soil (Lavigne et al., 1971). Belowground food webs also have important effects on soils and plant productivity (Moore et al., chapter 11, this volume) and, in turn, aboveground feeding interactions. Food web studies are needed on the multichannel feeding linkages that include aboveground herbivores and soil detritivores in grasslands and terrestrial ecosystems in general (Polis and Strong, 1996).

Ecosystem Processes

Insects are significant in the processing of energy, in soil modification, and in disturbance and patch dynamics in the shortgrass steppe (Lauenroth and Milchunas, 1992). These processes operate at different spatial and temporal scales, and are closely tied to population dynamics and community interactions.

Energy and Material Flows

Arthropods comprise less than 1% of the total heterotrophic biomass in the shortgrass steppe. Between 96% to 99% of the biomass in heterotrophs is estimated to occur in the belowground microflora (Lauenroth and Milchunas, 1992). Among arthropods, most biomass (84%) and energy flow (79%) occur below ground (Lauenroth and Milchunas, 1992). However, the fraction of the ANPP consumed by arthropods (17% with cattle present, 79% without cattle) is much larger than the fraction of NPP consumed by belowground arthropods (5%). Although most biomass and energy flow occur belowground, arthropods have a more important role as direct regulators of aboveground production (e.g., grasshoppers). Except for root feeders such as white grubs, arthropods primarily have an indirect effect on belowground production through the decomposition of organic matter. Therefore, from a whole-system view, aboveground arthropod energetics are primarily driven by biophagic feeding (consumption of live biomass) and belowground energetics are mostly saprophagic (consumption of dead biomass) in the shortgrass steppe (Lauenroth and Milchunas, 1992).

Beyond this generalization, there are three poorly known but potentially important aspects of the roles of insects in energy and nutrient flows. First, insect groups such as ground-dwelling beetles and ants have both above- and belowground activities that may accelerate the rate at which aboveground organic matter becomes incorporated into soil. Second, the redistribution and processing of materials by animals are spatially heterogeneous, which contributes to the patch dynamics of energy and nutrients within ecosystems (Whicker and Detling, 1988). Finally, the redistribution and processing of materials differ among insect groups, depending on their life history, vagility, and population size and distribution. Although it would be impractical to examine these differences on a species-by-species basis, one can focus on keystone species or representatives of functional groups (Lawton and Brown, 1994). Here I contrast the roles of insect species to illustrate how material redistribution and processing might vary among insect groups and across different spatial scales.

Fine-Scale Processes

Fine-scale patchiness is most obvious at the level of the individual plant. In the shortgrass steppe, *B. gracilis* accounts for around 90% of the total basal cover, which ranges from 20% to 40% (Milchunas et al., 1989). Plants have average basal areas of 394 cm^2 separated by bare ground (Aguilera and Lauenroth, 1993). Individual *B. gracilis* plants tend to form small hummocks surrounded by bare-ground depressions. This microtopography affects spatial patterns of organic matter and nutrient accumulation, which are concentrated in plant patches (Burke et al., chapter 13, this volume; Hook et al., 1991). Patches of cacti, shrubs, and forbs also form islands of nutrient and organic matter, but these patch types are more widely spaced among the dominant *B. gracilis*. Fine-scale patchiness of individual plants impinges on the spatial patterns of insect redistribution and processing of materials: Movement patterns of insects among plant patches affect redistribution, and residency of insects within patches affects the quantity of material processed.

The fine-scale patterns of insect movement and distribution described earlier are germane to these processes. For grasshoppers, patch residency is related to dietary preferences or thermoregulation (Anderson et al., 1979; Joern, 1987), and movements are related to body size or the spatial distribution of microhabitats (Joern, 1983b; With and Crist, 1995). Because grasshopper grazing can significantly reduce aboveground biomass, this reduction should vary spatially according to patterns of movement and microhabitat use. Patch-specific grasshopper feeding affects fine-scale material flows in three ways: assimilation of plant material that may leave the patch when grasshoppers move, unassimilated feces that are deposited in the feeding patch or in other patches, and uneaten clippings that remain within the patch. An estimated 75% to 85% of the plant biomass destroyed by grasshoppers goes to uneaten clippings and feces (Mitchell, 1975; Mitchell and Pfadt, 1974). Thus, grasshoppers are much more important in the processing of plants to litter than in energy storage and trophic conversion (Mitchell and Pfadt, 1974). Patterns of grasshopper movements and feeding among plant patches could substantially contribute to fine-scale heterogeneity in nutrients and organic matter.

Tenebrionid beetles also differ in their microhabitat use among plant cover types. Movement rates are greatest in bare ground and grass cover, and residence times are higher in cactus than in shrub patches (Crist and Wiens, 1995; Crist et al., 1992). Because tenebrionids feed primarily on plant litter and seeds (Allsopp, 1980), the net effect of tenebrionid beetles should be to redistribute litter (and seeds) from bare ground and grass to cactus and shrub patches. The rate of consumption and redistribution of litter have not been measured, but they could be significant because tenebrionid populations have a biomass that is comparable with grasshoppers (Lauenroth and Milchunas, 1992). Tenebrionids also spend a significant amount of time in burrows or soil crevices, so that some of the redistribution occurs belowground.

Dung beetles (Scarabaeidae) are another group that are important to fine-scale redistribution of nutrients, especially in grazed systems where fecal production by livestock is an important nutrient pathway (Lauenroth and Milchunas, 1992). The ball-rolling dung beetle, *Canthon pilularius*, is a seasonally common species in the shortgrass steppe that specializes on cattle feces. Mated pairs of beetles roll dung into large spheres and bury them several centimeters deep into the soil (Guertin, 1993). Their larvae then feed on these dung balls and pupate to the surface. Adult activity occurs primarily under warm, wet conditions in June and July (Guertin, 1993), so that belowground transport of feces by dung beetles likely occurs in distinct seasonal pulses.

Harvester ants retrieve seeds, litter, and insects from foraging areas and accumulate these organic materials in their nests. Some of these materials collected by ants may also be deposited in refuse piles on, or in areas away from, nests. Seeds are most likely to be removed from bare-ground areas (Crist and Wiens, 1994), but because ants are selective in both removing and discarding seeds, patterns of seed redistribution are dissimilar to that produced by wind transport (T. O. Crist, unpublished data). Harvester ants also clip aboveground vegetation immediately surrounding the nest to create a cleared disk. This clearing may result in a substantial number of plants removed from disks (Clark and Comanor, 1975), and the resulting litter becomes incorporated into soil within or near ant nests. Although we lack estimates on the amount of organic matter transferred into nests, the cumulative effects of concentrating organic matter into nests are clearly evident (see "Roles in Soils," this chapter).

Arid-land subterranean termites, *R. tibialis*, are another social insect that may have considerable importance in the belowground transfer of organic matter, especially woody litter. The considerable abundance of termites in areas with woody *Atriplex* shrubs (Fig. 10.5E) suggests that this species deserves further study regarding the importance of its role in the nutrient dynamics of shrub patches.

Broad-Scale Processes

Material flows at the level of the plant patch are closely linked to insect movement and microhabitat use. Variability in the flows of energy and nutrients across topographic gradients, soils, and plant assemblages should be more closely linked to the population distributions of insects. Insects often exhibit localized movements, so that long-distance transport of materials by insects is probably less important than patch-level movement. Heterogeneity in the effects of insects on energy and nutrient flows across landscapes should therefore stem primarily from variation in population abundance.

Some of the broad-scale variation in insect population abundance can be linked to topographic variation (Fig. 10.3). If ecosystem effects are related to insect abundance, then the detritus processing and movement by *E. extricata* should be highest on mid slopes. Similarly, the removal of seeds and processing of plant material by *P. occidentalis* is likely to be greatest in uplands (Crist and Wiens, 1996). In contrast, *F. obscuripes* uses woody debris in nest construction, preys extensively on arthropods, and tends aphids in shrubs; these patch-level processes should be more important in lowland areas where *F. obscuripes* is common. *Reticulitermes tibialis* is also more common in lowland areas, where patches of woody litter occur (Fig. 10.5).

Past studies of broad-scale patterns of grasshopper abundance have focused primarily on grazing rather than on topographic variation (Capinera and Sechrist, 1982; Welch et al., 1991). Spatial variation in grasshopper abundances may be considerable, however, and at least some of this variation is the result of vegetation (Przybyszewski and Capinera, 1990). This suggests that grasshopper densities might also vary substantially across a toposequence along with changes in vegetation (e.g., Milchunas et al., 1989), and these spatial distribution patterns could translate into topographic variation in the rates of herbivory and litter production in grassland ecosystems.

Roles in Soils

The effects of *Pogonomyrmex* ants on the chemical and physical properties of soils have been studied in the shortgrass steppe (Rogers and Lavigne, 1974) and in several other grassland ecosystems (reviewed by MacMahon et al., 2000). Substantial enrichment of soil nitrogen and phosphorus occurs in nest sites compared with surrounding areas (Mandel and Sorensen, 1982; Rogers and Lavigne, 1974; Whitford and DiMarco, 1995). In the shortgrass steppe, an estimated

2.8 kg soil·ha^{-1}·y^{-1} is moved from lower horizons and deposited on the surface (Rogers and Lavigne, 1974), as well as considerable redistribution and sorting of particle sizes (Mandel and Sorensen, 1982). Increased porosity and accumulation of organic matter result in a reduction in soil bulk density (Mandel and Sorensen, 1982; Rogers and Lavigne, 1974). Soil water may also be significantly higher in ant nests and disks compared with surrounding soils (Laundré, 1990).

Biotic enrichment of soils in ant nests occurs in response to increased levels of water, nutrients, and organic matter. In a shrub steppe ecosystem, Friese and Allen (1993) found roots within *P. occidentalis* nests to have a greater spore density of mycorrhizal fungi compared with surrounding soils. At the CPER, levels of mycorrhizae were consistently higher in ant nests compared with surrounding soils across topography and grazing (Snyder et al., 2002). Several other functional groups of microorganisms were enriched in ant nests compared with surrounding soils (Friese et al., 1997). Soils sampled from ant nests and off-mound areas in the shrub steppe and shortgrass steppe showed greater numbers of colony-forming units of microorganisms in ant nests compared with surrounding soils in both environments where *P. occidentalis* occurred (Fig. 10.9). The heightened activity of decomposer functional groups suggests that enriched levels of organic matter decompose more rapidly in nests than in surrounding soils. In contrast to nest mounds, the cleared disk becomes depleted of organic matter, nutrients, and microorganisms. Microbial biomass and mineralization of carbon and nitrogen

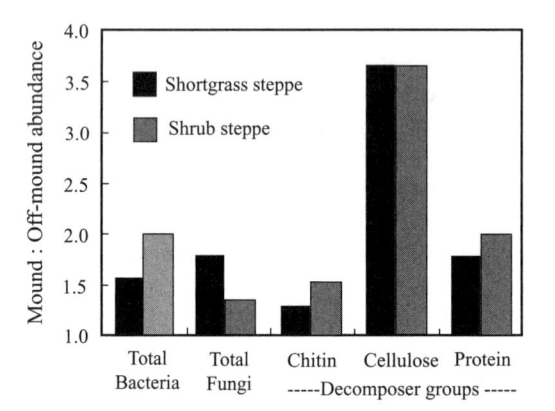

Figure 10.9 The relative increase in abundance (colony-forming units) of bacteria, fungi, and microbial functional groups in harvester ant mounds (*Pogonomyrmex occidentalis*) over nonmound areas in two rangeland ecosystems. The increase is expressed as the ratio of the means of mound and nonmound soil samples (three in each location). Off-mound samples were taken 3 m from nests in soils underneath the dominant plants at each site (blue grama [*B. gracilis*] in the shortgrass steppe and big sagebrush [*Artemisia tridentata*] in the shrub steppe).

were lower in soil from the cleared disk compared with soil under surrounding plants (Kelly et al., 1996).

Disturbance Agents

Harvester ants have long been recognized as disturbance agents in rangelands (reviewed by Rogers, 1987), but the various components of the ant disturbance regime—area, frequency, and turnover rates—are only more recently recognized (Coffin and Lauenroth, 1988). Clearings around ant nests, fecal pat deposition by cattle, and pocket gopher mounds are responsible for most of the animal-induced disturbances in blue grama grassland (Peters et al., chapter 6, this volume). An estimated 2.5 $m^2 \cdot ha^{-1} \cdot y^{-1}$ are cleared by ants in high-density areas (Coffin and Lauenroth, 1988, 1990). Rates of colony mortality and establishment can increase with ant density (Wiernasz and Cole, 1995), however, so that the frequency and turnover rate of disturbance may be greater in high-density areas (as in upland plains) than in low-density areas (as in lowlands). Vegetation surrounding ant clearings can have greater cover and altered species composition compared with unaffected areas (Coffin and Lauenroth, 1990; Nowak et al., 1990).

The patch dynamics of ant disturbances after nest abandonment are not well understood. Initially, after the cessation of aboveground clipping by ants, large numbers of annual plants are found, as well as scattered perennial grasses that resprout from roots (Coffin and Lauenroth, 1990). Plant colonization of abandoned ant mounds might differ from other more ephemeral animal disturbances such as pocket gophers (Carlson and Crist, 1999; Huntly and Inouye, 1988; Wu and Levin, 1994) because of long-term occupancy and alteration of nest sites (MacMahon et al., 2000; Whitford, 1997). Although increased water and nutrient availability may facilitate plant establishment, it is less clear how biotic enrichment affects the trajectory of patch dynamics. The greatly increased levels of mycorrhizal fungi, for example, could favor late-successional mycotrophic plants over ruderal nonmycotrophic species (Friese and Allen, 1993; Friese et al., 1997).

White grubs impose a somewhat different disturbance regime than harvester ants. Grub kills substantially reduce plant cover and create openings for new plant establishment (Coffin et al., 1998; Milchunas et al., 1990; Rottman and Capinera, 1983; Ueckert, 1979), but soils are not substantially modified as with harvester ants. The size of disturbance can be considerably larger (up to 1 ha or more) than ant clearings, but the frequency is generally lower and in different locations within the landscape (Peters et al., chapter 6, this volume). Ant disturbances are more common in upland soils, whereas grub densities and disturbances are greater near cacti or shrubs where soils are more penetrable (Wiener and Capinera, 1980). On a regional scale, grub kill areas may affect 100 to 1000 ha of grassland (Watts et al., 1982; Wiener and Capinera, 1980). The appearance of grub kill areas is often associated with drought years (Watts et al., 1982), but it is unclear whether grub densities increase in dry years or whether grub-induced plant mortality is greater with drought stress. The distribution and dynamics of *Phyllophaga* populations are not well understood, and their importance in grassland disturbance may be underestimated (Watts et al., 1982).

Linking Population, Community, and Ecosystem Processes

Insect Populations and Ecosystem Processes

I have focused on grasshoppers, beetles, and ants in the shortgrass steppe to illustrate how patterns of distribution and abundance have important ecosystem consequences. The flows of energy and nutrients among ecosystem compartments and the redistribution of materials by insects depend on their spatial distributions. Likewise, the frequency of disturbance and the roles in soils depend on insect density and distribution. I emphasized two spatial scales at which processes are likely to differ. First, the processing and transport of materials at the level of the plant patch is primarily a consequence of insect movements, patch residence times, and the dispersion of insects among patches. Second, the overall effects of a particular species or functional group on the flows of energy and nutrients vary with population abundance according to broad-scale effects of soils, topography, and grazing.

Conceptually, at least two scales of heterogeneity are therefore important in considering the effects of insect species on ecosystem processes. Differences in insect processing or transport of materials among patch types may reinforce or decrease the fine-scale heterogeneity in carbon and nitrogen (Hook et al., 1991), depending on the net effect produced by the movement and patch residency of the insect. The same can be said for the movements of propagules such as seeds or spores. At broader scales, population-level differences in rates of consumption, litter production, and decomposition by insects should be considered in concert with variation in soil nutrients (carbon, nitrogen, and phosphorus) and plant cover (or productivity), which generally increase from uplands to lowlands (Clark and Woodmansee, 1992; Milchunas et al., 1989; Schimel et al., 1985). Furthermore, the boundary contrast in plant cover, soil nutrients, and organic matter between plant patches and bare ground is greater in uplands than in lowlands, so that the collective effects of insects on patch-level heterogeneity will depend on topographic position.

Insect Species Diversity and Ecosystem Processes

What is the role of insect species diversity in ecosystem processes in the shortgrass steppe? Although a diverse insect fauna is described in the shortgrass steppe (Kumar et al., 1976), few studies of insect diversity have been conducted in relation to ecosystem function. Kirchner (1977) measured arthropod numbers, species diversity, and guild structure in the ecosystem stress experiment at the CPER. Plots with water or water-plus-nitrogen amendments showed significant increases in overall arthropod biomass and numbers. Arthropod diversity was also initially higher in these plots, but treatments showed highly variable patterns of diversity in subsequent years. There were significant positive relationships between the change in diversity of plants and arthropods across years. Diversity of feeding guilds increased in watered treatments compared with nitrogen-amended and control plots as a result of increased numbers of arthropod predators relative to herbivorous guilds. Kirchner (1977) attributed some of the variability in diversity to unmeasured structural changes in the community. Further efforts are needed to link insect diversity with functional attributes of rangeland ecosystems (Watts et al., 1982; West, 1994).

The relationship between species diversity and ecosystem function is a topic of ongoing interest and importance (Jones and Lawton, 1995; Lambers et al., 2004; Naeem and Li, 1997; Tilman et al., 1996). Some investigators focus on keystone species or ecosystem engineers that have effects in ecosystems disproportionately greater than that suggested by their biomass or abundance; others examine how species redundancy within functional groups affects ecosystem processes. The relative value of these and other species-based approaches to analyzing ecosystem function is likely to vary among systems or the particular function under investigation. In the shortgrass steppe, for example, we might view white grubs or harvester ants as important disturbance agents that have no functional equivalent (cf. MacMahon et al., 2000; Whitford, 1997). In contrast, ant species with smaller and less conspicuous nests have a greater role in scavenging and nutrient redistribution (Bestelmeyer and Wiens, 2003). Grasshopper effects on primary production or plant community structure might be viewed as functionally redundant, where several species can have similar feeding preferences (e.g., within a subfamily) and where some species deletion might not influence overall rates of herbivory. Similarly, tenebrionid beetles are primarily detritivores, and several abundant species can have substitutable roles in the processing of plant litter.

Spatial Scaling of Insect Abundance

Ecologists are increasingly concerned with questions of how small-scale process-oriented studies can be extrapolated to broad-scale landscape patterns, or the degree to which broad-scale processes constrain those operating at fine scales (Wiens, 1989). Linear extrapolation across scales is often conducted without reference to the form of variability (King et al., 1991) or without the recognition that different processes operate at different scales (i.e., processes are scale dependent [Wiens, 1989]). A variety of approaches are used to measure scale dependence in patterns and processes (Milne, 1991; Turner et al., 1991). One approach to measuring scale-dependent patterns includes the analysis of scaling exponents derived from log–log transformations (Johnson et al., 1992a; Milne, 1991; Schneider, 1994).

Here I provide an example of scaling exponents using colony densities of harvester ants, which differ in their spatial distribution across spatial scales. Semivariance measures the variation among points separated by different distances (lag length). A log–log regression of semivariance against lag length can be used to detect scales of spatial dependence. A linear relationship (on a log–log scale) in spatial dependence often occurs over a limited range of scales. As the lag length increases, however, the slope of this relationship may change as processes operating at broader scales produce a different form of spatial dependence. The fractal dimension, D, of the semivariance relationship is derived from the slope, m, of the regression line by $D = (4-m)/2$, and can be used to index shifts in

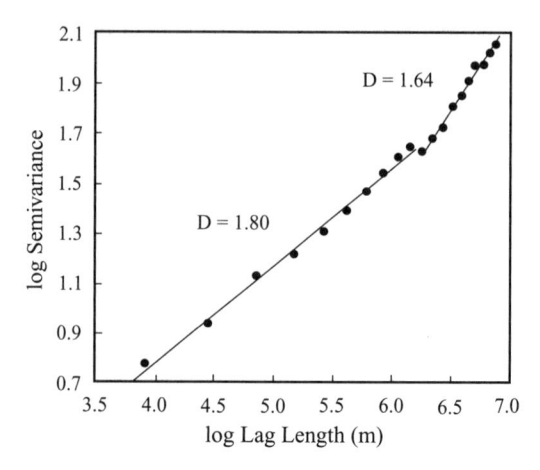

Figure 10.10 Semivariance in colony densities of harvester ants obtained from grid counts of the mapped area in Figure 10.4B. The log–log relationship between semivariance and lag length indicates two distinct regions of variability in colony density, which were determined by piecewise linear regression. The fractal dimension, D, is determined from the slope, m, of the regression line by $D = (4 - m)/2$.

spatial patterning across scales (Milne, 1991). If semivariance is constant across lag length (random noise), then $m = 0$ and $D = 2$.

The semivariance of colony densities of harvester ants (Fig. 10.4B) was calculated by overlaying a 16 × 32 grid of 50 × 50-m (0.25-ha) cells (total, 128 ha). A log–log plot of semivariance with lag length shows two distinct regions that were determined by piecewise linear regression (Fig. 10.10). The first region, from 50 to 450 m, has a fractal dimension of $D = 1.80$, which indicates a more homogenous distribution of colony density at scales less than 450 m. The second region, from 450 to 900 m, shows greater heterogeneity in colony density with a shift to $D = 1.64$. Broad-scale patterns of ant distribution can therefore be described by two domains of scale: one that is controlled by local variation within a topographic position and another that is controlled by broad-scale changes in topography (Fig. 10.10). A more detailed spatial analysis suggests that ant distributions are influenced by colony establishment and interactions at a local scale, and changes in soils and grazing regime at broader scales (Crist and Wiens, 1996).

Conclusions and Future Directions

A great deal is known about the common insects in the shortgrass steppe. A large number of species have been recorded, the major contributors to overall biomass quantified, and the population dynamics and distribution of some species have been measured. A considerable amount is also known about the impacts of cattle grazing on insect communities. However, there are still major gaps in our understanding of insects in shortgrass steppe ecosystems. We lack information on the long-term dynamics of important insect species. Few data are available on broad-scale spatial distributions of insects. There is only sparse information on the structure of aboveground food webs. There is also very little known about the relationships between insect diversity and ecosystem functioning. Nonetheless, the knowledge base of past studies will make possible future advances in these areas. Because of the important relationships among soils, nutrient dynamics, and plant community structure, future linkages of insect population distribution with ecosystem processes would be facilitated if measurements are taken along topographic gradients.

Studies of insects in the shortgrass steppe can serve as a model for understanding animal roles in other terrestrial ecosystems. First, heterotrophs have important roles in grasslands because of the large fraction of consumable biomass and relatively rapid turnover of organic matter (Lauenroth and Milchunas, 1992). Second, most aboveground insects can be readily observed and measured in the low, open plant canopy in the shortgrass steppe. Lastly, their tremendous abundance and diversity make them good candidates to examine the various relationships among population, community, and ecosystem processes.

Acknowledgments My thinking about insects in the shortgrass steppe has been influenced by discussions with John Wiens, Carl Friese, Kim With, Paul Stapp, Nancy McIntyre, and Brandon Bestelmeyer. John Capinera, Mike Vanni, Deb Peters, Indy Burke, and Becky Riggle provided several valuable suggestions on earlier versions of the manuscript. I gratefully acknowledge research support from the NSF (BSR-8805829 and DEB-9207010 to J. A. Wiens) and the USDA (National Research Initiative competitive grant 95031420 to T. O. Crist and C. F. Friese).

References

Aguilera, M. O., and W. K. Lauenroth. 1993. Neighborhood interactions in a natural population of the perennial bunchgrass, *Bouteloua gracilis*. *Oecologia* **94**:595–602.

Allsopp, P. G. 1980. The biology of false wireworms and their adults (soil-inhabiting Tenebrionidae) (Coleoptera): A review. *Bulletin of Entomological Research* **70**: 343–379.

Anderson, R. V., C. R. Tracy, and Z. Abramsky. 1979. Habitat selection in two species of short horned grasshoppers: The role of thermal and hydric stress. *Oecologia* **38**:359–374.

Andrews, R. M. 1979. *A shortgrass prairie macroarthropod community: Organization, energetics and effects of grazing by cattle.* U.S. IBP Grassland Biome technical report no. 208. Colorado State University, Fort Collins, Colo.

Barnes, P. W., and A. T. Harrison. 1982. Species distribution and community organization in a Nebraska sandhills mixed prairie as influenced by plant/soil–water relationships. *Oecologia* **52**:192–201.

Belovsky, G. E., and A. Joern. 1995. The dominance of different regulating factors for rangeland grasshoppers, pp. 359–386. In: N. Cappuccino and P. W. Price (eds.), *Population dynamics: New approaches and synthesis.* Academic Press, San Diego, Calif.

Belovsky, G. E., and J. B. Slade. 1993. The role of vertebrate and invertebrate predators in a grasshopper community. *Oikos* **68**:193–201.

Belovsky, G. E., and J. B. Slade. 1995. Dynamics of two Montana grasshopper populations: Relationships among weather, food abundance, and intraspecific competition. *Oecologia* **101**:383–396.

Bestelmeyer, B. T., and J. A. Wiens. 2001. Ant biodiversity in semiarid landscape mosaics: The consequences of grazing vs. natural heterogeneity. *Ecological Applications* **11**:1123–1140.

Bestelmeyer, B. T., and J. A. Wiens. 2003. Scavenging ant foraging behavior and variation in the scale and nutrient redistribution among semiarid grasslands. *Journal of Arid Environments* **53**:373–386.

Bossenbroek, J. M., H. H. Wagner, and J. A. Wiens. 2004. Taxon-dependent scaling: Beetles, birds, and vegetation at four North American grassland sites. *Landscape Ecology* **20**:675–688.

Capinera, J. L. 1987. Population ecology of rangeland grasshoppers, pp. 162–182. In: J. L. Capinera (ed.), *Integrated pest management on rangeland*. Westview Press, Boulder, Colo.

Capinera, J. L., and D. R. Horton. 1989. Geographic variation in effects of weather on grasshopper infestation. *Environmental Entomology* **18**:8–14.

Capinera, J. L., and T. S. Sechrist. 1982. Grasshopper (Acrididae)–host plant associations: Response of grasshopper populations to cattle grazing intensity. *Canadian Entomologist* **114**:1055–1062.

Capinera, J. L., and D. C. Thompson. 1987. Dynamics and structure of grasshopper assemblages in shortgrass prairie. *Canadian Entomologist* **119**:567–575.

Cappuccino, N., and P. W. Price (eds.). 1995. *Population dynamics: New approaches and synthesis*. Academic Press, San Diego, Calif.

Carlson, J. M., and T. O. Crist. 1999. Plant responses to pocket-gopher disturbances across pastures and topography. *Journal of Range Management* **52**:637–645.

Clark, W. H., and P. L. Comanor. 1975. Removal of annual plants from the desert ecosystem by Western harvester ants, *Pogonomyrmex occidentalis*. *Environmental Entomology* **4**:52–56.

Clark, F. E., and R. G. Woodmansee. 1992. Nutrient cycling, pp. 137–146. In: R. T. Coupland (ed.), *Natural grasslands*. Vol. 8A, *Ecosystems of the world*. Elsevier Science, Amsterdam.

Coffin, D. P., and W. K. Lauenroth. 1988. The effects of disturbance size and frequency on a shortgrass plant community. *Ecology* **69**:1609–1617.

Coffin, D. P., and W. K. Lauenroth. 1990. Vegetation associated with nest sites of the Western harvester ant (*Pogonomyrmex occidentalis*) in a semiarid grassland. *American Journal of Botany* **76**:53–58.

Coffin, D. P., W. A. Laycock, and W. K. Lauenroth. 1998. Disturbance intensity and above- and belowground herbivory effects on long term (14y) recovery of a semiarid grassland. *Plant Ecology* **139**:221–233.

Crist, T. O. 1998. The spatial distribution of subterranean termites in shortgrass steppe: A geostatistical approach. *Oecologia* **114**:410–416.

Crist, T. O., and C. F. Friese. 1994. The use of ant nests by subterranean termites in two semiarid ecosystems. *American Midland Naturalist* **131**:370–373.

Crist, T. O., D. S. Guertin, J. A. Wiens, and B. T. Milne. 1992. Animal movement in heterogeneous landscapes: An experiment with *Eleodes* beetles in shortgrass prairie. *Functional Ecology* **6**:536–544.

Crist, T. O., and J. A. MacMahon. 1992. Harvester ant foraging and shrub-steppe seeds: Interactions of seed resources and seed use. *Ecology* **73**:1768–1779.

Crist, T. O., and J. A. Wiens. 1994. Scale effects of vegetation on forager movement and seed harvesting by ants. *Oikos* **69**:37–46.

Crist, T. O., and J. A. Wiens. 1995. Individual movements and estimation of population size in darkling beetles (Coleoptera: Tenebrionidae). *Journal of Animal Ecology* **64**:733–746.

Crist, T. O., and J. A. Wiens. 1996. The distribution of ant colonies in a semiarid landscape: Implications for community and ecosystem processes. *Oikos* **76**:301–311.

Dyer, M. I., and U. G. Bokhari. 1976. Plant–animal interactions: Studies of the effects of grasshopper grazing on blue grama grass. *Ecology* **57**:762–772.

Friese, C. F., and M. F. Allen. 1993. The interaction of harvester ants and vesicular–arbuscular mycorrhizal fungi in a patchy semiarid environment: The effects of mound structure on fungal dispersion and establishment. *Functional Ecology* **7**:13–20.

Friese, C. F., S. J. Morris, and M. F. Allen. 1997. Disturbance in natural ecosystems: Scaling from fungal diversity to ecosystem functioning, pp. 47–65. In: D. T. Wicklow and B. Söderstrom (eds.), *The Mycota IV. Environmental and microbial relationships*. Springer-Verlag, Berlin.

Guertin, D. S. 1993. *Trade-offs between feeding and reproduction in a ball-rolling dung beetle,* Canthon pilularius *(L.).* PhD diss., Colorado State University, Fort Collins, Colo.

Hart, R. H. 2001. Plant biodiversity on shortgrass steppe after 55 years of grazing of zero, light, moderate, or heavy cattle grazing. *Plant Ecology* **155**:111–118.

Hewitt, G. B., and J. A. Onsager. 1983. Control of grasshoppers on rangeland in the United States: A perspective. *Journal of Range Management* **36**:202–207.

Hoffman, A. L., and J. A. Wiens. 2004. Scaling of the tenebrionid beetle community and its environment on the Colorado shortgrass steppe. *Ecology* **85**:629–636.

Hook, P. B., I. C. Burke, and W. K. Lauenroth. 1991. Heterogeneity of soil and plant N and C associated with individual plants and openings in North American shortgrass steppe. *Plant and Soil* **138**:247–256.

Huntly, N., and R. Inouye. 1988. Pocket gophers in ecosystems: Patterns and mechanisms. *BioScience* **38**:786–793.

Joern, A. 1982. Vegetation structure and microhabitat selection in grasshoppers (Orthoptera: Acrididae). *Southwestern Naturalist* **27**:197–209.

Joern, A. 1983a. Host plant utilization by grasshoppers (Orthoptera: Acrididae) from a sandhills prairie. *Journal of Range Management* **36**:793–797.

Joern, A. 1983b. Small-scale displacements of grasshoppers (Orthoptera: Acrididae) within arid grasslands. *Journal of the Kansas Entomological Society* **56**:131–139.

Joern, A. 1986. Experimental study of avian predation on coexisting grasshopper populations (Orthoptera: Acrididae) in a sandhills grassland. *Oikos* **46**:243–249.

Joern, A. 1987. Behavioral responses underlying ecological patterns of resource use in rangeland grasshoppers, pp. 137–161. In: J. L. Capinera (ed.), *Integrated pest management on rangeland*. Westview Press, Boulder, Colo.

Johnson, A. R., B. T. Milne, and J. A. Wiens. 1992a. Diffusion in fractal landscapes: Simulations and experimental studies of tenebrionid beetle movements. *Ecology* **73**:1968–1983.

Johnson, A. R., B. T. Milne, J. A. Wiens, and T. O. Crist. 1992b. Animal movements and population dynamics in heterogeneous landscapes. *Landscape Ecology* **7**:63–75.

Jones, C. G., and J. H. Lawton (eds.). 1995. *Linking species and ecosystems*. Chapman and Hall, New York.

Keeler, K. H. 1993. Fifteen years of colony dynamics in *Pogonomyrmex occidentalis*, the Western harvester ant, in western Nebraska. *Southwestern Naturalist* **38**:286–289.

Kelly, R. H., I. C. Burke, and W. L. Lauenroth. 1996. Soil organic matter and nutrient availability responses to reduced plant inputs in shortgrass steppe. *Ecology* **77**:2516–2527.

Kemp, W. P. 1992. Rangeland grasshopper (Orthoptera: Acrididae) community structure: A working hypothesis. *Environmental Entomology* **21**:461–470.

Kemp, W. P., and B. Dennis. 1993. Density dependence in rangeland grasshoppers (Orthoptera: Acrididae). *Oecologia* **96**:1–8.

Kemp, W. P., S. J. Harvey, and K. M. O'Neill. 1990. Patterns of vegetation and grasshopper community composition. *Oecologia* **83**:299–308.

King, A. W., A. R. Johnson, and R. V. O'Neill. 1991. Transmutation and functional representation of heterogeneous landscapes. *Landscape Ecology* **5**:239–253.

Kirchner, T. B. 1977. The effects of resource enrichment on the diversity of plants and arthropods in a shortgrass prairie. *Ecology* **58**:1334–1344.

Kumar, R., R. J. Lavigne, J. E. Lloyd, and R. E. Pfadt. 1976. *Insects of the Central Plains Experiment Range, Pawnee National Grassland*. Agricultural Experiment Station, Laramie, Wyo.

Lambers, J. H. R., W. S. Harpole, D. Tilman, J. Knops, and P. B. Reich. 2004. Mechanisms responsible for positive diversity–productivity relationships in Minnesota grasslands. *Ecology Letters* **7**:661–668.

Lauenroth, W. L., and D. G. Milchunas. 1992. Shortgrass steppe, pp. 183–226. In: R. T. Coupland (ed.), *Natural grasslands*. Vol. 8A, *Ecosystems of the world*. Elsevier Science, Amsterdam.

Laundré, J. W. 1990. Soil moisture patterns below mounds of harvester ants. *Journal of Range Management* **43**:10–12.

Lavigne, R. J., and M. K. Campion. 1978. The effect of ecosystem stress on the abundance and biomass of carabidae (Coleoptera) on the shortgrass prairie. *Environmental Entomology* **7**:88–92.

Lavigne, R. J., and R. Kumar. 1974. *Population densities and biomass of above-ground arthropods subjected to environmental stress treatments on the Pawnee site, 1973*. U.S. IBP Grassland Biome, technical report no. 268. Colorado State University, Fort Collins, Colo.

Lavigne, R. J., L. E. Rogers, and J. Chu. 1971. *Data collected on the Pawnee site relating to Western harvester ant and insect predators and parasites, 1970*. U.S. IBP Grassland Biome, technical report no. 107. Colorado State University, Fort Collins, Colo.

Lawton, J. H., and V. K. Brown. 1994. Redundancy in ecosystems, pp. 255–270. In: E. D. Schulze and H. A. Mooney (eds.), *Biodiversity and ecosystem function*. Springer-Verlag, New York.

Lockwood, J. A., and D. R. Lockwood. 1991. Rangeland grasshopper (Orthoptera: Acrididae) population dynamics: Insights from catastrophe theory. *Environmental Entomology* **20**:970–980.

MacMahon, J. A., J. F. Mull, and T. O. Crist. 2000. Harvester ants (*Pogonomyrmex* spp.): Their community and ecosystem influences. *Annual Review of Ecology and Systematics* **31**:265–291.

Mandel, R. D., and C. J. Sorensen. 1982. The role of the Western harvester ant (*Pogonomyrmex occidentalis*) in soil formation. *Soil Science Society of America Journal* **46**:785–788.

McIntyre, N. E. 1995. Methamidophos application effects on *Pasimachus elongatus* (Coleoptera: Carabidae): An update. *Environmental Entomology* **24**:559–563.

McIntyre, N. E. 1997. Scale-dependent habitat selection by the darkling beetle *Eleodes hispilabris* (Coleoptera: Tenebrionidae). *American Midland Naturalist* **138**:230–235.

McIntyre, N. E. 2000. Community structure of *Eleodes* beetles (Coleoptera: Tenebrionidae) in the shortgrass steppe: Scale-dependent uses of heterogeneity. *Western North American Naturalist* **60**:1–15.

Milchunas, D. G., W. K. Lauenroth, P. L. Chapman, and M. K. Kazempour. 1989. Plant communities in relation to grazing, topography, and precipitation in a semiarid grassland. *Vegetatio* **80**:11–23.

Milchunas, D. G., W. K. Lauenroth, P. L. Chapman, and M. K. Kazempour. 1990. Community attributes along a perturbation gradient in shortgrass steppe. *Journal of Vegetation Science* **1**:375–384.

Milchunas, D. G., O. E. Sala, and W. K. Lauenroth. 1988. A generalized model of the effects of grazing by large herbivores on grassland community structure. *American Naturalist* **132**:87–106.

Miller, R. H., and J. A. Onsager. 1991. Grasshopper (Orthoptera: Acrididae) and plant relationships under different grazing intensities. *Environmental Entomology* **20**:807–814.

Milne, B. T. 1991. Lessons from applying fractal models to landscape patterns, pp. 199–235. In: M. G. Turner and R. H. Gardner (eds.), *Quantitative methods in landscape ecology*. Springer-Verlag, New York.

Mitchell, J. E. 1975. Variation in food preferences of three grasshopper species (Acrididae: Orthoptera) as a function of food availability. *American Midland Naturalist* **94**:267–283.

Mitchell, J. E., and R. E. Pfadt. 1974. A role of grasshoppers in a shortgrass prairie ecosystem. *Environmental Entomology* **3**:358–360.

Mulkern, G. B. 1967. Food selection by grasshoppers. *Annual Review of Entomology* **12**:59–78.

Mull, J. F., and J. A. MacMahon. 1996. Factors determining the spatial variability of seed densities in a shrub-steppe ecosystem: The role of harvester ants. *Journal of Arid Environments* **32**:181–192.

Naeem, S., and S. Li. 1997. Biodiversity enhances ecosystem reliability. *Nature* **390**:507–509.

Nowak, R. S., C. L. Nowak, T. DeRocher, T. Cole, and M. A. Jones. 1990. Prevalence of *Oryzopsis hymenoides* near harvester ant mounds: Indirect facilitation by ants. *Oikos* **58**:190–198.

Onsager, J. A. 1987. Integrated management of rangeland grasshoppers, pp. 196–204. In: J. L. Capinera (ed.), *Integrated pest management on rangeland*. Westview Press, Boulder, Colo.

Parmenter, R. R., and J. A. MacMahon. 1988a. Factors influencing species composition and population sizes in a ground beetle community (Carabidae): Predation by rodents. *Oikos* **52**:350–356.

Parmenter, R. R., and J. A. MacMahon. 1988b. Factors limiting populations of arid-land darkling beetles (Coleoptera: Tenebrionidae): Predation by rodents. *Environmental Entomology* **17**:280–286.

Pfadt, R. E., and D. M. Hardy. 1987. A historical look at rangeland grasshoppers and the value of grasshopper control programs, pp. 183–185. In: J. L. Capinera (ed.), *Integrated pest management on rangeland*. Westview Press, Boulder, Colo.

Polis, G. A., and D. R. Strong. 1996. Food web complexity and community dynamics. *American Naturalist* **147**:813–846.

Porter, S. D., and C. D. Jorgensen. 1988. Longevity of harvester ant colonies in southern Idaho. *Journal of Range Management* **41**:104–107.

Przybyszewski, J., and J. L. Capinera. 1990. Spatial and temporal patterns of grasshopper (Orthoptera: Acrididae) phenology and abundance in shortgrass prairie. *Journal of the Kansas Entomological Society* **63**:405–413.

Rodell, C. F. 1977. A grasshopper model for a grassland ecosystem. *Ecology* **58**:227–245.

Rogers, L. E. 1974. Foraging activity of the Western harvester ant in the shortgrass plains ecosystem. *Environmental Entomology* **3**:420–424.

Rogers, L. E. 1987. Ecology and management of harvester ants in the shortgrass plains, pp. 261–267. In: J. L. Capinera (ed.), *Integrated pest management on rangeland*. Westview Press, Boulder, Colo.

Rogers, L. E., and R. J. Lavigne. 1974. Environmental effects of Western harvester ants on the shortgrass plains ecosystem. *Environmental Entomology* **3**:994–997.

Rottman, R. J., and J. L. Capinera. 1983. Effects of insect and cattle-induced perturbations on shortgrass prairie arthropod community. *Journal of the Kansas Entomological Society* **56**:241–252.

Schimel, D. S., M. A. Stillwell, and R. G. Woodmansee. 1985. Biogeochemistry of C, N, and P in a soil catena of the shortgrass steppe. *Ecology* **66**:276–282.

Schmitz, O. J. 1993. Trophic exploitation in grassland food chains: Simple models and a field experiment. *Oecologia* **93**:327–335.

Schmitz, O. J., P. A. Hamback, and A. P. Beckerman. 2000. Trophic cascades in terrestrial systems: A review of the effects of carnivore removals on plants. *American Naturalist* **155**:141–153.

Schneider, D. C. 1994. *Quantitative ecology*. Academic Press, San Diego, Calif.

Snyder, S. R., T. O. Crist, and C. F. Friese. 2002. Variability in soil chemistry and arbuscular mycorrhizal fungi in harvester ant nests: The influence of topography, grazing and region. *Biology and Fertility of Soils* **35**:406–413.

Stapp, P. T. 1996. *Determinants of habitat use and community structure of rodents in northern shortgrass steppe*. PhD diss., Colorado State University, Fort Collins, Colo.

Stapp, P. T. 1997. Microhabitat use and community structure of darkling beetles (Coleoptera: Tenebrionidae) in shortgrass prairie: Effects of season, shrub cover, and soil type. *American Midland Naturalist* **137**:298–311.

Tilman, D., D. Wedin, and J. Knops. 1996. Productivity and sustainability influenced by biodiversity in grassland ecosystems. *Nature* **379**:718–720.

Torell, L. A., and E. W. Huddleston. 1987. Factors affecting the economic threshold for control of rangeland grasshoppers, pp. 377–396. In: J. L. Capinera (ed.), *Integrated pest management on rangeland*. Westview Press, Boulder, Colo.

Tscharntke, T., and H. J. Greiler. 1995. Insect communities, grasses, and grasslands. *Annual Review of Entomology* **40**:535–558.

Turchin, P. 1998. *Quantitative analysis of movement*. Sinauer Associates, Sunderland, Mass.

Turchin, P., F. J. Odendaal, and M. D. Rausher. 1991. Quantifying insect movement in the field. *Environmental Entomology* **20**:955–963.

Turner, S. J., R. V. O'Neill, W. Conley, M. R. Conley, and H. C. Humphries. 1991. Pattern and scale: Statistics for landscape ecology, pp. 17–49. In: M. G. Turner and R. H. Gardner (eds.), *Quantitative methods in landscape ecology*. Springer-Verlag, New York.

Ueckert, D. N. 1979. Impact of a white grub (*Phyllophaga crinata*) on a shortgrass community and evaluation of selected rehabilitation practices. *Journal of Range Management* **32**:445–448.

Uvarov, B. 1977. *Grasshoppers and locusts: A handbook of general acridoidology.* Vol. 2. Centre for Overseas Pest Research, London, UK.

Van Horn, D. 1970. *Grasshopper population numbers and biomass dynamics on the Pawnee site from fall 1968 through 1970.* U.S. IBP Grassland Biome, technical report no. 148. Colorado State University, Fort Collins, Colo.

Watts, J. G., E. Huddleston, and J. C. Owens. 1982. Rangeland entomology. *Annual Review of Entomology* **27**:283–311.

Weeks, Jr., R. D. 1996. *Habitat associations of ground-dwelling spiders (Arachnida, Araneae) on shortgrass steppe in northeastern Colorado: Implications for community organization and distribution.* Masters thesis, Colorado State University, Fort Collins, Colo.

Weeks, Jr., R. D., and T. O. Holtzer. 2000. Habitat and season in structuring ground-dwelling spider communities (Araneae) in a shortgrass steppe ecosystem. *Environmental Entomology* **29**:1164–1172.

Welch, J. L., R. Redak, and B. C. Kondratieff. 1991. Effect of cattle grazing on the density and species of grasshoppers (Orthoptera: Acrididae) of the Central Plains Experimental Range, Colorado: A reassessment after two decades. *Journal of the Kansas Entomological Society* **64**:337–343.

West, N. E. 1994. Biodiversity of rangelands. *Journal of Range Management* **46**: 2–13.

Whicker, A. D., and J. K. Detling. 1988. Ecological consequences of prairie dog disturbances. *BioScience* **38**:778–785.

Whitford, W. G. 1997. The importance of biodiversity of soil biota in arid ecosystems. *Biodiversity and Conservation* **5**:185–195.

Whitford, W. G., and R. DiMarco. 1995. Variability in soils and vegetation associated with harvester ant (*Pogonomyrmex rugosus*) nests on a Chihuahuan Desert watershed. *Biology and Fertility of Soils* **20**:169–173.

Wiener, L. F., and J. L. Capinera. 1980. Preliminary study of the biology of the white grub *Phyllophaga fimbripes* (LeConte) (Coleoptera: Scarabaeidae). *Journal of the Kansas Entomological Society* **53**:701–710.

Wiens, J. A. 1989. Spatial scaling in ecology. *Functional Ecology* **3**:385–397.

Wiens, J. A., T. O. Crist, and B. T. Milne. 1993a. On quantifying insect movements. *Environmental Entomology* **22**:709–715.

Wiens, J. A., N. C. Stenseth, B. Van Horne, and R. A. Ims. 1993b. Ecological mechanisms and landscape ecology. *Oikos* **66**:369–380.

Wiens, J. A., and M. I. Dyer. 1975. Rangeland avifaunas: Their composition, energetics, and role in the ecosystem, pp. 146–181. In: D. R. Smith (tech. coord.), *Management of forest and range habitats for nongame birds.* USDA Forest Service general technical report WO-1.

Wiernasz, D. C., and B. J. Cole. 1995. Spatial distribution of *Pogonomyrmex occidentalis*: Recruitment, mortality, and overdispersion. *Journal of Animal Ecology* **64**:519–527.

With, K. A., and T. O. Crist. 1995. Critical thresholds in species' responses to landscape structure. *Ecology* **76**:2446–2459.

Wu, J., and S. A. Levin. 1994. A spatial dynamic modeling approach to pattern and process in an annual grassland. *Ecological Monographs* **64**:447–464.

11

Trophic Structure and Nutrient Dynamics of the Belowground Food Web within the Rhizosphere of the Shortgrass Steppe

John C. Moore
Jill Sipes
Amanda A. Whittemore-Olson
H. William Hunt
Diana H. Wall
Peter C. de Ruiter
David C. Coleman

Belowground organisms are key components of the trophic structure and they mediate the dynamics of nutrients of all terrestrial ecosystems. The interactions among assemblages of belowground microorganisms and their consumers mediate the cycling of plant-limiting nutrients, influence aboveground plant productivity, affect the course of plant community development, and affect the dynamic stability of aboveground communities following natural and anthropogenic disturbances (Clarholm, 1985; Ingham et al., 1985; Laakso and Setälä, 1999; Naeem et al., 1994; Tilman et al., 1996; Wall and Moore, 1999). The influence of belowground organisms on the aboveground plant community is heightened in systems such as the shortgrass steppe (Blair et al., 2000), given the relatively high percentage of plant production that is diverted belowground through plant roots.

Many of the human-induced changes that the shortgrass steppe has been subjected to during the past 150 years fall outside the scope of the natural variations in climate and grazing. This conflict between the natural history of the shortgrass steppe and the more recent human legacy forms the backdrop of this chapter. First we present a detailed description of the belowground food web for the native shortgrass steppe and present its structure in terms of the patterns of trophic interactions, the distribution of biomass, the flow of energy, and the strengths of interactions. Second, we explore how three disturbances—managed grazing, agricultural practices, and climate change (altered precipitation and temperature, and elevated CO_2)—have altered the structure of the belowground community. We conclude with a synthesis of the common patterns that we observed in the grassland's response to these disturbances, and speculate on their consequences.

The Belowground Food Web

Aboveground plant parts provide from 20% to 40% cover with exposed soil between them (Lauenroth and Milchunas, 1991). Much of the aboveground production remains in place as standing dead, rather than falling to the soil surface as litter. The ratio of shoot production to root production is roughly 1:1, contrasting sharply with forests, where far more production is allocated aboveground (Jackson et al., 1996; Milchunas and Lauenroth, 1993, 2000). Hence, in the shortgrass steppe, plant roots provide the major input of carbon to soil. As such, plant roots are the focal point of biological activity in soils (Coleman et al., 1983).

The belowground food web of the shortgrass steppe consists of bacteria, fungi, protozoa, nematodes, arthropods, annelids, small mammals, and their interactions. Numerous studies of the soil biota of the shortgrass steppe have been conducted, dating back to the U.S. IBP of the late 1960s to early 1970s. During the IBP era, much of the work on soil biota cataloged the species diversity of various sites and began the process of determining population densities under native and manipulated conditions such as cattle grazing.

The IBP studies of soil biota led to a body of work that linked various attributes of the soil (texture and structure), grass roots, nitrogen availability, and the feeding relationships among the taxa (Kumar et al., 1975; Coleman, 1976; Elliott and Coleman, 1977). These studies not only developed and refined the concepts of the rhizosphere, and the importance of soil biota to nutrient cycling and plant growth, but revealed three points that would change how we viewed communities and ecosystems. First, the studies used biomass estimates rather than densities, allowing for a common currency of carbon, nitrogen, or both. Second, the great diversity of soil biota made it necessary to develop a standard grouping scheme for taxa that shared similar attributes. Last, inorganic nitrogen was viewed as an integral component of the system, allowing it to be manipulated, modeled, and interpreted in the same vein as the soil biota.

A Model Based on the Rhizosphere

The food web diagrams and models we present are pictorial and mathematical representations of a conceptual model of the rhizosphere. Research of the shortgrass steppe led to the development of a widely used conceptual representation of the rhizosphere that emphasizes interaction (Coleman et al., 1983; Elliott, 1978; Moore et al., 2003; Trofymow and Coleman, 1982; Wall and Moore, 1999). From this perspective, the rhizosphere is defined as the narrow zone of soil that is influenced by plant roots and their products, and the trophic interactions that are affected by these products (Coleman et al., 1983; Moore et al., 2003; Trofymow and Coleman, 1982).

Trofymow and Coleman (1982) proposed that a growing root could be viewed as a continuum of overlapping zones of activity, from the root tip to the crown, where different microbial populations have access to a continuous flow of organic substrates derived from the root. The first zone is the root tip. The tip is the site of root growth characterized by rapidly dividing cells of the rootcap and a low carbon-to-nitrogen ratio, or labile secretions or exudates that lubricate the tip as it passes through the soil. The exudates and sloughed root cells provide carbon for bacteria and fungi, which in turn immobilize nitrogen and phosphorus. Farther up the root is the region of nutrient exchange. This region is characterized by a higher carbon-to-nitrogen ratio or resistant products such as root hairs, and to a lesser extent the labile products of exudation. The birth and death of root hairs stimulate additional microbial growth (Bringhurst et al., 2001). The upper zones of the rhizosphere include regions of remineralization of nutrients by predators, symbiotic–mutualistic relations (mycorrhizae), and structure (Coleman et al., 1983; Moorman and Reeves, 1979; Wall and Moore, 1999).

We present the connectedness, energy (nutrient) flow, and functional categories of descriptions that were proposed by Paine (1980) to describe the belowground food web of the shortgrass steppe. The connectedness food web describes the feeding relationships among groups. The energy flux food web describes the distribution of energy (usually in terms of biomass carbon or nitrogen) within and among the groups. The functional web describes the impacts that one group has upon the dynamics of the other groups.

The Connectedness Food Web Based on Functional Groups

The connectedness description of the belowground food web of the shortgrass steppe is based on functional groups rather than species (Fig. 11.1) (Hunt et al., 1987; Moore et al., 1988). We forego a discussion of the major taxa, as they have been discussed at length in several venues (Moore et al., 1988; Wallwork, 1970; Walter and Proctor, 1999); however, representative genera are presented in Table 11.1. Functional groups are collections of species that are similar in terms of the following: (1) food sources, (2) feeding modes, (3) life history, and (4) habitat use. These criteria draw from niche theory (food, habitat, and time axes [sensu Schoener, 1974]) and from the dimensions of the differential equations that define the dynamics of the populations (mass, area, and time). The physiological attributes of the functional groups and the densities of each functional group are used to estimate feeding rates and interaction strengths, which are defined as the effects of one group's dynamics on the dynamics of the other groups (Table 11.2).

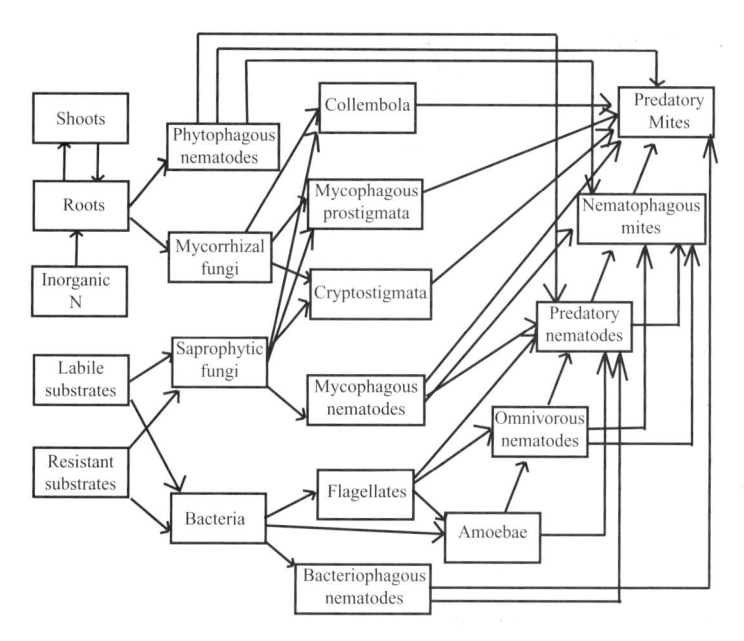

Figure 11.1 The connectedness description of the belowground food web of the shortgrass steppe based on functional groupings of soil biota, plants, detritus, and soil nitrogen (Hunt et al., 1987; Moore et al., 1988).

The Energy Flow Food Web

We have developed an energy flow description of the belowground food web, indexed by carbon flow (Fig. 11.2). The nutrient fluxes within the food web were estimated from the feeding rates between functional groups, the egestion rates of unassimilated consumption (feces, orts, and leavings), and the mineralization rates of carbon and nitrogen (metabolic release of CO_2 and NH_4^+/NO_3^-). We present these formulations using units of carbon.

Feeding rates are derived from population sizes and data on death rates and energy conversion efficiencies. The basic assumption underlying the calculation of feeding rates is that the annual (equilibrium) feeding rates should balance the annual death rate through natural death and predation (Hunt et al., 1987; O'Neill, 1969):

$$F_j = \frac{d_j B_j + D_j}{a_j P_j} \tag{11.1}$$

where F_j is the feeding rate (measured in kilograms carbon per hectare per year), d_j is the specific death rate per year, B_j is the average annual (equilibrium)

Table 11.1 Examples of Genera within the Functional Groups of the Shortgrass Steppe Belowground Food Web

Functional Group	Description	Examples (Genera)
Predatory mites	Attack most soil invertebrates small enough to overcome	Hypoaspis, Asca, Amblyseius, Rhodacarus, Gamasellodes, Macrocheles, Spinibdella, Cyta, Stigmaeus, Cocorhagidia
Nematophagous mites	Attack only nematodes	Alliphis, Eviphis, Alycus, Alicorhagia, Ololaelaps, Veigaia
Predatory nematodes	Attack nematodes and bacteria (minimal)	Discolaimium, Mononchus
Omnivorous nematodes	Consume bacteria and protozoa	Mesodiplogaster
Fungivorous nematodes	Feed on fungal cytoplasm	Aphelenchus, Aphelenchoides
Bacteriophagous nematodes	Consume bacteria	Acrobeloides, Pelodera, Rhabditis
Collembola	Consume fungal hyphae and spores, algae, pollen	Folsomia, Isotoma, Isotomides, Hypogastura, Tullbergia, Deuterosminthurus, Sminthurus
Mycophagous prostigmata	Pierce fungal hyphae and consume fungal cytoplasm	Tydeus, Eupodes, Tarsonemus, Bakerdania, Pediculaster, Scutacarus, Speleorchestes
Cryptostigmata	Consume fungal hyphae and spores	Haplozetes, Passalozetes, Zygoribatula, Pilogalumna, Tectocepheus, Oppiella, Ceratozetes
Amoebae	Consume bacteria	Acanthamoeba, Hartmanella, Tricamoeba, Mayorella, Varella
Flagellates	Consume bacteria	Pleuromonas, Bodo, Mastigamoeba
Phytophagous nematodes	Feed on plant roots	Helicotylenchus, Tylenchorhynchus, Xiphinema

From Kumar et al. (1975), Smolik and Dodd (1983), Hunt et al. (1987), and Moore et al. (1988).

population size (measured in kilograms carbon per hectare), D_j is the death rate resulting from predation (measured in kilograms carbon per hectare per year), a_j is the assimilation efficiency, and p_j is the production efficiency. For polyphagous predators, the feeding rate per prey type (F_{ij}) is based on the relative abundances of the prey types and on prey preference:

$$F_{ij} = \frac{w_{ij} B_i}{\sum_{k=1}^{n} w_{kj} B_k} F_j \qquad (11.2)$$

where F_{ij} is the feeding rate by predator j on prey i, and w_{ij} is the preference of predator j for prey i over its other prey types. The calculations of feeding rates

Table 11.2 Estimates of the Constants Needed to Calculate Carbon and Nitrogen Fluxes, F_i (Eq. 11.1), Egestion Rates, E_i (Eq. 11.3) and Rates of Mineralization, M_i (Eq. 11.4) for Each Functional Group Presented in Figure 11.1

Functional Group	C:N	Turnover Rate per y	Assimilation Efficiency, %	Production Efficiency, %	Biomass, kg C·ha⁻¹
Predatory mites	8	1.84	60	35	0.160
Nematophagous mites	8	1.84	90	35	0.160
Predatory nematodes	10	1.60	50	37	1.080
Omnivorous nematodes	10	4.36	60	37	0.650
Fungivorous nematodes	10	1.92	38	37	0.410
Bacteriophagous nematodes	10	2.68	60	37	5.800
Collembola	8	1.84	50	35	0.464
Mycophagous prostigmata	8	1.84	50	35	1.360
Cryptostigmata	8	1.20	50	35	1.680
Amoebae	7	6.00	95	40	3.780
Flagellates	7	6.00	95	40	0.160
Phytophagous nematodes	10	1.08	25	37	2.900
AM-mycorrhizal fungi	10	1.20	100	30	7.000
Saprobic fungi	10	2.00	100	30	63.000
Bacteria	4	1.20	100	30	304.000
Detritus	10	0.00	100	100	3000.000
Roots	10	1.00	100	100	300.000

AM, arbuscular mycorrhizae. From Hint et al., 1987.

started with the top predators, which suffer only from natural death, and proceeded working backward to the lowest trophic levels.

The egestion rate (measured in kilograms carbon per hectare per year) is defined as the amount of the prey that is killed that is not assimilated by the predator per time step:

$$E_j = (1 - a_j)F_j \tag{11.3}$$

where E_j (measured in kilograms carbon per hectare per year) is the egestion rate by predator j of unassimilated prey and a_j is the assimilation efficiency of predator j.

The mineralization rate (measured in kilograms carbon per hectare per year) is the amount of the assimilated prey that is released in an inorganic form resulting from maintenance or access per time step:

$$M_j = (1 - p_j)a_j F_j \tag{11.4}$$

where M_j is the mineralization rate (measured in kilograms carbon per hectare per year) by predator j and p_j is the production efficiency of predator j.

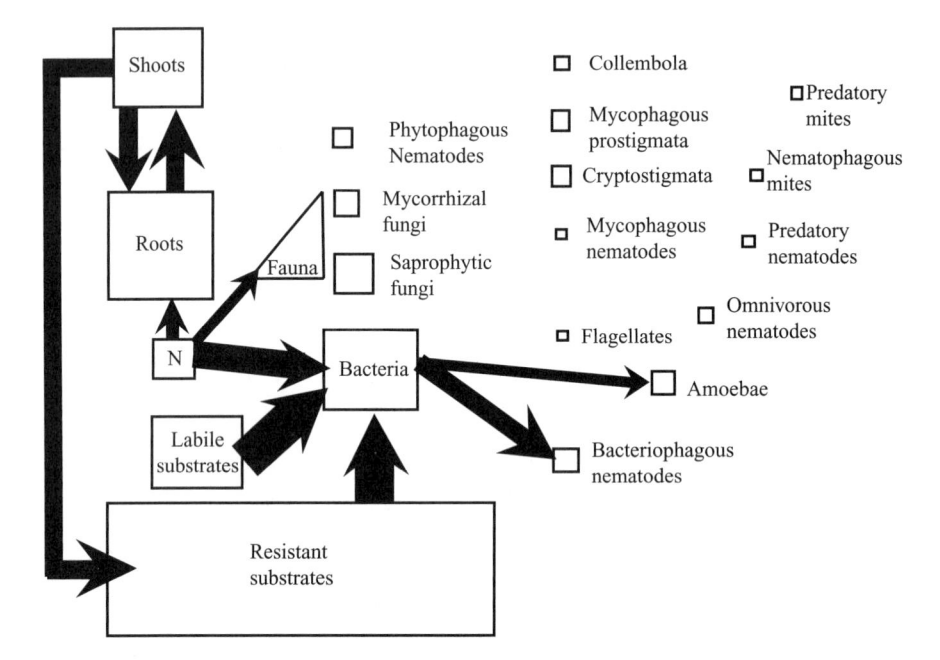

Figure 11.2 The energy flow description of the belowground food web presented in Figure 11.1. The boxes and arrows represent the relative biomasses of the functional groups and the estimates of the flow of energy (carbon) among functional groups (Eqs. 11.1 and 11.2, respectively). The flow to fauna represents all remaining flows not present in the diagram. (Adapted from Moore et al. [1988].)

The energy flux description reveals that the shortgrass steppe has a pyramidal structure that is compartmentalized into a root, bacterial, and fungal energy channel (Table 11.3) (Moore and Hunt, 1988). The pyramidal structure repeats itself within each energy channel. The majority of energy passes through the bacterial energy channel.

The Functional Food Web

We define the functional food web (Fig. 11.3) in terms of the interaction strengths. Interactions strengths are the elements of the community or Jacobian matrices (May, 1972, 1973), and represent the *per capita*—in this case, *per biomass*— effects of one group upon another at equilibrium. We estimated the interaction strengths from the population sizes and energy flow rates using Lotka-Volterra equations for the dynamics of the functional groups (de Ruiter et al., 1994; Moore et al., 1993):

$$\frac{dX_i}{dt} = X_i \left[b_i + \sum_{j=1}^{n} c_{ij} X_j \right] \qquad (11.5)$$

Table 11.3 Depiction of Energy Channels as Defined by the Proportion of Energy (Index by Carbon) Potentially Derived from Bacteria, Fungi and/or Plant Roots by the Different Functional Groups in the Belowground Food Web of the Shortgrass Steppe (adapted from Moore and Hunt [1988])

Functional Group	Bacteria	Fungi	Root
Protozoa			
Flagellates	100	0	0
Amoebae	100	0	0
Ciliates	100	0	0
Nematodes			
Phytophagous nematodes	0	0	100
Mycophagous nematodes	0	90	10
Omnivorous nematodes	100	0	0
Bacteriophagous nematodes	100	0	0
Predatory nematodes	68.67	3.50	27.83
Microarthropods			
Collembola	0	90	10
Cryptostigmata	0	90	10
Mycophagous prostigmata	0	90	10
Nematophagous mites	66.70	3.78	29.52
Predatory mites	39.54	38.56	21.91

Estimates were based on the carbon fluxes obtained using Eqs. 11.1 and 11.2.

where X_i and X_j represent the population sizes of group i and j, respectively; b_i is the specific rate of increase or decrease of group I; and c_{ij} is the coefficient of interaction between group i and group j. The matrix elements (α_{ij}) are defined as the partial derivatives near equilibrium: $\alpha_{ij} = (\partial \frac{dX_i}{dt} / \partial X_j)^*$. Values for the interaction strengths are derived from the equilibrium descriptions by equating the death rate of group i resulting from predation by group j in equilibrium, $c_{ij} X_i^* X_j^*$, to the average annual feeding rate, F_{ij} (Eq. 11.2) and the production rate of group j resulting from feeding on group i, $c_{ji} X_j^* X_i^*$, to $a_j P_j F_{ij}$. We assume that the long-term seasonal average population sizes of functional groups (B_i) approximate the theoretical steady-state densities (X_i^*). Hence, the effect of predator j on prey i is

$$\alpha_{ij} = c_{ij} X_j^* = -\frac{F_{ij}}{B_j} \tag{11.6}$$

and the effect of prey i on predator j is

$$\alpha_{ij} = c_{ij} X_j^* = -\frac{a_j \, p_j \, F_{ij}}{B_j} \tag{11.7}$$

(A)	(B)	
	Interaction Strength (yr^{-1})	
	-20 -10 0 0.1 0.2	
Resource	*Top Predators*	*Consumer*
Nematophagous Mites		Predaceous mites
Predaceous Nematodes		Predaceous mites
Predaceous Nematodes		Nematophagous mites
Omnivorous Nematodes		Predaceous mites
Ominivorous Nematodes		Nematophagous mites
Omnivorous Nematodes		Predaceous nematodes
Amoebae		Predaceous nematodes
Amoebae		Omnivorous nematodes
Bacteriophagous Nematodes		Predaceous mites
Fungivorous Nematodes		Predaceous mites
Mycophagous Prostigmata		Predaceous mites
Cryptostigmata		Predaceous mites
Collembola		Predaceous mites
Bacteriophagous Nematodes		Nematophagous mites
Fungivorous Nematodes		Nematophagous mites
Bacteriophagous Nematodes		Predaceous nematodes
Fungivorous Nematodes		Predaceous nematodes
Flagellates		Predaceous nematodes
Flagellates		Omnivorous nematodes
Flagellates		Amoebae
Phytophagous Nematodes		Predaceous mites
Phytophagous Nematodes		Nematophagous mites
Phytophagous Nematodes		Predaceous nematodes
Bacteria		Predaceous nematodes
Bacteria		Omnivorous nematodes
Bacteria		Amoebae
Bacteria		Bacteriophagous nematode
Bacteria		Flagellates
Fungi		Fungivorous nematodes
Mycorrhizal Fungi		Fungivorous nematodes
Fungi		Mycophagous prostigmata
Mycorrhizal Fungi		Mycophagous prostigmata
Fungi		Cryptostigmata
Mycorrhizal Fungi		Cryptostigmata
Fungi		Collembola
Mycorrhizal Fungi		Collembola
Detritus		Bacteria
Detritus		Fungi
Roots		Mycorrhizal fungi
Roots		Phytophagous nematodes
	Basal Resources	

10^{-1} 10^{1} 10^{3} 10^{5}

Feeding Rate
(kg ha-1 yr-1)

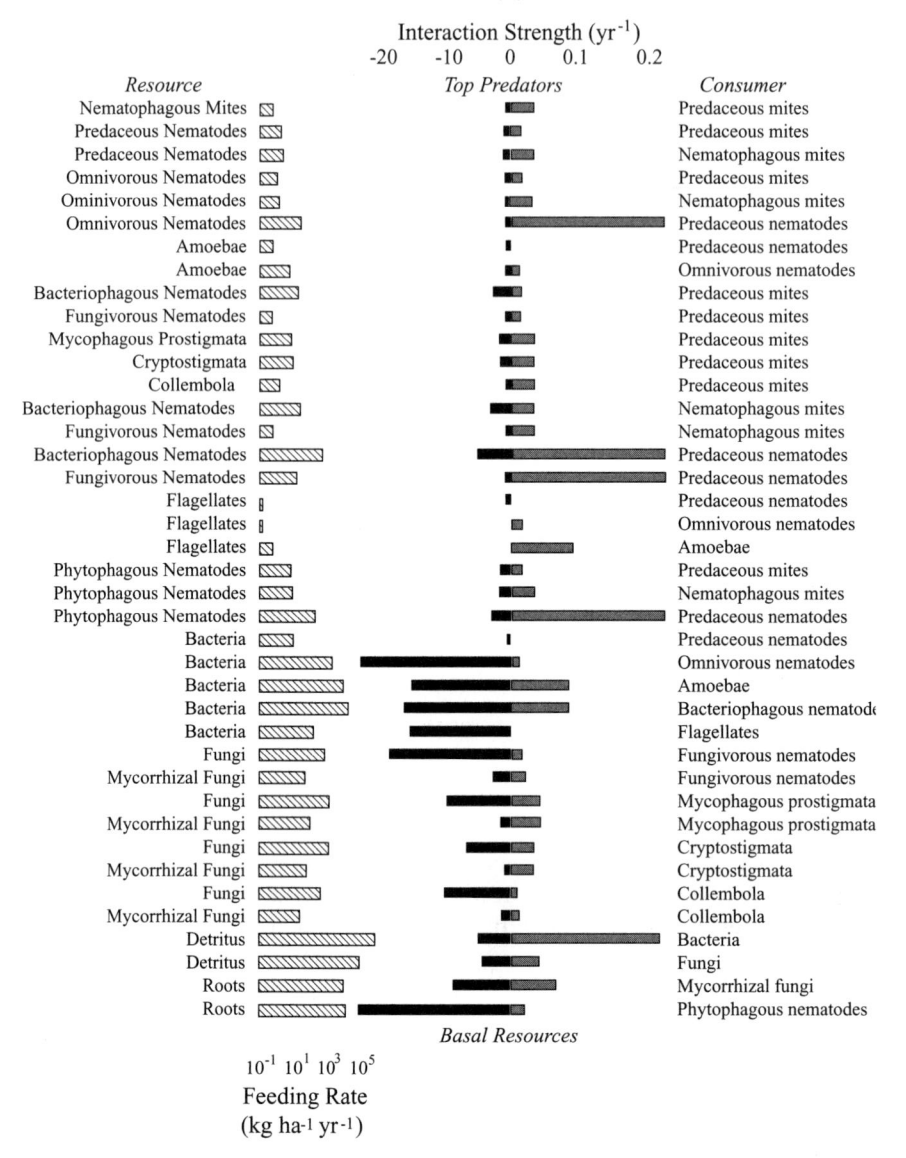

Figure 11.3 (A, B) The functional description of the belowground food web for the short-grass steppe, Nunn, Colorado (de Ruiter et al., 1995; Moore et al., 1986). The bars correspond to the levels of feeding rates and interaction strengths for pairwise trophic interactions in a food web (see Fig. 11.1) arranged from resources (A) to consumers (B), starting with interactions at the base of the food web (basal resources) to the top of the food web (top predators). The estimates for the feeding rates were obtained using Eqs. 11.1 and 11.2. The pairwise interaction strengths represent the elements of the Jacobian matrix for the system of equations (Eq. 11.5) describing the trophic interactions among functional groups. The elements include the impact of a consumer on a resource (predator on prey), and the impact of the resource on the consumer (prey on predator), obtained from Eqs. 11.6 and 11.7, respectively. The asymmetry in the feeding rates (measured in kilograms nitrogen per

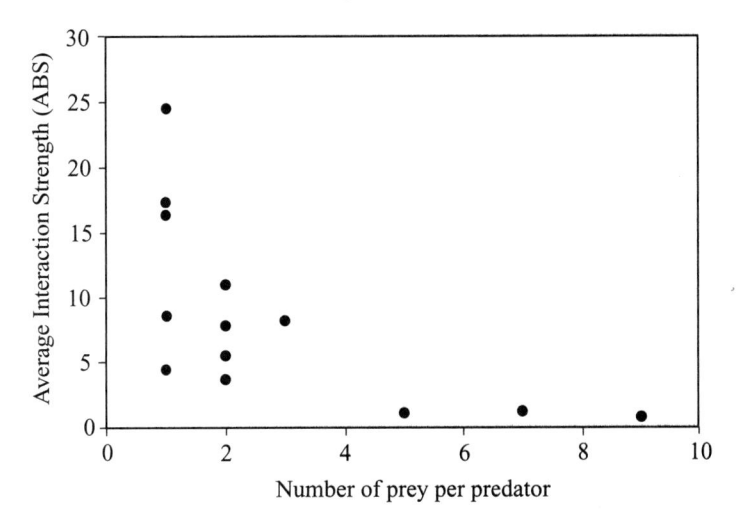

Figure 11.4 The relationship between the effect of a predator on prey, y, presented as the average of the absolute values of the estimates obtained from Eq. 11.6, and the number of prey per predator, x, for fauna within the belowground food web ($y = 14.1e^{-0.347x}$, $r^2 = .679$).

From the eigenvalues of the Jacobian matrix we can determine the dynamic stability of the community, and hence relate dynamic stability to the dynamics of nutrients.

Stable Patterns of Interaction

In the functional food web, we arranged the estimates of carbon flow and interaction strengths among the functional groups by trophic position (Fig. 11.3). Three sets of observations are evident. The first set deals with the distribution of biomass and feeding rates among functional groups with trophic position. The belowground food web of the shortgrass steppe forms a pyramid of biomass, with the majority of biomass positioned at the base of the food web and smaller amounts at the upper levels (Table 11.2). The feeding rates follow a similar pattern as the majority of nitrogen flows through the interactions at the base of the food web and declines with increased trophic position (Fig. 11.3).

The second set of observations deals with the strengths of the trophic interactions. For the shortgrass steppe, the average strength of a predator's impact on its prey follows a pattern that is important to stability (Fig. 11.4). May (1972) hypothesized that the average interaction strength between predator and prey

---

hectare per year) (A) and patterning of interaction strengths per year (B) within the Jacobian matrix with increased trophic level (bottom to top) were observed for food webs from the shortgrass steppe in Colorado (presented here), and agricultural sites in The Netherlands, Sweden, and Georgia, USA (de Ruiter et al., 1995; Moore et al., 1996).

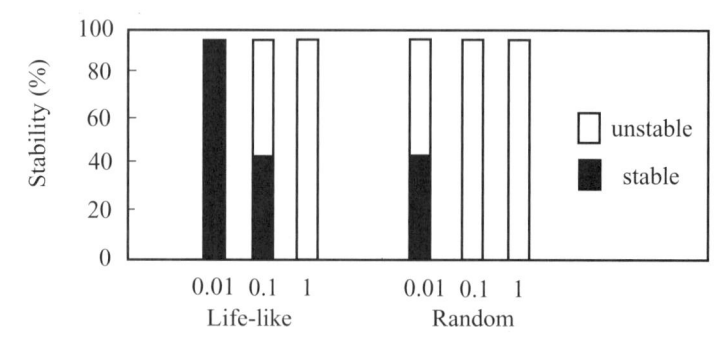

Figure 11.5 Monte Carlo trials comparing the lifelike food web of the shortgrass steppe with disturbed counterparts reveal that the asymmetrical patterning of interaction strength within the Jacobian matrix of the food web confers stability (de Ruiter et al., 1995). The black fraction in the bars denotes the percentage of stable matrices based on 1000 runs. For the lifelike matrices, the element values were sampled randomly from the uniform distributions $[0, 2\alpha_{ij}]$, in which α_{ij} is the element of the Jacobian matrix as derived from observations using Eq. 11.6 and Eq. 11.7. In the disturbed matrices, the values of the matrix elements were obtained by randomly permuting the nonzero pairs of elements after Yodzis (1981). The diagonal matrix elements referring to intragroup interference (α_{ii}) were set proportional at three levels of magnitude—0.01, 0.1, 1.0—to the specific death rates for that group.

should decrease as the number of prey species increased. Predators with few prey species would interact more strongly with each species than predators with many prey items.

The third observation is that the two patterns discussed earlier—the patterning of nutrient flows and the patterning of interaction strengths with trophic position—are evident within the root, bacterial, and fungal energy channels (Fig. 11.5). Taken together, these observations lend strong support to the idea that the shortgrass steppe food web is compartmentalized, and that the compartmentalized structure is important to the stability of the ecosystem (May, 1972; Moore and de Ruiter, 1997; Moore and Hunt, 1988).

Conceptual Model of Community Structure, Nutrient Dynamics, and Stability

These studies lead to the general conclusion that the energetic organization of communities forms the basis of ecosystem stability. We suggest that the changes in the trophic structure and material flow through the food webs brought about by disturbances should alter the distribution of biomass, the pattern of nutrient flow, and ultimately the patterning of interaction strength (Fig. 11.6). These changes in turn induce instabilities within the community much like those observed in the

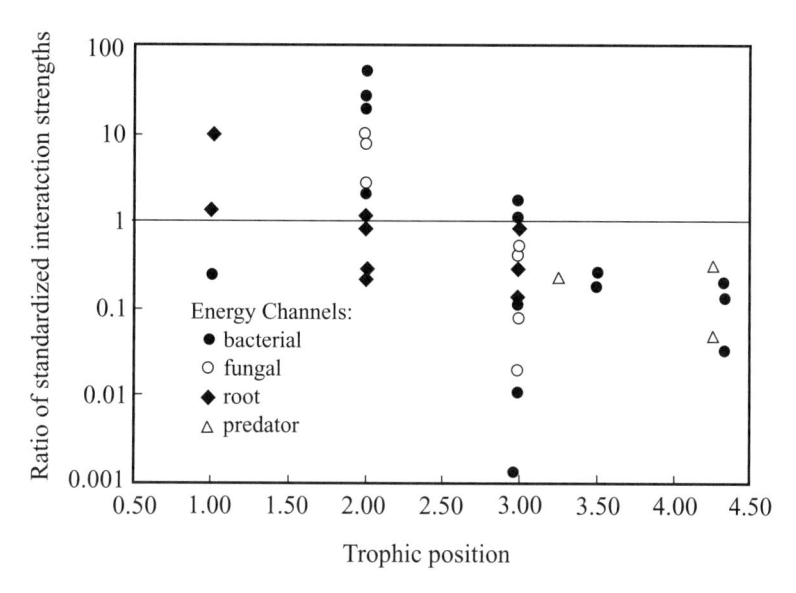

Figure 11.6 The asymmetry in the interaction strengths with increased trophic position for the belowground food web of the shortgrass steppe that was observed in Figure 11.3 can also be found within each energy channel (bacterial, fungal, and root). Each symbol represents the ratio of the pairwise interaction strengths between a consumer and a resource. Prior to taking the ratio, the interaction strengths were standardized by dividing each α_{ij} and α_{ji} by the average interaction strengths for all α_{ij} and α_{ji}, respectively. Because predators obtain a significant portion of their energy from more than one energy channel (Table 11.3), they were considered separately.

random webs (Fig. 11.5), or push the community away from its predisturbance steady state. A product of these periods of instability in an otherwise stable system (disturbance and recovery) are changes in material flow through the food web, and changes in the relationship between mineralization and immobilization of nitrogen. To illustrate these points we present the results of studies on the effects of three disturbances on the belowground food web: grazing by livestock, winter wheat agriculture, and elevated temperature and CO_2.

Impact of Grazing

The plant communities of the shortgrass steppe evolved along with large grazing ungulates that included bison and pronghorn antelope, small mammals, arthropods, and nematodes (Milchunas et al., 1988; Milchunas et al., chapter 16, this volume). Although there has been some debate regarding whether grazing should be referred to as a disturbance (Milchunas and Lauenroth, 1993), we consider grazing to be a frequent and natural disturbance on the shortgrass steppe.

The USDA Forest Service initiated a series of long-term studies in 1939 whereby exclosures were erected at the CPER to isolate areas from grazing by cattle. In 1991, the fences of several of the 1939 exclosures were moved to expose a portion of the exclosure to cattle. A new exclosure was built to prevent cattle grazing of an area that had been grazed each year since 1939. This new arrangement of fencing created four grazing treatments that combined historical treatments with new ones (historical/new): ungrazed, ungrazed/grazed, grazed, and grazed/ungrazed. Beginning in 1993, soil bacteria, saprophytic fungi, arbuscular mycorrhizal fungi, and protozoa were sampled at the end of July of each growing season through 1998. Nematodes were collected from 1992 through 1994.

Long-term grazing had mixed effects on soil bacteria, saprobic fungi, protozoa, and nematodes. Early during the study, bacteria densities were higher in the historically nongrazed plots (ungrazed and ungrazed/grazed) than in the grazed plots (grazed and grazed/ungrazed). However, by 1994, no significant differences could be detected between long-term grazed and ungrazed plots. Long-term grazing had no significant effect on the densities of individuals of nematodes, propagules of fungi, or infection potential of arbuscular mycorrhizal fungi. Significant year-to-year variation in densities and higher densities under plants than between plants were observed.

When viewed in a multivariate context, a different pattern emerged (Fig. 11.7). Wall-Freckman and Huang (1998) found that grazed plots (grazed and grazed/ungrazed) had

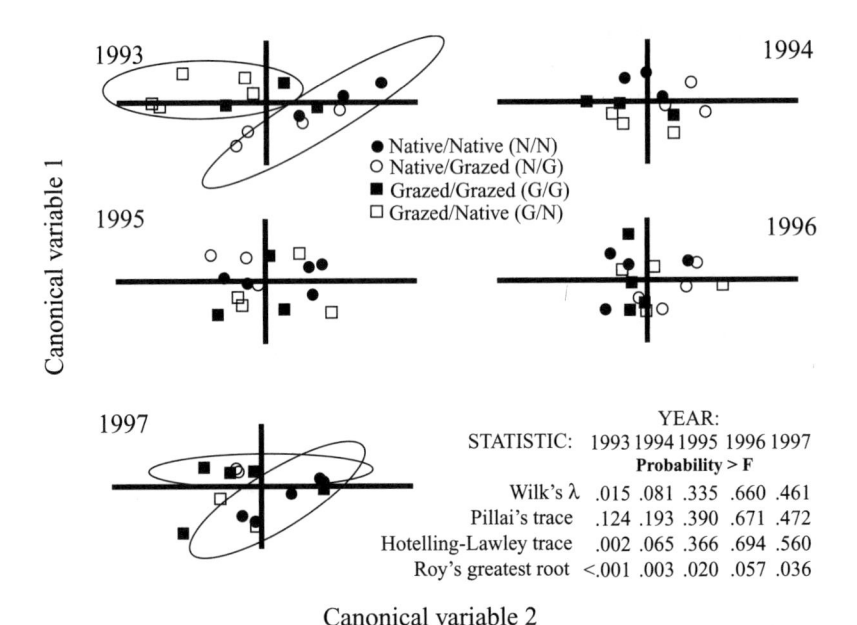

Figure 11.7 Canonical discriminant analysis of the long-term study on grazing at the shortgrass steppe comparing plots that had been under long-term grazing (G/G and G/N) and from those that had not been grazed (N/N and N/G). The analysis used bacteria (measured in colony-forming units per gram of soil) and fungi (measured in colony-forming units per gram of soil), and the densities of protozoa (measured in numbers per gram of soil) as input variables. G, grazed; N, native.

a higher proportion of fungal-feeding nematodes. Canonical discriminant analysis of the treatments, using the colony-forming units of bacteria and fungi, and the densities of protozoa as input variables, revealed that the plots that had been under long-term grazing (grazed and grazed/ungrazed) differed significantly from those that had not been grazed (ungrazed and ungrazed/grazed). During five growing seasons, the plots that had been exposed to grazing (ungrazed/grazed) shifted to resemble those that had been under long-term grazing (grazed), whereas those that had not been removed from long-term grazing (grazed/ungrazed) shifted to resemble those that been excluded from grazing (ungrazed).

Impacts of Agricultural Practices on Shortgrass Steppe Food Webs

The soils of the shortgrass steppe were first tilled at the end of the 19th century (Hart, chapter 4, this volume). Since that time, there has been a steady decline in the levels of soil organic carbon (SOC) and nitrogen in tilled soils; some estimates place losses at nearly 50% of the original soil carbon within the plow layer (Aguilar et al., 1988; Burke et al., 1989; Haas et al., 1957). These losses have been attributed to the adverse effects of tillage. Tillage alters the physical and chemical nature of soil in two ways. First, the O and A horizons contain the majority of soil organic material. These horizons are homogenized with tillage. Organic material that had been at the soil surface is buried, leaving the soil surface prone to wind and water erosion. Second, soils are a matrix of aggregates of sand, silt, and clay particles that are held together by partially decomposed organic material. Under natural conditions, soil aggregates are quite stable, with the organic material being shielded from further decomposition for decades to centuries. Tillage disrupts soil aggregates, thereby releasing the stored organic material at an accelerated rate (Elliott, 1986). This material is mineralized by microbes and eventually lost.

Studies conducted at Akron, Colorado, on winter wheat established clear connections among management practices, the densities and dynamics of soil organisms, and nutrient retention (Elliott et al., 1984; Moore, 1986). Dryland winter wheat management includes a crop-and-fallow rotation. The fallow rotation is needed for water storage in areas of low precipitation and no irrigation. Much of the tillage and other forms of management happen during the fallow period. The studies at Akron compared the stubble mulch and no-till management practices. Stubble mulch uses mechanical means (rod-weeder, sweep or tandem disks) to control weeds and to incorporate a portion of the crop residue into the soil. No-till uses herbicides to control weeds. Both practices use tillage for planting.

Tillage significantly altered the densities, dynamics, and distributions of soil biota (Elliott et al., 1984; Holland and Coleman, 1987; Moore, 1986). No-till management supports significantly higher levels of microbial and faunal biomass and biotic activity, and retains more nitrogen in the fallow rotation than stubble mulch (Tables 11.4 and 11.5). These results revealed that the loss of nitrogen and soil organic matter was as much a function of the timing of the levels of activity or biomass, as of the activity or biomass alone (Moore, 1986). Within the no-till management practice, the bacterial energy channel was more active early during

Table 11.4 Microbial Biomass Carbon, Mineralizable Carbon, and mineralizable nitrogen in the Top 10 cm of No-Till and Stubble Mulch Fallow Winter Plots at Akron, Colorado, for Five Dates in 1982 (from Elliott et al. [1984])

Variable	Treatment	8 Jun	6 Jul	2 Aug	23 Aug	13 Sep	ANOVA[a]
Microbial biomass, kg C·ha^{-1}	Stubble Mulch	245	204	271	186	255	$P = .019$ (D)
	No-till	329	273	299	194	256	$P = .054$ (T)
Respired Carbon, (kg CO$_2$·ha^{-1})	Stubble Mulch	49	42	56	53	24	$P = .075$ (D)
	No-till	80	69	84	57	24	$P = .042$ (T)
Mineralizable nitrogen, kg NO3·ha^{-1}	Stubble Mulch	10.63	6.39	5.79	2.62	1.95	$P = .026$ (D×T)
	No-till	12.68	5.79	8.90	−3.31	−1.38	

[a]Level of highest significant interaction for analysis of variance (ANOVA) of five dates (D)×two treatments (T)×three field replicates.
Values obtained using the chloroform method of Jenkinson and Powlson (1976). The estimates of mineralizable C and N were obtained from the 20-day incubations for the nonchloroformed controls.

Table 11.5 Soil Animal Biomass of Carbon in the Top 10 cm of No-Till and Stubble Mulch Fallow Winter Wheat Plots at Akron, Colorado, for Five Dates in 1982 (from Elliott et al. [1984])

Variable	Treatment	8 Jun	6 Jul	2 Aug	23 Aug	13 Sep	ANOVA[a]
Collembola × 100	Stubble mulch	6.69	1.03	2.18	0.61	1.49	$P = .019$
	No-till	5.58	8.29	4.01	2.60	1.49	(T×D)
Acari × 10	Stubble mulch	3.77	0.47	0.91	0.40	0.92	$P = .020$
	No-till	3.60	3.31	1.24	1.32	0.54	(T×D)
Holophagous nematodes × 100	Stubble mulch	0.88	0.46	1.12	0.51	0.56	$P = .002$
	No-till	0.48	1.69	0.96	0.57	0.32	(T×D)
Protozoa × 1	Stubble mulch	1.76	0.67	0.74	1.96	1.92	$P = .001$(D)
	No-till	2.20	0.96	0.93	2.29	1.61	

[a]Level of highest significant interaction for analysis of variance (ANOVA) of five dates (D) × two treatments (T) × three field replicates.
Arthropods were collected using Tullgren funnels (Merchant and Crossley, 1970), nematodes were collected using standard Baermann funnels (Goodell, 1982), and protozoa were estimated using the most probable number methods (Darbyshire et al., 1974).

the season, whereas the fungal pathway was more active later during the season. Under the stubble mulch management, the activities of both energy channels had shifted to mid or late season (Fig. 11.8).

Impact of Climate Change and Elevated CO$_2$

Climatic patterns in the area of the SGS LTER site and within the region have been affected by irrigation, urbanization, and air pollution (Pielke et al., 1997).

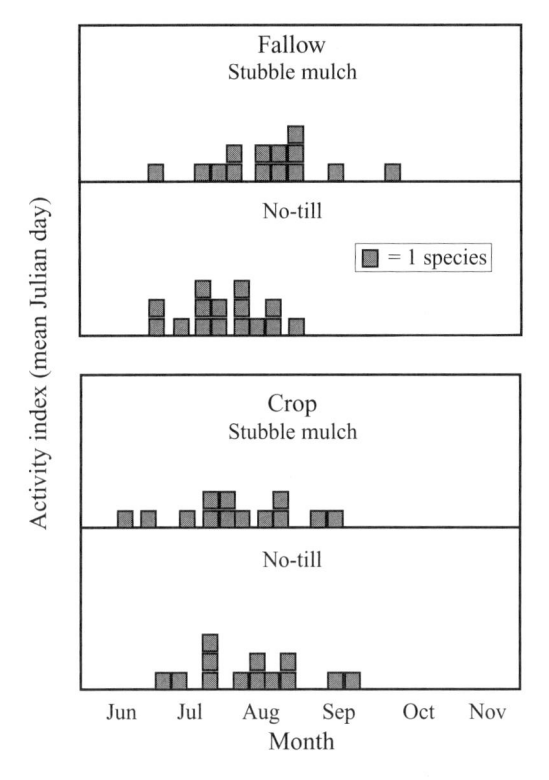

Figure 11.8 The seasonal distributions of the activities of different genera of soil arthropods as measured by the mean Julian day of activity (Moore, 1986) from crop and fallow fields of winter wheat in Akron, Colorado. Each box represents one species. The mean Julian day index is the weighted average of the species density by sample date (Julian day).

Irrigation, for example, may have resulted in cooler and wetter summers along the Front Range and in the adjacent mountain areas (Stohlgren et al., 1998). During the 1990s, nighttime minimum temperatures during the summers at the LTER site were 3 to 5 °C higher than they were in previous decades. Moreover, the frequency and intensity of thunderstorms during the summer months increased in later decades of the 20th century (Pielke et al., 1997). How human-induced changes in climate resulting from elevated concentrations of atmospheric CO_2 might affect the ecosystem is unclear, because temperature, precipitation, and CO_2 levels all affect the competitive balance among plant species and functional types.

In 1997, nine experimental plots were established to study the long-term impact of elevated CO_2 on plant production, species composition, and nutrient dynamics. Open-top chambers (4.5 m in diameter × 3 m in height) were placed over six of the plots, three of which were enriched with CO_2 to a level of 700 µmol·mol^{-1} (elevated), and three were not enriched (ambient). The three remaining plots were maintained as native plots without chambers (controls), as the chambers elevated temperature an average 3 °C. Soils were sampled during the fall after the growing season.

After four growing seasons under elevated CO_2, two trends suggest that the flow of matter and energy within the soil food web may have shifted from the

Figure 11.9 The responses of soil fungi (millimeters of saprophytic and arbuscular mycorrhizal hyphae per gram dry soil) and bacteria (log colony-forming unit [CFU] per gram dry soil) to treatments simulating elevated CO_2 and temperature. The treatments included elevated CO_2 (700 ppm) and temperature under open-top chambers, ambient CO_2, elevated temperature under open-top chambers, and native shortgrass steppe (controls). Letters that are different within a year differ at the $P \leq .05$ level.

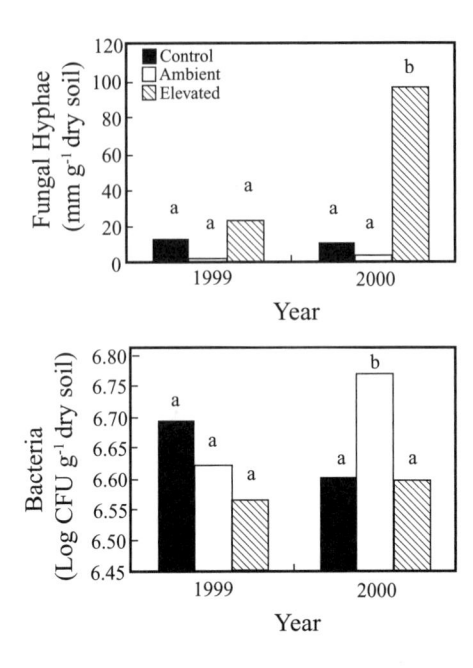

bacterial energy channel to the fungal energy channel. First, the soils exhibited a significant decline in bacteria and significant increases in fungi (Fig. 11.9). Soil fauna responded in a mixed manner (Table 11.6). Protozoa (consumers of bacteria) were unaffected by the treatments, but arthropods (consumers of fungi) and top predators increased under the chambers, with the cryptostigmata and mesostigmata significantly higher under the chambers with elevated CO_2.

Conclusions

The belowground food web of the shortgrass steppe is a dynamic system that is simple enough to study in detail, yet complex enough to elucidate patterns that can be generalized to other ecosystems. Beginning with the detailed inventories and studies of the early IBP through the landscape and regional efforts currently in vogue, several patterns about the structure, function, and responsiveness of the belowground food web to different disturbances have emerged.

Field and modeling studies have revealed that the utilization of resources by species within the belowground food web is compartmentalized into a series of quasi-independent interacting pathways of energy and material flows that originate from plant roots and detritus, with the detritus pathway divided into a bacterial pathway and a fungal pathway. Species within a pathway share similar habitat requirements and complementary life history traits. As a result of this, matter is processed and energy is transferred within the pathways at distinct rates.

Table 11.6 Soil Protozoa and Arthropod Population Densities to 60 cm Depth from Control (no open top-chamber), Ambient (open-top chamber ambient CO_2 levels) and Elevated (open-top chambers CO_2 at 700 ppm) at the CPER, Nunn, Colorado

Taxon	Treatment	1998	1999	2000
Protozoa,	Control	NA	311.06 a[a]	454.86 a
no./g soil	Ambient	NA	252.14 a	595.86 a
	Elevated	NA	383.75 a	692.29 a
Collembola	Control	927.75 b[b]	814.65 a	557.11 a
no./m²	Ambient	666.98 a	831.11 a	640.83 ab
	Elevated	673.68 b	660.34 a	767.21 b
Cryptostigmata,	Control	2900.86 a	1868.26 a	673.68 a
no./m²	Ambient	2598.69 a	1640.51 a	909.38 b
	Elevated	1440.52 a	2170.61 a	1484.39 c
Prostigmata,	Control	1191.25 a	2872.00 a	1088.72 a
no./m²	Ambient	1356.63 a	3049.59 a	1868.26 b
	Elevated	1303.44 a	4688.00 a	2127.63 b
Mesostigmata,	Control	1264.92 a	1426.19 a	822.84 a
no./m²	Ambient	1440.52 a	1484.39 a	798.52 a
	Elevated	1303.44 a	2423.00 b	1499.31 b

[†]Letters that differ within a year indicate a significant treatment effect at the $P \le .05$ level using the SNK test, unless otherwise stated.
[‡]Significant treatment effect at the $P \le .10$ level using the SNK test.
Arthropods were collected using modified Tullgren funnels (Merchant and Crossley, 1970), and protozoa were estimated using the most probable number method (Darbyshire et al., 1974).
NA, not applicable.

Manipulative studies of grazing, agricultural management, and elevated CO_2 have demonstrated that disturbances alter the pattern of energy flow, or shift activity from one pathway to another in ways that are important to stability. These disturbances have affected compartments within the food web in a manner consistent with other studies. For example, at other agricultural sites, the intensive conventional management practices that use deep tillage, pesticides, and inorganic fertilizer applications induce shifts in biomass, and carbon and nitrogen flow from the fungal compartment to the bacterial compartment (Adrén et al., 1990; Hendrix et al., 1986; Moore et al. 1991). Ten years of nitrogen plus phosphorus additions to the arctic tundra designed to simulate the effects of the melting of permafrost under global warming has resulted in significant shifts in biomass toward the bacterial compartment (Doles, 2000). Studies on the effects of nitrogen addition at the Konza tallgrass prairie resulted in an increase in the ratio of bacteria to fungi thought to have been precipitated by a decrease in the carbon-to-nitrogen ratio of the plant detritus (Rice et al., 1998).

Studies of the belowground food web of the shortgrass steppe illustrate the need to study systems in a multivariate context. We know that nutrient availability to plants and the retention of plant-limiting nutrients are affected by changes in soil properties and the activity of soil fauna. As our modeling efforts have demonstrated, many of the biota make small contributions to the overall energy budget of the systems (de Ruiter et al., 1993; Hunt et al., 1987). To echo a caution raised by Milchunas and Lauenroth (1993, page 346) and shared by others (Collins et al., 1998; McNaughton, 1983): "...[e]valuating any one variable can provide misleading perceptions concerning the potential for shifting a system to an alternative state" (Milchunas and Lauenroth, 1993). Nutrient availability and retention are a function of the combined activities of the root, bacterial, and fungal pathways, and how synchronous these activities are with plant growth. Disturbances have altered the activities of these pathways. Whether these changes constitute a change to an alternative state in soils remains to be seen, but the approaches developed at the SGS LTER and across the LTER network offer our best hope for understanding the response of systems to human-induced changes of the future.

References

Aguilar R. E., F. Kelly, and R. D. Heil. 1988. Effects of cultivation on soils in northern Great Plains rangeland. *Soil Science Society of America Journal* **52**:1081–1085.

Andrén, O., T. Lindberg, U. Bostrom, M. Clarholm, A.-C. Hansson, G. Johansson, J. Lagerlof, K. Paustian, J. Persson, R. Pettersson, J. Schnurer, B. Sohlenius, and M. Wivstad. 1990. Organic carbon and nitrogen flows. *Ecological Bulletins* **40**:85–126.

Blair, J. M., T. C. Todd, and M. A. Callaham, Jr. 2000. Responses of grassland soil invertebrates to natural and anthropogenic disturbances, pp. 43–71. In: D. C. Coleman and P. F. Hendrix (eds.), *Invertebrates as webmasters in ecosystems.* CABI Publishing, Oxford, UK.

Bringhurst, R. M., Z. G. Cardon, and D. J. Gage. 2001. Galactosides in the rhizosphere: Utilization by *Sinorhizobium meliloti* and development of a biosensor. *Proceedings of the National Academy of Sciences USA* **98**:4540–4545.

Burke I. C., C. M. Yonker, W. J. Parton, C. V. Cole, K. Flach, and D. S. Schimel. 1989. Texture, climate, and cultivation effects on soil organic matter content in U.S. grassland soils. *Soil Science Society of America Journal* **53**:800–805.

Clarholm, M. 1985. Possible roles of roots, bacteria, protozoa, and fungi in supplying nitrogen to plants, pp. 355–365. In: H. Fitter, D. Atkinson, D. J. Read, and M. B. Usher (eds.), *Ecological interactions in soils: Plants, microbes, and animals.* Blackwell Scientific, Oxford, UK.

Coleman, D. C. 1976. A review of root production processes and their influence on soil biota in terrestrial ecosystems, pp. 417–434. In: J. M. Anderson and A. MacFadyen (eds.), *The role of terrestrial and aquatic organisms in decomposition processes.* Blackwell, Oxford, UK.

Coleman, D. C., C. P. P Reid, and C. V. Cole. 1983. Biological strategies of nutrient cycling in soil systems, pp. 1–55. In: A. MacFadyen and E. D. Ford (eds.), *Advances in ecological research.* Vol. 13. Academic Press, London, UK.

Collins, S. L., A. K. Knapp, D. C. Hartnett, and J. M. Briggs. 1998. The dynamic tallgrass prairie: Synthesis and research opportunities, pp. 301–315. In: A. K. Knapp, J. M. Briggs, D. C. Hartnett, and S. L. Collins (eds.), *Grassland dynamics: Long-term ecological research in the tallgrass prairie.* Oxford University Press, New York.

de Ruiter, P. C., A. Neutel, and J. C. Moore. 1994. Modelling food webs and nutrient cycling in agro-ecosystems. *Tree* **9**:378–383.

de Ruiter, P. C., A. Neutel, and J. C. Moore. 1995. Energetics, patterns of interaction strengths, and stability in real ecosystems. *Science* **269**:1257–1260.

de Ruiter, P. C., J. A. Van Veen, J. C. Moore, L. Brussaard, and H. W. Hunt. 1993. Calculation of nitrogen mineralization in soil food webs. *Plant Soil* **157**:263–273.

Doles, J. 2000. *A survey of soil biota in the arctic tundra and their role in mediating terrestrial nutrient cycling.* Masters thesis, Department of Biological Sciences, University of Northern Colorado, Greeley, Colo.

Elliott, E. T. 1978. *Carbon, nitrogen, and phosphorus transformations in gnotobiotic soil microcosms.* Masters thesis, Department of Agronomy, Colorado State University, Fort Collins, Colo.

Elliott, E. T. 1986. Aggregate structure and carbon, nitrogen, and phosphorus in native and cultivated soils. *Soil Science Society of America Journal* **50**:518–524.

Elliott, E. T., and D. C. Coleman. 1977. Soil protozoan dynamics in a shortgrass prairie. *Soil Biology and Biochemistry* **9**:113–118.

Elliott, E. T., K. Horton, J. C. Moore, D. C. Coleman, and C. V. Cole. 1984. Mineralization dynamics in fallow dryland wheat plots, Colorado. *Plant and Soil* **79**:149–155.

Haas, H. J., C. E. Evans, and E. R. Miles. 1957. *Nitrogen and carbon changes in Great Plains soils as influenced by cropping and soil treatments.* USDA, U.S. Government Printing Office, Washington, D.C.

Hendrix, P. F., Parmelee, R. W, Crossley, D. A., Jr., Odum, E. P., and Groffman, P. 1986. Detritus foodwebs in conventional and non-tillage agroecosystems. *Bioscience* **36**:374–380.

Holland, E. A., and D. C. Coleman. 1987. Litter placement effects on microbial communities and organic matter dynamics. *Ecology* **68**:425–433.

Hunt, H. W., D. C. Coleman, E. R. Ingham, R. E. Ingham, E. T. Elliott, J. C. Moore, C. P. P. Reid, and C. R. Morley. 1987. The detrital food web in a shortgrass prairie. *Biology and Fertility of Soil* **3**:57–68.

Ingham, R. E., J. A. Trofymow, E. R. Ingham, and D. C. Coleman. 1985. Interactions of bacteria, fungi, and their nematode grazers: Effects on nutrient cycling and plant growth. *Ecological Monographs* **55**:119–140.

Jackson, R. B., J. Canadell, J. R. Ehleringer, H. A. Mooney, O. E. Sala, and E. D. Schulze. 1996. A global analysis of root distributions for terrestrial biomes. *Oecologia* **108**:389–411.

Kumar, R., R. J. Lavigne, J. E. Lloyd, and R. E. Pfadt. 1975. *Macroinvertebrates of the Pawnee site.* Technical report no. 278. Grassland Biome, U.S. International Biological Program, Colorado State University, Fort Collins, Colo.

Laakso, J., and H. Setälä. 1999. Sensitivity of primary production to changes in the architecture of belowground food webs. *Oikos* **87**:57–64.

Lauenroth, W. K., and D. G. Milchunas. 1991. Shortgrass steppe, pp. 397–403. In: R. T. Coupland (ed.), *Grasslands: Introduction and western hemisphere.* Elsevier, New York.

May, R. M. 1972. Will a large complex system be stable? *Nature* **238**:413–414.

May, R. M. 1973. *Stability and complexity of model ecosystems.* Princeton University Press, Princeton, N.J.

McNaughton, S. J. 1983. Serengeti grassland ecology: The role of composite environmental factors and contingency in community organization. *Ecological Monographs* **53**:291–320.

Milchunas, D. G., and W. K. Lauenroth. 1993. Quantitative effects of grazing on vegetation and soil over a global range of environments. *Ecological Monographs* **63**:327–366.

Milchunas D. G., O. E. Sala, W. K. Lauenroth. 1988. A generalized model of the effects of grazing by large herbivores on grassland community structure. *The American Naturalist* **132**:87–106.

Moore, J. C. 1986. *Micro-mesofauna dynamics and functions in dryland wheat-fallow agroecosystems*. PhD diss., Department of Zoology and Entomology, Colorado State University, Fort Collins, Colo.

Moore, J. C., and P. C. de Ruiter. 1997. Compartmentalization of resource utilization within soil ecosystems, pp. 375–393. In: A. Gange and V. Brown (eds.), *Multitrophic interactions in terrestrial systems*. Blackwell, Oxford, UK.

Moore, J. C., P. C. de Ruiter, and H. W. Hunt. 1993. The influence of ecosystem productivity on food web stability. *Science* **261**:906–908.

Moore, J. C., H. W. Hunt, and E. T. Elliott. 1991. Interactions between soil organisms and herbivores, pp. 105–140. In: P. Barbosa, V. Kirschik, and C. Jones (eds.), *Multitrophic level interactions among microorganisms, plants and insects*. John Wiley, New York.

Moore, J. C., K. McCann, H. Setälä, and P. C. de Ruiter. 2003. Top down is bottom up: Does predation in the rhizosphere regulate aboveground dynamics? *Ecology* **84**:846–857.

Moore J. C., D. E. Walter, and H. W. Hunt. 1988. Arthropod regulation of micro- and mesobiota in belowground detrital food webs. *Annual Review of Entomology* **33**:419–439.

Moorman, T., and F. B. Reeves. 1979. The role of endomycorrhizae in revegetation practices in the semi-arid west. II. A bioassay to determine the effects of land disturbance and endomycorrhizal populations. *American Journal of Botany* **66**:14–18.

Naeem, S., L. J. Thompson, S. P. Lawler, J. H. Lawton, and R. M Woodfin. 1994. Declining biodiversity can alter performance of ecosystems. *Nature* **368**:734–737.

O'Neill, R. V. 1969. Indirect estimation of energy fluxes in animal food webs. *Journal of Theoretical Ecology* **22**:284–290.

Paine, R. T. 1980. Food webs: Linkage, interaction strength and community infrastructure. *Journal of Animal Ecology* **49**:667–685.

Pielke, R. A., T. J. Lee, J. H. Copeland, J. L. Eastman, C. L. Ziegler, and C. A. Finley. 1997. Use of USGS-provided data to improve weather and climate simulations. *Ecological Applications* **7**:3–21.

Rice, C. W., T. C. Todd, J. M. Blair, T. R. Seastedt, R. A. Ramundo, and G. W. Wilson. 1998. Belowground biology and processes, pp. 244–264. In: A. K. Knapp, J. M. Briggs, D. C. Hartnett, and S. L. Collins (eds.), *Grassland dynamics: Long-term ecological research in the tallgrass prairie*. Oxford University Press, New York.

Schoener, T. W. 1974. Resource partitioning in ecological communities. *Science* **185**:27–39.

Smolik, J. D., and J. L. Dodd. 1983. Effect of water and nitrogen, and grazing on nematodes in a shortgrass prairie. *Journal of Range Management* **36**:744–748.

Stohlgren, T. J., T. N. Chase, R. A. Pielke, T. G. F. Kittel, and J. S. Baron. 1998. Evidence that local land use practices influence regional climate, vegetation, and stream flow patterns in adjacent natural areas. *Global Change Biology* **4**:495–504.

Trofymow, J. A., and D. C. Coleman. 1982. The role of bacterivorous and fungivorous nematodes in cellulose and chitin decomposition in the context of a root [rhizosphere] soil conceptual model, pp. 117–137. In: D. W. Freckman (ed.), *Nematodes in soil ecosystems*. University of Texas Press, Austin, Texas.

Tilman, D., D. Wedin, and J. Knops. 1996. Productivity and sustainability influenced by biodiversity in grassland ecosystems. *Nature* **379**:718–720.

Wall, D. W., and J. C. Moore. 1999. Interactions underground: Soil biodiversity, mutualism and ecosystem processes. *BioScience* **49**:109–117.

Wall-Freckman, D., and S. P. Huang. 1998. Response of the soil nematode community in a shortgrass steppe to long-term and short-term grazing. *Applied Soil Ecology* **9**:39–44.

Wallwork, J. A. 1970. *Ecology of soil animals*. McGraw-Hill, London, UK.

Walter, D. E., and H. Proctor. 1999. *Mites: Ecology, evolution, and behavior.* CABI Publishing, Wallingford, Oxon, UK.

12

Net Primary Production in the Shortgrass Steppe

William K. Lauenroth
Daniel G. Milchunas
Osvaldo E. Sala
Ingrid C. Burke
Jack A. Morgan

Net primary production (NPP), the amount of carbon or energy fixed by green plants in excess of their respiratory needs, is the fundamental quantity upon which all heterotrophs and the ecosystem processes they are associated with depend. Understanding NPP is therefore a prerequisite to understanding ecosystem dynamics. Our objectives for this chapter are to describe the current state of our knowledge about the temporal and spatial patterns of NPP in the shortgrass steppe, to evaluate the important variables that control NPP, and to discuss the future of NPP in the shortgrass steppe given current hypotheses about global change. Most of the data available for NPP in the shortgrass steppe are for aboveground net primary production (ANPP), so most of our presentation will focus on ANPP and we will deal with belowground net primary production (BNPP) as a separate topic. Furthermore, our treatment of NPP in this chapter will ignore the effects of herbivory, which will be covered in detail in chapter 16.

Our approach will be to start with a regional-scale view of ANPP in shortgrass ecosystems and work toward a site-scale view. We will begin by briefly placing ANPP in the shortgrass steppe in its larger context of the central North American grassland region. We will then describe the regional-scale patterns and controls on ANPP, and then move to the site-scale patterns and controls on ANPP. At the site scale, we will describe both temporal and spatial dynamics, and controls on ANPP as well as BNPP. We will then discuss relationships between spatial and temporal patterns in ANPP and end the chapter with a short, speculative section on how future global change may influence NPP in the shortgrass steppe.

Aboveground Net Primary Production of the Shortgrass Steppe in the Context of the Central North American Grassland Region

Temperate grasslands in central North America are found over a range of mean annual precipitation from 200 to 1200 mm·y^{-1} and mean annual temperatures from 0 to 20 °C (Lauenroth et al., 1999). The widely cited relationship between mean annual precipitation and average annual ANPP allows us to convert the precipitation gradient into a production gradient (Lauenroth, 1979; Lauenroth et al., 1999; Noy-Meir, 1973; Rutherford, 1980; Sala et al., 1988b). Therefore, the position of the central North American grasslands on the production gradient is from less than 100 g·m^{-2}·y^{-1} to more than 600 g·m^{-2}·y^{-1} (Lauenroth et al., 1999). The shortgrass steppe contains a large proportion of the least productive sites in the region. Paruelo and Lauenroth (1995) used normalized difference vegetation index (NDVI) data, which are correlated with ANPP (Paruelo et al., 1997), to compare sites within the grassland region. They found large differences among grassland types, supporting the idea that the shortgrass steppe is the least productive of the grassland types (Fig. 12.1).

Lane et al. (1998, 2000) sampled a transect across the precipitation gradient from the northern shortgrass steppe through the northern mixed prairie and into the tallgrass prairie. Their results also support the idea of low NPP in the shortgrass steppe compared with the remainder of the grassland region. They estimated annual ANPP and made measurements of leaf area index (LAI), and found that both increased significantly with increasing mean annual precipitation (Fig. 12.2A). Aboveground NPP ranged from 40 to 75 g·m^{-2} at the three shortgrass steppe sites, from 165 to 300 g·m^{-2} at three northern mixed prairie sites,

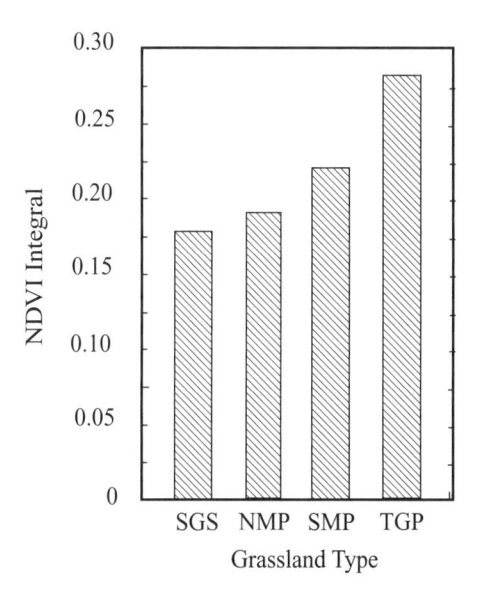

Figure 12.1 Normalized difference vegetation index (NDVI) integrated over the growing season for the shortgrass steppe (SGS), northern mixed prairie (NMP), southern mixed prairie (SMP), and tallgrass prairie (TGP). Each bar represents the average over several sites. (Adapted from Paruelo and Lauenroth [1995].)

272 Ecology of the Shortgrass Steppe

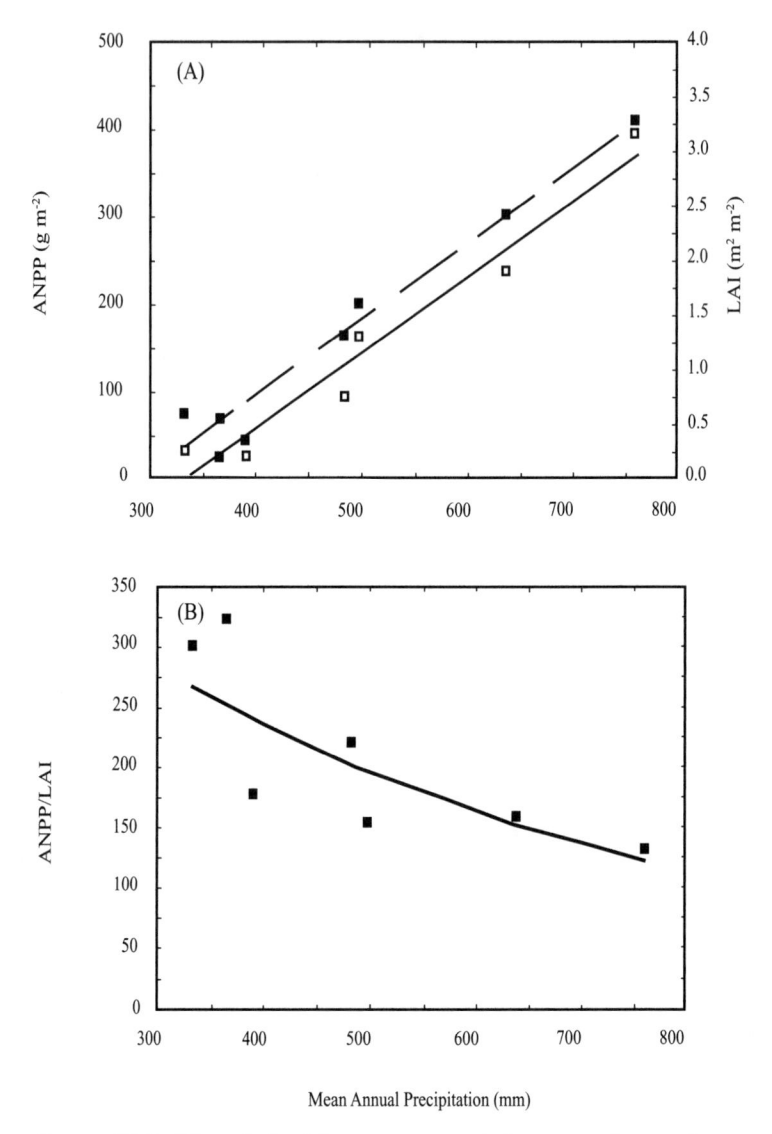

Figure 12.2 The relationship among annual aboveground net primary production (ANPP), leaf area index (LAI), mean annual precipitation (A); and ANPP per unit of LAI as a function of mean annual precipitation across the central grassland region (B). (Data from Lane et al. [1998, 2000].)

and was 410 g·m^{-2} for a tallgrass prairie site. Leaf area index averaged approximately 0.25 at the shortgrass sites, ranged from 0.75 to 1.9 at the northern mixed prairie sites, and was 3.15 at the tallgrass site. The relationship between LAI and ANPP suggests that the amount of NPP produced per unit LAI decreases along the precipitation gradient from the shortgrass steppe to the tallgrass prairie (Fig. 12.2B). Lane et al. (2000) estimated the amount of light reaching the soil

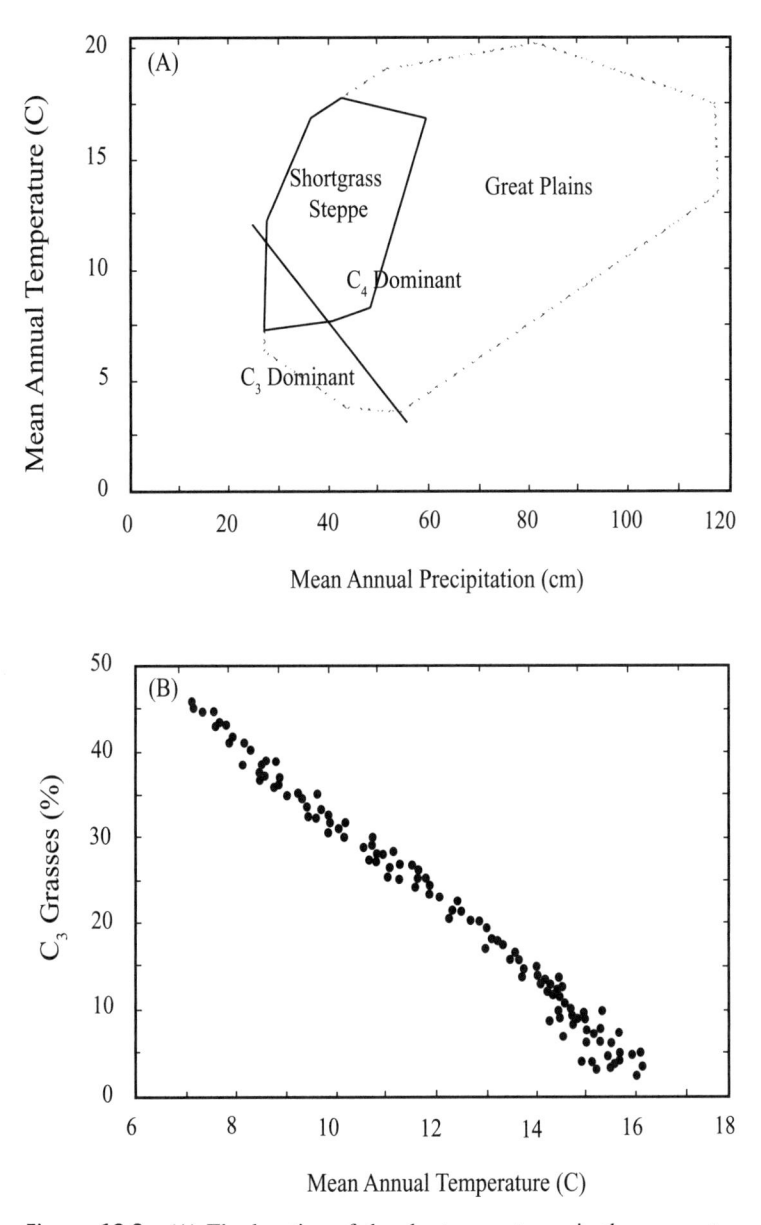

Figure 12.3 (A) The location of the shortgrass steppe in the parameter space defined by mean annual precipitation and mean annual temperature relative to the entire Great Plains. The solid diagonal line divides the space into portions dominated by C_3 or C_4 plants. (Adapted from Epstein et al. [1997b].) (B) The relationship between the percentage contribution of C_3 grasses to aboveground net primary production and mean annual temperature in the shortgrass steppe. (Data from Fan [1993].)

surface at maximum canopy development and reported that it was greater than 90% for the shortgrass sites, between 0% and 10% for the mixed prairie sites, and 0% for the tallgrass site. The picture of the shortgrass steppe that emerges from this work is one of a low-productivity environment with a small amount of the soil surface shaded by the leaves of plants.

The shortgrass steppe combines low precipitation with high temperatures. It is found in areas with mean annual temperatures more than 7 °C and mean annual precipitation less than 625 mm (Fig. 12.3A). Based upon the work of Epstein et al. (1997b) a very small portion of the environmental space of the shortgrass steppe (Fig. 12.3A) favors the dominance of C_3 species. Regression relationships from this work suggest that the percentage that C_3 grasses contribute to ANPP in the shortgrass steppe on sandy loam soils should range from more than 40% in the north to 0% in the south (Epstein et al., 1997b). The decrease in C_3 grasses from north to south is linear with increasing mean annual temperature (Fig. 12.3B).

Thus, the shortgrass steppe is a dry grassland that is dominated by short-stature C_4 grasses that develop a sparse canopy with a low LAI. These characteristics combine to result in its having the lowest annual ANPP in the central North American grassland region.

Regional-Scale Spatial Patterns and Controls on Net Primary Production

Effects of Precipitation and Temperature

Average ANPP throughout the shortgrass steppe ranges from 50 $g \cdot m^{-2}$ to more than 300 $g \cdot m^{-2}$ (Fig. 12.4 [USDA, 1967]). The mean of estimates for average years is 178 $g \cdot m^{-2}$, with a spatial coefficient of variation (CV) of 38%. Predictably, the distribution during unfavorable years shifts toward smaller values; and in favorable years, toward larger values (Fig. 12.4). During unfavorable years, 10% of the sites have ANPP ≤50 $g \cdot m^{-2}$, whereas during favorable years 10% of the sites have ANPP more than 400 $g \cdot m^{-2}$. The mean for unfavorable years is 103 $g \cdot m^{-2}$ (CV, 50%), and for favorable years is 250 $g \cdot m^{-2}$ (CV, 35%).

The key environmental variables that explain the differences among locations within the shortgrass steppe are precipitation and temperature (Fig. 12.5). Soil texture also has a small but significant influence on the regional distribution of ANPP. The sites with the lowest ANPP occur in the warmest and driest parts of the region (western Texas and southeastern New Mexico; Fig. 12.5B). Sites with the greatest ANPP occur at a mean annual precipitation of more than 400 $mm \cdot y^{-1}$ in three areas: (1) northeastern Colorado, southeastern Nebraska, and northeastern Kansas; (2) the northeastern portion of the panhandle of Texas and northern New Mexico; and (3) southern Colorado adjacent to the mountains. These are the sites with the greatest amounts of effective precipitation. As mean annual temperature (MAT) increases, mean annual potential evapotranspiration also increases (MAPET [measured in centimeters] = $94 + 5 \times MAT$; $r^2 = 0.71$ [Lauenroth and Burke, 1995]), resulting in a requirement for greater precipitation to maintain

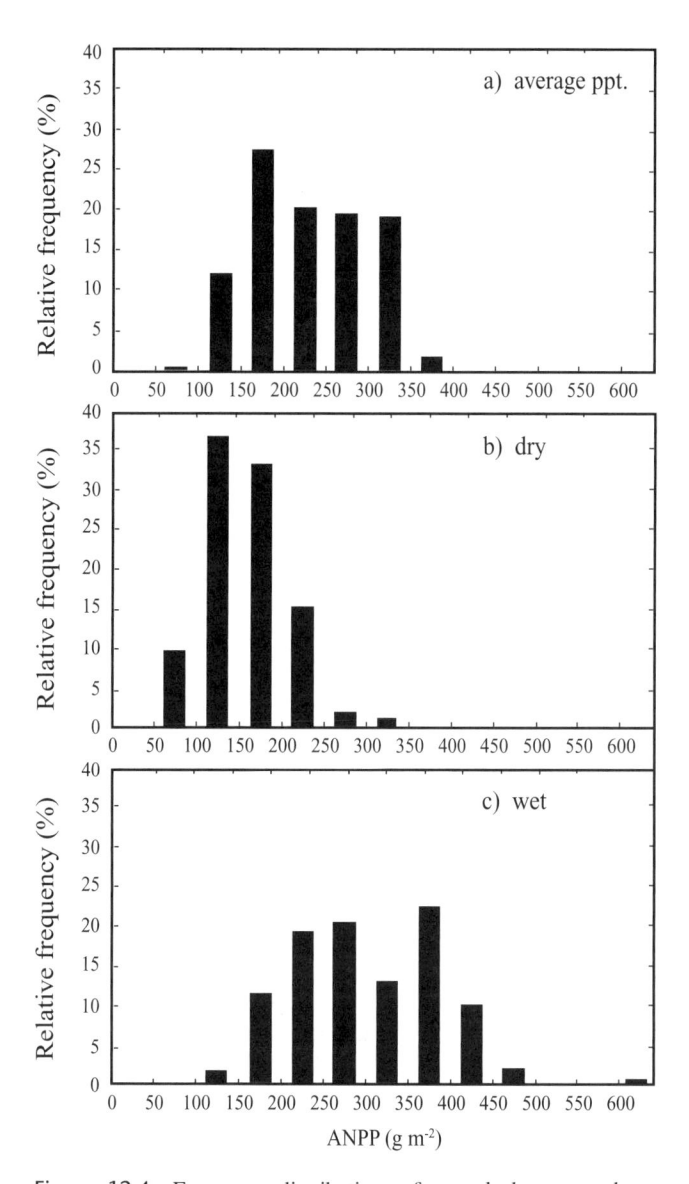

Figure 12.4 Frequency distributions of annual aboveground net primary production (ANPP) in the shortgrass steppe for normal (average) (A), unfavorable (dry) (B), and favorable (wet) (C) years. These data were collected by the Natural Resources Conservation Service as part of their Range Site database (USDA, 1967). The data represent ANPP for different sites throughout the shortgrass region.

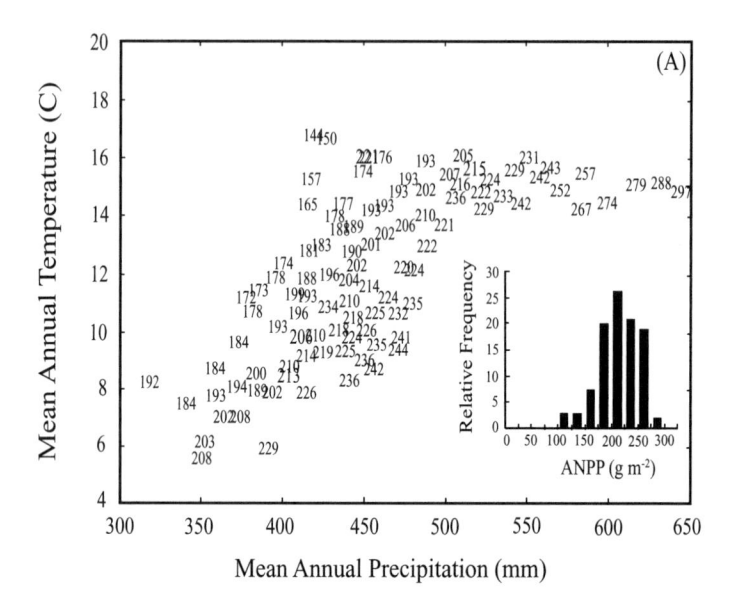

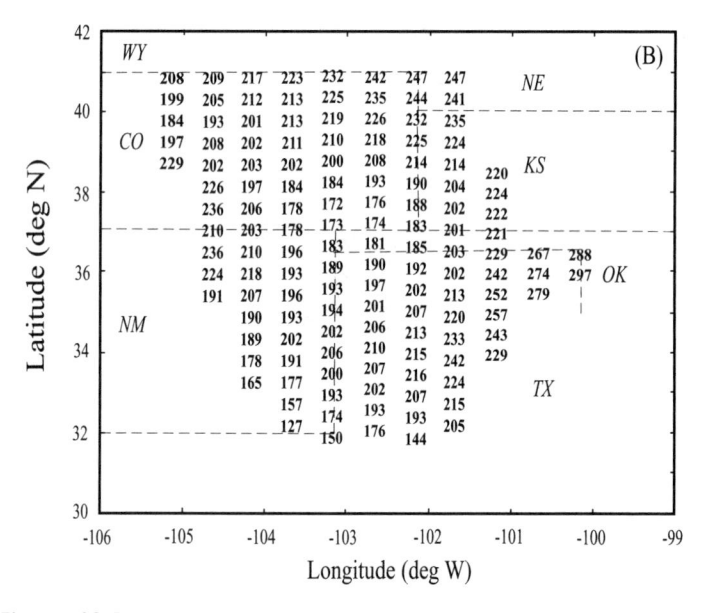

Figure 12.5 Distribution of annual aboveground net primary production (ANPP) in the shortgrass steppe. (A) Distribution of ANPP (measured in grams per square meter) as a function of mean annual precipitation and mean annual temperature. (B) Distribution of ANPP (measured in grams per square meter) as a function of longitude and latitude. See Figure 12.4 for description of the database.

a similar amount of ANPP. Vertical slices through Figure 12.5A provide examples of this effect. For instance, at a mean annual precipitation of 450 mm, ANPP ranges from 240 g·m^{-2}·y^{-1} at 9 °C mean annual temperature to 175 g·m^{-2}·y^{-1} at 16 °C.

Effects of Soil Texture

The role that soil texture plays in influencing ANPP can be illustrated by an analysis of data collected by the NRCS in north–central Colorado (Fig. 12.6A). Aboveground NPP decreases as water-holding capacity increases. Water-holding

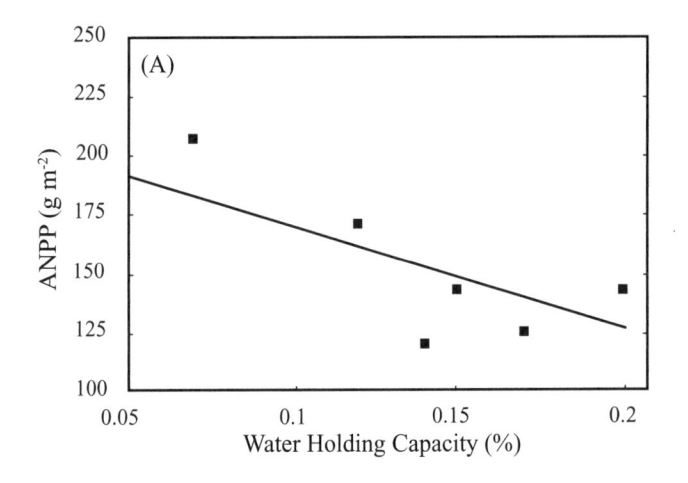

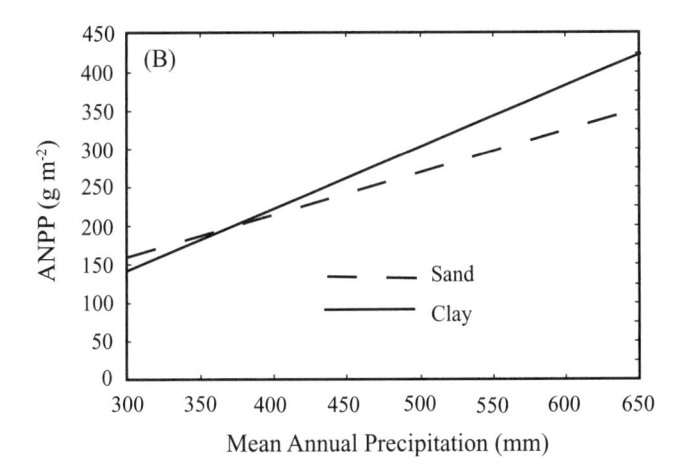

Figure 12.6 Relationship between (A) waterholding capacity of soils and average annual aboveground net primary production (ANPP) and (B) mean annual precipitation and ANPP for two soil texture classes from sites in Weld County, Colorado. See Figure 12.4 for a description of the database.

capacity is largely determined by soil texture and has a large influence on soil water availability and water balance (Cosby et al., 1984; Lauenroth and Bradford, 2006). A decrease in ANPP in dry areas as water-holding capacity increases is consistent with the inverse texture effect (Noy-Meir, 1973), which states that in dry areas, coarse-textured soils should be more productive than fine-textured soils and the reverse should be true for wet areas (Fig. 12.6B). The explanation lies in the relative roles of bare soil evaporation and percolation below the root zone. Coarse-textured soils have low water-holding capacity and low losses to bare soil evaporation. This results in more water available for plant growth in dry areas and less in wet areas. This effect can be best illustrated by evaluating the two extreme cases of clay and sand (Fig. 12.6B). At the wettest sites ($\approx$650 mm·y^{-1}), ANPP on clay soils averages 70 g·m^{-2} more than that on sandy soils; at the dry end of the gradient (300 mm·y^{-1}), sandy soils have an ANPP that is approximately 20 g·m^{-2} more than that found on clay. The inverse texture effect (Fig. 12.6B) provides an explanation for why the relationship between ANPP and water-holding capacity for sites in north–central Colorado decreases as water-holding capacity increases (Fig. 12.6A). These sites have mean annual precipitation that is less than 375 mm, which puts them on the dry side of the crossover point for the inverse texture effect (Fig. 12.6B).

Another source of evidence for the existence of an inverse texture effect can be found in the observation that deep sandy soils in the shortgrass region tend to support taller vegetation than adjacent fine-textured soils. Küchler (1964) mapped the potential natural vegetation of the conterminous United States and indicated that on deposits of deep sand associated with major rivers in the shortgrass region, tallgrasses (*Andropogon* sp.) are dominant, whereas the adjacent fine-textured soils are dominated by the shortgrasses *Bouteloua gracilis* and *Buchloë dactyloides*. Fan (1993) evaluated the distribution of species and functional groups in eastern Colorado and found that production of C$_4$ grasses was significantly and positively related to sand content of the soil ($r^2 = 0.56$; $P = .0001$). Investigation of the details of this relationship revealed that the largest proportion of ANPP at the highest sand contents was contributed by tallgrass species. Average ANPP at low sand contents (<60%) was 65 g·m^{-2}, and at high sand contents was 120 g·m^{-2}. Most of the evidence supports the idea of an inverse texture effect at both the shortgrass steppe and the central grassland region scales. The single contrasting study is that of Lane et al. (1998), and it is possible that the lack of support in these data for an inverse texture effect was the result of insufficient sample size to deal with quadrat-to-quadrat variability.

Effects of Land Use on Aboveground Net Primary Production

A current reality for the shortgrass steppe and all grassland areas worldwide is that only a portion of the original area of grassland remains in native vegetation (Lauenroth et al., 1999). More than half the area within the shortgrass steppe remains in native vegetation (Lauenroth and Milchunas, 1992; Lauenroth et al., 1999; Hart, chapter 4, this volume). The remainder is split between dryland and irrigated crops. The major dryland crop is winter wheat and the key irrigated crop is corn. Although other crops are grown under dryland and irrigated conditions, these two account for the largest part of the area harvested each year.

Net Primary Production in the Shortgrass Steppe 279

It is difficult to estimate NPP for crops, but by making some assumptions we can get approximate numbers for corn and wheat. If we assume that all corn produces the same amount of aboveground biomass as the corn grown for silage, then on average, production is 4250 to 4700 g·m^{-2}. This is under irrigated and fertilized conditions, and is approximately 20 times the average ANPP of the native shortgrass steppe. This level of production is restricted to those areas with either surface irrigation water, such as along the major rivers, or where water can be pumped to the surface.

Dryland wheat production can be statistically related to environmental variables, so that we can make comparisons with native grasslands. In the central Great Plains, including the shortgrass steppe, winter wheat is grown using the summer-fallow rotation system (USDA, 1974). This rotation system produces one crop in 2 years. Lauenroth et al. (2000) analyzed summer-fallow wheat production for the central Great Plains and found the following relationship between ANPP of winter wheat and mean annual precipitation (MAP):

$$\text{ANPP (g·m}^{-2}) = 197 + 0.2 \times \text{MAP}$$
$$(\text{mm}) \quad (r^2 = 0.67)$$

This equation incorporates the fact that at a particular location, only one wheat crop is produced in 2 years.

Over the range of mean annual precipitation found in the shortgrass steppe, grassland ANPP increases more rapidly with increasing precipitation than does wheat (Fig. 12.7). Between 325 and 575 mm mean annual precipitation, ANPP of winter wheat is greater than that of the native shortgrass steppe. At a mean annual precipitation more than 575 mm, native steppe production is greater than wheat. The likely explanation for greater production of wheat in the driest locations is that the

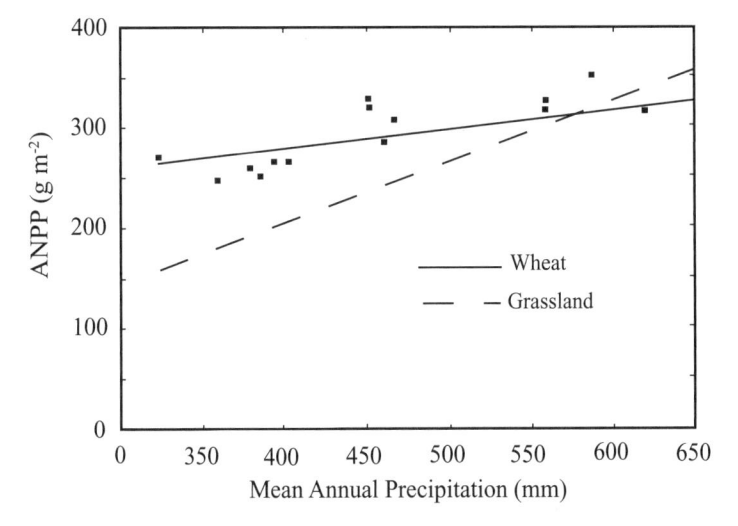

Figure 12.7 Relationship between mean annual precipitation and annual aboveground net primary production (ANPP) for native grasslands and summer-fallow winter wheat in the shortgrass steppe. (Adapted from Lauenroth et al. [2000].)

summer-fallow rotation system on average produces more favorable soil water conditions for plant growth and less variability among years than in the adjacent grasslands. The crossover point at 575 mm is the point at which there is no advantage to the summer-fallow system, and continuous cropping would be more desirable.

The decreased variability of winter wheat ANPP relative to the native shortgrass steppe is very apparent from a spatial analysis (Fig. 12.8). Although the

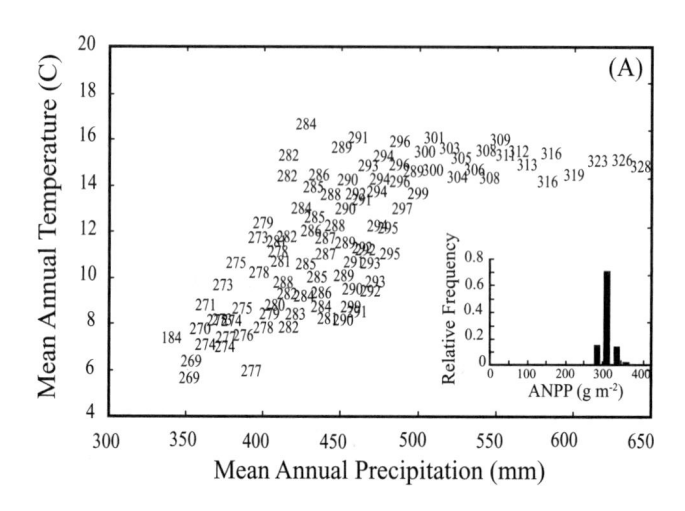

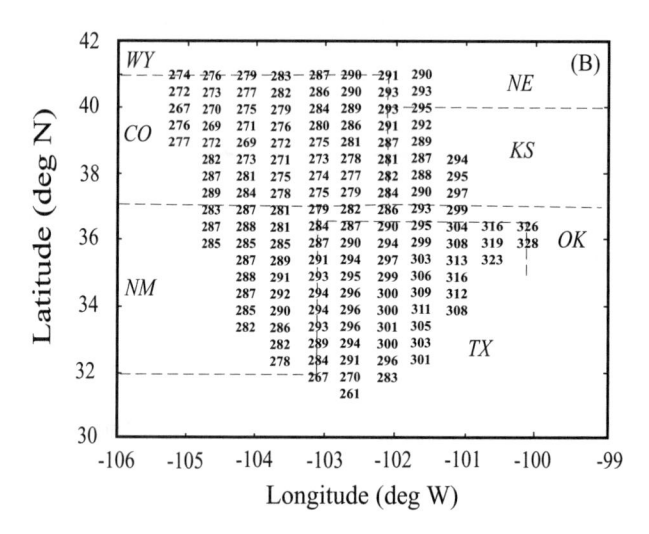

Figure 12.8 Distribution of annual aboveground net primary production (ANPP) for summer-fallow winter wheat in the shortgrass steppe. (A) Distribution of ANPP as a function of mean annual precipitation (MAP) and mean annual temperature. (B) Distribution of ANPP as a function of longitude and latitude. Aboveground NPP was calculated from the regression equation ANPP (g·m^{-2}) = 197 + 0.2 × MAP (mm) (r^2 = .67) (Lauenroth et al., 2000).

spatial distribution of average grassland ANPP ranges from 125 g·m^{-2} in the extreme southwestern portion of the steppe to almost 300 g·m^{-2} in the southeast (Fig. 12.5), winter wheat ANPP only varies from 261 to 328 g·m^{-2}. One caution about these results is that the relationship between wheat ANPP and mean annual precipitation was developed with data from the northern portion of the steppe and may not do as good a job representing the southern portion. The intercepts for the two equations relating ANPP to mean annual precipitation lead to additional speculation about the differences. The y-intercept for the grassland relationship is negative, leading to a positive x-intercept. Noy-Meir (1973) interpreted the x-intercept of such relationships as the amount of precipitation required to maintain zero production. In the case of grasslands, 57 mm is required to maintain zero production. The winter wheat equation had a positive y-intercept, suggesting that if precipitation is zero, ANPP would still be greater than zero. This is consistent with the idea of water storage during the fallow period. Average winter wheat ANPP over the shortgrass steppe precipitation gradient is 294 g·m^{-2}. The y-intercept (197 g·m^{-2}) suggests that, on average, 67% of ANPP is attributable to the fallow period.

Site-Scale Spatial Patterns and Controls on Net Primary Production

Controls, or at least correlates, of the spatial pattern of ANPP are much clearer at the regional scale than at the site scale. Climatic influences seem to be the key controls at the regional scale. At the site scale, climatic conditions are either constant or vary only a small amount. All the work on site-scale spatial variability of ANPP in the shortgrass steppe has been conducted at the Central Plains Experimental Range (CPER).

Effects of Soil Texture

One of the techniques that holds the greatest promise for assessing ANPP at the square kilometer scale is satellite-based remote sensing. Analysis of imagery for the CPER indicates the spatial variability in the vegetation at the scale of tens of square kilometers (Anderson, 1991; Todd, 1994). Anderson (1991) used a Landsat Thematic Mapper scene to create a spectral classification of a 42-km^2 area in the northeast portion of the CPER into greenness classes. Data for green biomass from harvested plots ranged from 43 to 190 g·m^{-2} over the range of greenness. He reported an excellent relationship between the average NDVI of the greenness classes and green biomass. His equation for the August 1989 data is

$$\text{Green Biomass (g·m}^{-2}) = 30.8 + 434.5 \times \text{NDVI} \quad (r^2 = 0.95)$$

Regressions for two other periods produced coefficients of determination of 0.71 and 0.95 (Anderson, 1991). Although Anderson (1991) did not attempt to relate green biomass to environmental variables, a qualitative analysis of the relationship between his spatial distribution of green biomass and a soil texture map for the CPER (not shown) suggests that landscape position and surface horizon soil texture both may influence green biomass.

The soils at the CPER, similar to most other sites in the shortgrass steppe that have remained in native vegetation, are predominantly coarse in texture. The most common texture is fine sandy loam. Todd (1994) used Landsat Thematic Mapper data for the entire CPER (65 km^2) to investigate relationships between spectral indices and standing crop biomass. She found a number of significant relationships between indices derived from satellite data and standing crop of grazed sites, but fewer for ungrazed sites. Reanalysis of the data presented by Todd (1994) revealed a significant relationship between soil texture (clay content) and standing crop (Fig. 12.9), but it is possible that the key controlling variable is landscape position rather than clay (see the next section on landscape effects). One of the things that argues against the importance of clay in this relationship is that the majority of our field data indicate that ANPP is greatest on sandy soils.

The data from the two Thematic Mapper images used by Todd (1994) produced relationships with different y-intercepts but otherwise similar forms. The June 22, 1984, image produced standing crop values that were, on average, 32 g·m^{-2} more than the July 12, 1991, image. Precipitation in 1991 was very near the long-term average, and in 1984 was 30% greater than the average. This likely accounts for the difference in standing crop between the 2 years.

Seven years of data from the CPER collected on a loamy sand and a clay loam site approximately 5 km apart support the idea that ANPP on coarse-textured soils is greater than on fine-textured soils (Fig. 12.10). Six of the seven ratios

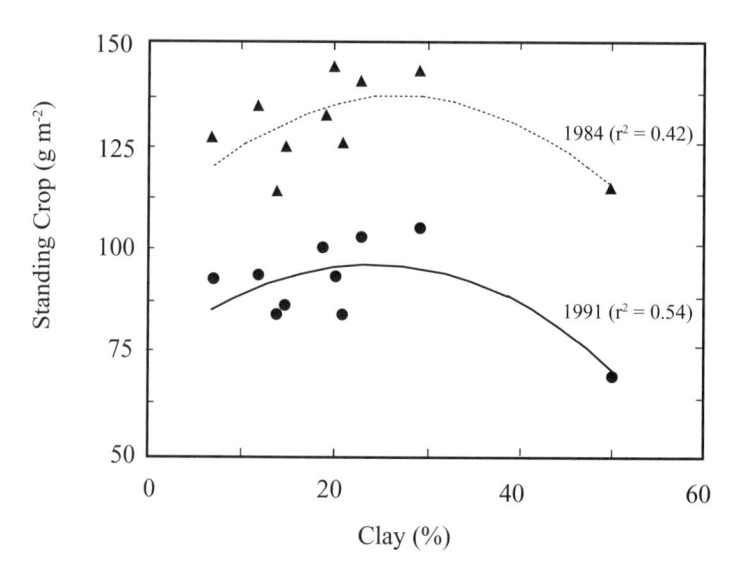

Figure 12.9 Relationship between standing crop biomass for the CPER predicted from Thematic Mapper data and clay content. (Data from Todd [1994].)

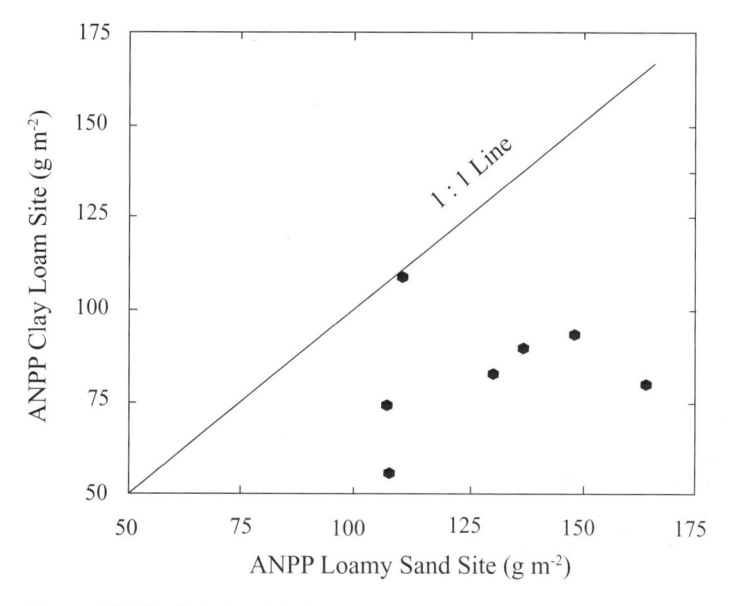

Figure 12.10 Relationship between annual aboveground net primary production (ANPP) on a loamy sand and a clay loam site at the CPER.

of ANPP fine to ANPP coarse were between 0.45 and 0.80 and were not significantly related to annual precipitation. In 1 year, the ratio was approximately one. Dodd and Lauenroth (1997) used a soil water simulation model to compare these two sites and found that, on average, over 50 years, a greater fraction of soil water was lost via transpiration from the loamy sand than from the clay loam site. Conversely, a greater fraction of the water was lost to bare soil evaporation on the clay loam than the loamy sand site. These results are consistent with greater ANPP for the coarse- versus the fine-textured site.

Landscape Effects

In addition to the effects of soil texture, landscape position has a complex effect on ANPP that is not easy to partition into individual controlling factors. Sixteen years of data collected from three landscape positions at the CPER indicate that there are small differences in ANPP between upland and midslope positions, but large differences between swales and either of the other two landscape positions (Fig. 12.11). A simple soil texture explanation does not account for these differences because in this example both the midslope and swale positions have the same soil textures. The upland position has a clay loam soil and the midslope and swale positions both are on sandy clay loam soils. The most likely explanation for these differences lies in a combination of effects of soil texture, soil water, and nitrogen. Swale positions can receive additional water via runon or interflow and in some cases have higher soil carbon and nitrogen contents (Burke et al., 1999; Yonker et al., 1988).

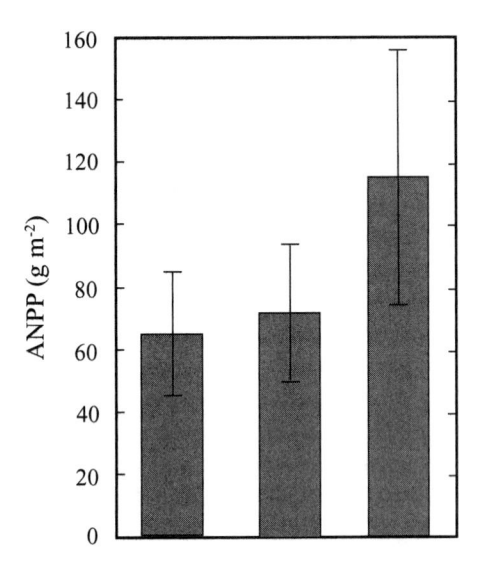

Figure 12.11 Annual aboveground net primary production (ANPP) for three landscape positions at the CPER. Each bar represents the average of 17 years. The vertical lines are 1 SE.

Temporal Patterns and Controls

Ecophysiology of Shortgrass Plants

The physiological responses of shortgrass steppe plants have often been studied within the context of functional types (e.g., morphological groups, such as shrubs, grasses, forbs, succulents; or photosynthetic groups, such as C_3, C_4, and crassulacean acid metabolism [CAM]), and generally in response to variation in water and temperature, the two most important environmental constraints to plant production in this region (Detling et al., 1978; Dickinson and Dodd, 1976; Kemp and Williams, 1980; Sala et al., 1992). These functional types have frequently been discussed in regard to plant distribution, but they have significance for temporal variation in primary production as well. For instance, shrubs can be important on soils (Dodd and Lauenroth, 1997; Lane et al., 1998; Lauenroth and Milchunas, 1992; Sala et al., 1997) or in climates (Knight, 1994; Sala et al., 1997) that encourage deep percolation of water. Deep-rooted shrubs and forbs can take advantage of water stored below surface layers where roots of grasses and succulents dominate (Coffin and Lauenroth, 1990). Similarly, variation in seasonal ANPP of a site and of individual species depends to a large extent on the interactions among seasonality and amount of precipitation, soils, and plant types (Epstein et al., 1997a). On sites with a significant shrub component, large precipitation events will likely lead to substantial production responses of both grasses and shrubs, whereas small precipitation events will enhance primarily grass production (Dodd and Lauenroth, 1997). Because deep soil water is less variable and ephemeral than shallow soil water, the seasonal dynamics of shrub production should be less variable than for shallow-rooted grasses (Sala and Lauenroth, 1982).

Similar kinds of morphological differences in rooting apply as well to the majority of the shortgrass steppe, which is grass dominated (Lauenroth and Milchunas, 1992). Cool-season grasses such as *Agropyron smithii* tend to be more deep rooted than the warm-season, dominant *B. gracilis* (Coupland and Johnson, 1965; Weaver and Albertson, 1956), and these differences in rooting characteristics can result in different soil–plant water dynamics (Dodd and Lauenroth, 1997; Sala et al., 1992). Furthermore, leaf water potential and leaf conductance of *B. gracilis* sometimes appear to be more sensitive to changes in soil water compared with *A. smithii* (Morgan et al., 1998; Sala and Lauenroth, 1982; Sala et al., 1982), responses that could be related to differences in rooting characteristics.

The other important functional typing that has often framed physiological studies on the shortgrass steppe involves the photosynthetic pathway—specifically, the categorization of plants into C_3, C_4, and CAM (Black, 1971, 1973). Of the three types, CAM plants are the smallest group represented in the shortgrass steppe and are especially noted for their adaptations to xeric environments. *Opuntia* is the most prominent CAM genus of the region (Lauenroth and Milchunas, 1992). An abundance of both C_3 and C_4 perennial grasses dominates the shortgrass steppe, although the majority of physiological studies have focused on contrasting two important, representative species: the C_3 *A. smithii* and the C_4 *B. gracilis*. *Bouteloua gracilis* has a higher temperature optimum for photosynthesis (Brown and Trlica, 1977; Kemp and Williams, 1980; Monson et al., 1983; Read et al., 1997), requires more light to saturate photosynthesis (Brown and Trlica, 1977; Williams and Kemp, 1978), and has greater water use efficiency (Monson et al., 1986; Morgan et al., 1998) compared with the C_3 native *A. smithii*. These traits partly explain the dominance of *B. gracilis* in this semiarid steppe; they have also been used to explain the temporal variation in production between these two grasses as a mechanism for niche separation (Monson et al., 1983).

At the site level, seasonal and yearly variation in the dynamics of precipitation and temperature account for most of the temporal variation in physiological activity. The period when both temperature and precipitation create favorable conditions is, to a large extent, limited to May, June, and July (Lauenroth and Milchunas, 1992; Sala et al., 1992). The early part of this period is most favorable for photosynthesis and growth of cool-season C_3 grasses and sedges (Kemp and Williams, 1980; Monson et al., 1983; Read and Morgan, 1996; Read et al., 1997). Furthermore, the distribution of soil water tends to be deeper in the profile early during the growing season than at other times (Sala et al., 1992), a factor that may be important for deeply rooted C_3 grasses.

As the growing season progresses and temperatures warm, growth of the dominant C_4 species *B. gracilis* and *B. dactyloides* commence, about a month to a month and a half after C_3 grasses and sedges first begin growth (Dickinson and Dodd, 1976; Monson and Williams, 1982). Precipitation amounts peak, and photosynthetic and growth rates reach seasonal maxima in May and June, resulting in rapid increases in biomass (Brown and Trlica, 1977; LeCain et al., 2002).

After the first of July, increased evaporative demand means that recharge of the soil water stores is unlikely, and water from precipitation is held only briefly in the surface soil layers before being transpired or evaporated (Sala et al., 1992).

These drier and warmer conditions in mid to late summer strongly favor growth of C_4 grasses. Rates of community photosynthesis tend to be less this time of year because of water stress (Brown and Trlica, 1977; LeCain et al., 2002). Continued photosynthesis and growth of C_4 grasses in late summer is due in large part to their warm-season C_4 metabolism, which can maintain photosynthetic activity up to relatively high temperatures (Brown and Trlica, 1977; Detling et al., 1978; Kemp and Williams, 1980; Monson et al., 1983; Read and Morgan, 1996; Reed et al., 1997), plus their shallow root system that is able to utilize soil water efficiently that is available mostly near the soil surface. The characteristic high water use efficiency (Monson et al., 1986; Morgan et al., 1998) and stomatal sensitivity (Morgan et al., 1998; Sala and Lauenroth, 1982; Sala et al., 1982) of the C_4 grass *B. gracilis* may be important adaptations that allow continued growth of this species at this time of year. High light saturation of photosynthesis in *B. gracilis* (Brown and Trlica, 1977; Williams and Kemp, 1978) may also be important in the dominance and functioning of this species during summer months in a region characterized by low leaf area and few cloudy days.

LeCain et al. (2002) illustrated how the seasonality of CO_2 fluxes on the shortgrass steppe is tied to soil water. They used chambers to measure CO_2 exchange every 2 to 3 weeks during the 1995 to 1997 growing seasons, and measurements were adjusted to subtract soil respiration, giving an estimate of community net photosynthesis. Precipitation in 1996 was close to long-term norms for the site (Fig. 12.12). In contrast, spring in 1995 was unusually wet, followed by a dry summer, whereas 1997 had below-average precipitation early in the year, followed by greater than normal precipitation late during the second half of the growing season. The effects of these different precipitation patterns on soil water content, and consequently on net photosynthesis of the plant community in this semiarid steppe, are evident (Fig. 12.12), and illustrate how seasonal and year-to-year variability in precipitation are, in large part, the basis for differences in NPP.

In summary, the temporal variation in productivity of the shortgrass steppe occurs primarily in response to variations in water and temperature, and involves the coexistence of different plant types in this grassland that are capable of growing and utilizing resources at different times of the year. The cool-season C_3 grasses take advantage of early-season cool temperatures and water stores that are deepest at that time of year. Both C_3 and C_4 grasses typically exhibit their highest growth rates in mid to late spring and on into early summer, but the typically warm, dry conditions of midsummer clearly favor the warm-season and shallow-rooted C_4 grasses that tend to have higher water use efficiency. The actual rates of NPP and the contribution made by different species can vary dramatically and depend on seasonal and year-to-year variability in soil water and temperature (LeCain et al., 2002).

Aboveground Net Primary Production

The combined patterns in the seasonality of precipitation and temperature in the central grassland region result in all the grasslands having similar temporal patterns of NPP. Average NDVI data from 1991 for a number of sites in each of the

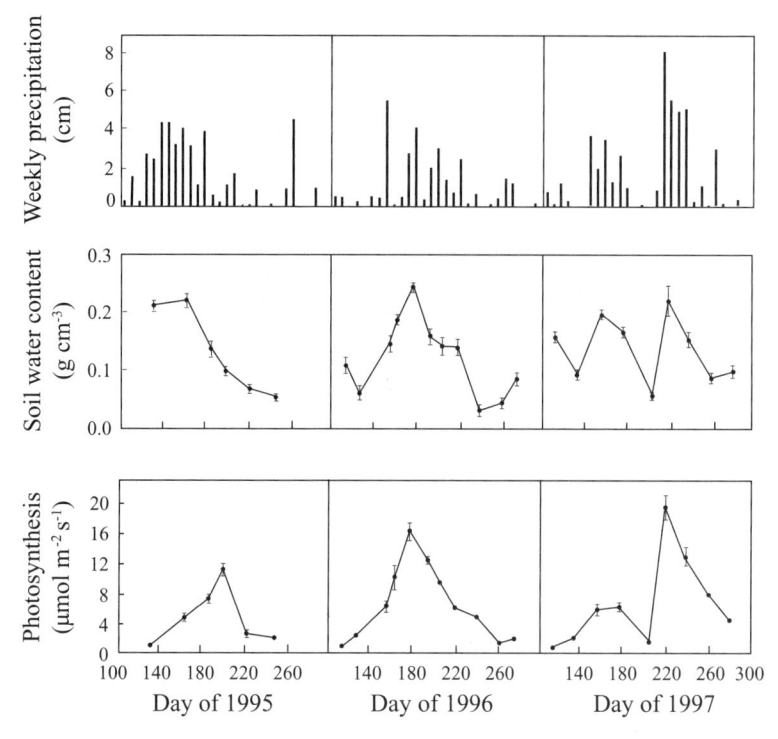

Figure 12.12 Weekly precipitation, soil water content, and chambered-determined photosynthesis rates of the shortgrass steppe 1995 to 1997 at the CPER. Photosynthesis measurements are gross rates, corrected for static chamber measurements of soil respiration. Photosynthesis data represent 5 means ± SEs. Soil water contents are 3 means ± SEs. For more details, see LeCain et al. (2002).

four grassland types suggests that the peak in green biomass for all types occurs between June 1 (day 152) and August 1 (day 213; Fig. 12.13A). Although average shortgrass steppe greenness is clearly lower than other types and shows strong seasonality, it is highly variable across sites (Fig. 12.13B). The peak in green biomass for these sites occurs between May 1 (day 121) and September 1 (day 244). Because of the important influence of the amount and timing of water availability in promoting ANPP, we should not be surprised to see such site-to-site variability in both the magnitude and timing of maximum green biomass. In addition to site-to-site variability in the same year, year-to-year variability in water availability results in differences in the timing and amount of green biomass. Lauenroth et al. (1980) found three distinct patterns of green biomass dynamics for 3 years at the CPER. Three peaks in green biomass occurred in 1973: one in early June, another in early August, and another in early September. In 1974, a single broad peak lasted from early July until early August, and 1975 was characterized by a single sharp peak in late June.

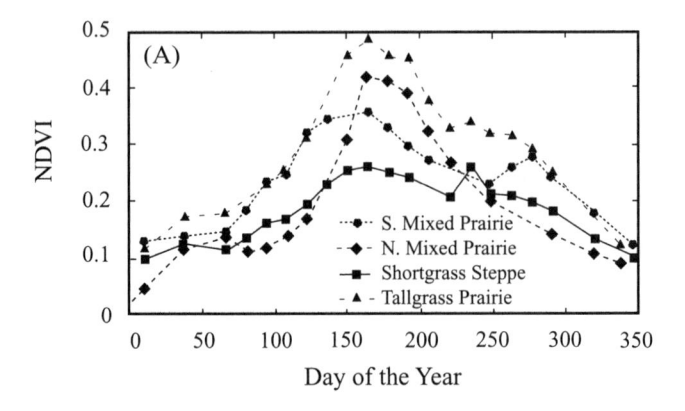

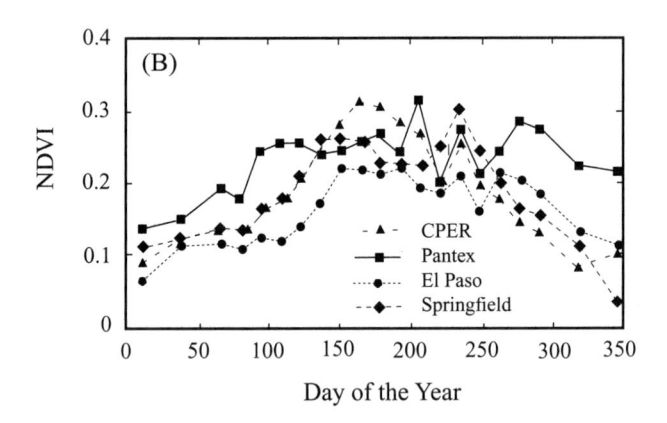

Figure 12.13 (A) Season dynamics of NDVI for the four grassland types in the central grassland region. (Adapted from Paruelo and Lauenroth [1995].) (B) Season dynamics of NDVI for four shortgrass steppe sites. (Adapted from Paruelo and Lauenroth [1995].)

Effects of Precipitation

Interannual variability in precipitation results in variability in water availability, and subsequent variability in annual NPP. Lauenroth and Sala (1992) analyzed 52 years of forage production data for the CPER and found a significant positive relationship between forage production and ANPP, making it possible to express the data in terms of ANPP. The 52-year average ANPP was 97 g·m^{-2} and deviated positively (maximum, 45 g·m^{-2}) and negatively (minimum, −36 g·m^{-2}) from the average (Fig. 12.14). During this time period, annual precipitation deviated positively (maximum, 267 mm) and negatively (minimum, −214 mm) from the 52-year mean of 321 mm·y^{-1}. Although there was a relatively good correspondence between the wet years and high ANPP, and dry years and low ANPP ($r=0.63$, $n=52$, $P < .01$), there was substantial residual variability not explained by annual precipitation. Using growing season precipitation improved the correlation a small

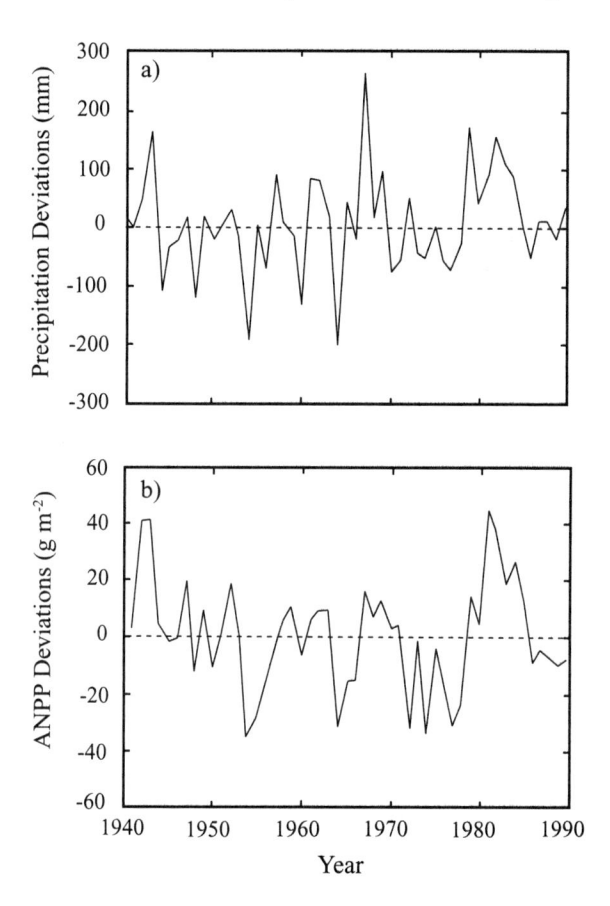

Figure 12.14 Deviations of annual precipitation (A) and deviations of annual aboveground net primary production (ANPP) (B) from their long-term means for the period 1939 to 1990 at the CPER. (Adapted from Lauenroth and Sala [1992].)

amount ($r=0.66$), but using different-size precipitation events did not result in any additional improvement ($r=0.67$).

Landscape Effects

The results found by Lauenroth and Sala (1992) were from a number of sites representing a variety of landscape positions. In at least some cases, individual landscape positions can have substantially different temporal dynamics of ANPP. Three landscape positions from a single catena were sampled at the CPER from 1983 to 1997 (Fig. 12.15). Interannual fluctuations of ANPP were similar for the midslope and upland positions, but quite different for the swale. All three

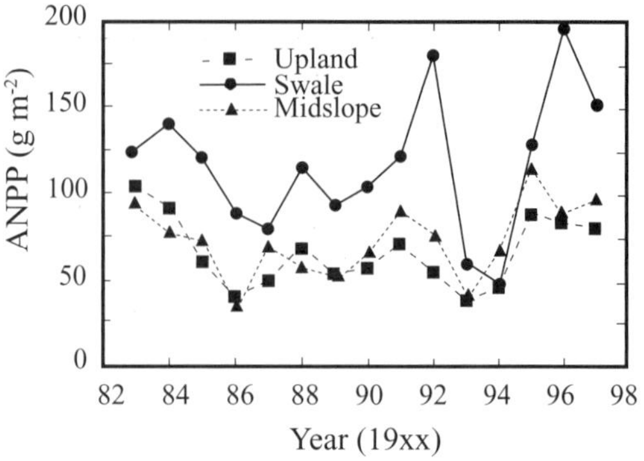

Figure 12.15 Interannual fluctuations in aboveground net primary production (ANPP) for three landscape positions on a single catena. (Data from Lauenroth and Sala [1992].)

landscape positions followed the same large-scale trends, but ANPP on the swale fluctuated more widely between low-production and high-production years. Minimum ANPP was similar for all three landscape positions, but maximum ANPP for the swale position was almost 200 g·m^{-2}, whereas for the upland and midslope positions it was between 100 and 125 g·m^{-2}. Aboveground NPP for all landscape positions was related to annual precipitation (Fig. 12.16).

The greater slope for the swale compared with the other two landscape positions suggests that the controlling factors may be different among the landscape positions. The classic catena model (Gerrard, 1981) is based upon the downslope movement of material as a result of the movement of water. Although many hill slopes at the CPER do not fit the classic catena model because of the importance of wind movement of material, it is possible that during certain years there is movement of water either aboveground or belowground from the upland and midslope positions to the swale position (Singh et al., 1998). Another characteristic of swales is that many of them have higher soil organic carbon and nitrogen, as well as higher nitrogen availability (Burke et al., 1999; Hook and Burke, 2000; Yonker et al., 1988). Higher nitrogen availability in the swale position would explain the greater slope.

Nitrogen Effects

Nitrogen additions increase ANPP in most nonforested terrestrial ecosystems, and the shortgrass steppe is not an exception (Dodd and Lauenroth, 1979; Hart et al., 1995; Lauenroth et al., 1978). Burke et al. (1997) reported that the shortgrass steppe has a low nitrogen use efficiency (ANPP/annual net nitrogen mineralization) compared with other locations in the central North American grassland region. One of many possible interpretations of this is that ANPP is more limited by water than by nitrogen. Several experiments have been conducted at the CPER that provide additional information about this issue, although only two provide direct information about the response of ANPP to nitrogen addition.

Hart et al. (1995) reported the results of an experiment that was carried out between 1979 and 1985, and consisted of the addition of 2.2 g N·m^{-2} annually

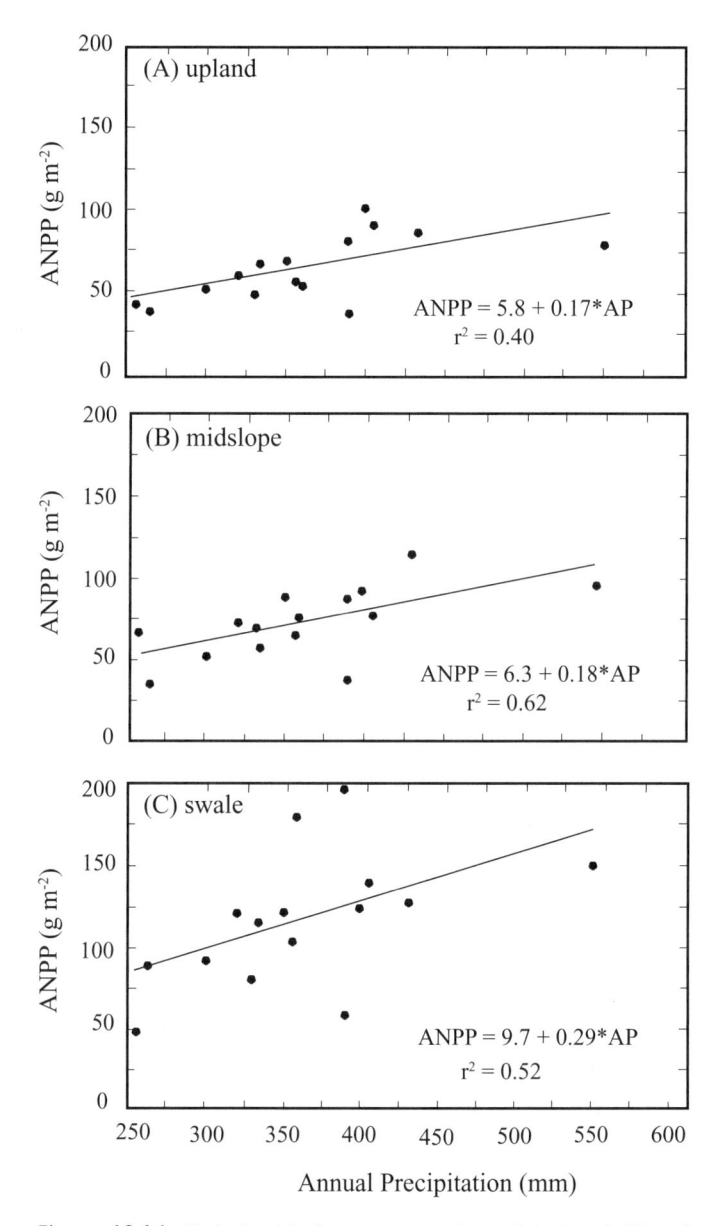

Figure 12.16 Relationship between annual precipitation (AP) and annual aboveground net primary production (ANPP) for upland (A), midslope (B), and swale (C) landscape positions at the CPER.

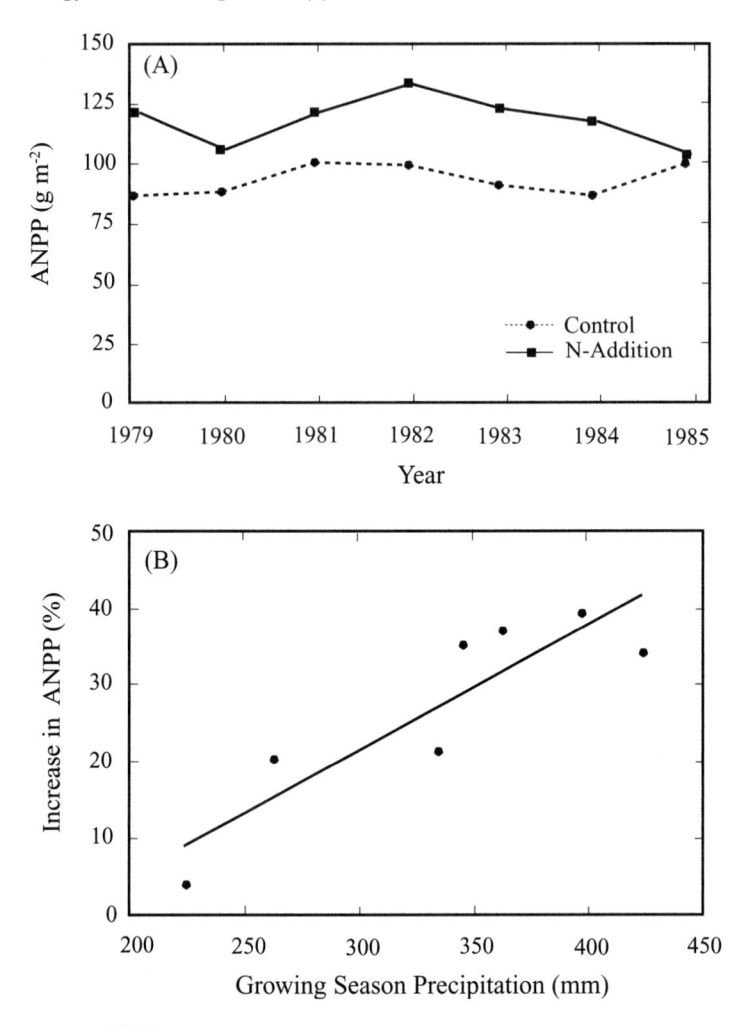

Figure 12.17 (A) The effect of nitrogen addition on annual aboveground net primary production (ANPP) at the CPER. (Data from Hart et al. [1995].) (B) Relationship between growing season precipitation and the amount by which ANPP was increased by nitrogen fertilization at the CPER. (Data from Hart et al. [1995].)

in the fall to two 32.4-ha pastures. Two similar-size pastures were used as controls. Nitrogen addition increased ANPP in each of the years of the experiment (Fig. 12.17A). The average increase in ANPP compared with the control was 27%, with a range of 4% to 39%. The variability in the response of ANPP to nitrogen addition was related to growing season precipitation (Fig. 12.17B). The largest increases occurred in years with high growing season precipitation, and the smallest increases occurred in years with low growing season precipitation. *Bouteloua gracilis* was the dominant species in terms of contributions to total ANPP in both

Table 12.1 Average Annual ANPP for Species and Species Groups with and without Nitrogen Fertilization

Species or Group	Control, kg·ha⁻¹	Nitrogen Addition, kg·ha⁻¹
Bouteloua gracilis	450 a	680 b
Cool-season grasses	40 c	80 d
Other grasses	80 e	60 e
Forbs	40 f	130 g

ANPP within a species or group not followed by the same letter are significantly different at $P \leq .05$. (Data from Hart et al. [1995].)

the control and nitrogen addition pastures, increasing by 50% as a result of nitrogen addition (Table 12.1). The forb group had the greatest percentage increase to nitrogen addition of 225%. Forbs contributed only 7% of total ANPP under control conditions, but increased to 14% with nitrogen addition. These results suggest an interaction between water and nitrogen availability in controlling ANPP at the CPER.

One of the most important experiments for understanding the long-term effects of resource availability on ANPP, as well as community structure of the shortgrass steppe, was started during the International Biological Program (IBP). (Results of this long-term experiment are reported in many other chapters in this book.) The interactive effects of water and nitrogen on ANPP were clearly illustrated in this experiment, which consisted of a factorial combination of nitrogen and water additions (Dodd and Lauenroth, 1979; Lauenroth et al., 1978). The unwatered nitrogen treatment consisted of an addition of inorganic nitrogen as ammonium nitrate at the rate of 5 g N·m⁻² over the level of the control plots. This resulted in the addition of 35 g N·m⁻² over the 5 years of the experiment to the nitrogen addition plots. The watered nitrogen treatment consisted of annual additions of 10 g N·m⁻² for a total of 50 g N·m⁻² to the water-plus-nitrogen plots in 5 years. The objective of the water addition treatment was to maintain soil water potential in the top 10 cm of the soil greater than–0.08 Mpa. This resulted in an average addition to the water-only plots of 550 mm per growing season and to the water-plus-nitrogen plots of 600 mm per growing season.

The addition of water and nitrogen increased ANPP in all but the first year of the experiment, when the treatments were not begun early enough to influence production (Fig. 12.18). Nitrogen on average increased ANPP by 100%, water addition increased ANPP by 250%, and water-plus-nitrogen increased ANPP by 700% (Fig. 12.18). The decreases in ANPP for the water and water-plus-nitrogen treatments during the last year of the experiment were the result of limited water additions. In both the Hart et al. (1995) nitrogen addition and Lauenroth et al. (1978) water and nitrogen experiments, the group of plants that had the greatest relative response to the addition of nitrogen was the forbs. In the results of Hart et al. (1995), grasses increased 44% and forbs increased 225%. The dominant grass, *B. gracilis*, increased by 50%. Hyder et al. (1975) suggested that on average during the 7 years of their experiment, the addition of 2.2 g N·m⁻² had negative

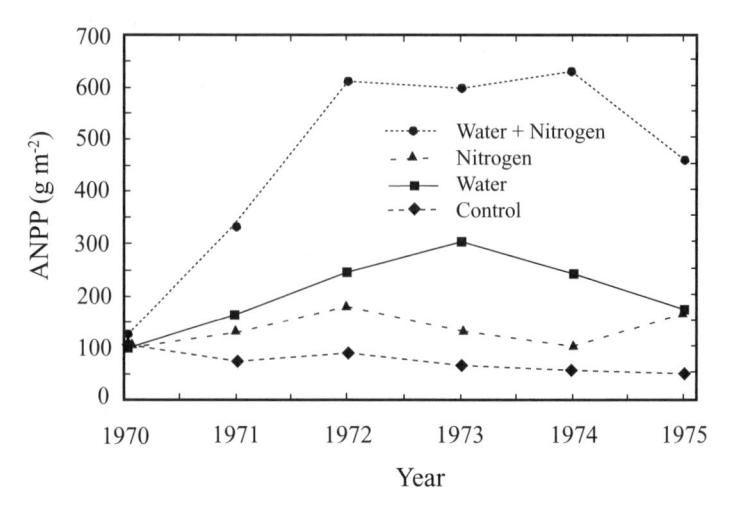

Figure 12.18 Effects of water, nitrogen, and water–nitrogen additions on annual aboveground net primary production (ANPP) at the CPER. (Adapted from Dodd and Lauenroth [1979].)

effects on the frequency of occurrence of perennial species and positive effects on annual species. In the water and nitrogen experiment by Lauenroth et al. (1978), nitrogen addition increased grasses by 70%, forbs by 425%, and the dominant grass by 50%. Water increased grasses by 175%, forbs by 725%, and the dominant grass by 175%. Water-plus-nitrogen increased grasses by 575%, forbs by 1600%, and the dominant grass by 650%. These results suggest two important conclusions. First, that it is not possible to speak of water or nitrogen as the single factor that controls ANPP at the CPER. Water addition had the greatest absolute and relative effect on ANPP, but the effects of nitrogen addition were also substantial. Second, there is a clear, positive interaction between water and nitrogen availability. With a similar amount of nitrogen available in the soil, increasing water availability had a huge positive influence. Average ANPP of the water-plus-nitrogen treatment was 3.7 times greater than ANPP of the nitrogen addition treatment. It is also true that with similar amounts of water availability, the addition of nitrogen had a huge effect on ANPP. Average ANPP of the water-plus-nitrogen treatment was 2.3 times that of the water treatment. Interestingly, many of the effects of nitrogen have persisted decades after the nitrogen additions were halted (Burke et al., chapter 13, this volume; Lauenroth, chapter 5, this volume; Milchunas and Lauenroth, 1995).

Effects of Carbon Dioxide

The CO_2 concentration of the atmosphere has increased 30% during the past 150 years and is expected to continue to increase into the foreseeable future (IPCC, 2007; Körner, 2000). Because the atmosphere is the source of carbon to plants, there are good theoretical reasons to expect that increasing CO_2 concentrations will result in increases in NPP in many ecosystems, and especially those dominated by C_3 species (Bowes, 1993). There are also good theoretical reasons for ecosystems such as the shortgrass steppe that are dominated by C_4 species to experience minimal increases in NPP (Bowes, 1993). In contrast to theoretical predictions, many C_4-dominated ecosystems, including the shortgrass steppe, have shown positive responses of NPP to elevated CO_2 (Ghannoum et al., 2000; Hunt et al., 1996; Morgan et al., 2001; Wand et al., 1999). We have conducted two CO_2 enhancement experiments for the shortgrass steppe, one in the greenhouse (Hunt et al., 1996; LeCain and Morgan, 1998; Morgan et al., 1998) and another in large open-top field chambers experiencing ambient weather conditions (LeCain et al., 2003; Morgan et al., 2001, 2004).

Under a greenhouse treatment of 700 $\mu L \cdot L^{-1}$ CO_2, biomass responses of both *B. gracilis* (C_4) and *A. smithii* (C_4) were greater than the control, but not different from each other (Hunt et al., 1996). For both species, their response to elevated CO_2 was enhanced with water addition. The response of *B. gracilis* was greatest at ambient temperatures, but the response of *A. smithii* was increased by elevated temperature (Hunt et al., 1996).

In open-top field chambers and 720 $\mu L \cdot L^{-1}$ CO_2, ANPP was increased an average of 42% over 2 years, both of which were wetter than average (Morgan et al., 2001, 2004). The increase in ANPP was spread proportionately over C_4 and C_3 species, suggesting an equal response of plants of the two photosynthetic pathway types. The chamber effect resulted in an average increase of 57% in ANPP, suggesting a positive effect of elevated temperature. The available information suggests that the shortgrass steppe will have a positive response of ANPP to increasing CO_2 concentrations and perhaps to warming.

Belowground Net Primary Production

Belowground NPP has been a major hurdle in our understanding of shortgrass steppe ecosystem structure and function. The challenge has been finding a method that will produce reliable results. This is a problem that is common to the study of all ecosystems, not just the shortgrass steppe (Lauenroth, 2000), but the magnitude of belowground allocation and its importance for consumers in this ecosystem (see Moore et al., chapter 11, this volume) makes it especially important that we understand BNPP. We have used two methods in the shortgrass steppe. The first, sequential harvesting of soil cores, has been used at two sites: the CPER (Colorado) and Pantex (located in the panhandle of Texas near Amarillo). The second method, ^{14}C turnover, has been used at the CPER.

One of the key difficulties in estimating BNPP is separating the annual increment from the standing crop of belowground plant biomass. A defining characteristic of semiarid grasslands is that a large fraction of the total plant biomass occurs belowground. In the shortgrass steppe, 70% to more than 80% of the total plant biomass (live+dead) occurs belowground (Table 12.2) (Sims et al., 1978). There are two components to belowground biomass: crowns and roots (including rhizomes). Crowns represent the transition zone between above- and belowground

Table 12.2 Above- and Belowground Biomass Components for Two Sites in the Shortgrass Steppe

		CPER		Pantex	
	Year	Ungrazed	Grazed	Ungrazed	Grazed
Aboveground					
Live					
	1970	63	48	25	62
	1971	58	82	78	103
	1972	89	57	107	93
	Mean	70	62	70	86
Recent dead					
	1970	—	—	20	19
	1971	—	—	70	71
	1972	35	28	10	10
	Mean	35	28	33	33
Old dead					
	1970	—	—	52	45
	1971	—	—	101	84
	1972	39	37	136	109
	Mean	39	37	96	79
Litter					
	1970	180	164	182	185
	1971	140	91	273	214
	1972	210	128	242	247
	Mean	177	128	232	215
Total					
	1970	243	212	279	311
	1971	198	173	522	472
	1972	373	250	495	459
	Mean	271	212	432	414
Belowground					
Crown					
	1970	—	—	123	186
	1971	340	322	289	302
	1972	258	308	349	410
	Mean	299	315	254	299
Roots					
	1970	1631	1701	1212	1631
	1971	1014	996	344	402
	1972	803	1290	580	562
	Mean	1149	1329	620	725
Total					
	1970	1631	1701	1335	1817
	1971	1354	1318	633	704
	1972	1061	1598	929	972
	Mean	1349	1539	966	1164

Net Primary Production in the Shortgrass Steppe 297

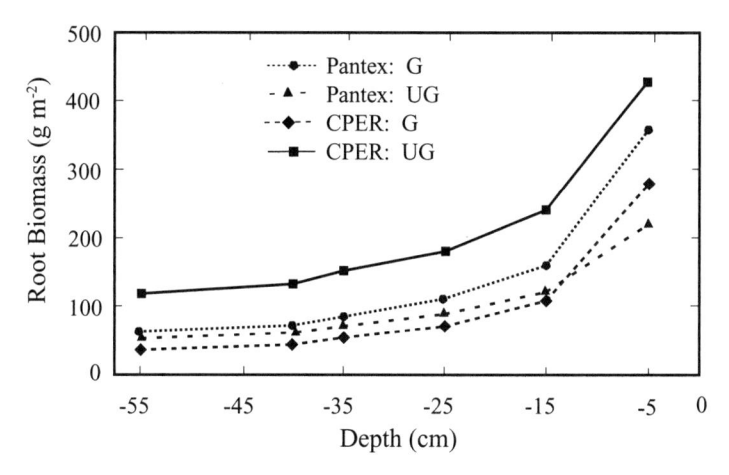

Figure 12.19 Relationship between root biomass and depth for two shortgrass steppe sites. Pantex is the Texas Tech University Research Farm in Amarillo, Texas, and CPER is the Central Plains Experimental Range. G, grazed; UG, ungrazed treatments. (Data from Sims and Singh [1978].)

organs in grasses, and in the shortgrass steppe, this includes all the perennial aboveground material. At the CPER and Pantex sites, crowns represent 20% to 26% of belowground biomass. Roots and rhizomes represent the remaining 74% to 80%.

Root biomass is distributed as a negative exponential function with depth (Fig. 12.19). Both sites have more than 30% of their roots in the top 10 cm of the soil regardless of whether they are grazed. The top 30 cm of the soil contains 70% to 80% of the root biomass and, because all the crowns are at or just below the soil surface, the 0 to 30-cm-depth increment includes more than 70% to 80% of the total belowground biomass.

Sims and Singh (1978) reported estimates of BNPP using sequential harvesting and summation of significant increments in biomass (Table 12.3) (Singh et al., 1975). Estimates of BNPP ranged from approximately 400 to almost 1000 g·m^{-2}. These values are associated with ANPP-to-BNPP ratios of 0.39 to 0.22, suggesting that BNPP is 2.5 to 4.5 times greater than ANPP. However, analysis of methods of estimating BNPP from harvest data has suggested that there is a high likelihood that all estimates based upon harvest data have a positive bias (Singh et al., 1984; Sala et al., 1988a).

Carbon isotope labeling resulted in values of BNPP substantially smaller than those reported by Sims and Singh in 1978 (Milchunas and Lauenroth, 1992). This method requires a pulse label with a carbon tracer (either ^{14}C or ^{13}C) and relies on an estimate of the time required for all of the tracer to be lost from the belowground organs (turnover time). Milchunas and Lauenroth (1992) estimated values of BNPP ranging from 202 to 262 g·m^{-2} and ANPP-to-BNPP ratios ranging from

Table 12.3 Estimates of Aboveground and Belowground ANPP for Two Sites in the Shortgrass Steppe

	CPER		*Pantex*	
	Ungrazed	Grazed	Ungrazed	Grazed
Aboveground				
1970	160	123	155	155
1971	218	108	289	218
1972	138	77	327	302
Mean	172	103	257	225
Belowground				
1970	411	471	417	410
1971	686	722	606	987
1972	607	429	876	968
Mean	568	541	633	788

Data from Sims and Singh (1978, Tables 2 and 4).

0.42 to 0.72. These results suggested that BNPP is 1.4 to 2.4 times greater than ANPP in the shortgrass steppe. With further sampling of these plots, we found some complications with the method as a result of contamination of roots by soil (Milchunas and Lauenroth, 2001). Correction for contamination resulted in a 22% increase in the 10-year average annual root production from 183 to 223 $g \cdot m^{-2}$. Because the 10-year uncorrected average production for roots is essentially identical to the original 4-year average (183 $g \cdot m^{-2}$ vs. 175 $g \cdot m^{-2}$), it seems reasonable to assume that the average for crowns will be similar. If that is true, the average ANPP-to-BNPP ratio is approximately 0.36, suggesting that BNPP is 2.8 times greater than ANPP. This falls within the lower portion of the range reported by Sims and Singh (1978) of 2.5 to 4.5.

Comparison of Spatial and Temporal Patterns

Shortgrass scientists have conducted a great deal of research on the spatial patterns in ANPP across grassland regions (e.g., Epstein et al., 1997a, b; Sala et al., 1988b), as well as conducted long-term monitoring and experimental analyses (Lauenroth and Sala, 1992; Lauenroth et al., 1978). These experiments have all demonstrated that precipitation is a key control over NPP. However, are the relationships gained from analysis of large spatial gradients the same as those from long-term, site-level analyses? Conceptually, the relationships differ. In the temporal case, the relationship is between ANPP in a specific year and the amount of precipitation received during that year; in the spatial case, the relationship is between mean annual ANPP and mean annual precipitation. The temporal model reflects the ability of the vegetation to capitalize on the amount of precipitation and nitrogen made available as a result of the amount and timing of water inputs in a specific year.

Because water availability is a key control on NPP in the shortgrass region, one of the ways to compare spatial and temporal patterns is to ask whether the slopes of regressions between ANPP and precipitation are similar for spatial and temporal data sets. Lauenroth and Sala (1992) summarized 52 years of ANPP data for the CPER and developed a regression between ANPP and annual precipitation:

$$\text{ANPP}_i = 56 + 0.131 \times \text{APPT}_i$$

where ANPP_i is ANPP in year i (measured in grams per square meter) and APPT_i is annual precipitation in year i (measured in millimeters). They then compared this temporal model with a spatial model developed for the entire grassland region by Sala et al. (1988b):

$$\text{ANPP}_m = -34 + 0.60 \text{ MAP}$$

where ANPP_m is mean ANPP (measured in grams per square meter) and MAP is mean annual precipitation (measured in millimeters). It is clear from this comparison that the temporal relationship is different from the spatial relationship (Fig. 12.20). That is, the response of ANPP to precipitation for a given year and at a single site is not what we would predict, based upon relationships generated from regional average precipitation and production data. In dry years ($\text{APPT}_i <$ MAP), the vegetation (the number, identity, and size of plants) and the biogeochemistry (the size and quality of organic matter pools) reflect the average wetter conditions of the site, and ANPP exceeds the value one would expect for a site

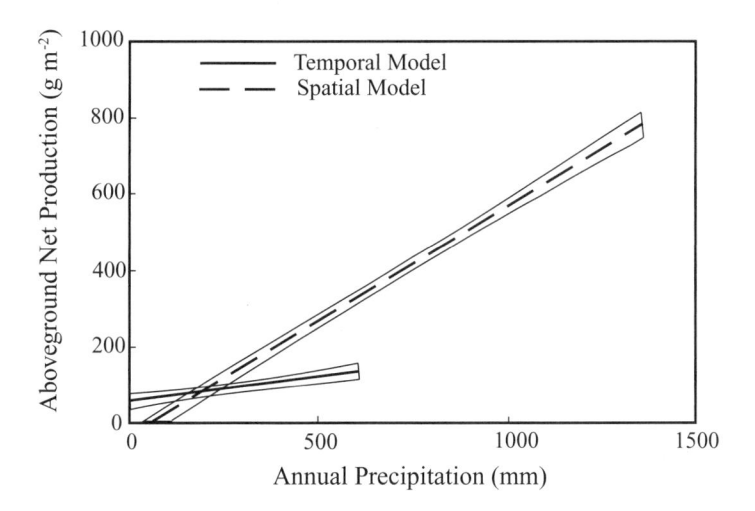

Figure 12.20 Relationships between annual aboveground net primary production (ANPP) from a spatial model for the central grassland region (Sala et al., 1988b) and a temporal model for the CPER (Lauenroth and Sala, 1992). The shaded areas are the 95% confidence interval. The spatial model is $\text{ANPP} = 0.60 \times \text{AP} - 34$ and the temporal model is $\text{ANPP} = 0.13 \times \text{AP} + 56$. AP, annual precipitation. (Adapted from Lauenroth and Sala [1992].)

that had $APPT_i = MAP$. In a wet year, ($APPT_i > MAP$), the vegetation and biogeochemical processes at a single site reflect drier conditions, and cannot capitalize on the extra water available, so that ANPP is lower than the spatial model would predict for a site with $APPT_i = MAP$.

Our results from the shortgrass steppe have substantial implications for understanding the response of ecosystems to changes in climate—short term or long term. For instance, simulation models that utilize such long-term spatial relationships to reflect the relationship between climate and production cannot adequately predict how a single site will change.

Furthermore, our results (Lauenroth and Sala, 1992) have been validated at other locations. For instance, Paruelo et al. (1999) evaluated relationships between spatial and temporal models for the entire grassland region precipitation gradient (200 mm < MAP < 1200 mm). They found that temporal precipitation use efficiency ($PUE_t = \Delta ANPP/\Delta APPT$) was low at both the dry (shortgrass steppe) and wet (tallgrass prairie) ends of the gradient. Their interpretation was that at the dry end of the precipitation gradient, vegetation constraints were the dominant influence (that is, that the ability of vegetation to respond to changes in precipitation limited ANPP response). At the wet end of the gradient, they inferred that biogeochemical constraints dominated (that is, that the ability of ANPP to respond was limited by nitrogen or other nutrient availability). These results suggest that the low slope of the shortgrass steppe temporal model reported by Lauenroth and Sala (1992) is the result of the low productive potential of the dominant shortgrass species (*B. gracilis* and *B. dactyloides*).

Our results have been further validated in the tallgrass prairie (Briggs and Knapp, 1995). The slope of the relationship between ANPP and annual precipitation was substantially lower than that reported by Sala et al. (1988b) for the grassland region and very close to the similar slope for the shortgrass steppe.

Summary

The shortgrass steppe occupies the warmest and driest portion of the central grassland region and as a consequence is the least productive of the grassland types. More than half the shortgrass steppe remains in native grasslands, whereas the remainder has been converted to croplands. Native grasslands are dominated by short-stature C_4 grasses, and ANPP ranges from 50 to more than 300 $g \cdot m^{-2}$. Major crops include corn on irrigated fields and wheat under dryland conditions. Average annual ANPP for corn ranges from 4250 to 4700 $g \cdot m^{-2}$ and average annual ANPP for wheat ranges from 260 to 330 $g \cdot m^{-2}$. The key variables influencing the spatial pattern of ANPP for both crops and native grasslands are water and soil texture. The major effects of soil texture are through its effects on the storage and availability of both water and nutrients.

The seasonality of temperature and precipitation account for an early- to midsummer peak in ANPP of native grasslands. Interannual variability in precipitation is the key determinant of interannual availability of soil water, which is the most frequent control on annual ANPP. Other factors that can affect temporal variability of ANPP include landscape effects, soil nitrogen, and atmospheric CO_2. Landscape position influences water and nutrient availability through topography (runoff and runon) and soil texture. Although connections between annual variability in nitrogen supply and ANPP are difficult to find, additions of nitrogen as fertilizer have consistently resulted in increases in ANPP. The increases are largest in the wettest years, or when nitrogen addition is combined with water addition. Effects of CO_2 have been demonstrated with open-top chambers and resulted in an average 42% increase over 2 years.

Annual belowground NPP is one of the most difficult quantities to estimate for any ecosystem, including the shortgrass steppe, and is particularly important for this system. We have estimated BNPP for shortgrass ecosystems using biomass harvest and carbon isotope turnover techniques. Biomass harvest estimates range from 400 to almost 1000 g·m^{-2}, and carbon isotope turnover estimates of BNPP range from 180 to 225 g·m^{-2}. The carbon isotope turnover estimates seem to have the fewest problems and therefore represent the best estimates of annual BNPP for shortgrass ecosystems.

References

Anderson, G. L. 1991. *Using simulation models and spatial information systems for evaluating rangeland ecosystems.* PhD diss., Colorado State University, Fort Collins, Colo.

Black, C. C. 1971. Ecological implications of dividing plants into groups with distinct photosynthetic production capacities. *Advances in Ecological Research* **7**:87–114.

Black, C. C., Jr. 1973. Photosynthetic carbon fixation in relation to net CO2 uptake. *Annual Review of Plant Physiology* **24**:253–286.

Bowes, G. 1993. Facing the inevitable: Plants and increasing atmospheric CO_2. *Annual Review of Plant Physiology and Plant Molecular Biology* **44**:309–332.

Briggs J. M. and A. K. Knapp. 1995. Interannual variability in primary production in tallgrass prairie—climate, soil-moisture, topographic position, and fire as determinants of aboveground biomass. *American Journal of Botany* **82**:1024–1030.

Brown, L. F., and M. J. Trlica. 1977. Interacting effects of soil water, temperature and irradiance on CO_2 exchange rates of two dominant grasses of the shortgrass prairie. *Journal of Applied Ecology* **14**:197–204.

Burke, I. C., W. K. Lauenroth, and W. J. Parton. 1997. Regional and temporal variation in net primary production and net nitrogen mineralization in grasslands. *Ecology* **78**:1330–1340.

Burke, I. C., W. K. Lauenroth, R. Riggle, P. Brannen, B. Madigan, and S. Beard. 1999. Spatial variability of soil properties in the shortgrass steppe: The relative importance of topography, grazing, microsite, and plant species in controlling spatial patterns. *Ecosystems* **2**:422–438.

Coffin, D. P., and W. K. Lauenroth. 1990. A gap dynamics simulation model of succession in a semiarid grassland. *Ecological Modelling* **49**:229–266.

Cosby, B. J., G. M. Hornberger, R. B. Clapp, and T. R. Ginn. 1984. A statistical exploration of the relationships of soil moisture characteristics to the physical properties of soils. *Water Resources Research* **20**:682–690.

Coupland, R. T., and R. E. Johnson. 1965. Rooting characteristics of native grassland species in Saskatchewan. *Ecology* **53**:475–507.

Detling, J. K., W. J. Parton, and H. W. Hunt. 1978. An empirical model for estimating CO_2 exchange of *Bouteloua gracilis* (H.B.K.) Lag. in the shortgrass prairie. *Oecologia* **33**:137–147.

Dickinson, C. E., and J. L. Dodd. 1976. Phenological pattern in the shortgrass prairie. *American Midland Naturalist* **96**:367–378.

Dodd, J. L., and W. K. Lauenroth. 1979. Analysis of the response of a grassland ecosystem to stress, pp. 43–58. In: N. R. French (ed.), *Perspectives in grassland ecology*. Springer-Verlag, New York.

Dodd, M. B., and W. K. Lauenroth. 1997. The influence of soil texture on the soil water dynamics and vegetation characteristics of a shortgrass steppe ecosystem. *Plant Ecology* **133**:13–28.

Epstein, H. E., W. K. Lauenroth, and I. C. Burke. 1997a. Effects of temperature and soil texture on aboveground net primary production in the U.S. Great Plains. *Ecology* **78**:2628–2631.

Epstein, H. E., W. K. Lauenroth, I. C. Burke, and D. P. Coffin. 1997b. Productivity patterns of C_3 and C_4 functional types in the U.S. Great Plains. *Ecology* **78**:722–731.

Fan, W. 1993. *Regional analysis of plant species and environmental variables in eastern Colorado*. PhD diss., Colorado State University, Fort Collins, Colo.

Ghannoum, O., S. van Caemmerer, L. H. Ziska, and J. P. Conroy. 2000. The growth response of C_4 plants to rising atmospheric CO_2 partial pressure: A reassessment. *Plant, Cell, and Environment* **23**:931–942.

Gerrard, A. 1981. *Soils and landforms*. George Allen and Unwin, Boston, Mass.

Hart R. H., M. C. Shoop, and M. M. Ashby. 1995. Nitrogen and atrazine on shortgrass: Vegetation, cattle and economic responses. *Journal of Range Management* **48**:165–171.

Hook, P. B., and I. C. Burke. 2000. Biogeochemistry in a shortgrass landscape: Control by topography, soil texture, and microclimate. *Ecology* **81**:2686–2703.

Hunt, H. W., E. T. Elliot, J. K. Detling, J. A. Morgan, and D.- X. Chen. 1996. Responses of a C_3 and a C_4 perennial grass to elevated CO_2 and temperature under different water regimes. *Global Change Biology* **2**:35–47.

Hyder, D. N., R. E. Bement, E. E. Remmenga, and D. F. Hervey. 1975. *Ecological responses of native plants and guidelines for management of shortgrass range*. USDA technical bulletin no. 1503.U. S. Government Printing Office, Washington, D.C.

IPCC. 2007. Climate change 2007: Mitigation. Contribution of working group III to the fourth assessment report of the Intergovernmental Panel on Climate Change. Cambridge University Press, Cambridge, UK.

Kemp, P. R., and G. J. Williams, III. 1980. A physiological basis for niche separation between *Agropyron smithii* (C_3) and *Bouteloua gracilis* (C_4). *Ecology* **61**:846–858.

Knight, D. H. 1994. *Mountains and plains*. Thomson-Shore, Dexter, Mich.

Korner, C. 2000. Biosphere responses to CO2 enrichment. *Ecological Applications* **10**:1590–1619.

Küchler, A. W. 1964. *Potential natural vegetation of the conterminous United States*. American Geographical Society, New York.

Lane, D. R., D. P. Coffin, and W. K. Lauenroth. 1998. Effects of soil texture and precipitation on above-ground net primary productivity and vegetation structure across the central grassland region of the United States. *Journal of Vegetation Science* **9**:239–250.

Lane, D. R., D. P. Coffin, and W. K. Lauenroth. 2000. Changes in grassland canopy structure across a precipitation gradient. *Journal of Vegetation Science* **11**:359–368.

Lauenroth, W. K. 1979. Grassland primary production: North American grasslands in perspective, pp. 3–24. In: N. R. French (ed.). *Perspectives in grassland ecology*. Springer-Verlag, New York.

Lauenroth, W. K. 2000. Methods of estimating belowground net primary production, pp. 58–71. In: Sala, O. E., R. B. Jackson, H. Mooney and R. W. Howarth (eds.). *Methods in ecosystem science*. Springer-Verlag, New York.

Lauenroth, W. K., and J. B. Bradford. 2006. Ecohydrology and the partitioning AET between transpiration and evaporation in a semiarid steppe. *Ecosystems* **9**:756–767.

Lauenroth, W. K., and I. C. Burke. 1995. Great Plains, climate variability. *Encyclopedia of Environmental Biology* **2**:237–249.

Lauenroth, W. K., I. C. Burke, and M. P. Gutmann. 1999. The structure and function of ecosystems in the central North American grassland region. *Great Plains Research* **9**:223–259.

Lauenroth, W. K., I. C. Burke, and J. M. Paruelo. 2000. Patterns of production and precipitation use efficiency in winter wheat and native grasslands in the central Great Plains of the United States. *Ecosystems* **3**:344–351.

Lauenroth, W. K., J. L. Dodd, and P. L. Sims. 1978. The effects of water and nitrogen-induced stresses on plant community structure in a semiarid grassland. *Oecologia* **36**:211–222.

Lauenroth, W. K., J. L. Dodd, and C. E. Dickinson. 1980. Biomass dynamics of blue grama in a shortgrass prairie and evaluation of a method for separating live and dead material. *Journal of Range Management* **33**:210-212.

Lauenroth, W. K., and D. G. Milchunas. 1992. Shortgrass steppe, pp. 183–226. In: R. T. Coupland (ed.), *Natural grasslands: Introduction and western hemisphere*. Elsevier, Amsterdam.

Lauenroth, W. K., and O. E. Sala. 1992. Long term forage production of North American shortgrass steppe. *Ecological Applications* **2**:397–403.

LeCain, D. R., J. A. Morgan. 1998. Growth, gas exchange, leaf nitrogen and carbohydrate concentrations in NAD-ME and NADP-ME C-4 grasses grown in elevated CO_2. *Physiologia Plantarum* **102**:297–306.

LeCain, D. R., J. A. Morgan, A. R. Mosier, and J. A. Nelson. 2003. Soil and plant water relations determine photosynthetic responses of C-3 and C-4 grasses in a semi-arid ecosystem under elevated CO_2. *Annals of Botany* **92**(1):41–52.

LeCain, D. R., J. A. Morgan, G. E. Schuman, J. D. Reeder, and R. H. Hart. 2002. Carbon exchange and species composition of grazed pastures and exclosures in the shortgrass steppe of Colorado. *Agriculture, Ecosystems and the Environment* **93**:421–435.

Milchunas, D. G., and W. K. Lauenroth. 1992. Carbon dynamics and estimates of primary production by harvest, C-14 dilution and C-14 turnover. *Ecology* **73**:1593–1607.

Milchunas, D. G., and W. K. Lauenroth. 1995. Inertia in plant community structure-state changes after cessation of nutrient-enrichment stress. *Ecological Applications* **5**:452–458.

Monson, R. K., R. O. Littlejohn, Jr., and G. J. Williams, III. 1983. Photosynthetic adaptation to temperature in four species from the Colorado shortgrass steppe: A physiological model for coexistence. *Oecologia* **58**:43–51.

Monson, R. K., M. R. Sackschewsky, and G. J. Williams, III. 1986. Field measurements of photosynthesis, water-use efficiency, and growth in *Agropyron smithii* (C_3) and *Bouteloua gracilis* (C_4) in the Colorado shortgrass steppe. *Oecologia* **68**:400–409.

Monson, R. K., and G. J. Williams, III. 1982. A correlation between photosynthetic temperature adaptation and seasonal phenology patterns in the shortgrass prairie. *Oecologia* **54**:58–62.

Morgan, J. A., D. R. LeCain, A. R. Mosier, and D. G. Milchunas. 2001. Elevated CO_2 enhances water relations and productivity and affects gas exchange in C_3 and C_4 grasses of the Colorado Shortgrass Steppe. *Global Change Biology* **7**:1–16.

Morgan, J. A., D. R. LeCain, J. J. Read, H. W. Hunt, and W. G. Knight. 1998. Photosynthetic pathway and ontogeny affect water relations and the impact of CO_2 on *Bouteloua gracilis* (C_4) and *Pascopyrum smithii* (C_3). *Oecologia* **114**:483–493.

Morgan, J. A., A. R. Mosier, D. G. Milchunas, D. R. LeCain, J. A. Nelson, and W. J. Parton. 2004. CO2 enhances productivity, alters species composition, and reduces digestibility of shortgrass steppe vegetation. *Ecological Applications* **14**:208–219.

Noy-Meir, I. 1973. Desert ecosystems: Environment and producers. *Annual Review of Ecology and Systematics* **4**:25–51.

Paruelo, J. M., H. E. Epstein, W. K. Lauenroth, and I. C. Burke. 1997. ANPP estimates from NDVI for the central grassland region of the United States. *Ecology* **78**:953–958.

Paruelo, J. M., and W. K. Lauenroth. 1995. Regional patterns of normalized difference vegetation index in North American shrublands and grasslands. *Ecology* **76**:1888–1898.

Paruelo, J. M., W. K. Lauenroth, I. C. Burke, and O. E. Sala. 1999. Grassland precipitation use efficiency varies across a resource gradient. *Ecosystems* **2**:64–67.

Read, J. J., and J. A. Morgan. 1996. Growth and partitioning in *Pascopyrum smithii* (C_3) and *Bouteloua gracilis* (C_4) as influenced by carbon dioxide and temperature. *Annals of Botany* **77**:487–496.

Read, J. J., J. A. Morgan, N. J. Chatterton, and P. A. Harrison. 1997. Gas exchange and carbohydrate and nitrogen concentrations in leaves of *Pascopyrum smithii* (C_3) and *Bouteloua gracilis* (C_4) at different carbon dioxide concentrations and temperatures. *Annals of Botany* **79**:197–206.

Rutherford, M. C. 1980. Annual plant production precipitation relations in arid and semi-arid regions. *South African Journal of Science* **76**:53–56.

Sala, O. E., M. E. Biondini, and W. K. Lauenroth. 1988a. Bias in estimates of primary production: An analytical solution. *Ecological Modelling* **44**:43–55.

Sala, O. E., and W. K. Lauenroth. 1982. Small rainfall events, an ecological role in semi-arid regions. *Oecologia* **53**:301–304.

Sala, O. E., W. K. Lauenroth, and R. A. Gollucio. 1997. Plant functional types in temperate semiarid regions, pp. 217–233. In: T. M. Smith, H. H. Shugart, and F. I. Woodward (eds.), *Plant functional types*. Cambridge University Press, Cambridge, UK.

Sala, O. E., W. K. Lauenroth, and W. J. Parton. 1992. Long term soil water dynamics in the shortgrass steppe. *Ecology* **73**:1175–1181.

Sala, O. E., W. K. Lauenroth, and C. P. P. Reid. 1982. Water relations: A new dimension for niche separation between *Bouteloua gracilis* and *Agropyron smithii* in North American semi-arid grasslands. *Journal of Applied Ecology* **19**:647–657.

Sala, O. E., W. J. Parton, L. A. Joyce, and W. K. Lauenroth. 1988b. Primary production of the central grassland region of the United States. *Ecology* **69**:40–45.

Sims, P. L., and J. S. Singh. 1978. The structure and function of ten western North American grasslands: III. Net primary production, turnover and efficiencies of energy capture and water use. *Journal of Ecology* **66**:573–597.

Sims, P. L., J. S. Singh, and W. K. Lauenroth. 1978. The structure and function of ten western North American grasslands: I. Abiotic and vegetational characteristics. *Journal of Ecology* **66**:251–285.

Singh, J. S., W. K. Lauenroth, and R. K. Steinhorst. 1975. Review and assessment of various techniques for estimating net aerial primary production in grasslands from harvest data. *Botanical Review* **41**:181–232.

Singh, J. S., W. K. Lauenroth, H. W. Hunt, and D. M. Swift. 1984. Bias and random errors in estimators of net root production: A simulation approach. *Ecology* **65**:1760–1764.

Singh, J. S., D. G. Milchunas, and W. K. Lauenroth. 1998. Soil water dynamics and vegetation patterns in a semiarid grassland. *Plant Ecology* **134**:77–89.

Todd, S. W. 1994. Spatiotemporal estimation of biomass on the shortgrass steppe using Landsat TM vegetation and soil indices, field data, and simulation models. PhD diss., Colorado State University, Fort Collins, Colo.

USDA. 1967. *National handbook for range and related grazing lands.* Soil Conservation Service, U.S. Department of Agriculture, Washington, D.C.

USDA. 1974. *Summer fallow in the western United States.* Conservation research report no. 17. U.S. Department of Agriculture, Agricultural Research Service, Washington, D.C.

Wand, S. J. E., G. F. Midgley, M. H. Jones, and P. S. Curtis. 1999. Responses of wild C_4 and C_3 grass (Poaceae) species to elevated atmospheric CO_2 concentration: A meta analytic test of current theories and perceptions. *Global Change Biology* **5**:723–741.

Weaver, J. E., and F. W. Albertson. 1956. *Grasslands of the Great Plains: Their nature and use.* Johnsen, Lincoln, Neb.

Williams, G. J., III, and P. R. Kemp. 1978. Simultaneous measurements of leaf and root gas exchange of shortgrass prairie species. *Botanical Gazette* **150**:150–157.

Yonker, C. M., D. S. Schimel, E. Paroussis, and R. D. Heil. 1988. Patterns of organic carbon accumulation in a semiarid shortgrass steppe. *Soil Science Society of America Journal* **52**:478–483.

13

Soil Organic Matter and Nutrient Dynamics of Shortgrass Steppe Ecosystems

Ingrid C. Burke
Arvin R. Mosier
Paul B. Hook
Daniel G. Milchunas
John E. Barrett
Mary Ann Vinton
Rebecca L. McCulley
Jason P. Kaye
Richard A. Gill
Howard E. Epstein
Robin H. Kelly
William J. Parton
Caroline M. Yonker
Petra Lowe
William K. Lauenroth

Since the days of the IBP, there has been a strong emphasis on research about the biogeochemistry of shortgrass steppe ecosystems (e.g., Clark, 1977; Woodmansee, 1978). A major theme has been seeking to understand spatial and temporal patterns and controls of biogeochemical pools and fluxes at scales that span from several centimeters to hundreds of kilometers, and from hours to millennia. The synthesis of this work has resulted in a conceptual framework regarding the biogeochemical dynamics of the shortgrass steppe, with two key components:

1. Spatial and temporal patterns are controlled by five major factors: climate, physiography, natural disturbance, human use, and biotic interactions. Plants are the most important biotic component. The interaction of these factors as they change in time and space determines the distribution and size of biogeochemical pools and the rates of biogeochemical processes.

2. Carbon (C), nitrogen (N), and other associated biologically active elements are overwhelmingly located belowground, with more than 90% found in soils (Burke et al., 1997a). This distribution determines the biogeochemical sensitivity of the shortgrass steppe to perturbations.

These ideas have been synthesized in the development of the CENTURY ecosystem simulation model, originally developed for grasslands and agroecosystems in the shortgrass steppe region of the western Great Plains (Parton et al., 1987, and chapter 15, this volume). The model represents complex interactions among the five controlling factors to simulate C and N cycling, and has served as an organizing framework for developing hypotheses and for evaluating questions that are difficult to address in the field (Parton et al., chapter 15, this volume).

The objectives of this chapter are to describe how nutrient pools and fluxes are distributed in the shortgrass steppe, to characterize how the five controlling factors interact to create spatial and temporal patterns, and to evaluate the potential future changes to which the biogeochemistry of the shortgrass steppe may be particularly vulnerable.

Distribution and Fluxes of Ecosystem Nutrient Pools

Carbon

The dominant concentration of organic matter belowground in the shortgrass steppe is the result of the high allocation of primary production to roots (Kelly et al., 1996; Lauenroth et al., chapter 12, this volume), the fast turnover rates of plant biomass producing above- and belowground litter, and the slow decomposition rates that are characteristic of semiarid systems (Gill et al., 1999; Meentemeyer, 1984). Current estimates of C pool sizes suggest that aboveground C (as an index of organic matter) averages approximately 68 g $C \cdot m^{-2}$, whereas belowground C is more than 100 times higher, averaging 7946 g $C \cdot m^{-2}$. Note that belowground C combines live roots (901 g $C \cdot m^{-2}$) and all pools of SOC (7045 g $C \cdot m^{-2}$; Fig. 13.1).

A common analytical framework for SOC divides this detrital material into three kinetic fractions, characterized by their turnover time (Jenkinson and Rayner, 1977): active SOC (turning over every few years), an intermediate pool (every several decades), and recalcitrant SOC (thousand-year timescales for turnover). Although the assignment of these kinetic pools to physical substances is largely conceptual, we have conducted sufficient organic matter fractionations to have found a good correspondence between these kinetic pools and empirical fractions in the shortgrass steppe. We have estimates of the sizes as well as the turnover times of these pools (Gill et al., 1999; Kelly et al., 1996). We conceive of the active pool as that portion mineralizable over a 6-month time period (Gill et al., 1999), the intermediate pool as the coarse and fine particulate organic matter (POM) (Cambardella and Elliott, 1992), and the recalcitrant soil organic matter as the C that is not in either of the other two pools. In the soil profile to a 1-m depth, the approximate proportions of these pools are 0.2% active, 34% intermediate, and ≈65% recalcitrant (Table 13.1).

308 Ecology of the Shortgrass Steppe

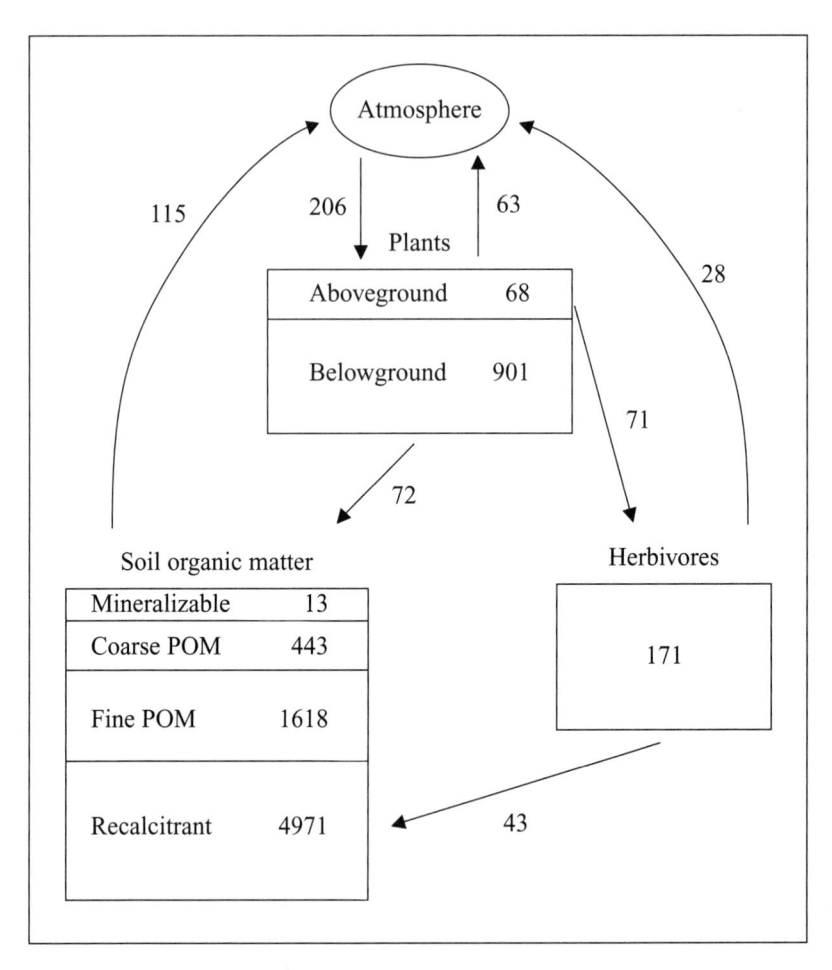

Figure 13.1 Shortgrass steppe carbon budget. Units of pools are grams per square meter to a 1-m depth, and fluxes are in grams per square meter per day. POM, particulate organic matter. (Synthesized from data presented in Gill et al. [1999] and Lauenroth and Milchunas [1992].)

Although root biomass is largely concentrated in surface soils, with a large proportion of root biomass in the surface 10 cm (Lauenroth, chapter 5, this volume), soil organic matter is more evenly distributed throughout the profile, with only 34% in the surface 10 cm (Fig. 13.2, Table 13.1). In much of the literature, considerable focus has been placed on surface pools of C, as an indicator of C storage or biologically active material (e.g., Burke et al., 1989, 1995b; Ihori et al., 1995a,b; Parton et al., 1987). However, it is important to realize that the surface soil contains a small proportion of the C in soils to a depth of a meter, with some soils having considerably more C in older, buried horizons (Kelly et al., chapter 3, this volume; Yonker et al., 1988). This represents substantial storage of organic

Table 13.1 Site Characteristics and C Mass for Three SOC Fractions and Bulk Soil Collected from ^{14}C-labeled Plots at the CPER

Depth, (cm)	Bulk Density, (g·cm^{-3})	Texture Sand, %	Clay, %	Root Biomass, g·m^{-2}	6-Month Mineralized C, mg·m^{-2}	Fine-POM C, g·m^{-2}	Coarse-POM C, g·m^{-2}	C, g·m^{-2}
0–5	0.88	67	16	752.6 (0.389)	1189.2 (0.378)	202.1 (0.271)	37.5 (0.181)	701.2 (0.199)
5–10	1.09	68	17	227.6 (0.417)	860.5 (0.401)	86.6 (0.170)	35.5 (0.229)	494.5 (0.115)
10–15	1.16	64	18	158.1 (0.413)	850.0 (0.414)	53.1 (0.342)	27.3 (0.259)	412.8 (0.281)
15–20	1.17	63	21	134.0 (0.784)	751.1 (0.351)	42.4 (0.165)	22.4 (0.349)	394.4 (0.102)
20–35	1.32	59	25	217.1 (0.351)	1936.9 (0.365)	202.0 (0.291)	100.0 (0.299)	1429.5 (0.158)
35–50	1.32	58	26	194.4 (0.367)	1734.81 (0.349)	153.7 (0.371)	68.2 (0.310)	1221.9 (0.293)
50–75	1.34	69	21	210.4 (0.565)	3055.34 (0.421)	360.1 (0.467)	72.3 (0.368)	1318.4 (0.196)
75–100	1.34	66	21	107.4 (0.432)	3070.4 (0.325)	517.9 (0.244)	80.2 (0.655)	1074.4 (0.142)

Bulk density was measured as the total core mass/core volume. Roots were removed from soil by sieving soil through a 2-mm sieve (n = 16). Six-month mineralized C was measured using laboratory incubations and represent potentially-mineralizable C. Coarse POM-C is organic C larger than 0.5 mm but smaller than 2 mm. Fine-POM C is organic C smaller than coarse-POM C but larger than 53mμ. Total C is C found in bulk soil, after removing carbonates. Coefficient of variation is expressed as standard deviation/mean in parentheses (n = 16). (From Gill et al. [1997].)

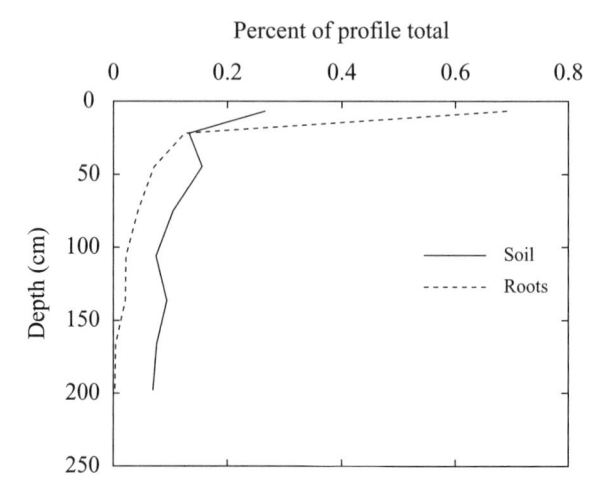

Figure 13.2 The depth distribution of roots versus soil C, shown as a percent of the profile total to 2 m. (From Gill et al. [1999]. Reprinted with kind permission of Springer Science and Business Media.)

matter from an ecosystem capital perspective, as well as potentially significant pools for nutrient cycling.

To date, we do not have adequate information to determine the significance of these deep soil layers to ecosystem functioning. Fractionation of organic matter pools (Gill et al., 1999) suggests that the highest concentrations of recent organic matter inputs are in surface soils (0–5 cm), but that decomposition rates, k (Fig. 13.3), are highest in subsurface soils (10–15-cm depth), which are most frequently moist (Sala et al., 1992). Lowest decomposition rates occur at depth (75–100 cm). Interestingly, although belowground primary production is lower at depth than in surface horizons (Lauenroth et al., chapter 12, this volume), a higher proportion of the organic matter at depth is in active or intermediate soil organic matter pools, likely because decomposition rates are slow at depth (Fig. 13.3). Dodd et al. (2002) showed that a significant amount of N is mineralized at depth; surprisingly, mineralization rates at 150 cm were 25% of the surface soil mineralization rates in a sandy soil. Future work on deep soil layers will be important to our understanding of shortgrass steppe biogeochemistry.

Nitrogen

Nitrogen is distributed similarly to carbon, with only a very small proportion of the total N capital cycling each year (Fig. 13.4, Table 13.2 [Holland et al., 1999; Hook and Burke, 1995, 2000; National Atmospheric and Deposition Program, 1998]). Much of the research that has been conducted on the N cycle has focused on elaborating the rates of input and output fluxes such as wet deposition and trace gas losses, and internal cycling (including gross and net mineralization).

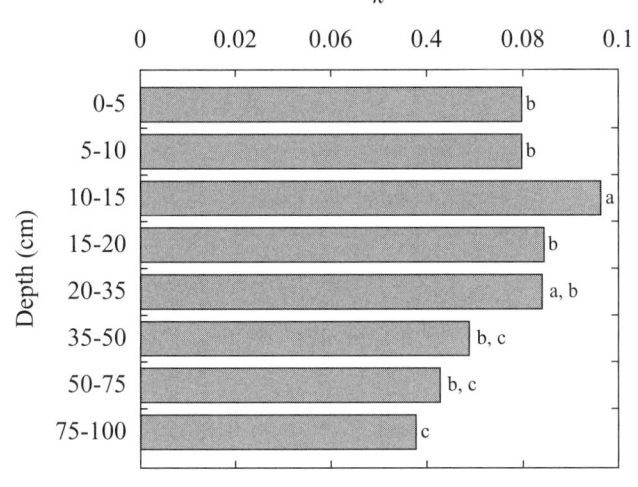

Figure 13.3 Mass specific decomposition rate constant (k) for total particulate organic matter. Depths marked by different letters are significantly different ($P < .05$). This constant was determined using a flux over pool size approximation, averaging over 7.5 years. (From Gill et al. 1999. Reprinted with kind permission of Springer Science and Business Media.)

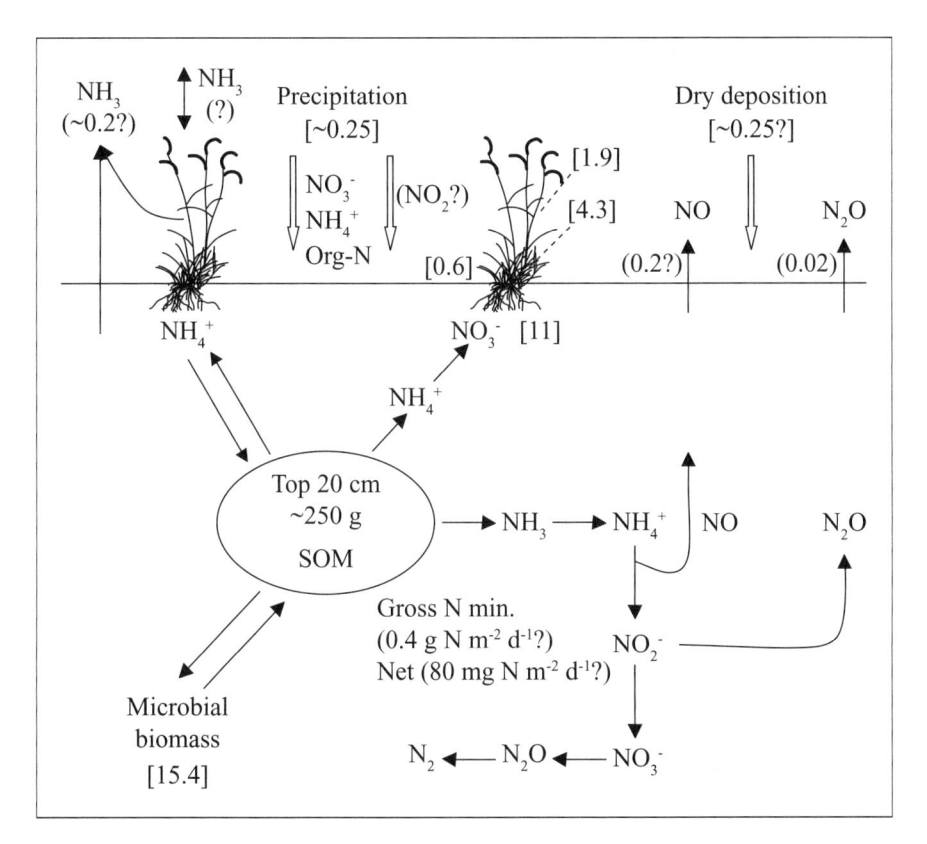

Figure 13.4 Nitrogen budget for the shortgrass steppe. SOM, soil organic matter. Units are g m^{-2} for pods and g m^{-2} y for processes except as shown. (?) denotes high uncertainty in estimate.

Table 13.2 Estimated Fluxes of Nitrogen

Internal recycling		
	Gross N mineralization, $g \cdot m^{-2} \cdot d^{-1}$	0.40
	Net N mineralization, $g \cdot m^{-2} \cdot d^{-1}$	0.08
	Net N mineralization and litterfall, $g \cdot m^{-2} \cdot y^{-1}$	1.0–2.5
	Plant N uptake and litterfall, $g \cdot m^{-2} \cdot y^{-1}$	1.0–2.5
Inputs		
	Wet deposition, $g \cdot m^{-2} \cdot y^{-1}$	0.25
	Dry deposition, $g \cdot m^{-2} \cdot y^{-1}$	0.25
	N fixation, $g \cdot m^{-2} \cdot y^{-1}$	??
Outputs		
	NH_3 volatilization, $g \cdot m^{-2} \cdot y^{-1}$	0.20
	NO flux, $g \cdot m^{-2} \cdot y^{-1}$	0.20
	N_2O flux, $g \cdot m^{-2} \cdot y^{-1}$	0.02
	N leaching	?

Values for internal recycling estimated from Hook and Burke (1995, 2000) and Mosier (chapter 13, this volume,). Wet deposition estimated from the National Atmospheric and Deposition Program (1998); dry deposition likely to be a similar value (Holland et al., 1999). Nitrogen fixation and N leaching have not been measured; leaching is likely to be low. See Mosier et al. (chapter 14, this volume) for a detailed description and citations for the gaseous output fluxes.

A recent synthesis of the N cycle (Burke et al., 2002) found that thus far, we cannot close the N budget. Although our site has long been a member of the National Atmospheric Deposition Program, and we have measured trace gas fluxes on many of our long-term plots (Mosier et al., chapter 14, this volume), there are still several key areas of element balance that we do not understand. Key gaps are in understanding total N inputs, volatile ammonia (NH_3) losses from plants, and NO_2 and NH_3 absorption by plants.

Experimental alterations of N availability result in substantial and long-lasting effects on nutrient cycling in the shortgrass steppe (Vinton and Burke, 1995). As described by Lauenroth et al. (chapter 12, this volume), NPP is strongly controlled by water availability, with interesting interactions with N availability. Increases in N with and without water additions resulted in increases in NPP and changes in plant community structure that have persisted more than 25 years (Milchunas and Lauenroth, 1995). Increased N availability, alone and in combination with water, increased the importance of annual forbs, which have lower nutrient use efficiencies. Consequently, litter quality improved (higher percent N), potentially leading to a positive feedback for maintaining high N availability (Vinton and Burke, 1995). Twenty years after N additions ceased, N mineralization remained high on N addition plots (Vinton and Burke, 1995), and mycorrhizal fungi infection was lower in plants from N addition plots (Corkidi et al., 2002). Clearly, N availability is a significant factor in structuring shortgrass steppe ecosystems.

Climatic Controls

Climate (Pielke and Doesken, chapter 2, this volume) is a primary control over biogeochemical pools and processes. Because available soil water is the resource most frequently limiting to NPP in the steppe (Lauenroth and Sala, 1992; Lauenroth et al., 1978; Sala et al., 1988), patterns of precipitation and their influence over both ecosystem structure and functioning have received a great deal of attention (Barrett et al., 2002a; Burke et al., 1991, 1997a, b; Hook and Burke, 2000; Lauenroth and Sala, 1992; McCulley, 2002; McCulley and Burke, 2004; McCulley et al., 2005; Sala et al., 1988). These studies demonstrate that water availability exerts a strong influence over both the spatial and temporal variability associated with pools and turnover of C and N.

Temperature also influences broad-scale patterns in biogeochemistry in shortgrass steppe ecosystems, as well as across the entire central grassland region. For example, soil organic matter storage, N immobilization, and potential N mineralization increase with decreasing gradients in mean annual temperature in shortgrass steppe and mixed-grass prairie ecosystems (Barrett and Burke, 2000; Burke et al., 1989; Epstein et al., 2002). Hence, an important generalization emerging from the research conducted in the shortgrass steppe is the recognition that biogeochemical cycling responds differentially to spatial and temporal variability in both precipitation and temperature (Burke et al., 1997a, b; Lauenroth and Sala, 1992). In this section we discuss the influence of spatial gradients and temporal variability in climate on pools and turnover of C and N.

Spatial Variability of Temperature and Precipitation

Regional patterns in precipitation and temperature strongly influence biogeochemical cycling in the central grassland region at large scales (Barrett and Burke, 2002; Barrett et al., 2002a; Burke et al., 1989, 1997a, b; Epstein et al., 1996, 1998a, b; Lane et al., 1998; Sala et al., 1988). Together with local-scale variation in soil properties and topography, these climate patterns control both organic matter content and turnover in the shortgrass steppe (Hook and Burke, 1995, 2000; Schimel and Parton, 1986; Schimel et al., 1985a). Soil organic matter storage is particularly sensitive to broad-scale climate gradients (Barrett and Burke, 2002; Barrett et al., 2002a, b; Burke et al., 1989), whereas C and N turnover may respond to finer scale variability resulting from the influences of soil texture and topography on microclimate (Burke et al., 1999; Hook and Burke, 2000).

Early work across spatial gradients of the entire central grassland region (Burke et al., 1989) indicated that total soil surface (0–20 cm) organic matter pools (C and N) increase linearly with precipitation, and decrease linearly with temperature, within the range of mean annual temperature and precipitation of the shortgrass steppe. Increasing precipitation clearly influences NPP (Epstein et al., 1997; Lauenroth et al., chapter 12, this volume; Sala et al., 1988), the eventual input into soil organic matter. Our original interpretation was that soil organic matter declined with increasing temperature because of the positive influence of temperature on decomposition rates (Burke et al., 1989). However, in a recent analysis, Epstein et al. (2002) separated the effects of temperature on decomposition and NPP. Their analysis indicated that within the shortgrass steppe region, decomposition rates have a weak positive relationship with mean annual temperature, but concurrently there is a significant decline in NPP with increasing temperatures. In summary, the decrease in soil organic matter with increasing temperature within the shortgrass steppe is more likely the result of the effects of increased water stress on plant inputs, rather than the effects of temperature on decomposition rates.

Temporal Variability of Temperature and Precipitation

Resource-limited ecosystems are often dominated by pulse availability of the limiting resources, which creates significant but often predictable spatial and temporal variability (Burke et al., 1997b; Noy-Meir, 1973; Sala and Lauenroth, 1982; Sala et al., 1992; Schlesinger et al., 1990). Pulses of water regulate important nutrient fluxes. For example, Hook and Burke (2000) showed that seasonal patterns of net N mineralization were highly dependent upon individual precipitation events, and Schimel and Parton (1986) reported that net N mineralization responded significantly to pulses of water availability in the form of dew. These pulses of mineral N may represent important sources of available N to plants and microbes, particularly because they accompany discrete pulses of water availability. Similarly, there could be increased atmospheric deposition during the spring, when fertilizers are applied to adjacent cultivated lands (Burke et al., 2002).

Pulses occur over several temporal scales (Austin et al., 2004; Huxman et al., 2004). Plant phenology and production coincide with temporal pulses of precipitation over seasonal timescales, whereas soil microbial and invertebrate activity are responsive to diel-scale variation in soil water availability associated with summer convective storms and dew. Augustine and McNaughton (2004) found that in semiarid savanna systems, there is an important asynchrony in plant production and N mineralization. Burke et al. (unpublished data) found that in situ soil respiration immediately after a July thunder storm was twice as high as average rates for that part of the growing season (Fig. 13.5), but only for C-amended soils, suggesting both a substrate and a moisture limitation. In the nearby mixed-grass prairie, Chimner and Welker (2005) showed increases in midsummer soil respiration in response to both summer and winter precipitation increases. Because the shortgrass steppe is dominated by warm-season, or C_4, plants, shifts in seasonality of precipitation may have special significance for the relationships among microbial dynamics, plant nutrient uptake, and litterfall. Simulation analyses (Epstein et al., 1999) have suggested that shifts in rainfall seasonality change the relationship between moisture and the season of maximum growth of the warm-season plants, and that this may have a large influence on N turnover and soil organic matter pools.

Although climate exerts a strong influence over C cycling (i.e., NPP, soil respiration, decomposition, and soil C storage), its influence over N dynamics is less clear. For example, several data sets collected over a range of temporal and spatial scales have revealed no clear trends in the influence of monthly climate

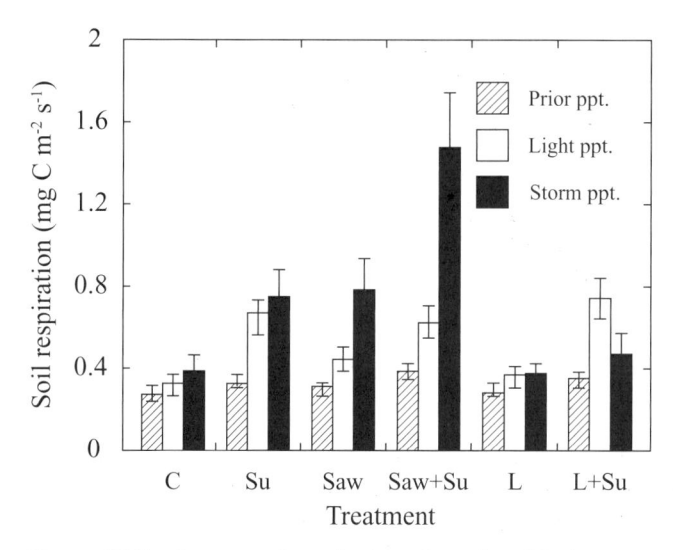

Figure 13.5 Average soil respiration after prior nighttime precipitation, light precipitation, and heavier storm precipitation on the shortgrass steppe. The data were collected on soils that are carbon amended. The treatments are control (C), sugar (Su), sawdust (Saw), sawdust plus sugar (Saw+Su), lignin (L), and lignin plus sugar (L+Su). ppt., precipitation. (From Burke et al. [manuscript in preparation].)

variables on net N mineralization in uncovered cores (Fig. 13.6 [Barrett et al., 2002a; McCulley, 2002; McCulley et al., 2005]). Although these data may suggest a parabolic influence of temperature over N mineralization with an optimal mean temperature near 18 °C, a polynomial fit to these data is weak ($r^2 = .13$). No single climate driver has emerged from these studies as an effective predictor of N mineralization in the field. Rather, N mineralization seems to be controlled by a suite of interacting microclimatic, physiographic, and microbial influences. This apparent unresponsiveness of the N cycle to climate variables may be a feature of the threshold/pulse behavior of dry ecosystems to multiple driving variables.

Ultraviolet Radiation

Recent results indicate that ultraviolet radiation may be a significant control over decomposition dynamics in arid and semiarid ecosystems (Austin and Vivanco, 2006). Analysis of long-term decomposition data from a continental-scale LTER study including the SGS LTER (Bontti et al., submitted; Parton et al., 2007) indicates that photodegradation may hasten decomposition in shortgrass steppe systems. Changes in ultraviolet radiation may alter the rate of C and nutrient cycling in these systems (see chapter 19).

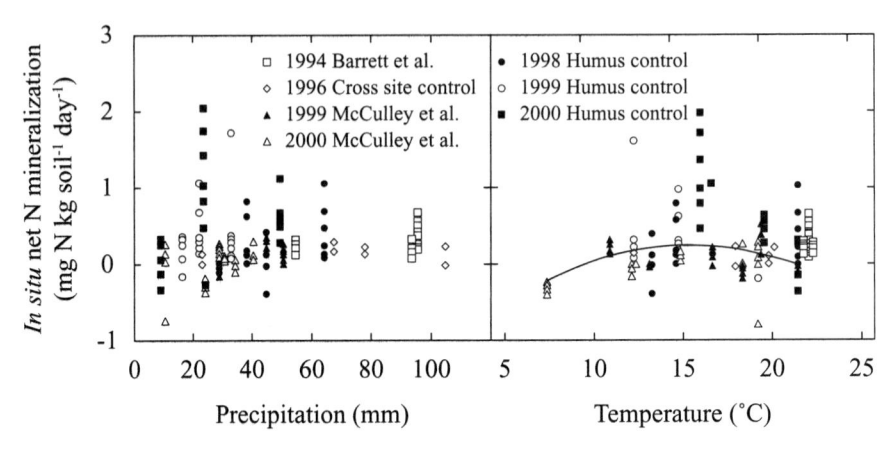

Figure 13.6 Thirty day net N mineralization as estimated by the intact core procedure in shortgrass steppe ecosystems for multiple years and sites versus observed precipitation during the incubation and mean monthly soil temperature during the incubation. (Data are modified from Barrett et al. [2002a], McCulley [2002], McCulley et al. [2005], and part of the SGS LTER database.)

Physiographic Control

Topography and soil texture interact with water, wind, and animals to control landscape-scale biogeochemical patterns in the shortgrass steppe (Fig. 13.7). Although topography and texture are dictated mainly by geology and long-term geomorphic processes (see Kelly et al., 1998; and chapter 3, this volume), they influence biogeochemistry directly and indirectly via processes operating at both short and long timescales. Soil parent material and texture influence organic matter accumulation at any point in the landscape directly through effects on aggregate formation and indirectly through effects on soil water availability, which in turn influences primary production and decomposition. The combined effects of topography, parent material, and texture on water dynamics create areas that are transiently drier or wetter than average. Runoff is limited compared with more humid regions (Parton et al., 1981; Singh et al., 1998; unpublished IBP and LTER data), but when it occurs, it likely results in some transport of soil and nutrients downslope. Effects of runoff on topographic patterns of soil development may be either offset or reinforced by eolian processes (Blecker et al., 1997; Kelly et al., chapter 3, this volume; Madole, 1994; Muhs, 1985). Finally, differential use of the landscape by livestock and wildlife can redistribute nutrients and alter vegetation structure, primary production, and organic matter turnover rates (Milchunas et al., 1989; Senft et al., 1985, 1987).

Most of our knowledge of landscape biogeochemistry comes from studies comparing sites arrayed along individual hill slopes, often using the catena concept (Gerrard, 1992) to integrate geomorphology, soil science, and ecosystem ecology (Schimel et al., 1985a,b; Woodmansee and Adamsen, 1983). According to the catena concept, plant production and soil development are typically greater in

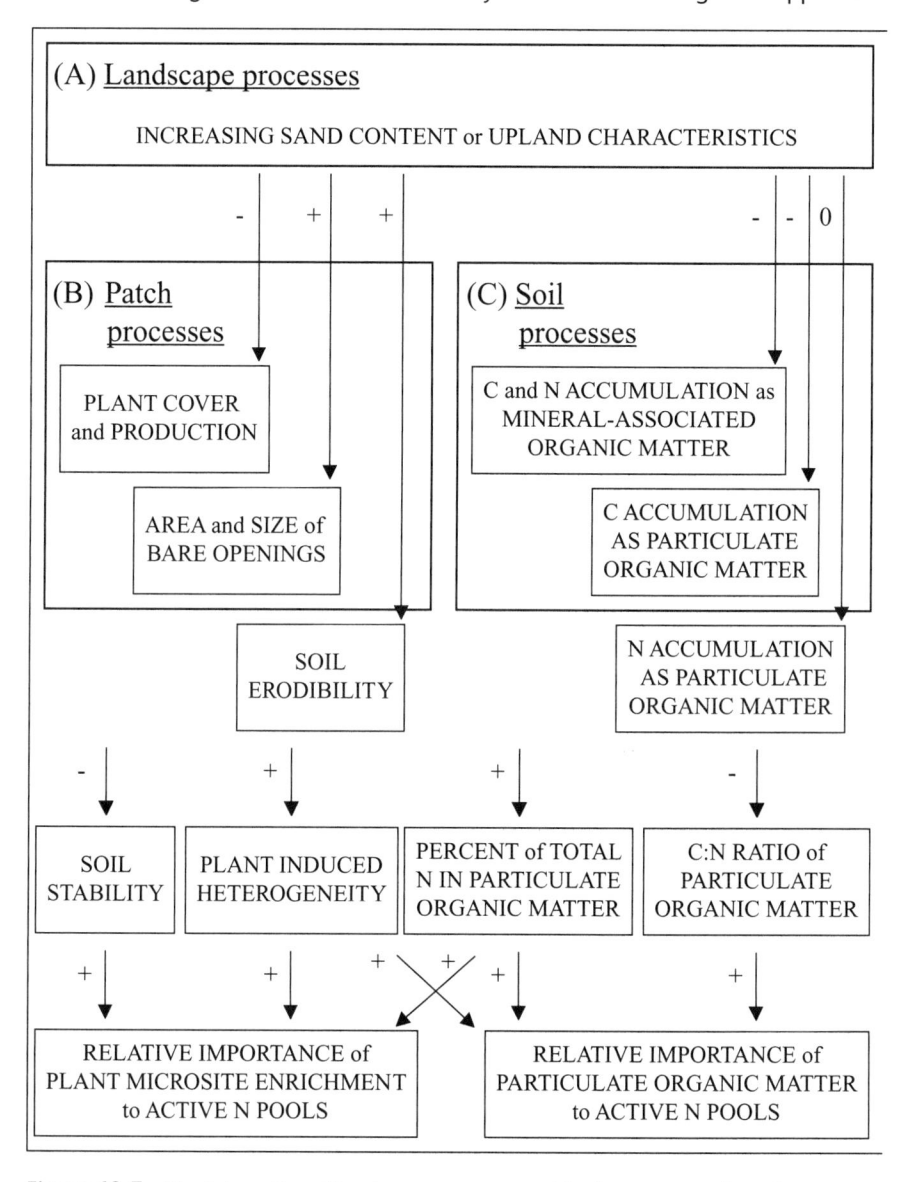

Soil Organic Matter and Nutrient Dynamics in the Shortgrass Steppe 317

Figure 13.7 The interaction of landscape processes with plant-generated patch processes and soil processes. (From Hook and Burke [2000].)

lower slope positions. This is presumed to result mainly from runoff, erosion, and deposition, although topographic patterns may be modified by other processes.

Catenary patterns are illustrated by an intensively studied hill slope at the CPER that featured strongly contrasting soils at three slope positions (Schimel et al., 1985a). On this site, organic C, N, and phosphorus (P) increase downslope as a result of increases in both soil depth and element concentrations. Available

N and P as well as root biomass also increase downslope. Although depositional sites make up only one third of this study area, they contain about two thirds of the organic C and N in this landscape (Schimel et al., 1985a). Soil morphology indicates that on this site, summit and midslope soils have developed in a different, coarser parent material than the footslope soil, and that the summit has experienced episodic erosion whereas the footslope has been subject to substantial, episodic deposition (Schimel et al., 1985b). Elevated C, N, and P in the depositional footslope site are attributed to downslope redistribution of organic matter, clay, and associated nutrients; enhancement of primary production by higher water and nutrient availability in lowlands; protection of organic matter by aggregation and adsorption to clays; and soil organic matter burial by deposition.

The topographic differences at this site are reinforced by greater deposition of N in urine and feces in lowland than upland sites (Schimel et al., 1986); topographic differences in soil in turn affect the short- and long-term fates of the N deposited. When ^{15}N-labeled urea was applied to simulate cattle urine, three times as much ^{15}N was recovered in soil organic matter at the footslope compared with the midslope (Table 13.3 [Delgado et al., 1996]). Volatile losses of NH_3 after urea application were negligible at the footslope but significant at the midslope, which also may have lost additional nitrate (NO_3) by leaching. These large topographic differences in ^{15}N retention persisted for at least a decade and likely reflect higher clay content, organic matter, and microbial activity in the footslope (Delgado et al., 1996). The footslope also exhibited higher N_2O emissions and lower methane (CH_4) consumption than the midslope (Mosier et al., 1991, and chapter 14, this volume).

Table 13.3 Nitrogen Content, and Percent of the Added ^{15}N Recovered in Different Components of Footslope and Backslope Positions in 1992, 10 Years after its Addition

	Footslope		Backslope	
Compartment	g N·m^{-2}	%[a]	g N·m^{-2}	%[a]
Crowns	11	8	8	2
Roots	11	6	7	2
Shoots	5	3	2	1
Litter	8	4	4	1
SON[b]	1108	64	754	24
POM[c]	379	37	200	10
MAON[d]	729	27	554	14

[a]Amount of N recovered at a percentage of the amount applied (53.5 g N·m^{-2}).
[b]Soil organic N (SON), 0 to 60 cm. Inorganic N was less than 1 mg·kg–1. SON = POM + MAON.
[c]Particulate organic matter (POM), 0 to 60 cm.
[d]Organic N associated with the soil mineral particles (MAON), 0 to 60 cm.
Nitrogen was added as urea to simulate cattle additions (Delgado et al., 1996).

Catenary patterns are not consistent across the shortgrass steppe region, however, and apparently depend more on differences in parent material, texture, or erosion and deposition than on either topographic position per se or predictable topographic patterns in soil water (Yonker et al., 1988). Topographic patterns of soil organic matter vary considerably from one hill slope to another and may feature either increases or decreases in organic C and N downslope (Woods and Schuman, 1988; Yonker et al., 1988).

Studies that control for differences in parent material have shown weaker topographic variation than those that confound topography and parent material (Woods and Schuman, 1988; Yonker et al., 1988). These studies support the conclusion that low-lying, depositional topographic positions generally differ from upland, erosional positions, but they cast doubt on the existence of strong, repeatable patterns of variation along hill slopes. Comparing five topographic positions (summit, shoulder, backslope, footslope, and toeslope) across 24 hill slopes, Yonker et al. (1988) (Fig. 13.8) found that organic C concentrations in the surface horizon were usually higher in toeslope than summit soils, but otherwise did not show consistent topographic patterns. Average SOC concentrations were 11.9 g·kg^{-1} soil in toeslope soils and 7.5 to 8.3 g·kg^{-1} in all other slope positions (weighted averages based on Yonker et al. [1988]), suggesting broad differences between depositional lower slope positions and other locations, but not systematic trends from upper to lower slopes (Fig. 13.8). Similarly, Hook and Burke (2000) contrasted surface soils from eight paired uplands (erosional) and lowlands (depositional), representing a wide range of parent materials, soil textures, and geomorphic settings. As expected, depositional lowlands generally had finer soils, more continuous plant cover, and substantially higher levels of C and N in pools representing active, intermediate, and passive fractions (Fig. 13.9). From this broad

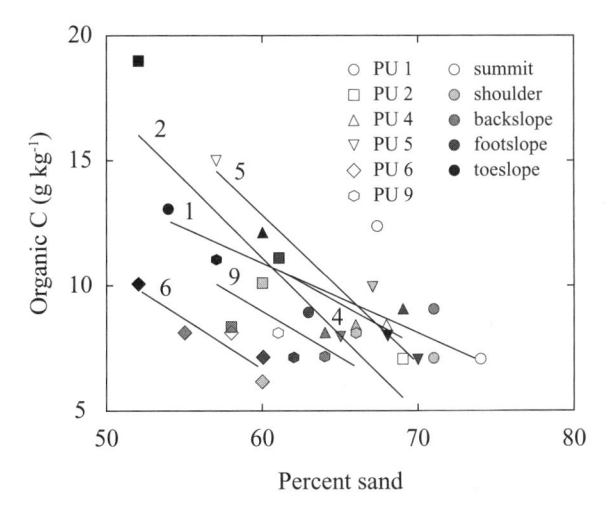

Figure 13.8 Comparison of landscape positions and soil textures across six different physiographic units (PU). (Data from Yonker et al. [1988].)

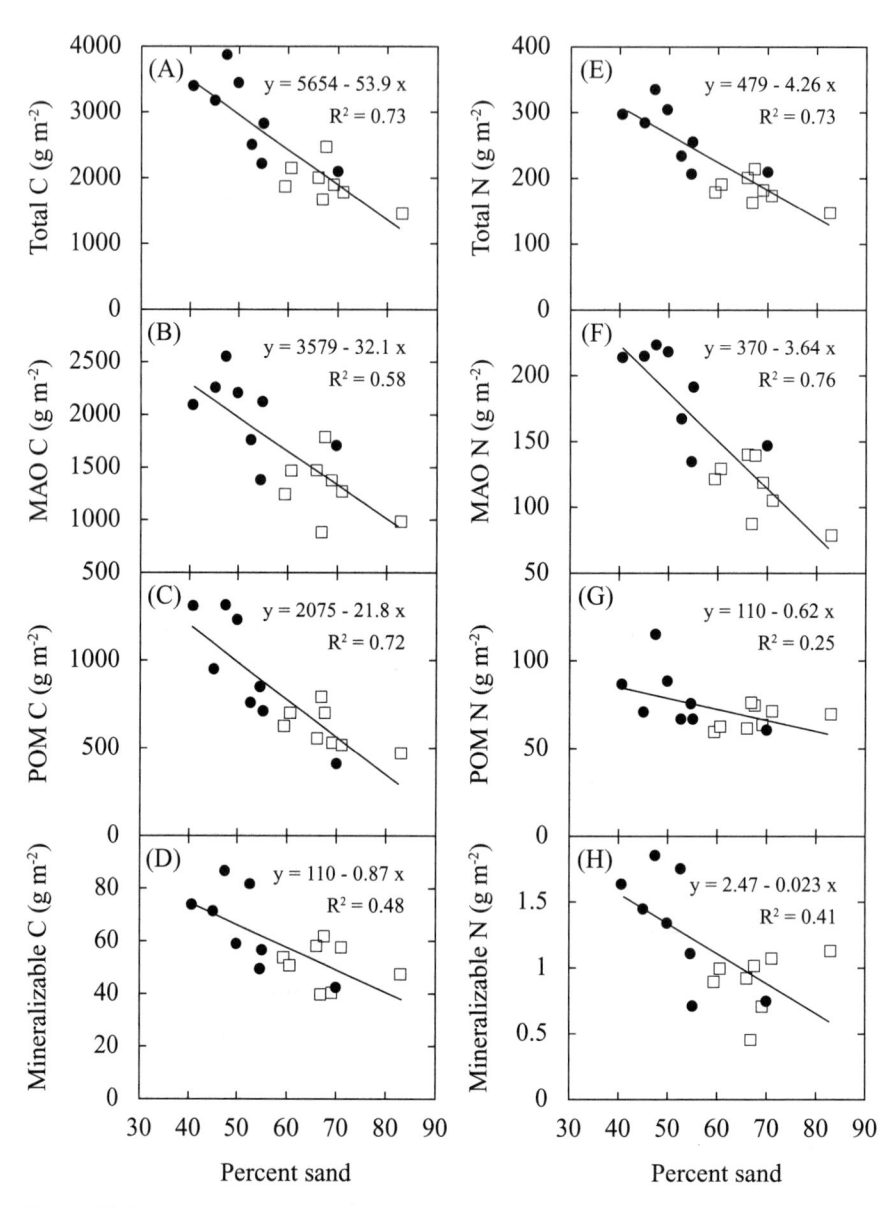

Figure 13.9 (A–H) Influence of sand content (measured as a percentage) on total soil organic C and N, mineral-associated organic (MAO) C and N, particulate organic matter (POM) C and N, and mineralizable C and N. (From Hook and Burke [2000].)

perspective, hydrology appears to promote topographic differences in soil C and N despite low frequency of runoff and minimal evidence for significant current eolian erosion and deposition.

Soil texture appears to account for much of the topographic variation in C and N (Hook and Burke, 2000) (Fig. 13.9). Plant cover and most C and N pools are negatively correlated with sand content, and relationships between texture and pool sizes are generally consistent across uplands and lowlands. Most C and N pools increase by 2 to 4.5 times across the range of textures encountered (40% to 83% sand). The negative effect of sand content on SOC is consistent across landscape-scale data sets, although specific quantitative relationships vary among physiographic units (Yonker et al., 1988) and land uses (Moore et al., 1993). Although topography and texture often vary together, texture is likely the more important proximal control of soil organic matter.

There is limited information about variation of in situ C and N dynamics with topography and texture, but landscape-scale differences are weaker than those for most other measures of C and N pools, including laboratory indices of potential N mineralization (Hook and Burke, 2000). Topographic differences in in situ N availability and C respiration appear to be weak or absent most of the year and develop mainly during moist spring periods (Burke et al., 1999).

Biotic Control: Plants

Vegetation type can have large effects on ecosystem development and biogeochemical cycling. Because plant litter is the major input into soil organic matter, it is not surprising that both the quantity and the type of plants present are key factors affecting biogeochemical pools and processes.

Spatial Variability in Biotic Controls

Although water is a key factor controlling NPP and thus the quantity of plant litter available for belowground processes, there is a whole suite of individual plant characteristics that control the spatial pattern and turnover of organic matter. Among these individual plant characteristics is the vertical placement of biomass. The shortgrass community typically has a majority of biomass belowground (Lauenroth et al., chapter 12, this volume). This, combined with a water-limited turnover of the organic matter, contributes to a soil organic matter level that is higher than one might anticipate, given the relatively low precipitation and plant production.

The horizontal placement of biomass is also a key factor affecting litter inputs and biogeochemical cycling (Fig. 13.10). Low water availability results in discontinuous plant cover aboveground, and a relatively high proportion of bare ground (48% to 62% [Hook et al., 1991]). Most bare-ground openings are 5 to 10 cm across and are not associated with disturbances, but rather with normal spacing between individual plants (Aguilera and Lauenroth, 1993; Coffin and Lauenroth, 1988; Hook et al., 1994). As demonstrated by numerous studies in the shortgrass steppe (Hook et al., 1991; Vinton and Burke, 1995) and other arid to semiarid systems (Bolton et al., 1990; Burke, 1989; Charley and West, 1977), C and N pool sizes and turnover under plants are higher than those from bare soils between plants, forming resource islands. The prominence of resource islands results in a strong connection between plant population dynamics and small spatial-scale biogeochemical processes (Lauenroth et al., 1997). Turnover in the population

```
                Water                              Plant establishment
                  │                                         ▲
                  ▼         ┌─────────────┐                 │
        discontinuous  ───► physical soil ───►      soil nutrient
        aboveground cover   redistribution          heterogeneity
                      plant functional type
                            time
                            wind
                            soil type
                        concentration of
                            biomass
```

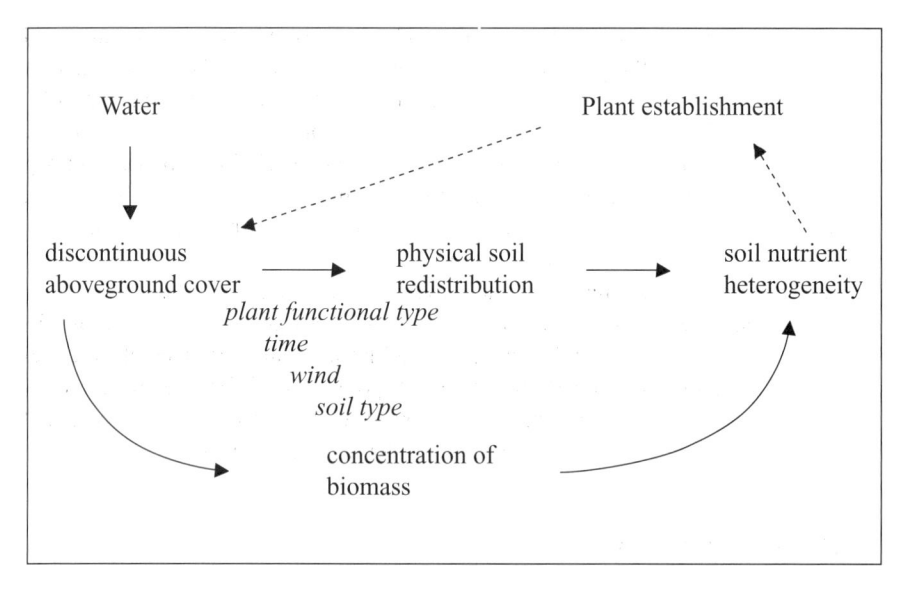

Figure 13.10 Plant–soil interactions that lead to plant–interplant heterogeneity in shortgrass steppe ecosystems. (From Burke et al. [1998]. Reprinted with kind permission of Springer Science and Business Media.)

of the long-lived dominant bunchgrass *Bouteloua gracilis* causes redistribution of the C and N associated with the affected individuals and a change in the spatial distribution of biogeochemical processes (Kelly and Burke, 1997). Because *B. gracilis* is a clonal species and therefore does not have a well-defined life span, the most common cause of population turnover is small-scale disturbance (Coffin and Lauenroth, 1988, 1990). In addition to the dominant bunchgrass, shrubs such as *Artemisia frigida* and *Gutierrezia sarothrae* also foster this patchy distribution of soil organic matter and activity (Vinton and Burke, 1997).

There are at least three sets of processes that influence the concentration of organic matter and nutrients beneath individual plants relative to openings (Burke et al., 1998, 1999; Hook et al., 1991) (Figs. 13.10 and 13.11). The first is the biological concentration of organic matter beneath individual plants, through above- and belowground litter production. The second is the physical redistribution of soil materials by wind and water, from between individual plants to under them. Substantial microtopography is associated with individual plants, with an average height of hummocks beneath *B. gracilis* of about 3 cm (Burke et al., 1999; Hook et al., 1991; Martinez-Turanzas et al., 1997). Microtopographic relief is higher in disturbed sites, on landscape positions with highest potential for erosion (summit and mid slopes), and under plants with the largest physical resistance to wind and water (e.g., *Opuntia polyacantha*; Fig. 13.12), or those with the longest life span. The third is the demography of the plant species or functional types comprising shortgrass communities (Lauenroth et al., 1997).

Soil Organic Matter and Nutrient Dynamics in the Shortgrass Steppe 323

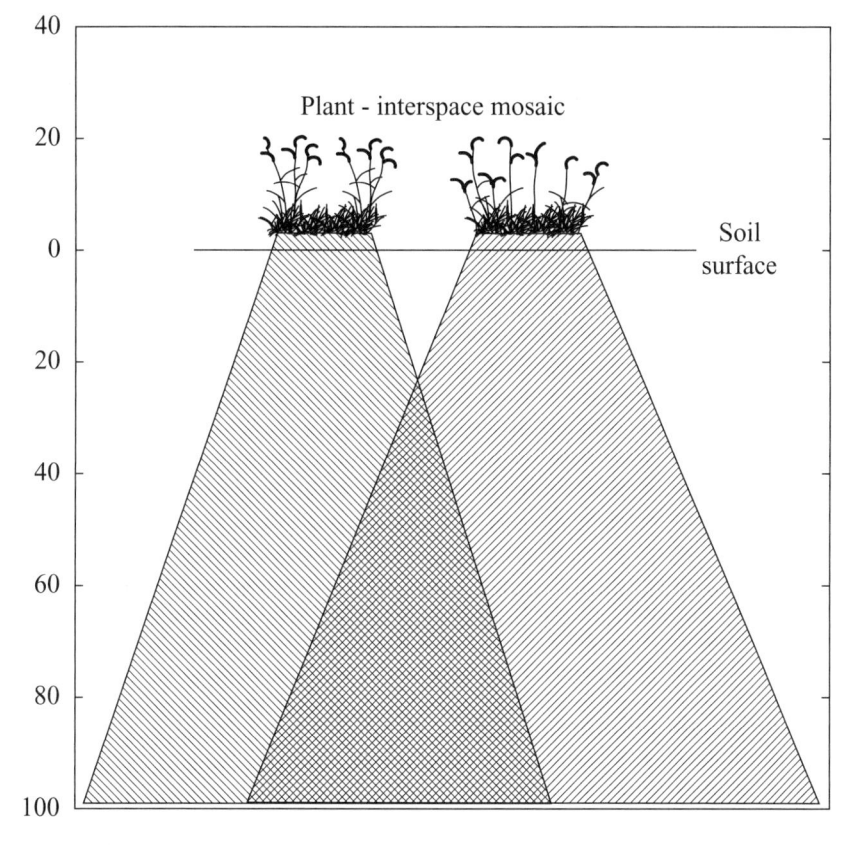

Figure 13.11 The influence of perennial plant location is strongest at the surface, because of gaps in biomass accumulation. (From Burke et al. [1998]. Reprinted with kind permission of Springer Science and Business Media.)

Plant functional-type characteristics also determine the quality of litter entering the decomposer pathway, and the subsequent biogeochemical dynamics. Using ^{15}N studies, Clark (1977) demonstrated a tight plant N-to-soil N linkage in the case of *B. gracilis*. Wedin and Tilman (1990) showed that plant species affected N cycling in the tallgrass prairie and that the key plant characteristic was the quality of litter, indexed by C-to-N ratio or lignin content. Litter quality can affect the rate of soil respiration and N mineralization. For example, the nonnative annual *Kochia scoparia*, which is locally abundant in some disturbed areas in the shortgrass steppe, had lower C-to-N ratios and lignin content and higher N mineralization rates in the soil beneath it than other species (Vinton and Burke, 1995) (Fig. 13.13). This species became abundant in a fertilized and irrigated area (Milchunas and Lauenroth, 1995), and persists today, more than 30 years since the addition of fertilizer and water stopped. The enhanced N cycling and N availability under *K. scoparia* may have allowed it to outcompete natives and to persist in a landscape normally dominated by shortgrass species.

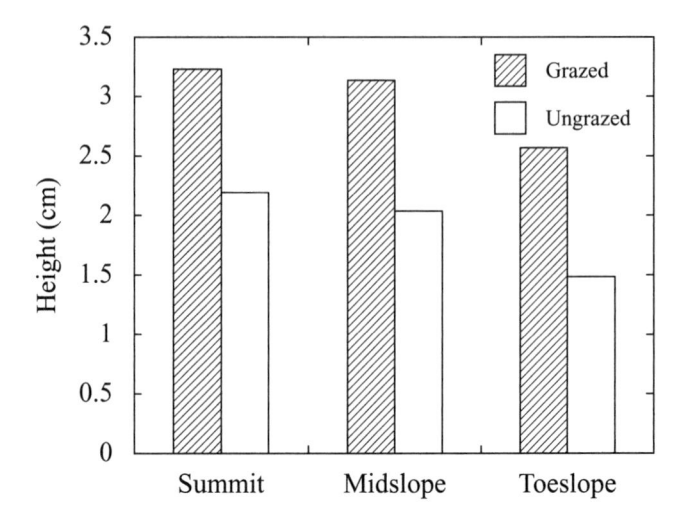

Figure 13.12 Height of soil mounds (indirect indication of organic matter accumulation) beneath patches of *Opuntia polyacantha* by landscape position and grazing regime. (From Burke et al. [1999]. Reprinted with kind permission of Springer Science and Business Media.)

Photosynthetic pathway differences between plant functional types (C_3 vs. C_4) also influence biogeochemical cycling. Epstein et al. (1998a, 2001) showed that a higher proportion of ^{15}N added in the spring was retained in soils and plants in C_3-dominated communities than C_4, indicating that both spatial and temporal patterns of nutrient cycling are influenced by plant functional type. In addition, the presence of shrubs may alter the depth distribution of soil organic matter and N dynamics (Dodd et al., 2000; Gill and Burke, 1999).

Although plant functional types have a significant influence on biogeochemistry, litter quality is less important as a controlling variable in the shortgrass steppe than in more mesic grasslands for at least two reasons. One reason is that species differences in tissue quality may not be expressed during decomposition because water is a more important limitation on organic matter breakdown than tissue chemistry. Another reason is that the low precipitation in the shortgrass steppe likely limits tissue C-to-N ratios to a fairly narrow range at the low (high-quality) end of the scale (Burke et al., 1998; Murphy et al., 2002; Vinton and Burke, 1997). Water-limited plants fix less C per unit N than plants in more mesic systems. For instance, Murphy et al. (2002) reported an increased C-to-N ratio in *B. gracilis* across a precipitation gradient from 300 to 800 mm·y^{-1}. A striking illustration of this phenomenon in mesic tallgrass prairie is the very high C-to-N ratios (100 or greater) attained by season-end tissues of tallgrass species such as *Sorgastrum nutans* (indian grass) and *Andropogon gerardi* (big bluestem) (Vinton and Burke, 1997). Finally, as mentioned earlier, there is recent evidence that ultraviolet radiation influences litter decomposition in the shortgrass steppe and other semiarid ecosystems (Austin and Vivanco, 2006; Parton et al., 2007).

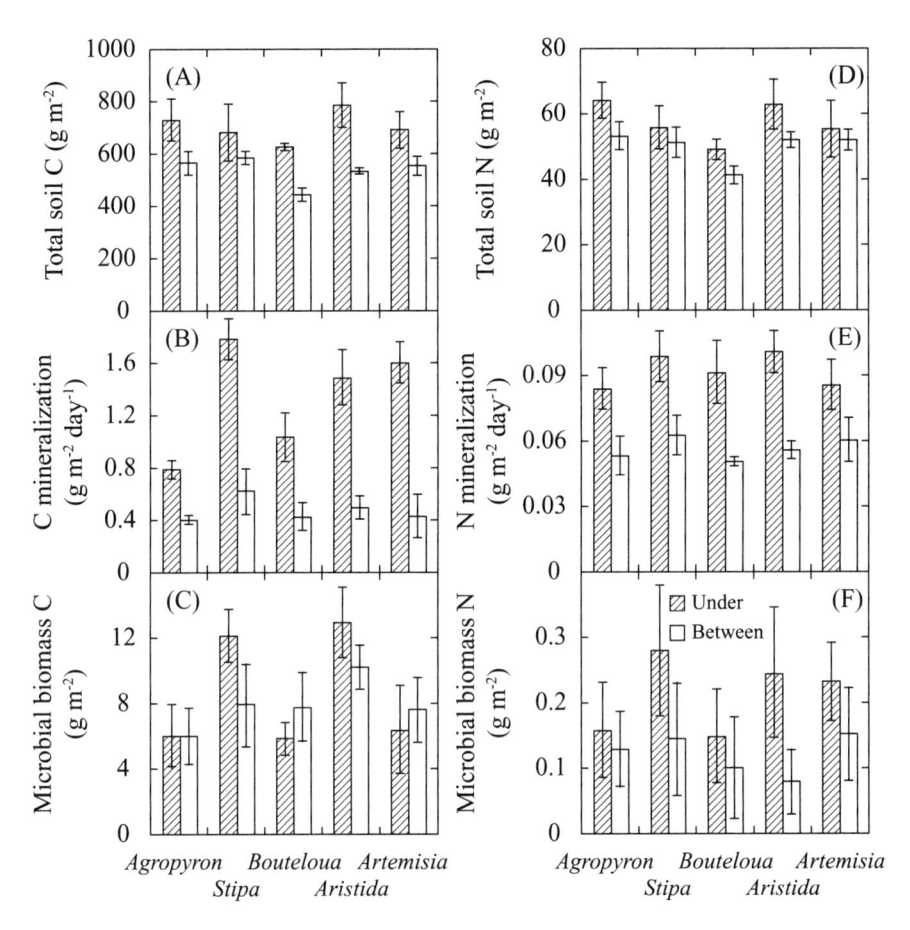

Figure 13.13 The effect of plant presence (under) or absence (between) for five different species on total soil carbon and nitrogen (A, D), potential C mineralization (B, E), and microbial biomass C and N (C, F). (From Vinton and Burke [1995].).

Biotic Control: Fauna

Small-scale disturbances mediated by fauna represent a major influence on biogeochemistry (Crist, chapter 10, this volume; Moore et al., chapter 11, this volume; Peters et al., chapter 6, this volume; Stapp et al., chapter 8, this volume). Burrowing activities by small mammals, ant mounds, and white grub kills are the most frequent disturbances (Peters et al., chapter 6, this volume). To date, there has been relatively little work on the effect of these disturbances on soil organic matter pools and turnover. To the extent that these organisms affect the recruitment, survival, and mortality of individuals of *B. gracilis*, they potentially have a large influence on at least the active pools of organic matter (Lauenroth et al., 1997). Kelly and Burke (1997) found that immediately after the death of individual *B. gracilis* plants, active and total soil organic matter increased, with subsequent decreases as the inputs to soil organic matter declined and decomposition proceeded. Thus, death of individual plants resulted in the disappearance of resource islands after just several months.

In empirical and simulation studies in mixed-grass ecosystems, many investigators have found increases in N mineralization, and increases in plant N concentration on prairie dog towns relative to off-town locations (Coppock et al., 1983; Holland and Detling, 1990; Holland et al., 1992). Such increases in N turnover may occur because of physical soil perturbations, and increases in N cycling resulting from herbivory and excretion. Initial studies in the shortgrass steppe have shown similar results (Dempsey, unpublished data; Farrar, 2002), and long-term studies are underway.

Human Use Control: Grazing

Cattle grazing directly influences the quantity and quality of plant litter entering the soil organic matter pools by the removal and processing of plant tissue (Milchunas and Lauenroth, 1993; Milchunas et al., chapter 16, this volume). Additionally, grazing can influence the distribution of limiting resources, such as N, on the landscape via urine and fecal deposition, and can potentially increase cycling and losses. Cattle may also affect soils by changing erosion patterns resulting in the alterations in soil texture and bulk density associated with grazing activity.

Grazing Effects on Erosion

One of the potential direct effects of grazing on biogeochemistry is the physical effect of cattle on soil redistribution. Studies of the effects of grazing on erosional processes in the shortgrass steppe are limited and have only been conducted at small spatial scales. Average wind speeds at the CPER are high (Pielke and Doesken, chapter 2, this volume), and eolian movement of soil was a major force shaping the current-day landscape (Kelly et al., chapter 3, this volume). The potential for fluvial erosion is higher on grazed sites. Striffler (1972) estimated runoff in small watersheds, and found that although the quantity of runoff was small, it was greater on heavily grazed (0.35 mm) compared with ungrazed (0.17 mm) or lightly grazed (0.09 mm) treatments. Wind erosion may be enhanced by grazing. Microtopography associated with *B. gracilis* clumps compared with bare-ground interspaces suggests a degree of small-scale local movement of soil (Martinez-Turanzas et al., 1997). Burke et al. (1999) measured soil texture between plants and in hummocks under *B. gracilis* and *O. polyacantha* (plains prickly pear cactus) plants, and found greater sand and silt content in the hummocks, especially under *O. polyacantha* clumps. Grazing influenced the degree of hummock microrelief only in the case of *O. polyacantha*, but this was significant for total N and coarse POM C and N. Although cacti can act as accumulators of moving soil, the amounts removed from the larger, surrounding grazed area are small on a unit-area basis.

Soil bulk densities are also increased as a result of cattle trampling (Burke et al., 1999). Bulk density between plants in moderately grazed pastures was

1.58 g·cm⁻³ compared with 1.45 g·cm⁻³ in ungrazed, and under plants was 1.52 g·cm⁻³ compared with 1.32 g·cm⁻³, respectively. This could explain the greater runoff on the grazed treatment described earlier, and may also influence infiltration rates and soil water dynamics. Infiltration rates in a heavily grazed treatment have been found to be lower than those in lightly and moderately grazed treatments for a loamy Aridic Argiustoll in swales (Rauzi and Smith, 1973). Contrary to these infiltration results, Leetham and Milchunas (1985) found a 19% greater soil water content in a heavily grazed treatment compared with a lightly grazed treatment when averaged over six sample dates and to a depth of 60 cm. Lewis (1971) reported slightly greater early growing season soil water content in ungrazed compared with heavily grazed shortgrass steppe. Overall, the impacts of cattle grazing on soil texture, microtopography, and microclimate appear to be small.

Grazing Effects on Nitrogen

Animals generally do not graze uniformly across a landscape, and one of the major possible effects of grazing in the shortgrass steppe is the redistribution of N. Preference for consumption in a particular community and disproportionate fecal and urine deposition in other communities resulting from grazing, traveling, resting, and bedding behaviors can lead to a redistribution of nutrients across the landscape. Dry matter intake and fecal excretion per unit body weight by cattle grazing on the shortgrass steppe have been observed to be greatest in September and October (Senft et al., 1984, 1987). Nitrogen intake, urine N output, and fecal N concentration have been shown to be associated with seasonal forage N concentrations, peaking in June. However, total fecal N excretion in these studies peaked in October, several months after intake and urine output peaked. Total N deposited as excreta was 0.2 g N·m⁻²·y⁻¹, with 44% as urine and 56% as fecal N.

Although grazing might redistribute N, Senft et al. (1984, 1987) found that movement among landscape positions was generally small (Table 13.4). The exceptions were for fence lines and watering areas, which were both large sinks for N. Seasonal differences were apparent. Lowlands and south-facing slopes had a negative N balance during the growing season, but became N sinks during the dormant season. On an annual basis, all topographic positions showed N losses to fence lines and watering areas, ranging from 70% of consumptive removal for ridgetops to 10% for lowlands.

Senft (1983) calculated that spatial coverage by excreta within a pasture ranged from 0.5% to 2%, depending on the plant community type, with higher percentages for areas where cattle congregate. The short-term effect of fecal pat deposition, however, is to kill the vegetation covered by the pat, and this represents a significant, patchy disturbance to the plant community (Peters et al., chapter 6, this volume). The effect of urine deposition on plants is that of immediate fertilization. A complete hydrolysis of urea occurred within 4 days on lowlands and mid slopes, but proceeded for a couple of weeks on drier ridgetops (Stillwell, 1983; Stillwell and Woodmansee, 1981). From 9% to 55% of the ¹⁵N urea added could be found in inorganic N in the soil after 120 days, depending on topographic position (Stillwell, 1983). The fertilization effect of the urine patches increased both plant production

Table 13.4 Nitrogen Removal, Deposition in Excreta, and Balance (measured in grams N per square meter per season) after Transport among Communities for Various Topographic Zones and Fence Lines and Watering Areas

Period	Cattle-Mediated Process	Ridgetops	South-Facing Slopes	North-Facing Slopes	Draws and Lowlands	Fence Lines	Watering Area
Growing season (April–October)	N removal	0.209	0.150	0.149	0.332	0.239	0.214
	N deposition	0.144	0.096	0.134	0.253	0.384	1.324
	N balance	−0.065	−0.054	−0.015	−0.079	0.145	1.110
Dormant season (November–March)	N removal	0.113	0.034	0.048	0.038	0.042	0.067
	N deposition	0.036	0.053	0.025	0.080	0.102	0.415
	N balance	−0.077	0.019	−0.023	0.042	0.060	0.348
Annual Total	N removal	0.322	0.184	0.197	0.370	0.281	0.281
	N deposition	0.180	0.149	0.159	0.333	0.486	1.739
	N balance	−0.142	−0.035	−0.038	−0.037	0.205	1.458

Adapted from Senft (1983).

and N concentrations, generally to a distance of 45 cm from the center of urine application. However, root biomass did not increase in the urine-affected areas.

Volatilization of NH_3 from fertilizer urea N can be as much as 50% of that applied when soils are coarse textured or calcareous, and this magnitude of loss can occur from pastures as well (Mosier et al., chapter 14, this volume; Mulvaney and Bremner, 1981). The magnitude of volatile loss can depend on many edaphic and abiotic factors, including temperature, humidity, soil pH, soil moisture, the soil drying regime, soil texture, and depth of the urea in the soil profile (Ernst and Massey, 1960). Ammonia losses from urine under field conditions range from 1.5% on a fine-textured footslope to 14% on a coarse-textured backslope in summer, with an 8% loss from the same backslope in winter (Milchunas et al., 1988a). When we extrapolated these measurements to a pasturewide basis using spatially and temporally explicit cattle urine deposition rates, we estimated losses to be 0.016 g $N \cdot m^{-2} \cdot y^{-1}$ from lowlands, with negligible losses from uplands (Schimel et al., 1986). We estimated losses from a typical pasture from NH_3 volatilization, animal biomass removal, and N_2O production to total 0.07 g $N \cdot m^{-2} \cdot y^{-1}$. This loss rate is approximately 25% of wet deposition inputs to the system. Long-term experiments on grazing (Burke, unpublished data; Reeder et al., 2004) have shown variable and small effects on N availability. Thus, the overall effect of cattle on the shortgrass steppe N budget appears small relative to other inputs and outputs of N and internal N cycling (Table 13.2).

Grazing Effects on Total Soil Organic Matter

What is the net effect of grazing on soil organic matter pools in shortgrass steppe ecosystems? As described by Milchunas et al. (1988b, 1989, 1998, and chapter 16, this volume), grazing the shortgrass steppe has small effects on NPP, crown biomass, and root biomass. Aboveground litter is consistently lower in grazed treatments (Milchunas et al., 1989), and one might expect that the integrated effect of decreased litter inputs would result in decreased soil organic matter and available nutrients (Milchunas et al., 1998). However, comparisons of long-term heavily grazed pastures with ungrazed plots have shown only very slight differences. In a long-term SGS LTER study that extended across three grazing treatments and three landscape positions (Burke et al., 1999), we found that although ungrazed plots have slightly higher levels of more active pools, including coarse and fine particulate soil organic matter, only the most active pools show effects of grazing. In two other studies (Derner et al., 2006; Reeder and Schuman, 2002), we found slightly higher levels of C in the surface soils of heavily grazed areas. Perhaps, as suggested by our conceptual framework, grazing only directly affects the smallest pool of organic matter in the system: aboveground biomass.

Human Use Control: Cultivation and Recovery

Since the early 1900s, much of the land area in the shortgrass steppe has been managed as row–crop agriculture (Hart, chapter 4, this volume). Early efforts at dryland agriculture, following the Homestead Act, resulted in substantial conversion of native shortgrass steppe to wheat-fallow agriculture. Areas located where irrigation water was available, either from surface flow or from groundwater, were converted to irrigated corn–bean rotations or similar systems. Through periods of drought in the 1930s and 1950s, as well as government support programs such as the Soil Bank and the CRP, much of the area in dryland wheat has been returned to perennial grasses (Skold, 1989). In some cases, pastures have gone through several cycles of plow-out and recovery.

The effects of dryland wheat-fallow agriculture on native shortgrass steppe have been well documented (Campbell et al., 2005; Haas et al., 1957; Lauenroth et al., 1999; and many others; reviewed in Paustian et al., 1997). Historical cultivation practices in dryland systems reduced soil organic matter by both decreasing the rate of soil organic matter inputs and increasing the rates of outputs (Burke et al., 1997a). Removal of plant residues and decreased root production characteristic of annual crops led to decreased effective NPP. Plowing increased decomposition and N mineralization rates through mixing, breaking soil aggregates, and increasing the amount of contact between litter and soil organisms (Doran and Werner, 1990; Elliott, 1986). Over several decades, soil organic matter contents were reduced between 20% and 50%—a very rapid rate of decline relative to the pace at which soils form in these semiarid systems. Losses of active organic matter eventually reduced N mineralization and nutrient supply capacity (Burke et al., 1995b). Increased C oxidation was sufficient to represent a significant redistribution from terrestrial to atmospheric pools (Burke et al., 1991). More current land-use management practices, including reduced intensity of plowing or no-till agriculture, fertilization, and addition of residues to soils may result in slow rates of organic matter recovery in some of these systems (Burke et al., 1995a; Campbell et al., 2005; Paustian et al., 1997; Wood et al., 1990, 1991).

The effects of irrigated agriculture on ecosystem biogeochemistry within the shortgrass steppe region are much less well studied. Mosier et al. (chapter 14, this volume) have shown substantial increases in denitrification associated with irrigated corn systems, as well as decreased CH_4 oxidation. Some recent initial work (Sinton, 2001) has indicated small losses in soil C and N throughout the soil profile in fields that have been irrigated for four decades. These results are somewhat counterintuitive, given the high levels of productivity and potential organic matter inputs, and potentially reduced decomposition resulting from anaerobic conditions. However, a very large proportion of that enhanced production is removed and exported with harvest, and plowing is well known to increase decomposition rates.

Although there are currently no geographic databases on historical cropland abandonment, it is clear from surveys of aerial photographs that as much as 30% of the "native" rangelands in northeastern Colorado have been cultivated at one time. Because cultivation in many of these areas has been abandoned for as long as 70 years, they provide a good opportunity for evaluating the potential recovery rate of soil organic matter after cultivation (Burke et al., 1995b; Ihori et al., 1995a,b), with potential implications for the CRP. A study of 12 paired native and abandoned fields indicated that recovery rates of soil organic matter are relatively slow, with an estimated 25 g C·m^{-2} recovering after 50 years, relative to a

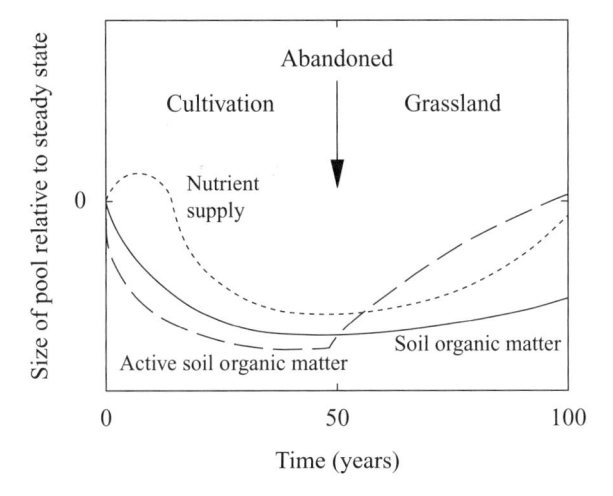

Figure 13.14 The recovery of total soil organic matter, active soil organic matter, and nutrient supply after 50 years of cultivation of the native shortgrass steppe, followed by 50 years of recovery (Burke et al., 1995b).

loss rate of approximately 250 g·m^{-2} during 20 years of cultivation (Burke et al., 1995b). However, active soil organic matter pools, those with turnover times of several years, including microbial biomass and nutrient supply capacity, and the resource islands characteristic of native fields do seem to recover after 50 years (Fig. 13.14). Such recovery is likely to be dependent upon the recovery of perennial bunchgrasses. Our studies of currently managed CRP fields, in perennial grasses for a decade, show very slight increases in organic matter, primarily in active pools (Robles and Burke, 1997, 1998). The rates of recovery of N mineralization are slightly higher in fields in which legumes are part of the seeding mixture (Robles and Burke, 1997).

Human Use Control: Urbanization

A portion of the western shortgrass steppe lies in the most rapidly urbanizing region of the Great Plains. In the 1990s, six of the country's fastest growing cities were in central Colorado (U.S. Bureau of the Census, 1999) and urban development is replacing agricultural areas at a rapid rate. Some of the converted area was cultivated agriculture, but uncultivated rangeland is also being urbanized. What are the biogeochemical consequences of native steppe conversion to urban use? Our research suggests that urban land management increases water, C, and N fluxes compared with uncultivated shortgrass steppe (Kaye et al., 2004, 2005) (Fig. 13.15).

Four main urban management strategies (Fig. 13.15) cause urban and steppe ecosystems to differ: (1) species composition changes, (2) irrigation, (3) N

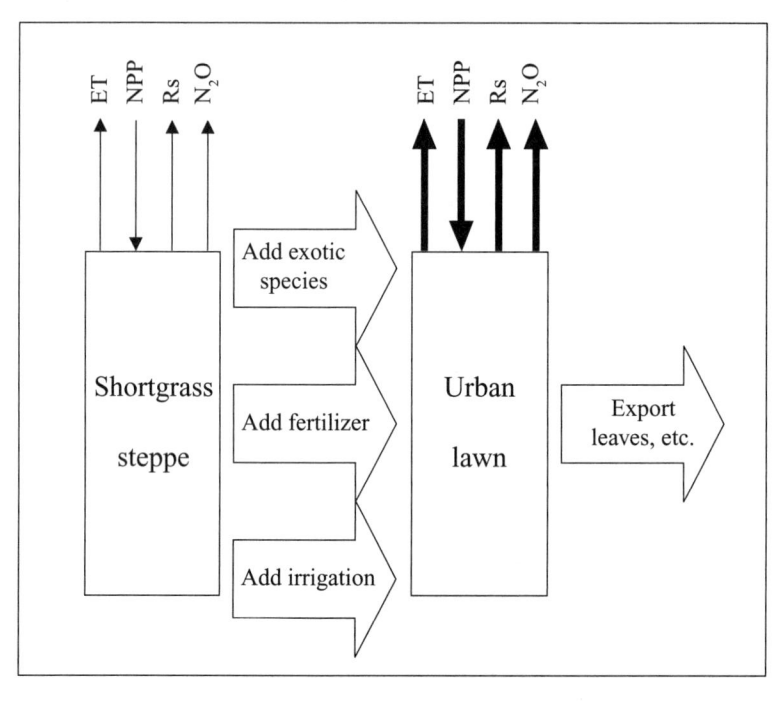

Figure 13.15 When the shortgrass steppe is converted to urban lawn, changes in land management (horizontal open arrows) cause large changes in water, carbon, and nitrogen fluxes (vertical filled arrows) from the biologically active component of the landscape. The size of the vertical arrows reflects the hypothesis that management inputs of fertilizer and irrigation should increase evapotranspiration (ET), net primary productivity (NPP), soil respiration (Rs), and soil nitrous oxide fluxes (N_2O) relative to the shortgrass steppe.

fertilization, and (4) biomass manipulations. The conversion of the shortgrass steppe to urban use represents a complete plant species composition shift from a mixture of 15 to 30 species of drought-tolerant grasses, forbs, shrubs, and cacti, to a monoculture of *Poa pratensis* (Kentucky bluegrass) in a matrix of mostly exotic deciduous trees. Maintaining bluegrass lawns and trees in this semiarid environment requires 1 to 5 cm of irrigation water per week in the summer (Koski and Skinner, 2001). If lawns occupy half the urban landscape, irrigation at this rate quadruples natural growing season precipitation per unit area. Nitrogen fertilization is also common for bluegrass lawns at rates of 5 to 20 g $N \cdot m^{-2} \cdot y^{-1}$, which is several times larger than N deposition or biological fixation. Last, mowing, grass removal, leaf raking, and pruning transfer biomass and nutrients within or from urban ecosystems at rates that likely exceed analogous transfers (e.g., overgrazing) in the shortgrass steppe.

We have been comparing the biogeochemistry of urban lawns with uncultivated grassland ecosystems in Fort Collins, Colorado. Our results (Kaye et al., 2005) show that ANPP in urban ecosystems ($\approx$380 g $C \cdot m^{-2} \cdot y^{-1}$) is approximately five times larger than in unmanaged grasslands ($\approx$75 g C·m^{-2}·y^{-1}). Soil respiration in urban ecosystems (2800 g C·m^{-2}·y^{-1}) is three times greater than fluxes from uncultivated grasslands (920 g C·m^{-2}·y^{-1}). Although soil respiration in unmanaged grasslands is highest in spring when soils are both wet and warm, urbanization decouples soil respiration from the natural moisture regime, and high rates are maintained from April to October. Fertilization, and perhaps increased atmospheric N deposition and N retention, have increased mineral soil N content (0–15 cm) by approximately 50% (to 450 g N·m^{-2}) in 100-year-old urban ecosystems. Net N mineralization potential and N$_2$O fluxes (Kaye et al., 2004) are often an order of magnitude larger in urban ecosystems than in unmanaged grasslands, although annual N flux budgets are not yet available. Our simulation results (Bandaranayake et al., 2003; Qian et al., 2003) indicate that C sequestration may increase as much as 11% to 59%, and N sequestration by 14% to 78%, depending upon management strategies. These simulations are supported by Golubiewski (2006), based upon measurements in lawns and adjacent shortgrass areas of Denver, Colorado.

All the data discussed here are measurements from the biologically active component of urban ecosystems that occupy an unknown proportion of the Fort Collins landscape. In addition, large fluxes of water, N, and C occur through human consumption (as opposed to lawn applications) that we have not measured. Nevertheless, it is clear that urban fluxes of water, N, and C are several times larger than in unmanaged grasslands. In the western shortgrass steppe, where land is rapidly being converted to urban uses, a complete understanding of regional biogeochemistry is not possible without more research in urban areas.

Vulnerability to Future Changes

Which of the likely environmental changes (increased atmospheric CO$_2$ and N deposition rates, global climate change, or land-use change) are most likely to have large influences on shortgrass steppe biogeochemistry? Which pools are most responsive to which major controls? In a recent study (Burke et al., 1999), we compared the overall importance of topography, microsite, plant species, and grazing as they influence the spatial distribution of active, intermediate, and recalcitrant soil organic matter pools (Fig. 13.16). We found that, first, these four factors account for 70% to 90% of the variance in soil organic matter pools and dynamics, and second, that topography and microsite together account for most of the variance. Neither grazing nor plant species (*B. gracilis* and *O. polyacantha*) could account for more than 2% of the variance. The influence of topography was dominant for the recalcitrant pools, indicating that the influence of geomorphology is exerted over centuries or longer on pools that are turning over slowly. Topographic controls are not likely to change over short timescales, unless the landscape is subjected to large-scale erosion and redistribution events, such as those that occurred during the Dust Bowl. The influence of microsite or resource islands was most important for the active and intermediate pools (mineralizable and POM), suggesting that resource islands form and grow over decadal timescales.

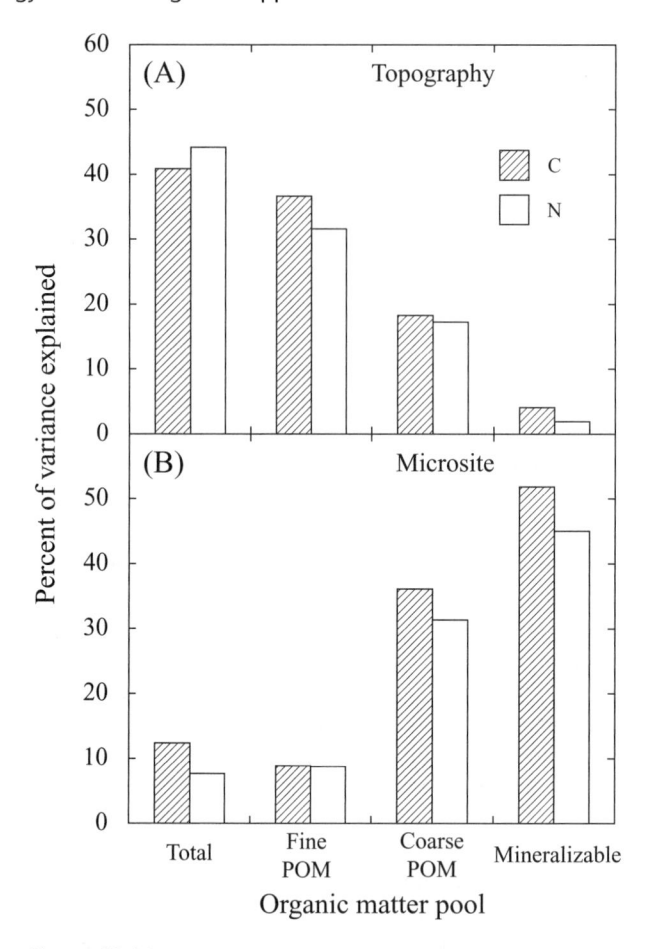

Figure 13.16 The influence of microsite (between and under plants)-level spatial variability versus topographic scale variability on four pools of soil organic carbon and nitrogen, with varying turnover times. POM, particulate organic matter. (From Burke et al. [1999]. Reprinted with kind permission of Springer Science and Business Media.)

Our conceptual framework asserts that disturbances affecting the belowground components of the shortgrass steppe are likely to have the largest impact. Clearly, these include cultivation management and urbanization. However, organic matter pools and processes are also sensitive to changes in water availability, temperature, and N availability (Holland et al., 1999), all of which are likely to change in the future. To address the interacting responses of the system to changes in these factors, we have initiated two types of studies. First, we are conducting long-term experiments in which we are manipulating temperature, moisture, and N availability alone and in combination, as well as increasing CO_2 concentrations and ultraviolet radiation (chapter 19, this volume). Because the long-term results of these experiments will not be evident for some time, we are also conducting simulation experiments (Parton et al., chapter 15, this volume). Such sensitivity analyses can provide us with useful hypotheses and guidance for these and future experiments.

References

Aguilera, M. O., and W. K. Lauenroth. 1993. Neighborhood interactions in a natural population of the perennial bunchgrass *Bouteloua gracilis*. *Oecologia* **94**:595–602.

Augustine, D. J., and S. J. McNaughton. 2004. Temporal asynchrony in soil nutrient dynamics and plant production in a semiarid ecosystem. *Ecosystems* **7**:829–840.

Austin, A. T., and L. Vivanco. 2006. Plant litter decomposition in a semi-arid ecosystem controlled by photodegradation. *Nature* **442**:555–558.

Austin, A. T., L. Yahdjian, J. M. Stark, J. Belnap, A. Porporato, U. Norton, D. A. Ravetta, and S. M. Schaeffer. 2004. Water pulses and biogeochemical cycles in arid and semiarid ecosystems. *Oecologia* **141**:221–235.

Bandaranayake, W., Y. L. Qian, W. J. Parton, D. S. Ojima, and R. F. Follett. 2003. Estimation of soil organic carbon changes in turfgrass systems using the CENTURY model. *Agronomy Journal* **95**:558–563.

Barrett, J. E., and I. C. Burke. 2000. N immobilization in grassland soils: Controls by soil organic matter. *Soil Biology and Biochemistry* **32**:1707–1716.

Barrett, J. E., and I. C. Burke. 2002. Nitrogen retention in semiarid ecosystems across a soil organic-matter gradient. *Ecological Applications* **12**:878–890.

Barrett, J. E., R. L. McCulley, D. R. Lane, I. C. Burke, and W. K. Lauenroth. 2002a. Influence of climate variability on plant production and N mineralization in central U.S. grasslands. *Journal of Vegetation Science* **13**:383–394.

Barrett, J. E., D. W. Johnson, and I. C. Burke. 2002b. Abiotic nitrogen uptake in semiarid grassland soils of the US Great Plains. *Soil Science Society of America Journal* **66**:979–987.

Blecker, S. W., C. M. Yonker, C. G. Olson, and E. F. Kelly. 1997. Paleopedologic and geomorphic evidence for Holocene climate variation, shortgrass steppe, Colorado, USA. *Geoderma* **76**:113–130.

Bolton, H., Jr., J. L. Smith, and R. E. Wildung. 1990. Nitrogen mineralization potentials of shrub-steppe soils with different disturbance histories. *Soil Science Society of America Journal* **54**:887–891.

Burke, I. C. 1989. Control of nitrogen mineralization in a sagebrush steppe landscape. *Ecology* **70**:1115–1126.

Burke, I. C., E. T. Elliott, and C. V. Cole. 1995a. Influence of macroclimate, landscape position, and management on soil organic matter in agroecosystems. *Ecological Applications* **5**:124–131.

Burke, I. C., T. G. F. Kittel, W. K. Lauenroth, P. Snook, C. M. Yonker, and W. J. Parton. 1991. Regional analysis of the central Great Plains, sensitivity to climate variability. *BioScience* **41**:685–692.

Burke, I. C., W. K. Lauenroth, and D. P. Coffin. 1995b. Soil organic matter recovery in semiarid grasslands: Implications for the Conservation Reserve Program. *Ecological Applications* **5**:793–801.

Burke, I. C., W. K. Lauenroth, G. Cunfer, J. E. Barrett, A. R. Mosier, and P. Lowe. 2002. Nitrogen in the central grasslands region of the U.S.: A regional perspective. *BioScience* **52**:813–823.

Burke, I. C., W. K. Lauenroth, and D. G. Milchunas. 1997a. Biogeochemistry of managed grasslands in the central grasslands of the U.S., pp. 85–101. In: E. Paul, K. Paustian, and E. T. Elliott (eds.), *Organic matter in U.S. agroecosystems*. CRC Press, Boca Raton, Fla.

Burke, I. C., W. K. Lauenroth, and W. J. Parton. 1997b. Regional and temporal variation in net primary production and nitrogen mineralization in grasslands. *Ecology* **78**:1330–1340.

Burke, I. C., W. K. Lauenroth, R. Riggle, P. Brannen, B. Madigan, and S. Beard. 1999. Spatial variability of soil properties in the shortgrass steppe: The relative importance of topography, grazing, microsite, and plant species in controlling spatial patterns. *Ecosystems* **2**:422–438.

Burke, I. C., W. K. Lauenroth, M. A. Vinton, P. B. Hook, R. H. Kelly, H. E. Epstein, M. R. Aguiar, M. D. Robles, M. O. Aguilera, K. L. Murphy, and R. A. Gill. 1998. Plant–soil interactions in temperate grasslands. *Biogeochemistry* **42**(1–2):121–143.

Burke, I. C., C. M. Yonker, W. J. Parton, C. V. Cole, K. Flach, and D. S. Schimel. 1989. Texture, climate, and cultivation effects on soil organic matter content in U.S. grassland soils. *Soil Science Society of America Journal* **53**:800–805.

Cambardella, C. A., and E. T. Elliott. 1992. Particulate soil organic-matter changes across a grassland cultivation sequence. *Soil Science Society of America Journal* **56**:777–783.

Campbell, C. A., H. H. Janzen, K. Paustian, E. G. Gregorich, L. Sherrod, B. C. Liang, and R. P. Zentner. 2005. Carbon storage in soils of the North American Great Plains: Effect of cropping frequency. *Agronomy Journal* **97**:349–363.

Charley, J. L., and N. E. West. 1977. Micro-patterns of nitrogen mineralization activity in soils of some shrub-dominated semi-desert ecosystems of Utah. *Soil Biology and Biochemistry* **9**:357–365.

Chimner, R. A., and J. M. Welker. 2005. Ecosystem respiration responses to experimental manipulations of winter and summer precipitation in a mixed-grass prairie, WY, USA. *Biogeochemistry* **73**:257–270.

Clark, F. E. 1977. Internal cycling of 15-nitrogen in shortgrass prairie. *Ecology* **58**:1322–1333.

Coffin, D. P., and W. K. Lauenroth. 1988. The effects of disturbance size and frequency on a shortgrass plant community. *Ecology* **69**:1609–1617.

Coffin, D. P., and W. K. Lauenroth. 1990. A gap dynamics simulation model of succession in a semiarid grassland. *Ecological Modelling* **49**:229–266.

Coppock D. L., J. K. Detling, J. E. Ellis, and M. I. Dyer. 1983. Plant–herbivore interactions in a North American mixed-grass prairie. I. Effects of prairie dogs on intraseasonal aboveground plant biomass, nutrient dynamics and plant species diversity. *Oecologia* **56**:1–9.

Corkidi, L., D. L. Rowland, N. C. Johnson, and E. B. Allen. 2002. Nitrogen fertilization alters the functioning of arbuscular mycorrhizas at two semiarid grasslands. *Plant and Soil* **240**:299–310.

Delgado, J. A., A. R. Mosier, D. W. Valentine, D. S. Schimel, and W. J. Parton. 1996. Long term ^{15}N studies in a catena of the shortgrass steppe. *Biogeochemistry* **32**:41–52.

Derner, J. D., T. W. Boutton, and D. D. Briske. 2006. Grazing and ecosystem carbon storage in the North American Great Plains. *Plant and Soil* **280**:77–90.

Dodd, M. B., W. K. Lauenroth, and I. C. Burke. 2000. Nitrogen availability through a coarse-textured soil profile in the shortgrass steppe. *Soil Science Society of America Journal* **64**:391–398.

Dodd, M., W. K. Lauenroth, I. C. Burke, and P. Chapman. 2002. Associations between vegetation patterns and soil texture in the shortgrass steppe. *Plant Ecology* **158**:127–137.

Doran, J. W., and M. R. Werner. 1990. Management and soil biology, pp. 205–230. In: C. A. Francis, C. B. Flora, and L. D. King (eds.), *Sustainable agriculture in temperate zones*. Wiley, New York.

Elliott, E. T. 1986. Aggregate structure and carbon, nitrogen, and phosphorus in native and cultivated soils. *Soil Science Society of America Journal* **50**(3):627–632.

Epstein, H. E., I. C. Burke, and W. K. Lauenroth. 1999. Response of the shortgrass steppe to changes in rainfall seasonality. *Ecosystems* **2**:139–150.

Epstein, H. E., I. C. Burke, and W. K. Lauenroth. 2002. Regional patterns of decomposition and primary production rates in the US Great Plains. *Ecology* **83**:320–327.

Epstein, H. E., I. C. Burke, and A. R. Mosier. 2001. Plant effects on nitrogen retention in shortgrass steppe 2 years after N-15 addition. *Oecologia* **128**:422–430.

Epstein, H. E., I. C. Burke, A. R. Mosier, and G. L. Hutchinson. 1998a. Plant functional type effects on trace gas fluxes in the shortgrass steppe. *Biogeochemistry* **42**:145–168.

Epstein H. E., W. K. Lauenroth, and I. C. Burke. 1997. Effects of temperature and soil texture on aboveground net primary production in the U.S. Great Plains. *Ecology* **78**:2628–2631.

Epstein H. E., W. K. Lauenroth, I. C. Burke, and D. P. Coffin. 1996. Ecological responses of dominant grasses along two climatic gradients in the Great Plains of the United States. *Journal of Vegetation Science* **7**:777–788.

Epstein, H. E., W. K. Lauenroth, I. C. Burke, and D. P. Coffin. 1998b. Regional productivities of plant species in the Great Plains of the United States. *Plant Ecology* **134**:173–195.

Ernst, J. W., and H. F. Massey. 1960. The effects of several factors on volatilization of ammonia formed from urea in the soil. *Soil Science Society Proceedings* **24**:87–90.

Farrar, J. P. 2002. Effects of prairie dog mound-building and grazing activities on vegetation in the central grasslands. Masters thesis, Colorado State University, Fort Collins, Colo.

Gerrard, J. 1992. Soil geomorphology: An integration of pedology and geomorphology. Chapman and Hall, London, UK.

Gill, R. A, and I. C. Burke. 1999. Ecosystem consequences of plant life form changes at three sites in the semiarid United States. *Oecologia* **121**:551–563.

Gill, R. A., I. C. Burke, D. G. Milchunas, and W. K. Lauenroth. 1999. Relationship between root biomass and soil organic matter pools in the shortgrass steppe of eastern Colorado. *Ecosystems* **2**:226–236.

Golubiewski, N. E. 2006. Urbanization increases grassland carbon pools: Effects of landscaping in Colorado's Front Range. *Ecological Applications* **16**:555–571.

Haas, H. J., C. E. Evans, and E. R. Miles. 1957. Nitrogen and carbon changes in Great Plains soils as influenced by cropping and soil treatments. USDA U.S. Government Printing Office, Washington, D.C.

Holland, E. A., and J. K. Detling. 1990. Plant response to herbivory and belowground nitrogen cycling. *Ecology* **71**:1040–1049.

Holland, E. A., F. J. Dentener, B. H. Braswell, and J. M. Sulzman. 1999. Contemporary and pre-industrial global reactive nitrogen budgets. *Biogeochemistry* **46**:7–43.

Holland, E. A., W. J. Parton, J. K. Detling, and D. L. Coppock. 1992. Physiological responses of plant populations to herbivory and their consequences for ecosystem nutrient flow. *The American Naturalist* **140**:685–706.

Hook, P. B., and I. C. Burke. 1995. Evaluation of methods for estimating net nitrogen mineralization in a semiarid grassland. *Soil Science Society of America Journal* **59**:831–837.

Hook, P. B., and I. C. Burke. 2000. Biogeochemistry in a shortgrass steppe landscape: Control by topography, soil texture, and microclimate. *Ecology* **81**:2686–2703.

Hook, P. B., I. C. Burke, and W. K. Lauenroth. 1991. Heterogeneity of soil and plant N and C associated with individual plants and openings in North American shortgrass steppe. *Plant and Soil* **138**:247–256.

Hook, P. B., W. K. Lauenroth, and I. C. Burke. 1994. Spatial patterns of plant cover and roots in a semiarid grassland: Abundance of canopy openings, root gaps, and resources gaps. *Journal of Ecology* **82**:485–494.

Huxman, T. E., K. A. Snyder, D. Tissue, A. J. Leffler, K. Ogle, W. T. Pockman, D. R. Sandquist, D. L. Potts, and S. Schwinning. 2004. Precipitation pulses and carbon fluxes in semiarid and arid ecosystems. *Oecologia* **141**:254–268.

Ihori T., I. C. Burke, and P. B. Hook. 1995a. Nitrogen mineralization in native, cultivated, and abandoned fields in shortgrass steppe. *Plant and Soil* **171**:203–208.

Ihori T., I. C. Burke, W. K. Lauenroth, and D. P. Coffin. 1995b. Effects of cultivation and abandonment on soil organic matter in northeastern Colorado. *Soil Science Society of America Journal* **59**:1112–1119.

Jenkinson, D. S., and J. H. Rayner. 1977. Turnover of soil organic-matter in some of the Rothamsted classical experiments. *Soil Science* **123**:298–305.

Kaye, J. P., I. C. Burke, A. R. Mosier, and J. P. Guerchman. 2004. Methane and nitrous oxide fluxes from urban soils to the atmosphere. *Ecological Applications* **14**(4):975–981.

Kaye, J. P., R. L. McCulley, and I. C. Burke. 2005. Carbon fluxes, nitrogen cycling and soil microbial communities in adjacent urban, native and agricultural ecosystems. *Global Change Biology* **11**:575–587.

Kelly, E. F., S. W. Blecker, C. M. Yonker, C. G. Olson, E. E. Wohl, and L. C. Todd. 1998. Stable isotope composition of soil organic matter and phytoliths as paleoenvironmental indicators. *Geoderma* **82**:59–81.

Kelly, R. H., and I. C. Burke. 1997. Heterogeneity of soil organic matter following death of individual plants in shortgrass steppe. *Ecology* **78**:1256–1261.

Kelly, R. H., I. C. Burke, and W. K. Lauenroth. 1996. Soil organic matter and nutrient availability responses to reduced plant inputs in shortgrass steppe. *Ecology* **77**:2516–2527.

Koski, T., and Skinner, V. 2001. *Lawn care*. Colorado State University Cooperative Extension. Fact sheets. Online. Available at www.ext.colostate.edu/pubs/garden/pubgard.html.

Lane, D. R., D. P. Coffin, and W. K. Lauenroth. 1998. Effects of soil texture and precipitation on above-ground net primary productivity and vegetation structure across the central grassland region of the United States. *Journal of Vegetation Science* **9**(2):239–250.

Lauenroth, W. K., I. C. Burke, and M. P. Gutmann. 1999. The structure and function of ecosystems in the central North American grassland region. *Great Plains Research* **9**:223–259.

Lauenroth, W. K., D. P. Coffin, I. C. Burke, and R. A. Virginia. 1997. Effects of plant mortality on population dynamics and ecosystem structure: a case study, pp. 234–254. In: T. M. Smith and H. H. Shugart (eds.), *Plant functional types*. Cambridge University Press, Cambridge, Mass.

Lauenroth, W. K., J. L. Dodd, and P. L. Simms. 1978. The effects of water- and nitrogen-induced stresses on plant community structure in a semiarid grassland. *Oecologia* **36**:211–222.

Lauenroth, W. K., and O. E. Sala. 1992. Long-term forage production of North American shortgrass steppe. *Ecological Applications* **2**:397–403.

Leetham, J. W., and D. G. Milchunas. 1985. The composition and distribution of soil microarthropods in the shortgrass steppe in relation to soil water, root biomass, and grazing by cattle. *Pedobiologia* **28**:311–325.

Lewis, J. K. 1971. The grassland biome: A synthesis of structure and function, pp. 317–387. In: N. R. French (ed.), P*reliminary analysis of structure and function in grasslands.* Range Science Department Science series no. 10. Colorado State University, Fort Collins, Colo.

Madole, R. F. 1994. Stratigraphic evidence of desertification in the west–central Great Plains within the last 1000 yr. *Geology* **22**:483–486.

Martinez-Turanzas, G. A., D. P. Coffin, and I. C. Burke. 1997. Development of microtopographic relief in a semiarid grassland: Effects of disturbance size and soil texture. *Plant and Soil* **191**:163–171.

McCulley, R. L. 2002. Biogeochemical response of U.S. Great Plains grasslands to regional and interannual variability in precipitation. PhD diss., Colorado State University, Fort Collins, Colo.

McCulley, R. L., and I. C. Burke. 2004. Microbial community composition across the Great Plains: Landscape versus regional variability. *Soil Science Society of America Journal* **68**:106–115.

McCulley, R. L., I. C. Burke, J. A. Nelson, W. K. Lauenroth, A. K. Knapp, and E. F. Kelly. 2005. Regional and temporal patterns in carbon cycling across the Great Plains of North America. *Ecosystems* **8**:106–121.

Meentemeyer, V. 1984. The geography of organic decomposition rates. *Annals of the Association of American Geographers* **74**:551–560.

Milchunas, D. G., and W. K. Lauenroth. 1993. Quantitative effects of grazing on vegetation and soils over a global range of environments. *Ecological Monographs* **63**:327–366.

Milchunas, D. G., and W. K. Lauenroth. 1995. Inertia in plant community structure: State changes after cessation of nutrient-enrichment stress. *Ecological Applications* **5**(2):452–458.

Milchunas, D. G., W. K. Lauenroth, and I. C. Burke. 1998. Livestock grazing: Animal and plant biodiversity of shortgrass steppe and the relationship to ecosystem function. *Oikos* **83**(1):65–74.

Milchunas, D. G., W. K. Lauenroth, P. L. Chapman, and M. K. Kazempour. 1989. Effects of grazing, topography, and precipitation on the structure of a semiarid grassland. *Vegetatio* **80**:11–23.

Milchunas, D. G., W. J. Parton, D. S. Bigelow, and D. S. Schimel. 1988a. Factors influencing ammonia volatilization from urea in soils of the shortgrass steppe. *Journal of Atmospheric Chemistry* **6**:323–340.

Milchunas, D. G., O. E. Sala, and W. K. Lauenroth. 1988b. A generalized model of the effects of grazing by large herbivores on grassland community structure. *The American Naturalist* **132**:87–106.

Milchunas, D. G., and W. K. Lauenroth. 1992. Carbon dynamics and estimates of primary production by harvest, C-14 dilution and C-14 turnover. *Ecology* **73**:1593–1607.

Moore, I., P. Gessler, G. Nielsen, and G. A. Peterson. 1993. Soil attribute prediction using terrain analysis. *Soil Science Society of America Journal* **57**:443–452.

Mosier, A. R., D. Schimel, D. Valentine, K. Bronson, and W. Parton. 1991. Methane and nitrous oxide fluxes in native, fertilized, and cultivated grasslands. *Nature* **350**:330–332.

Muhs, D. R. 1985. Age and paleoclimatic significance of Holocene sand dunes in northeastern Colorado. *Annals of the Association of American Geographers* **75**(4):566–582.

Mulvaney, R. L., and J. M. Bremner. 1981. Control of urea transformation in soils. *Soil Biochemistry* **5**:153–196.

Murphy, K. L., I. C. Burke, M. A. Vinton, W. K. Lauenroth, M. R. Aguiar, D. A. Wedin, R. A. Virginia, and P. Lowe. 2002. Regional analysis of litter quality in the central grassland region of North America. *Journal of Vegetation Science* **13**:395–402.

National Atmospheric and Deposition Program. 1998. *Isopleth maps.* Online. Available at http://nadp.sws.uiuc.edu/isopleths/maps1998/.

Noy-Meir, I. 1973. Desert ecosystems: Environment and producers. *Annual Review of Ecology and Systematics* **4**:25–51.

Parton, W. J., W. K. Lauenroth, and F. M. Smith. 1981. Water loss from a shortgrass steppe. *Agricultural Meteorology* **24**:97–109.

Parton, W. J., D. S. Schimel, D. S. Ojima, and C. V. Cole. 1987. Analysis of factors controlling soil organic matter levels in Great Plains grasslands. *Soil Science Society of America* **51**:1173–1179.

Parton, W., W. L. Silver, I. C. Burke, L. Grassens, M. E. Harmon, W. S. Currie, J. Y. King, E. C. Adair, L. A. Brandt, S. C. Hart, and B. Fasth. 2007. Global-scale similarities in nitrogen release patterns during long-term decomposition. *Science* **315**:361–364.

Paustian, K. H., H. P. Collins, and E. A. Paul. 1997. Management controls on soil carbon, pp. 15–49. In: E. A. Paul, K. Paustian, E. T. Elliott, and C. V. Cole (eds.), *Soil organic matter in temperate agroecosystems.* CRC Press, Boca Raton, Fla.

Qian, Y. L., W. Bandaranayake, W. J. Parton, B. Mecham, M. A. Harivandi, and A. R. Mosier. 2003. Long-term effects of clipping and nitrogen management in turfgrass on soil organic carbon and nitrogen dynamics: The CENTURY model simulation. *Journal of Environmental Quality* **32**:1694–1700.

Rauzi, F., and F. M. Smith. 1973. Infiltration rates: Three soils with three grazing levels in northeastern Colorado. *Journal of Range Management* **26**:126–129.

Reeder, J. D., and G. E. Schuman. 2002. Influence of livestock grazing on C sequestration in semi-arid mixed-grass and short-grass rangelands. *Environmental Pollution* **116**:457–463.

Reeder, J. D., G. E. Schuman, J. A. Morgan, and D. R. LeCain. 2004. Response of organic and inorganic carbon and nitrogen to long-term grazing of the shortgrass steppe. *Environmental Management* **33**:485–495.

Robles, M. D., and I. C. Burke. 1997. Legume, grass, and conservation reserve program effects on soil organic matter recovery. *Ecological Applications* **7**:345–357.

Robles, M. D., and I. C. Burke. 1998. Soil organic matter recovery on Conservation Reserve Program fields in southeastern Wyoming. *Soil Science Society of America Journal* **62**:725–730.

Sala, O. E., and W. K. Lauenroth. 1982. Small rainfall events: An ecological role in semiarid regions. *Oecologia (Berl.)* **53**:301–304.

Sala, O. E., W. K. Lauenroth, and W. J. Parton. 1992. Long-term soil water dynamics in the shortgrass steppe. *Ecology* **73**:1175–1181.

Sala, O. E., W. J. Parton, L. A. Joyce, and W. K. Lauenroth. 1988. Primary production of the central grassland region of the United States. *Ecology* **69**:40–45.

Schimel, D. S., and W. J. Parton. 1986. Microclimate controls of nitrogen mineralization and nitrification in shortgrass prairie steppe soils. *Plant and Soil* **93**:347–357.

Schimel, D. S., W. J. Parton, F. J. Adamsen, R. G. Woodmansee, R. L. Senft, and M. A. Stillwell. 1986. The role of cattle in the volatile loss of nitrogen from a shortgrass steppe. *Biogeochemistry* **2**:39–52.

Schimel, D., M. A. Stillwell, and R. G. Woodmansee. 1985a. Biogeochemistry of C, N, and P in a soil catena of the shortgrass steppe. *Ecology* **66**:276–282.

Schimel, D. S., M. A. Stillwell, and R. G. Woodmansee. 1985b. Effects of erosional procession nutrient cycling in semiarid landscapes, pp. 571–580. In: D. E. Coldwell, J. A. Brierly, and C. L. Brierly (eds.), *Planetary ecology.* Van Nostrand Reinhold, New York.

Schlesinger, W. H., J. F. Reynolds, G. L. Cunningham, L. F. Huenneke, W. M. Jarrell, R. A. Virginia, and W. G. Whitford. 1990. Biological feedbacks in global desertification. *Science* **247**:1043–1048.

Senft, R. L. 1983. *The redistribution of nitrogen by cattle.* PhD diss., Colorado State University, Fort Collins, Colo.
Senft, R., M. Coughenour, D. Bailey, L. Rittenhouse, O. Sala, and D. Swift. 1987. Large herbivore foraging and ecological hierarchies. *BioScience* **37**:789–799.
Senft, R. L., L. R. Rittenhouse, and M. A. Stillwell. 1984. Seasonal changes in nitrogen and energy budgets of range cattle. *Proceedings of the Western Section of the American Society of Animal Science* **35**:200–203.
Senft, R., L. Rittenhouse, and R. G. Woodmansee. 1985. Factors influencing patterns of cattle grazing behavior on shortgrass steppe. *Journal of Range Management* **38**:82–87.
Singh, J. S., D. G. Milchunas, and W. K. Lauenroth. 1998. Soil water dynamics and vegetation patterns in a semiarid grassland. *Plant Ecology* **134**:77–89.
Sinton, P. J. 2001. *Effects of irrigated and dryland cultivation on soil carbon, nitrogen and phosphorus in northeastern Colorado.* Masters thesis, Colorado State University, Fort Collins, Colo.
Skold, M. D. 1989. Crop retirement policies and their effects on land use policies in the Great Plains. *Journal of Production Agriculture* **2**:197–201.
Stillwell, M.A. 1983. *The effects of bovine urine on the nitrogen cycle of a shortgrass prairie.* PhD diss., Colorado State University, Fort Collins, Colo.
Stillwell, M. A., and R. G. Woodmansee. 1981. Chemical transformations of urea-nitrogen and movement of nitrogen in a shortgrass prairie soil. *Soil Science Society of America Journal* **45**:893–898.
Striffler, W. D. 1972. *Hydrologic data at Pawnee site, 1971.* U.S. International Biological Program. Grassland Biome, 1972. Technical report no. 196. Natural Resource Ecology Laboratory, Colorado State University, Fort Collins, Colo.
U.S. Bureau of the Census 1999. *Population estimates program, population estimates for cities with populations of 100,000 and greater, SU-98–2.* U.S. Government Printing Office, Washington, D.C.
Vinton, M. A., and I. C. Burke. 1995. Interactions between individual plant species and soil nutrient status in shortgrass steppe. *Ecology* **76**:1116–1133.
Vinton, M. A., and I. C. Burke. 1997. Plant effects on soil nutrient dynamics along a precipitation gradient in Great Plains grasslands. *Oecologia* **110**:393–402.
Wedin, D. A., and D. Tilman. 1990. Species effects on nitrogen cycling: A test with perennial grasses. *Oecologia* **4**:433–441.
Wood, C. W., D. G. Westfall, and G. A. Peterson. 1991. Soil carbon and nitrogen changes on initiation of no-till cropping systems. *Soil Science Society of America Journal* **55**:470–476.
Wood, C. W., D. G. Westfall, G. A. Peterson, and I. C. Burke. 1990. Impacts of cropping intensity on C and N mineralization in no-till dryland agroecosystems. *Agronomy Journal* **82**(6):1115–1120.
Woodmansee, R. G. 1978. Additions and losses of nitrogen in grassland ecosystems. *BioScience* **28**(7):448–453.
Woodmansee, R. G., and F. J. Adamsen. 1983. Biogeochemical cycles and ecological hierarchies, pp. 497–516. In: R. R. Lowrance, R. L. Todd, L. F. Asmussen, and R. A. Leonard (eds.), *Nutrient cycling in agricultural ecosystems.* College of Agriculture Experiment Station special publication no. 23. University of Georgia, Athens, Ga.
Woods, L. E., and G. E. Schuman. 1988. Cultivation and slope position effects on soil organic matter. *Soil Science Society of. America Journal* **52**(5):1371–1376.
Yonker, C. M., D. S. Schimel, E. Paroussis, and R. D. Heil. 1988. Patterns of organic carbon accumulation in a semiarid shortgrass steppe, Colorado. *Soil Science Society of America Journal* **52**:478–483.

14

Soil–Atmosphere Exchange of Trace Gases in the Colorado Shortgrass Steppe

Arvin R. Mosier
William J. Parton
Roberta E. Martin
David W. Valentine
Dennis S. Ojima
David S. Schimel
Ingrid C. Burke
E. Carol Adair
Stephen J. Del Grosso

During the past half century, atmospheric concentrations of important greenhouse gases including carbon dioxide (CO_2), methane (CH_4), and nitrous oxide (N_2O) have been increasing at unprecedented rates (IPCC, 1996, 2007). Trace gases such as methane (CH_4), nitric oxide (NO), and nitrous oxide (N_2O) are exchanged regularly between the soil and atmosphere, playing important roles in the greenhouse effect, in atmospheric chemistry, and in the redistribution of ecosystem nitrogen (N). Soils can be important sources of greenhouse gases, commonly contributing up to two thirds of atmospheric N_2O and more than one third of atmospheric CH_4 (Monson and Holland, 2001; Smith et al., 2003). Recent extensive changes in land management and in cultivation, which can stimulate N_2O production and/or decrease CH_4 uptake, could be contributing to the observed increases of both CH_4 and N_2O in the atmosphere (IPCC, 2007). Although the absolute amount of trace gases (such as CH_4, NO, and N_2O) released into the atmosphere from soils may be small, these gases are extremely effective at absorbing infrared radiation (Smith et al., 2003). Methane, for example, is 20 to 30 times more efficient than CO_2 as a greenhouse gas (LeMer and Roger, 2001).

As a result, even small changes in the production or consumption of these gases by soils could dramatically influence climate change.

Of the gases exchanged between the soil and atmosphere, the major reactive ones are oxides of N (NO and NO_2, collectively referred to as NO_x). Combustion is a major source of NO_x, but native and N-fertilized soils also contribute significant amounts of NO_x to the atmosphere (Williams et al., 1992). Nitric and nitrous oxide play a complex role in atmospheric chemistry. At low concentrations, it catalyzes the breakdown of ozone. At higher concentrations it can interact with carbon monoxide (CO), hydroxyl radicals (OH·), and hydrocarbons to produce ozone. Atmospheric NO_x is converted within days to nitric acid, which is an important component (30% to 50%) of acidity in precipitation (Williams et al., 1992).

Methane has important effects on the oxidative capacity of the troposphere. By reacting with OH radicals, it reduces the ability of the lower atmosphere to eliminate pollutants such as chlorofluorocarbons (CFCs) and it forms methyl peroxy radicals (LeMer and Roger, 2001; Monson and Holland, 2001). These radicals react with NO and oxygen and can ultimately contribute to the formation of ozone (Monson and Holland, 2001). Changes in land use and increases in N fertilization may alter both soil uptake and soil production of CH_4 (Monson and Holland, 2001).

The N trace gases described earlier clearly constitute one pathway for ecosystem N loss and redistribution. Ammonia (NH_3) is also emitted from soils to the atmosphere. In the atmosphere, NH_3 can be dissolved or adsorbed by aerosols and/or water droplets, or be taken up directly by vegetation. Most of the NH_3 that enters the troposphere is returned as NH_4^+ (ammonium) in bulk precipitation or captured aerosols. Ammonia in aerosols or dissolved in cloud droplets can accept a proton to form NH_4^+, thereby reducing the acidity of precipitation (Schlesinger and Hartley, 1992).

Developing a working knowledge of the controls on soil–atmosphere exchange of these trace gases is clearly important. Until we began the studies discussed in this chapter, little information existed on CH_4 uptake in temperate grasslands or on the impact of changes of land use and management on CH_4 uptake or nitrogen oxides (NO_x, connoting both NO and NO_2) and N_2O emissions. We began these studies to understand and clarify the relationships between soil trace gas fluxes and climatic factors, soil properties, landscape position, and land-use management in temperate grasslands.

An Approach for Estimating Trace Gas Fluxes at a Long-Term Site

Historical Perspective of Trace Gas Measurements in the Shortgrass Steppe

During the late 1980s and early 1990s, the idea that the land–atmosphere exchange of trace gases could significantly affect local and global atmospheric conditions was just being given serious consideration (Andreae and Schimel,

1989; Crutzen, 1983). Such considerations led to the development of components of the International Geosphere Biosphere Program. As a result of these discussions, in early 1990 we initiated measurement of the soil–atmosphere exchange of CH_4, and revived N_2O flux studies to ascertain the impact of land-use change and nutrient additions on the shortgrass steppe (Mosier et al., 1991).

Rationale for Measurement

A number of factors influence soil biogeochemistry and should therefore affect the soil–atmosphere exchange of trace gases. We knew that N_2O and NO_x fluxes are characteristically highly spatially and temporally variable (both between and within years), but we did not know how variable CH_4 uptake rates were in time and space. Based upon studies of soil nutrient pools (Burke et al., chapter 13, this volume), we expected that climate, topography, soil properties, and nutrient and water availability would be important, and that long-term land-use change and management would, in turn, affect soil properties and nutrient availability. Long-term research study sites were needed to be able to address these issues.

The USDA–ARS CPER, located within the SGS LTER site, provided a long-term research site that met our needs. These needs included sites with known history, typical topography, accessibility, different land uses and management, and the ability to use the sites over many years.

In designing our field sampling program, it was important to consider all the known major sources of variability in biogeochemical pools and processes, including plant–interplant location, soils, and landscape position (Burke et al., chapter 13, this volume). The measurement unit needed to be within the dominant vegetation (Lauenroth, chapter 5, this volume), and contain the main source of spatial variability: bunchgrass hummocks and the empty spaces between them. We also needed to encompass the variation in land use and management found in the shortgrass steppe and the CPER.

Within the shortgrass steppe region and the CPER, a variety of changes in land use and management had taken place during the 20th century. Early during the century much of the land was cultivated, but most areas under cultivation reverted back to grassland after the Dust Bowl days of the late 1930s (Hart, chapter 4, this volume). Throughout the years at the CPER, different management studies were implemented, which included N fertilization and grazing intensity trials. On agricultural fields adjacent to the CPER, dryland wheat-fallow cropping was common. To encompass this variability, we set up several series of long-term sites: a sand catena, a clay catena, and a series of fertilized and unfertilized, plowed and unplowed sites in native pasture and in cultivated areas.

Variability of gas fluxes over time was also of concern. We knew from limited studies within the shortgrass steppe and broader studies in other ecosystems that N_2O and NO_x fluxes vary greatly with time, depending primarily upon soil moisture conditions, temperature, and substrate (Andreae and Schimel, 1989). To quantify fluxes of these gases properly, continuous measurements are needed (Brumme et al., 1999), but instrumentation to conduct such measurements is prohibitively expensive. To accommodate resource limitations we decided to take weekly measurements at each site on a year-round basis. The idea was that such a measurement scheme would allow cross-site comparisons to be made under the same weather conditions. When conducted over at least 2 years, the measurement periods can be expected to encompass a wide variety of different soil conditions, which are needed to provide information for simulation model parameterization and verification. Such models are used to simulate trace gas exchange at large temporal and spatial scales. Maintaining a rigorous sampling schedule over years provided the possibility of comparing gas fluxes from sites having different topography, soil properties, moisture conditions, nutrient input, land use, seasonal climate, and year-to-year climate variations. We made measurements including weekly fluxes of CH_4, CO_2, N_2O, and NO_x (after 1994), soil water content (0–15 cm), soil temperature (5 cm), air temperature, and soil nitrate (NO_3^-) and NH_4^+, for 2 to 5 years at 28 different locations.

Research Approach

We established a set of sites (Table 14.1) to address the following four major questions:

1. How do gaseous fluxes vary across landscape positions, and how does this topographic effect vary across soils of different textures?
2. What is the effect of fertilization on trace gas fluxes?
3. What is the effect of current and past cultivation management and abandonment on trace gas fluxes?
4. What is the effect of enhanced atmospheric CO_2 concentration on these fluxes?

Gas Flux Measurements

Methane and Nitrous Oxide We used a chamber method to sample trace gas fluxes in the field. We installed chamber anchors about 5 m apart, randomly within each location, with three to five replicate locations for each treatment we were testing (Table 14.1). The permanently placed anchors were made from PVC pipe, with an inside diameter of 20 cm, driven 8 cm into the soil. Anchors within wheat fields were removed before each tillage operation and were reinstalled later in new, nearby locations. We measured soil–atmosphere exchange of CH_4 and N_2O within each location by fitting 7.5-cm high, closed, vented chambers (Hutchinson and Mosier, 1981) onto the anchors. Gas samples from inside the chambers were removed at 0, 15, and 30 minutes after installation using 60-mL polypropylene syringes fitted with nylon stopcocks. We analyzed the samples by gas chromatography, usually within 6 hours. These flux measurements were made mid morning of each sampling day, and generally weekly at each site (details of sampling methods in Mosier et al., 1991, 1993, 1996).

Nitrogen Oxides (NO_x) We used a slightly different technique for NO_x, involving in situ measurements. We developed a system that flowed outside air

Table 14.1 Site Soil Properties (0–15 cm)

	Flux Measurement Period, years (19xx)	Sand, %	Clay, %	Bulk Density, g/cm^{-3}	Total N, %	Total C, %	pH	Gravimetric Soil Water at WFPS 0.2 kg·kg^{-1}	0.6 kg·kg^{-1}	Annual N Min., g N·m^{-2}·y^{-1}	Microbial Biomass, (μg N·g^{-1} × 100)
Sand catena											
Midslope native (MN)	1990–1995	76	13	1.42	0.14	1.25	6.5	6.5	19.6	4.1	ND
Midslope fertilized (MF)	1990–1995	76	13	1.42	0.14	1.22	5.6	ND	ND	ND	ND
Swale native (SN)	1990–1995	58	24	1.34	0.19	1.84	6	7.4	24.7	ND	ND
Swale fertilized (SF)	1990–1995	58	24	1.34	0.22	2.22	5.6	ND	ND	ND	ND
Clay catena											
Top clay	1991–1995	50	24	1.32	0.09	0.80	6.9	7.6	22.8	ND	ND
Midslope clay	1991–1995	70	13	1.33	0.14	1.3	5.5	7.5	22.5	0.9	290
Swale –clay	1991–1995	47	27	1.32	0.17	1.80	5.7	7.6	22	1.2	1140
Pasture A											
Pasture native (PN)	1990–1995	74	13	1.41	0.12	1.07	6.2	6.6	19.9	4.3	310
Pasture fertilized (PF)	1990–1995	74	15	1.34	0.14	1.18	5.6	7.4	22.1	8.5	580
Pasture plowed (PP)	1992–1995	74	14	1.29	0.11	0.89	6	7.5	22.5	ND	ND
Pasture B											
Field native (FN)	1992–1993	70	12	1.4	0.12	1.10	6.5	6.7	20.2	ND	ND
Field tilled 39 (FT)	1992–1993	70	12	1.4	0.10	0.86	6.5	6.7	20.2	ND	ND
W-F-CRP											
Wheat west (WW)	1992–1995	66	15	1.27	0.08	0.54	7.3	8.2	24.6	ND	600
Wheat east (WE)	1992–1995	67	13	1.32	0.08	0.65	6.5	7.6	22.8	ND	270
	1992–1995	64	16	1.44	0.09	0.74	7.3	6.3	18.9	ND	ND
Pasture C	1996–1997	74	11	1.36	0.13	1.21	7.1	7.2	21.6	ND	ND
Pasture D	1996–1997	21	39	0.91	0.3	3.09	7.3	14.4	43.2	ND	ND
CO$_2$ study site	1997–2002	76	10	1.28	0.1	0.89	7.3	7.2	14.4	ND	ND

CRP, Conservation Research Program.; Min., minimum; ND, not determined.

through a 6-L chamber at the rate of 1 L·min^{-1} for 6 minutes after the chamber was placed on the anchor. The change in concentration of NO and NO$_2$ in the air flowing through the chamber was determined using a Scintrex LNC-3 convertor and LMA-3 analyzer to measure a chemiluminescent reaction with Luminol (Scintrex Ltd., Concord, ON, Canada) (Martin et al.,1998).

Major Controls over Trace Gas Exchange

A number of factors generally control soil biogeochemistry and therefore the soil–atmosphere exchange of CH$_4$, N$_2$O, and NO$_x$. Our studies, discussed next, show that weather and climate, topography and soil texture, nutrients, plant species composition, and land-use management all influence trace gas exchange in predictable ways.

Under the semiarid conditions of the shortgrass steppe, soils are seldom wet enough to generate net production of CH$_4$ in the soil, which requires very low oxidation–reduction conditions (Reeburgh et al., 1993). However, because shortgrass steppe soils are typically sandy and have a significant fraction of the soil pore space filled with air rather than water (Schimel et al., 1986), they are well suited for oxidation of atmospheric CH$_4$. Methane-oxidizing microorganisms appear to be ubiquitous in nature and generally oxidize atmospheric CH$_4$ at rates that relate to gas diffusion and microbial activity.

Nitrous oxide is produced both through nitrification and denitrification. Nutrient supply, soil water, and temperature typically are the main regulators of its production (Firestone and Davidson, 1989). Emission from the soil to the atmosphere is dependent mainly upon soil properties and moisture (Mosier et al., 1996). The largest production and emission events are generally related to denitrification rather than nitrification (Parton et al., 1988). Nitric oxide is produced during both nitrification and denitrification, but emissions from the soil are most directly related to nitrification (Tortoso and Hutchinson, 1990).

Impact of Weather and Climatic Variability: Control by Soil Temperature and Moisture

Our-long term weekly measurements of N$_2$O and CH$_4$ fluxes indicate that, across all landscape positions, gas fluxes follow the same temporal pattern through the seasons. As soil moisture increases, N$_2$O emissions increase and CH$_4$ uptake decreases. Nitrous oxide fluxes are generally higher during warmer wetter periods, very low during dry periods, and generally low but occasionally spiking during the cold seasons. Fluxes of both gases do not stop during the winter (Fig. 14.1A). The 4 months of November through February typically represent 15% to 30% of the annual flux. The 4-year average monthly flux maxima occur in February and August for N$_2$O (Fig. 14.1A). Methane fluxes typically follow an inverse relationship with soil water content until soils have been very dry for some time (Valentine et al., 1993). Methane uptake is highest in April through June and August through October, and lowest in July when soils are warm and dry. The interactions

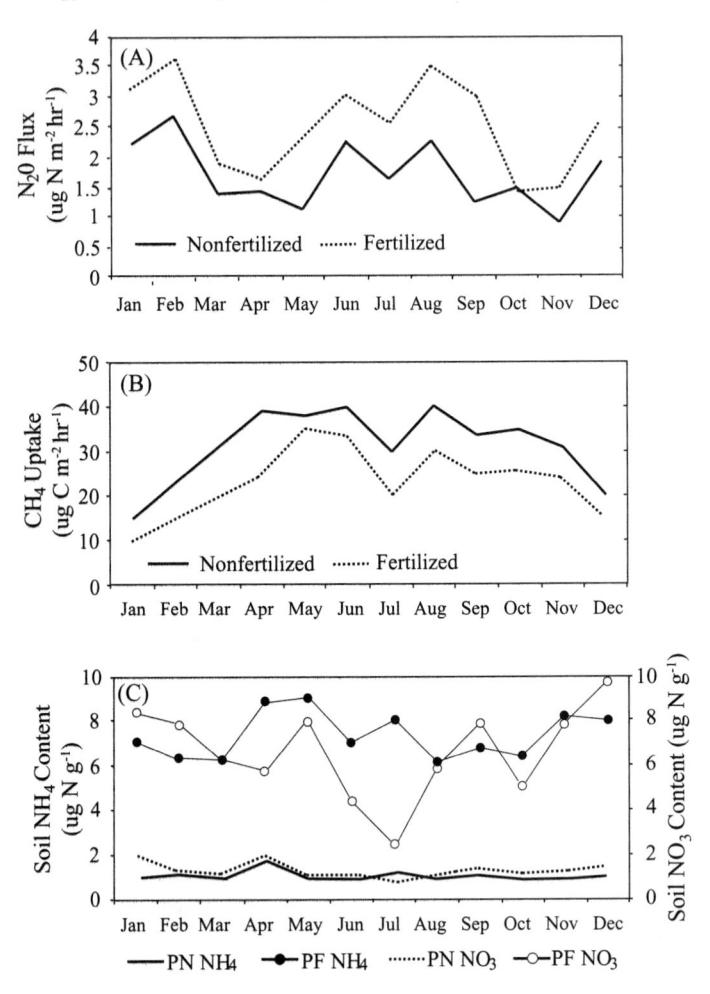

Figure 14.1 (A) Mean monthly N_2O flux. (B) Methane uptake based on 4-year means (1991–1994) at three unfertilized sites and three fertilized sites. (C) Mean monthly soil nitrate and ammonium content based on 4-year means at two sites. PF, native pasture fertilized; PN, native pasture nonfertilized.

among water, temperature, and substrate concentration determine N_2O production (Li et al., 1992; Parton et al., 1988), so that direct correlation between N_2O flux and any one of these variables is typically very low in the shortgrass steppe.

The high N_2O fluxes during the winter (>5 μg N·m^{-2}·hr^{-1}) were most common during January and February (Fig. 14.1A) and were associated with snowmelt periods (80% of the high N_2O fluxes during the winter occurred during periods of snowmelt). The winter N_2O flux data show that significant fluxes occurred despite low soil temperature when microbial activity is assumed to be low. Because N_2O may be produced from two processes—one aerobic (nitrification) and one anaerobic (denitrification)—we deduce that the high fluxes are caused by denitrification during the wet periods associated with snowmelt. The four months of November through February typically represent 15% to 30% of the annual flux at each site. In some years, maximum mean monthly N_2O flux rates were observed in January and February. The second peak in N_2O emissions generally occurred during the May through September (Fig. 14.1A) period as a result of increased N turnover at the warmer temperatures, even though water is frequently limiting during the summer months. The high fluxes of N_2O during the summer are primarily a result of nitrification because the observed soil water contents rarely are high enough to promote denitrification (Li et al., 1992; Parton et al., 1988, 1996). The large winter N_2O emissions occurred only when soil, snow, and snowmelt fluctuations induced appropriate conditions.

We attempted to quantify the effects of soil water and soil temperature on CH_4 and N_2O fluxes by aggregating data from all the treatments and sites we had studied on the SGS LTER. In the summer, at soil temperature more than 15 °C, N_2O fluxes tend to increase exponentially with increasing soil temperature (Fig. 14.2), and increase linearly as soil water-filled pore space (WFPS) increases from 10% to 50% (Fig. 14.3). We evaluated the soil temperature effect by using N_2O flux data for cases when water was not limiting (WFPS was more than 30% and N_2O fluxes were relatively high) and then fit an exponential curve to the data. We stratified the data into two groups: fertilized sites (Fig. 14.2A) and unfertilized sites (Fig. 14.2B).

Similarly, we evaluated the effect of soil water on N_2O flux by aggregating all the data from cases when soil temperature was not limiting (>15 °C) and used a linear regression to represent the effect of WFPS (Fig. 14.3). The data were stratified into the same groups used for the temperature effect on N_2O flux. The results (Figs. 14.2 and 14.3) show that N_2O fluxes increase with increasing WFPS and soil temperature, that the fluxes are highest for the fertilized sites, and that they are similar for the unfertilized sites in sandy loam and swales.

We evaluated the effect of soil temperature on CH_4 uptake (Figs. 14.4 and 14.5) by aggregating the data for WFPS values from 15% to 25% (highest CH_4 consumption) and calculating a linear regression of CH_4 fluxes to soil temperature (Fig. 14.4). The data were further aggregated into two categories according to surface soil texture (Fig. 14.4). Although CH_4 consumption increased linearly with increasing soil temperature (Fig. 14.4A) in the sandy sites ($r^2 = .23$), in the fine-textured sites soil temperature had little or no effect on CH_4 consumption ($r^2 = 0.0$; Fig. 14.4B).

The effect of WFPS on CH_4 consumption was determined by using all the data with soil temperature more than 15 °C (higher CH_4 consumption) and then fitting a beta function to the data (Fig. 14.5). We used the beta function because CH_4 consumption was highest for WFPS values from 15% to 25%, and decreased with WFPS values more than 25% or less than 15%. Again, we aggregated data into fertilized sites (Fig. 14.5A), nonfertilized sand (Fig. 14.5B), and nonfertilized clay sites (Fig. 14.5C). The results showed that peak CH_4 consumption occurred at 15% WFPS for the sandy soils and 20% WFPS for the fine-textured soils. The unfertilized sandy soils had the highest CH_4 consumption rates, followed by fertilized sandy sites and then by fine-textured sites ($r^2 = .11, .2, .19$, respectively).

350   Ecology of the Shortgrass Steppe

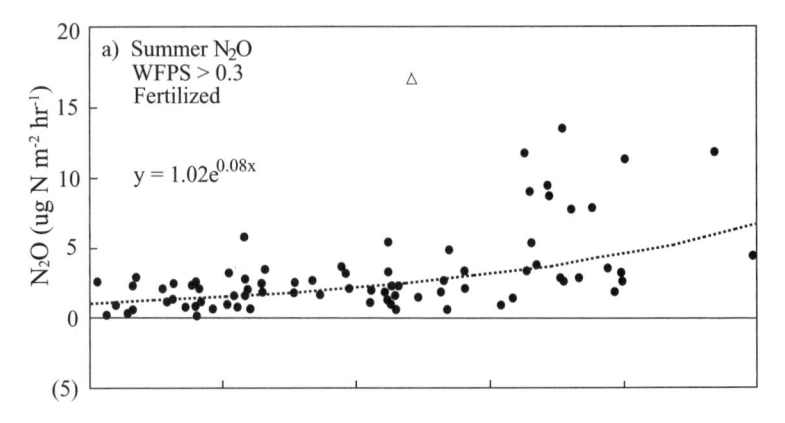

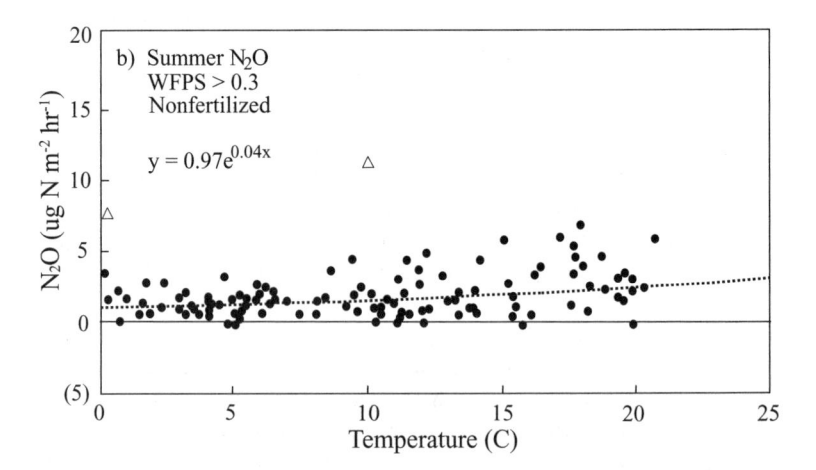

Figure 14.2   Regression analysis of the effect of soil temperature on N_2O flux at soil water-filled pore space (WFPS) more than 0.3 in fertilized sites (A) (n=3; r^2=.30) and unfertilized sites (B) (n=3; r^2=.11). Data points represented as triangles were considered outliers and were not included in the regression analysis.

Seasonal NO emissions appear to be dependent on temperature (r^2= .36), with the highest fluxes occurring during the summer and the lowest during the winter (Fig. 14.6). It has been observed in both tropical forest and savanna ecosystems that the largest NO emissions are associated with precipitation events that follow long dry periods (Davidson et al., 1993; Johansson and Sanhueza, 1988). Spring and early summer were unusually wet in 1995, when May through July precipitation was double that received in 1994, which may account for the relatively small increases in flux after precipitation events in 1995.

Zachariassen and Schimel (1991) measured CH_4 fluxes on 8 days during four different parts of the growing season: green-up (mid June), peak biomass

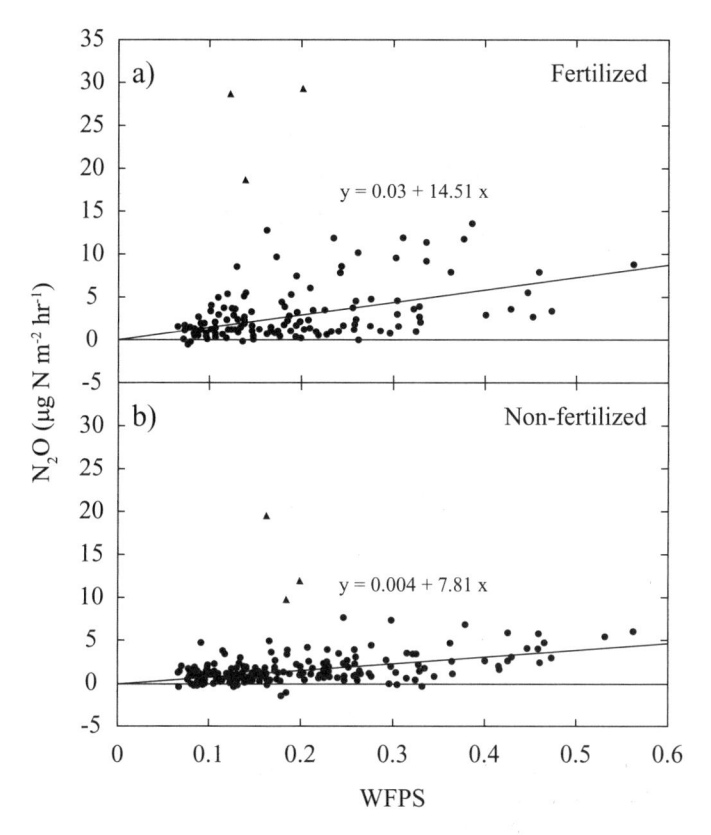

Figure 14.3 Regression analysis of the effect of soil water-filled pore space (WFPS) on N_2O flux when soil temperatures were more than 15 °C in fertilized sites (A) (n=3; r^2=.23) and unfertilized sites (B) (n=3; r^2=.29). Data points represented as triangles were considered outliers and were not included in the regression analysis.

(late June), dry-down/senescence (early July), and postsenescence (mid to late September). The 24-hour mean flux rates observed for these periods were 0.004, −0.012, 0.007, and 0.034 µg $N \cdot m^{-2} \cdot s^{-1}$ respectively (Langford et al., 1992). Using these values, we estimated an annual plant-generated CH_3 emission rate of 0.1 g $N \cdot m^{-2} \cdot y^{-1}$ for the shortgrass steppe.

Impact of Landscape Position and Soil Texture

We have evaluated the effects of soil texture and landscape position on trace gas fluxes using several sites (Mosier et al., 1998). Our results indicate that when landscape position is held constant, soil texture significantly affects both NO_x and CH_4 fluxes. Nitric oxide emissions and CH_4 oxidation rates were higher in soils with the lowest clay content (Fig. 14.7). During a July through August sampling

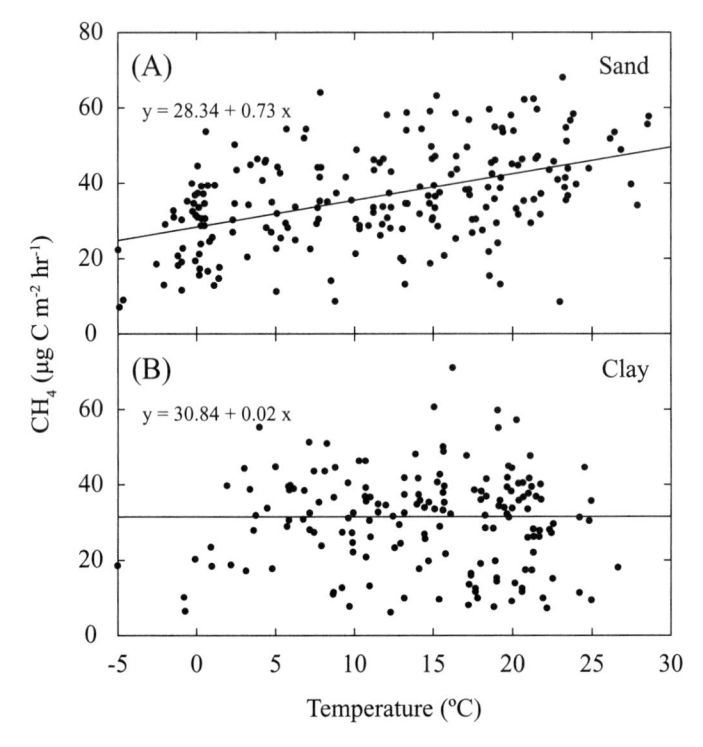

Figure 14.4 Regression analysis of the effect of soil temperature on CH_4 uptake at soil water-filled pore space between 0.15 and 0.25 in coarse-textured soils (A) (n=2; r^2=.23) and finer textured soils (B) (n=2; r^2=.0).

period, NO emission from the undisturbed sandy loam soil was more than double that observed from the clay loam soil ($P < .05$; Fig. 14.7). At the same time, N_2O emissions were lower from the soil having the lowest clay content.

The effects of landscape position on trace gas fluxes of the shortgrass steppe are strong, but difficult to sort out from the direct effects of soil texture, as seen earlier. Our results from both sandy and clay catenas indicate that N_2O fluxes from swales are typically higher than from mid slopes (e.g., on sandy catena, 54-month mean of 1.2 μg N·m^{-2}·hr^{-1} for mid slopes and 2.0 μg N·m^{-2}·hr^{-1} for swales [$P< .05$]; Fig. 14.8A). Conversely, CH_4 uptake is higher in the summits and mid slopes than in swales for both sandy and clay catenas (36.7 μg C·m^{-2}·hr^{-1} on sandy mid slopes compared with 21.3 μg N·m^{-2}·hr^{-1} in the swales [$P < .05$]; Fig. 14.8B).

Impact of Soil Nitrogen Additions

Nitrous Oxide

Annual N_2O emissions from shortgrass steppe soils that have only natural N addition through atmospheric N deposition average about 100 g N·ha^{-1}·y^{-1}, which

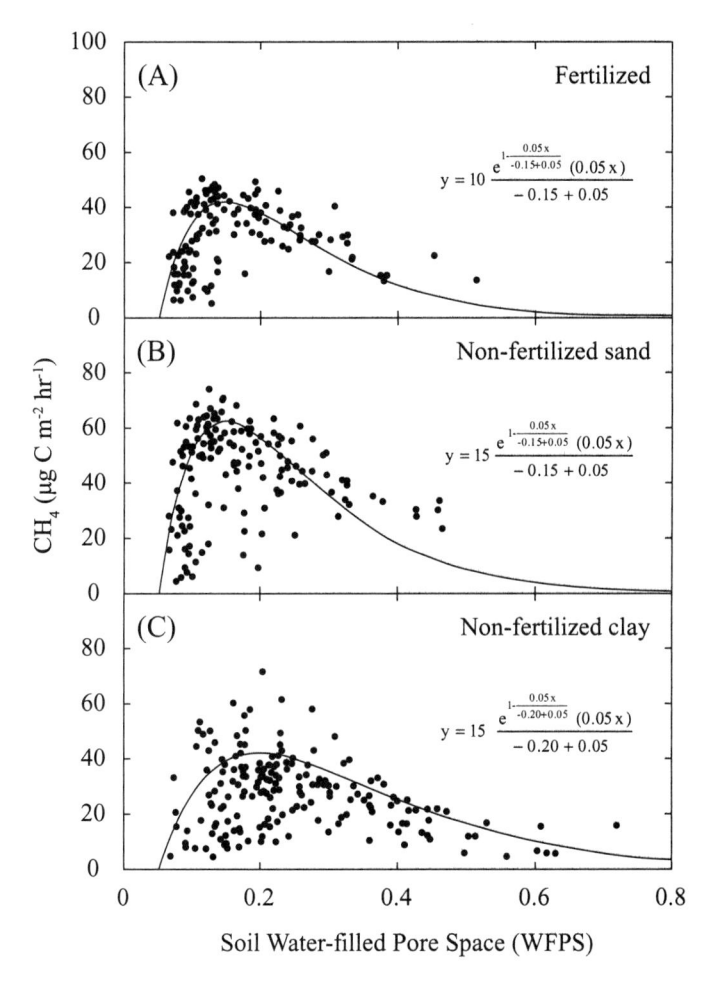

Figure 14.5 Effect of soil water-filled pore space (WFPS) on CH$_4$ uptake with soil temperature more than 15°C in fertilized sites (A) (n=2; r^2=0.11), unfertilized coarse-textured sites (B) (n=2; r^2=0.20), and unfertilized fine-textured sites (C) (n=2).

is 1.5% to 5% of the annual N input estimated from wet and dry deposition (Parton et al., 1988). We have conducted several fertilization studies to evaluate the effects of increased N availability on N$_2$O emissions (Mosier and Schimel, 1991) by adding N to several landscape positions, using several different levels of fertilization.

In studies simulating cattle urine deposition, we found that during the year after N application (Mosier and Parton, 1985; Mosier et al., 1982), N losses were small (0.5% to 1% of N added). In other studies with low N addition, we found a residual effect of fertilization on N$_2$O emissions still evident after 12 to 13 years,

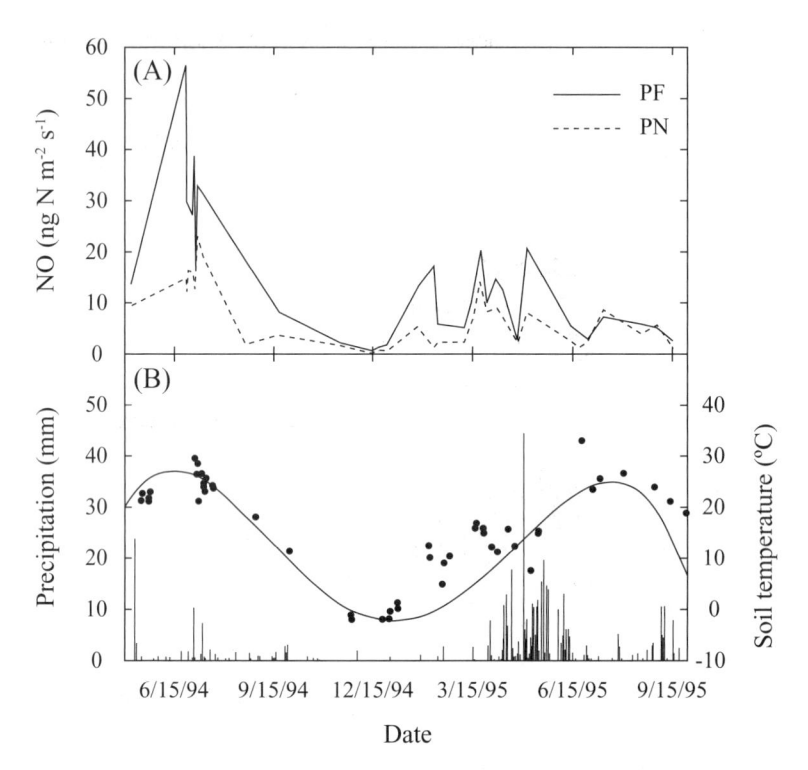

Figure 14.6 (A) Fluxes of NO from pasture sites PF (native fertilized) and PN (native nonfertilized). (B) Precipitation and soil temperature measured at the CPER weather station (solid line), and soil temperature measured at the time of flux measurements (circles).

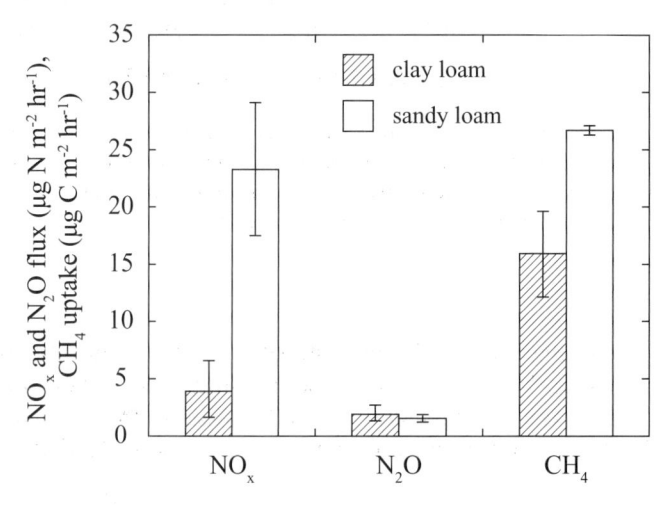

Figure 14.7 Impact of soil texture on mean NO_x and N_2O emission rates, and CH_4 uptake rates in a pasture with sandy loam soil (SL) and in a pasture with clay loam soil (CL). Gas flux data are means of weekly measurements over 2 years.

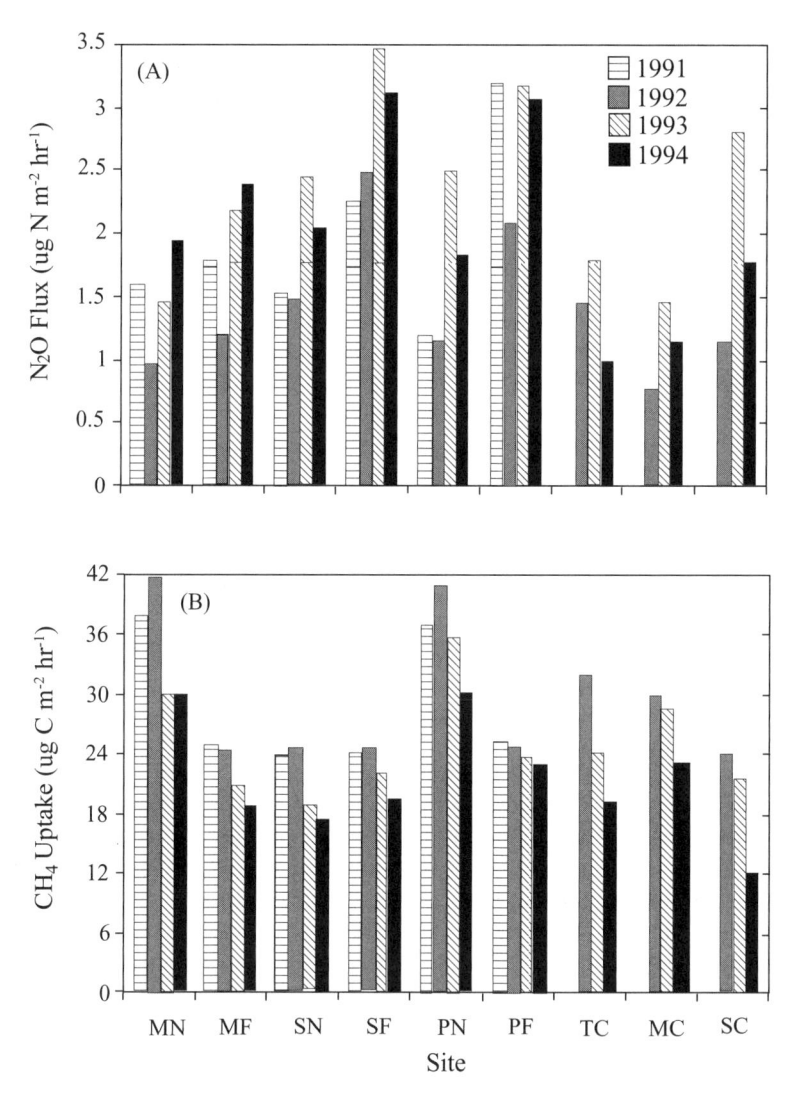

Exchange of Trace Gases in the Shortgrass Steppe 355

Figure 14.8 (A) Mean annual N_2O flux from nine sites at the CPER. (B) Mean annual CH_4 uptake rates at nine sites in the CPER. Flux rates were calculated from mean monthly rates of weekly measurements from 1991 through 1994. Sand catena: TC, top clay; MF, midslope fertilized; MN, midslope-native; PF, pasture fertilized; PN, pasture native; SF, swale fertilized; SN, swale native. Clay catena: MC, midslope clay; SC, swale clay.

even though soil mineral N values were no longer elevated (Table 14.1, for sites that were fertilized in 1981 and 1982).

Higher levels of N addition create a large response in N_2O flux. Application of 22 kg $N \cdot ha^{-1} \cdot y^{-1}$ from 1976 through 1989 increased N_2O emissions by an average of 73% during 1991 to 1994 (Figs. 14.1 and 14.8A). It is interesting to note that even though soil mineral N content was much higher in 1991 than in later years (an average of 10 µg N·g^{-1} of both NO_3^- and NH_4^+ in 1991 compared with an average of 1 µg N·g^{-1} of each in 1994), N_2O fluxes were not lower in 1994. This observation suggests that it is the N turnover rate (coupled with precipitation patterns) that is controlling N_2O emission, rather than bulk soil mineral N content. Bulk soil NH_4^+ concentration may not regulate nitrification rates (Davidson and Hackler, 1994), because the process may be limited by NH_4^+ supply on the microsite scale (Davidson et al., 1990). In the years immediately after fertilization, when bulk soil NH_4^+ and NO_3^- concentrations were high, NH_4^+ supply may have exceeded nitrifier demand.

Methane

Steudler et al. (1989) showed that mineral N additions to forest soils in the northeastern United States significantly decreased soil consumption of atmospheric CH_4. Our work in the shortgrass steppe has shown that CH_4 consumption decreases by 80% to 90% with the addition of 25 µg NH_4^+ N·g^{-1} soil and about 30% with the same amount of NO_3^- N (Bronson and Mosier, 1994). When we returned, after a number of years, to sites that had been fertilized in 1981, we were surprised to find that CH_4 consumption rates were still much lower than at unfertilized locations (Mosier et al., 1991).

Nitrogen Oxides (NO_x)

We measured NO_x flux periodically from June 1994 through September 1995 to assess the effect of N fertilization last applied about 5 years earlier (Fig. 14.6A). During this 16-month period, the mean annual fluxes of NO (calculated by averaging across seasons for each observation location) were 24 and 11.5 µg NO N·m^{-2}·hr^{-1} for fertilized and unfertilized sites, respectively. On an annual basis, about 0.21 and 0.10 g N·m^{-2}·y^{-1} were emitted from these sites. Generally, fluxes at both sites followed the same temporal patterns, with emissions from the previously fertilized site exceeding those from the native site.

Ammonia

After the application of a urea solution, we measured NH_3 volatilization continuously for 4 days, then rewet the soils and measured for another 4 days. Total NH_3 volatilized during these periods was 1.5% and 14.1% of the N applied, respectively (Milchunas et al., 1988).

Influence of Plant Species on Trace Gas Exchange

We conducted a 2-year study to determine the impact of C_3 and C_4 plant composition on trace gas exchange (Epstein et al., 1998). We hypothesized that in the spring, during the time that C_3 plants are most actively assimilating inorganic N, N gas fluxes would be lower within C_3 communities than in C_4 communities.

Conversely, we hypothesized that C_4 communities would assimilate more N during the warmest part of the growing season, thus N gas emissions would be lower from C_4-dominated communities than from C_3-dominated ones during that time. We collected gas samples weekly from two sites differing in soil texture (one site with sandy clay loam and the other with clay), throughout two growing seasons. Plant functional type had no apparent influence on trace gas exchange in the clay soil. However, several differences were observed among plant communities on the sandy clay loam. We observed that CH_4 uptake was greater ($P < .05$) on C_4 plots than on C_3 plots. Nitric oxide fluxes were significantly greater from C_4 plots than from C_3 plots during the drier of the 2 years of observation. The study indicated that under certain environmental conditions, particularly when factors such as moisture and temperature are not limiting, plant community composition can play an important role in regulating trace gas exchange.

Impact of Land-Use Change in the Shortgrass Steppe

Grasslands occupy large areas of the earth and likely influence the global biosphere. The soil–atmosphere exchange of trace gases within grasslands, and perturbations of this exchange by changing management and use of grasslands, has probably affected local, regional, and global atmospheres during the past century. Cultivation of grassland soils, for example, results in depletion of soil organic matter and reduces site fertility over time (Aguilar et al., 1988; Burke et al., chapter 13, this volume; Elliot and Cole, 1989; Ihori et al., 1995; Russel, 1929; Tieszen et al., 1982). Tillage increases both erosion and decomposition, leading to decreased soil organic matter and therefore a decrease in the ability of soil to retain mineral nutrients. These disturbances have long-term effects. Formerly cropped areas within the shortgrass steppe that were abandoned and have since reverted to grasslands appear to require several decades to return to typical native vegetation types and distribution (Coffin et al., 1996), and much longer to recover their soil organic matter (Burke et al., 1995).

The most common agricultural conversion of Colorado grasslands has been to a dryland wheat-fallow cropping system (Hart, chapter 4, this volume). In this system, land is planted to wheat in alternate years and in the intervening years the land is not cropped but is kept weed free either by tillage or, in recent years, by herbicides. To study the effects of land use, we monitored gas fluxes from September 1992 through September 1995 in a wheat-fallow system (both phases), an undisturbed native pasture, a plowed pasture, and in an adjacent CRP site (Figs. 14.9 and 14.10). Methane uptake was always highest in the native pasture, relative to any of the other systems (Figs. 14.9A and 14.10a). Generally, the wheat-fallow sites exhibited CH_4 consumption rates similar to those at the CRP site. However, CH_4 uptake rates were generally lower in the recovering grassland (CRP) than in undisturbed grasslands (Figs. 14.9A and 14.10A). This may have been because soil moisture content was frequently higher in the CRP site during winter because of the inclusion of *Agropyron cristatum* in the plant community. Being taller than the native *Bouteloua gracilis*, this grass tends to collect more winter snow. During the 4 years of the study, CH_4 uptake averaged 18.9 μg CH_4

358 Ecology of the Shortgrass Steppe

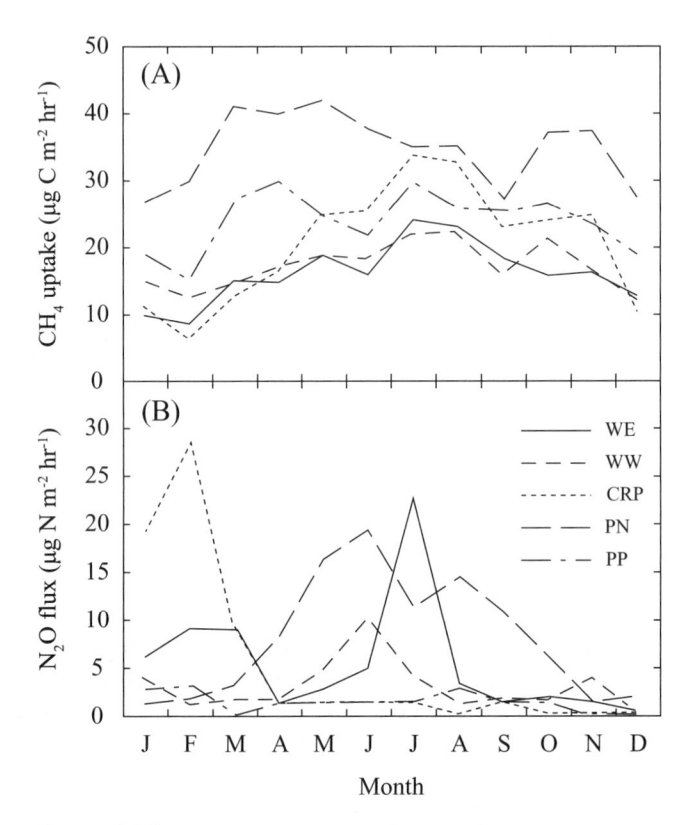

Figure 14.9 (A) Mean monthly CH_4 uptake rate. (B) Mean monthly N_2O emission rate for five sites during the 3-year period September 1992 to September 1995. CRP, Conservation Reserve Program; PN, pasture native; PP, pasture plowed; WE, wheat east; WW, wheat west.

$C \cdot m^{-2} \cdot hr^{-1}$ at the CRP site, compared with 35.2 µg CH_4 $C \cdot m^{-2} \cdot hr^{-1}$ at the native grassland site (Fig. 14.10A). Thus, the 5 to 8 years since the last cultivation of the CRP fields was apparently insufficient to return the microbial and physical processes responsible for CH_4 oxidation to the native rate. Higher uptake rates in the CRP fields during the summer compared with those in wheat fields could be attributed to lower soil water content and lower N turnover rates, because the wheat fields were biannually fertilized with sewage sludge.

The patterns in N_2O emissions were generally opposite those of CH_4 uptake, as we have observed before in grassland sites (Mosier et al., 1996). Thus, they were lowest in native fields and highest in plowed fields (Fig. 14.9B). All sites showed wintertime flux maxima in 1993, but in 1994 only the CRP site exhibited peak February flux. The following winter (1994–1995) was very warm and dry, thus eliminating the frozen soil N_2O emission phenomenon frequently observed (Mosier et al., 1996). Nitrous oxide emissions were significantly lower in the

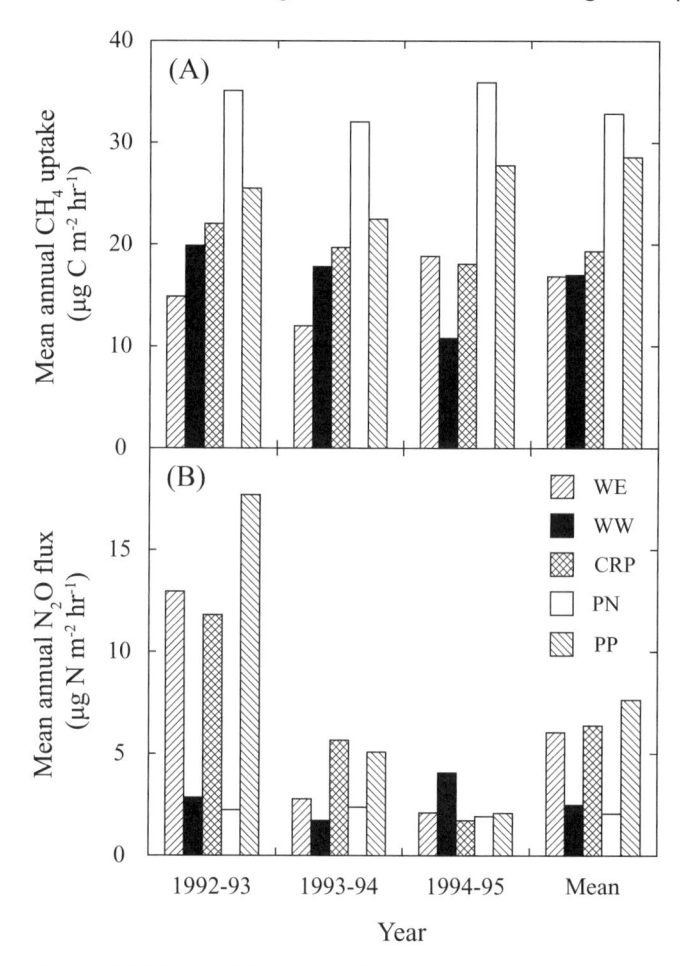

Figure 14.10 (A) Mean annual CH_4 uptake rate. (B) Mean annual N_2O emission rate for five sites during 1992 through 1995. CRP, Conservation Reserve Program; PN, pasture native; PP, pasture plowed; WE, wheat east; WW, wheat west. (Bars for the wheat-fallow fields are means of one complete cycle.)

native sites than in wheat or CRP sites (Fig. 14.10B). Reversion of the CRP to grassland did not lead to an observable decrease in N_2O emissions after 8 years. Surprisingly, in the wheat sites, increased N_2O emissions could not be linked with the timing of sewage sludge applications. Because the fields were merely harrowed after application of the sludge, much of the organic material remained on the soil surface, possibly allowing NH_3 volatilization to continue as N was mineralized. Site management influences emissions as well, with a large N_2O emission event occurring (Fig. 14.9B, site WE) coincident with combined surface sewage sludge application and rain.

Effect of Time on the Return of Trace Gas Fluxes to Native Levels after Plowing

To observe the changes in trace gas flux after plowing, we plowed a small area of native pasture and maintained the area free from plants for the following 2 years (Table 14.1, Fig. 14.10, site PP). Methane uptake rates declined to about 65% of the native site immediately after plowing and remained consistently lower throughout the following 3 years (Fig. 14.10A). The plowed site typically remained wetter through the year because the plot contained little vegetation. The combination of wetter soils and increased N cycling after plowing likely explain the reduced CH_4 consumption (Mosier et al., 1996). Methane uptake rates were typically lowest during the winter and highest during the spring at both native and plowed sites (Fig. 14.9A). The uptake rates in the newly plowed site were generally higher than those in the wheat sites (Figs. 14.9A and 14.10A). Apparently, continued tillage and agronomic use tended to maintain CH_4 oxidation at reduced rates.

Nitrous oxide emissions increased immediately after plowing and continued to be higher for the next 3 years (Fig. 14.10B), although they did begin to decline markedly the third year after plowing (Fig. 14.10B). Unlike the native, wheat, and CRP fields, no large winter N_2O emission events were observed at the newly plowed site. Because the major N and C sources that stimulate nitrification–denitrification are derived from decomposing plant material, after plowing the N_2O release patterns tended to follow those of plant residue decomposition. The large increase in N_2O after the single tillage event in this study was relatively short-lived. In 1994 and 1995, N_2O fluxes were similar to those at the CRP site, with both sites having higher emissions in those years than the native grassland (Fig. 14.10B).

To observe the long-term effect of reverting plowed fields to grasslands, we studied fields that were last cultivated 0 to 53 years before sampling and were allowed to revert back gradually to native vegetation (Fig. 14.11). During the 15-month measurement period (April 1992–July 1993), CH_4 uptake (based on monthly means) averaged 36.0 and 36.3 $\mu g \cdot cm^{-2} \cdot hr^{-1}$, and N_2O emissions averaged 1.9 and 1.3 μg $N \cdot m^{-2} \cdot hr^{-1}$ at native and old replanted sites, respectively. Neither CH_4 nor N_2O fluxes differed significantly between native and replanted sites ($P < .05$), and gas fluxes at both sites followed similar weekly and seasonal trends (Fig. 14.11). These results are an important indicator that 50 years is sufficient time for recovery of trace gas fluxes after cessation of cultivation. Burke et al. (1995; chapter 13, this volume) found that labile pools of organic matter had recovered on 50-year-old replanted fields, but that the total soil organic matter was still as much as 30% lower than that of never-cultivated sites. Trace gas fluxes follow the patterns of the labile soil pools.

Effect of Elevated Carbon Dioxide on Trace Gas Exchange

During the past few decades, the atmospheric concentration of CO_2 has increased at historically unprecedented rates (IPCC, 2007). Increasing CO_2 concentration has a direct effect on plant production and plant communities, but indirectly feeds back into a number of soil biotic systems that influence long-term ecosystem

Exchange of Trace Gases in the Shortgrass Steppe   361

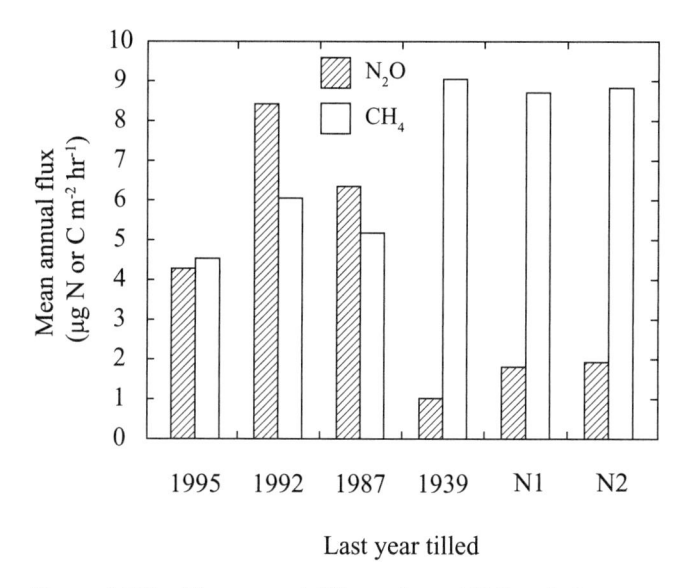

Figure 14.11 Mean annual CH_4 uptake and N_2O emission rates, calculated from 3 years of weekly flux measurements, for sites that were last tilled and allowed to revert gradually to native vegetation 0, 2, 7, and 53 years before flux measurements were made, and from two native grassland sites (N1 and N2).

viability (Hungate et al., 1997a,b,c; Owensby et al., 1993). These interactive feedbacks on the soil C and N cycles, and their influence on trace gas fluxes, have potentially important impacts on the global atmospheric budgets of these gases and on the long-term sustainability of the grassland. Our studies of trace gas fluxes in the shortgrass steppe demonstrate that such grasslands play an important role as consumers of atmospheric CH_4, and producers of N_2O (Mosier et al., 1991, 1996, 1997). The impact of elevated CO_2 on the production and consumption of other trace gases (NO_x, N_2O, and CH_4) is not well understood and has not yet been assessed in semiarid grasslands. The few short-term measurements of NO_x, N_2O, CH_4, and CO_2 fluxes in CO_2 enrichment studies in other ecosystems give contradictory results, and long-term measurements had not been made within any ecosystem before the shortgrass steppe studies reported in Mosier et al. (2002a).

To quantify the effects of doubling CO_2 on shortgrass steppe ecosystem dynamics and trace gas exchange, we monitored the soil–atmosphere exchange of CO_2, NO_x, N_2O, and CH_4 weekly, year-round, April 1997 through November 2001, on treatments that included unchambered control, chambered with ambient CO_2 (≈365 µmol·mol^{-1}), and chambered with enriched CO_2 (≈720 µmol·mol^{-1} CO_2) plots in the Colorado shortgrass steppe (Lauenroth et al., chapter 12, this volume).

Even though both C_3 and C_4 plant biomass increased under elevated CO_2, and soil moisture content was typically higher than under ambient CO_2 conditions,

362 Ecology of the Shortgrass Steppe

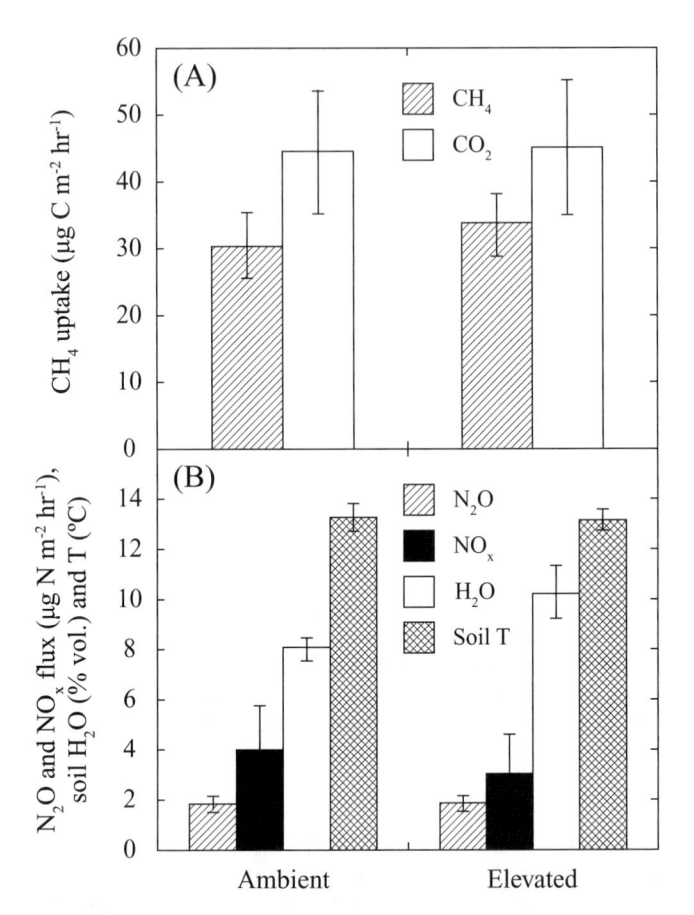

Figure 14.12 Effect of doubling atmospheric CO_2 on trace gas exchange, soil water, and soil temperature. Data presented are averages of flux measurements, and soil water and temperature measurements made at each sampling day weekly between April 1997 and November 2001. Error bars represent the SD (n=3). (A) Methane uptake and system respiration (CO_2 flux). (B) Nitrous oxide and NO_x flux, soil water, and temperature (T).

none of the trace gas fluxes were significantly altered by CO_2 enrichment. During the 55 months of observation, NO_x flux averaged 4.3 µg N·m^{-2}·hr^{-1} in ambient CO_2 And 3.1 µg N·m^{-2}·hr^{-1} in enriched CO_2 (Fig. 14.12A), and it was negatively correlated to plant biomass production. Under ambient and elevated CO_2, N_2O emission rates averaged 1.8 and 1.7 µg N·m^{-2}·hr^{-1}, CH_4 flux rates averaged –31 and –34 µg C·m^{-2}·hr^{-1}, and ecosystem respiration averaged 43 and 44 mg C·m^{-2}·hr^{-1}, respectively (Fig. 14.12) (Mosier et al. 2002a,b).

Methane oxidation tended to be higher in elevated CO_2 than in control plots, whereas NO_x and N_2O tended to be lower, but not significantly so in either case.

Aboveground biomass production was higher under elevated CO_2 (Morgan et al., 2001), and utilized more soil N (King et al., 2004). However, soil N mineralization was probably somewhat enhanced under elevated CO_2 because of moister soils (Hungate et al., 1997a,b,c). The two opposing processes apparently offset each other, because NO_x and N_2O emissions, which reflect system N mineralization and nitrification, did not differ. Ecosystem respiration, which included soil and aboveground plant respiration, was not generally higher under elevated CO_2.

By analyzing the concentration of soil CO_2 at different depths in the open-top chambers and calculating soil respiration, Pendall et al. (2003) found that elevated CO_2 increased soil respiration by about 25% in a moist growing season and by about 85% in a dry season.

Significant increases in soil respiration rates occurred only during dry periods. Some $\delta^{13}C$ analyses of soil CO_2 revealed that soil organic matter decomposition rates were more than doubled under elevated CO_2, whereas rhizosphere respiration rates were not changed. Estimates of net ecosystem production, which account for both inputs and losses of C, suggest that soil C sequestration is not increased under elevated CO_2 during dry years, but may be in wet years (Morgan et al., 2004; Pendall et al., 2004).

Residual Effects of Carbon Dioxide

We were interested to determine whether residual effects on microbial processes persisted after CO_2 enrichment. The open-top chambers were removed at the end of the long-term study, and we conducted a short-term study to determine whether soil microbial processes were altered in soils that had been exposed to double-ambient CO_2 concentrations during the previous five growing seasons. We measured the response of CO_2, NO_x, and N_2O emissions and atmospheric CH_4 uptake to water addition and water-plus-N fertilization to soils that had and had not (ambient) been exposed to elevated CO_2. These responses are detailed in Mosier et al. (2003).

Nitric oxide emissions from the ambient CO_2 soils were higher than the elevated CO_2 treatment when the soils were irrigated, and N_2O emissions from elevated CO_2 soils were significantly lower than the ambient CO_2 soils ($P<.05$) when irrigated (Fig. 14.13A). This reflects the N-depleted state of the soils under elevated CO_2 that occurs as a result of enhanced plant production (King et al., 2004). During the measurement period, NO-to-N_2O ratios ranged between 2 and 5, with the highest in the elevated CO_2 soils.

When NH_4NO_3 was added, N_2O flux tripled in elevated CO_2 soils ($P<.05$), but increased only slightly in ambient CO_2 soils ($P>.05$; Fig. 14.13B). Nitric oxide emissions and N_2O emissions in N-fertilized soils increased markedly after each irrigation and precipitation event. Nitric oxide fluxes increased almost 10-fold with N addition in ambient CO_2 soils, but only about fivefold in elevated CO_2 soils. Nitric oxide emissions were significantly lower from elevated CO_2 soils than from ambient CO_2 soils, again indicating the N-depleted state of the elevated CO_2 soils. Hungate et al. (1997a, c) found that, during wet-up, NO emissions were depressed by 55% in high nutrient conditions under elevated CO_2 (ambient + 360 µmol·mol^{-1}),

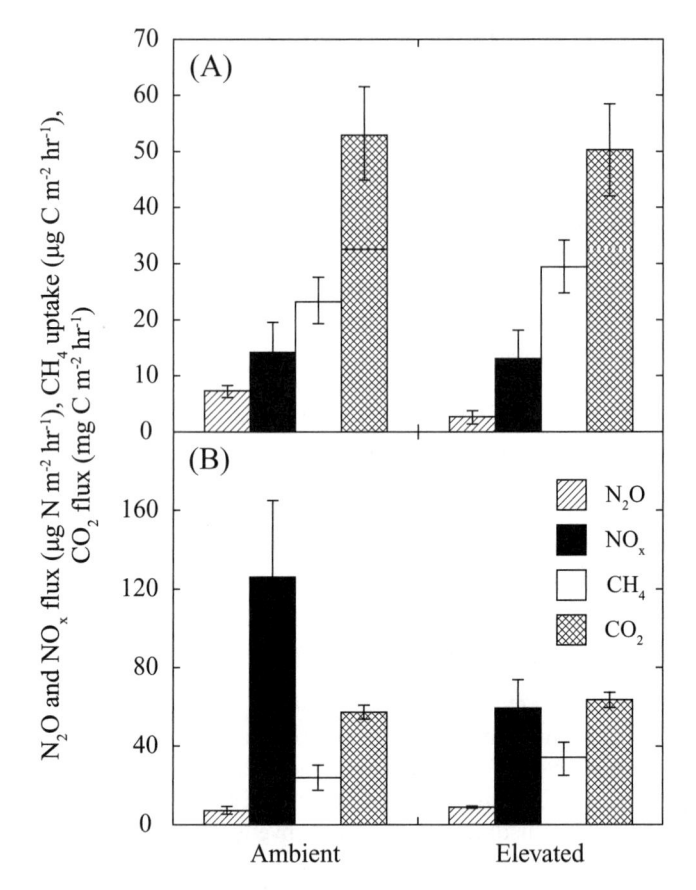

Figure 14.13 The effect of 5 years of CO_2 enrichment on the flux of N_2O, NO_x, and CO_2 and CH_4 uptake after irrigation (A) and irrigation plus ammonium nitrate addition (10 g N·m^{-2}) (B). Values are averages of 11 flux measurements made over a month-long observation period; error bars represent SD (n = 3).

whereas there was no difference among treatments in N_2O emissions. They attributed the decreased NO emissions under elevated CO_2 to increased N immobilization. Increased utilization of added N by soil microbes, thus a decrease in NO emissions, appears to be the case in this study as well.

Plant growth during the time of the study was virtually nonexistent because of the very low amount of precipitation that had fallen the preceding year. Increases in ecosystem CO_2 flux (dark chamber respiration, which includes plant, root, and soil microbial respiration) after water addition were similar in all soils. Only with the water-plus-N addition did CO_2 fluxes from elevated CO_2 soils exceed those from control and ambient soils ($P < .05$). Microbial respiration appears to be enhanced under elevated CO_2 (Pendall et al., 2003), especially when microbes are not limited by water or N availability. Nitrogen addition appeared to stimulate soil microbial respiration while decreasing NO emissions because of increased microbial immobilization of added N.

After irrigation and irrigation plus fertilization, the rate of uptake of atmospheric CH_4 was significantly greater ($P < .05$) in elevated CO_2 soils than ambient CO_2 soils (Fig. 14.13). Methane uptake rates were not measurably enhanced with N addition in ambient CO_2 soils, but tended to be greater in elevated CO_2 soils ($P > .05$). During the 5 years of CO_2 enrichment, CH_4 uptake rates tended to be higher under elevated CO_2. This short-term study suggests that a microbial population developed under elevated CO_2 capable of increasing utilization of atmospheric CH_4. This is in contrast to the study by Ineson et al. (1998), who observed lower CH_4 uptake rates under elevated CO_2 within a free-atmosphere CO_2 enrichment study in Switzerland. They also observed lower CO_2 respiration rates and increased N_2O emissions under elevated CO_2. McLain et al. (2002) also observed lower CH_4 consumption rates under elevated CO_2 in a pine plantation. The decrease in CH_4 consumption was attributed, in part, to wetter soils under elevated CO_2. Soil conditions in the pine forest were likely much more comparable with the grassland soils in Switzerland (Ineson et al., 1998) than with the much drier conditions in the Colorado shortgrass steppe. The wetter soil conditions under elevated CO_2 in the semiarid grassland likely produced more favorable conditions for methanotrophic activity, rather than limiting CH_4 diffusion into the soil as in the Swiss grassland (Ineson et al., 1998) and in the pine forest (McLain et al., 2002). Hu et al. (2001) suggested that over the long term, soil microbial decomposition is slowed under elevated CO_2 because of N limitation. Conversely, Hungate et al. (1997a) found that higher soil water content under elevated CO_2 in an annual grassland stimulated soil N mineralization and resulted in greater plant N uptake. The N addition study confirms the observations of Pendall et al. (2003) that soil respiration is enhanced under elevated CO_2 and N immobilization is increased, thereby decreasing NO emissions, despite the fact that we observed no general CO_2-induced effect on NO_x and N_2O flux during the 5-year observation period (Mosier et al., 2002a,b).

Annual Nitrogen Trace Gas Losses for a Shortgrass Steppe

Although cattle grazing contributes both to N redistribution within a pasture and potential losses through NH_3 volatilization, the N loss rates are relatively small. Schimel et al. (1986) measured NH_3 emissions from sets of microplots on midslope and swale positions of a sand catena in 1981 and 1982. They found that 27% of summer-applied N was lost through NH_3 volatilization from the midslope soil during the month after application, whereas less than 1% was lost from swale soils during the same time period. They extrapolated NH_3 losses observed in the direct measurements for the two catena positions to the pasture scale. Using seasonal rates of urine and feces decomposition, landscape position stratification data from Stillwell (1983), and measured NH_3 rates, Schimel et al. (1986) estimated that 0.01 g $N \cdot m^{-2} \cdot y^{-1}$ was lost from the shortgrass steppe through NH_3 volatilization. They assumed that pastures contained 70% upland (coarse-textured soil) and

30% lowland (fine-textured soil). They also suggested that NH_3 loss from senescing plant material is an important loss vector, emitting about 0.2 g $N \cdot m^{-2} \cdot y^{-1}$. Zachariassen and Schimel (1991) estimated from seasonal direct flux measurements that about 0.07 g NH_3 $N \cdot m^{-2}$ was lost from a shortgrass pasture during this time The estimates of Schimel et al. (1986) were based on observed differences in total N content (1.5%) of peak standing biomass during the summer, N content of standing biomass (0.9% to 1.1%) in the late autumn, and estimated N translocation to roots in the autumn.

Mosier and Parton (1985) and Parton et al. (1988) show that significant N losses through N_2 emissions from denitrification are generally unlikely because the conditions for denitrification to occur in the system are rare and brief. More recent studies (Mosier et al., 1996, 1997) do suggest, however, that winter denitrification may result in significant N losses when conditions are conducive (frozen soil, snow accumulation, and brief warm temperatures that puddle the surface soil where NO_3^- accumulated after the growing season). We can see from data in Figure 14.1A and Figure 14.9B that N_2O emissions were, on average, highest in January and February. The relative ratios of N_2 and N_2O and NO produced during these events have not yet been quantified. We estimate that about 0.04 g $N \cdot m^{-2} \cdot y^{-1}$ is emitted as N_2 annually.

Periodic measurements of NO_x emissions from one site March through July 1988, were conducted by Stocker et al. (1993) using micrometeorological eddy correlation, and by Williams and Fehsenfeld (1991) using chambers. Using the information presented in Stocker et al. (1993), we estimate an annual NO emission of 0.11 g $N \cdot m^{-2} \cdot y^{-1}$. This estimate was made from their flux estimates of midsummer (7.2 ng $N \cdot m^{-2} \cdot s^{-1}$ for 30 days), summer (5.7 ng $N \cdot m^{-2} \cdot s^{-1}$ for 90 days), and long term (2.2 ng $N \cdot m^{-2} \cdot s^{-1}$ for 265 days), and summing the three periods to obtain an annual flux estimate.

To improve this NO emission estimate and to observe how fluxes vary temporally as well as across the landscape, we began studies in 1994 to measure emissions throughout the year (e.g., Fig. 14.6) in a number of different sites (Martin et al., 1998). These new studies generally agree with the previous NO emission studies, in that the N lost by this mechanism represents an important part of the annual N budget of the system. The year-round measurements do suggest that fluxes are probably higher than the earlier studies indicate, with an estimated annual emission of 0.2 g NO $N \cdot m^{-2} \cdot y^{-1}$. This estimate comes from observations at two coarse-textured sites and two finer textured sites where fluxes averaged 7.6 $ng \cdot m^{-2} \cdot s^{-1}$ and 5.2 $ng \cdot m^{-2} \cdot s^{-1}$, respectively. Using the array of Schimel et al. (1986) of 70% coarse-textured upland soil and 30% finer textured lowland soil across the shortgrass steppe, we estimate an integrated annual emission of 0.22 g NO $N \cdot m^{-2} \cdot y^{-1}$. We do not, however, have an adequate assessment of interannual variation needed to compare fluxes from year to year.

Summary

Our long-term, year-round measurements of CH_4 and N_2O fluxes show that gas fluxes vary across the shortgrass steppe landscape, and that these grasslands serve as both an important sink for atmospheric CH_4 and a source of N_2O. Our studies also provide information for an increased understanding of the seasonal dynamics and interannual variation of CH_4 uptake and N_2O flux, and an appreciation of the importance of emissions of N gases (NH_3, NO_x, N_2O) as losses from the ecosystem. These data have been used in the development and refinement of simulation models that can be used to describe fluxes of CH_4, NO_x, and N_2O across the shortgrass steppe in response to variations of climate, landscape, and land-use and management change (Del Grosso et al., 2000a,b; Parton et al., 2001).

Important findings from the field studies include the following:

1. Emissions of NO_x and N_2O are important N-loss processes, and volatile N emissions likely regulate many aspects of the shortgrass steppe soil–plant system. We are lacking, however, a complete understanding of the impacts of annual NO_x deposition, NH_3 emissions and uptake from plants, and N losses through winter denitrification on the N budgets of the system.
2. Climate and weather are important controls over trace gas fluxes. Wintertime fluxes of both CH_4 and N_2O constitute a large portion of the annual flux of these gases. Because of climatic variability, there are interannual variations of the annual mean fluxes of both N_2O and CH_4.
3. Land use is an important control over trace gas flux. Pastures having similar land-use history and soil texture show similar seasonal and annual N_2O emission and CH_4 uptake rates and patterns, but there are strong differences among land-use types. Most important, conversion of grasslands to croplands promotes changes in CH_4 and N_2O flux. Immediately after tillage, CH_4 consumption decreased by about 35% and remained at the decreased rate for at least 3 years. Nitrous oxide emissions were about eight times higher from tilled soil for about 18 months after tillage. Two to 3 years after tillage, N_2O emissions averaged only 25% to 50% higher in the tilled soils than the native grassland. Long-term cropping (winter wheat) resulted in lower CH_4 uptake. Recovery of cultivated soils back to perennial grasslands eventually leads to soils having CH_4 oxidation and N_2O emissions similar to those of native soils of the same texture and parent material. Complete recovery after tillage requires longer than 8 years but less than 50 years.
4. Under elevated CO_2, none of the trace gas fluxes appear to be significantly altered over a 5-year period. However, addition of water and NH_4^+ NO_3^- to soils that had been exposed to double-ambient CO_2 concentrations for five growing seasons increased ecosystem respiration and atmospheric CH_4 oxidation, and decreased NO emissions. These observations suggest that methanotrophic populations were enhanced under elevated CO_2 whereas soil N supply was depleted by increased plant growth (King et al., 2004; Morgan et al., 2001).
5. Soil respiration was higher in previously elevated CO_2 soils after irrigation and N addition, suggesting that microbes were becoming N limited. Although midway through the 5-year experiment decomposition rates were twice as high under elevated as under ambient CO_2 (Pendall et al.,

2003), the N fertilization response observed here suggests that eventually microbial decomposition rates will slow, as predicted by Hu et al. (2001), leading to increased C sequestration potential.

Our data confirm that fluxes of N_2O and CH_4 within grasslands represent an important part of the global atmospheric budget. Temperate grasslands comprise about 8% ($\approx 11.5 \times 10^{12}$ m^2) of global land surface area (Bouwman, 1990). Assuming a spatial distribution in all temperate grasslands of 70% upland and 30% lowland (Schimel et al., 1986), our estimated average N_2O emission and CH_4 uptake rates are 1.6 µg $N \cdot m^{-2} \cdot hr^{-1}$ and 31.2 µg $C \cdot m^{-2} \cdot hr^{-1}$, or 14.2 mg $N \cdot m^{-2} \cdot y^{-1}$ and 314 mg $C \cdot m^{-2} \cdot y^{-1}$, respectively. Assuming a global temperate grassland area of 11.5×10^{12} m^2, and that our grassland sites are typical of the rest of the world, N_2O emissions averaged 0.16 $Tg \cdot y^{-1}$. This amount represents about 1.1% of annual global production (IPCC, 1992). Tropical grasslands are expected to emit N_2O at several times higher annual rates (Keller et al., 1993; Matson and Vitousek, 1991; Mosier et al., 1993). Consumption of atmospheric CH_4 by grassland soils averaged 3.2 Tg CH_4 $C \cdot y^{-1}$ (4.2 Tg CH_4), about 1% of CH_4 production (IPCC, 1992). As Ojima et al. (1993) point out, without the aerobic soil sink for atmospheric CH_4, the rate of CH_4 concentration increase in the atmosphere would be about two times higher than the rate observed in the 1980s.

Acknowledgments We appreciate the invaluable assistance of Anita Kear for conducting the majority of the gas flux measurements. We also thank Brian Newkirk for his help with graphics and statistical analyses, and Becky Riggle, Chris Hegdahl, David Jensen, and Larry Tisue for their capable technical assistance. Financial support for these studies came principally from USDA–ARS and SGS LTER NSF grant DEB 0217631.

References

Aguilar, R., E. F. Kelly, and R. D. Heil. 1988. Effects of cultivation on soils in northern Great Plains rangelands. *Soil Science Society America Journal* **52**:1076–1081.

Andreae, M. O., and D. S. Schimel (eds.). 1989. *Exchange of trace gases between terrestrial ecosystems and the atmosphere.* John Wiley, New York.

Bouwman, A. F. 1990. Exchange of greenhouse gases between terrestrial ecosystems and the atmosphere, pp. 61–127. In: A. F. Bouwman (ed.), *Soils and the greenhouse effect.* John Wiley, New York.

Bronson, K. F., and A. R. Mosier. 1994. Suppression of methane oxidation in aerobic soil by nitrogen fertilizers, nitrification inhibitors, and urease inhibitors. *Biology and Fertility of Soils* **17**:263–268.

Brumme, R., W. Borken, and S. Finke. 1999. Hierarchical control on nitrous oxide emission in forest ecosystem. *Global Biogeochemical Cycles* **13**:1137–1148.

Burke, I. C., E. T. Elliott, and C. V. Cole. 1995. Influence of macroclimate, landscape position, and management on soil organic matter in agroecosystems. *Ecological Applications* **5**:124–131.

Coffin, D. P., W. K. Lauenroth, and I. C. Burke. 1996. Recovery of vegetation in a semiarid grassland 53 years after disturbance. *Ecological Applications* **6**:538–555.

Crutzen, P. J. 1983. Atmospheric interactions in homogeneous gas reactions of C, N and S containing compounds, pp. 67–112. In: B. Bolin and R. B. Cook (eds.), *The major biogeochemical cycles and their interactions.* SCOPE vol. 21. John Wiley, New York.

Davidson, E. A., and J. L Hackler. 1994. Soil heterogeneity can mask the effects of ammonium availability on nitrification. *Soil Biology Biochemistry* **26**:1449–1453.

Davidson, E. A., P. A. Matson, P. M. Vitousek, R. Riley, K. D. Dunkin, G. Garcia-Mendez, and J. M. Maass. 1993. Processes regulating soil emissions of NO and N_2O in a seasonally dry tropical forest. *Ecology* **74**:130–139.

Davidson, E. A., J. M. Stark, and M. K. Firestone. 1990. Microbial production and consumption of nitrate in an annual grassland. *Ecology* **71**:1968-1975.

Del Grosso, S. J., W. J. Parton, A. R. Mosier, D. S. Ojima, A. E. Kulmala, and S. Phongpan. 2000a. General model for N_2O and N_2 gas emissions from soils due to denitrification. *Global Biogeochemical Cycles* **14**:1045–1060.

Del Grosso, S .J., W. J. Parton, A. R. Mosier, D. S. Ojima, C. S. Potter, W. Borken, R. Brumme, K. Butterbach-Bahl, P. M. Crill, K. Dobbie, and K. A. Smith. 2000b. General CH_4 oxidation model and comparisons of CH_4 oxidation in natural and managed systems. *Global Biogeochemical Cycles* **14**:999–1019.

Elliott, E. T., and C. V. Cole. 1989. A perspective on agroecosystem science. *Ecology* **70**:1597–1602.

Epstein, H. E., I. C. Burke, A. R. Mosier, and G. L. Hutchinson. 1998. Plant functional type effects on trace gas fluxes in the shortgrass steppe. *Biogeochemistry* **42**:145–168.

Firestone, M. K., and E. A. Davidson. 1989. Microbiological basis of NO and N_2O production and consumption in soil, pp. 7–21. In: M. O. Andreae and D. S. Schimel (eds.), *Exchange of trace gases between terrestrial ecosystems and the atmosphere.* John Wiley, New York.

Hu, S., F. S. Chapin, III, M. K. Firestone, C. B. Field, and N. R. Chiariello. 2001. Nitrogen limitation of microbial decomposition in a grassland under elevated CO_2. *Nature* **409**:188–191.

Hungate, B. A., F. S. Chapin, H. Zhong, E. A. Holland, and C. B. Field. 1997a. Stimulation of grassland nitrogen cycling under carbon dioxide enrichment. *Oecologia* **109**:149–153.

Hungate, B. A., E. A. Holland, R. B. Jackson, F .S. Chapin, H. A. Mooney, and C. B. Fields. 1997b. The fate of carbon in grasslands under carbon dioxide enrichment. *Nature* **388**:576–579.

Hungate, B. A., C. P. Lund, H. L. Pearson, and F. S. Chapin. 1997c. Elevated CO_2 and nutrient addition alter soil N cycling and N trace gas fluxes with early season wet-up in a California annual grassland. *Biogeochemistry* **37**:89–109.

Hutchinson, G. L., and A. R. Mosier. 1981. Improved soil cover method for field measurement of nitrous oxide fluxes. *Soil Science Society of America Journal* **45**:311–316.

Ihori, T., I. C. Burke, W. K. Lauenroth, and D. P. Coffin. 1995. Effects of cultivation and abandonment on soil organic matter in northeastern Colorado. *Soil Science Society of America Journal* **59**:1112–1119.

Ineson, P., P. A. Coward, and U. A. Hartwig. 1998. Soil gas fluxes of N_2O, CH_4 and CO_2 beneath *Lolium perenne* under elevated CO_2: The Swiss free air carbon dioxide enrichment experiment. *Plant and Soil* **198**:89–95.

IPCC. 1992. Climate change 1992. The supplementary report to the IPCC Scientific Assessment, Intergovernmental Panel on Climate Change. Cambridge University Press, Cambridge, UK.

IPCC. 1996. *Climate change 1995. The science of climate change. Intergovernmental panel on climate change.* Cambridge University Press, Cambridge, UK.

IPCC. 2007. Intergovernmental panel on climate change. Climate change: The physical basis. Summary for Policy makers. Contribution of Working Group I to the fourth assessment report of the IPCC. Online. Available at www.ipcc.ch.

Johansson, C., and E. Sanhueza. 1988. Emission of NO from savanna soils during rainy season. *Journal Geophysical Research* **93**:14193–14198.

Keller, M., E. Veldkamp, A. M. Weltz, and W. A. Reiners. 1993. Effect of pasture age on soil trace gas emissions from a deforested area of Costa Rica. *Nature* **365**:244–246.

King, J. Y., A. R. Mosier, J. A. Morgan, D. R. LeCain, D. G. Milchunas, and W. J. Parton. 2004. Plant nitrogen dynamics in shortgrass steppe under elevated atmospheric carbon dioxide. *Ecosystems* **7**:147–160.

Langford, A. O., F. C. Fehsenfeld, J. Zachariassen, and D. S. Schimel. 1992. Gaseous ammonia fluxes and background concentrations in terrestrial ecosystems of the United States. *Global Biogeochemical Cycles* **6**:459–483.

LeMer, J., and P. Roger. 2001. Production, oxidation, emission and consumption of methane by soils: A review. *European Journal of Soil Biology* **37**:25–50.

Li, C., S. Frolking, and T. A. Frolking. 1992. A model of nitrous oxide evolution from soil driven by rainfall events, 1: Model structure and sensitivity. *Journal Geophysical Research* **97D**:9759–9776.

Martin, R. E., M. C. Scholes, A. R. Mosier, D. S. Ojima, E. A. Holland, and W. J. Parton. 1998. Controls on annual emissions of nitric oxide from soils of the Colorado shortgrass steppe. *Global Biogeochemical Cycles* **12**:81–91.

Matson, P. A., and P. M. Vitousek. 1990. Ecosystem approach to a global nitrous oxide budget. *BioScience* **40**:667–672.

McLain, J. E. T., T. B. Kepler, and D. M. Ahmann. 2002. Belowground factors mediating changes in methane consumption in a forest soil under elevated CO2. *Global Biogeochemical Cycles* **16**(23):1–14.

Milchunas, D. G., W. J. Parton, D. S. Bigelow, and D. S. Schimel. 1988. Factors influencing ammonia volatilization from urea in soils of the shortgrass steppe. *Journal Atmospheric Chemistry* **6**:323–340.

Monson, R. K., and E. A. Holland. 2001. Biospheric trace gas fluxes and their control over tropospheric chemistry. *Annual Review of Ecology and Systematics* **32**:547–576.

Morgan, J. A., D. R. LeCain, A. R. Mosier, and D. G. Milchunas. 2001. Elevated CO_2 enhances water relations and productivity and affects gas exchange in C3 and C4 grasses of the Colorado shortgrass steppe. *Global Change Biology* **7**:451–466.

Morgan, J. A., A. R. Mosier, D. G. Milchunas, D. R. LeCain, J. A. Nelson, and W. J. Parton. 2004. CO2 enhances productivity, alters species composition, and reduces digestibility of shortgrass steppe vegetation. *Ecological Applications* **14**:208–219.

Mosier, A. R., G. L. Hutchinson, B. R. Sabey, and J. Baxter. 1982. Nitrous oxide emissions from barley plots treated with ammonium nitrate or sewage sludge. *Journal of Environmental Quality* **11**:78–81.

Mosier, A. R., J. A. Morgan, J. Y. King, D. LeCain, and D. G. Milchunas. 2002a. Soil atmosphere exchange of CH4, CO2, NOx and N2O in the Colorado shortgrass steppe under elevated CO2. *Plant and Soil* **240**:200–211.

Mosier, A. R., J. A. Morgan, J. Y. King, D. LeCain, and D. G. Milchunas. 2002b. *Soil atmosphere exchange of CH4, CO2, NOx, and N2O in the Colorado shortgrass steppe under elevated CO2.* Presented at the 7th Scientific Conference of the International Committee on Global Atmospheric Chemistry. Herakliion, Crete, Greece, September 18–25.

Mosier, A. R., and W. J. Parton. 1985. Denitrification in a shortgrass prairie: A modeling approach, pp. 441–451. In: D. E. Caldwell, J. A. Brierley, and C. L. Brierley (eds.), *Planetary Ecology.* Van Nostrand Reinhold, New York.

Mosier, A. R., W. J. Parton, and S. Phongpan. 1998. Long term large N and immediate small N addition effects on trace gas fluxes in the Colorado shortgrass steppe. *Biology & Fertility of Soils* **28**:44–50.

Mosier, A. R., W. J. Parton, D. W. Valentine, D. S. Ojima, D. S. Schimel, and J. A. Delgado. 1996. CH$_4$ and N$_2$O fluxes in the Colorado shortgrass steppe: I. Impact of landscape and nitrogen addition. *Global Biogeochemical Cycles* **10**:387–399.

Mosier, A. R., W. J. Parton, D. W. Valentine, D. S. Ojima, D .S. Schimel, and O. Heinemeyer. 1997. CH$_4$ and N$_2$O fluxes in the Colorado shortgrass steppe: 2. Long term impact of land use change. *Global Biogeochemical Cycles* **11**:29–42.

Mosier, A. R., E. Pendall, and J. A. Morgan. 2003. Soil–atmosphere exchange of CH4, CO2, NOx, and N2O in the Colorado shortgrass steppe following five years of elevated CO2 and N fertilization. *Atmospheric Chemistry and Physics Discussions* **3**:2691–2706.

Mosier, A. R., and D. S. Schimel. 1991. Influence of agricultural nitrogen on atmospheric methane and nitrous oxide. *Chemistry & Industry* **23**:874–877.

Mosier, A. R., D. S. Schimel, D. W. Valentine, K. F. Bronson, and W. J. Parton. 1991. Methane and nitrous oxide fluxes in native, fertilized, and cultivated grasslands. *Nature* **350**:330–332.

Mosier, A., D. W. Valentine, D. S. Schimel, W. J. Parton, and D. S. Ojima. 1993. Methane consumption in the Colorado short grass steppe. *Mitteilungen der Deutschen Bodenkundlichen Gesellschaft* **69**:219–226.

Ojima, D., A. Mosier, S. Del Grosso, and W. J. Parton. 2000. TRAGNET analysis and synthesis of trace gas fluxes. *Global Biogeochemical Cycles* **14**:995–997.

Owensby, C. E., P. I. Coyne, and L. M. Auen. 1993. Nitrogen and phosphorus dynamics of a tall grass prairie ecosystem exposed to elevated carbon dioxide. *Plant Cell Environment* **16**:843–850.

Parton, W. J., E. A. Holland, S. J. Del Grosso, M. D. Hartman, R. E. Martin, A. R. Mosier, D. S. Ojima, and D. S. Schimel. 2001. Generalized model for NO$_x$ and N$_2$O emissions from soils. *Journal of Geophysical Research* **106**(D15):17403–17420.

Parton, W. J., A. R. Mosier, D. S. Ojima, D. W. Valentine, D. S. Schimel, K. Weier, and A. E. Kulmala. 1996. Generalized model for N$_2$ and N$_2$O production from nitrification and denitrification. *Global Biogeochemical Cycles* **10**:401–412.

Parton, W. J., A. R. Mosier, and D. S. Schimel. 1988. Rates and pathways of nitrous oxide production in a shortgrass steppe. *Biogeochemistry* **6**:45–48.

Pendall, E., S. Del Grosso, J. Y. King, D. R. LeCain, D. G. Milchunas, J. A. Morgan, A. R. Mosier, D. S. Ojima, W. A. Parton, P. P. Tans, and J. W. C. White. 2003. Elevated atmospheric CO2 effects and soil water feedbacks on soil respiration components in a Colorado grassland. *Global Biogeochemical Cycles* **17**:1–13.

Pendall, E., A. R. Mosier, and J. A. Morgan. 2004. Rhizodeposition stimulated by elevated CO2 in a semiarid grassland. *New Phytologist* **162**:447–458.

Reeburgh, W. S., S. C. Whalen, and J. J. Alpern. 1993. The role of methylotrophy in the global methane budget, pp. 1–14. In: J. C. Murrell and D. P. Kelly (eds.), *Microbial growth on C1 compounds. Proceedings of the 7th International Symposium*. Intercept, Andover, UK.

Russel, J. C. 1929. Organic matter problems under dry-farming conditions. *Agronomy Journal* **21**:960–969.

Schimel, D. S., W. J. Parton, F. J. Adamsen, R. G. Woodmansee, R. L. Senft, and M. A. Stillwell. 1986. The role of cattle in the volatile loss of nitrogen from a shortgrass steppe. *Biogeochemistry* **2**:39–52.

Schlesinger, W. H., and A. E. Hartley. 1992. A global budget for atmospheric NH_3. *Biogeochemistry* **15**:191–211.

Smith, K. A., T. Ball, F. Conen, K. E. Dobbie, J. Massheder, and A. Rey. 2003. Exchange of greenhouse gases between soil and atmosphere: Interactions of soil physical factors and biological processes. *European Journal of Soil Science* **54**:779–791.

Steudler, P. A., R. D. Bowden, J. M. Melillo, and J. D. Aber. 1989. Influence of nitrogen fertilization on methane uptake in temperate forest soils. *Nature* **341**:314–331.

Stillwell, M. A. 1983. *The effect of bovine urine on the nitrogen cycle in a shortgrass prairie*. PhD diss., Colorado State University, Fort Collins, Colo.

Stocker, D. W., D. H. Stedman, K. F. Zeller, W. J. Massman, and D. G. Fox. 1993. Fluxes of nitrogen oxides and ozone measured by eddy correlation over a shortgrass prairie. *Journal Geophysical Research* **98**:12619–12630.

Tieszen, H., J. W. B. Stewart, and J. R. Bettany. 1982. Cultivation effects on the amounts and concentration of carbon, nitrogen, and phosphorus in grassland soils. *Agronomy Journal* **74**:831–835.

Tortoso, A. C., and G. L. Hutchinson. 1990. Contributions of autotrophic and heterotrophic nitrifiers to soil NO and N_2O emissions. *Applied Environmental Microbiology* **56**:1799–1805.

Valentine, D. W., A. R. Mosier, R. W. Lober, and J. W. Doran. 1993. Methane uptake in grassland soils: Concentration profiles, flux measurements, and modeling. *Bulletin Ecological Society of America* **74**:466.

Williams, E. J., and F. C. Fehsenfeld. 1991. Measurement of soil nitrogen oxide emissions at three North American ecosystems. *Journal Geophysical Research* **96**:1033–1042.

Williams, E. J., G. L. Hutchinson, and F. C. Fehsenfeld. 1992. NO_x and N_2O emissions from soil. *Global Biogeochemical Cycles* **6**:359–388.

Zachariassen, J., and D. S. Schimel. 1991. Ammonia exchange above grassland canopies. EOS transactions. *American Geophysical Union* **72**:110.

15

The Shortgrass Steppe and Ecosystem Modeling

William J. Parton
Stephen J. Del Grosso
Ingrid C. Burke
Dennis S. Ojima

Ecological modeling has played a key role in scientific investigations of the SGS LTER during the past several decades. The SGS LTER site, focused initially on the Central Plains Experimental Range (CPER), was the main grassland research site for the Grassland Biome component of the U.S. IBP effort (Lauenroth et al., this volume, chapter 1). Initial development of ecosystem models occurred from 1970 to 1975 as part of the IBP. All the U.S. IBP projects (grassland, tundra, desert, deciduous forest, and coniferous forest biomes) included research on the development of ecosystem models, with the goals of using models to help formulate and interpret field experiments, and of projecting the impact of changes in management practices on ecosystem dynamics. Models were developed as part of the Grassland Biome project (Bledsoe et al., 1971; Innis, 1978), and included modeling specialists who worked with research biologists on the development and formulation of the ecosystem models.

The modeling activities of the U.S. IBP Grassland Biome project included developing the ELM Grassland model (Innis, 1978). The ELM model was a complex process-oriented model that was intended to be used at all the Grassland Biome sites in the United States. This model was developed by postdoctoral fellows who were to formulate the different submodels, and then link the submodels using software that was developed as part of the program. The submodels included a plant production submodel, a cattle production submodel, a linked nutrient cycling and soil organic matter submodel, a grasshopper dynamics submodel, and a soil temperature and water submodel. Biophysical and biological data from the different sites were collected to develop and test the model. Model development was constrained by lack of knowledge about the biological processes that control ecosystem behavior, and by lack of appropriate data to test the ability of the model to simulate ecosystem responses to changes in grazing and fertility management practices. However, the ELM Grassland model was quite successful at investigating the interactions of different components of the ecosystem, and at helping to formulate new research efforts. Unfortunately, the model was very difficult to use because of the amount of time required to make computer runs, and the large amount of data required to verify the response of the model to changes in the environment. Ultimately, the knowledge developed as part of building the ELM Grassland model was successfully used to develop the next generation of grassland ecosystem models.

The SGS LTER program and its IBP Grassland Biome predecessor have had a major impact on the development of grassland ecosystem models since the ELM model was formulated in the 1970s. Important models that have been developed during the last 30 years using data from the SGS LTER site include the ROOT, PHOENIX, CENTURY, Foodweb, Grassland Ecosystem Model (GEM), SPUR, STEPPE, and DAYCENT models. Here, we will briefly review these models.

The ROOT model (Parton et al., 1978) used detailed root growth data (root window data) from the SGS LTER site to formulate a mechanistic root growth model that simulated the response of roots to changes in soil temperature, water, and fertility. The PHOENIX model (McGill et al., 1981) simulated soil nutrient cycling and carbon (C) dynamics for grasslands and was developed using many of the concepts included in the ELM nutrient cycling submodel. The CENTURY model (Parton et al., 1987) is a generalized grassland ecosystem model that was developed to simulate plant production, nutrient cycling, and soil organic matter dynamics for grasslands and agroecosystems in the Great Plains. The CENTURY model used experience gained from the development of the PHOENIX, ROOT, and ELM models to develop a simplified ecosystem model that could simulate grassland and crop ecosystem dynamics at the regional scale for the Great Plains. Hunt et al. (1984) developed the Foodweb model to simulate nutrient cycling, C dynamics, and soil animal dynamics using extensive data sets collected at the SGS LTER site during the 1980s. The gap dynamics model, STEPPE (Coffin and Lauenroth, 1990), was developed to represent shortgrass steppe plant community dynamics; this model is described at length in Peters and Lauenroth (chapter 7, this volume). Hunt et al. (1991) developed the GEM, and Hanson et al. (1988) developed the SPUR grassland ecosystem model. Both of these models were developed using data sets from the SGS LTER, in addition to experimental data, and were designed to simulate the impact of climatic change and increases in atmospheric CO_2 on grassland ecosystem dynamics. Most recently, detailed process-oriented nutrient cycling, trace gas flux, and soil organic matter data collected during the 1990s have been used to develop and test the DAYCENT model. DAYCENT has been used to simulate trace gas fluxes and ecosystem dynamics for grasslands and agroecosystems in the Great Plains and the Midwest, and to estimate nitrous oxide (N_2O) emissions from agricultural soils for the U.S. National Greenhouse Gas Inventory (Del Grosso et al., 2001, 2006; Kelly et al., 2000; Parton et al., 1998). The detailed process-oriented plant production, nutrient cycling, soil organic matter dynamics, and trace gas flux data collected at the SGS LTER site have played a critical role in the development of most of the grassland ecosystem models used in the United States during the past 30 years.

Ecosystem models developed in conjunction with the SGS LTER site have been used extensively to evaluate the impacts of changes in grassland and cropping management practices, climate, and atmospheric CO_2 level. The models have been used for impact assessment at the site-specific level as well as at regional and global scales. Ecosystem models (e.g., CENTURY and SPUR) developed at the SGS LTER site were among the first models used to extrapolate climate change impacts on C cycling at regional and global scales (Parton et al., 1994, 1995; Schimel et al., 1991). Model results from Pepper et al. (2005) suggested that increasing atmospheric CO_2 levels and air temperature will increase plant growth in shortgrass steppe, tallgrass prairie, and boreal forest ecosystems. However, decomposition rates also increased and net C uptake was close to neutral unless nitrogen (N) inputs were increased, in which case all three systems became net sinks of C for the duration of the simulations (2000–2100). Likewise, more recent model simulations from Parton et al. (2007) predict that increasing atmospheric CO_2 levels and air temperature will increase plant growth in semiarid grasslands in southeast Wyoming, particularly during wet years. In agreement with previous model results, increasing atmospheric CO_2 levels and air temperature also increased decomposition rates, and hence soil organic matter declined slightly during the period simulated (2000–2015).

In this chapter we will first present a short review of how models developed for the shortgrass steppe have been used for testing our understanding of grassland ecosystem processes, and for impact assessment during the past 30 years. Here, our goal is to illustrate how the models have been used to advance our understanding of grasslands dynamics. Second, we will apply the DAYCENT model to assess the vulnerabilities of the shortgrass steppe to potential climate change during the next 100 years. This analysis will consider the impacts of changing precipitation, temperature, atmospheric CO_2, and N deposition levels.

Site-Specific Impact Assessment

Models developed originally for the shortgrass steppe have been used for numerous environmental impact studies. Grassland ecosystem models have been used to evaluate the environmental impacts of insect outbreaks, range management schemes, and climatic patterns on soil organic matter dynamics, grassland production, nutrient cycling, and secondary production. Some of the earliest work included the use of the ELM model (Innis, 1978) to simulate grassland ecosystem responses to potential climate change associated with both cloud seeding for hail suppression and the deployment of a fleet of supersonic transports (Parton and Smith, 1974a, b). The ELM model was also used to assess the ecosystem impact of overgrazing in the Sahel region of Africa during the drought of the 1970s (Parton and Schnell, 1975). The ELM model and other grassland ecosystem models have been used to evaluate grazing and fire management implications for an Oklahoma tallgrass prairie (Holland et al., 1992; Ojima et al., 1990; Parton and

Risser, 1980; Parton et al., 1980), to develop an impact assessment of management practices used for strip-mine reclamation (Parton et al., 1979), and to evaluate the impact of grasshopper and caterpillar grazing on plant production and nutrient cycling (Capinera et al., 1983a, b). Numerous studies have applied CENTURY to scenarios of cultivation management to evaluate ecosystem responses at a site level, or across multiple sites in the Great Plains (e.g., Campbell et al., 2005).

During the early 1990s, there was substantial interest in evaluating the impact of potential climatic change associated with increasing atmospheric CO_2. The GEM model (Hunt et al., 1991), CENTURY model (Schimel et al., 1990), and SPUR model (Hanson et al., 1988) were used to simulate the impact of increased atmospheric CO_2-induced potential climate changes on plant production, nutrient cycling, and animal production for sites in the Great Plains. Results indicated that plant production was most sensitive to changes in precipitation, whereas soil organic matter dynamics, decomposition, and nutrient cycling were sensitive to changes in both precipitation and temperature (increasing temperature and precipitation increased both nutrient cycling and decomposition of litter and soil organic matter). The GEM model results (Hunt et al., 1991) suggested that increasing atmospheric CO_2 levels would increase plant production and soil organic matter levels, and reduce the negative impact of potential decreases in precipitation.

Regional Impact Assessment

The CENTURY model was one of the first ecosystem models to be applied at a regional scale (Parton et al., 1987) or to be linked to spatially explicit input data organized in a geographic information system to simulate regional ecosystem dynamics (Burke et al., 1990). This structure takes information from spatial and temporal data of soils, climate, and land use to simulate regional responses of ecosystems to changes in management and environmental variables. Burke et al. (1991) and Schimel et al. (1991) used the CENTURY model to simulate ecosystem dynamics at the regional scale with a primary focus on the impact of potential climate changes on ecosystem dynamics within the Great Plains. The results from these studies suggested that increasing temperature may result in substantial loss of soil C, and that plant production was positively correlated to changes in precipitation. The paper by Schimel et al. (1991) also showed that the simulated regional patterns in grassland production were well correlated with the observed regional patterns in the NDVI (remotely sensed index of plant production). Burke et al. (1997) used CENTURY-simulated regional patterns of N mineralization and other ecosystem parameters to evaluate the relative importance of the factors that control plant production at the regional scale.

Burke et al. (1991) have found that the regional C consequences of land-use management may be greater than those of climate change. Parton et al. (2005) recently used a regional application of CENTURY to simulate the large-scale responses of grasslands to cultivation management. Their results showed large-scale losses of soil C with early cultivation practices. Plowing, which enhances decomposition rates, and low crop residue inputs associated with dryland wheat-fallow cropping both contributed to soil C losses. However, Great Plains soils converted from dryland to irrigated cropping since the 1970s have become a C sink—estimated at 21.3 Tg C for the region. Del Grosso et al. (2006) applied DAYCENT to cropped soils to evaluate national-scale consequences of cropping on NO_2 production. Last, CENTURY has been incorporated into economic models to estimate the regional-scale economic potentials for agricultural C sequestration (Antle et al., 2007).

Land–Atmosphere Interactions

Our understanding of the interactions between the atmosphere and the land surface (referring to the soil, vegetation, water system) is critical to estimating the vulnerability of key natural resources to climate and land-use changes (Pielke et al., 1997). Changes in these interactions affect mesoscale physical and chemical climate, water basin hydrology, and ecological properties, such as vegetation composition, disturbance regime, and biogeochemical cycles. Terrestrial biospheric processes respond strongly to atmospheric temperature, humidity, CO_2 levels, N deposition, precipitation, and radiative transfers. Grassland simulation models have been important for understanding the biophysical feedbacks that couple the land surface to the atmosphere through processes controlling energy and water exchanges (Eastman et al., 1998; Vidale et al., 1997; Walko et al., 2000). These feedbacks operate rapidly and are estimated many times each hour. Biogeochemical and ecosystem interactions with atmospheric processes have been implemented using the CENTURY-RAMS model (Lu et al., 2001; Pielke et al., 1997). The coupled CENTURY biogeochemical and RAMS atmospheric model showed that seasonal vegetation growth patterns strongly influence surface water and energy exchanges, and regional climate patterns. These research efforts have been directed toward developing a better understanding of how biophysical coupling to the atmosphere changes over time, as the land surface and ecosystem processes change the constraints on water, C, and energy fluxes, and on hydrological and ecological processes (Ojima et al., 1991; Schimel et al., 1990, 1994).

Long-Term Ecosystem Response to Environmental Change: A Vulnerability Assessment

In this section we use a model developed from long-term data at the SGS LTER to assess the vulnerability of the shortgrass steppe ecosystem to possible future long-term changes in the environment. Our analysis focuses on CO_2 increases, associated climate change, and increased N deposition.

Atmospheric concentration of CO_2 has increased from a preindustrial level of ≈280 ppm to ≈380 ppm at present, and is expected to exceed ≈650 ppm by 2100 unless anthropogenic CO_2 emissions fall below 1990 levels by the end of this century (IPCC, 2001, 2007). Current predictions are that temperatures will increase between 1.8 and 4.0 °C (IPCC, 2007). Increased atmospheric CO_2 concentration has direct and indirect effects on many important ecosystem parameters. In addition to climate change resulting from increased radiative forcing from elevated greenhouse gases, CO_2 levels also influence plant growth. Three mechanisms are responsible for the CO_2 fertilization effect observed in both laboratory and field experiments (Körner, 2006; Woodward, 2002). Photosynthesis rates can increase, stomatal conductance can decrease, and C-to-nutrient ratios may widen (Gifford et al., 2000; Morgan et al., 2004b). Climate change related to increased greenhouse gases in the atmosphere has the potential to alter net primary production (NPP) by changing water and temperature stress. Climate change also can alter soil C levels and N mineralization rates (Hu et al., 2001), and thus influence NPP indirectly.

Open-top chamber experiments performed at the shortgrass steppe show significant effects of elevated CO_2 on NPP and soil water content (Morgan et al., 2001). Net primary production increased by 25% to 50% under elevated CO_2 (720 ppm), and soil water content was higher during the 4 years of the experiment. The primary mechanism was likely increased water use efficiency, with minor increases in photosynthesis rates and in C-to-N ratios for some species. Morgan et al. (2001) conclude that drier systems will likely show higher NPP increases under elevated CO_2 than less water-stressed systems. With elevated CO_2 levels, leaf stomata may close to some extent, but will still maintain some CO_2 diffusion, thus resulting in higher photosynthetic rates than in low CO_2 conditions. However, the long-term effects of increased CO_2 on plant growth rates, N availability, and soil organic matter levels are uncertain. Climate data show a trend of increasing temperature since approximately 1950 for the Great Plains region, particularly nighttime minimum temperatures (Alward et al., 1999; Pielke et al., 2000). Rainfall was variable during the 20th century, and future changes in rainfall patterns are uncertain. Model simulations are required to quantify the sensitivity of ecosystem responses to these uncertain drivers and to simulate long-term changes that cannot be addressed using field experiments.

DAYCENT and CENTURY Model Descriptions

DAYCENT (Del Grosso et al., 2001; Kelly et al., 2000; Parton et al., 1998) is the daily time step version of the CENTURY model (Parton et al., 1994). DAYCENT and CENTURY simulate exchanges of C, nutrients, and trace gases among the atmosphere, soil, and vegetation (Fig. 15.1). Both models simulate decomposition and nutrient mineralization of plant litter and soil organic matter, plant growth and senescence, and soil water and temperature fluxes. CENTURY has a monthly time step, and uses monthly maximum/minimum temperature and precipitation data as input. Site-specific data such as soil texture and hydraulic properties (one value for all soil layers), and vegetation type are also required as input data. DAYCENT requires daily maximum/minimum temperature and precipitation data, and soil texture, bulk density, and hydraulic properties must be specified for each soil layer. Both models simulate three pools of soil organic matter (active, slow, passive) that have different maximum turnover rates and two mineral N

DAYCENT Model

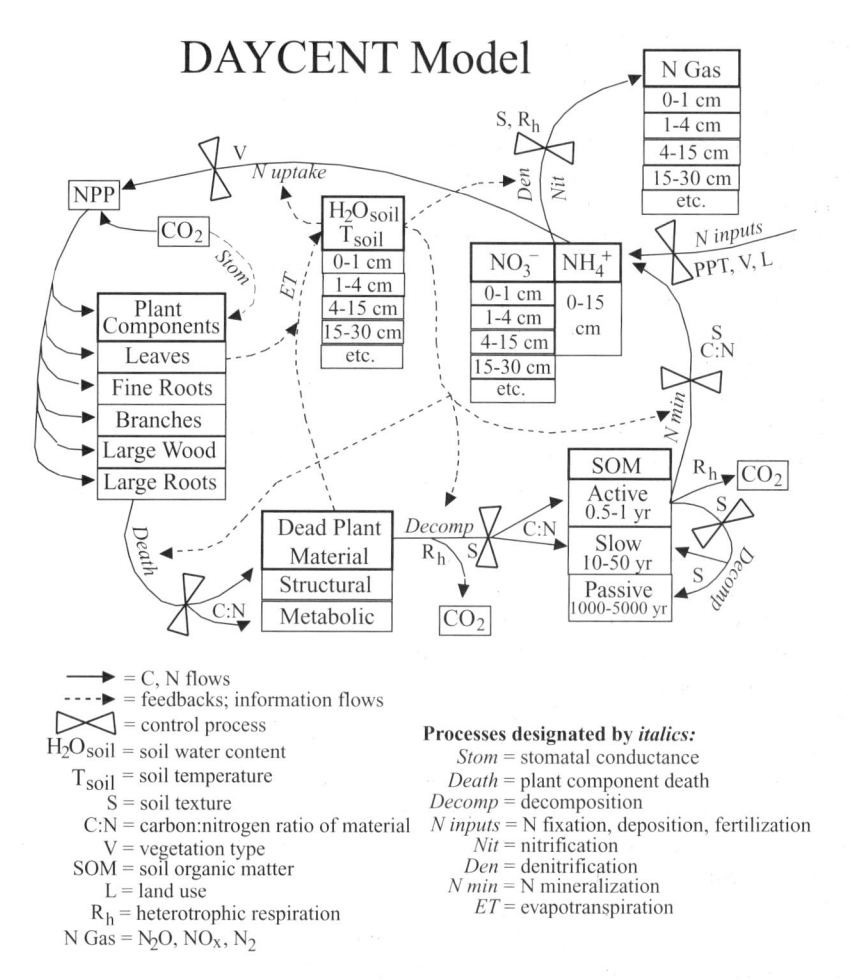

Figure 15.1 Flow diagram for the DAYCENT biogeochemical model. (From Parton et al. [2007].)

pools (ammonium and nitrate). Ammonium (NH_4^+) is assumed to be confined to the top 15 cm of soil whereas nitrate (NO_3^-) is mobile and can leach out the bottom of the profile. In addition to having finer temporal resolution, soil layers are more refined in DAYCENT. Typically, soil layers are 15 or 30 cm thick in CENTURY but are 15 cm or less (for surface layers) in DAYCENT. In both models, users can change the soil layer structure and the depth of the profile. Typical management and disturbance events (plowing, burning, harvesting, fertilizing, irrigating, and so forth) can be readily implemented in both models. In terms of processes, the major difference between the models is that DAYCENT explicitly represents the processes (nitrification and denitrification) that lead to N_2O, NO_x, and N_2 emissions, whereas CENTURY assumes that a constant proportion of available N in each time step is lost as N gas without distinguishing between the different N gas species. Unlike CENTURY, DAYCENT includes a submodel to represent uptake of atmospheric methane (CH_4) by soil microbes. The soil water and temperature submodels are also much more detailed in DAYCENT.

The primary motivation for developing DAYCENT was reliable simulation of N gas emissions from soils. The response of N gas emissions to soil water content and texture is nonlinear, and N gas emissions are highly pulse driven. It is not possible to correctly simulate the short-term (daily timescale) conditions that result in pulses of N gas flux using monthly scale driving variables. The shorter time step allows DAYCENT to include submodels for nitrification and denitrification to simulate N gas emissions.

Although denitrification is generally not an important part of the annual N budget in the shortgrass steppe, there are events that may lead to significant N losses through denitrification, particularly in high clay content soils (Mosier et al., 1996; Parton et al., 1988). Del Grosso et al. (2000) developed the DAYCENT denitrification submodel for nitrous oxide (N_2O) and NO_x gas emissions from soils from observations of N gas losses from incubations of intact and disturbed soil cores. Nitrous oxide emissions from denitrification are a function of soil NO_3^- concentration, water-filled pore space (WFPS), heterotrophic respiration, and texture. The model assumes that denitrification rates are controlled by the availability of soil NO_3^- (electron acceptor), labile C compounds (electron donor), and oxygen (competing electron acceptor). Heterotrophic soil respiration is used as a proxy for labile C availability, whereas oxygen availability is a function of soil physical properties that influence gas diffusivity, soil WFPS, and oxygen demand. Oxygen demand (indicated by respiration rates) varies inversely with a soil gas diffusivity coefficient, which is regulated by soil porosity and pore size distribution. Model inputs include soil heterotrophic respiration rate, texture, NO_3^- concentration, and WFPS. The model selects the minimum of the NO_3^- and CO_2 functions to establish a maximum potential denitrification rate for particular levels of electron acceptor and C substrate, and accounts for limitations of oxygen availability, to estimate daily $N_2 + N_2O$ flux rates. The output of the ratio function is combined with the estimate of total N gas flux rate to infer N_2O emission.

DAYCENT Simulations and Results

We used DAYCENT to investigate the effects of changes in precipitation, temperature, CO_2 concentration, and N deposition on some key ecosystem properties and processes. Soil and climate data from the CPER were used to drive DAYCENT for these simulations. We simulated a sandy loam soil and used long-term (1948–2000) weather data to initialize the C and nutrient pools in the model. To validate the parameters in the model used to simulate the CO_2 effect, we compared simulated and observed NPP data for ambient and elevated (720 ppm) CO_2. DAYCENT showed an approximate 30% increase in NPP for the shortgrass steppe, which agrees with data collected from open-top chamber elevated CO_2 experiments performed at the CPER (Morgan et al., 2001).

To establish a control to compare with altered CO_2, climate, and N deposition, we simulated ambient CO_2 and nonmodified climate from 2001 until 2100 by recycling 10 years (1991–2000) of actual weather data. We then simulated five alternative scenarios (see Table 15.1 for input variables changed) to represent the effects of elevated CO_2, increased temperature, increased precipitation, decreased precipitation, and increased N deposition. We included moderate grazing during the growing season (April–October) in all scenarios. The input variables were linearly ramped up (or down) in 10-year increments. For example, with the water treatment, precipitation was increased 2% above the control during 2001 to 2010, 4% from 2011 to 2020, and 20% from 2091 to 2100. We did not alter the number of precipitation events, but increased or decreased the size of the existing events by the appropriate percentage. To represent the effect of increased CO_2, transpiration was assumed to decrease and the C-to-N ratio of aboveground biomass was assumed to increase. These effects are based on data from the shortgrass steppe showing that soil water contents were higher and aboveground biomass N concentrations tended to be lower in plots from elevated CO_2 chambers compared with plots from ambient CO_2 chambers (Morgan et al., 2001). Increased air temperature was simulated, with minimum temperature being increased to a greater extent than maximum temperature, because data show that temperatures have been increasing in the Great Plains, and that nighttime minimum temperatures have increased to a greater extent than daily maximum temperatures (Alward et al., 1999; Pielke et al., 2000).

We summarized critical ecosystem output variables, including annual and 10-year average values for aboveground net primary production (ANPP), belowground net primary production (BNPP), system C (defined as the sum of C in soil organic matter, surface organic matter, litter, and live biomass), N_2O emissions, NO_x emissions, and CH_4 uptake for each scenario (Table 15.1). Increased N, elevated CO_2, and increased precipitation all increased total NPP and ANPP (Fig. 15.2A, B). Decreased precipitation led to decreased NPP, and increased temperature had little effect on NPP. Elevated CO_2, increased temperature, and increased precipitation each led to slightly lower system C compared with the control (Fig. 15.2C). Increased N deposition led to large increases in system C, whereas decreased precipitation led to a small increase in system C. Elevated

Table 15.1 Control and Alternative Scenarios Simulated for the Shortgrass Steppe

Scenario	Transpiration, %	Aboveground C:N[a], %	T_{min}, °C	T_{max}, °C	PPT, %	N, g·m^{-2}
Control	—	—	—	—	—	—
CO_2	↓ 30	↑ 10	—	—	—	—
+T	—	—	↑ 2	↑ 1	—	—
+H_2O	—	—	—	—	↑ 20	—
−H_2O	—	—	—	—	↓ 20	—
+N	—	—	—	—	—	1

[a]carbon-to-nitrogen ratio for aboveground biomass.
Each row represents a different simulation and the columns show how the model inputs were changed relative to the control. The inputs were linearly ramped up (or down) from 2001 to 2100. PPT, precipitation; T, temperature.

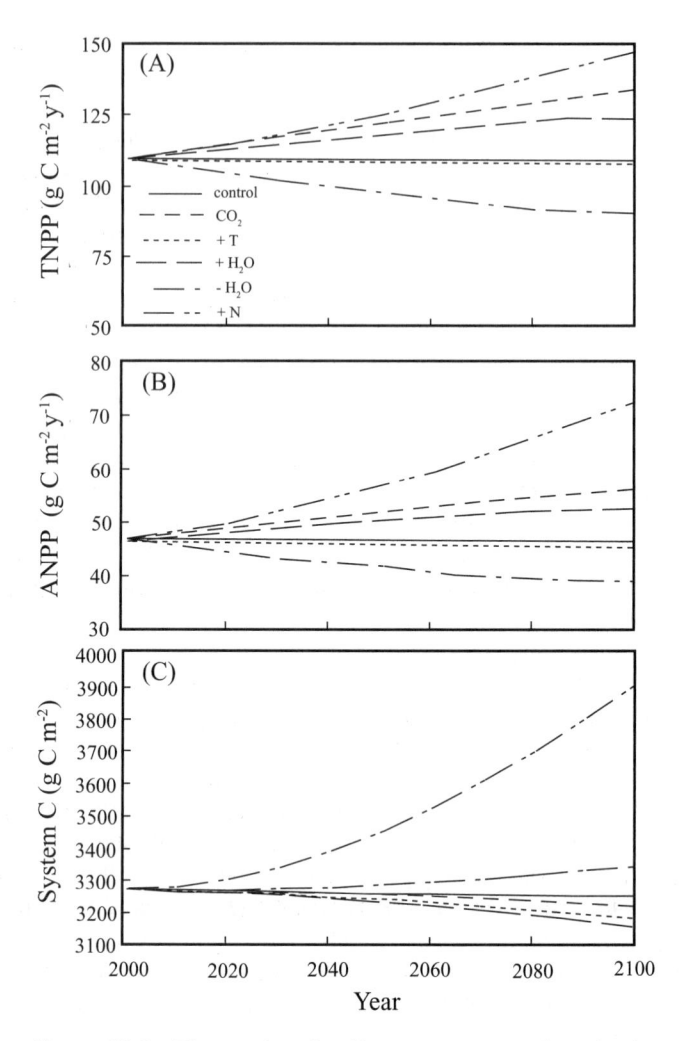

Figure 15.2 Time series of a 10-year average total productivity, aboveground productivity, and system C for shortgrass steppe simulations under elevated CO_2, climate change, and increased N deposition. System C is the sum of C in SOC, surface organic matter, litter, and live biomass. TNPP, total NPP.

CO_2 and increased precipitation both substantially increased NPP (Fig. 15.3A), but temperature had little impact. Decreased precipitation resulted in decreased NPP. Increased N deposition resulted in the largest ANPP enhancement and also increased BNPP. Increased N, elevated CO_2, and increased precipitation each led to higher N_2O and NO_x emissions (Fig. 15.3B). Increased temperature was found to increase NO_x emissions to a much greater extent than N_2O emissions. Decreased precipitation decreased both N_2O and NO_x emissions substantially. Methane uptake rates were not strongly affected by any of the treatments except for N addition, which decreased CH_4 uptake.

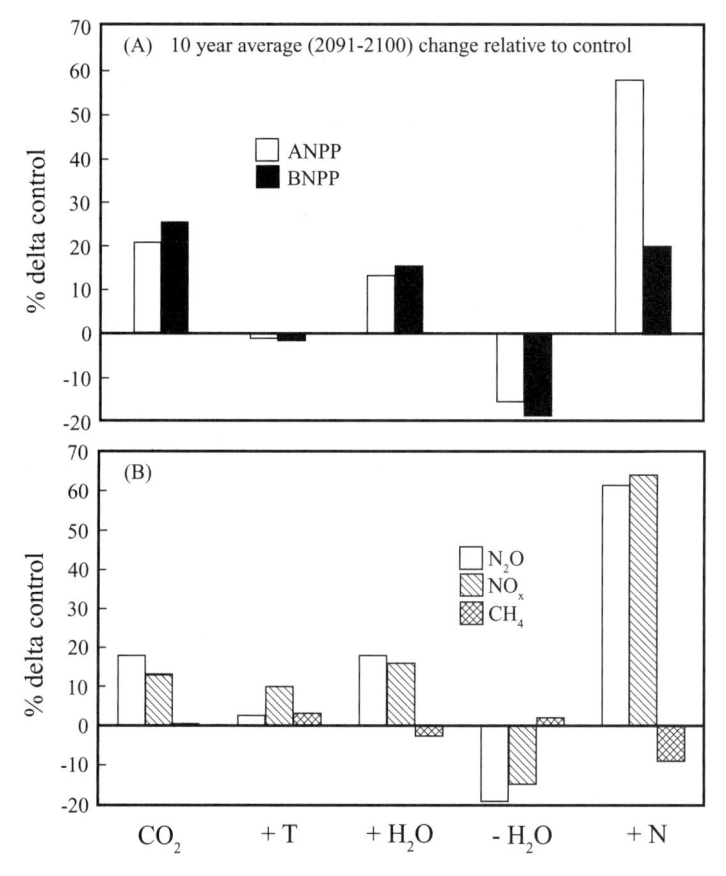

Figure 15.3 Effects of DAYCENT simulations for the shortgrass steppe on NPP and trace gas fluxes. ANPP, aboveground net primary production; BNPP, belowground net primary production, T, temperature.

Vulnerability of the Shortgrass Steppe

Most of the simulated effects exhibited by the different scenarios can be explained in the context of how the model simulates nutrient and water limitation. Treatments that reduced nutrient stress (N) or water stress (CO_2, +H_2O) increased NPP. Increased temperature had little effect on simulated NPP because higher water stress is compensated for by lower nutrient stress resulting from enhanced N mineralization. Nitrogen addition decreased simulated belowground allocation as a fraction of total NPP because the model assumes that as N and water become more available, belowground allocation will decrease. However, the scenarios that decreased water stress (CO_2, +H_2O) did not lead to a decrease in belowground allocation (Fig. 15.3A). This occurred because the model assumes that roots must forage to supply additional N needed to support increased aboveground biomass that results from decreased water stress. These results suggest that the shortgrass steppe ecosystem is influenced by both water and N limitations, which interact on various scales to determine total NPP and biomass allocation.

Increased N deposition was the only scenario to increase simulated system C substantially, even though the CO_2 and increased precipitation treatments increased BNPP much the same as the increased N treatment. Contrary to our expectations, some scenarios showed an inverse relationship between NPP and system C. For example, decreased precipitation, the only scenario with lower NPP, also resulted in increased system C, whereas the CO_2 and plus-water treatments both lost C (Fig. 15.2). This trend is a function of generally dry conditions in the shortgrass steppe, and the model assumption that microbial activity and NPP have different sensitivities to soil water content.

The results of the DAYCENT simulations suggest that although increased water availability from elevated CO_2 will increase NPP significantly ($\approx$25%), the effect on soil C levels will be minor because increased decomposition rates approximately balance higher C inputs. If N deposition increases, DAYCENT predicts that NPP will increase by approximately 35% and soil C levels will increase significantly.

One factor not considered in these simulations is the effect of elevated CO_2 on species composition. Results from open-top chamber experiments performed at the shortgrass steppe (Morgan et al., 2004a) suggest that elevated CO_2 favors drought-sensitive species with lower forage quality. Another limitation of these simulation experiments is that we varied only the amount of precipitation, but not the frequency of precipitation events. Recent DAYCENT simulations performed by Burke et al. (2002) show that decreasing the number of rainfall events while increasing the size of the events leads to substantially lower NPP, lower aboveground to belowground biomass and production ratios, and lower mineralization rates, but higher N gas emissions.

Our DAYCENT simulations provide an excellent example of using a model to guide future experimental research. Our results indicate that key ecosystem processes in the shortgrass steppe are sensitive to climate change and elevated greenhouse gas levels. Future observations and experiments will be required to assess reliably how the shortgrass steppe will respond to changes in climate, CO_2 levels, and N deposition.

Acknowledgments This research was supported with funds from the NSF (LTER BSR9011 659, DEB 9632852), National Institute of Child Health and Human Development (#1 R01 HD33554), National Aeronautics and Space Administration (EOS NAGW2662), and Environmental Protection Agency (EPA Regional Assessment R824a39–01–0).

References

Alward, R. D., J. K. Detling, and D. G. Milchunas.1999. Grassland vegetation changes and nocturnal global warming. *Science* **283**:229–231.

Antle, J. M., S. M. Capalbo, K. Paustian, and M. K. Ali. 2007. Estimating the economic potential for agricultural soil carbon sequestration in the central United States using an aggregate econometric–process simulation model. *Climatic Change* **80**:145–171.

Bledsoe, J. L., R. C. Francis, G. L. Swartzman, and J. D. Gustafson.1971. PWNEE: A grassland ecosystem model, pp. 179. In: U.S. IBP Grassland Biome technical report no. 64.Colorado State University, Fort Collins.

Burke, I. C., E. C. Adair, R. L. McCulley, P. Lowe, S. Del Grosso, and W. K. Lauenroth. 2002. *The importance of pulse dynamics in nutrient availability and ecosystem functioning: Evidence and questions.* Presented at the Responses to Resource Pulses in Arid and Semi-arid Ecosystems Symposium, Tucson, Ariz., August 2002.

Burke, I. C., T. G. F. Kittel, W. K. Lauenroth, P. Snook, C. M. Yonker, and W. J. Parton. 1991. Regional analysis of the central Great Plains: Sensitivity to climate variability. *BioScience* **41**:685–692.

Burke, I. C., W. K. Lauenroth, and W. J. Parton. 1997. Regional and temporal variation in aboveground net primary productivity and net N mineralization in grasslands. *Ecology* **78**(5):1330–1340.

Burke, I. C., D. S. Schimel, C. M. Yonker, W. J. Parton, L. A. Joyce, and W. K. Lauenroth. 1990. Regional modeling of grassland biogeochemistry using GIS. *Landscape Ecology* **4**:45–54.

Campbell, C. A., H. H. Janzen, K. Paustian, E. G. Gregorich, L. Sherrod, B. C. Liang, and R. P. Zentner. 2005. Carbon storage in soils of the North American Great Plains: Effect of cropping frequency. *Agronomy Journal* **97**:349–363.

Capinera, J. L., J. K. Detling, and W. J. Parton. 1983a. Assessment of range caterpillar (Lepidoptera: Saturniidae) effects with a grassland simulation model. *Journal of Economic Entomology* **76**:1088–1094.

Capinera, J. L., W. J. Parton, and J. K. Detling. 1983b. Application of a grassland simulation model to grasshopper pest management on the North American shortgrass prairie, pp. 335–344. In: W. K. Lauenroth, G. V. Skogerboe, and M. Flug (eds.), *Analysis of ecological systems: State-of-the-art in ecological systems.* Elsevier, New York.

Coffin, D. P., and W. K. Lauenroth. 1990. A gap dynamics simulation model of succession in a semiarid grassland. *Ecological Modeling* **49**:229–266.

Del Grosso, S. J., W. J. Parton, A. R. Mosier, M. D. Hartman, J. Brenner, D. S. Ojima, and D. S. Schimel. 2001. Simulated interaction of carbon dynamics and nitrogen trace gas fluxes using the DAYCENT model, pp. 303–332. In: M. Schaffer, L. Ma, and S. Hansen (eds.), *Modeling carbon and nitrogen dynamics for soil management.* CRC Press, Boca Raton, Fla.

Del Grosso, S. J., W. J. Parton, A. R. Mosier, D. S. Ojima, A. E. Kulmala, and S. Phongpan. 2000. General model for N_2O and N_2 gas emissions from soils due to denitrification. *Global Biogeochemical Cycles* **14**:1045–1060.

Del Grosso, S. J., W. J. Parton, A. R. Mosier, M. K. Walsh, D. S. Ojima, and P. E. Thornton. 2006. DAYCENT national-scale simulations of nitrous oxide emissions from cropped soils in the United States. *Journal of Environmental Quality* **35**:1451–1460.

Eastman, J. L., R. A. Pielke, and D. J. McDonald. 1998. Calibration of soil moisture for large eddy simulations over the FIFE area. *Journal of Atmospheric Science* **55**:1131–1140.

Gifford, R. M., D. J. Barrett, and J. L. Lutze. 2000. The effects of elevated [CO2] on the C:N and C:P mass ratios of plant tissues. *Plant and Soil* **224**:1–14.

Hanson, J. D., J. W. Skiles, and W.J. Parton. 1988. A multi-species model of rangeland plant communities. *Ecological Modeling* **44**:89–123.

Holland, E. A., W. J. Parton, J. K. Detling, and D. L. Coppock. 1992. Physiological responses of plant populations to herbivory and their consequences for ecosystem nutrient flow. *American Naturalist* **140**:685–706.

Hu, S., F. S. Chapin, III, M. K. Firestone, C. B. Field, and N. R. Chiariello. 2001. Nitrogen limitation of microbial decomposition in a grassland under elevated CO_2. *Nature* **409**:188–191.

Hunt, H. W., D. C. Coleman, C. V. Cole, R. E. Ingham, E. T. Elliott, and L. E. Woods. 1984. Simulation model of a food web with bacteria, amoebae, and nematodes in soil, pp. 346–352. In: M. J. Klug and C. A. Reddy (eds.), *Current perspectives in microbial ecology*. American Society for Microbiology, Washington, D.C.

Hunt, H. W., M. J. Trlica, E. F. Redente, J. C. Moore, J. K. Detling, T. G. F. Kittel, D. E. Walter, M. C. Fowler, D. A. Klein, and E. T. Elliott. 1991. Simulation model for the effects of climate change on temperate grassland ecosystems. *Ecological Modeling* **53**:205–246.

Innis, G. S. 1978. *Grassland simulation model: Ecological studies, 26*. Springer-Verlag, New York.

IPCC. 2001. *Climate change 2001: The scientific basis–Contribution of Working Group I to the IPCC Third Assessment Report*. Cambridge University Press, Cambridge, UK.

IPCC. 2007. *Intergovernmental panel on climate change. Climate change: The physical basis. Summary for policy makers. Contribution of Working Group I to the fourth assessment report of the IPCC*. Online. Available at www.ipcc.ch.

Kelly, R. H., W. J. Parton, M. D. Hartman, L. K. Stretch, D. S. Ojima, and D. S. Schimel. 2000. Intra- and interannual variability of ecosystem processes in shortgrass steppe. *Journal of Geophysical Research: Atmospheres* **10**:20093–20100.

Körner, C. 2006. Plant CO_2 responses: An issue of definition, time and resource supply. *New Phytologist* **172**:393–411.

Lu, L., R. Pielke, G. Liston, W. J. Parton, D. Ojima, and M. Hartman. 2001. Implementation of a two-way interactive atmospheric and ecological model and its application to the central United States. *Journal of Climate* **14**:900–919.

McGill, W. B., H. W. Hunt, R. G. Woodmansee, J. O. Reuss, and K. H. Paustian. 1981. Formulation, process controls, parameters and performance of PHOENIX: A model of carbon and nitrogen dynamics in grassland soils, pp. 171–191. In: M. Frissel and J. Van Heem (eds.), *Simulation of nitrogen behavior of soil–plant systems*. Pudoc, Centre for Agricultural Publishing and Documentation, Wageningen, the Netherlands.

Morgan, J. A., D. R. LeCain, A. R. Mosier, and D. G. Milchunas. 2001. Elevated CO_2 enhances water relations and productivity and affects gas exchange in C_3 and C_4 grasses of the Colorado shortgrass steppe. *Global Change Biology* **7**:451–466.

Morgan, J. A., A. R. Mosier, D. G. Milchunas, D. R. LeCain, J. A. Nelson, and W. J. Parton. 2004a. CO2 enhances productivity, alters species composition, and reduces digestibility of shortgrass steppe vegetation. *Ecological Applications* **14**:208–219.

Morgan, J. A., D. E. Pataki, C. Körner, H. Clark, S. J. Del Grosso, J. M. Grünzweig, A. K. Knapp, A. R. Mosier, P. C. D. Newton, P. A. Niklaus, J. B. Nippert, R. S. Nowak, W. J. Parton, H. W. Polley, and M. R. Shaw. 2004b. Water relations in grassland and desert ecosystems exposed to elevated atmospheric CO_2. *Oecologia* **140**:11–25.

Mosier, A. R., W. J. Parton, D. W. Valentine, D. S. Ojima, D. S. Schimel, and J. A. Delgado. 1996. CH_4 and N_2O fluxes in the Colorado shortgrass steppe, 1. Impact of landscape and nitrogen addition. *Global Biogeochemical Cycles* **10**:387–399.

Ojima, D. S., T. G. F. Kittel, T. Rosswall, and B. H. Walker. 1991. Critical issues for understanding global change effects on terrestrial ecosystems. *Ecological Applications* **1**(3):316–325.

Ojima, D. S., W. J. Parton, D. S. Schimel, and C. E. Owensby. 1990. Simulated impacts of annual burning on prairie ecosystems, pp. 118–132. In: S.L. Collins and L. L. Wallace (eds.), *Fire in North American tallgrass prairies.* University of Oklahoma Press, Norman, Okla.

Parton, W. J., J. E. Ellis, and D. M. Swift. 1979. The impacts of strip mine reclamation practices: A simulation study, pp. 584–591. In: M. K. Wali (ed.), *Ecology and coal resource development, vol. 12.* Pergamon Press, New York.

Parton, W. J., M. P. Gutmann, S. A. Williams, M. Easter, and D. Ojima. 2005. Ecological impact of historical land-use patterns in the great plains: A methodological assessment. *Ecological Applications* **15**:1915–1928.

Parton, W. J., M. D. Hartman, D. S. Ojima, and D. S. Schimel. 1998. DAYCENT: Its land surface submodel: Description and testing. *Global Planetary Change* **19**:35–48.

Parton, W. J., J. A. Morgan, G. Wang, and S. J. Del Grosso. 2007. Projected ecosystem impact of the prairie heating and CO_2 enrichment experiment. *New Phytologist.***174**:823–834.

Parton, W. J., A. R. Mosier, and D. S. Schimel. 1988. Rates and pathways of nitrous oxide production in a shortgrass steppe. *Biogeochemistry* **6**:45–58.

Parton, W. J., and P. G. Risser. 1980. Impact of management practices on the tallgrass prairie. *Oecologia* **46**(2):223–234.

Parton, W. J., P. G. Risser, and R. G. Wright. 1980. Grazing responses in a prairie National Park. *Environmental Management* **4**(2):165–170.

Parton, W. J., D. S. Schimel, C. V. Cole, and D. S. Ojima. 1987. Analysis of factors controlling soil organic matter levels in Great Plains grasslands. *Soil Science Society of America Journal* **51**:1173–1179.

Parton, W. J., D. S. Schimel, and D. S. Ojima. 1994. Environmental change in grasslands: Assessment using models. *Climatic Change* **28**:111–141.

Parton, W. J., and R. C. Schnell. 1975. Use of an ecosystem model to study potential interactions between the atmosphere and the biosphere, pp. 721–726. In: *Proceedings of the 1975 Summer Computer Simulation Conference, vol. I.* Simulation Councils, La Jolla, Calif.

Parton, W. J., J. M. O. Scurlock, D. S. Ojima, D. S. Schimel, D. O. Hall, M. B. Coughenour, E. Garcia Moya, T. G. Gilmanov, A. Kamnalrut, J. I. Kinyamario, T. Kirchner, T. G. F. Kittel, J. C. Menaut, O. E. Sala, R. J. Scholes, and J. A. van Veen. 1995. Impact of climate change on grassland production and soil carbon worldwide. *Global Change Biology* **1**:13–22.

Parton, W. J., J. S. Singh, and D. C. Coleman. 1978. A model of production and turnover of roots in shortgrass prairie. *Journal of Applied Ecology* **47**:515–542.

Parton, W. J., and F. M. Smith. 1974a. Exploring some possible effects of potential SST-induced weather modification in a shortgrass prairie ecosystem, pp. 255–258. In: *6th Conference on Aerospace and Aeronautical Meteorology.* American Meteorological Society, Boston, Mass.

Parton, W. J., and F. M. Smith. 1974b. Simulating the effects of growing season rainfall enhancement and hail suppression on the production, consumption, and decomposition functions of a native shortgrass prairie ecosystem, pp. 523–528. In: *4th Conference on Weather Modification.* American Meteorological Society, Boston, Mass.

Pepper, D. A., S. J. Del Grosso, R. E. McMurtrie, and W. J. Parton. 2005. Simulated carbon sink response of shortgrass steppe, tallgrass prairie and forest ecosystems to rising [CO_2], temperature and nitrogen input. *Global Biogeochemistry Cycles* **19**: GB1004, doi:10.1029/2004GB002226. .

Pielke, R. A, Sr., G. E. Liston, L. Lu, P. L. Vidale, R. L. Walko, T. G. F. Kittel, W. J. Parton, and C. B. Field. 1997. *Coupling of land and atmospheric models over the GCIP area: CENTURY, RAMS, and SiB2C.* Presented at the 13th annual Conference on Hydrology. 77th AMS annual meeting, Long Beach, Calif., February 2–7.

Pielke, R. A., T. Stohlgren, W. Parton, J. Moeny, N. Doesken, L. Schell, and K. Redmond. 2000. Spatial representativeness of temperature measurements from a single site. *American Meteorological Society* **81**(4):826–830.

Schimel, D. S., B. H. Braswell, E. A. Holland, R. McKeown, D. S. Ojima, T. H. Painter, W. J. Parton, and A. R. Townsend. 1994. Climatic, edaphic, and biotic controls over storage and turnover of carbon in soils. *Global Biogeochemical Cycles* **8**:279–293.

Schimel, D. S., T. G. F. Kittel, and W. J. Parton. 1991. Terrestrial biogeochemical cycles: Global interactions with the atmosphere and hydrology. *Tellus* **43AB**:188–203.

Schimel, D. S., W. J. Parton, T. G. F. Kittel, D. S. Ojima, and C. V. Cole. 1990. Grassland biogeochemistry: Links to atmospheric processes. *Climatic Change* **17**:13–25.

Vidale, P. L., R. A. Pielke, A. Barr, and L. T. Steyaert. 1997. Case study modeling of turbulent and mesoscale fluxes over the BOREAS region. *Journal of Geophysical Research* **102**:29167–29188.

Walko, R. L., L. E. Band, J. Baron, T. G. F. Kittel, R. Lammers, T. J. Lee, D. S. Ojima, R. A. Pielke, C. Taylor, C. Tague, C. J. Tremback, and P. L. Vidale. 2000. Coupled atmosphere–biophysics–hydrology models for environmental modeling. *Journal of Applied Meteorology* **39**:931–944.

Woodward, F. I. 2002. Potential impacts of global elevated CO_2 concentrations on plants. *Current Opinion Plant Biology* **5**:207–211.

16

Effects of Grazing on Vegetation

Daniel G. Milchunas
William K. Lauenroth
Ingrid C. Burke
James K. Detling

Evolutionary History of Grazing and Semiaridity

Grazing by large native ungulates and semiaridity are the two main forces that have had a large influence in shaping the current-day structure of the shortgrass steppe ecosystem (Milchunas et al., 1988). With the uplift of the Rocky Mountain chain during the Miocene (approximately one million years ago), forests of the Great Plains were gradually replaced by grasslands (Axelrod, 1985). Large grazing and browsing animals inhabited the Great Plains during the middle to late Pleistocene, as did grasses of the genera *Stipa*, *Agropyron*, *Oryzopsis*, and *Elymus* (Axelrod, 1985; Stebbins, 1981). Bison occurred both east and west of the Rockies during the Wisconsin glacial period in the latter part of the Pleistocene (Wilson, 1978). During the early Holocene, approximately 10,000 years ago, bison and grasses of the genera *Bouteloua*, *Buchloë*, *Andropogon* or *Schizachyrium*, and *Sorghastrum* concomitantly increased throughout the Great Plains (Stebbins, 1981), but bison did not proliferate west of the continental divide (Mack and Thompson, 1982; Van Vuren, 1987). The natural shift in fauna from horses, pronghorn, and camels to bison and wild sheep from Eurasia is thought to have favored the spread of shortgrasses such as *Bouteloua* and *Buchloë* (Stebbins, 1981). Furthermore, grassland flora east and west of the Rocky Mountains probably had separate origins (Leopold and Denton, 1987).

The shortgrass steppe is unique from other North American semiarid ecosystems in having bison play an important role. Bison did not proliferate west of the Rocky Mountains as they did on the Great Plains to the east. This is due in part to a lack of coincidence in timing of bison lactation and the phenological development of C_3 grasses in the more Mediterranean–like climate west of the Rockies, in contrast to the mix of C_3 and C_4 grasses and pattern of spring–summer precipitation on the Great Plains (Mack and Thompson, 1982). Other explanations for the low numbers of bison west of the Rocky Mountains include physiographic barriers restricting immigration (Kingston, 1932), low protein content of forage (Daubenmire, 1985; Johnson, 1951), heavy snowfall as a cause of mortality (Daubenmire, 1985), and low aboveground primary production coupled with disjunct suitable habitat (Van Vuren, 1987). Bison also did not prosper in the southwestern United States, nor did a large herbivore fauna develop in South America (Stebbins, 1981). The evolution of plains grasses was rapid in North America compared with South America (Stebbins, 1981). Evolutionary lines of bison in North America included six species, three of which were endemic, and the extant *Bison bison* subsp. *bison* of the plains that developed about 4000 to 5000 years ago (McDonald, 1981).

The shortgrass steppe occupies the driest portion of the Great Plains, because of its location in the rain shadow created by the Rocky Mountains (Pielke and Doeskin, chapter 2, this volume). Precipitation ranges from 300 mm in the west to 550 mm in the east (Lauenroth and Milchunas, 1991). This trend of increasing precipitation continues from west to east and from shortgrass steppe to mixed-grass prairie to tallgrass prairie to eastern deciduous forest. Across this gradient, there is increased allocation to aboveground plant production, and a shift in the relative importance of belowground competition for soil water on the dry end to more intense competition for nitrogen and light in a dense canopy on the wet end. Although approximately 90% of plant biomass in the shortgrass steppe is belowground, BNPP is only 67% of total ANPP (Milchunas and Lauenroth, 1992, 2001). Grazed or not, the plant canopy in this dry environment is short and sparse relative to many other grasslands (Milchunas et al., 1988).

Plants of the shortgrass steppe are, therefore, unique in the western hemisphere in terms of having evolved both with a long evolutionary history of grazing by large herbivores and with semiaridity. Selection for individuals in a plant population that have characteristics providing a greater tolerance or avoidance of grazing can be rapid. Various studies have indicated phenotypic changes within 4 months (Brougham and Harris, 1967) and apparent genotypic changes in population structure in 10 to 15 years (Detling and Painter, 1983; Jaramillo and Detling, 1988), 13 years (Peterson, 1962), 25 years (Briske and Anderson, 1992), and 35 years (Kemp, 1937). Thus, although the Great Plains flora is considered young and there are few endemic plants, insects, and birds (Mengel, 1970; Ross, 1970; Stebbins, 1981), the 9000- to 10,000-year association between grasses and herbivores has been sufficient to separate it distinctly from other ecosystems in the western hemisphere, particularly in terms of its capacity to withstand herbivory (Larson, 1940; Mack and Thompson, 1982; Milchunas et al., 1988). Although fire has been an important force maintaining and shaping the structure of the more productive eastern grasslands of the Great Plains, it has not been as important in the shortgrass steppe with its low fuel loads. Weaver and Clements (1938) considered the shortgrass steppe to be a disclimax community, caused by the disturbance of grazing by domestic animals. Although grazing can shift mixed-grass prairie to resemble shortgrass steppe more closely (Lauenroth et al., 1994), Larson (1940) argued that bison maintained the shortgrass steppe, and that it should, therefore, not be considered a disturbance community. However, early classifications considering all of the Great Plains as mixed-grass prairie (Weaver and Albertson, 1956) may have led to confusion; responses to grazing differ across shortgrass, mixed-grass, and tallgrass types (Lauenroth et al., 1994). Although all grasslands of the Great Plains coevolved with *Bison*, we now know that the shortgrass steppe is climatically maintained (Lauenroth, chapter 5, this volume; Lauenroth and Milchunas, 1991). Grazing and semiaridity can be convergent, complementary selection forces (Milchunas et al., 1988), contributing to the unique responses to grazing that we will examine in this chapter.

Recent History of Land Use in the Shortgrass Steppe

Estimates of the numbers of bison that occupied the plains prior to their extinction as a commercially, socially, and ecologically important species in the mid 1870s range from 5.5 to 60 million, based upon extrapolation of observations by early explorers and hide market numbers or various combinations of potential carrying capacity and early observations (Hart, chapter 4, this volume; Shaw, 1995). Some single herds were estimated to have totaled four million (Hornaday, 1889), and others covered 3500 sq. km. (Farnham, 1839 [as cited in McHugh, 1972]). The demise of the bison coincided with a period (1870–1890) of large increases in the numbers (tripling) of cattle in the United States (U.S. Department of Commerce, Bureau of Census, 1935–1982 [as cited in Joyce, 1989]).

Up to about 1900, livestock grazed freely over the shortgrass steppe and received little or no supplementation from crop or hay meadows (Cook and Redente, 1993). Most ownership boundaries were established by fencing by 1900, and permits for grazing of public lands were being issued in the early 1900s. By 1930, cropland had become an important component of the landscape, and supplementation increased to about 20% of feed supply. Grazing of public lands was further regulated with the passing of the Taylor Grazing Act in 1934 (Bement, 1993). Cultivated cropland reached a peak prior to the Dust Bowl drought of 1934 to 1937, after which much of the abandoned land reverted back to public ownership.

Current land use in the shortgrass steppe region is approximately 70% rangeland and 30% cropland (Lauenroth et al., 1994; Parton et al., 2003, Hart, chapter 4, this volume). All public and private grassland is grazed by domestic livestock, except for small areas set aside as experimental or "nature" areas. Some grassland is former cropland that was abandoned in the 1930s; approximately 20% to 30% of the Pawnee National Grassland (PNG) and the Central Plains Experimental Range (CPER) is in old-field succession following cessation of cultivation, and is still ecologically (and visually) discernible from native grassland (Peters et al., chapter 6, this volume). Fluctuations in land use have been driven by economics, government programs, and weather cycles (Joyce, 1989). However, basic environmental constraints have not allowed cropland area expansion since the 1950s (Parton et al., 2003). Much of the shortgrass steppe is marginal for sustaining crop agriculture, is highly susceptible to erosion when the native perennial grass cover is destroyed, and recovers very slowly from disturbance.

Herbivory and Grazing in the Shortgrass Steppe

The Herbivore Pathway of Energy Flow and Grazing Intensities

For moderately stocked shortgrass steppe, aboveground consumption by all trophic levels is dominated by cattle (Table 16.1 [Lauenroth and Milchunas, 1991]). Although cattle are the most obvious herbivore on the shortgrass steppe, their contribution to energy flow accounts for only 13% of total nonsaprophytic consumption. Consumption in terms of herbivory is approximately equally divided among cattle (aboveground), arthropods (both above- and belowground), and nematodes (belowground), with lagomorphs, rodents, and birds each totaling less than 1%. Although cattle do not dominate total herbivory (because of the large amount of belowground herbivory by arthropods and nematodes), cattle are the dominant aboveground herbivore. Herbivory by cattle is a dominant force on plant communities from aboveground, but their role as a dominant herbivore diminishes when placed in the context of the total plant.

Herbivory in grasslands and shrublands is large in comparison with most other systems. In forests, folivory is typically 10% or less of annual foliage standing crop of trees, except during periods of insect outbreaks (Detling, 1989; Schowalter et al., 1986). Herbivory by native ungulates on Serengeti grasslands can reach 92% removal (McNaughton, 1985), which is greater than a heavily grazed shortgrass steppe. Light, moderate, and heavy grazing of a shortgrass steppe is generally considered to be 20%, 40%, and 60% removal of ANPP, respectively (Klipple and Costello, 1960). What is considered heavy grazing in some grasslands of the Great Plains may be considered moderate in the shortgrass steppe. For example, Heitschmidt et al. (1985) considered heavy grazing of mixed-grass prairie in Texas to average approximately 40% to 50% consumption of ANPP.

Table 16.1 The Importance of Different Groups of Consumers in Terms of Herbivorous Processing of Aboveground, Belowground, and Total Plant Energy in the Northern Shortgrass Steppe

Consumer Group	Total NPP,	Aboveground NPP, %	Belowground NPP, %
Ruminants	4.77	29.90	—
Lagomorphs	0.06	0.39	—
Rodents	0.04	0.22	—
Birds	0.01	0.06	—
Arthropods	5.05	5.11	5.04
Macro	4.20	5.11	4.03
Micro	0.85	—	1.01
Nematodes	5.11	—	6.08
Total	15.04	35.68	11.12

Values show the proportion of total (NPP), aboveground (ANPP), and belowground (BNPP) net primary production that is consumed by the group of herbivores indicated. (Adapted from Lauenroth and Milchunas [1991].)

Stocking rates in different systems are generally adjusted to reflect the capacity of the vegetation to withstand the grazing pressure, and there are large differences across the world between the qualitative descriptors of grazing intensity and the quantitative estimates of consumption (Milchunas and Lauenroth, 1993). Average forage production for the shortgrass steppe is approximately 70 g·m^{-2}·y^{-1} in moderately grazed level uplands (Fig. 16.1; see Lauenroth et al., chapter 12, this volume) compared with an average of 285 g·$^{-2}$·y^{-1} in the moderately grazed

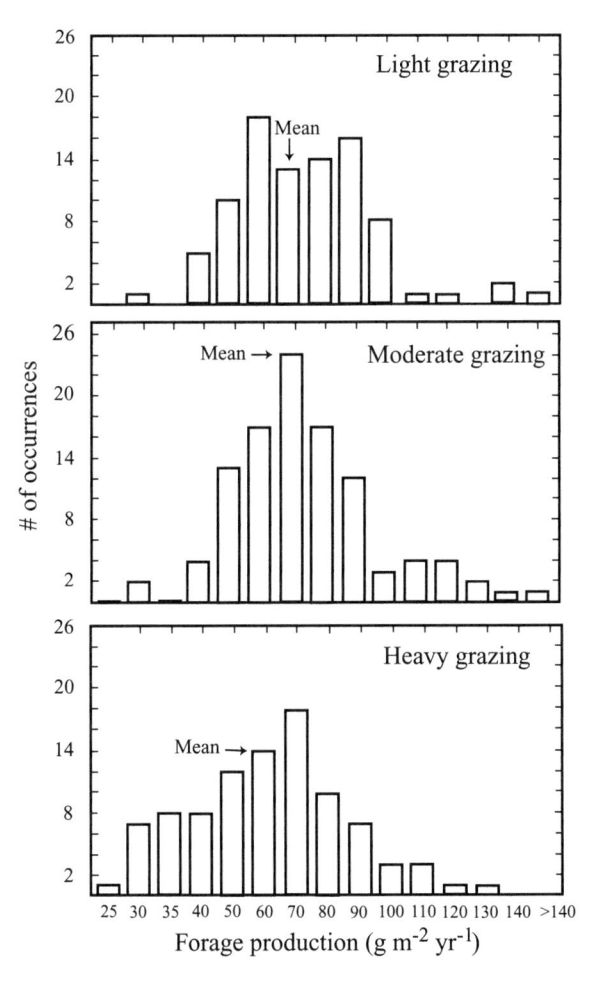

Figure 16.1 Distributions (number of annual occurrences) of forage production (measured in grams per square meter per year) in long-term lightly, moderately, and heavily grazed treatments in the northern shortgrass steppe. Data are for 89, 97, and 88 site samplings for the aforementioned respective treatments, and span the years 1940 to 1990. (From Milchunas et al. [1994].)

sites of the Texas mixed-grass prairie (Heitschmidt et al. 1985). Across grasslands of the Great Plains, there is an inverse relationship between grazing intensities and ANPP within a particular qualitative description of moderate or heavy grazing (Fig. 16.2). Impacts of grazing on plant species composition and ANPP generally increase with increasing ANPP of a vegetation type (Milchunas and Lauenroth, 1993).

Most of the studies we report on here were conducted at the CPER and the PNG. Grazing at the CPER is primarily on a summer–winter pasture basis.

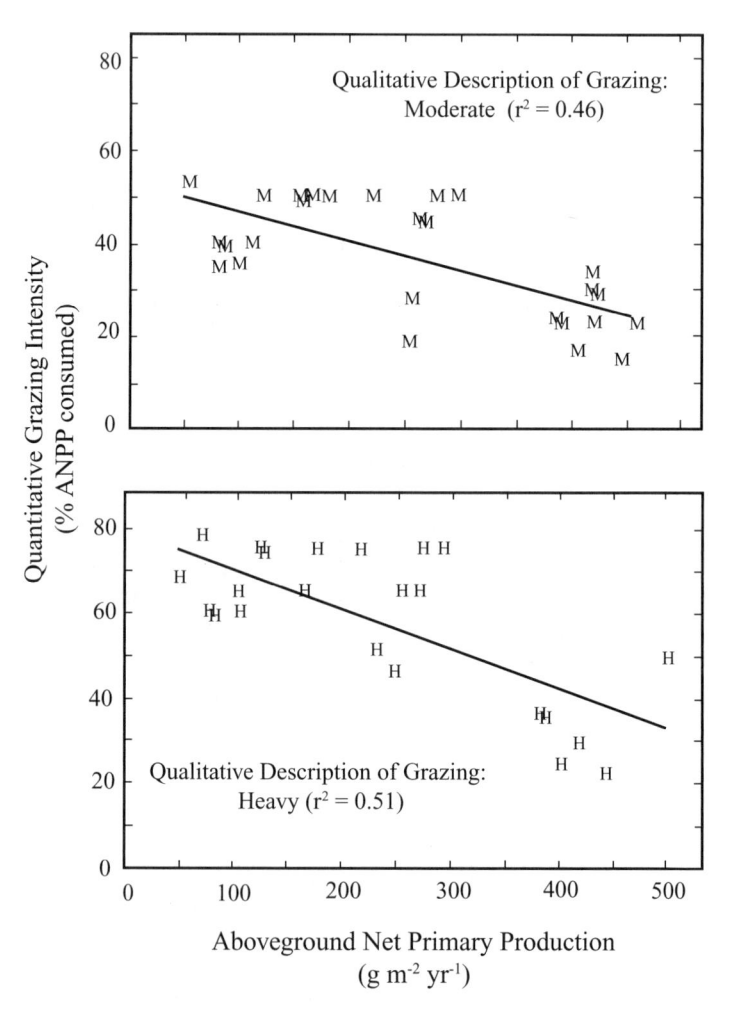

Figure 16.2 Quantitative grazing intensity (measured percentage of aboveground net primary production [ANPP] consumed) in relation to ANPP (measured in grams per square meter per year) for various authors' qualitative description of moderate and heavy grazing intensities. Studies are from grasslands of the Great Plains. (Compiled from data found in Appendix I of Milchunas and Lauenroth [1993].)

The summer grazing season begins in late May, approximately 1 month after the start of the cool-season growth period, and extends for a maximum of 184 days (late November). The length of the grazing period depends upon the time necessary to achieve the intensity criteria, and has been as short as 48 days under the heavy grazing regime (Ashby et al., 1993; Hart and Ashby, 1998). The frost-free dates are from May 15 through September 15, but vegetation is often brown by August and short green-ups and periods of growth depend upon fall precipitation. Since 1939, numbers of animals on the half-section pastures have ranged from 6 to 22, 11 to 29, and 14 to 45 yearling heifers under the light, moderate, and heavy stocking regimes, respectively (Ashby et al., 1993). Ungrazed exclosures were established in 1939 (Fig. 16.3 a, b).

Although an average removal of 20%, 40%, and 60% of ANPP for lightly, moderately, and heavily grazed conditions, respectively, has been maintained since 1939, the means of achieving these removal rates has changed. From 1939 until 1964, intensity criteria were based upon the percent of ANPP removed, but since 1965 the intensity criteria have been based upon leaving a certain amount of biomass. Residual biomass of 22, 34, and 50 $g \cdot m^{-2}$ has been the goal for lightly, moderately, and heavily grazed areas, respectively (Ashby et al., 1993; Hart and Ashby, 1998). Across years, the average percent removal of ANPP remains approximately the same under the two methods, but there are several important differences between them. From a management standpoint, it is easier for both researchers and ranchers to estimate residual biomass than to predict actual production to allow consumption of a certain percentage of that amount. From an ecological standpoint, the two methods are opposite in terms of conditions under which year-to-year grazing pressures on plant communities are most severe. In years of low productivity, removing a percentage of a small amount of growth may have deleterious effects; whereas during highly productive years, removing a percentage still leaves a large residual. The opposite occurs when managing based on residual. Defoliation pressure will be lessened under conditions of drought stress. Under extreme drought, for example, the required amount of residual to be left may never be produced, so animals would not even be allowed to graze. During years of high plant production, grazing pressures are higher when managing based on residuals. Thus, leaving a certain amount compared with consuming a certain amount lessens the grazing pressure in years of low productivity, but increases the grazing pressure in years of high productivity. This tends to lower year-to-year variability in the impact of unfavorable growing conditions on the system, but it also means greater year-to-year fluctuation in potential numbers of animals allowed on the range.

Plant Communities, Grazing Behavior, Diet Selection, and Forage Quality

In this section we briefly describe the shortgrass steppe landscape and the manner in which domestic livestock forage at various scales. The interactions between wildlife, other consumers, and domestic livestock are addressed separately by Milchunas and Lauenroth in chapter 18, and the livestock responses to the grazing

Figure 16.3 The shortgrass steppe is dominated by prostrate C_4 grasses across the gently rolling topography. (A) Loamy uplands ridgetop site at the CPER that has been heavily grazed since 1939 (left of fence) or ungrazed since 1939 (right of fence). (Photo by

treatments and effects of grazing management systems are discussed by Hart and Derner in chapter 17. As described by Lauenroth in chapter 5, most plant communities in the shortgrass steppe are dominated by *B. gracilis,* but there are exceptions. The dominant species and other warm-season (C$_4$) grasses comprise a large proportion of the basal cover and of the ANPP in upland communities. Within the upland communities, Senft et al. (1985) identified six community types based upon the relative proportions of the important grass, half-shrub, and cactus species, as well as soil texture and topographic (catenary) position. In lowland areas of the CPER, *Atriplex canescens* shrublands occupy bands of sandy soils (overflow sites) along larger ephemeral stream channels (see Fig. 18.1B, Milchunas and Lauenroth, chapter 18, this volume), but understory species are similar to those of level upland habitat (Liang et al., 1989). Because low abundance of cool-season (C$_3$) species is a limiting factor in animal production in the shortgrass steppe region (Hart, chapter 4, this volume), the C$_4$ shrub *A. canescens* habitat is often used as winter pasture.

Herbivore foraging patterns are based upon decisions that the animal must make at several levels of spatial resolution, and these can vary through time (Senft et al., 1987). Landscape-scale foraging patterns in the shortgrass steppe can be categorized according to topographic or soil characteristics, which are associated with differences in plant communities and primary productivity (Lauenroth, chapter 5, this volume). Diet selection is a function of the chemical and physical characteristics of plant species and plant parts (palatability), and selection of feeding areas is often related to diet preference (Senft, 1989). Diet selection by cattle can directly affect plant community composition resulting from differences in species' abilities to tolerate or avoid herbivory, or indirectly through altering competitive relationships between neighboring plants. Differential habitat use can affect the distributions of communities as well as the degree of compositional change that may occur in both primary producer and consumer populations.

Although considered to be large generalist herbivores, cattle display a reasonable degree of diet selectivity (Table 16.2). In some studies, the dominant grass (*B. gracilis*) appears to be consumed in proportion to its cover (Senft et al., 1984a), but other cattle diet studies have found its consumption to be smaller than its proportion of cover (Hanson and Gold, 1977; Shoop et al., 1985; Vavra et al., 1977). *Agropyron smithii*, a major component of mixed-grass prairie, has been found to be highly preferred (Lauenroth and Milchunas, 1991); it is also an important component of small-mammal diets. The dominant upland subshrub (*Artemisia frigida*) and forb (*Sphaeralcea coccinea*) are usually consumed in lesser amounts

---

Daniel G. Milchunas.) (B) Loamy uplands swale site at the CPER that has been heavily grazed (left of fence) or ungrazed since 1939 (right of fence). (Photo by Daniel G. Milchunas.) (C) Grazing lawn-type physiognomy in heavily grazed swale with taller vegetation inside cactus clumps where cattle do not graze. (Photo by Mark Lindquist.) Note that in (A) and (B), long-term ungrazed vegetation is not dramatically taller than long-term heavily grazed vegetation, but differences inside and outside the exclosure are greater in the more productive swales than in the ridgetops.

Table 16.2 Percentages of Major Plant Species Available to and Consumed by Cattle, and Their Preference Rankings in Long-Term Moderately Stocked Pastures in the Northern Shortgrass Steppe

Species	Bites, %	Cover, %	Preference, %[a]
Psoralea tenuiflora	0.19	0.03	6.63
Stipa comata	3.13	0.51	6.08
Agropyron smithii	4.96	0.86	5.77
Chenopodium album	1.05	0.23	4.50
Eriogonum effusum	0.33	0.09	3.79
Gutierrezia sarothrae	0.72	0.21	3.42
Sporobolus cryptandrus	3.96	1.32	3.01
Astragalus gracilis	0.08	0.03	2.84
Leucocrinum montanum	0.08	0.03	2.84
Carex eleocharis	9.95	5.24	1.90
Gaura coccinea	0.06	0.03	1.90
Lepidium densiflorum	2.36	1.35	1.75
Sitanion hystrix	4.55	3.96	1.15
Bouteloua gracilis	59.27	59.34	1.00
Aristida longiseta	1.16	1.78	0.65
Chrysothamnus nauseosus	0.25	0.49	0.51
Astragalus spatulatus	0.33	0.67	0.49
Sphaeralcea coccinea	1.91	3.88	0.49
Buchloë dactyloides	4.77	10.67	0.45
Plantago patagonica	0.14	0.61	0.23
Artemisia frigida	0.22	1.10	0.20
Lappula redowskii	0.03	0.20	0.14
Vulpia octoflora	0.03	0.41	0.07
Chenopodium leptophyllum	0.03	0.50	0.06
Chrysopsis villosa	0.00	0.25	0.00
Euphorbia glyptosperma	0.00	0.20	0.00
Mirabilis linearis	0.00	0.12	0.00
Picradeniopsis oppositifolia	0.00	0.29	0.00

[a]Percent bites/percent cover.
Values based upon 3603 bites from belt transects in six sites over 3 years (Milchunas, unpublished data).

than their abundance. Seven grass or *Carex* species comprised the greatest proportion of cattle diets in moderately stocked grassland. However, other cattle diet studies at the CPER site found much less consumption of *B. gracilis* (Hanson and Gold, 1977; Shoop et al., 1985; Vavra et al., 1977). Not unexpectedly, dietary preferences of cattle change with time of year. Schwartz and Ellis (1981) observed that C_4 warm-season grasses increased and C_3 cool-season grasses decreased in importance in cattle diets from spring through late fall. Forbs were important items in the diet in summer whereas shrubs were selected primarily in spring. The major shrub *Atriplex* was an important food source in winter and early spring both in terms of quantity (up to 55% in March) and nutrient content (Cook et al., 1977; Shoop et al., 1985).

An intensive study of grazing behavior in moderately grazed upland plains indicated that cattle may select foraging areas by plant community type or topographic zones, although these two classifications are not independent in many areas (Senft et al., 1985). Preference for plant communities and topographic zones changed seasonally. Lowland swales and draws were heavily utilized during the growing season, as were fence lines and areas adjacent to watering tanks (Table 16.3). In contrast, ridgetops and south-facing slopes were more heavily utilized during the dormant season. Senft et al. (1985) hypothesized that the seasonal shift was the result of movement to less preferred areas after growth ceased and forage became depleted in the preferred areas.

Simulations have suggested that increasing stocking rates would increase use of less productive communities because of a more rapid depletion of preferred areas (Senft, 1989). Increasing stocking rates was predicted to have a much greater effect on species preferences in the diet than on preferences for plant communities/topographic zones (Fig. 16.4). Utilization of lowland swales under heavy stocking rates decreases plant species diversity that is normally associated with gradients in soil quality (Milchunas et al., 1989). Relative preferences for topographic zones/plant communities may also vary over long timescales of grazing pressure. Varnamkhasti et al. (1995) found greater quantities of forage removal from lowland swales in long-term lightly grazed pastures than in heavily grazed pastures, and could not attribute this pattern to differences in size of swales or distances to water in the two treatments.

Forage quality may also interact with topographic location and grazing intensity. Nitrogen availability (standing nitrogen) was greater in the lowland swales of lightly grazed than of heavily grazed treatments (Milchunas et al., 1995),

Table 16.3 Grazing Preferences for Plant Communities and Topographic Zones for Two Seasons in Upland Plains Sites in the Northern Shortgrass Steppe

	Plant Community					
Period	Buda-Bogr	Buda-Agsm-Carex	Agsm-Dist	Bogr-Oppo	Bogr-Eref-Oppo	Bogr-Eref
Growing season (Apr–Oct)	1.10	1.68	1.11	0.60	0.72	1.39
Dormant season (Nov–Mar)	1.55	0.94	0.70	1.05	1.38	0.40

	Topographic Zone					
Period	Ridgetops	South-Facing Slopes	North-Facing Slopes	Draws and Swales	Fence Lines	Watering Area
Growing season (Apr–Oct)	0.43	0.93	0.86	1.39	1.05	1.95
Dormant season (Nov–Mar)	1.61	1.16	1.09	0.80	0.20	1.53

Values represent the amount of time spent grazing compared with the area available. (Adapted from Senft et al. [1985].).

Agsm, *Agropyron smithii*; Bogr, *Bouteloua gracilis*; Buda, *Buchloë dactyloides*; Dist, *Distichlis stricta*; Oppo, *Opuntia polyacantha*; Eref, *Eriogonum effusum*.

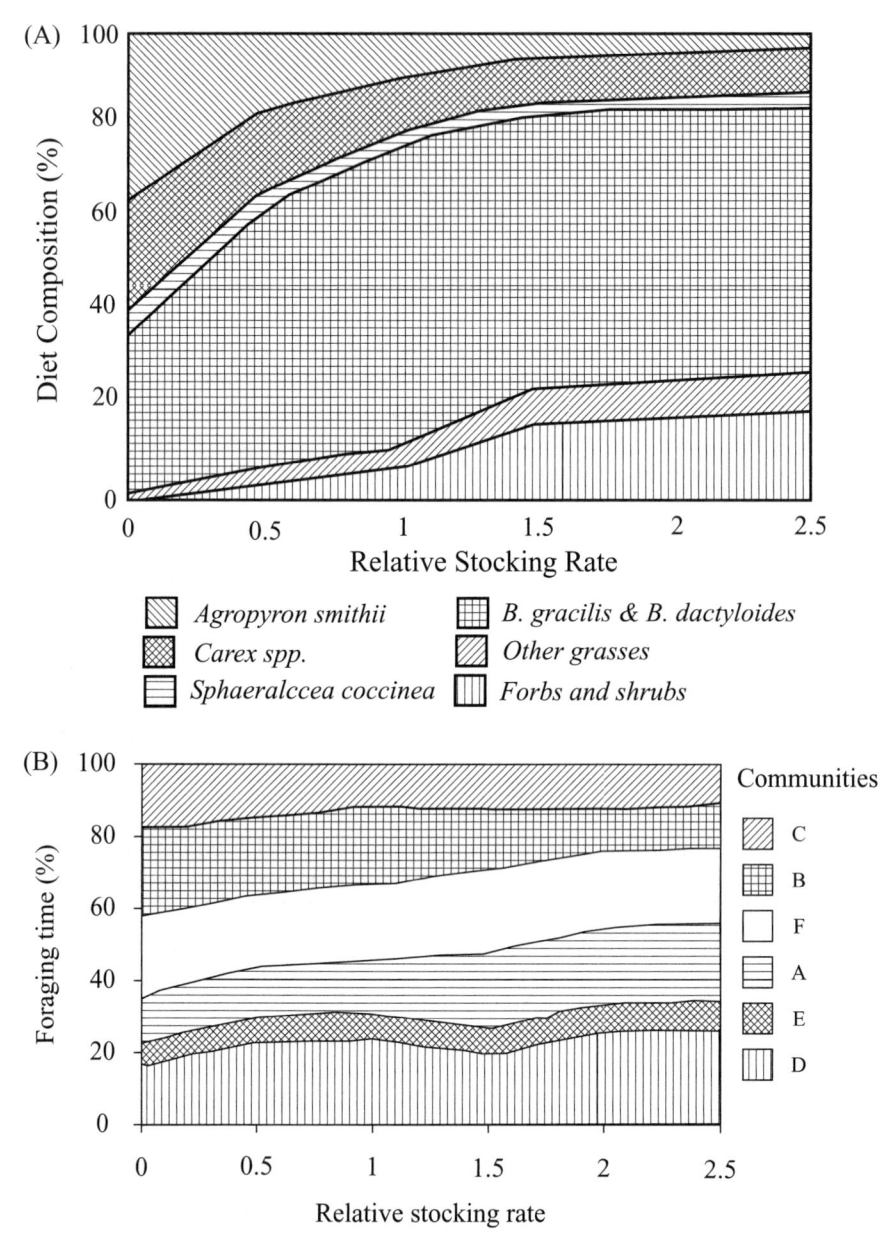

Figure 16.4 (A–B) Simulations of the effect of stocking rate on grazing behavior of cattle at plant community and plant species levels of selection in the shortgrass steppe. Relative stocking rate effects on species composition of diets (percent) (A), percent of time spent foraging in different plant communities (B), relative preferences for species or plant groups.

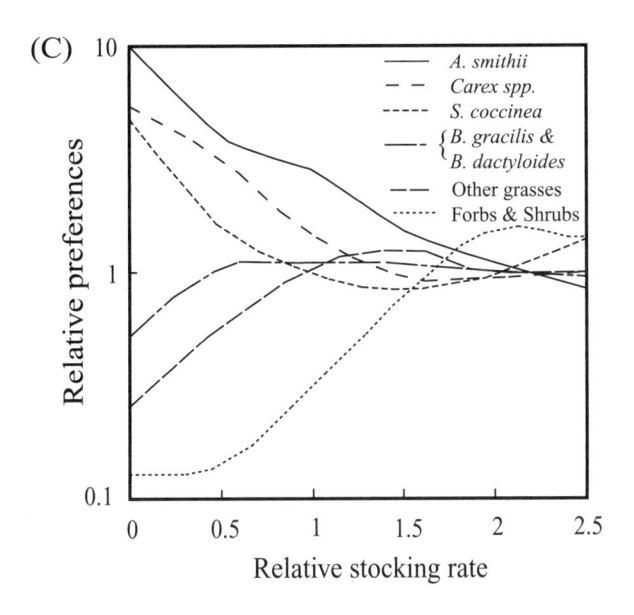

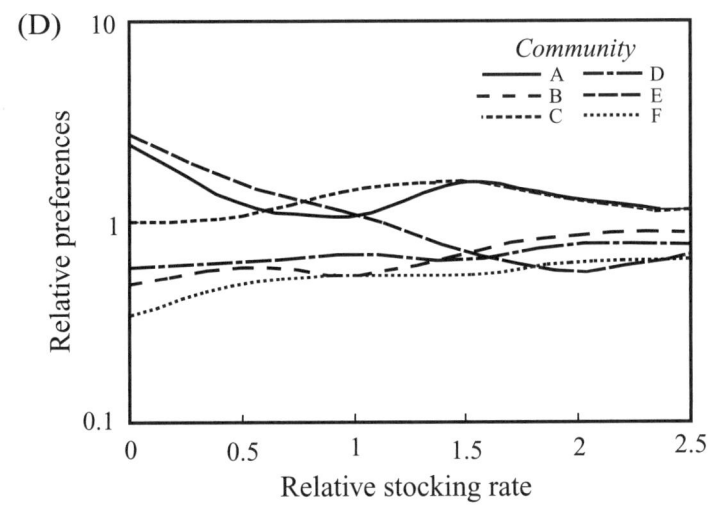

Figure 16.4 (C–D) Simulations of the effect of stocking rate on grazing behavior of cattle at plant community and plant species levels of selection in the shortgrass steppe. Relative stocking rate effects on relative preferences for species (C), and relative preferences for six plant communities in the pasture (D). The plant communities are A, *Buchloë dactyloides–Bouteloua gracilis*; B, *B. dactyloides–Pascopyrum smithii–Carex* spp.; C, *P. smithii–Distichlis spicata*; D, *B. gracilis–Opuntia polyacantha*; E, *B. gracilis–Eriogonum effusum–O. polyacantha*; F, *B. gracilis–E. effusum*. (From Senft [1989].)

suggesting that long-term preferential grazing of swales may have reduced time spent in these areas in the heavily grazed treatments relative to those in the lightly grazed treatments. Senft et al. (1985) found that standing nitrogen in preferred species ($r=.75$), biomass of preferred species ($r=.71$), standing nitrogen in live plants ($r=.71$), and standing live biomass ($r=.69$) were all correlated with community/topographic zone preferences, but that total aboveground biomass was only weakly correlated ($r=.45$) with preference. Nutrient density (quality × quantity) appears to be an important factor in plant community preference by cattle, particularly in the shortgrass steppe, where sparse canopy cover and prostrate growth forms mean that overall nutrient density is low and bite sizes are small.

Similar to differences in plant communities, forage or diet quality display less seasonal variation in the shortgrass steppe than in many other systems. This may be due in part to the high-quality, dormant-season curing characteristics of the dominant species (*B. gracilis*) and to the low stem-to-leaf ratios of this very short-stature grass. These seasonal dynamics are evident even within Colorado shortgrass–mixed-grass types (Fig. 16.5). Tall-stature grasses generally display greater declines in nutritive value with advancing phenology (Fig. 16.6). On the other hand, although shrubs and forbs maintain greater protein, carotene, and phosphorus levels with advancing maturity (Cook et al., 1977), these are not large components of most shortgrass steppe plant communities. Crude protein and digestible dry matter of cattle diets vary little more with season of the year than between plant communities on the shortgrass steppe (Senft et al., 1984a). Both nitrogen and energy budgets of cattle can fall below maintenance requirements by late in the year when supplementation is not provided (Table 16.4) (Senft et al., 1984b). These nutritional factors have implications for cattle weight gains and

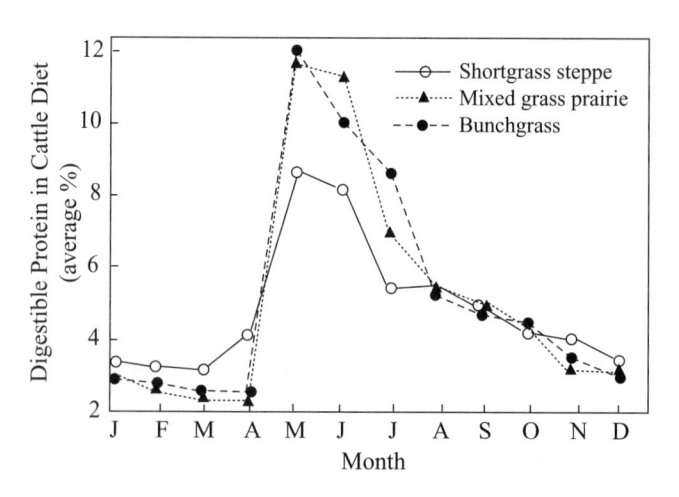

Figure 16.5 Seasonal digestible protein (average percent) in cattle diets for the shortgrass steppe in southeastern Colorado, a mixed-grass type in sandy soils (Akron, Colorado), and bunchgrass range in northwestern Colorado. (Data from Cook et al. [1977].)

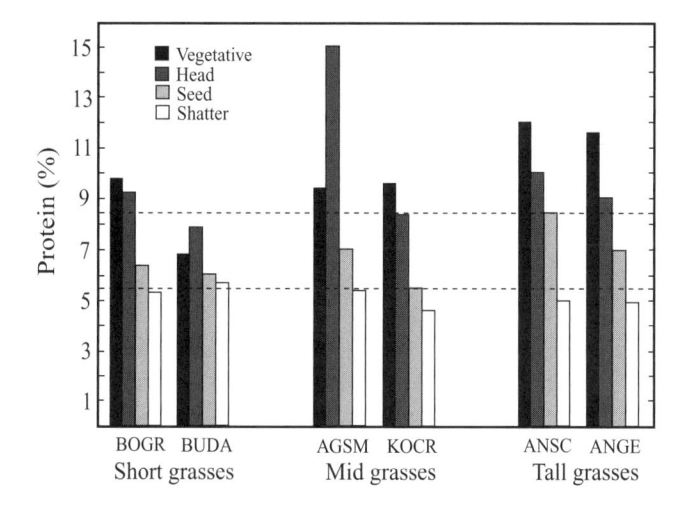

Figure 16.6 Protein concentrations of short-, mid-, and tall-grass species at vegetative, head, seed, and shatter phenological stages of growth (generally spring through early winter periods). AGSM, *Agropyron smithii;* ANGE, *Andropogon gerardi;* ANSC, *A. scoparius;* BOGR, *Bouteloua gracilis;* BUDA, *Buchloë dactyloides;* KOCR, *Koeleria cristata.* These are important species in the shortgrass steppe, mixed-grass prairie, and tallgrass prairie. The upper dashed line is the mean for the vegetative phenological stage of shortgrasses and the lower dashed line is the mean for the shatter phenological stage for shortgrasses. (Data from Cook et al. [1977].)

Table 16.4 Seasonal Nitrogen and Metabolizable Energy Required and Eaten (Intake) for Cattle Grazing at Moderate Stocking Rates in Upland Plains Sites in the Northern Shortgrass Steppe

Month	Nitrogen, g·d⁻¹ Required[a]	Intake	ME, Mcal·d⁻¹ Required[a]	Intake	Body Weight, kg
May	56	89	8.2	9.6	245
June	56	139	8.2	14.8	252
July	64	97	9.4	14.7	287
August	64	112	9.4	16.8	292
September	64	102	9.4	14.9	291
October	74	97	10.6	18	331
November	74	72	10.6	16.1	326
January	74	69	10.6	9.8	323
March	74	71	10.6	6.7	317

[a]National Research Council (1976).
Adapted from Senft et al. (1984b).

management practices (Hart, chapter 4, this volume), as well as for the impacts of livestock grazing on wild herbivores (Milchunas and Lauenroth, chapter 18, this volume).

Effects of Grazing on Vegetation

Plant Community Structure

It is well known that different methods of sampling vegetation may bias results with respect to species or life-form abundance and composition (Milchunas, 2006). This is especially true for grazing studies in which animals are selectively removing plant parts through the season or year. Measures such as canopy cover or biomass of species, unless they are from temporarily caged plots, confound current-year removal by grazing with potential long-term effects on population dynamics (mortality, establishment) and plant growth capacity (potential productivity). Current-year removal always has an impact on community structure, which may or may not manifest in long-term effects if the removals are terminated. Other measures such as species density, frequency, or basal cover do not confound current-year removal with long-term population dynamics, but do have other drawbacks. It is important to keep in mind what the technique measures when interpreting studies that examine the effects of grazing on plant communities.

Species, Life-Forms, and Functional Groups

The effects of grazing on plant community structure have been studied at several northern and southern shortgrass steppe sites. The longest and most intensively studied controlled experiments are at the CPER in the northern shortgrass steppe. Based upon periodic sampling from 1940 through the early 1950s, researchers concluded that the shortgrass steppe shows little response to grazing (Klipple and Costello, 1960). Changes in species abundances resulting from fluctuations in weather were much greater than those resulting from grazing. This prompted the authors to comment that with heavy grazing "some changes in the vegetation, such as the sod-type growth of *B. gracilis* (blue grama) and the disappearance of highly palatable species, would go unnoticed if similar ranges subject to other grazing treatments were not nearby for comparison" (p. 78). The shift to prostrate growth forms of *B. gracilis*, and the sodlike appearance that Harper (1969) and McNaughton (1984) would later term grazing lawns (Fig. 16.3C), was considered an undesirable consequence of heavy grazing. Change in morphology greatly reduced available forage to the grazing animals, cattle weight gains were lower, and animal mortality increased significantly (Fig. 16.7). Impacts of heavy grazing on plant community structure were considerably lower than impacts on cattle, however. This has potentially important ramifications for the shortgrass region, where much of the land is owned by private ranchers, and economics play an important role in management decisions. Loss of profits resulting from

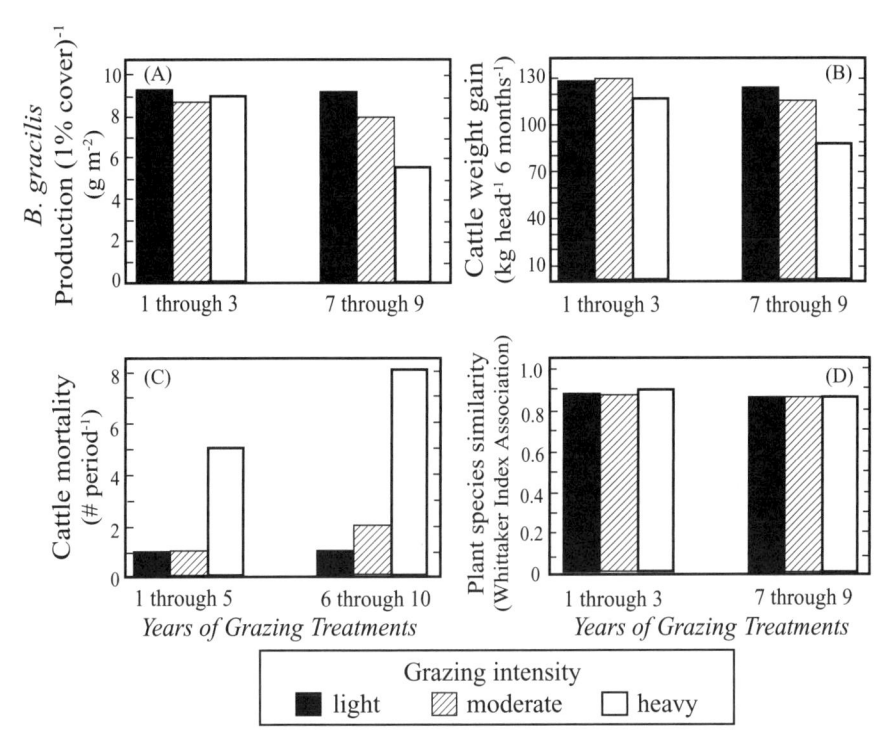

Figure 16.7 Vegetation and cattle responses to grazing intensity treatments in the northern shortgrass steppe averaged for the first 3 years of treatment (1940–1942) and years 7 through 9 (1945–1949). (A) *Bouteloua gracilis* yield per unit area of ground cover. (B) Cattle weight gain. (C) Cattle mortality. (D) Plant community species similarity (ungrazed compared with light, moderate, and heavily grazed treatments calculated using the Whittaker [1952] index of community association with cover data). (Compiled from data in Klipple and Costello [1960].)

overgrazing are important compared with shifts in plant community composition, and an animal/profitability threshold may be reached before the system moves to an alternative state. In other words, the effects on the animals are greater than the effects on plant community species composition. Furthermore, the increased vegetative cover with heavy compared with light grazing has effects on plant community structure that are discussed later.

Still, some changes in shortgrass steppe species abundances do occur with changes in grazing intensity, and our understanding of the regulating factors and specifics of population dynamics has progressed with additional studies and years of treatment. Results from early years did not show much difference between grazing treatments in cover of the dominant *B. gracilis* (Table 16.5). Cover of *B. dactyloides*, a stoloniferous shortgrass, increased with grazing, as did the annual *Festuca octoflora*. Midheight grasses such as *S. comata* and the palatable *A. smithii* decreased with grazing.

Table 16.5 Species Abundance or Composition of Long-Term Grazing Treatments in the Northern Shortgrass Steppe

	Klipple & Costello (1960), Cover												Milchunas et al. (1989), Density										Ashby et. al. (1993), Frequency			
	1940–1942				1952–1953								1984						1986					1992			
													Ridge		Swale				Ridge		Swale						
Species	H	M	L	U		H	M	L	U				H	U	H	U			H	U	H	U		H	M	L	U
Grasses																											
Bouteloua gracilis	7.41	6.98	5.90	8.40		5.48	5.82	5.34	6.75				1569	1409	1921	1491			1340	1174	1748	1272		81	76	57	16
Buchloë dactyloides	0.97	1.38	1.23	0.62		1.14	1.37	1.03	0.54				—	—	—	—			—	—	—	—		14	17	1	7
Aristida longiseta	0.16	0.39	0.45	0.18		0.13	0.33	0.62	0.30				3.1	1.5	3.0	0.2			1.6	1.7	1.0	1.1		31	32	72	31
Sporobolus cryptandrus	0.03	0.04	0.04	0.03		0.02	0.05	0.02	0.04				0.1	0.2	1.1	0.4			T	T	T	T		22	9	37	10
Agropyron smithii	0.13	0.09	0.23	0.20		0.01	0.02	0.07	0.07				1.8	36.0	10.0	193.6			2.5	46.8	8.7	132.8		1	8	6	29
Stipa comata	0.03	0.03	0.04	0.01		0.01	0.03	0.06	0.15				T	T	T	T			T	0.5	T	0.1		11	0	27	25
Sitanion hystrix	T	0.01	0.01	T		T	0.01	0.01	0.04				0.6	1.9	0.1	0.4			1.33	2.84	0.09	1.11		5	2	10	20
Carex eleocharis	0.08	0.06	0.02	0.01		0.05	0.04	0.03	0.07				298.0	194.5	474.9	316.9			337.4	202.6	368.6	358.6		72	77	46	59
Muhlenbergia torreyi	0.09	0.54	0.58	0.29		0.07	0.21	0.16	0.11				T	T	T	T			T	T	T	T					
Festuca octoflora	0.07	0.09	0.10	0.04		0.01	0.01	T	T				32.3	28.8	27.3	10.4			9.8	4.1	27.2	1.2					
Forbs																											
Sphaeralcea coccinea	0.22	0.20	0.27	0.09		0.02	0.02	0.06	0.05				11.0	13.4	9.2	10.0			10.3	16.4	6.4	20.1		58	54	57	74
Chenopodium album	0.02	0.03	0.03	0.03		T	T	T	T				T	T	T	T			T	0.1	0.1	0.4		3	1	4	10
Lepidium densiflorum													0.4	0.3	0.1	0.2			0.5	0.9	0.2	0.2		20	8	20	27
Salsola iberica	0.04	0.04	0.02	0.02		T	T	T	T				0.3	1.6	0.6	1.8			0.1	1.0	0.2	1.2		2	0	1	10
Thelesperma spp.	T	T	0.02	0.02		T	T	0.01	0.01				T	T	T	T			T	T	T	0.1		6	0	16	10
Astragalus gracilus	0.04	0.07	0.06	0.05		0.02	0.01	0.03	T				0.1	1.2	0.3	0.2			0.2	1.5	T	0.2					

Psoralea tenuiflora	0.02	0.06	0.04	0.03	T	T	0.02	0.01	0.0	2.1	0.1	1.1	T	T	T	1.5
Sophora sericea	0.05	0.14	0.12	0.08	0.01	0.04	0.02	T	0	0	0	0	0	0	0	0
Shrubs																
Artemisia frigida	0.02	0.02	0.03	0.19	0.09	0.14	0.15	0.24	0.1	7.6	1.6	19.8	T	T	T	0.8
Gutierrezia sarothrae	0.18	0.07	0.06	0.14	0.05	0.03	0.07	0.02	3.8	16.6	0.2	6.7	0.1	0.7	T	0.1
Chrysothamnus nauseosus	0.06	0.44	0.14	0.16	0.07	0.17	0.05	0.07	0.5	2.5	0.4	9.8	T	T	T	0.4
Eriogonum effusum	0.22	0.21	0.19	0.11	0.11	0.11	0.12	0.08	6.1	6.1	6.5	10.6	0.1	0.3	0.4	1.0
Cacti																
Opuntia polyacantha	0.86	0.52	0.74	0.97	0.51	0.40	0.44	0.62	37.4	27.2	14.1	12.4	28.9	29.6	11.4	7.7	56	46	76	60

H, heavily grazed; L, lightly grazed; M, moderately grazed; U, ungrazed.

A study of these same treatments in 1962 through 1963, stratifying replicates according to soil type, reported only three species with significant, strong correlations with grazing intensities (Hyder et al., 1966). These and other species varied in the degree and direction of response among soil types. The effects of grazing on species composition were small, and Hyder et al. (1966, 1975) suggested this as a reason to question the usefulness of vegetation-based range condition classification on these grasslands. Hyder et al. (1975) suggested that a management goal for the shortgrass steppe must be to maintain a good stand of *B. gracilis*, which is thinned after drought but has not been observed to be negatively affected by grazing.

More recent assessments of these grazing treatments (Ashby et al., 1993; Hart and Ashby, 1998; Milchunas et al. 1989) show more distinct relationships between grazing intensities and changes in species composition, although conclusions concerning the relatively small magnitude of the overall differences remains similar (Table 16.5). The dominant *B. gracilis* and other shortgrasses clearly increase with grazing, and less abundant midgrasses generally decrease. This is commonly observed in mixed-grass and tallgrass prairie ecosystems, where grazing can often shift classifications between mixed-grass prairie/shortgrass steppe or tallgrass prairie/mixed-grass prairie (Lauenroth et al., 1994). Convergent, long-term selection forces of semiaridity and grazing have produced plant communities in which large changes in community physiognomy do not occur even after more than 50 years of very large differences in grazing intensity (Milchunas et al., 1988).

Early studies of shortgrass steppe plant population dynamics in response to grazing intensity attempted to define general, longer term responses at spatial scales of the whole-pasture management unit. Even though loamy upland sites have only gently rolling topography (Fig. 16.3A,B), grazing by cattle is more intense in swales than on ridgetops, and primary production is often greater in swales (Lauenroth et al., chapter 12, this volume). Differences in the amounts and seasonal patterns of precipitation can favor different species and plant functional types. Interactions among grazing, topography, and short-term weather can be important controls on plant population dynamics (Milchunas et al., 1989). For example, *S. coccinea* densities were found to decrease from a wet year to a dry year in heavily grazed treatments, but to increase in ungrazed exclosures (Table 16.5). Sensitivity of *S. coccinea* to grazing during the dry year was in the more favorable swale habitat. Other species display greater sensitivity to grazing during wet-year conditions or in ridgetop communities. Whether a species can be classified as an increaser or a decreaser with grazing depends upon its interactions with other species in a particular community matrix and upon the abiotic conditions.

The time of the year that a pasture is grazed can also affect species composition. Repeated heavy grazing during any particular month in the growing season results in approximately three times as many responses in key species as does grazing during any particular month when plants are senescent (Hyder et al., 1975). Heavy grazing in April, May, and June has been found to affect cool-season species most negatively, whereas heavy grazing in September favors cool-season species and best controls annuals. However, weather conditions have been observed to exert greater control over species composition than grazing during any single month of the year at intensities that "could not have been more severe without endangering the lives of the cattle" (Hyder et al., 1975).

Grasses are the most abundant life-form of the shortgrass steppe, accounting for approximately 98% of plant density and 80% of aboveground biomass. As a group, grasses generally track increases of the dominant *B. gracilis* in response to grazing (Fig. 16.8, Table 16.5, and Fig. 16.9). Forbs and shrubs display opposite responses to those of grasses. Both decrease with grazing either because of less tolerance to grazing or because of greater competition with grasses under grazing. Forbs, particularly annuals, are dynamically interactive with respect to

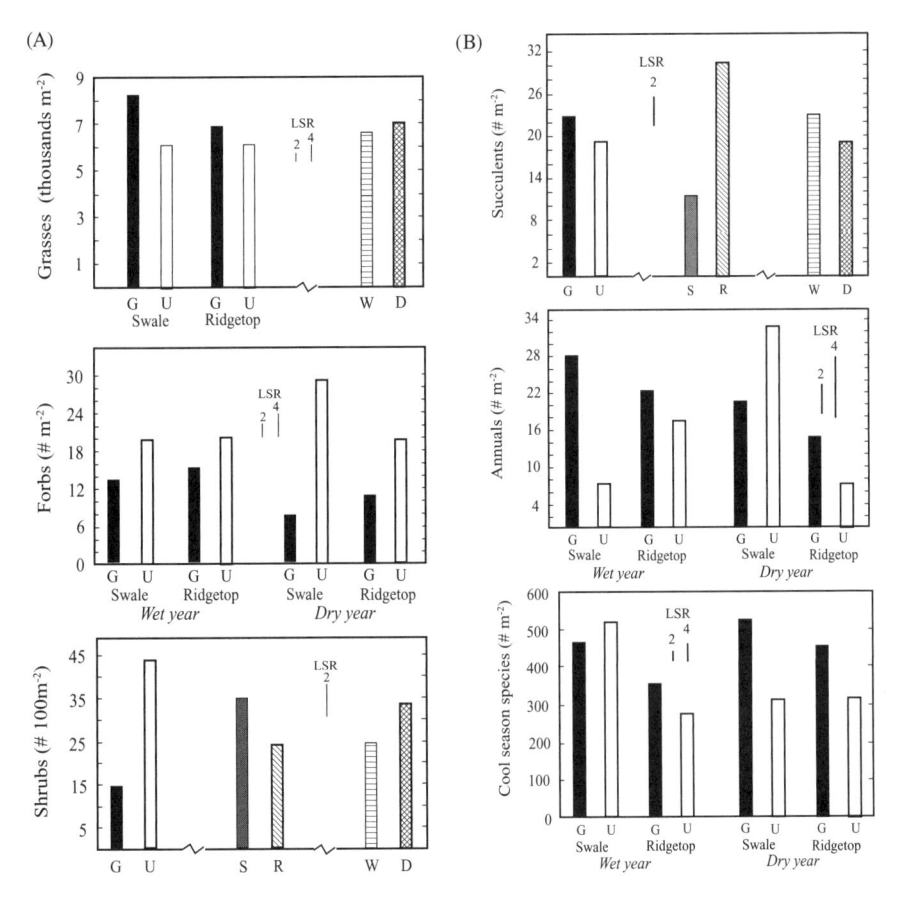

Figure 16.8 Density of functional groups (A) and species or life-forms (B) in relation to long-term grazing treatments (46 and 48 years after initiation) in the northern shortgrass steppe. D, dry year; G, grazed at heavy intensity; R, ridgetop or summit; S, swale or toe slope topographic position; U, ungrazed; W, wet year. Use LSR$_2$ for significance test when crossing any one treatment within the other two treatments and LSR$_4$ when crossing any two treatment categories within a third. A broken x-axis represents a two-way interaction followed by a main effect, or three main effects. (Data from Milchunas et al. [1989].)

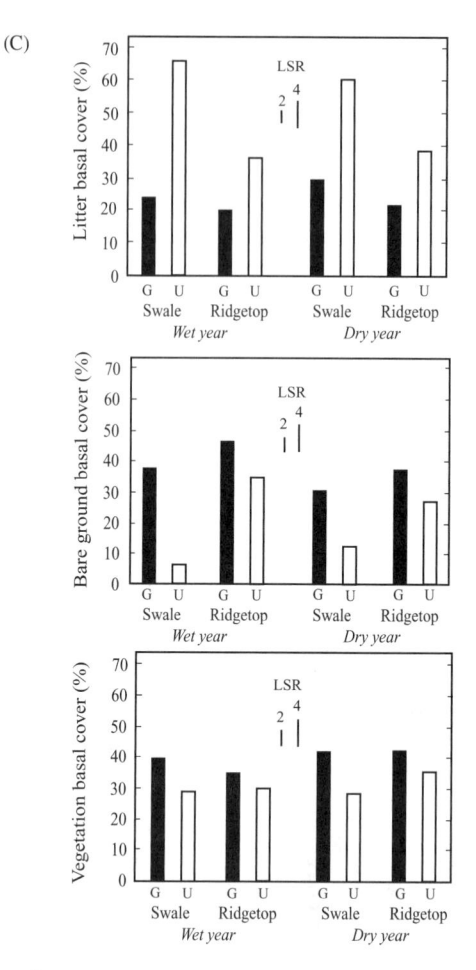

Figure 16.8 (*continued*) Basal cover (C) in relation to long-term grazing treatments (46 and 48 years after initiation) in the northern shortgrass steppe. D, dry year; G, grazed at heavy intensity; R, ridgetop or summit; S, swale or toe slope topographic position; U, ungrazed; W, wet year. Use LSR$_2$ for significance test when crossing any one treatment within the other two treatments and LSR$_4$ when crossing any two treatment categories within a third. A broken x-axis represents a two-way interaction followed by a main effect, or three main effects. (Data from Milchunas et al. [1989].)

grazing, topography, and short-term, annual wet–dry cycles. Succulents do not respond either to grazing or to short-term fluctuations in weather, but are more abundant on ridgetops than swales (Table 16.5). Cool-season species are a small proportion of the plant community, are more abundant in swales than ridgetops, and can increase or decrease with grazing depending upon topographic position or the species comprising the group. *Agropyron smithii* is a cool-season grass that is one of the most palatable species in the community and shows very large

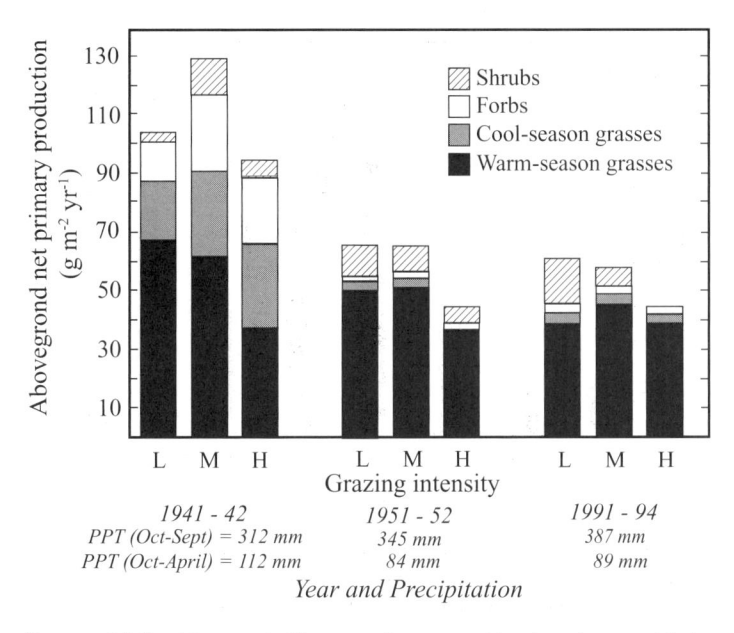

Figure 16.9 Aboveground net primary production (measured in grams per square meter per year) of warm- and cool-season grasses, forbs, and shrubs for lightly, moderately, and heavily grazed treatments in the northern shortgrass steppe averaged for years 1941 to 1942, 1951 to 1952, and 1991 to 1994. PPT, precipitation. (Data from Ashby et al. [1993] and Ashby [unpublished data].)

declines with grazing under all situations. The warm-season group is dominated by *B. gracilis*.

The response of annuals to long-term grazing treatments is highly dependent upon topography (Fig. 16.8). Topography influences the interaction effects among soils, water availability, and nutrient budgets (Singh et al., 1998). We have found annuals to have more than four times greater density in heavily grazed than in ungrazed swale communities, but with no significant difference between grazing treatments on ridgetops. Densities of annuals in swale compared with ridgetop communities are very different in wet and dry years. The most abundant annuals are *F. octoflora* and *Plantago patagonica*. Both species are unpalatable, very small-stature, short-lived, cool-season species that are possibly favored by grazing as a result of lower litter cover and warmer soil temperatures in grazed than in ungrazed sites during the period of high precipitation in spring.

Although annuals are generally more abundant in grazed communities than ungrazed, we found that both native and exotic opportunistic weedy species are more abundant in ungrazed than in heavily grazed communities (Fig. 16.10 [Milchunas et al., 1989, 1990]). Richness of exotic species is also greater in ungrazed compared with grazed communities (see later sections in this chapter).

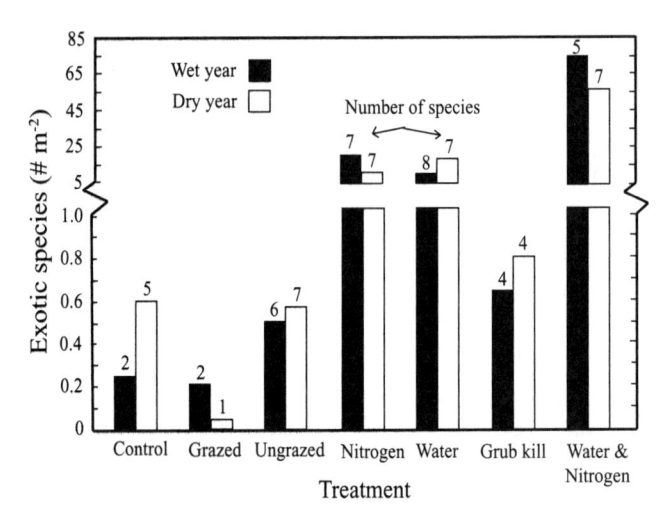

Figure 16.10 Density (number of individual plants per square meter) and richness (number of species [indicated by number over bars]) of exotic species in long-term heavily grazed and ungrazed treatments in wet and dry years, and compared with nutrient enrichment and grub kill disturbances in the northern shortgrass steppe. The Control is the control for the nutrient enrichment and the grub kill treatments, and is a short-term exclosure compared with long-term exclosure treatment represented by "ungrazed" treatment. (Data from Milchunas et al. [1989, 1990].)

Differences in community structure between long-term heavily grazed and ungrazed treatments are small (Fig. 16.11), even though some individual species and groups of species respond. However, comparisons between topographic positions indicate that differences between swale and ridgetop communities were greater in ungrazed than in heavily grazed treatments (Fig. 16.3). Segregation of plant communities along the edaphic and microclimatic gradient imposed by landscape topography is expressed in ungrazed treatments, but grazing tends to create a more homogeneous distribution of species across this environmental gradient (Fig. 16.11). Grazers in this particular system also smooth the distributions of soil nutrients and plant biomass across the landscape, but can also create small-scale heterogeneity in species composition by killing long-lived grasses through fecal pat deposition (Peters et al., chapter 6, this volume), by grazing outside of cactus clumps, and by influencing microerosion patterns (see later sections in this chapter). The dual role of grazers in creating both heterogeneity and homogeneity in this system is counter to the common perception of grazers as primarily creating heterogeneity in many other systems. Grazers have often been observed to create heterogeneity by overgrazing particular patches of vegetation (Bakker et al., 1984; Mott, 1987; Vinton et al., 1993), by mediating fire intensity (Hobbs et al., 1991; Vesey-Fitzgerald, 1972), by regulating populations of other consumers

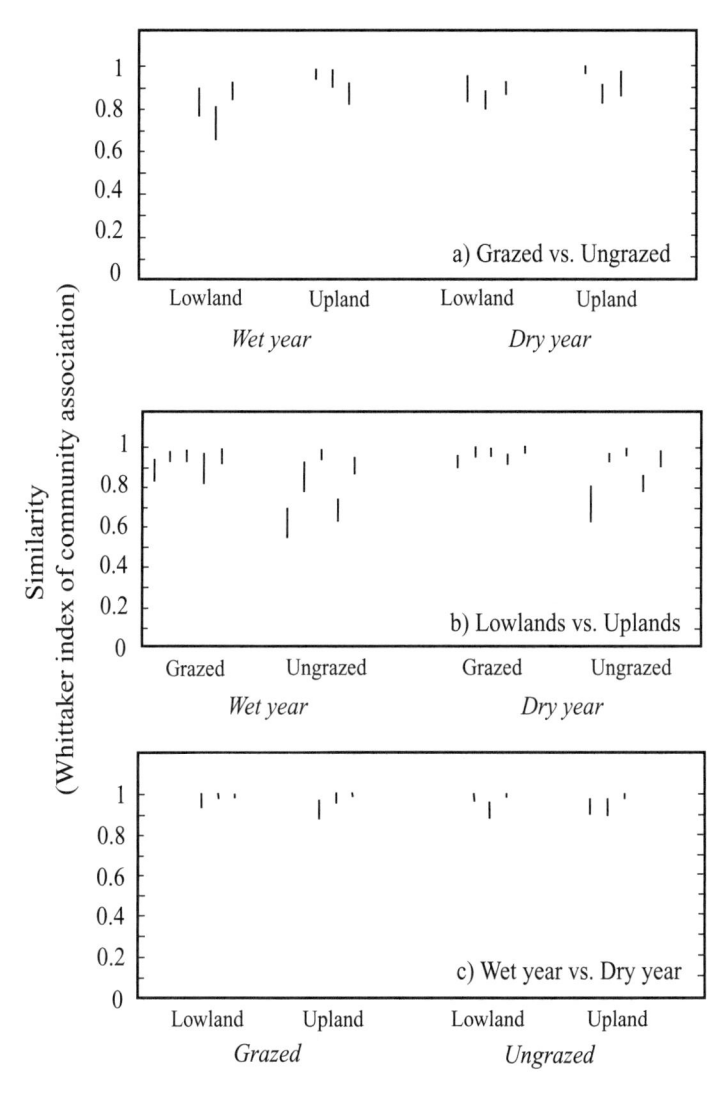

Figure 16.11 Plant community species similarity of heavily grazed compared with ungrazed communities within year and topographic position (A), swales (i.e., Lowland) compared with ridgetops (i.e., Upland) within grazing treatment and year (B), and a wet (1984) compared with a dry (1986) year within grazing treatment and topographic position (C). The long-term treatments were initiated in 1939. Similarity was calculated using the Whittaker (1952) index of community association. A value of one means all species are found in common to both communities and occur in the same proportional abundances in the two communities that are contrasted. A value of zero indicates that there are no species common to both communities being contrasted. Values represent confidence interval range based upon bootstrap method for each replicate–site comparison. (From Milchunas et al. [1989].)

that affect vegetation (Noy-Meir, 1988), and by promoting redistribution of soil and soil nutrients (Schlesinger et al., 1990).

The relatively small differences in plant community similarities with grazing in the shortgrass steppe can only be appreciated by comparing them with responses in other systems. Species similarity between grazing intensity treatments in the shortgrass steppe are consistently greater than averages for other grasslands in North and South America, Australia, Asia, Europe, and Africa (Fig. 16.12). Changes in species similarities with grazing are most strongly influenced by increasing productivity of the vegetation (Milchunas and Lauenroth,

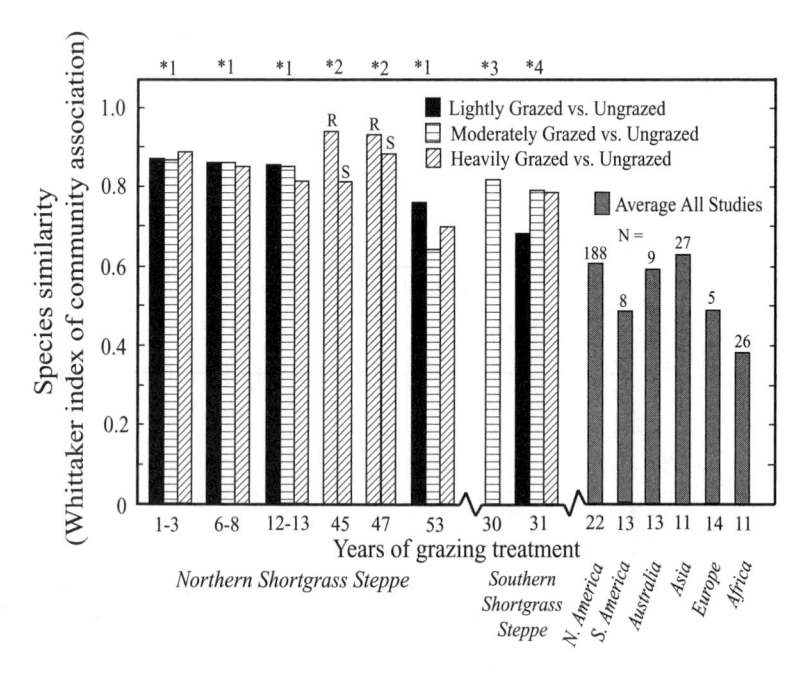

Figure 16.12 Plant community species similarities (Whittaker [1952] index of community association) for northern and southern shortgrass steppe sites, and the years and intensities of grazing treatments compared with average similarities between ungrazed and grazed sites for six other continents. Data for northern shortgrass steppe site are from Klipple and Costello (1960), Milchunas et al. (1989), and Ashby et al. (1993) (all from same treatments at CPER, Colorado); for the southern shortgrass steppe site are from Grant (1971) and Sims et al. (1978), reported in Milchunas and Lauenroth (1993) (Pantex site near Amarillo, Texas) and Vokhiwa (1994) (Comanche Grasslands near Springfield, Colorado); and for the six continents from studies and data reported in Milchunas and Lauenroth (1993). N, number of site comparisons from studies cited in Appendix I of Milchunas and Lauenroth (1993); R, ridgetop community; S, swale community; *1, quadrat cover; *2, density; *3, biomass; *4, line–transect–point cover. See legend for Figure 16.11 for explanation of similarity index.

1993; Milchunas et al., 1988), which is consistent with the small changes observed in the low-production shortgrass steppe. Productive plant communities with long evolutionary histories of grazing often display the capacity to shift species composition with changes in grazing pressure (Milchunas and Lauenroth, 1993). The relatively small response of shortgrass steppe communities with a long evolutionary history of grazing may be the result of convergent selection pressures of long-term grazing and semiaridity through evolutionary time. Furthermore, the dominant species increases in abundance with grazing in the shortgrass steppe.

Diversity and Dominance

The relative and absolute increase of the dominant plant species (*B. gracilis*) with grazing, as well as the lack of influence of grazing on the abundant *Opuntia polyacantha*, have a large influence on plant species diversity. In other systems, dominant species most often decline with increasing intensities of grazing (Milchunas and Lauenroth, 1993), and the release in competitive dominance allows for the coexistence of a greater diversity of other species (Milchunas et al., 1988). Concomitant with the increase in *B. gracilis* with grazing in the shortgrass steppe is a decrease in plant diversity in swales but not on ridgetops (Table 16.6). Within a pasture, swales are grazed more intensively than ridgetops. However, richness is greater in both swales and ridgetops of ungrazed compared with grazed communities. The slight increase in *B. gracilis* with grazing on upland communities may be countered by the role *O. polyacantha* plays in maintaining diversity in grazed grassland. This cactus is more abundant in ridgetop than

Table 16.6 Diversity, Richness, Evenness, and Dominance Indices for Long-Term Heavily Grazed and Ungrazed Swales and Ridgetops in Wet and Dry Years for the Northern Shortgrass Steppe

Community Attribute	Year	Grazed Swale $\bar{X}$	SD	Grazed Ridgetop $\bar{X}$	SD	Ungrazed Swale $\bar{X}$	SD	Ungrazed Ridgetop $\bar{X}$	SD
Diversity	Wet	1.33	(0.06)	1.31	(0.08)	1.50	(0.10)	1.31	(0.04)
	Dry	1.26	(0.03)	1.27	(0.05)	1.46	(0.07)	1.28	(0.03)
Richness	Wet	18.7	(5.0)	18.0	(2.7)	25.7	(4.9)	27.0	(1.7)
	Dry	14.7	(8.3)	15.0	(3.6)	18.0	(1.7)	19.7	(4.9)
Evenness	Wet	0.10	(0.02)	0.09	(0.02)	0.13	(0.02)	0.08	(0.01)
	Dry	0.09	(0.01)	0.09	(0.01)	0.11	(0.02)	0.08	(0.01)
Dominance	Wet	0.88	(0.03)	0.89	(0.05)	0.83	(0.04)	0.90	(0.02)
	Dry	0.90	(0.01)	0.90	(0.03)	0.84	(0.02)	0.91	(0.02)

Values based upon density data. Diversity calculated as Shannon-Weaver (1949) H'. Richness calculated as the total number of species sampled using 48 0.25·m² quadrats per each treatment topographic year, and SD based on three replicates each treatment, location, year. Evenness calculated using Pielou (1966), and dominance using Simpson (1949). (Adapted from Milchunas et al. [1989].)

in swale communities (Table 16.5), and individuals act as microrefugia from grazing (Bayless et al., 1996; Rebollo et al., 2002, 2005). Cattle avoid grazing in the thorny clumps, creating patches of vegetation in heavily and moderately grazed treatments that may be more diverse than outside the clumps (Bayless et al., 1996; Rebollo et al., 2002) (Fig. 16.3C). Cactus protection in long-term moderately grazed treatments has positive effects on some groups of species (Table 16.7). The refuge effect of cactus does not translate into greater plant richness or communities more similar to those that are ungrazed. Most of the effects of cactus are the result of indirect effects of grazing such as litter cover, rather than direct effects of protection from defoliation. Grazing would likely have a greater effect on the plant community if cacti did not create a mosaic of ungrazed patches across the landscape.

The unusual response of the shortgrass steppe to grazing (increases in basal cover of total vegetation and of the dominant species, and declines in opportunistic weedy species common in a variety of other disturbed communities) raised questions for shortgrass steppe scientists concerning whether grazing was a disturbance in this system. The predation hypothesis (Paine, 1966, 1971), the intermediate disturbance hypothesis (Fox, 1979; Grime, 1973), and the Huston hypothesis (Huston, 1979, 1985) predict low diversity in undisturbed communities that are dominated by a few superior competitors, highest diversity at intermediate levels of disturbance resulting from suppression of competitive dominants, and low diversity at high levels of disturbance where only a few species are adapted to the harsh conditions. However, in the shortgrass steppe, diversity does not show the predicted bell-shaped (or humpbacked) relationship predicted with increasing intensity of grazing. Diversity (exp. H') calculated from the frequency data of Ashby et al. (1993) (Table 16.5) shows values of 2.5, 2.4, 2.0, and 2.2 for the ungrazed, lightly, moderately, and heavily grazed long-term treatments at the CPER, respectively. Hart (2001a) calculated H' values of 0.71, 0.76, 0.84, and 0.73, respectively, based on biomass at peak standing crop, with values heavily weighted by differences in the large biomass of cactus compared with herbaceous material. However, diversity (exp. H') calculated from frequency data collected the first year of grazing treatment (Klipple and Costello, 1960), before grazing effects could be manifested, are 3.2, 3.3, 3.2, and 3.2, respectively. Similar values of 3.0, 3.2, 3.1, and 3.1, respectively, are obtained from data collected 12 to 13 years after initiating the grazing treatments. In general, the diversity response across grazing intensities appears flat and generally unresponsive through time.

It would be reasonable to hypothesize that grazing may not be a disturbance in a system with a long evolutionary history of grazing. However, both the intermediate disturbance and Huston hypotheses were able to predict grazing–diversity relationships in African grasslands, which also have a long evolutionary history of grazing (Milchunas et al., 1988). Evolutionary history alone does not appear to be a good predictor of grazing–diversity relationships. We proposed a model to explain the very different responses in diversity to increasing grazing intensity, based upon interactions with precipitation or productivity, and evolutionary history of grazing (Fig. 16.13)

Table 16.7 Mean Canopy Cover (percent) of Plant Functional Groups inside and outside Cactus Clumps in Ungrazed and Moderately Grazed Treatments since 1939, and Significant Positive and/or Negative Effects of Grazing and Cactus Presence

	Ungrazed In Cactus	Ungrazed Out Cactus	Moderately Grazed In Cactus	Moderately Grazed Out Cactus	Grazing Effects In Cactus	Grazing Effects Out Cactus	Cactus Effects Ungrazed	Cactus Effects Grazed
Grasses	15.28	14.92	10.81	10.11		(−)		
Forbs	4.36	5.04	4.39	3.54		(−)		
Shrubs	3.31	5.91	1.35	2.38	(−)	−	−	
Barrel cacti	0.08	0.12	0.06	0.03		−		
Annuals	0.48	0.97	0.69	1.04			(−)	
Perennials	22.54	25.03	15.92	15.02	−	−	(−)	
Cool season	18.16	19.69	9.70	7.33	−	−		(+)
Warm season	4.79	6.19	6.85	8.70		(+)		
Cool-season annual grasses	0.00	0.00	0.006	0.05		(+)		
Cool-season annual forbs	0.19	0.30	0.11	0.24				−
Cool-season perennial grasses	11.86	11.15	5.51	3.44	−	−		+
Cool-season perennial forbs	2.99	2.87	2.91	1.78		−		+
Warm-season annual forbs	0.29	0.67	0.57	0.74	(+)			
Warm-season perennial grasses	3.42	3.77	5.29	6.61		(+)		
Warm-season perennial forbs	3.42	3.77	5.29	6.61				
Cool-season shrubs	3.12	5.36	1.15	1.80	(−)	−	−	
Warm-season shrubs	0.19	0.55	0.20	0.57				
Exotics	0.18	0.34	0.14	0.10	−	(−)		
Weeds	6.77	10.65	7.62	7.19	(−)	−		
Species without asexual reproduction	18.79	21.33	14.33	11.94	−	−	(−)	(+)
Selected for by cattle	13.02	12.53	6.25	4.71	−	−		
Not selected for by cattle	9.40	12.67	9.47	10.64			−	
Increasers	0.16	0.04	0.20	0.37	+			−
Decreasers	13.02	14.80	4.31	3.15	−	−	(−)	
Indifferents	9.80	10.97	12.12	12.47				

Positive or negative signs within a bracket represent $P < .1$, and those unbracketed represent $P < .05$. (Adapted from Rebollo et al. [2002].).

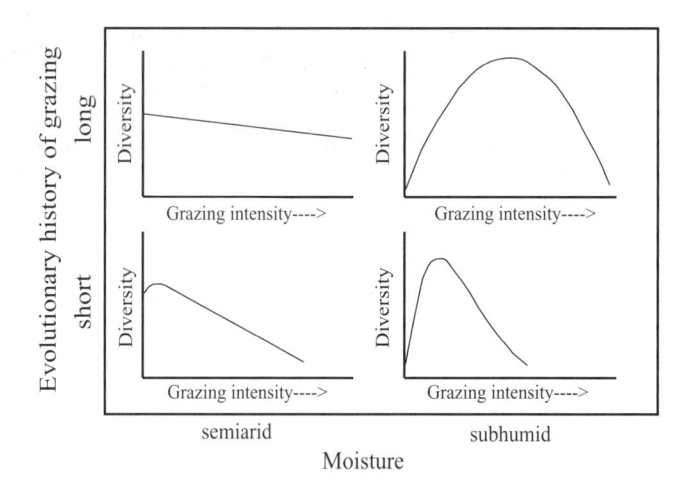

Figure 16.13 Theoretical plant species diversity in relation to grazing intensity for communities in semiarid to subhumid moisture environments and with short and long evolutionary histories of grazing. (From Milchunas et al. [1988].)

Along a precipitation–productivity gradient, competition changes from primarily belowground, for soil water, to increasingly greater competition for light in the aboveground canopy. Short-stature grasslands with large belowground allocation develop in semiarid climates, whereas taller growth forms with greater aboveground allocation develop in grasslands in subhumid environments. Because tall growth forms are generally more susceptible to grazing (Díaz et al., 2007; Milchunas and Lauenroth, 1993), adaptations to grazing and competition for light in a well-developed canopy are divergent selection forces (Milchunas et al., 1988). Short growth forms with relatively greater allocation belowground are adaptations to both herbivory and drought (i.e., convergent selection forces).

The evolutionary history of grazing axis represents the time that plants have been exposed to grazing pressures and, thus, the potential for resistance and avoidance mechanisms to have evolved. In subhumid environments, divergent selection forces result in communities with species adapted for either withstanding grazing pressure or competing in a canopy. We expect rapid switches in community composition with changes in grazing pressure (see examples in Milchunas et al., 1988). The shortgrass steppe evolved under convergent selection forces, where removal or imposition of grazing pressure alone is not expected to produce large effects on community composition. Differences between systems with short or long evolutionary histories of grazing within the same precipitation–productivity range result from different capacities to survive and regrow after defoliation. This results in differences in survivorship and the potential for invasion by opportunistic weedy species. There are very few systems such as the shortgrass steppe that have developed under both semiaridity and a long evolutionary history of grazing (Milchunas and Lauenroth, 1993).

Invasibility and Comparison with Other Disturbed Communities

The greater abundances of weedy species associated with disturbed areas in ungrazed rather than grazed treatments, and the lack of agreement with the intermediate disturbance hypothesis in diversity responses with increasing grazing intensities, raised the question for shortgrass scientists of whether grazing should be considered a disturbance in the shortgrass steppe with its long evolutionary history of grazing. We contrasted communities of both the long-term heavily grazed and the ungrazed treatments with a variety of other disturbances at the CPER (Milchunas et al., 1990). The other disturbances included areas killed by an outbreak of white grubs (Peters et al., chapter 6, this volume) and nutrient enrichment stress plots subjected to large amounts of water, nitrogen, and water-plus-nitrogen. When the ungrazed versus disturbed community comparisons were plotted against the grazed versus disturbed community comparisons, data points fell above the line of equality, in the direction of the ungrazed versus disturbed community comparisons (Fig. 16.14). This indicated that ungrazed communities

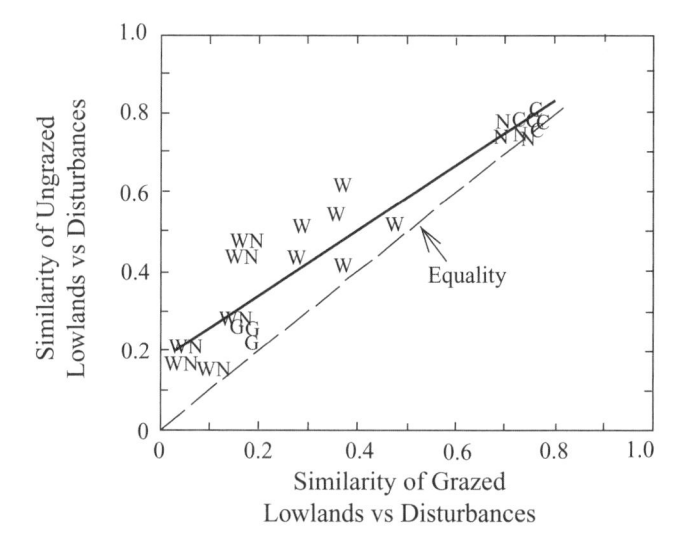

Figure 16.14 Plant community similarity of long-term ungrazed treatments compared with a variety of other disturbed northern shortgrass steppe communities plotted against similarity of long-term heavily grazed treatments compared with the same variety of disturbed communities. Points falling above the line of equality (1:1 relationship) indicate the ungrazed communities are more similar to the disturbed communities than the grazed communities. The disturbed communities were W+N, water plus nitrogen; W, water; and N, nitrogen enrichment applied as stress treatments; G, white grub-killed areas; and C, control versus other control site. $r^2 = .88$. Data are for swales that are more heavily grazed than ridgetops, calculated from density data, and for 2 years and three replicate sites. See legend for Figure 16.11 for an explanation of the similarity index. (From Milchunas et al. [1990].)

were more similar to the disturbed communities than were grazed communities. The relationship was more pronounced for the more heavily grazed swale communities than for ridgetops. Although not exactly alike, cattle are a close surrogate for bison grazing, whereby the removal of cattle results in plant communities characteristic of disturbed areas.

Although any community can be disturbed by excessive grazing, and grazing by native bison may have occurred in different temporal patterns (Milchunas and Lauenroth, chapter 18, this volume), grazing by domestic livestock appears closer to historical conditions than no grazing at all. Grazing in the shortgrass steppe is similar to the role of fire, flooding, and so forth, in many systems, where these forces are integral, endogenous (*sensu* Margalef, 1968) components of the system. Establishing "natural" fire regimes for forests, where timing of historical fires is recorded in tree rings, is less subjective than in the case of grasslands. Similarly, there are no good surveys or scientific studies of the numbers or migration patterns of the large herds of bison that inhabited the plains for thousands of years, but Hart (2001b) concluded from journals of early travelers that grazing by bison and other ungulates was heavy and frequent on the Great Plains. Nevertheless, establishing nominal, natural grazing regimes for grasslands of the Great Plains would be difficult and subjective.

Responses to grazing discussed thus far suggest that levels of competition and the potential for invasions by exotic species may be greater in ungrazed than grazed communities of the shortgrass steppe. We have examined this indirectly by assessing spatial patterns in root distributions, and directly by seeding plots with five different opportunistic weedy species (Milchunas et al., 1992). We followed seedling establishment and phenological development of individuals in five treatments: (1) long-term heavily grazed, currently grazed during the growing season of study (GG); (2) long-term heavily grazed, currently ungrazed (GU); (3) long-term ungrazed, currently ungrazed (UU); (4) vegetation killed the previous summer with herbicide and structure left intact, currently ungrazed (KU); and (5) vegetation disturbed by blading with a tractor and bare soil hoed, currently ungrazed (DU). This design permitted us to make several comparisons: the direct effects of current-year defoliation compared with the long-term indirect effects of grazing (GG vs. GU), the effects of long-term grazing versus long-term absence of grazing (in the absence of current grazing [i.e., GU vs. UU]); and long-term grazing treatments (GU, UU) compared with initially competition-free plots that were otherwise structurally undisturbed (KU) or disturbed (DU).

Kochia scoparia and *Salsola iberica* established successfully, but emergence of *Sisymbrium altissimum*, *Cirsium arvense*, and *Lepidium densiflorum* was low on all treatments. Of the 500 seeds sown in each quadrat, very few but similar numbers of *K. scoparia* and *S. iberica* seedlings established in the GG and GU treatments, whereas very large numbers of seedlings were found in the DU and KU treatments (Fig. 16.15). Nearly four times as many seedlings of *S. iberica* emerged on the UU treatment than on either GG or GU treatments. Numbers of *K. scoparia* on UU quadrats were even greater than those emerging on KU, suggesting that the microenvironment created by living plants amplified

Effects of Grazing on Vegetation   421

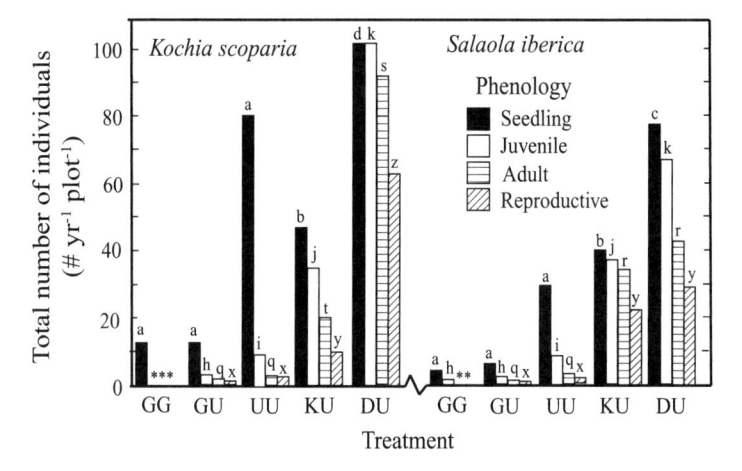

Figure 16.15 Numbers of *Kochia scoparia* and *Salsola iberica* individuals reaching seedling, juvenile, adult, and reproductive phenological stages in treatments of GG, long-term heavily grazed, grazed during the experiment; GU, long-term heavily grazed, ungrazed during the experiment; UU, long-term ungrazed, ungrazed during the experiment; KU, vegetation previously killed by herbicide and left intact, ungrazed during experiment, but previously grazed; DU, vegetation previously disturbed by blading and hoeing, ungrazed during experiment. Means within a species and phenological stage not sharing a common letter are significantly different with respect to treatment. Asterisks indicate groups left out of the analysis because of a large number of zeros. Five hundred seeds were sown to each plot, with one species per plot. Species are exotic weeds commonly found along roadsides and other disturbed sites. (From Milchunas et al. [1992].)

germination. Very young seedlings emerging in favorable moisture conditions of early spring probably do not experience competition from established neighbors unless they are in very close proximity. Microenvironmental conditions for germination were more favorable in the UU treatment than the GG and GU treatments.

The effects of competition from neighbors probably increases as seed stores are depleted, as root systems of seedlings expand, and as drought periods of summer become more intense. The proportions of the populations that were dead by the following month were high in both the long-term grazed treatments, were intermediate in the long-term ungrazed treatment, and were very low in both competition-free treatments (Table 16.8). By the end of the season, no or very few individuals were found on the GG and GU treatments, and few of these had progressed beyond the seedling stage (Fig. 16.15). A significant number of individuals did survive in the UU treatment, and some of these reproduced. Densities the following spring were 0.4, 0.1, 2.6, 25.9, and 428.0 individuals per quadrat for *K. scoparia* in the GG, GU, UU, KU, and DU

Table 16.8 Monthly Proportion of Deaths for June Cohort of *Kochia scoparia* and *Salsola iberica* Individuals on Plots Sown with 500 Seeds per Plot of One of the Species

	\multicolumn{6}{c}{Monthly[a] Proportion of Deaths of June Cohort, % of previous month's population}
	\multicolumn{3}{c}{Kochia scoparia}	\multicolumn{3}{c}{Salsola iberica}
Treatment	Jul	Aug	Sep	Jul	Aug	Sep
GG	88 a	100[c]	—	84 a	91 a	100[c]
GU	71 a	46 a	69 a	28 b	23 b	67 a
UU	43 b	27 b	40 b	26 b	23 b	32 c
KU	12 c	7 c	8 d	6 c	2 c	1 d
DU	13 c	13 d	51 b[b]	12 c	7 c	52 b[b]

[a] From previous month to month indicated.
[b] High value a result of senescence after flowering rather than premature death.
[c] Not included in statistical analyses because of large number of zeros.
Treatments: GG, long-term heavily grazed and grazed during the experiment; GU, long-term heavily grazed but ungrazed during the experiment; UU, long-term ungrazed and ungrazed during the experiment; KU, vegetation previously killed by herbicide but left intact and ungrazed during experiment but previously grazed; DU, vegetation previously disturbed by blading and hoeing and ungrazed during experiment. Means within a date and species not sharing a common letter are significantly different with respect to treatment. (Adapted from Milchunas et al. [1992].).

treatments, respectively; and 0.0, 0.0, 2.4, 32.6, and 27.4 individuals per quadrat, respectively, for *S. iberica*. It was not possible to assess clearly the differences in levels of competition between UU and GU treatments because of large differences in seedling emergence. However, it was clear that more favorable microenvironmental conditions for germination of these weedy species existed in long-term ungrazed than in grazed communities, and that the indirect effects of grazing on establishment were more important than the direct effects of current-year defoliation. A greater number of small-mammal disturbances are often observed in ungrazed compared with grazed shortgrass steppe, and this is sometimes thought to be the reason for greater numbers of weeds in ungrazed areas (Ashby et al., 1993; Hart and Ashby, 1998). The very large number of seed-producing individuals in the plots with soil disturbance illustrates that the patches disturbed by small mammals may play a large role in seed availability to surrounding areas, but the emergence in long-term ungrazed, undisturbed plots implicates additional factors in the greater susceptibility to invasion of ungrazed compared with grazed shortgrass steppe. A prolonged cool, record-wet spring occurred in 1995, and these conditions were highly favorable for emergence of *Bromus tectorum* (annual cheatgrass). Only isolated individuals of this species were previously observed in exclosures, but the large numbers that grew in exclosures in 1995 compared with very few in adjacent grazed grassland substantiate findings from the seed addition experiment that was performed earlier during a year of average precipitation.

Distribution of Biomass

An important characteristic of grassland plants is the large proportion of biomass in near-surface and belowground organs (Fig. 16.16). Crowns and roots are inaccessible to large herbivores such as cattle, and provide a storage reserve for regrowth after defoliation. In general, the ratio of aboveground to belowground biomass

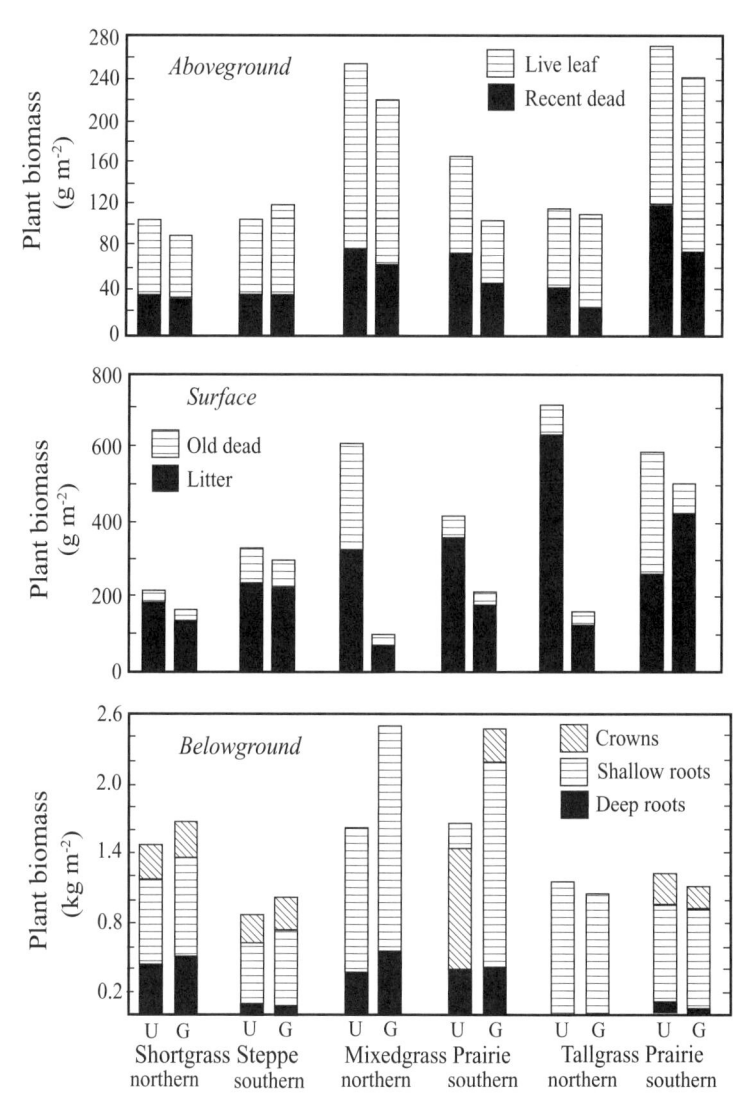

Figure 16.16 Plant biomass in aboveground (live leaf and recent dead), surface (old dead and litter), and belowground (crown, shallow root, and deep root) categories for ungrazed (U) and grazed (G) treatments at northern and southern shortgrass steppe, northern and southern mixed-grass prairie, and tallgrass prairie sites in the North American Great Plains. (Data from Sims and Singh [1978].)

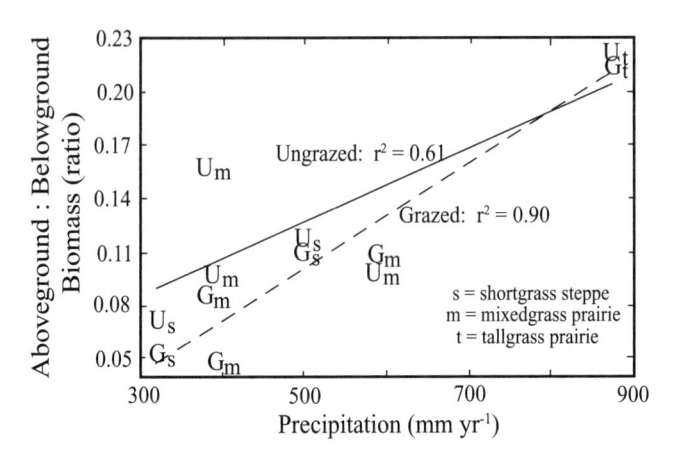

Figure 16.17 Aboveground-to-belowground biomass ratios in relation to precipitation (measured in millimeters per year) for ungrazed and grazed treatments in shortgrass steppe, mixedgrass prairie, and tallgrass prairie communities in the North American Great Plains, with trend lines. G, grazed; m, mixedgrass prairie; s, shortgrass steppe; t, tallgrass prairie; U, ungrazed. Aboveground biomass includes live plus recent dead, and belowground is crowns plus roots. (Data from Sims and Singh [1978].)

increases with increasing precipitation (Fig. 16.17). As this ratio increases, defoliation of the canopy has a greater effect on individual plants because a given percentage of aboveground herbivory removes a greater percentage of the total plant biomass. Furthermore, increasing aboveground plant biomass with increasingly productive communities results in increasing levels of competition for light in the canopy. Therefore, alterations in canopy structure (physiognomy) by defoliation will have a greater impact on plant–plant competitive interactions with increasing productivity (Milchunas et al., 1988). We have synthesized grazing studies from around the world (Milchunas and Lauenroth, 1993) and found increasingly greater changes in plant species composition with grazing, and increasingly greater negative effects of grazing on ANPP with increasing aboveground primary production. Comparing results from eight North American grasslands, Sims and Singh (1978) observed that cooler sites had a greater amount of roots in grazed compared with ungrazed treatments, and that warmer sites had no differences or slight reductions of roots in grazed treatments.

Several studies have examined root biomass in relation to grazing intensity treatments in the shortgrass steppe. Averaged over seasonal sampling for 3 years, Sims et al. (1978) found no effect of moderate grazing on root biomass at a southern shortgrass steppe site in Texas, and a slightly greater amount in moderately grazed than in ungrazed sites at the CPER in northern Colorado (Fig. 16.16). Samples taken in increments to 80 cm in depth through one growing season indicated no differences in depth distribution of roots at the CPER between ungrazed,

Effects of Grazing on Vegetation 425

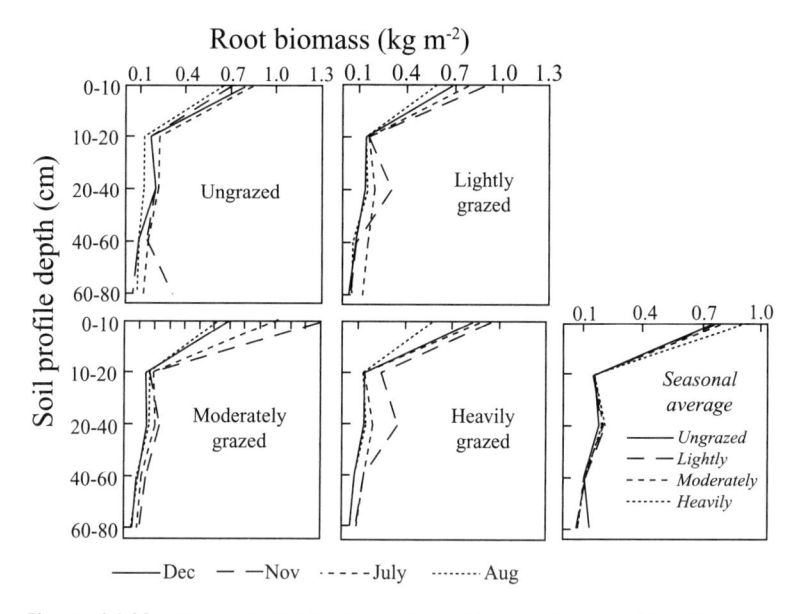

Figure 16.18 Seasonal distributions of root biomass (measured in kilograms per square meter) by depth in the soil profile (measured in centimeters) for long-term ungrazed, lightly, moderately, and heavily grazed treatments, and seasonal averages for the four grazing treatments, in northern shortgrass steppe communities. (Data from Sims et al. [1971].)

lightly, moderately, and heavily grazed treatments, but somewhat greater seasonal variability at higher grazing intensities (Fig. 16.18 [Sims et al., 1971]). Similarly, Leetham and Milchunas (1985) found no difference in the total biomass or in the depth distribution of roots between the same lightly and heavily grazed treatments at the CPER. Milchunas and Lauenroth (1989) observed slight reductions in 0 to 10-cm root biomass in the heavily grazed compared with ungrazed swales and ridgetops at the CPER, and the same effect of grazing on 10 to 20-cm-deep roots only in swales. For both depths, there were greater differences between topographic positions than there were between grazing treatments. Vokhiwa (1994) found no difference in root biomass or depth distributions between ungrazed and moderately grazed treatments at a southern shortgrass steppe site near Springfield, Colorado. These studies together suggest that there is little to no effect of grazing on root biomass in the shortgrass steppe. In contrast, increases in root biomass with grazing were found at northern mixed-grass prairie sites but not at southern mixed-grass and tallgrass prairie sites (Fig. 16.16).

A large effect of grazing on roots was, however, observed in the shortgrass steppe for a distributional characteristic that is seldom examined in root studies. Horizontal spatial distributions of roots and crowns were found to be much more homogeneous in heavily grazed than ungrazed treatments at the CPER (Fig. 16.19). Data from plots that were completely cored in a grid fashion indicated

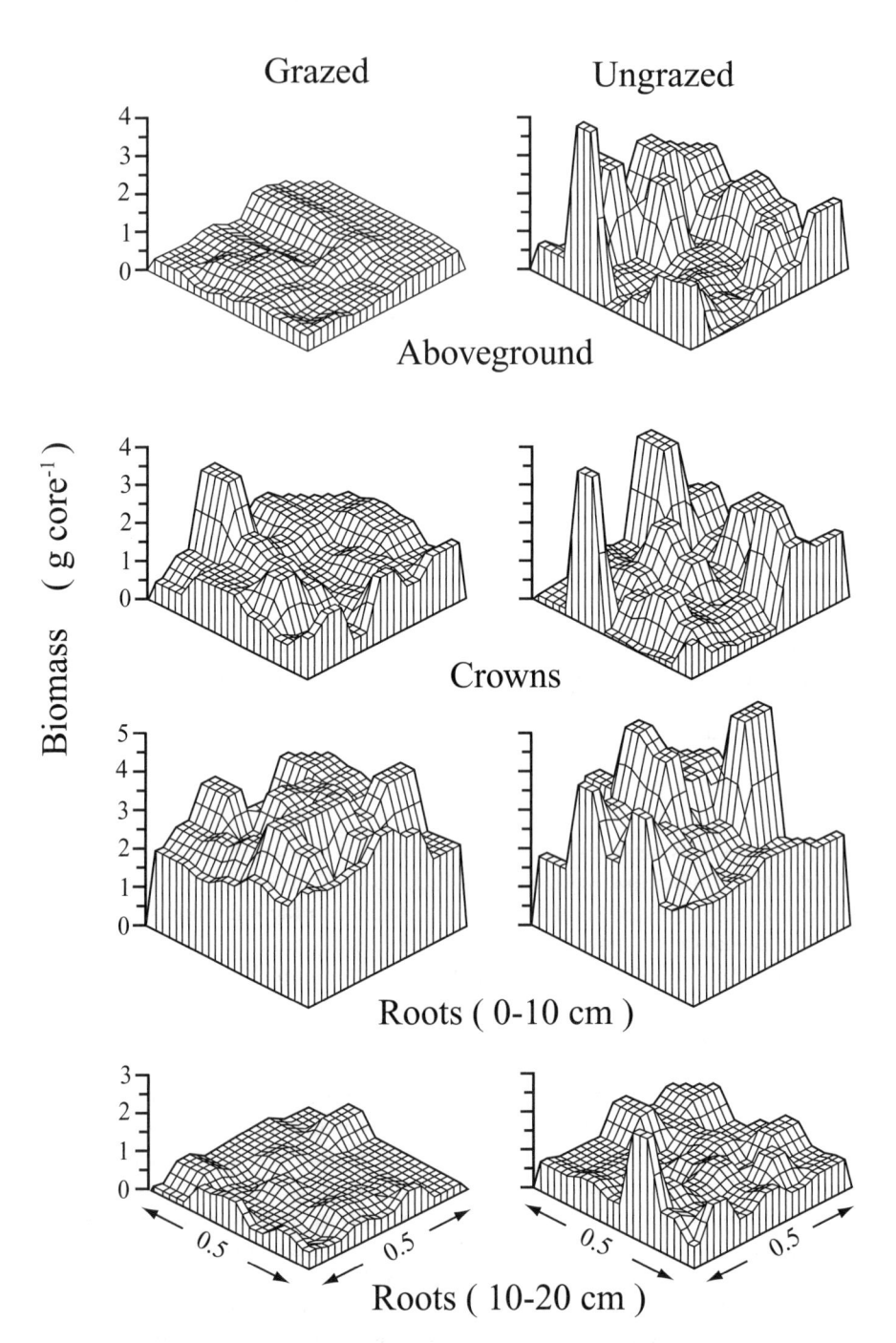

Figure 16.19 Horizontal- and vertical-plane spatial distribution of plant biomass (measured in grams per core) in long-term grazed and ungrazed treatments in the northern shortgrass steppe. Data are for one representative plot in a swale topographic position. (From Milchunas and Lauenroth [1989], and see this chapter for means for all plots.)

that mean absolute differences between adjacent first-neighbor or second-neighbor cores were sometimes as much as two times greater in ungrazed than in heavily grazed grassland (Milchunas and Lauenroth, 1989). The grazing lawn structure (Fig. 16.3C) of aboveground basal cover and crown biomass under grazed conditions translated to a more uniform exploitation of the soil volume than that found in ungrazed sites. This has potential implications for plant–plant interactions. Microsites favorable for the establishment of invasive opportunistic weedy species may be less available in heavily grazed than in ungrazed shortgrass steppe. Very few weed seedlings established and survived within or directly adjacent to existing plants compared with the numbers found in bare ground or litter (Milchunas et al., 1992).

Litter and standing old dead material play important roles in ecosystem function, plant population dynamics, and herbivory. Consumption of plant material, and the return of a portion in the form of urine and feces, reduces the amount of plant material that would otherwise senesce and fall to the surface as litter. Litter affects soil temperature, water dynamics, and the physical structure of the soil surface (Facelli and Pickett. 1991). The large accumulation of litter in ungrazed or unburned highly productive grasslands such as the tallgrass prairie can limit germination and future productivity (Kelting, 1954; Knapp and Seastedt, 1986). The lesser amounts of litter in semiarid regions can still have effects on germination and plant community composition (Milchunas and Lauenroth, 1995; Milchunas et al., 1992), and may play an important role in productivity of grazed versus ungrazed shortgrass steppe (see the next section). Standing dead material from previous-year production not only shades photosynthetically active material, but can also deter grazing by large herbivores as a result of lower quality of forage bites compared with clumps with less dead material. Herbivores often return to areas previously grazed, sometimes leading to patch overgrazing in a heterogeneous mosaic of ungrazed patches in both semiarid (Bridge et al., 1983; Fuls, 1992) and subhumid environments (Bakker et al., 1984; Hunter, 1962; Mott, 1987; Vinton et al., 1993). However, differences between grazed and ungrazed treatments in both litter and standing old dead in the shortgrass steppe are relatively small compared with those in mixed- or tallgrass prairie (Figs. 16.16, 16.20). There is a greater negative effect of grazing on litter biomass with increasing productivity ($r^2 = 0.69$), ranging from –11% in the southern shortgrass steppe to –84% in a North Dakota mixed-grass prairie (Fig. 16.20).

Plant Community Productivity

Estimates of NPP are subject to a variety of biases and errors (Lauenroth et al., 1986), particularly for belowground components (crown and root) (Milchunas and Lauenroth, 1992, 2001; Singh et al., 1984). Because of serious problems inherent in maxima–minima (peak–trough) methods and the lack of isotope or minirhizotron work, the effects of grazing intensities on roots and crowns will only be addressed here in terms of biomass. Comparisons of traditional methods of estimating ANPP with those obtained using ^{14}C turnover methodology indicate that a single harvest at peak standing crop (in temporarily caged or ungrazed plots) is the best estimate of ANPP in shortgrass steppe communities (Milchunas and Lauenroth, 1992, 2001). The short growing season and dominance by warm-season species

428 Ecology of the Shortgrass Steppe

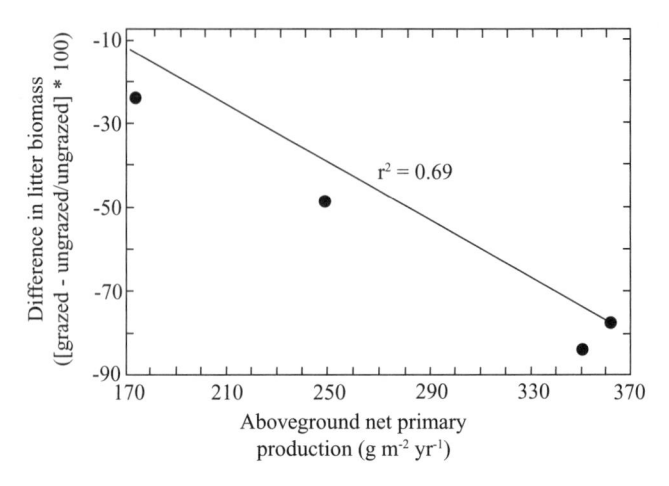

Figure 16.20 Difference in litter biomass between grazed and ungrazed treatments (measured as a percentage) in relation to aboveground net primary production (measured in grams per square meters per year) for shortgrass steppe and mixed-grass prairie sites in the Great Plains of North America. (Data from Appendix II in Sims and Singh [1998].)

contribute to a modest underestimate (approximately 16%) of ANPP using this method in the shortgrass steppe compared with greater underestimates that may be obtained in communities with multiple seasonal peaks in the various components of biomass. Estimates of ANPP in grazed systems have the additional problem that the animals are continually consuming plants as they grow. Estimates obtained by clipping within temporary, year-long-placed cages do not account for the potential positive or negative effects of current-year defoliation (McNaughton et al., 1996), but do account for long-term cumulative effects of grazing treatments. We will first examine ANPP in relation to grazing intensities based upon peak standing crop estimates from temporarily caged plots, and then explicitly address the potential for compensatory regrowth resulting from defoliation of the community.

Long-Term Effects on Primary Productivity and Seed Production

Forage production, excluding cactus and other inedible species, was the most usual method of sampling through the early 1960s. Forage production of the long-term grazing intensity treatments at the CPER spanning a 50-year period averaged 75, 71, 68, and 57 $g \cdot m^{-2} \cdot y^{-1}$ for the ungrazed, lightly, moderately, and heavily grazed treatments (n = 15, 89, 97, and 88), respectively (Fig. 16.1). Multiple regression analysis using the long-term forage production data indicated that productivity was most sensitive to amount of cool-season precipitation, followed by amount of warm-season precipitation, then soil fertility, and, last, grazing intensity (Milchunas et al., 1994). Grazing at 20% to 35% annual removal did not alter production. For pastures of average fertility, grazing at a level of 60% removal was predicted to decrease production 3% in wet years and 12% in dry years. The forage available to consumers across the grazing intensity treatments displayed a similar degree of year-to-year variability (standard deviations). This is not consistent with Le Houerou's (1988) prediction of increasing variability in productivity with increasing grazing intensity. The lack of a grazing treatment-by-precipitation interaction in the regression analyses further suggests that there may not be a decreasing capacity for response to favorable years with increasing grazing intensity.

In a study conducted 50 years after initiating the grazing treatments, plots watered to simulate a very wet year compared with unsupplemented plots in a year of average precipitation showed greater differences in ANPP in ungrazed than in lightly or heavily grazed treatments (Varnamkhasti et al., 1995). The long-term ungrazed treatment was more productive than the heavily grazed treatment in the simulated wet year, but the opposite was true under the drier conditions. We conducted this study in swale communities, and mechanically applied defoliation treatments to simulate grazing. Possibly because of these factors and the seasonal distribution of the applied water, the results from this study and the long-term data set analyses do not correspond. However, differences in the response of grazing treatments to large precipitation events versus the usual small events characteristic of drier years suggest a mediating role of litter cover on utilization of precipitation. Higher litter cover in ungrazed treatments may act as an insulator to evaporative loss after large, penetrating events, but may intercept much of the moisture that falls during small precipitation events.

Unusual abiotic conditions appear to alter productivity responses of the long-term grazing treatments. This is apparent not only in response to additions of very large precipitation events, but possibly also during recovery from drought. A drought in 1954 resulted in precipitation 62% below average for years in which productivity was monitored on the grazing treatments. Precipitation was +3%,–24%, +30%, and +4% for years 1 through 4 postdrought, respectively. Precipitation use efficiency is a convenient term for comparing postdrought years because it normalizes across different amounts of precipitation by placing productivity on a per-unit-of-precipitation basis. Precipitation use efficiency for 4 years after the drought was reduced in all treatments compared with that for other years of similar precipitation (Fig. 16.21). However, the precipitation use efficiencies were 60% of comparable years in lightly grazed compared with only 41% in moderately and heavily grazed treatments the first year after the drought. Values 4 years after the drought were 75%, 62%, and 52% of those in similar years for the lightly, moderately, and heavily grazed treatments, respectively, and returned to the usual pattern by the fifth year after the drought. Although these data represent only one case, the heavily grazed treatment appeared to be most negatively affected by extreme drought. However, precipitation use efficiencies the second year postdrought were greater in the moderately and heavily treatments than in the lightly grazed treatment, when precipitation was 24% below average.

Estimates of ANPP for grazing treatments in the southern shortgrass steppe are limited. Peak standing crops were 59 and 66 g·m^{-2} in the ungrazed and moderately grazed treatments, respectively, at a Texas site (Lewis, 1971). This compared with 83 and 56 g·m^{-2} for the respective ungrazed and heavily grazed

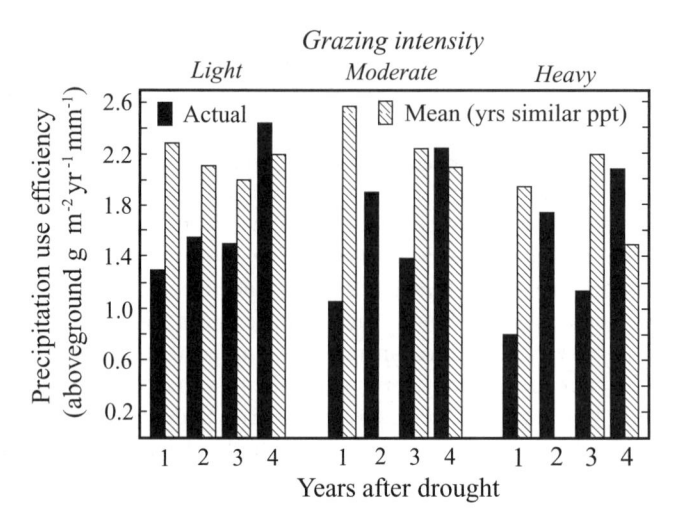

Figure 16.21 Precipitation use efficiency for aboveground plant production (aboveground grams per square meter per year per millimeter precipitation) in relation to long-term light, moderate, and heavy grazing treatments for 1, 2, 3, and 4 years after the severe drought of 1954, and for years of similar precipitation (ppt) to those after the drought. (Data from Milchunas et al. [1994].)

treatments at a northern site. Three years of average standing crop data tend to support differences in the response to grazing of southern versus northern shortgrass sites (Fig. 16.16).

Aboveground NPP by functional groups for three periods through the 55-year history of the grazing intensity treatments at the CPER show that weather has a larger effect on ANPP and its botanical composition than grazing. Forbs, shrubs, and cool-season grasses were generally a larger proportion of ANPP, and total ANPP was greater in all grazing treatments in the early 1940s than in the 1950s or 1990s (Fig. 16.9) (Ashby, unpublished data; Ashby et al., 1993; Hart and Ashby, 1998). Cool-season precipitation was lower during the latter two periods, although the total annual amount was greater. Cool-season precipitation has a greater effect on productivity than warm-season precipitation, and particularly on the productivity of forbs and cool-season grasses, both of which are sensitive to grazing intensity. We found greater differences in ANPP between ungrazed and heavily grazed treatments in swales than in upland communities (Milchunas et al., 1992).

Grazing also has an effect on the spatial distribution of *B. gracilis* seed production. Soil texture has a significant influence on seed production of *B. gracilis* in ungrazed shortgrass steppe (greatest on sandy soils), but not in grazed sites (Coffin and Lauenroth, 1992). Ungrazed sites have been found to have twice as much viable seed production per flowering culm, and greater seed production per area as well. In contrast, Rebollo et al. (2002) found greater overall seedhead production in moderately grazed compared with ungrazed treatments. This was primarily because of the increase in seedhead production of *B. gracilis* inside cactus clumps in the moderately grazed treatment. Grazing also has positive effects on the number of inflorescents per unit of cover for shrubs and cool-season annual grasses and forbs, and negative effects on cool-season perennial grasses.

Compensatory Regrowth: Interactions with Grazing History

McNaughton (1979) found greater productivity where plants were defoliated during the current-year growing season than where they were not. Williamson et al. (1989) grazed grasshoppers for short periods, caged at different densities, in level upland communities that had not been grazed by cattle for 10 to 11 years. When grazing season precipitation was higher than average, current-year defoliation by grasshoppers increased ANPP in only a few cases. However, when grazing season precipitation was lower than average, increases in ANPP were found across a range of grasshopper grazing intensities. These authors concluded that compensatory regrowth was most likely to occur during short-term recovery from dry periods. In a study not designed to assess compensatory regrowth, Milchunas et al. (1992) observed that end-of-season biomass was the same in long-term heavily grazed treatments regardless of whether they were grazed during that particular year. This suggested that compensatory regrowth in grazed areas was as great as the amount that had been removed by the cattle.

The study by Williamson et al. (1989) and the observation by Milchunas et al. (1992) led to a more detailed study of interactions among grazing history, precipitation, and current-year-defoliation in compensatory regrowth responses. There are two methods to assess the effects of current-year grazing by large herbivores on ANPP (McNaughton et al., 1996) that mimic the actual pattern and frequency of grazing by the animals: (1) moving cages through the growing season and clipping caged and grazed plots, and (2) observing the pattern of bites taken from grazed reference plots and then clipping caged plots in the same manner. The latter method was developed and applied in long-term heavily and lightly grazed areas to which either supplemental water was added to simulate a very wet year or no water was added in a year of average precipitation (Varnamkhasti et al., 1995). Water treatment had a greater overall effect on ANPP than did either long-term grazing history or current-year defoliation, although all three factors interactively determined ANPP (Fig. 16.22). Slightly higher ANPP and precipitation use efficiency of the unwatered, long-term lightly grazed treatment with defoliation resulted in different relationships between grazing and watering treatments. These data and the results of Williamson et al. (1989) suggest that defoliation may increase ANPP during periods of water stress, but that some of the potential for regrowth may diminish with heavy grazing. Other studies have reported conservation of soil water or improved plant–water relations in response to grazing or defoliation (Archer and Detling, 1986; Baker and Hunt, 1961; Coughenhour et al., 1990; Day and Detling, 1994; Eck et al., 1975; Hodgkinson, 1976; McNaughton et al., 1983; Reed and Dwyer, 1971; Simoes and Baruch, 1991), although only a few observed increased ANPP. Young leaves can have a lower transpiration loss per unit of growth than older leaves removed by defoliation.

432 Ecology of the Shortgrass Steppe

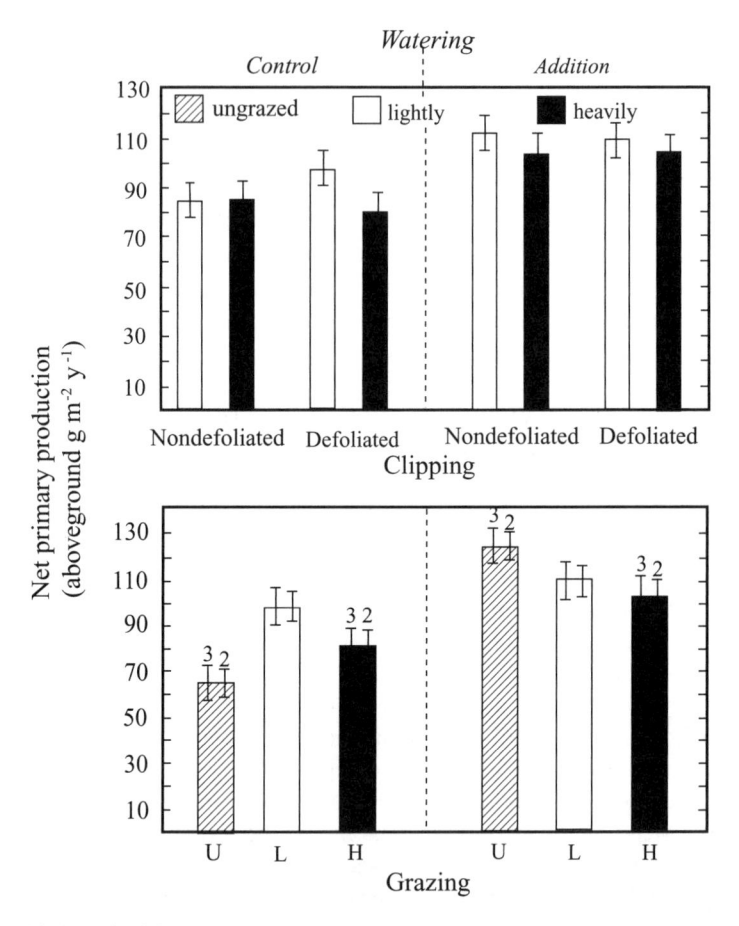

Figure 16.22 Aboveground net primary production (measured in grams per square meter per year) in long-term lightly and heavily grazed, clipped or not clipped, and control or watered treatments (A); and ungrazed, and lightly and heavily grazed treatments (with clipping), and control or watered treatments (B). Clipping treatments (D, defoliated; N, nondefoliated) simulated natural patterns and intensities observed in adjacent uncaged reference plots. Watering treatments were an unwatered control during a year of average precipitation and a watered treatment equivalent to very wet years in the northern shortgrass steppe. Confidence intervals (Tukey's Honest Significant Difference) are for differences between means of any one type of treatment, holding the level of the other treatments constant. H, heavily grazed; L, lightly grazed; U, ungrazed. (From Varnamkhasti et al. [1995].)

Small effects of defoliation on compensatory regrowth were observed in the previously mentioned studies at the CPER. At the same research site but in more productive sandy soils, very large differences between mid growing season-defoliated and -nondefoliated treatments were observed in two separate studies. In a study of the effects of elevated levels of atmospheric CO_2, Morgan et al. (2001)

found an ANPP of 78 g·m^{-2}·y^{-1} on nondefoliated compared with 134 g·m^{-2}·y^{-1} on defoliated treatments, and no interaction with CO$_2$ treatment. Milchunas et al. (2005) observed 38 versus 50 g·m^{-2}·y^{-1} of digestible forage production for nondefoliated and defoliated treatments averaged over 4 years and all CO$_2$ treatments, and no interaction with CO$_2$ treatment. This site had been lightly to moderately grazed in the past. Differences between the responses observed in this study and the smaller compensatory responses discussed earlier could be the result of the greater ANPP at the CO$_2$ study site or because there was a single defoliation during the season of entire (but small and alternating) quadrats. Similar conditions of greater ANPP than at sites reported earlier, previously long-term moderately grazed, and clipping of entire quadrats once at mid growing season were used in a study of ultraviolet radiation and defoliation effects on ANPP, and a very wet and a very dry year were simulated (Milchunas et al., 2004). Aboveground NPP of nondefoliated treatments averaged 170 g·m^{-2}·y^{-1} compared with 290 g·m^{-2}·y^{-1} for defoliated treatments in the very wet year, and there was no defoliation effect for the dry year treatment when ANPP averaged 50 g·m^{-2}·y^{-1}. The interaction of current-year defoliation with level of precipitation differed in this study compared with the studies reported earlier. This may be the result of the severe drought imposed in the ultraviolet study or any of the other differences among the studies, but suggests that the potential for compensatory regrowth after current-year defoliation may be affected by many factors and is not easily defined.

Forage Quality and Consumption

Conclusions concerning forage quality in relation to grazing intensity treatments in the northern shortgrass steppe differ depending upon the study and/or the methods of collecting samples. Uresk and Sims (1975) found no difference in nitrogen concentrations of live or dead *B. gracilis* hand-plucked from long-term ungrazed, lightly, moderately, and heavily grazed treatments for 11 sample dates in each of 2 years. Similarly, Milchunas (unpublished data) observed no differences in nitrogen concentrations of *B. gracilis*, *S. coccinea*, or *G. sarothrea* (grass, forb, and shrub, respectively) hand-plucked from moderately grazed and 13- to 17-year-old exclosures, averaged over 5 years and 17 sampling dates. However, differences between the grazed and the ungrazed treatment were sometimes positive and sometimes negative, depending upon the sample date and year.

In a more controlled experiment, Milchunas et al. (1995) assessed nitrogen content and digestibility of vegetation from entire plots that were subjected to defoliation and/or water supplementation in long-term ungrazed, lightly, and heavily grazed treatments. Effects of grazing on nitrogen concentration and on percent in vitro digestible dry matter (Fig. 16.23) were similar to those observed for productivities (Fig. 16.22), with the effects being more pronounced on quality than on quantity of vegetation. Multiplying the biomass productivities by the concentrations of nitrogen and digestible dry matter to estimate nitrogen and digestible forage yield accentuated differences between the treatments. We observed large increases in nitrogen yield with water addition in the ungrazed treatment, but not in the lightly grazed treatment (Fig. 16.24B), probably because of the already large stimulation in nitrogen yield with defoliation (Milchunas et al., 1995). Similar

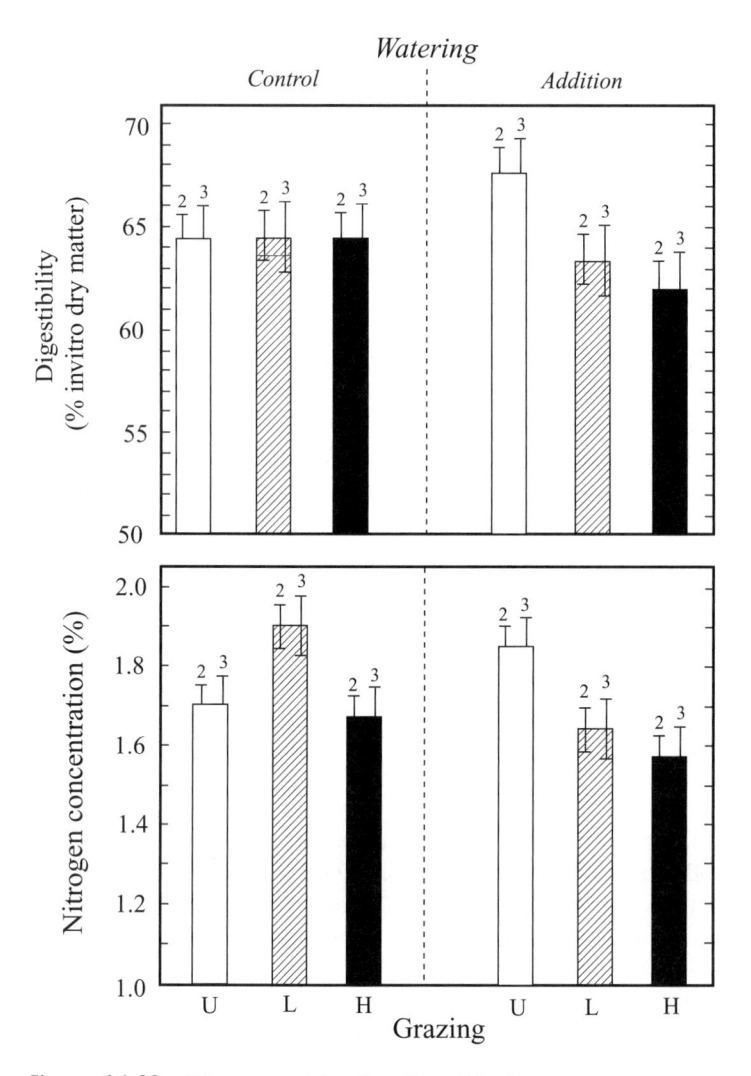

Figure 16.23 Nitrogen and in vitro digestible dry matter concentrations (measured as a percentage) of vegetation in ungrazed, lightly, and heavily grazed treatments, with or without water addition. The grazed treatments were clipped to simulate natural patterns and intensities observed in adjacent uncaged reference plots. Watering treatments were an unwatered control during a year of average precipitation and a watered treatment equivalent to very wet years in the northern shortgrass steppe. Confidence intervals (Tukey's Honest Significant Difference) are for differences between means of any one type of treatment holding the level of the other treatments constant. H, heavily grazed; L, lightly grazed; U, ungrazed. (From Milchunas et al. [1995].)

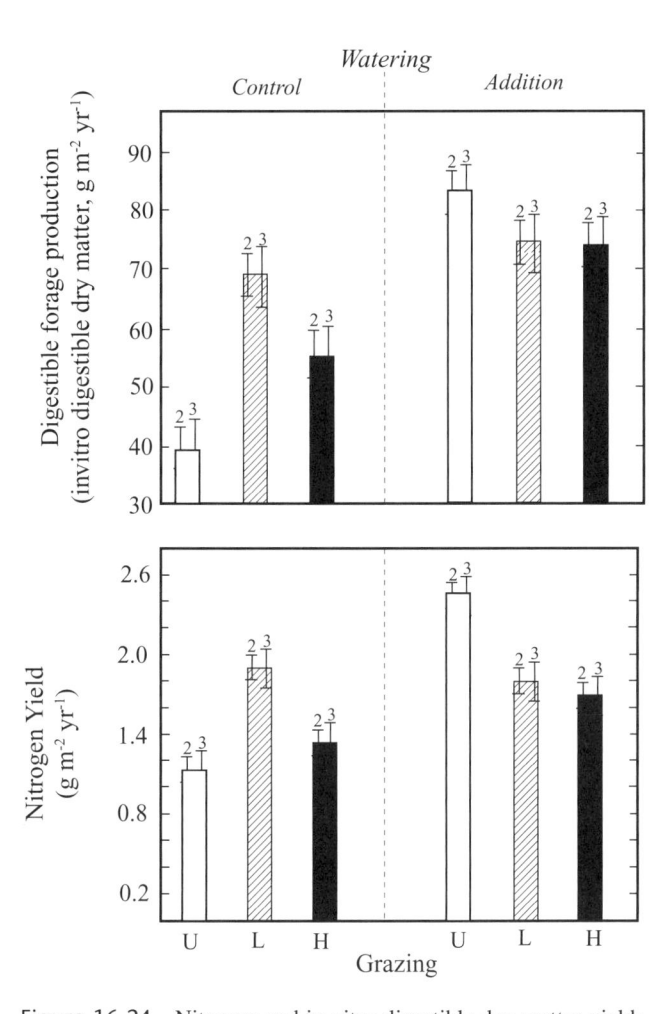

Figure 16.24 Nitrogen and in vitro digestible dry matter yields (measured in grams per square meter per year) of vegetation in ungrazed, lightly, and heavily grazed treatments, with or without water addition. The grazed treatments were clipped to simulate natural patterns and intensities observed in adjacent uncaged reference plots. Watering treatments were an unwatered control during a year of average precipitation and a watered treatment equivalent to very wet years in the northern shortgrass steppe. Confidence intervals (Tukey's Honest Significant Difference) are for differences between means of any one type of treatment holding the level of the other treatments constant. H, heavily grazed; L, lightly grazed; U, ungrazed. (From Milchunas et al. [1995].)

relationships between the ungrazed, lightly, and heavily grazed treatments were found for digestible forage production (Fig. 16.24A), but the effects of defoliation on digestible forage production were less than for nitrogen yield.

Because of the importance of forage quality to consumers, ANPP may not be the most important indicator for assessing the effects of grazing. For instance, cattle weight gain on the heavily grazed treatment is more negatively affected than ANPP (Hart, chapter 4, this volume), and pronghorn are able to maintain similar crude protein levels in their diets throughout the year when grazing the lightly or heavily grazed pastures, even though the quantity of forage was lower on the heavily grazed treatment (Schwartz et al., 1977). Increased forage quality with current-year defoliation may partially compensate for declining productivity and biomass availability, but the capacity for this to occur is diminished with long-term heavy grazing intensities. The lightly grazed pastures provided more stable year-to-year yields of forage quality than either the ungrazed or heavily grazed treatments, and provided greater quality yields in the year of average precipitation.

Comparison of Response with That of Other Ecosystems

We found that the average reduction in ANPP with grazing for all shortgrass steppe studies and treatments (Milchunas and Lauenroth, 1993) was–10%, at an average grazing intensity of 44% (n=10). For similar grazing intensities, northern and southern mixed-grass prairie sites display average reductions in ANPP of–22% and–23%, respectively. This agrees with the statistical model derived from a worldwide data set that shows increasingly greater negative effects of grazing on ANPP with increasing productivity of a site. The model by Milchunas and Lauenroth (1993) suggests that declines in ANPP with grazing are more than twice as sensitive to increasing site productivity than they are to increasing intensities of grazing. An anomaly is the potential for grazing to increase productivity in very productive grasslands where litter buildup can inhibit germination or emergence of vegetation. For instance, tallgrass sites increased 18% with grazing of unburned sites (average grazing intensity, 36%; n=4).

Similar to the relatively small ANPP response, grazing may be expected to have relatively less impact on forage quality in the shortgrass steppe compared with other more productive grasslands (Milchunas et al., 1995). We base this conclusion on the relatively minor impact of grazing on community physiognomy and on species composition in this system. Increases in less palatable species with grazing can decrease overall forage quality (Cook and Harris, 1968; Pieper et al., 1959; Westoby, 1985, 1986), but this, as well as high stem-to-leaf and dead-to-live ratios can more readily develop in communities with taller growth forms (Coppock et al., 1983; Fryxell, 1991; McNaughton, 1984).

Summary and Overview

Plant species composition in the shortgrass steppe is very resistant to grazing, more so than other grasslands (Fig. 16.25). Changes that take place in response

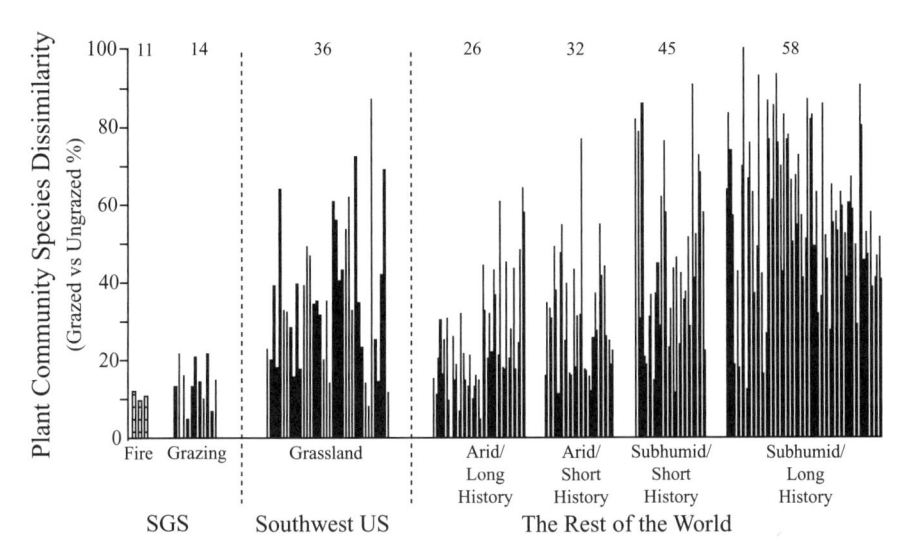

Figure 16.25 Plant community species dissimilarity of burned versus unburned and grazed versus ungrazed treatments in the northern shortgrass steppe compared with species dissimilarities of grazed versus ungrazed grasslands in the adjacent southwestern United States (with a short evolutionary of grazing and similar productivity) and grazed versus ungrazed rangelands around the world. (Fire data are from Milchunas [unpublished data]; shortgrass steppe grazing data are from Klipple and Costello [1960], Grant [1971], Milchunas et al. [1989], and Ashby et al. [1993]; data for the southwestern United States is from Milchunas [2006], and data from around the world from 276 grazed versus ungrazed comparisons are from Milchunas and Lauenroth [1993].) Dissimilarity is converse of similarity, as calculated and described in the legend for Figure 16.11.

to long-term grazing indicate that ungrazed, rather than grazed, plant communities are more likely to be invaded by exotic and native weedy species. Ungrazed communities, then, are more representative of disturbed plant communities than are grazed communities. The mechanism for this appears to be the increased spread (basal cover) by native prostrate growth forms in grazed areas, resulting in a more uniform exploitation of the soil volume and reductions in safe sites for weedy colonization.

This resistance of the shortgrass plant community to grazing pressure includes components of both avoidance and tolerance. A long evolutionary history of grazing by large native herbivores (bison) that are ecologically similar to cattle produced plant species tolerant of defoliation. The semiarid environment produced a plant community with a large proportion of biomass belowground that is inaccessible to large mammalian herbivores, and one with a sparse canopy in which the potential for herbivore-mediated alteration of competition for light is minimal. The long history of grazing and semiaridity act as convergent selection forces providing resistance to current-day grazing pressure (Milchunas et al., 1988).

These attributes are exemplified in the dominant species *B. gracilis*. Because a large proportion of *B. gracilis* biomass is in its roots it is not severely affected by grazing by large herbivores. Furthermore, its dominance in the community may give it some protection from grazing. Although a palatable and nutritious plant, it is consumed less than its proportional abundance in the community, as grazers seem to prefer a mixed diet, possibly to balance their nutritional needs.

Grazing of livestock on public lands has been and continues to be a controversial issue. Proponents of the removal of livestock from public lands often fail to distinguish between systems with long versus short evolutionary histories of grazing and the potential for this to result in very different outcomes in responses to grazing (Milchunas, 2006). Similarly, administration of grazing fees and conservation programs by federal agencies is generally applied uniformly across all rangelands. A conclusion of the global analyses of grazing impacts by Milchunas and Lauenroth (1993) was that, "within levels not considered abusive 'overgrazing,' the geographical location where grazing occurs may be more important than how many animals are grazed or how intensively an area is grazed" (p. 327). This obviously does not mean that good local grazing management is not important, but that the larger regional scale of grazing management, particularly with respect to the geography of evolutionary history, is important as well. When viewed at a regional scale, where we graze is more important than how we graze.

Traditional range management focuses on individual ranch-scale practices, and much progress has been made during the past century. Differential application of policy at larger scales based upon what we know of system response to grazing is more difficult, but offers potential for balancing environmental and production needs. It is interesting to note that the southwestern United States is considered to be a hotspot of species endangerment (Flather et al., 1994). This is an area where grazing is currently an important land use of native ecosystems that have only short histories of grazing by large herbivores. Similar to fire, grazing can be a devastating force in some systems but an integral part of others. Milchunas (2006) and Milchunas et al. (1998) suggest that Margalef's (1968) terminology of endogenous and exogenous disturbances may provide a useful conceptual framework for sociopolitical issues concerning utilization of rangelands by domestic livestock. Although any system can be overgrazed, moderate grazing of the shortgrass steppe by domestic livestock more closely resembles the endogenous disturbance of bison grazing than does protection from grazing in this particular system. Endogenous disturbances are a stabilizing force in a system in which the flora and fauna have evolved in concert with the disturbance.

Acknowledgments Rod Heitschmidt, Paul Stapp, and Dick Hart provided reviews of early versions of the manuscript. We thank LTER site managers Mark Lindquist, Don Hazlett, and Ray Souther, and the many students on the LTER field crews, for a constant source of help in field sampling, and Judy Hendryx for support and coordination in the laboratory.

References

Archer, S., and J. K. Detling. 1986. Evaluation of potential herbivore mediation of plant water status in a North American mixed-grass prairie. *Oikos* **47**:287–291.

Ashby, M. M., R. H. Hart, and J. R. Forewood. 1993. *Plant community and cattle responses to fifty years of grazing on shortgrass steppe*. USDA–ARS, Rangeland Resources Research Unit, RRRU-1, Fort Collins, Colo.

Axelrod, D. I. 1985. Rise of the grassland biome, central North America. *Botanical Review* **51**:163–201.

Baker, J. N., and O. J. Hunt. 1961. Effects of clipping treatments and clonal differences on water requirement of grasses. *Journal of Range Management* **14**:216–219.

Bakker, J. P., J. de Leeuw, and S. E. van Wieren. 1984. Micro-patterns in grassland vegetation created and sustained by sheep grazing. *Vegetatio* **55**:153–161.

Bayless, M., W. K. Lauenroth, and I. C. Burke. 1996. Refuge effect of the *Opuntia polyacantha* (plains prickly pear) on grazed areas of the shortgrass steppe. *Bulletin of the Ecological Society of America*.

Bement, R. E. 1993. Colorado rangelands: A land manager's historical perspective. *Rangelands* **15**:208–210.

Bridge, B. J., J. J. Mott, and R. J. Hartigan. 1983. The formation of degraded areas in the dry savanna woodlands of northern Australia. *Australian Journal of Soil Research* **21**:91–104.

Briske, D. D., and V. J. Anderson. 1992. Competitive ability of the bunchgrass *Schizachyrium scoparium* as affected by grazing history and defoliation. *Vegetatio* **103**:41–49.

Brougham, R. W., and W. Harris. 1967. Rapidity and extent of changes in genotypic structure induced by grazing in rye grass populations. *New Zealand Journal of Agricultural Research* **10**:56–65.

Coffin, D. P., and W. K. Lauenroth. 1992. Spatial variability in seed production of the perennial grass *Bouteloua gracilis* (H.B.K.) LAG. ex Griffiths. *American Journal of Botany* **79**:347–353.

Cook, C. W., R. D. Child, and L. L. Larson. 1977. *Digestible protein in range forages as an index to nutrient content and animal response*. Colorado State University, Range Science Department, science series no. 29, Fort Collins, Colo.

Cook, C. W., and L. E. Harris. 1968. *Nutritive value of seasonal ranges*. Utah State University, Agricultural Experiment Station Bulletin 472, Logan, Utah.

Cook, C. W., and E. F. Redente. 1993. Development of the ranching industry in Colorado. *Rangelands* **15**:204–207.

Coppock, D. L., J. K. Detling, J. E. Ellis, and M. I. Dyer. 1983. Plant–herbivore interactions in a North American mixed-grass prairie. I. Effects of black-tailed prairie dogs on intraseasonal aboveground plant biomass and nutrient dynamics and plant species diversity. *Oecologia* **56**:1–9.

Coughenour, M. B., J. K. Detling, I. E Bamberg, and M. M. Mugambi. 1990. Production and nitrogen responses of the African dwarf shrub *Indigofera spinosa* to defoliation and water limitation. *Oecologia* **83**:546–552.

Daubenmire, R. F. 1985. The western limits of the range of the American bison. *Ecology* **66**:622–624.

Day, T. A., and J. K. Detling. 1994. Water relations of *Agropyron smithii* and *Bouteloua gracilis* and community evapotransporation following long-term grazing by prairie dogs. *American Midlands Naturalist* **132**:381–392.

Detling, J. K. 1989. Grasslands and savannas: Regulation of energy flow and nutrient cycling by herbivores, pp. 131–148. In: L. R. Pomeroy and J. J. Alberts (eds.), *Concepts of ecosystem ecology: A comparative view*. Ecological studies 67. Springer-Verlag, New York.

Detling, J. K., and E. L. Painter. 1983. Defoliation responses of western wheatgrass populations with diverse histories of prairie dog grazing. *Oecologia* **57**:65–71.

Díaz., S., S. Lavorel, S. McIntyre, V. Falczuk, F. Casanoves, D. G. Milchunas, C. Skarpe, G. Rusch, M. Sternberg, I. Noy-Meir , J. Landsberg, Z. Wei, H. Clark, and B. D. Campbell. 2007. Plant trait responses to grazing: A global synthesis. *Global Change Biology* **13**:313–341.

Eck, H. V., W. G. McCully, and J. Stubbendieck. 1975. Response of shortgrass plains vegetation to clipping, precipitation, and soil water. *Journal of Range Management* **28**:194–197.

Facelli, J. M., and S. T. A. Pickett. 1991. Plant litter: Its dynamics and effects on plant community structure. *Botanical Review* **57**:1–32.

Flather, C. H., L. A. Joyce, and C. A. Bloomgarden. 1994. *Species endangerment patterns in the United States*. United States Department of Agriculture Forest Service, general technical report RM-241. Rocky Mountain Forest and Range Experiment Station, Fort Collins, Colo.

Fox, J. F. 1979. Intermediate-disturbance hypothesis. *Science* **204**:1344–1345.

Fryxell, J. M. 1991. Forage quality and aggregation by large herbivores. *The American Naturalist* **138**:478–498.

Fuls, E. R. 1992. Ecosystem modification created by patch-overgrazing in semi-arid grassland. *Journal of Arid Environments* **23**:59–69.

Grant, W. E. 1971. *Comparisons of aboveground plant biomass on ungrazed pastures vs. pastures grazed by large herbivores, 1970 season*. U.S. International Biological Program, Grassland Biome technical report no. 131. Natural Resource Ecology Laboratory, Colorado State University, Fort Collins, Colo.

Grime, J. P. 1973. Control of species density in herbaceous vegetation. *Journal of Environmental Management* **1**:151–167.

Hansen, R. M., and I. K. Gold. 1977. Blacktail prairie dogs, desert cottontails and cattle trophic relations on shortgrass range. *Journal of Range Management* **30**:210–214.

Harper, J. L. 1969. The role of predation in vegetational diversity, pp. 48–62. In: G. M. Woodwell and H. H. Smith (eds.) *Diversity and stability in ecological systems*. Brookhaven Symposia in Biology no. 22. National Technical Information Service, Springfield, Va.

Hart, R. H. 2001a. Plant biodiversity on shortgrass steppe after 55 years of zero, light, moderate, or heavy cattle grazing. *Plant Ecology* **155**(1):111–118.

Hart, R. H. 2001b. Where the buffalo roamed: Or did they? *Great Plains Research* **11**:83–102.

Hart, R. H., and M. M. Ashby. 1998. Grazing intensities, vegetation, and heifer gains: 55 years on shortgrass. *Journal of Range Management* **51**:392–398.

Heitschmidt, R. K., S. L. Dowhower, R. A. Gordon, and D. L. Price. 1985. *Response of vegetation to livestock grazing at the Texas Experimental Ranch*. Texas Agriculture Experiment Station, B-1515, College Station, Texas.

Hobbs, N. T., D. S. Schimel, C. E. Owensby, and D. J. Ojima. 1991. Fire and grazing in the tallgrass prairie: Contingent effects on nitrogen budgets. *Ecology* **72**:1374–1382.

Hodgkinson, K. C. 1976. The effects of frequency and extent of defoliation, summer irrigation, and fertilizer on the production and survival of the grass *Danthonia caespitosa* Gaud. *Australian Journal of Agricultural Research* **27**:755–767.

Hornaday, W. T. 1889. *The extermination of the American bison with a sketch of its discovery and life history.* Smithsonian report no. 1887, Smithsonian Institution, Washington, D.C.

Hunter, R. F. 1962. Hill sheep and their pasture: A study of sheep grazing in southeast Scotland. *Journal of Ecology* **50**:651–680.

Huston, M. 1979. A general hypothesis of species diversity. *The American Naturalist* **113**:81–101.

Huston, M. A. 1985. Patterns of species diversity on coral reefs. *Annual Review of Ecology Systematics* **16**:149–177.

Hyder, D. N., R. E. Bement, E. E. Remmenga, and D. F. Hervey. 1975. *Ecological responses of native plants and guidelines for management of shortgrass range.* United States Department of Agriculture–Agricultural Research Service, technical bulletin no. 1503. U.S. Government Printing Office, Washington, D.C.

Hyder, D. N., R. E. Bement, E. E. Remmenga, and C. Terwillager, Jr. 1966. Vegetation-soils and vegetation-grazing relations from frequency data. *Journal of Range Management* **19**:11–17.

Jaramillo, V. J., and J. K. Detling. 1988. Grazing history, defoliation, and competition: Effects on shortgrass population and nitrogen accumulation. *Ecology* **69**:1599–1608.

Johnson, C. W. 1951. Protein as a factor in the distribution of the American bison. *Geography Review* **41**:330–331.

Joyce, L. A. 1989. *An analysis of the range forage situation in the United States: 1989–2040.* United States Department of Agriculture, general technical report RM-180. Forest Service, Rocky Mountain Forest and Range Experiment Station. Fort Collins, Colo.

Kelting, R. W. 1954. Effects of moderate grazing on the composition and plant production of a native tallgrass prairie in central Oklahoma. *Ecology* **35**:200–207.

Kemp, W. B. 1937. Natural selection within plant species as exemplified in a permanent pasture. *Journal of Heredity* **28**:329–333.

Kingston, C. S. 1932. Buffalo in the Pacific Northwest. *Washington Historical Quarterly* **23**:163–172.

Klipple, G. E., and D. F. Costello. 1960. *Vegetation and cattle responses to different intensities of grazing on short-grass ranges on the central Great Plains.* United States Department Agriculture technical bulletin no. 1216. U.S. Government Printing Offfice, Washington D.C.

Knapp, A. K., and T. R. Seastedt. 1986. Detritus accumulation limits productivity of tallgrass prairie. *BioScience* **36**:662–668.

Larson, F. 1940. The role of the bison in maintaining the short grass plains. *Ecology* **21**:113–121.

Lauenroth, W. K., H. W. Hunt, D. M. Swift, and J. S. Singh. 1986. Estimating aboveground net primary production in grasslands: A simulation approach. *Ecological Modeling* **33**:297–314.

Lauenroth, W. K., and D. G. Milchunas. 1991. The shortgrass steppe, pp. 183–226. In: R. T. Coupland (ed.), *Natural grasslands, introduction and western hemisphere. Ecosystems of the world 8A.* Elsevier, Amsterdam.

Lauenroth, W. K., D. G. Milchunas, J. L. Dodd, R. H. Hart, R. K. Heitschmidt, and L. R. Rittenhouse. 1994. Grazing in the Great Plains of the United States, pp. 69–100. In: M. Vavra, W. A. Laycock, and R. D. Pieper (eds.), *Ecological implications of livestock herbivory in the West*. American Institute of Biological Sciences Symposium, San Antonio, Texas, 1991. Society for Range Management, Denver, Colo.

Leetham, J. W., and D. G. Milchunas. 1985. The composition and distribution of soil microarthropods in the shortgrass steppe in relation to the soil water, root biomass, and grazing by cattle. *Pedobiologia* **28**:311–325.

Le Houerou, H. N. 1988. Relationship between the variability of primary production and the variability of annual precipitation in world arid lands. *Journal of Arid Environments* **15**:1–18.

Leopold, E. B., and M. F. Denton. 1987. Comparative age of grassland and steppe east and west of the northern Rocky Mountains. *Annals of the Missouri Botanical Gardens* **74**:841–867.

Lewis, J. K. 1971. The grassland biome: A synthesis of structure and function, 1971, pp. 317–387. In: N. R. French (ed.), *Preliminary analysis of structure and function in grasslands*. Range Science Department science series no. 10. Colorado State University, Fort Collins, Colo.

Liang, Y. M., D. L. Hazlett, and W. K. Lauenroth. 1989. Biomass dynamics and water use efficiencies of five plant communities in the shortgrass steppe. *Oecologia* **80**:148–153.

Mack, R. N., and J. N. Thompson. 1982. Evolution in steppe with few large, hooved mammals. *The American Naturalist* **119**:757–773.

Margalef, R. 1968. *Perspectives in ecological theory*. University of Chicago Press, Chicago, Ill.

McDonald, J. N. 1981. *North American Bison: Their classification and evolution*. University of California Press, Berkeley, Calif.

McHugh, T. 1972. *The time of the buffalo*. A. A. Knopf, New York.

McNaughton, S. J. 1979. Grazing as an optimization process: Grass–ungulate relationships in the Serengeti. *The American Naturalist* **113**:691–703.

McNaughton, S. J. 1984. Grazing lawns: Animals in herds, plant form, and coevolution. *The American Naturalist* **124**:863–886.

McNaughton, S. J. 1985. Ecology of a grazing ecosystem: The Serengeti. *Ecological Monographs* **55**:259–294.

McNaughton, S. J., D. G. Milchunas, and D. A. Frank. 1996. How can productivity be measured in grazing ecosystems? *Ecology* **77**:974–977.

McNaughton, S. J., L. L. Wallace, and M. B. Coughenour. 1983. Plant adaptation in an ecosystem context: Effects of defoliation, nitrogen, and water on growth of an African C_4 sedge. *Ecology* **64**:307–318.

Mengel, R. M. 1970. The North American Central Plains as an isolating agent in bird speciation, pp. 279–340. In: W. Dort and J. K. Jones (eds.), *Pleistocene and recent environments of the central Great Plains*. University Kansas Department Geology special publication no. 3. University of Kansas Press, Lawrence, Kans.

Milchunas, D. G. 2006. *Responses of plant communities to grazing in the southwestern United States*. General technical report RMRS–GTR-169. USDA Forest Service, Rocky Mountain Research Station, Fort Collins, Colo.

Milchunas D. G., J. R. Forwood, and W. K. Lauenroth. 1994. Forage production across fifty years of grazing intensity treatments in shortgrass steppe. *Journal of Range Management* **47**:133–139.

Milchunas, D. G., J. Y. King, A. R. Mosier, J. C. Moore, J. A. Morgan, M. H. Quirk, and J. R. Slusser. 2004. UV radiation effects on plant growth and forage quality in a shortgrass steppe ecosystem. *Phytochemical Phytobiology* **79**:404–410.

Milchunas, D. G., and W. K. Lauenroth. 1989. Three-dimensional distribution of plant biomass in relation to grazing and topography in the shortgrass steppe. *Oikos* **55**:82–86.

Milchunas, D. G., and W. K. Lauenroth. 1992. Carbon dynamics and estimates of primary production by harvest, C^{14} dilution, C^{14} turnover. *Ecology* **73**:593–607.

Milchunas, D. G., and W. K. Lauenroth. 1993. A quantitative assessment of the effects of grazing on vegetation and soils over a global range of environments. *Ecological Monographs* **63**:327–366.

Milchunas, D. G., and W. K. Lauenroth. 1995. Inertia in plant community structure: State changes after cessation of nutrient-enrichment stress. *Ecological Applications* **5**(2):452–458.

Milchunas, D. G., and W. K. Lauenroth. 2001. Belowground primary production by carbon isotope decay and long-term root biomass dynamics. *Ecosystems* **4**:139–150.

Milchunas, D. G., W. K. Lauenroth, and I. C. Burke. 1998. Livestock grazing: Animal and plant biodiversity of shortgrass steppe and the relationship to ecosystem function. *Oikos* **83**:65–74.

Milchunas, D. G., W. K. Lauenroth, and P. L. Chapman. 1992. Plant competition, abiotic, and long- and short-term effects of large herbivores on demography of opportunistic species in a semiarid grassland. *Oecologia* **92**:520–531.

Milchunas, D. G., W. K. Lauenroth, P. L. Chapman, and M. K. Kazempour. 1989. Effects of grazing, topography, and precipitation on the structure of a semiarid grassland. *Vegetatio* **80**:11–23.

Milchunas, D. G., W. K. Lauenroth, P. L. Chapman, and M. K. Kazempour. 1990. Community attributes along a perturbation gradient in a shortgrass steppe. *Journal of Vegetation Science* **1**:375–384.

Milchunas, D. G., W. K. Lauenroth, and O. E. Sala. 1988. A generalized model of the effects of grazing by large herbivores on grassland community structure. *The American Naturalist* **132**:87–106.

Milchunas, D. G., A. R. Mosier, J. A. Morgan, D. LeCain, J. Y. King, and J. A. Nelson. 2005. CO_2 and grazing effects on a shortgrass steppe: Forage quality versus quantity for ruminants. *Agriculture Ecosystems and Environment* **111**:166–184.

Milchunas, D. G., A. S. Varnamkhasti, W. K. Lauenroth, and H. Goetz. 1995. Forage quality in relation to long-term grazing history, current-year defoliation, and water resource. *Oecologia* **101**:366–374.

Morgan, J. A., D. R. LeCain, A. R. Mosier, and D. G. Milchunas. 2001. Elevated CO_2 enhances water relations and productivity and affects gas exchange in C_3 and C_4 grasses of the Colorado shortgrass steppe. *Global Change Biology* **7**:451–466.

Mott, J. J. 1987. Patch grazing and degradation in native pastures of the tropical savannas in northern Australia, pp. 153–161. In: F. P. Horn, J. Hodgson, J. J. Mott, and R. W. Brougham (eds.), *Grazing-lands research at the plant–animal interface*. Winrock International, Morrilton, Alaska.

Noy-Meir, I. 1988. Dominant grasses replaced by ruderal forbs in a vole year in undergrazed Mediterranean grasslands in Israel. *Journal of Biogeography* **15**:579–587.

Paine, R. T. 1966. Food web complexity and species diversity. *The American Naturalist* **100**:65-75.

Paine, R. T. 1971. A short-term experimental investigation of resource partitioning in a New Zealand rocky intertidal habitat. *Ecology* **52**:1096–1106.

Parton, W. J., M. P. Gutmann, and W. R. Travis. 2003. Sustainability and historical land-use change in the Great Plains: The case of eastern Colorado. *Great Plains Research* **3**:97–125.

Peterson, R. A. 1962. Factors affecting resistance to heavy grazing in needle-and-thread grass. *Journal of Range Management* **15**:183–189.

Pieper, R., C. W. Cook, and L. E. Harris. 1959. The effect of intensity of grazing upon nutritive content of diet. *Journal of Animal Science* **18**:1031–1037.

Rebollo, S., D. G. Milchunas, and I. Noy-Meir. 2005. Refuge effects of a cactus in grazed shortgrass steppe under different productivity, grazing intensity and cactus clump structure. *Journal of Vegetation Science* **16**:85–92.

Rebollo, S., D. G. Milchunas, I. Noy-Meir, and P. L. Chapman. 2002. The role of a spiny plant refuge in structuring grazed shortgrass steppe plant communities. *Oikos* **98**:53–64.

Reed, J. L., and D. D. Dwyer. 1971. Blue grama response to nitrogen and clipping under two soil moisture levels. *Journal of Range Management* **24**:47–51.

Ross, H. H. 1970. The ecological history of the Great Plains: Evidence from grassland insects, pp. 225–240. In: W. Dort and J. K. Jones (eds.), *Pleistocene and recent environments of the central Great Plains*. University Kansas Department Geology special publication no. 3. University of Kansas Press, Lawrence, Kans.

Schlesinger, W. H., J. F. Reynolds, G. L. Cunningham, L. F. Huenneke, W. M. Jarrell, R. A. Virginia, and W. G. Whitford. 1990. Biological feedbacks in global desertification. *Science* **247**:1043–1048.

Schowalter, T. D., W. W. Hargrove, and D. A. Crossley, Jr. 1986. Herbivory in forested ecosystems. *Annual Review of Entomology* **31**:177–196.

Schwartz, C. C., and J. E. Ellis. 1981. Feeding ecology and niche separation in some native and domestic ungulates on the shortgrass prairie. *Journal of Applied Ecology* **18**:343–353.

Schwartz, C. C., J. G. Nagy, and R. W. Rice. 1977. Pronghorn dietary quality relative to forage availability and other ruminants in Colorado. *Journal of Wildlife Management* **41**:161–168.

Senft, R. L. 1989. Hierarchical foraging models: Effects of stocking and landscape composition on simulated resource use by cattle. *Ecological Modeling* **46**:283–303.

Senft, R. L., M. B. Coughenour, D. W. Bailey, L. R. Rittenhouse, O. E. Sala, and D. M. Swift. 1987. Large herbivore foraging and ecological hierarchies. *BioScience* **37**:789–799.

Senft, R. L., L. R. Rittenhouse, and M. A. Stillwell. 1984a. Diets selected by cattle from plant communities on shortgrass range. *Proceedings of the Western Section of the American Society of Animal Science* **35**:180–183.

Senft, R. L., L. R. Rittenhouse, and M. A. Stillwell. 1984b. Seasonal changes in nitrogen and energy budgets of range cattle. *Proceedings of the Western Section of the American Society of Animal Science* **35**:200–203.

Senft, R. L., L. R. Rittenhouse, and R. G. Woodmansee. 1985. Factors influencing patterns of cattle grazing behavior on shortgrass steppe. *Journal of Range Management* **38**:82–87.

Shaw, J. H. 1995. How many bison originally populated western rangelands? *Rangelands* **17**:148–150.

Shoop, M. C., R. C. Clark, W. A. Laycock, and R. M. Hansen. 1985. Cattle diets on shortgrass ranges with different amounts of fourwing saltbush. *Journal of Range Management* **38**:443–449.

Simoes, M., and Z. Baruch. 1991. Responses to simulated herbivory and water stress in two tropical C4 grasses. *Oecologia* **88**:173–180.

Sims, P. L., J. S. Singh, and W. K. Lauenroth. 1978. The structure and function of ten western North American grasslands. I. Abiotic and vegetational characteristics. *Journal of Ecology* **66**:251–285.

Sims, P. L., and J. S. Singh. 1978. The structure and function of ten western North American grasslands. II. Intra-seasonal dynamics in primary producer components. *Journal of Ecology* **66**:547–572.

Sims, P. L., D. W. Uresk, D. L. Bartos, and W. K. Lauenroth. 1971. *Herbage dynamics on the Pawnee site: Aboveground and belowground herbage dynamics on the four grazing intensity treatments; and preliminary sampling on the ecosystem stress site*. U.S. International Biological Program, Grassland Biome technical report no. 99. Natural Resource Ecology Laboratory, Colorado State University, Fort Collins, Colo.

Singh, J. S., W. K. Lauenroth, H. W. Hunt, and D. M. Swift. 1984. Bias and random errors in estimation of net root production: A simulation approach. *Ecology* **65**:1760–1764.

Singh, J. S., D. G. Milchunas, and W. K. Lauenroth. 1998. Soil water dynamics and vegetation patterns in a semiarid grassland. *Plant Ecology* **134**:77–89.

Stebbins, G. L. 1981. Coevolution of grasses and herbivores. *Annals of the Missouri Botanical Gardens* **68**:75–86.

Uresk, D. W., and P. L. Sims. 1975. Influence of grazing on crude protein content of blue grama. *Journal of Range Management* **28**:370–371.

Van Vuren, D. R. 1987. Bison west of the Rocky Mountains: An alternative explanation. *Northwest Science* **61**:65–69.

Varnamkhasti, A. S., D. G. Milchunas, W. K. Lauenroth, and H. Goetz. 1995. Production and rain use efficiency in short-grass steppe: Grazing history, defoliation, and water resource. *Journal of Vegetation Science* **6**:787–796.

Vavra, M., R. W. Rice, R. M. Hansen, and P. J. Sims. 1977. Food habits of cattle on the shortgrass range in northeastern Colorado. *Journal of Range Management* **30**:251–263.

Vesey-Fritzgerald, D. 1972. Fire and animal impact on vegetation in Tanzania national parks. *Tall Timbers Fire Ecology Conference* **11**:297–317.

Vinton, M. A., D. C. Hartnett, E. J. Finck, and J. M. Briggs. 1993. Interactive effects of fire, bison (*Bison bison*) grazing and plant community composition in tallgrass prairie. *American Midland Naturalist* **129**:10–18.

Vokhiwa, Z. M. 1994. *Carbon and nitrogen dynamics in grazed and protected semiarid shortgrass steppe*. PhD diss., Colorado State University, Fort Collins, Colo.

Weaver, J. W., and F. W. Albertson. 1956. *Grasslands of the Great Plains: Their nature and use*. Johnsen, Lincoln, Neb.

Weaver, J. W., and F. E. Clements. 1938. *Plant ecology*. McGraw-Hill, New York.

Westoby, M. 1985. Does heavy grazing usually improve the food resource for grazers? *The American Naturalist* **126**:870–871.

Westoby, M. 1986. Mechanisms influencing grazing success for livestock and wild herbivores. *The American Naturalist* **128**:940–941.

Whittaker, R. H. 1952. A study of foliage insect communities in the Great Smoky Mountains. *Ecological Monographs* **22**:1–44.

Williamson, S. C., J. K. Detling, J. L. Dodd, and M. I. Dyer. 1989. Experimental evaluation of the grazing optimization hypothesis. *Journal of Range Management* **42**:149–152.

Wilson, M. 1978. Archaeological kill site populations and the Holocene evolution of the genus *Bison*. *Plains Anthropology* **23**:9–22.

17

Cattle Grazing on the Shortgrass Steppe

Richard H. Hart
Justin D. Derner

Cattle are the primary grazers on the shortgrass steppe. For example, during the late 1990s, 21 shortgrass counties in Colorado reported about 2.36 million cattle compared with 283,000 sheep (National Agricultural Statistics Service, USDA, 1997a), 60,000 pronghorn antelope, and a few thousand bison (Hart, 1994). Assuming one bison or five to six sheep or pronghorn consume as much forage as one bovine (Heady and Child, 1994), cattle provide about 97% of the large-herbivore grazing pressure in this region. The ratio of cattle to other grazers is even greater in the remainder of the shortgrass steppe. In 1997, the three panhandle counties of Oklahoma reported 387,000 cattle and only 1300 sheep, whereas the 38 panhandle counties of Texas reported 4.24 million cattle and 14,000 sheep (National Agricultural Statistics Service, USDA, 1997b,c). However, only about half the cattle in the panhandle counties of Texas and Oklahoma graze on rangeland the remainder are in feedlots.

Grazing Research on the Shortgrass Steppe

Rangeland research on the shortgrass steppe (Table 17.1 describes the parameters of the major research stations in the shortgrass steppe) has included a long history of both basic ecology and grazing management. The responses of rangeland plant communities to herbivory are addressed by Milchunas et al. (chapter 16, this volume) and to disturbance are discussed by Peters et al. (chapter 6, this volume). Here we focus on research pertaining to three management practices important to cattle ranching on shortgrass steppe: stocking rates, grazing systems,

Table 17.1 Parameters of Major Research Stations in the Shortgrass Steppe

Location	Latitude, Longitude	Elevation, m	Precipitation, mm	Frost-Free Days
Central Plains Experimental Range, Nunn, Colorado	40°50'N, 104°40'W	1600–1700	325	135
Southeast Colorado Research Center, Springfield, Colorado	37°20'N, 102°40'W	1400	400	180
Pantex IBP Site, Amarillo, Texas	35°10'N, 101°50'W	1170	495	200
Texas Experimental Ranch, Lubbock, Texas	33°40'N, 101°50'W	1000	470	210

and extending the grazing season via complementary pastures and use of pastures dominated by *Atriplex canescens* [Pursh] Nutt (fourwing saltbush).

Stocking Rates

Stocking rate, defined as the number of animals per unit area for a specified time period, is the primary and most easily controlled variable in the management of cattle grazing. Cattle weight gain responses to stocking rate or grazing pressure (animal days per unit of forage produced) have been quantified in several grazing studies on the shortgrass steppe (Bement, 1969, 1974; Hart and Ashby, 1998; Klipple and Costello, 1960). Average daily gains per animal are better estimated as a function of grazing pressure, rather than stocking rate, as forage production is highly variable in this semiarid environment (Lauenroth and Sala, 1992; Milchunas et al., 1994). Average daily gain remains constant over a range of very low grazing pressures until a *critical grazing pressure* is reached; at this point, average daily gain declines linearly with increasing grazing pressure (Hart, 1972; Jones and Sandland, 1974). Decreases in average daily gains are the result of reduced nutrient intake as grazing pressure increases (Olson et al., 2002). The relation of gain to grazing pressure can be used to calculate profitability under a range of stocking rates, forage production levels, and cattle prices.

In contrast to average daily gains, gains per unit area exhibit a quadratic function with stocking rate (Fig. 17.1). Economic returns per unit area depend not only on the relationships between gains and stocking rate, but also on selling price, maintenance, and operating costs. Maximum economic returns per unit area occur at stocking rates lower than those needed to maximize gain per unit land area (e.g., Hart et al., 1988; Manley et al., 1997).

Relationships of cattle gains to stocking rate and grazing pressure for the shortgrass steppe have been developed from long-term experimental studies. The longest of these studies began in 1939 at the USDA–ARS Central Plains Experimental Range (CPER). This study encompasses light, moderate, and heavy

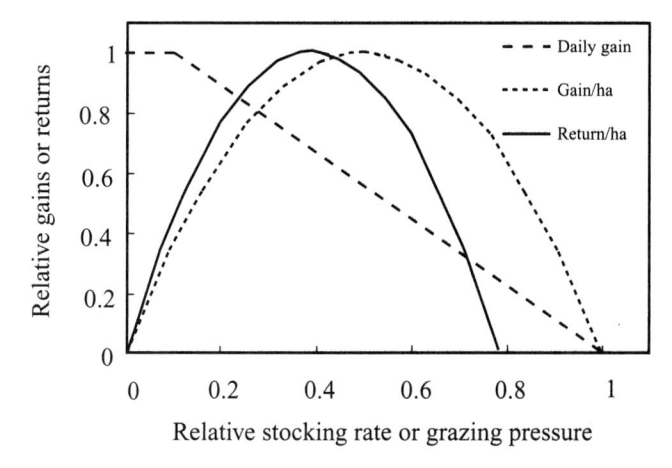

Figure 17.1 The classic stocking rate guide (modified from Bement, 1969).

grazing intensities, with four replications of 130-ha pastures at the beginning of the investigation. Replicates were removed from the study between 1950 and 1978, with only a single replicate pasture for each treatment remaining after 1978. From 1940 to 1964, light, moderate, and heavy grazing pastures were stocked with Hereford yearling heifers to remove 20%, 40%, and 60%, respectively, of the current year's growth of grasses during a 5-month grazing season (mid May–mid October). From 1965 to the present, light, moderate, and heavy grazing pastures have been stocked to leave forage residual levels of 500, 335, and 225 kg·ha^{-1}, respectively, at the end of the grazing season. The average stocking rates for the light, moderate, and heavy grazing treatments have been 15, 20, and 30 Hereford yearling heifers, respectively, in the 130-ha pastures over the 5-month grazing season. Of note, grazing pressure at each of the treatment intensity levels has nearly doubled from 1939 to 2006 (Derner et al., unpublished data), with this increase primarily a result of greater beginning grazing season weights of the Hereford yearling heifers, which are attributable to changes in calving seasons (from mid May–February and March), and genetic advances (e.g., artificial insemination).

Grazing season gains from 1940 to 1949 were 129, 123, and 100 kg·head^{-1} under light, moderate, and heavy stocking, respectively (Klipple and Costello, 1960). Gains per unit area for this same time period were 13.0, 18.9, and 25.7 kg·ha^{-1} under light, moderate, and heavy stocking, respectively (Klipple and Costello, 1960). Primarily using these data, the seminal stocking rate guide for the shortgrass steppe was developed by Bement in 1969 (Fig. 17.1). Predicted maximum economic returns per unit area, calculated using cattle prices for 1964, 1965, and 1966, occurred when 336 kg·ha^{-1} of ungrazed forage was left at the end of the grazing season (Bement, 1974).

Forage production, estimated as peak standing crop, was only determined in 17 of the years from 1940 through 1990. From 1991 to present, forage production has been determined annually. Using available data on forage production

450 Ecology of the Shortgrass Steppe

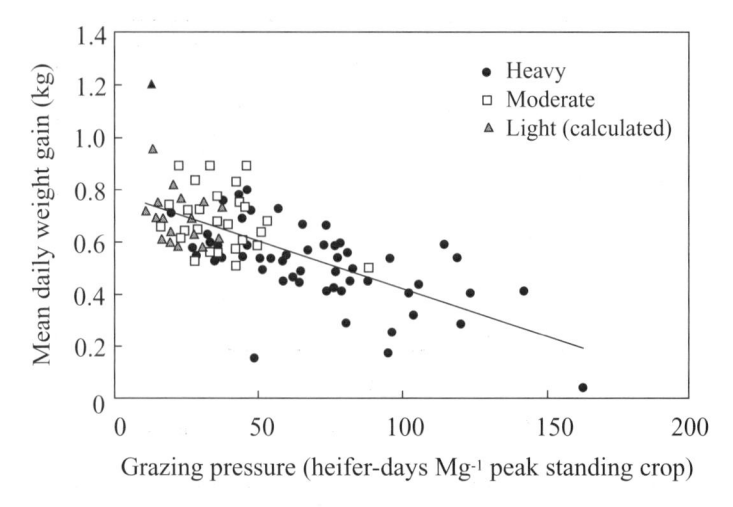

Figure 17.2 Daily gain of yearling heifers under three grazing intensities and a range of grazing pressures at the USDA–ARS CPER, near Nunn, Colorado. (From Hart and Ashby [1998].)

and cattle gains, Hart and Ashby (1998) determined that average daily gains decrease linearly with increasing grazing pressure (Fig. 17.2), a relationship that can be used to predict cattle weight gain using a spreadsheet approach (Hart, 2000).

Length of grazing season has a major effect on the response of cattle weight gains to grazing pressure (Hart and Hanson, 1993), because gains decrease rapidly as vegetative growth slows and stops, and nutrient concentrations decrease. Hyder et al. (1975) reported that gains declined sharply after the summer (June, July, August) maximum, even though forage availability remained high. Klipple and Costello (1960) found that heifers gained no weight in October under moderate stocking, and lost weight under heavy stocking rates.

Yearling steer average daily gains decreased as stocking rate increased in an investigation conducted at the Southeastern Colorado Research Center (SEREC). Average daily gains were 0.75 kg·head^{-1}·day^{-1} under moderate and 0.69 kg·head^{-1}·day^{-1} under heavy stocking rates, at forage allowances of 285 and 215 kg·head^{-1}·28 days^{-1}, respectively (Cook and Rittenhouse, 1988). In 1974, the first year of the study, aboveground biomass production was 1280 kg·ha^{-1} under moderate and 1390 kg·ha^{-1} under heavy stocking rates. In 1980, the last year of the study, biomass production was 1240 and 1130 kg·ha^{-1} under moderate and heavy stocking rates, respectively. Grazing season was 168 days, from early May to late October.

Grazing Systems

Grazing systems have been developed as an alternative to season-long or continuous grazing. They typically involve dividing a larger pasture into smaller paddocks, and these small paddocks are grazed according to a schedule of use and grazing plan. Heady and Child (1994) described a grazing plan as one that stipulates the order and time at which pastures are to be grazed and rested. They also list several possible objectives of a grazing plan:

1. Improve range condition to maintain and increase plant vigor, to promote seed production and seedling establishment, to ensure vegetational succession, and/or to provide fuel for prescribed fire.
2. Improve quality and quantity of forage, and provide reserve forage for emergencies.
3. Achieve improved, regular distribution of grazing animals.
4. Promote uniform forage use by reducing selectivity of grazing.
5. Increase livestock weight gains and reproductive success, per head and/or per unit area, to increase ranch income.
6. Increase flexibility and decrease biological and financial risk in ranch operations.
7. Coordinate domestic animal grazing with habitat needs of wildlife and other uses of the land.

Effects of time-controlled, short-duration rotational and season-long continuous grazing on yearling steer gains were evaluated from 1995 to 2003 at a moderate stocking rate (1.95 ha · animal unit^{-1}·month^{-1}) at the CPER. Steer average daily gains, grazing season gains, and beef production did not differ between grazing systems (Derner and Hart, 2007). Relationships between precipitation (annual or growing season) and average daily gain were not observed. In contrast, both grazing season gains (Fig. 17.3) and beef production exhibited a significant curvilinear response to both length of growing season and annual precipitation. Regression equations demonstrated that beef production is optimized when annual precipitation is 491 mm (24.8 kg·ha^{-1}) and growing season precipitation (May–September) is 368 mm (25.1 kg·ha^{-1}) (Derner and Hart, 2007).

Average daily gains of yearling steers has been shown to be higher under continuous grazing (0.75 kg · head^{-1} · day^{-1}) compared with a three-pasture rest–rotation grazing system (0.68 kg · head^{-1} · day^{-1}) at SEREC from 1969 through 1977 (Cook and Rittenhouse, 1988). In this study, pastures were stocked to allow approximately 285 kg forage·steer^{-1}· 28 days^{-1}. Grazing began about May 1 and ended about October 21, for a grazing season of 168 days. Mean aboveground biomass production was 1530 kg·ha^{-1} under continuous grazing and 1060 kg·ha^{-1} under rotation grazing. Eight-paddock rotation grazing and continuous season-long grazing were compared from 1983 to 1986 at SEREC using a stocking rate of 1 steer· 3.24 ha^{-1}. Pastures were grazed from mid May until mid October, and aboveground biomass production was not estimated. Average daily gains of yearling steers was higher under continuous grazing (1.23 kg · head^{-1}· day^{-1}) compared with the eight-paddock rotation grazing system (1.07 kg · head^{-1}· day^{-1}) (Cook and Rittenhouse, 1988).

Average daily gains for yearling steers were similar for continuous yearlong grazing and a rotation grazing system when stocking rates were the same at the Texas Experimental Ranch (Pitts and Bryant, 1987). When stocking rate was increased on the 16-paddock rotation grazing system, average daily gains

452 Ecology of the Shortgrass Steppe

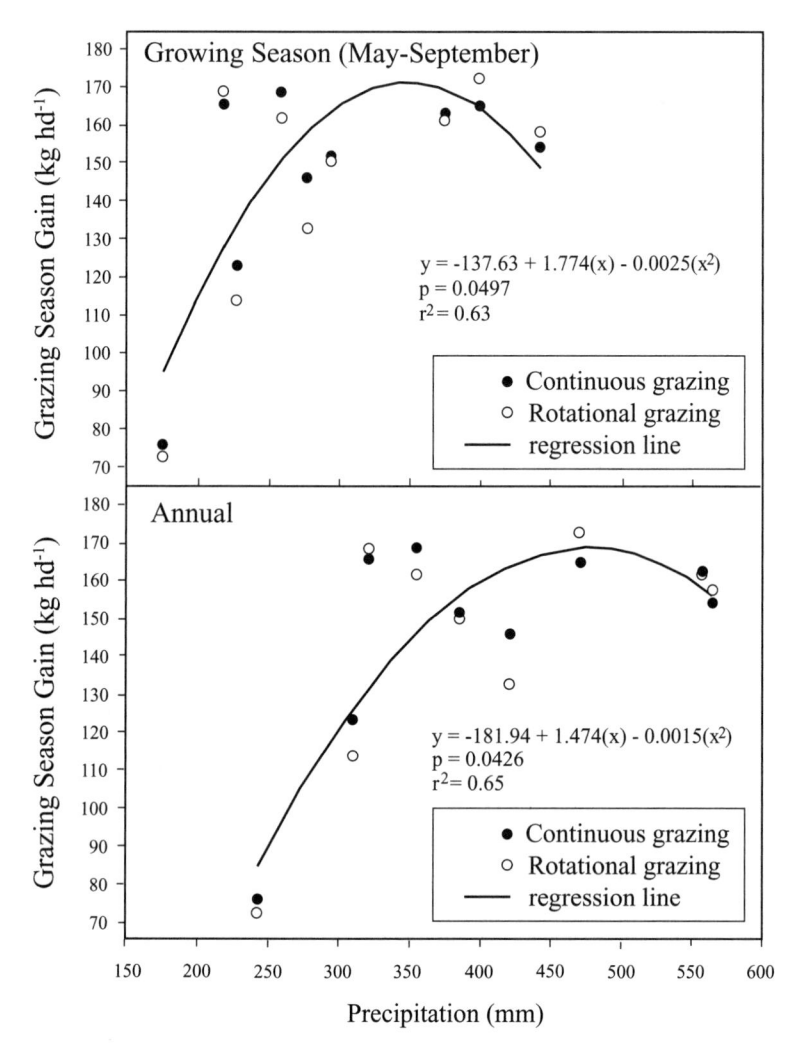

Figure 17.3 Relationships between grazing season gain and growing season precipitation (upper panel) and annual precipitation (lower panel) from 1995 to 2003 for season-long continuous and short-duration rotational grazing systems in the shortgrass steppe at the USDA–ARS CPER, near Nunn, Colorado. (From Derner and Hart [2007].)

decreased compared with continuous year-long grazing, but gains per unit land area increased (Table 17.2); this is the usual response to increased stocking rate or grazing pressure, as modeled by Hart (1978).

Extending the Grazing Season Using Complementary Forages

Supplemental hay, energy, and protein feedstuffs are fed to livestock in the shortgrass steppe from November until the next summer grazing season, because the

Table 17.2 Stocking Rates, Average Daily Gain, and Gain per Hectare under 16-Paddock Short-Duration Rotation and Continuous Grazing on Shortgrass Rangeland, Texas Experimental Ranch, Lubbock, Texas

Dates	Steers·ha^{-1} SDG	Steers·ha^{-1} CG	ADG, kg SDG	ADG, kg CG	Gain, kg·ha^{-1} SDG	Gain, kg·ha^{-1} CG
May 1979–Apr 1980	0.125	0.125	0.33	0.33	15.0	15.0
Apr 1980–Mar 1981	0.249	0.125	0.15	0.25	13.7	11.4
May 1981–Apr 1982	0.187	0.125	0.33	0.37	22.6	16.9
Apr 1982–Nov 1982	0.187	0.125	0.55	0.61	22.5	16.6

ADG, average daily gain; CG, continuous grazing; SDG, short-duration rotation grazing.
Pitts and Bryant (1987).

wheat-fallow cropping regime predominantly used in this semiarid environment does not provide opportunities such as those in the more mesic environments in the Great Plains region, where grazing of crop residues can provide complementary forage during this time. Therefore, the efficiency of livestock production in the shortgrass steppe may be increased by grazing pastures that are seeded to cool-season perennial grasses and/or rangeland dominated by *Atriplex canescens* ([Pursh] Nutt) (fourwing saltbush) in late fall and/or early spring. Such pastures can complement native shortgrass rangeland by supplying forage in greater abundance and with greater quality during late fall and early spring, concurrent with periods of lower nutrient quality of warm-season grasses. Grazing of the saltbush-dominated rangeland may be deferred until after the summer months to extend the grazing season and to provide opportunities for additional animal gain via grazing.

Atriplex canescens is adapted to semiarid conditions and produces palatable, digestible, high-protein forage (e.g., Cordova and Wallace, 1985; Garza and Fulbright, 1988; Rumbaugh et al., 1982). These shrubs typically begin rapid growth in May, flower during June, and complete seed set by the end of August. Trlica et al. (1977) determined that the most detrimental period for defoliating *A. canescens* is near maturity (i.e., flower production) compared with defoliation at quiescence, early growth, or rapid growth. Furthermore, Buwai and Trlica (1977) demonstrated that moderate defoliation (i.e., 60% removal of current year's growth) during periods of rapid growth, seed set, or quiescence stimulated twig growth, but heavy defoliation (i.e., 90% removal) could kill plants. These authors did suggest, however, that use of *A. canescens* during either early spring or late fall may not adversely affect plants. Previous findings suggest that periodic rest is needed to maintain stable *A. canescens* populations (Pieper and Donart, 1978), as studies have indicated that continuous grazing results in marginal plant regrowth (Price et al., 1989), in reduced population densities (Schuman et al., 1990), and in increased proportion of nonflowering plants (Cibils et al., 2003). Shoop et al. (1985) estimated that *A. canescens* provided 32% of cattle diets on a pasture when its frequency was 19%, and 14% of diets on a pasture when its frequency was 8%. Although *A. canescens* plants that are continuously grazed by cattle produce little regrowth, plants entirely protected from grazing may also appear sometimes to produce little regrowth, because regrowth progressively decreases with increasing years of protection (Price et al., 1989).

Weight gains of yearling heifers on *A. canescens*-dominated rangeland at the CPER are influenced by stocking rate for both the late fall (November–mid January) and early spring (April–mid May) grazing periods, but beef production is similar for both light and moderate stocking rates during each of the grazing periods (Derner and Hart, 2005). Similar beef production per unit land area between light and moderate stocking rates, both for late fall and early spring grazing periods, demonstrates that increased stocking rate does not compensate for lower individual animal gain that is achieved with the moderate stocking rate. Therefore, there is not an economic advantage to increase stocking rates from light to moderate levels when grazing *A. canescens* in the shortgrass steppe. These results, showing (1) adequate livestock gains with light stocking rates during both the late fall and early spring grazing periods and (2) that *A. canescens* plants clipped during either fall or spring exhibit equal recovery after the clipping just a few months later (Rumbaugh et al., 1982), suggest that incorporating light stocking rates not only provides reasonable beef production, but also is a sustainable land management practice in this ecosystem that does not degrade the rangeland resource. We have shown individual animal weight gains to be greater with light compared with moderate stocking rates for both of the grazing periods (Derner and Hart, 2005). Our findings suggest that land managers in the shortgrass steppe can effectively extend their grazing season by utilizing *A. canescens*-dominated rangeland both prior to and after the traditional summer grazing season using light stocking rates, which enhance individual animal gains without sacrificing gains per unit land area.

Combining summer grazing of native rangeland with grazing of *A. canescens*-dominated pastures and pastures seed to *Psathrostachys juncea* cv. Bozoisky (Bozoisky wildrye) and *Agropyron cristatum* × *desertorum* cv. Hycrest (Hycrest wheatgrass) results in a system in which cattle can graze for most of the calendar year, with the exception of the calving season, when cattle are commonly concentrated in small pastures near the ranch headquarters and fed hay or other stored feed. Research at the CPER demonstrates that weight gains of yearling heifers were 120 kg·head^{-1} on native shortgrass steppe during a 171-day summer grazing season, but 228 kg·head^{-1} when native rangeland was combined with seeded pastures and *A. canescens*-dominated pastures during a 312-day grazing period (start of April to end of January) (Hart, unpublished data).

Discussion and Conclusions

Research and more than a century of experience have demonstrated that cattle grazing on the shortgrass steppe at light to moderate stocking rates is biologically sustainable, just as wild herbivore grazing was sustainable for millennia (Larson, 1940; Milchunas et al., chapter 16, this volume). Proper stocking rates are essential for economic sustainability; complementary forages must be evaluated for their contribution to economic sustainability. Grazing systems seem to confer few benefits to grazing cattle if other good management practices are followed. Climatic, atmospheric, economic, and sociopolitical changes may require compensatory changes in grazing management to ensure continued profitability and sustainability.

Profitability of rangeland cattle grazing is threatened by economic, political, and social pressures outside the control of the cattle producer. Spitler (2001a) lists some of these factors: (1) the cyclical nature of the U.S. cattle industry; (2) high fixed costs and extreme price volatility; (3) until recently, decreasing per-capita demand for beef; (4) the beef industry's failure to develop new easy-to-prepare products; and (5) exacerbation of oversupply by small producers not concerned with profits. In addition, four meat packers slaughter 85% to 90% of the cattle in the United States and thus exert great control over market prices (Spitler, 2001b). Spitler (2001a) urges ranchers to find ways to diversify their operation and to develop income sources unaffected by fluctuations of the conventional cattle market. The noneconomic roles of rangeland cattle grazing in preserving open space and habitat for other species, and the aesthetic value of open space to humans are often recognized. Of the total animal species found in the United States, 84% of the mammals, 74% of the birds, and 58% of the amphibians are found in nonforested rangelands, including but not limited to the shortgrass steppe (Hart, 1994). Several key threatened or endangered species are present as well (see other chapters in this book). Lastly, many observers would agree that extensive areas of shortgrass steppe rangeland, punctuated by herds of cattle and pronghorn antelope, have more aesthetic appeal than small-acreage housing tracts associated with overgrazed horse pastures.

Land managers and livestock producers on the shortgrass steppe must balance profitability, stability, and sustainability, especially in light of high variability in precipitation amounts and resulting forage production (Derner and Hart, 2007; Lauenroth and Sala, 1992; Milchunas et al., 1994). A modeling approach for evaluating rangeland and livestock systems allows managers to assess the impacts of alternative management prior to actual implementation, thereby reducing risks in decision making. Models such as SMART (Hart, 1989), SPUR (Wright and Skiles, 1987), and SPUR2 (Hanson et al., 1992) have been developed to assist land managers. A new decision support system for the whole ranch–the Great Plains Framework for Agricultural Resource Management (GPFARM)—can serve as a tactical (short-term) and strategic (long-term) planning tool for production, economic, and environmental impact analysis, and site-specific database generation, from which alternative agricultural management systems can be tested and compared (Shafer et al., 2000). The GPFARM model has displayed excellent agreement in tracking growth and senescence trends in both warm- and cool-season perennial grasses, thereby demonstrating that it has functional utility for simulating forage production in semiarid environments such as the shortgrass steppe (Andales et al., 2005).

References

Andales, A. A., J. D. Derner, P. N. S. Bartling, L. R. Ahuja, G. H. Dunn, R. H. Hart, and J. D. Hanson. 2005. Evaluation of GPFARM for simulation of forage production and cow-calf weights. *Rangeland Ecology and Management* **58**:247–255.

Bement, R. E. 1969. A stocking-rate guide for beef production on blue-grama range. *Journal of Range Management* **22**:83–86.

Bement, R. E. 1974. Strategies used in managing blue-grama range on the Central Great Plains, pp. 160–166. In: K. W. Kreitlow and R. H. Hart (eds.), *Plant morphogenesis as the basis for scientific management of range resources.* USDA miscellaneous publication 1271. USDA, Washington, D.C.

Buwai, M., and M. J. Trlica. 1977. Multiple defoliation effects on herbage yield, vigor and total nonstructural carbohydrates of five range species. *Journal of Range Management* **30**:164–171.

Cibils, A. F., D. M. Swift, and R. H. Hart. 2003. Changes in shrub fecundity in fourwing saltbush browsed by cattle. *Journal of Range Management* **56**:39–46.

Cook, C. W., and L. R. Rittenhouse. 1988. *Grazing and seeding research at the Southeastern Colorado Experiment Station, Springfield, CO.* Colorado State University Agricultural Experiment Station technical bulletin LTB88–8. Colorado State University Agricultural Experiment Station, Fort Collins.

Cordova, F. J., and J. Wallace. 1985. Nutritive value of some browse and forb species. *Proceedings of American Society of Animal Science* **26**:160–162.

Derner, J. D., and R. H. Hart. 2005. Heifer performance under two stocking rates on fourwing saltbush-dominated rangeland. *Rangeland Ecology and Management* **58**:489–494.

Derner, J. D., and R. H. Hart. 2007. Livestock and vegetation responses to rotational grazing in shortgrass steppe. *Western North American Naturalist* **67**:359–367.

Garza, A., Jr., and T. E. Fulbright. 1988. Comparative chemical composition of armed saltbush and fourwing saltbush. *Journal of Range Management* **41**:401–403.

Hanson, J. D., B. B. Baker, and R. M. Bourdon. 1992. *SPUR2 documentation and user guide.* GPSR technical report no. 1. USDA–ARS, Fort Collins, Colo.

Hart, R. H. 1972. Forage yield, stocking rate, and beef gains on pasture. *Herbage Abstracts* **42**:345–353.

Hart, R. H. 1978. Stocking rate theory and its application to grazing on rangelands, pp. 550–553. In: D. N. Hyder (ed.), *Proceedings of the First International Rangeland Congress.* Society for Rangeland Management, Denver, Colo.

Hart, R. H. 1989. SMART: A simple model to assess range technology. *Journal of Range Management* **42**:421–424.

Hart, R. H. 1994. Rangeland. In: Charles Arntzen (ed.), *Encyclopedia of Agriculture Science* 3491–-501. Academic Press, Inc., San Diego, CA.

Hart, R. H. 2000. The right rate: New spreadsheets help producers estimate most profitable stocking rates. *Western Farmer-Stockman* **120**(11):WB4, WB13, WB14.

Hart, R. H., and M. M. Ashby. 1998. Grazing intensities, vegetation, and heifer gains: 55 years on shortgrass. *Journal of Range Management* **51**:392–398.

Hart, R. H., and J. D. Hanson. 1993. Managing for economic and ecological stability of range and range-improved grassland systems with the SPUR II model and the STEERISKIER spreadsheet. *Proceedings of the XVII International Grassland Congress*, Palmerston North. 8–21 February 1993:1593–1598.

Hart, R. H., M. J. Samuel, P. S. Test, and M. A. Smith. 1988. Cattle, vegetation and economic responses to grazing systems and grazing pressure. *Journal of Range Management* **41**:282–286.

Heady, H. F., and R. D. Child. 1994. *Rangeland ecology and management.* Westview Press, Boulder, Colo.

Hyder, D. N., R. E. Bement, E. E. Remmenga, and D. F. Hervey. 1975. *Ecological responses of native plants and guidelines for management of shortgrass range.* USDA technical bulletin 1503. U.S. Government Printing Office, Washington, D.C.

Jones, R. J., and R. L. Sandland. 1974. The relation between animal gain and stocking rate. Derivation of the relations from the results of grazing trials. *Journal of Agricultural Science* **83**:335–342.

Klipple, G. E., and D. F. Costello. 1960. *Vegetation and cattle responses to different intensities of grazing on shortgrass ranges on the Central Great Plains.* USDA technical bulletin 1216. U.S. Government Printing Office, Washington, D.C.

Larson, F. 1940. The role of bison in maintaining the shortgrass plains. *Ecology* **21**:113–121.

Lauenroth, W. K., and O. E. Sala. 1992. Long term forage production of North American shortgrass steppe. *Ecological Applications* **2**:397–403.

Manley, W. A., R. H. Hart, M. J. Samuel, M. A. Smith, J. W. Waggoner, Jr., and J. T. Manley. 1997. Vegetation, cattle and economic responses to grazing strategies and pressures. *Journal of Range Management* **50**:638–646.

Milchunas, D. G., J. R. Forwood, and W. K. Lauenroth. 1994. Productivity of long term grazing treatments in response to seasonal precipitation. *Journal of Range Management* **47**:133–139.

National Agricultural Statistics Service, USDA. 1997a. *1997 Census of agriculture.* Vol. 1, part 6, *Colorado.* State & county data. U.S. Government Printing Office, Washington, D.C.

National Agricultural Statistics Service, USDA. 1997b. *1997 Census of agriculture.* Vol. 1, part 36, *Oklahoma.* State & county data. U.S. Government Printing Office, Washington, D.C.

National Agricultural Statistics Service, USDA. 1997c. *1997 Census of agriculture.* Vol. 1, part 43, *Texas.* State & county data. U.S. Government Printing Office, Washington, D.C.

Olson, K. C., J. R. Jaeger, J. R. Brethour, and T. B. Avery. 2002. Steer nutritional response to intensive-early stocking on shortgrass rangeland. *Journal of Range Management* **55**:222–228.

Pieper, R. D., and G. B. Donart. 1978. Response of fourwing saltbush to periods of protection. *Journal of Range Management* **31**:314–315.

Pitts, J. S., and F. C. Bryant. 1987. Steer and vegetation response to short duration and continuous grazing. *Journal of Range Management* **40**:386–389.

Price, D. L., G. B. Donart, and G. M. Southward. 1989. Growth dynamics of fourwing saltbush as affected by different grazing management systems. *Journal of Range Management* **42**:158–162.

Rumbaugh, M. D., D. A. Jounson, and G. A. Van Epps. 1982. Forage yield and quality in a Great Basin shrub, grass, and legume pasture experiment. *Journal of Range Management* **35**:604–609.

Schuman, G. E., D. T. Booth, and J. W. Waggoner. 1990. Grazing reclaimed mined land seeded to native grasses in Wyoming. *Journal of Soil and Water Conservation* **45**:653–657.

Shafer, M. J., P. N. S. Bartling, and J. C. Ascough, II. 2000. Object-oriented simulation of whole farms: GPFARM framework. *Computers and Electronics in Agriculture* **28**:29–49.

Shoop, M. C., R. C. Clark, W. A. Laycock, and R. M. Hansen. 1985. Cattle diets on shortgrass ranges with different amounts of fourwing saltbush. *Journal of Range Management* **38**:443–449.

Spitler, J. 2001a. Spread your risk. *Western Farmer-Stockman* **121**(2):WB4–7.

Spitler, J. 2001b. Utah hard times. *Western Farmer-Stockman* **121**(2):WB1–2.

Trlica, M. J., M. Buwai, and J. Menke. 1977. Effects of rest following defoliations on the recovery of several range species. *Journal of Range Management* **30**:21–27.

Wright, J. R., and J. W. Skiles (eds.). 1987. *SPUR: Simulation of production and utilization of rangelands. Documentation and user guide.* USDA-ARS 63. National Technical Information Service, Springfield, Va.

18

Effects of Grazing on Abundance and Distribution of Shortgrass Steppe Consumers

Daniel G. Milchunas
William K. Lauenroth

Although livestock are the most obvious consumers on the shortgrass steppe, they are certainly not the only consumers. However, livestock may influence the other consumers in a number of different ways. They may directly compete for food resources with other aboveground herbivores. There is behavioral interference between livestock and some species of wildlife (Roberts and Becker, 1982), but not others (Austin and Urness, 1986). The removal of biomass by livestock alters canopy structure (physiognomy) and influences microclimate. Bird, small-mammal, and insect species can be variously sensitive to these structural alterations (Brown, 1973; Cody, 1985; MacArthur, 1965; Morris, 1973; Rosenzweig et al., 1975; Wiens, 1969). There are both short- and long-term effects of grazing on plant community species composition, primary production, and plant tissue quality. Belowground consumers can also be affected by the effects of grazing on soil water infiltration, nutrient cycling, carbon allocation patterns of plants, litter accumulation, and soil temperature. The overall effects of livestock on a particular component of the native fauna can be negative or can be positive through facilitative relationships (Gordon, 1988).

In this chapter we assess the effects of cattle grazing on other above- and belowground consumers, on the diversity and relative sensitivity of these groups of organisms, and on their trophic structure. We first present some brief background information on plant communities of the shortgrass steppe and on the long-term grazing treatments in which many of the studies reported herein were conducted. Details on the plant communities are presented by Lauenroth in chapter 5 (this volume), grazing effects on plant communities by Milchunas et al. in chapter 16 (this volume); and grazing effects on nutrient distributions and cycling by Burke et al. in chapter 13 (this volume).

Plant Communities and Grazing Intensities

The physiognomy of the shortgrass steppe is indicated in its name. The dominant grasses (*Bouteloua gracilis* and *Buchloë dactyloides*), forb (*Sphaeralcea coccinea*), and carex (*Carex eleocharis*) have the majority of their leaf biomass within 10 cm of the ground surface. A number of less abundant midheight grasses and dwarf shrubs are sparsely interspersed among the short vegetation, but usually much of their biomass is within 25 cm of the ground. Basal cover of vegetation typically totals 25% to 35%, and is greater in long-term grazed than in ungrazed grassland. Bare ground (more frequent on grazed sites) and litter-covered ground (more frequent on ungrazed sites) comprise the remainder of the soil surface (Milchunas et al., 1989). On sandy soils, the true shrub *Atriplex canescens* provides added structural complexity, although understory dominants remain similar (Fig. 18.1A). Consumer populations can differ substantially between the shrubland and grassland communities, probably because of differences in structure of canopy cover. Topography also influences consumer populations if they preferentially use either the uplands (catena ridgetops) or lowlands (catena toe slopes, also referred to as swales). Grazing by cattle is much greater in lowlands, and this tends to homogenize differences in plant community species composition that occur in the absence of grazing resulting from soil differences (Milchunas et al., chapter 16, this volume). Often, less extensive community types are associated with particular soils (Lauenroth, chapter 5, this volume).

The shortgrass steppe is in a semiarid environment, with annual precipitation averaging only 321 mm·y^{-1} at the Central Plains Experimental Range (CPER). Aboveground net primary production (ANPP) averages 95 g·m^{-2} in moderately grazed uplands, with slightly greater amounts in lowlands compared with uplands and in shrublands compared with grasslands (Lauenroth et al., chapter 12, this volume).

Long-term grazing treatments were established at the CPER in 1939. Light, moderate, and heavy grazing treatments result in approximately 20%, 40%, and 60% removal of ANPP when averaged over years (see Milchunas et al., chapter 16, this volume, for details of stocking rates and different annual removal rates). Pastures not in the more tightly regulated treatment pastures are generally grazed at moderate levels, whereas grasslands in the adjacent Pawnee National Grassland (PNG) or in private ownership are more often grazed at 50% to 65% removal of ANPP. The long-term grazing treatments at the CPER are grazed either summer or winter only. Other grazing systems may be used at other locations in the region.

Large Herbivores

Pronghorn (*Antilocapra americana*, sometimes referred to as antelope) are the most abundant large native herbivore in the shortgrass steppe (Fig. 18.1B). Numbers in eastern Colorado were estimated to be 10,000 in 1958 (Hoover et al., 1959) and 60,000 in 1992 (Hart, 1994), compared with two million presettlement

Figure 18.1 (A) Pronghorn antelope on a level uplands site. (Photo from Natural Resource Ecology Lab archives.) (B) An *Atriplex* shrubland community ungrazed since 1939 left of fence line and moderately grazed by cattle right of fence line. (Photo by Mark Vandever.) Note shrub abundance is lower in the ungrazed treatment.

and a low of 1000 in 1918 (Hoover et al., 1959). Lauenroth and Milchunas (1991) calculated a presettlement biomass of pronghorn to have been 16 kg·ha^{-1} compared with the current cattle biomass of 76 kg·ha^{-1} in heavily grazed pastures. Current numbers are regulated by hunting, and this sometimes depends upon problems associated with pronghorn grazing of winter wheat crops. Across their range, pronghorn numbers consistently increased from the early 1970s through the early 1990s (Langner and Flather, 1994). Although mule deer (*Odocoileus hemionus*) and white-tailed deer (*O. virginianus*) are not common on the open plains, they can often be found near ravines and riparian/shrubland areas where cover is available. There are no longer any free-ranging bison (*Bison bison*), but they are found in increasing numbers in some natural areas and parks where populations are being restored, and on bison ranches, which are also increasing in numbers. The long evolutionary history of grazing by these native ungulates in the shortgrass steppe is discussed by Hart in chapter 4 (this volume).

Bison and cattle are closely related and similar in size. This raises questions concerning the effect of the replacement of bison with cattle on other native consumers and the compatibility of the two in areas where bison are being reintroduced. Grasses and sedges are a large proportion of the diet of both species. Relatively minor differences in their diets have been found in the shortgrass steppe (Peden et al., 1974; Schwartz and Ellis, 1981), in the Great Basin shrub steppe in Utah (Van Vuren, 1982), and the mixed-grass prairie of South Dakota (Plumb and Dodd, 1993). In the shortgrass steppe, cattle consume more cool-season grasses, forbs, and shrubs than bison (Schwartz and Ellis, 1981). Bison diets have been found to be largely warm-season grasses, and they are somewhat less selective feeders than cattle. The low degree of selectivity by bison is reflected in the poor quality of their diets, and in higher intake as a proportion of their body weight (75 g·kg^{-1}wt$^{0.75}$) relative to cattle (52 g·kg^{-1} wt$^{0.75}$) or sheep (46 g·kg^{-1} wt$^{0.75}$) (Rice et al., 1974). There is some spatial niche separation between cattle and bison, in that cattle spend more time during the growing season grazing in swales (Peden et al., 1974; Van Vuren, 1982).

Bison and cattle are socially compatible down to distances of about 4 m, past which cattle are subordinate to bison (Van Vuren, 1982). Observation of a mixed herd in lightly and heavily grazed shortgrass steppe has shown that bison are dominant over cattle in selection of grazing and bedding areas and in obtaining water (Sparks, 1972). No differences are apparent in time spent grazing or resting, although both species spend more time grazing in the heavily than lightly grazed pasture. The high dietary overlap between cattle and bison, however, suggests a high degree of competition, but also a generally similar pressure on the plant community where cattle have replaced the native bison.

There are some minor differences between cattle and bison grazing. Cattle are a little more selective than bison for components much more common in pronghorn diets (forbs and shrubs), and these components are more susceptible to fluctuating abiotic conditions. This led Lauenroth and Milchunas (1991) to conclude that a bison–pronghorn assemblage was probably more stable than a pronghorn–cattle one. However, Schwartz et al. (1977) found that pronghorn were able to maintain similar diet quality on long-term lightly and heavily cattle-grazed pastures. This may be the result of a capacity for high selectivity in their foraging, differences in preference for topographic positions across the landscape (Schwartz and Ellis, 1981), and the tendency for cattle to be less mobile and restricted in their foraging areas (Van Vuren, 1982). However, competition between species may only occur during uncommon bottleneck periods, when all food becomes scarce. An example of this is a pronghorn die-off that occurred during a time of drought in Texas, and was attributed to overgrazing by domestic animals (Hailey et al., 1966).

Dietary overlap between pronghorn and sheep is much greater than between pronghorn and cattle (Clary and Holmgren, 1982; Schwartz and Nagy, 1976). Although social avoidance of sheep by pronghorn has not been observed, pronghorn in the Great Basin avoid areas grazed by sheep during winter until spring regrowth occurs, and favor areas temporarily rested from sheep use (Clary and Beale, 1983; Clary and Holmgren, 1982).

Small Mammals

Lagomorphs and rodents are an important component of the native mammalian fauna of the shortgrass steppe. However, Lauenroth and Milchunas (1991) estimated that consumption by lagomorphs and rodents combined was less than 1% of ANPP in the northern shortgrass steppe, even though each of their biomasses was greater than that of pronghorn antelope. Rodents process approximately 40,000 kJ·ha^{-1}·y^{-1} via herbivory compared with 96,000 kJ·ha^{-1}·y^{-1} as predators, whereas lagomorphs are 100% herbivores. The ratio of rodent to lagomorph biomass was much greater at the southern shortgrass site (Texas) compared with the northern site (Colorado). The ecology of lagomorphs and rodents in the shortgrass steppe is addressed by Stapp et al. in chapter 8 (this volume). The effects of livestock grazing on small mammals can manifest through direct dietary competition, by altering forage quality and seed or arthropod abundances, and by changing community physiognomy and predation rates.

Lagomorphs

Grasses and sedges represent more than 50% of the diet of black- and white-tailed jackrabbits (*Lepus californicus, L. townsendi*) and desert cottontails (*Sylvilagus auduboni*) (Flinders and Hansen, 1972; Hansen and Gold, 1977). *Agropyron smithii*, a grass species highly preferred by cattle, is also a very important component in the diet of all three lagomorphs. Both cattle and black-tailed jackrabbits have been found to graze swales preferentially during the growing season (Flinders and Hansen, 1975; Milchunas et al., chapter 16, this volume), but the white-tailed jackrabbit prefer uplands, and cottontails show no preference. There is potential for overlap between cattle and lagomorphs in both diet and habitat preference.

Flinders and Hansen (1975) found that black-tailed jackrabbits are more than twice as abundant in lightly and moderately summer-grazed than in heavily summer-grazed or moderately winter-grazed treatments. They noted that this is not the usual response of black-tailed jackrabbits to grazing, because in more productive habitats they tend to avoid dense vegetation and prefer heavily grazed or burned communities. This has been observed in the more productive grasslands of eastern Texas (Taylor and Lay, 1944), the sand hills of Colorado (Sanderson, 1959), southern Arizona (Taylor et al., 1935), and the mixed-grass (Smith, 1940) and tallgrass prairie (Phillips, 1936) of Oklahoma. Rabbit and rodent numbers become more abundant as range condition deteriorates in desert grassland, leading to increased competition with livestock (Schmutz et al., 1992). Heavy grazing of short-stature vegetation may reduce cover or forage to less than a critical minimum. White-tailed jackrabbits show no significant differences between grazing treatments in the shortgrass steppe, but average densities are highest in moderately grazed treatments (Flinders and Hansen, 1975). Cottontails are significantly more abundant in moderately grazed treatments compared with those either lightly or heavily grazed. Cottontails appear to be closely associated with shrubs, and moderate grazing of the shrubs seems to increase adventitious budding and the development of a denser shrub canopy.

Prairie Dogs

Prairie dogs (*Cynomys ludovicianus*) have an important influence on both the flora and fauna of Great Plains grasslands. Potentially 170 vertebrate species are closely associated with prairie dog colonies (Miller et al., 1994), and plant diversity is increased in some communities as a result of their burrowing and clipping activities (Bohnam and Lerwick, 1976; Hansen and Gold, 1977; Whicker and Detling, 1988). Millions of prairie dogs occupied the Plains before settlement, and a single colony in Texas was reported to have covered 6.5 million ha (Merriam, 1902). Large-scale poisoning throughout their range, because of perceived competition with cattle and their potential as disease vectors, has drastically reduced numbers and sizes of towns (McNulty, 1971).

Dietary overlap between cattle and prairie dogs is relatively high in the shortgrass steppe, ranging from a high of 69% in spring to a low of 41% in winter (Hansen and Gold, 1977). Hansen and Gold (1977) concluded that prairie dogs are not, however, influenced by cattle grazing in the shortgrass steppe. This is not the usual response to grazing. Prairie dogs prefer heavily grazed areas in more productive grasslands than the shortgrass steppe, and can decline in numbers if the cattle are removed (Allen and Osborn, 1954; Knowles, 1986; Koford, 1958). Prairie dogs may either positively or negatively impact cattle. In the shortgrass steppe, Hansen and Gold (1977) estimated that forage consumption and soil disturbances by prairie dogs and the associated cottontails reduced the amount of ANPP available to cattle by 24%. Cattle either gained no weight or lost weight during winter in pastures with prairie dog towns. O'Meilia et al. (1982) similarly reported that weight losses appeared to occur only during fall and winter in *B. gracilis*-dominated grassland in Oklahoma. The lack of weight gain differences during the growing season between pastures with and without prairie dog towns was attributed to potentially greater forage quality on the towns. Knowles (1986) observed that cattle occurred significantly more often, and Hassien (1976) found greater numbers of fecal pats, in areas with prairie dog colonies as opposed to those without colonies. This suggests that prairie dogs facilitate grazing by cattle in a manner similar to that reported for bison (Whicker and Detling, 1988). Greater mineralization and lower immobilization rates in prairie dog towns compared with uncolonized grassland resulted in increased availability of soil nitrogen and increased shoot nitrogen concentrations in mixed-grass prairie (Holland and Detling, 1990). The defoliation activities of the prairie dogs also lowered stem-to-leaf ratios and increased digestibilities of plants for bison in the mixed-grass prairie (Coppock et al., 1983).

Recent studies from the shortgrass steppe in Colorado show that the facilitation effect of prairie dogs on large herbivores does not necessarily occur as it does in more productive plant communities, and that weight gains of cattle are negatively affected by growing-season grazing. Observations of grazing behavior of cattle indicate no preference for prairie dog towns compared with adjacent off-town locations (Guenther and Detling, 2003). This suggests the quality of forage did not differ enough to alter grazing behavior. Grazing by prairie dogs in more productive communities is likely to have greater effects on leaf-to-stem ratios and on plant species composition that it does in short, less productive plant communities (Milchunas et al., 1988). Cattle weight gains declined 5.5% when the percentage of pastures occupied by prairie dogs was 20% compared with no towns, and was reduced 13.9% when the percentage occupied was 60% (Derner et al., 2006). The somewhat low weight gain losses compared with the percentage area occupied was attributed to the grazing tolerance of the dominant shortgrasses. This study was on areas recently experiencing large increases in prairie dog towns, and greater impacts may be expected in areas occupied for longer times.

Small Rodents

Rodent populations are often correlated with various structural attributes of plant communities, particularly cover (Brown, 1973; French et al., 1976; Grant and Birney, 1979; Rosenzweig et al., 1975; Stapp et al., chapter 8, this volume). Rodents may be important consumers of seeds and/or insects (Flake, 1971; French et al., 1976). Competitive interactions between rodent species influence the composition of rodent communities (Valone and Brown, 1995). Grazing can alter plant community structure, the abundance of insects, and seed production. These alterations may affect competition and/or social interactions between rodent species. In the shortgrass steppe, grazing results in large reductions in aboveground arthropod abundances (Crist, chapter 10, this volume), decreased seed production of *B. gracilis* (Coffin and Lauenroth, 1992), but increased seed production in some species (Rebollo et al., 2002). Possibly because of the relatively small changes in community structure and plant species composition with grazing in this system, there are only small effects of grazing on rodent communities in the shortgrass steppe compared with those found in more productive communities (Grant et al., 1982). For example, small-mammal production has been estimated to be 346 and 486 kcal·ha^{-1} in lightly and heavily grazed shortgrass steppe compared with 5812 and 1004 kcal·ha^{-1} in ungrazed and grazed tallgrass prairie. The primarily surface-dwelling, granivorous and omnivorous species of the shortgrass steppe are adapted to open habitat (Stapp et al., chapter 8, this volume).

Although rodent productivity was found to be somewhat greater, rodent biomass was slightly lower in heavily (160 g live weight·ha^{-1}) than in lightly (175 g live weight·ha^{-1}) grazed shortgrass steppe over a 4-year study (Grant et al., 1982). Only deer mice (*Peromyscus maniculatus*) increased with heavy grazing. In a 2-year study of three grazing intensity treatments, Flake (1971) found no consistent relationship across treatments in abundances of Ord's kangaroo rats (*Dipodomys ordii*), more northern grasshopper mice (*Onychomys leucogaster*) in moderately grazed than in either lightly or heavily grazed treatments, increases of deer mice with increasing grazing intensity, and decreases of thirteen-lined ground squirrels (*Spermophilus tridecemlineatus*) with increasing grazing intensity. Pocket gophers (*Thomomys bottae*) did not appear to select habitat based upon grazing treatments. Mounds averaged 6.5%, 2.5%, and 8% of land cover in the ungrazed, lightly, and heavily grazed treatments, respectively (Grant et al., 1980).

Birds

Similar to rodents, birds often respond greatly to alterations in vegetation canopy structure (Cody, 1985; MacArthur, 1965; Wiens, 1969). However, vertical layering of habitat structure in grasslands is not as great a potential control on bird diversity as it would be in forests, and birds do not appear to respond to differences in horizontal habitat heterogeneity across or within 15 sites and five community types (Wiens, 1974a,b; Wiens and McIntyre, chapter 9, this volume). Furthermore, with respect to food resources, data from four grassland and shrub steppe communities suggests that food may not usually limit bird populations, although niche overlap may increase in less productive habitats (Wiens and McIntyre, chapter 9, this volume; Wiens and Rotenberry, 1979). Therefore, one may expect the response of birds to grazing in the shortgrass steppe to be similar to that reported for small rodents. However, this is not the case.

A study of nesting birds in the long-term grazing treatments in the northern shortgrass steppe showed a complete shift in the dominant species from Lark Bunting (*Calamospiza melanocorys*) to Horned Lark (*Eremophila alpestris*), and very low similarity in species composition between lightly and heavily grazed treatments (Giezentanner, 1970). Utilization of heavily, moderately, and lightly summer-grazed, upland grassland communities averaged 3805, 2446, and 2789 bird-use days (40/ha) in summer, respectively, and 3680, 627, and 1945 bird-use days (40/ha) in winter. Estimates for winter-grazed, lowland shrubland communities were 2384, 2438, and 2114 bird-use days (40/ha) in summer; and 1171, 1263, and 541 bird-use days (40/ha) in winter for the heavily, moderately, and lightly grazed treatments, respectively. In a 5-year study, Ryder (1980) found that nesting by Mountain Plover (*Eupoda montana*), Horned Lark, and McCown's Longspur (*Rhynchophanes mccownii*) was greater on heavily than lightly grazed pastures. In contrast, Chestnut-collared Longspur (*Calcarius ornatus*), Western

Meadowlark (*Sturnella neglecta*), and Lark Bunting nesting were all greatest on the lightly grazed treatment.

A study at a southern shortgrass site in Texas, however, shows very little difference in bird species composition with grazing. Wiens (1973) calculated a 98% similarity between grazed and ungrazed bird communities, and 96% similarity between plant communities. Abundance of birds was, however, greater in heavily (7.2 kg · 100 ha^{-1}) than in lightly (2.4 kg · 100 ha^{-1}) grazed shortgrass steppe sites at Muleshoe National Wildlife Refuge in west Texas (Grzybowski, 1980, 1982). McCown's Longspurs were found only in the heavily grazed treatment. Horned Larks and Chestnut-collared Longspurs showed large increases with grazing, and Baird's Sparrow (*Ammodramus bairdii*) showed decreases. Similarity between the grazing treatments in species composition was 74%.

The response of a particular bird species to grazing is not necessarily consistent across sites of different productivities. Depending on the site, grazing may lead toward or away from some vegetation density seen as optimum by a given bird species (Bock and Webb, 1984; Kantrud and Kologiski, 1982). Endemic grassland birds that are closely associated with the shortgrass steppe often prefer open, sparse vegetation (Fig. 18.2 [Knopf, 1996]). Although endemic grassland birds of North America are a threatened group, this is more likely the result of loss of grassland to cultivation and development, and of conditions on the wintering grounds than a result of grazing by domestic livestock. All endemic grassland birds evolved with grazing by large native ungulates, with the exception of Cassin's Sparrow (*Aimophila cassinii*), whose southwest distribution is beyond the historical range of bison (Bock and Webb, 1984; Knopf, 1996). The Mountain Plover is a category I threatened species (major declines have been documented), and this species selects nesting sites that are in very sparse vegetation. Graul (1973, 1975) and Ryder (1980) found Mountain Plovers nesting almost exclusively on very heavily grazed shortgrass steppe. Plovers have been observed to nest on plowed ground (Shackford, 1991), in the middle of two-track roadways through

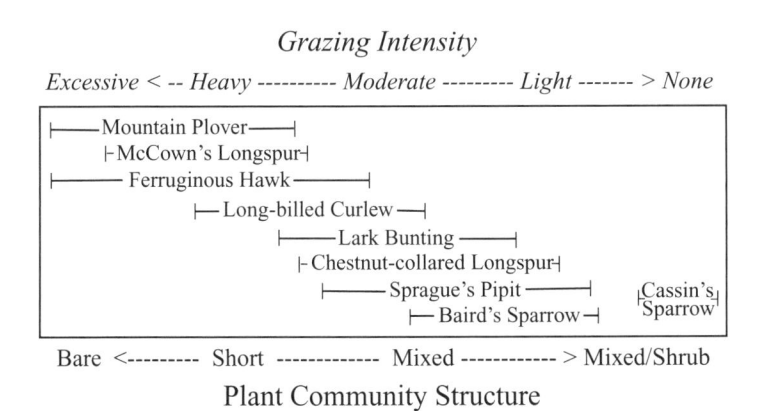

Figure 18.2 Distributions of endemic grassland bird species in relation to grazing intensity and plant community structure. (From Knopf [1996].)

pastures (Milchunas, personal observation), and alongside cattle fecal pats, which provide both protection from wind and a source of insect food. This illustrates the close association of Mountain Plovers with the large herds of bison that "... so completely consumed the herbage of the plains that detachments of the United States Army found it difficult to find sufficient grass for their mules and horses" (Hornaday, 1889 [quoted by Larson, 1940, p. 117]). Nighthawks (*Chordeiles minor*), Killdeer (*Charadrius vociferus*) (Ryder 1980), and McCown's Longspurs (Knopf, 1996) also prefer heavily grazed habitat throughout the PNG. Long-billed Curlews (*Numenius americanus*; category II threatened candidate) prefer grazed areas in Comanche National Grasslands in southern Colorado (King, 1978) and short vegetation throughout its range in western North America (Paton and Dalton, 1994). However, nest losses to predators were greater in heavily than in moderately grazed pastures (With, 1994).

The response of raptors to different grazing intensities is difficult to assess because of their wide-ranging habits. The influence of grazing is probably related to some extent to the impact on their primary food source of lagomorphs and rodents. Indirect effects can include cover for prey, nesting sites, and human management activities. The Ferruginous Hawk (*Buteo regalis*; category II threatened candidate) prefers hunting in grazed areas because the reduced cover affords better detection of prey (Wakeley, 1978). The Ferruginous Hawk is the only raptor to nest sometimes on the ground, and all raptors are nest-site limited in this primarily open grassland habitat (Olendorff and Stoddart, 1974). Nesting sites created by humans may have positive impacts on raptor populations, and percentages of these types of nesting situation were 0%, 4%, 34%, and 52% for Prairie Falcon (*Falco peregrinus*), Golden Eagle (*Aquila chrysatos*), Swainson's Hawk (*B. swainsoni*), and Ferruginous Hawk, respectively (Olendorff, 1972). Human control of prairie dog populations can negatively affect raptors by reducing a food source important to them. Raptor numbers have been closely tied to the presence or absence of active prairie dog colonies (Cully, 1991).

Arthropods

Approximately 90% of arthropod consumption occurs belowground in the shortgrass steppe (Crist, chapter 10, this volume; Lauenroth and Milchunas, 1991). Arthropods are an important herbivore group, accounting for approximately one third of total and 14% of aboveground herbivory. Herbivores comprise approximately 85% of all macroarthropods and 28% of all microarthropods.

Macroarthropods

Grazing has very large impacts on both aboveground and belowground macroarthropods in terms of both biomass and numbers. Aboveground macroarthropods averaged 42, 51, 32, and $32 \cdot m^{-2}$, and belowground macroarthropods averaged 183, 151, 87, and $119 \cdot m^{-2}$, in the ungrazed, lightly, heavily grazed, and current-year ungrazed (previously heavily grazed) treatments, respectively (Andrews,

Effects of Grazing on Abundance and Distribution of Consumers 469

1977). The lack of an aboveground response to current-year removal of grazers from the heavily grazed treatment suggests that the longer term effects of the grazers may be more important than the immediate loss of plant biomass to consumption.

Productivity of macroarthropods in relation to grazing treatments follows a pattern similar to that for density, and all trophic groups were somewhat similarly affected by ungrazed, lightly, and heavily grazed treatments (Fig. 18.3 [Milchunas et al., 1998]). The decline in the proportion of belowground omnivores and saprophages with increasing grazing intensity is probably not significant, because values for these relatively small groups belowground were quite variable. The lack of a proportionately greater decline in herbivores is surprising, given the potential for direct competition with cattle. Slight increases in aboveground herbivorous macroarthropods from ungrazed to lightly grazed treatments may be the result of greater quality of plant tissues in the lightly grazed treatment (Milchunas et al., 1995). Homoptera (primarily leafhoppers) are especially abundant in this treatment (Lavigne et al., 1972). Coleoptera (beetles, especially the most abundant family, Scarabaeidae) are generally relatively scarce in the heavily grazed treatment, and Chrysomelidae (leaf-feeding beetles) are found most commonly in the permanently ungrazed treatment. At the CPER,

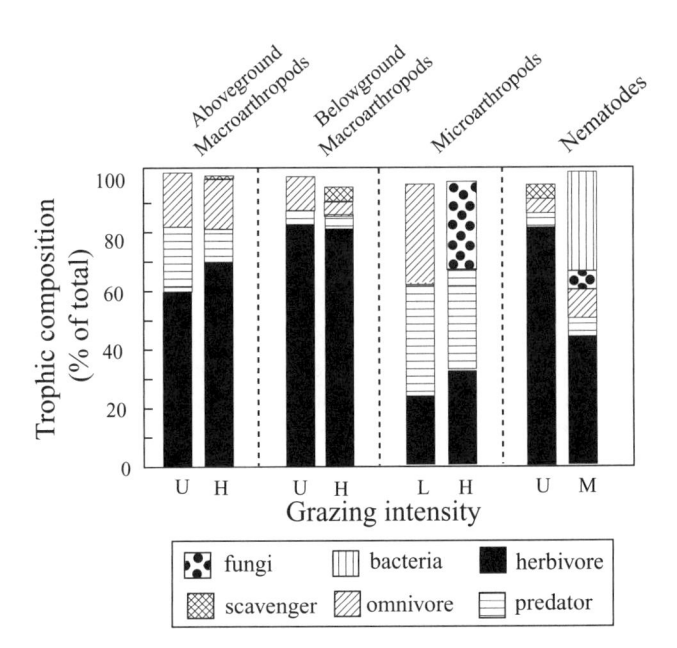

Figure 18.3 Trophic structure of above- and belowground macroarthropods, microarthropods, and nematodes in relation to long-term grazing treatments in the northern shortgrass steppe. H, heavily grazed; L, lightly grazed; M, moderately grazed; U, ungrazed. (From Milchunas et al. [1998].)

soil cores used to determine numbers of belowground macroarthropods included plant crowns and litter (Lloyd et al., 1973). Therefore, "soil" macroarthropods included some aboveground species, but did not include ants, which were dealt with separately. Two peaks, in May and then in August and September, in numbers and biomass of soil macroarthropods were observed in all treatments except the heavily grazed.

In a separate 1-year study, Rottman and Capinera (1983) found smaller differences between grazing treatments in macroarthropod composition and/or abundances than those reported earlier. Total biomass averaged 21, 35, and 17 mg·m^{-2} in the ungrazed, moderately, and heavily grazed treatments, respectively, with the moderately grazed value being significantly higher than the other two. Homoptera, Hemiptera, and Diptera followed patterns for abundance similar to that of biomass. However, total numbers of arthropods did not significantly differ among treatments.

Western harvester ants and grasshoppers are conspicuous arthropods in the shortgrass steppe (Crist, chapter 10, this volume). Harvester ants can move 2.8 kg/colony of soil to the surface, and the disk around the mound cleared of vegetation can be more than 1 m in diameter (Rogers and Lavigne, 1974). Western harvester ant colonies are more numerous at intermediate intensities of long-term grazing, averaging 23, 28, 31, and 3 colonies/ha in the ungrazed, lightly, moderately, and heavily grazed treatments, respectively, at the CPER (Rogers et al., 1972). Disk diameters average 1.2, 0.9, 0.7, and 0.9 m in the previously noted grazing treatments, representing a range of only 0.02% to 0.3% of the land area cleared (Rogers and Lavigne, 1974). Differences between colonies in lightly and heavily grazed treatments are not apparent in terms of rates of forage extraction, foraging distance, time per foraging trip, or availability of seed (Rogers, 1974). Seed production of *B. gracilis* is lower in heavily grazed than ungrazed pastures (Coffin and Lauenroth, 1992), but Rogers (1974) observed that ants in the heavily grazed pasture tend to occupy areas where seeds are more abundant as a result of grazing of those areas late during the growing season. Seed consumption by ants is estimated to be only 2% of production.

Grasshoppers normally constitute only 8% of total macroarthropod biomass (Lauenroth and Milchunas, 1991), but may exert significant pressure on plant communities during outbreak years, when populations are three to five times greater than average (Pfadt, 1977). Several studies have assessed grasshopper densities on the grazing treatments at the CPER. Van Horn et al. (1970) found no significant difference between lightly and heavily grazed treatments, but Capinera and Sechrist (1982) and Welch et al. (1991) found greater densities in lightly than in more heavily grazed treatments. Grasshoppers of different subfamilies responded differently to the grazing treatments. Catantopinae and Gomphocerinae were positively correlated with plant biomass (Capinera and Sechrist, 1982). Oedipodinae, generally associated with bare areas, were negatively correlated with plant biomass and thus increased with increasing grazing intensities, but were a smaller proportion of the total in all treatments. Studies conducted in mixed-grass and tallgrass prairies generally showed greater densities at higher grazing intensities (reviewed in Capinera and Sechrist, 1982).

Microarthropods

In contrast to macroarthropods, microarthropods display only minor responses to long-term grazing treatments at the CPER (Leetham and Milchunas, 1985). Microarthropod sampling in this study did not include surface litter and crowns, and soil was sampled to a depth of 60 cm. Total numbers of microarthropods were 135,000 and 150,000 $\cdot$ m^{-2}, and biomass totals were 62.3 and 68.7 mg $\cdot$ m^{-2}, in the lightly and heavily grazed treatments, respectively. Differences between treatments in orders or families were not consistent with respect to trophic classification. However, significant increases with grazing were observed for Linotetranidae and Pseudococcidae, and decreases observed for Tardigrade and Bdellidae. The only significant effect of grazing intensity on depth distribution of microarthropod families was that Linotetranidae was found at deeper depths in the heavily grazed treatment. In contrast to grazing treatments, soil water and root biomass had large influences on microarthropods. Soil water was 19% greater in the heavily than lightly grazed treatment, which may have compensated for any negative effects of grazing. Root biomass did not differ with grazing treatment in this study.

Crossley et al. (1975) sampled ungrazed, lightly, moderately, and heavily grazed treatments to a depth of only 5 cm and found no more than a "suggestion" of microarthropod abundance being greater in lightly and moderately grazed treatments. In this study also, grazing treatments had little effect on the composition of groups of microarthropods.

Nematodes and Microfungi Populations

Plant parasitic nematodes are one of the three major groups of herbivores in the shortgrass steppe along with arthropods and large ungulates. This group of nematodes has been estimated to be 37% by Leetham and Scott (unpublished data), and 38% to 43% by Wall-Freckman and Huang (1998) of all nematodes in this system. Bacterial feeders, fungal feeders, predators, and omnivores are important regulators of decomposition and nutrient cycling in ecosystems (Freckman and Caswell, 1985). There is some evidence to suggest that herbivory by nematodes may exert a greater influence on primary production than would be predicted based upon their levels of consumption (Detling et al., 1980; Stanton, 1983). Leaf photosynthetic rates increased 35% when *B. gracilis* roots were clipped (Detling et al., 1980). Furthermore, aboveground herbivory can influence root chemistry (Cook et al., 1958; Kinsinger and Hopkins, 1961) and exudation (Dyer and Bokhari, 1976; Vancura and Stanek, 1975). Therefore, interactions may occur between above- and belowground herbivores.

Nematodes show very little response to grazing at the CPER. No significant differences were found in abundances of nematodes between long-term ungrazed and moderately grazed treatments, or between long-term grazing treatments or those recently exposed to grazing after long-term exclosure and recently exclosed to grazing after long-term moderate grazing (Wall-Freckman and Huang, 1998). Samples were collected both under plants (*B. gracilis*) and in the bare interspaces between them. Respective means were 1400 and 8500 nematodes·kg⁻¹ dry soil in the ungrazed treatment and 1400 and 6800 nematodes·kg⁻¹ dry soil in the grazed treatment. *Bouteloua gracilis* and total plant basal cover is greater in grazed treatments, which may tend to bring the means even closer. A total of 81 (under plant) and 83 (interplant) taxa were found in the ungrazed treatment compared with 82 and 79 in the grazed. No significant differences between long-term grazing treatments were found for species richness, fungivore/bacterivore, trophic diversity, Shannon-Weiner diversity (H'), exponential H', Simpson's diversity, evenness of the latter two, dominance diversity curves, two similarity indices, maturity index, or plant–parasite index. The proportion of bacterial feeders was slightly greater in long-term ungrazed than grazed treatment, and the opposite was found for plant parasites.

Microfungal colonies averaged 84,500 and 93,000·g⁻¹ dry soil in the lightly and heavily grazed treatments, respectively (Christensen and Scarborough, 1969). An average of 64 species were identified in the lightly grazed treatment compared with 54 in the heavily grazed, and dominance of particular species increased with the decreasing species richness in the heavily grazed treatment. Species similarity was 73% between the two grazing treatments compared with 81% between lightly grazed replicates.

Diversity, Relative Sensitivity, and Trophic Structure of Consumer Groups

The effect of grazing by domestic animals on biodiversity has been, and will probably continue to be, an important sociopolitical and conservation issue. Individual studies generally focus on one group of consumers and on its relationship to plant community or soil environment. Even when reviewed together, as in the previous sections, it is often difficult to discern relative sensitivities clearly. In this section we examine the relative degree to which long-term grazing by cattle has affected abundances, composition, and diversity by directly comparing differences in responses among the groups of organisms.

Grazing generally has negative effects on the abundances of consumer groups of the shortgrass steppe (Fig. 18.4, [Milchunas et al., 1998]). The few cases in which increases occur are associated with lighter intensities of grazing. Groups of consumers respond very differently to grazing intensity. Aboveground macroarthropods have the greatest response, and rodents and microarthropods have the least.

Dissimilarities in community composition of the groups of organisms (Fig. 18.5) do not always correspond to changes in abundance (Fig. 18.4) in relation to grazing intensities. Birds show small responses in overall abundance to grazing intensity, but are second to aboveground arthropods with respect to changes in community composition. Aboveground macroarthropods display large reductions in abundance with grazing, as well as large changes in species composition. Plants, microarthropods, and nematodes change little with grazing in either abundance or community composition.

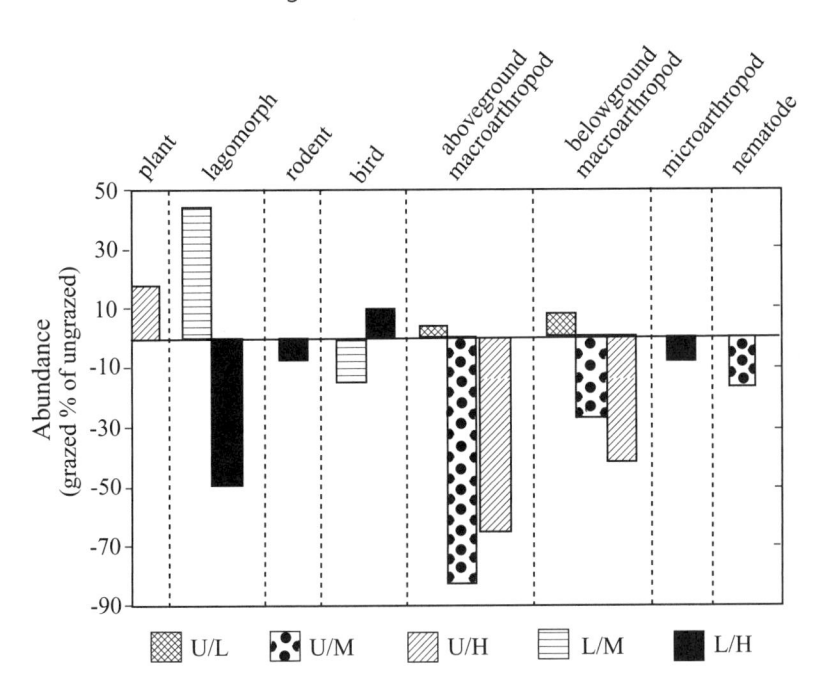

Figure 18.4 The abundance (measured as grazed percent of ungrazed) of plants, lagomorphs, rodents, birds, above- and belowground macroarthropods, microarthropods, and nematodes in relation to long-term grazing treatment contrasts in the northern shortgrass steppe. H, heavily grazed; L, lightly grazed; M, moderately grazed; U, ungrazed. (From Milchunas et al. [1998].)

Bird and plant diversity decline with increasing intensity of grazing (Fig. 18.6). Most other groups either increase in diversity with increasing grazing intensity and/or display highest diversity at intermediate intensities of grazing. Macroarthropods increase and birds decrease in diversity with increasing intensity of grazing, whereas both groups have large changes in species composition. Birds are the only group that change dominant species from the lightly to the heavily grazed treatment (Milchunas et al., 1998).

Over all indices, birds and aboveground macroarthropods appear to be the most sensitive to grazing in the northern shortgrass steppe. Changes in abundance and dissimilarity of communities are not related to responses in diversity, and there are no consistent responses in diversity between groups of organisms. Birds are the only group to display a similar response in diversity to that of plants.

That diversity of an assemblage increases or decreases with grazing does not necessarily indicate whether the changes are qualitatively positive or negative. Plants and birds are perhaps the most studied groups of organisms, and the ones to which quality attributes such as endemic/exotic, globally rare/abundant, can be most readily applied to species. Our work has shown (Milchunas et al., chapter 16, this volume) that plant species decline in diversity may not be considered a negative

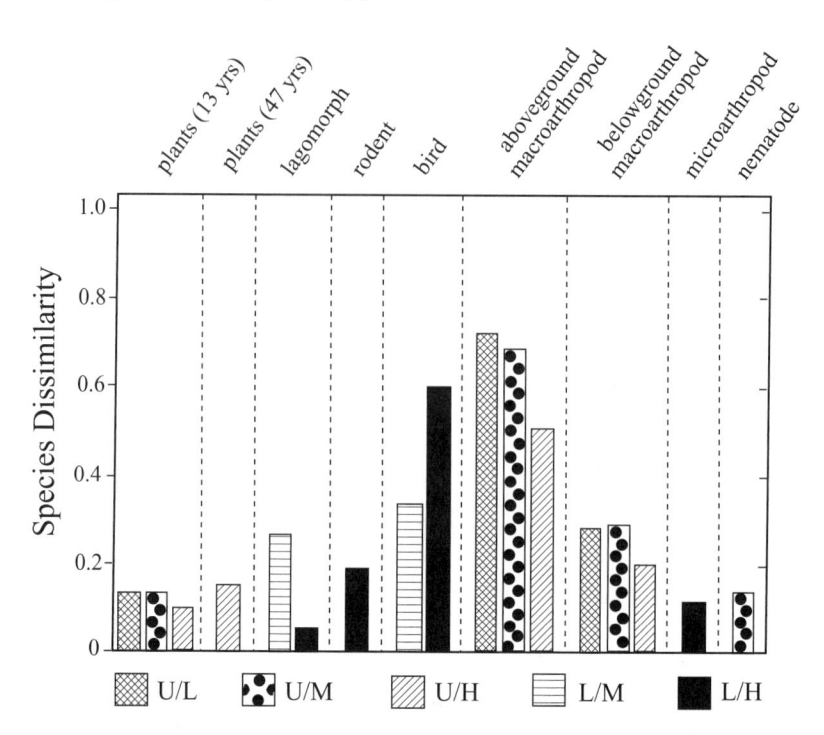

Figure 18.5 Species dissimilarity (Whittaker index of community association) of plants, lagomorphs, rodents, birds, above- and belowground macroarthropods, microarthropods, and nematodes in relation to long-term grazing treatment contrasts in the northern shortgrass steppe. H, heavily grazed; L, lightly grazed; M, moderately grazed; U, ungrazed. (From Milchunas et al. [1998].)

effect of grazing in this system that has evolved in the presence of large herbivores. Exotic and native weedy species contribute to the greater diversity of ungrazed compared with grazed communities. The situation for birds is somewhat less clear, possibly because migratory birds are subject to favorable or unfavorable conditions on both summer and winter ranges. Such factors as abundance, long-term population trends, and distributional characteristics, are used to determine vulnerability to extirpation (Carter and Barker, 1993) in bird species. In the shortgrass steppe, four out of six species of nesting birds have relatively high rankings of vulnerability (Table 18.1). Mountain Plovers, a federal category I species, are the most threatened grassland endemic species breeding in the shortgrass steppe. They nest primarily in heavily grazed areas. This is also true of McCown's Longspur (Table 18.1). Horned Larks, which are widespread in North America and not threatened, also increase with grazing. Two other endemics found in the shortgrass steppe but more closely associated with mixed-grass prairie (Lark Bunting, Chestnut-collared Longspur) prefer lightly grazed shortgrass steppe, but are more abundant in moderately and heavily grazed areas in the more productive mixed-grass prairie (Kantrud and Kologiski, 1982). Knopf (1996) commented

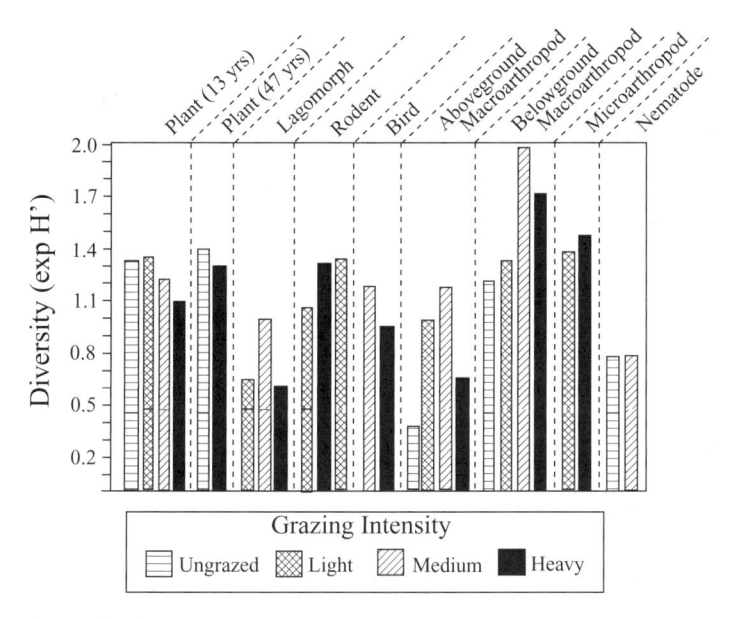

Figure 18.6 Species diversity (exp. H') of plants, lagomorphs, rodents, birds, above- and belowground macroarthropods, microarthropods, and nematodes in relation to long-term grazing treatments in the northern shortgrass steppe. (From Milchunas et al. [1998].)

Table 18.1 Breeding Bird Population Response to Long-Term Grazing Intensity Treatments in the Northern Shortgrass Steppe, with Population Abundance, Trend, Distribution, and Vulnerability Parameters

Species[d]	Breeding Pairs/20ac[a] Grazing Intensity L	M	H	BBS[b] no./Route N.Am.	BBS[b] Trend N.Am.	Distribution[c] Breed, Winter N.Am.	Vulnerable Index[c] Colorado
Horned Lark, S	2.5	4.5	7.3	24.7	−0.7[f]	11	1.43
Lark Bunting, E	5.3	3.1	0	27.3	−2.1[f]	43	3.29
W. Meadowlark, S	1.3	1.7	0	42.2	−0.5	21	2.14
McCown's Longspur, E	2.25	5	3.65	4.7	+7.3[g]	54	3.71
Mountain Plover, E,R	0	0	1.8	0.4	−3.7[g]	54	4
Chestnut-c. Longspur, E	0.3	0.5	0	9.9	+0.4	44	3.57

[a] Giezentanner (1970).
[b] Compiled from Breeding Bird Survey (BBS) data (1966–1993) by Knopf (1996).
[c] Carter and Barker (1993).
[d] E, endemic grassland; R, rare, federal category I candidate; S, Secondary grassland.
[e] Other studies indicate this specie prefers heavy grazed sites.
[f] Significant at P = .05.
[g] Significant at P = .001.
Breeding and winter distribution rank: 1, very widespread; 2, widespread; 3, intermediate; 4, local; 5, very local.
Vulnerability to extirpation: 1, very low; 2, low; 3, moderate; 4, high; 5, very high. H, heavy; L, light; M, moderate; N. Am., North America.
(From Milchunas et al. [1998].)

that all endemic grassland species of the Great Plains evolved within the context of an intensively grazed landscape, but that current management of domestic animals creates a uniformly grazed situation rather than the mosaic of differentially grazed sites that may have been characteristic of the grazing patterns of native ungulate herds.

Cattle, arthropods, and nematodes are the primary herbivores in the shortgrass steppe, each representing almost a third of the total herbivory. The presence or absence of cattle in this system has a large influence on the amount of plant material available to other herbivores, and potentially on the trophic structure of all consumers. Abundances of some groups of consumers are dramatically affected by grazing intensity treatments (Fig. 18.4). Although some statistically significant differences have been found for each particular group of consumers, different grazing intensities have very little overall effect on trophic structure of arthropods or nematodes (Fig. 18.3). This is also true for the herbivore component, which tends to increase rather than decrease with grazing by cattle.

Herbivore groups consuming relatively small amounts of plant material in the shortgrass steppe also display minor shifts in trophic structure. Small-mammal biomasses in herbivore, carnivore, omnivore, and granivore categories vary little among grazing treatments in the shortgrass steppe compared with very large differences found in differentially grazed tallgrass prairie (Grant et al., 1982). The proportion of insect and seed consumed by birds in the southern shortgrass steppe varies little between grazed and ungrazed treatments, but differences are evident in more productive mixed-grass prairie (Wiens and Dyer, 1975). Differences in trophic structure between grazed and ungrazed shortgrass steppe appear to be very small across all groups of organisms, and this is similar to the relatively small changes in plant species composition.

Summary and Overview

Grazing of public lands by domestic livestock is a controversial issue in the United States (Painter and Belsky, 1993) and will probably continue to be so. In a review of species endangerment patterns in the continental United States, Flather et al. (1994) listed grazing as a contributing factor to endangerment of 187 of a total of 667 threatened and endangered species. Grazing ranked high as a factor for plants (104 of 285 species), intermediate for birds (31 of 85) and insects (9 of 22), and low for reptiles (6 of 33), mammals (10 of 68), snails (4 of 13), and amphibians (4 of 11 species). Regions identified with high species endangerment are those with a short evolutionary history of grazing (southwestern United States and areas west of the Rocky Mountains).

This analysis of grazing effects on other consumers in the shortgrass steppe does not imply large negative impacts of livestock grazing. In contrast, the small effects of grazing on consumer trophic structure hints at a degree of stability. Birds are a particularly sensitive group to grazing. However, the response of the bird community suggests a conclusion similar to that for plants: The removal of the domestic large herbivore may be considered a disturbance in this system.

This may be the result of the long history of grazing by bison, and the relatively high similarity between diets of bison and cattle. However, the restricted movement of domestic stock compared with the unfettered wandering over long distances and wide areas by wild ungulates may be an ecologically important difference between the two (Werger, 1977). Research that contrasts rotational with season-long cattle grazing systems has found no differences with respect to plant community response in either the shortgrass steppe or nearby northern mixed-grass prairie (Derner and Hart, 2007). The influence of different grazing systems on wild consumers has not been explored in the shortgrass steppe. However, the degree to which any particular grazing system mimics the past pattern of free-ranging bison will always be speculative, because nominal conditions are not well documented.

In addition to the long history of grazing, the short stature of this semiarid grassland is an additional factor in the small effects of livestock grazing on both plant and consumer populations. Grazing by large herbivores directly affects canopy structure, but taller canopies are more affected by vegetation removal and trampling than shorter ones. Competition for light between grazed and ungrazed neighbors is more likely to be altered in a tall canopy than in a short one. Thus, grazing in productive systems with a long evolutionary history of grazing, such as the tallgrass prairie of North America or the Serengeti of Africa, often results in very large responses in consumer dynamics. In even greater contrast to the shortgrass steppe, grazing of domestic herbivores in productive systems with a short evolutionary history of grazing can be a massively destructive force, resulting in extinctions in and greatly altered functioning of systems such as those in Australia, New Zealand, and the southwestern United States. The shortgrass steppe appears to be among the most grazing-resistant ecosystems in the world.

References

Allen, P. F., and B. Osborn. 1954. Tallgrass defeats prairie dogs. *Soil Conservation* **20**:103–105.

Andrews, R. M. 1977. *A shortgrass prairie macroarthropod community: Organization, energetics and effects of grazing by cattle.* U.S. International Biological Program, Grassland Biome preprint no. 208. Natural Resource Ecology Laboratory, Colorado State University, Fort Collins, Colo.

Austin, D. D., and P. J. Urness. 1986. Effects of cattle grazing on mule deer diet and area selection. *Journal of Range Management* **39**:18–21.

Bock, C. E., and B. Webb. 1984. Birds as grazing indicator species in southeastern Arizona. *Journal of Wildlife Management* **48**:1045–1049.

Bohnam, C. D., and A. Lerwick. 1976. Vegetation changes induced by prairie dogs on shortgrass range. *Journal of Range Management* **29**:221–225.

Brown, J. H. 1973. Species diversity of seed-eating desert rodents in sand dune habitats. *Ecology* **54**:775–787.

Carter, M. F., and K. Barker. 1993. An interactive database for setting conservation priorities for western neotropical migrants, pp. 120–144. In: D. M. Finch and P. W. Stangel (eds.), *Status and management of neotropical migratory birds.* USDA Forest Service, general technical report RM-229. Rocky Mountain Forest and Range Experiment Station, Fort Collins, Colo.

Capinera, J. L., and T. S. Sechrist. 1982. Grasshopper (Acrididae)–host plant associations: Response of grasshopper populations to cattle grazing intensity. *Canadian Entomologist* **114**:1055–1062.

Christensen, M., and A. M. Scarborough. 1969. *Soil microfungi investigations Pawnee site.* U.S. International Biological Program, Grassland Biome technical report no. 23. Natural Resource Ecology Laboratory, Colorado State University, Fort Collins, Colo.

Clary, W. P., and D. M. Beale. 1983. Pronghorn reactions to winter sheep grazing, plant communities, and topography in the Great Basin. *Journal of Range Management* **36**:749–752.

Clary, W. P., and R. C. Holmgren. 1982. Observations of pronghorn distribution in relation to sheep grazing on the Desert Experimental Range, pp. 581–592. In: J. M. Peek and P. D. Dalke (eds.), *Wildlife–livestock relationships symposium.* University of Idaho, Forest, Wildlife, and Range Experiment Station, Moscow, Idaho.

Cody, M. L. 1985. *Habitat selection in birds.* Academic Press, Orlando, Fla.

Coffin, D. P., and W. K. Lauenroth. 1992. Spatial variability in seed production of the perennial grass *Bouteloua gracilis* (H.B.K.) LAG. ex Griffiths. *American Journal of Botany* **79**:347–353.

Cook, C. W., L. A. Stoddart, and F. Kinsinger. 1958. Responses of crested wheat grass to various clipping treatments. *Ecological Monographs* **38**:237–272.

Coppock, D. L., J. K. Detling, J. E. Ellis, and M. I. Dyer. 1983. Plant–herbivore interactions in a North American mixed-grass prairie, I. Effects of black-tailed prairie dogs on intraseasonal aboveground plant biomass and nutrient dynamics and plant species diversity. *Oecologia* **56**:1–9.

Crossley, D. A., Jr., C. W. Proctor, Jr., and C. Gist. 1975. Summer biomass of soil microarthropods of the Pawnee National Grassland, Colorado. *American Midland Naturalist* **93**:491–495.

Cully, J. F., Jr. 1991. Response of raptors to reduction of a Gunnison's prairie dog population by plague. *American Midland Naturalist* **125**:140–149.

Derner, J. D., J. K. Detling, and M. F. Antolin. 2006. Are livestock weight gains affected by black-tailed prairie dogs? *Frontiers in Ecology and Environment* **4**:459–464.

Derner, J. D., and R. H. Hart. 2007. Grazing-induced modifications to peak standing crop in northern mixed-grass prairie. *Rangeland Ecology and Management* **60**:270–276.

Detling, J. K., D. J. Winn, C. Procter-Gregg, and E. L. Painter. 1980. Effects of simulated grazing by belowground herbivores on growth, CO_2 exchange, and carbon allocation patterns of *Bouteloua gracilis. Journal of Applied Ecology* **17**:771–778.

Dyer, M. I., and U. G. Bokhari. 1976. Plant–animal interactions: Studies of the effects of grasshopper grazing on blue grama grass. *Ecology* **57**:762–772.

Flake, L. D. 1971. *An ecological study of rodents in a shortgrass prairie of northeastern Colorado.* U.S. International Biological Program, Grassland Biome technical report no. 100. Natural Resource Ecology Laboratory, Colorado State University, Fort Collins, Colo.

Flather, C. H., L. A. Joyce, and C. A. Bloomgarden. 1994. *Species endangerment patterns in the United States.* USDA Forest Service, general technical report RM-241. Rocky Mountain Forest and Range Experiment Station, Fort Collins, Colo.

Flinders, J. T., and R. M. Hansen. 1972. *Diets and habits of jackrabbits in northeastern Colorado.* Science series, 1. Range Science Department, Colorado State University, Fort Collins, Colo.

Flinders, J. T., and R. M. Hansen. 1975. Spring population responses cottontails and jackrabbits to cattle grazing shortgrass prairie. *Journal of Range Management* **28**:290–293.

Freckman, D. W., and E. P. Caswell. 1985. The ecology of nematodes in agroecosystems. *Annual Review Phytopathology* **23**:275–296.

French, N. R., W. E. Grant, W. Grodzinski, and D. M. Swift. 1976. Small mammal energetics in grassland ecosystems. *Ecological Monographs* **46**:201–220.

Giezentanner, J. B. 1970. *Avian distribution and population fluctuations on the shortgrass prairie of north central Colorado.* U.S. International Biological Program, Grassland Biome technical report no. 62. Natural Resource Ecology Laboratory, Fort Collins, Colo.

Gordon, I. J. 1988. Facilitation of red deer grazing by cattle and its impact on red deer performance. *Journal of Applied Ecology* **25**:1–10.

Grant, W. E., and E. C. Birney. 1979. Small mammal community structure in North American grasslands. *Journal of Mammalogy* **60**:23–36.

Grant, W. E., E. C. Birney, N. R. French, and D. M. Swift. 1982. Structure and productivity of grassland small mammal communities related to grazing-induced changes in vegetation cover. *Journal of Mammalogy* **63**:248–260.

Grant, W. E., N. R. French, and L. J. Folse, Jr. 1980. Effects of pocket gopher mounds on plant production in shortgrass prairie ecosystems. *Southwest Naturalist* **25**:215–224.

Graul, W. D. 1973. Adaptive aspects of the Mountain Plover social system. *Living Bird* **12**:69–94.

Graul, W. D. 1975. Breeding biology of the Mountain Plover. *Wilson Bulletin* **87**:6–31.

Grzybowski, J. A. 1980. *Ecological relationships among grassland birds during winter.* PhD diss., University of Oklahoma, Norman, Okla.

Grzybowski, J. A. 1982. Population structure in grassland bird communities during winter. *Condor* **84**:137–152.

Guenther, D. A., and J. K. Detling. 2003. Observations of cattle use of prairie dog towns. *Journal of Range Management* **56**:410–417.

Hailey, T. L., J. W. Thomas, and R. M. Robinson. 1966. Pronghorn die off in Trans-Pecos Texas. *Journal of Wildlife Management* **30**:488–496.

Hansen, R. M., and I. K. Gold. 1977. Blacktail prairie dogs, desert cottontails and cattle trophic relations on shortgrass range. *Journal of Range Management* **30**:210–214.

Hart, R. H. 1994. Rangeland, pp. 491–501. In: C. Arntzen (ed.). *Encyclopedia of agricultural science, vol. 3.* Academic Press, Los Angeles, Calif.

Hassien, F. D. 1976. *A search for black-footed ferrets in the Oklahoma Panhandle and adjacent area and an ecological study of black-tailed prairie dogs in Texas County, Oklahoma.* Masters thesis, Oklahoma State University, Stillwater, Okla.

Holland, E. A., and J. K. Detling. 1990. Plant responses to herbivory and belowground nitrogen cycling. *Ecology* **71**:1040–1049.

Hoover, R. L., C. E. Till, and S. Ogilvie. 1959. *The antelope of Colorado.* Colorado Game, Fish and Parks Department, technical bulletin no. 4. State of Colorado, Denver.

Hornaday, W. T. 1889. *The extermination of the American bison with a sketch of its discovery and life history.* Smithsonian report 1887. Smithsonian Institution, Washington, D.C.

Kantrud, H. A., and R. L. Kologiski. 1982. *Effects of soils and grazing on breeding birds of uncultivated upland grasslands of the northern Great Plains.* U.S. Department of Interior–Fish and Wildlife Service, Wildlife Research report 15.

King, R. 1978. *Habitat use and relative behaviors of breeding Long-billed Curlews.* Masters thesis. Colorado State University, Fort Collins, Colo.

Kinsinger, F. E., and H. H. Hopkins. 1961. Carbohydrate content of underground parts of grasses as affected by clipping. *Journal of Range Management* **14**:9–12.

Knopf, F. L. 1996. Prairie legacies: Birds, pp. 135–148. In: F. B. Samson and F. L. Knopf (eds.), *Prairie conservation: Preserving North America's most endangered ecosystem.* Island Press, Covelo, Calif.

Knowles, C. J. 1986. Some relationships of black-tailed prairie dogs to livestock grazing. *Great Basin Naturalist* **46**:198–203.

Koford, C. B. 1958. Prairie dogs, whitefaces and blue grama. *Wildlife Monograph* **3**:1–78.

Langner, L. L., and C. H. Flather. 1994. *Biological diversity: Status and trends in the United States.* USDA Forest Service, general technical report RM-244. Rocky Mountain Forest and Range Experiment Station, Fort Collins, Colo.

Larson, F. 1940. The role of the bison in maintaining the short grass plains. *Ecology* **21**:113–121.

Lauenroth, W. K., and D. G. Milchunas. 1991. The shortgrass steppe, pp. 183–226. In: R. T. Coupland (ed.), *Natural grasslands, introduction and western hemisphere. Ecosystems of the world 8A.* Elsevier, Amsterdam.

Lavigne, R. J., R. Kumar, J. Leetham, and V. Keith. 1972. *Population densities and biomass of aboveground arthropods under various grazing and environmental stress treatments on the Pawnee site, 1971.* U.S. International Biological Program, Grassland Biome technical report no. 204. Natural Resource Ecology Laboratory, Fort Collins, Colo.

Leetham, J. W., and D. G. Milchunas. 1985. The composition and distribution of soil microarthropods in the shortgrass steppe in relation to the soil water, root biomass, and grazing by cattle. *Pedobiologia* **28**:311–325.

Lloyd, J. E., R. Kumar, R. R. Grow, J. W. Leetham, and V. Keith. 1973. *Abundance and biomass of soil macroinvertebrates of the Pawnee site collected from pastures subjected to different grazing pressures, irrigation and/or water fertilization, 1970–1971.* U.S. International Biological Program, Grassland Biome technical report no. 239. Natural Resource Ecology Laboratory, Colorado State University, Fort Collins, Colo.

MacArthur, R. H. 1965. Patterns of species diversity. *Biological Review* **40**:510–533.

McNulty, F. 1971. *Must they die?* Doubleday, Garden City, N.J.

Merriam, C. H. 1902. The prairie dog of the Great Plains, pp. 257–270. In: *United States Department of Agriculture yearbook 1901.* U.S. Government Printing Office, Washington, D.C.

Milchunas, D. G., W. K. Lauenroth, and I. C. Burke. 1998. Livestock grazing: Animal and plant biodiversity of shortgrass steppe and the relationship to ecosystem function. *Oikos* **83**:65–74.

Milchunas, D. G., W. K. Lauenroth, P. L. Chapman, and M. K. Kazempour. 1989. Effects of grazing, topography, and precipitation on the structure of a semiarid grassland. *Vegetatio* **80**:11–23.

Milchunas, D. G., O. E. Sala, and W. K. Lauenroth. 1988. A generalized model of the effects of grazing by large herbivores on grassland community structure. *American Naturalist* **132**:87–106.

Milchunas, D. G., A. S. Varnamkhasti, W. K. Lauenroth, and H. Goetz. 1995. Forage quality in relation to long-term grazing history, current-year defoliation, and water resource. *Oecologia* **101**:366–374.

Miller, B., G. Ceballos, and R. Reading. 1994. The prairie dog and biotic diversity. *Conservation Biology* **8**:677–681.

Morris, M. G. 1973. The effects of seasonal grazing on the Heteroptera and Auchenorrhyncha (Hemiptera) of chalk grassland. *Journal of Applied Ecology* **10**:761–780.

Olendorff, R. R. 1972. *The large birds of prey of the Pawnee National Grassland: Nesting habits and productivity, 1969–1971*. U.S. International Biological Program, Grassland Biome technical report no. 151. Natural Resource Ecology Laboratory, Colorado State University, Fort Collins, Colo.

Olendorff, R. R., and J. W. Stoddart. 1974. The potential for management of raptor populations in western grasslands, pp. 44–88. In: F. N. Hammerstrom, B. E. Harrell, and R. R. Olendorff (eds.), *Management of raptors*. Raptor research report 2. Raptor Research Foundation, Vermillion, S.Dak.

O'Meilia, M. E., F. L. Knopf, and J. C. Lewis. 1982. Some consequences of competition between prairie dogs and beef cattle. *Journal of Range Management* **35**:580–585.

Painter, E. L., and A. J. Belsky. 1993. Application of herbivore optimization theory to rangelands of the western United States. *Ecological Applications* **3**:2–9.

Paton, P. W. C., and J. Dalton. 1994. Breeding ecology of Long-billed Curlews at Great Salt Lake, Utah. *Great Basin Naturalist* **54**:79–85.

Peden, D. G., G. M. Van Dyne, R. W. Rice, and R. M. Hansen. 1974. The trophic ecology of *Bison bison* L. on shortgrass plains. *Journal of Applied Ecology* **11**:489–497.

Pfadt, R. E. 1977. Some aspects of the ecology of grasshopper populations inhabiting the shortgrass prairie, pp. 73–79. In: H. M. Kulman and H. C. Chiang (eds.), *Insect ecology*. Papers presented in the A. C. Hodson lectures. University Minnesota Agricultural Experiment Station technical bulletin 310. University of Minnesota, St. Paul.

Phillips, P. 1936. The distribution of rodents in overgrazed and normal grasslands of central Oklahoma. *Ecology* **17**:673–679.

Plumb, G. E., and J. L. Dodd. 1993. Foraging ecology of bison and cattle on a mixed prairie: Implications for natural area management. *Ecological Applications* **3**:631–643.

Rebollo, S., D. G. Milchunas, I. Noy-Meir, and P. L. Chapman. 2002. The role of a spiny plant refuge in structuring grazed shortgrass steppe plant communities. *Oikos* **98**:53–64.

Rice, R. W., R. B. Dean, and J. E. Ellis. 1974. Bison, cattle, and sheep dietary quality and food intake. *Proceedings of the Western Section of the American Society of Animal Science* **25**:194–197.

Roberts, H. B., and K. G. Becker. 1982. Managing central Idaho rangelands for livestock and elk, pp. 537–543. In: J. M. Peek and P. D. Dalke (eds.), *Wildlife–livestock relationships symposium*. University of Idaho, Forest, Wildlife, and Range Experiment Station, Moscow, Idaho.

Rogers, L. E. 1974. Foraging activity of the Western harvester ant in the shortgrass plains ecosystem. *Environmental Entomology* **3**:420–424.

Rogers, L. E., and R. J. Lavigne. 1974. Environmental effects of Western harvester ants on the shortgrass plains ecosystem. *Environmental Entomology* **3**:994–997.

Rogers, L. E., R. J. Lavigne, and J. L. Miller. 1972. Bioenergetics of the Western harvester ant in the shortgrass plains ecosystem. *Environmental Entomology* **1**:763–768.

Rosenzweig, M. L., B. Smigel, and A. Kraft. 1975. Patterns of food, space, and diversity, pp. 241–268. In: I. Prakash and P. K. Ghosh (eds.), *Rodents in desert environments*. Dr. W. Junk, The Hague, the Netherlands.

Rottman, R. J., and J. L. Capinera. 1983. Effects of insect and cattle-induced perturbations on a shortgrass prairie arthropod community. *Journal of the Kansas Entomological Society* **56**:241–252.

Ryder, R. A. 1980. Effects of grazing on bird habitats, pp. 51–66. In: R. M. DeGraaf (ed.), *Management of western forests and grasslands for nongame birds.* USDA Forest Service general technical report INT-86. USDA Forest Service, Ogden, Utah.

Sanderson, R. H. 1959. *Relationship between jackrabbit use and availability of forage on native sandhills range.* Masters thesis, Colorado State University, Fort Collins, Colo.

Schmutz, E. M., E. L. Smith, P. R. Ogden, M. L. Cox, J. O. Klemmedson, J. J. Norris, and L. C. Fierro. 1992. Desert grassland, pp. 337–362. In: R. T. Coupland (ed.), *Natural grasslands, introduction and western hemisphere. Ecosystems of the world 8A.* Elsevier, Amsterdam.

Schwartz, C. C., and J. E. Ellis. 1981. Feeding ecology and niche separation in some native and domestic ungulates on the shortgrass prairie. *Journal of Applied Ecology* **18**:343–353.

Schwartz, C. C., and J. G. Nagy. 1976. Pronghorn diets relative to forage availability in northeastern Colorado. *Journal of Wildlife Management* **40**:469–478.

Schwartz, C. C., J. G. Nagy, and R. W. Rice. 1977. Pronghorn dietary quality relative to forage availability and other ruminants in Colorado. *Journal of Wildlife Management* **41**:161–168.

Shackford, J. S. 1991. Breeding ecology of the Mountain Plover in Oklahoma. *Bulletin of the Oklahoma Ornithological Society* **24**:9–13.

Smith, C. C. 1940. The effect of overgrazing and erosion upon the biota of the mixed-grass prairie of Oklahoma. *Ecology* **21**:381–397.

Sparks, K. L. 1972. *Grazing behavior of bison and cattle on a shortgrass prairie.* U.S. International Biological Program, Grassland Biome technical report no. 149. Natural Resource Ecology Laboratory, Colorado State University, Fort Collins, Colo.

Stanton, N. L. 1983. The effect of clipping and phytophagous nematodes on net primary production of blue grama, *Bouteloua gracilis. Oikos* **40**:249–257.

Taylor, W. P., and D. W. Lay. 1944. Ecologic niches occupied by rabbits in eastern Texas. *Ecology* **25**:120–121.

Taylor, W. P., C. T. Vorhies, and W. P. Lister. 1935. The relation of jack rabbits to grazing in southern Arizona. *Journal of Forestry* **33**:490–498.

Valone, T. J., and J. H. Brown. 1995. Effects of competition, colonization, and extinction on rodent species diversity. *Science* **267**:880–883.

Vancura, V., and M. Stanek. 1975. Root exudates of plants V. Kinetics of exudates from bean root as related to the presence of reserve compounds in cotyledons. *Plant and Soil* **43**:547–559.

Van Horn, D. H., R. Mason, P. Grove, D. Maujean, and G. Hemming. 1970. *Grasshopper population numbers and biomass dynamics on the Pawnee site.* U.S. International Biological Program, Grassland Biome technical report no. 148. Natural Resource Ecology Laboratory, Colorado State University, Fort Collins, Colo.

Van Vuren, D. 1982. Comparative ecology of bison and cattle in the Henry Mountains, Utah, pp. 449–457. In: J. M. Peek and P. D. Dalke (eds.), *Wildlife–livestock relationships symposium.* University of Idaho, Forest, Wildlife, and Range Experiment Station, Moscow, Idaho.

Wakeley, J. S. 1978. Factors affecting the use of hunting sites by Ferruginous Hawks. *Condor* **80**:316–326.

Wall-Freckman, D., and S. P. Huang. 1998. Response of the soil nematode community in a shortgrass steppe to long-term and short-term grazing. *Applied Soil Ecology* **9**:39–44.

Welch, J. L., R. Redak, and B. C. Kondratieff. 1991. Effects of cattle grazing on the density and species of grasshoppers (Orthoptera: Acrididae) of the Central Plains Experimental Range, Colorado: A reassessment after two decades. *Journal of the Kansas Entomological Society* **64**:337–343.

Werger, M. J. A. 1977. Effects of game and domestic livestock on vegetation in east and southern Africa, pp. 149–159. In: W. Krause (ed.), *Handbook of vegetation science, part XIII. Application of vegetation science to grassland husbandry.* Dr. W. Junk Publishers, The Hague, the Netherlands.

Whicker, A. D., and J. K. Detling. 1988. Ecological consequences of prairie dog disturbances. *BioScience* **38**:778–785.

Wiens, J. A. 1969. An approach to the study of ecological relationships among grassland birds. *Ornithology Monographs* **8**:1–93.

Wiens, J. A. 1973. Pattern and process in grassland bird communities. *Ecological Monographs* **43**:237–270.

Wiens, J. A. 1974a. Climatic instability and the "ecological saturation" of bird communities in North American grasslands. *Condor* **76**:385–400.

Wiens, J. A. 1974b. Habitat heterogeneity and avian community structure in North American grasslands. *American Midland Naturalist* **91**:195–213.

Wiens, J. A., and M. I. Dyer. 1975. Rangeland avifaunas: Their composition, energetics, and role in the ecosystem, pp. 146–182. In: D. R. Smith general technical report WO-1. USDA Forest Service, Washington, D.C.

Wiens, J. A., and J. T. Rotenberry. 1979. Diet niche relationships among North American grassland and shrubsteppe birds. *Oecologia* **42**:253–292.

With, K. A. 1994. McCown's Longspur. *Birds of North America* **96**:8–22.

19

The Future of the Shortgrass Steppe

Ingrid C. Burke
William K. Lauenroth
Michael F. Antolin
Justin D. Derner
Daniel G. Milchunas
Jack A. Morgan
Paul Stapp

Where lies the future of the shortgrass steppe? In prior chapters we have described the remarkable resilience of the shortgrass steppe ecosystem and its organisms to past drought and grazing, and their sensitivity to other types of change. Emerging from this analysis is the idea of vulnerability to two main forces: future changes in precipitation or water availability, and direct human impacts.

What are the likely changes in the shortgrass steppe during the next several decades? Which of the changes are most likely to affect major responses in the plants, animals, and ecosystem services of the shortgrass steppe? In this chapter we evaluate the current status of the shortgrass steppe and its potential responses to three sets of factors that will be driving forces for the future of the steppe: land-use change, atmospheric change, and changes in diseases.

Land-Use Change and Conservation

Conservation and Management Challenges in the Shortgrass Steppe: Traditional and Emerging Land-Use Practices

Paul Stapp, Justin D. Derner, Ingrid C. Burke,
William K. Lauenroth, Michael F. Antolin

Referring to the early 1900s, James Michener in his novel *Centennial* (1974) wrote the following:

> The old two-part system that had prevailed at the end of the nineteenth century—rancher and irrigator—was now a tripartite cooperation: the rancher used the rougher upland prairie; the irrigation farmer kept to the bottom lands; and the drylands gambler plowed the sweeping field in between, losing his seed money one year, reaping a fortune the next, depending on the rain. It was an imaginative system, requiring three different types of man, three different attitudes toward life.... (p. 1081)

Even today, because of the strong water limitation for cropping, the shortgrass steppe remains relatively intact, or at least unplowed, in contrast to other grassland ecosystems (Samson and Knopf, 1994). More than half of the shortgrass steppe remains in untilled, landscape-scale tracts, compared with only 9% of tallgrass prairie and 39% of mixed-grass prairie (The Nature Conservancy, 2003). These large tracts, including those in the national grasslands (Pawnee, Cimarron, Comanche, and Kiowa/Rita Blanca), provide the greatest opportunity for preserving key ecological processes and biological diversity.

The landscape of the 1900s has been rapidly changing during the past several decades. Increased habitat loss and fragmentation threaten biological diversity in the shortgrass steppe. Land-use changes in the shortgrass steppe are similar to those throughout the United States, where, during the latter half of the 20th century, most human population growth in lower density regions surrounds urban centers, contributing to land-use shifts from agricultural to exurban developments (Brown et al., 2005; Theobald, 2005). During recent years, exurban land development increased at a rate 25 times higher than overall U.S. population growth (Theobald, 2005). During the next decade and a half, it is estimated that exurban developments will expand to 14.3% of U.S. land area. The result for eastern Colorado may be a loss from production of as much as 35% of row–crop agriculture (Parton et al., 2003). Concomitantly, the composition of the rural population is changing, as the proportion of elderly people on agricultural lands increases (Hauteniemi and Gutmann, 2005; Parton et al., 2007a).

These changes are highly significant to agricultural ecosystem services (Millennium Ecosystem Assessment Series, 2003), including food, timber, and fiber production, as well as to cultural ecosystem services associated with open space, the social structure of rural communities, ecosystem processes including productivity, and biological diversity (Theobald, 2004). The interactions among land use, social processes and cultural values, and ecosystem services represent new frontiers in both ecological and social sciences (Gunderson et al., 2005), and pose many of the most important challenges for the future of the shortgrass steppe.

In the eastern plains of Colorado, population has grown exponentially during the past 50 years (2.47% increase per year; data from U.S. Census Bureau [2000]), especially in the 11 highly developed counties bordering the Front Range of the Rocky Mountains (Fig. 19.1). Between 1987 and 2002, the number of housing units in these counties increased by 30%, to more than 1.5 million, compared with only a 2% increase in the more rural counties of the plains (data from

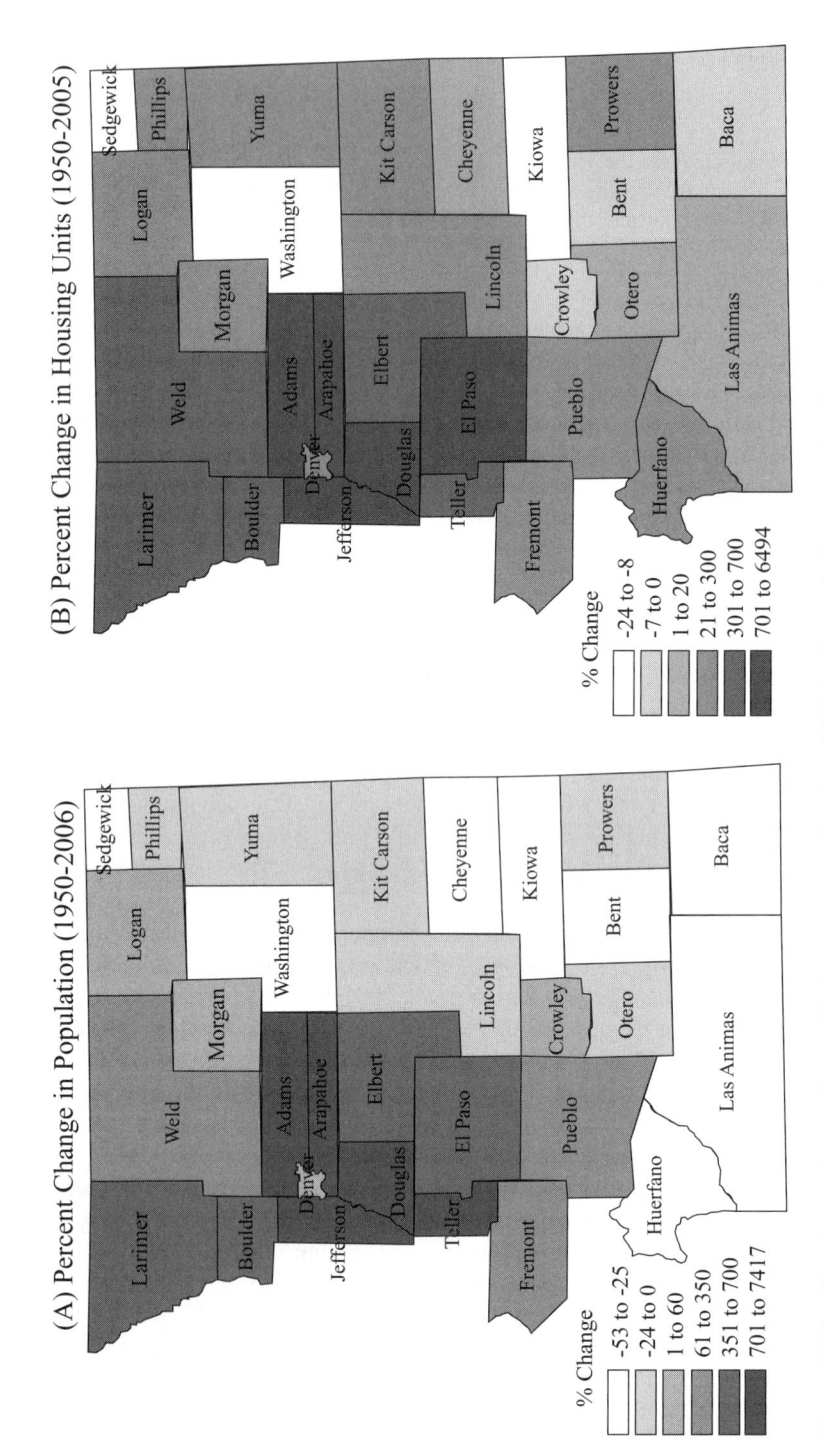

Figure 19.1 The changes in population (A), housing units (B), since 1950 for the available census data for eastern Colorado (U.S. Census).

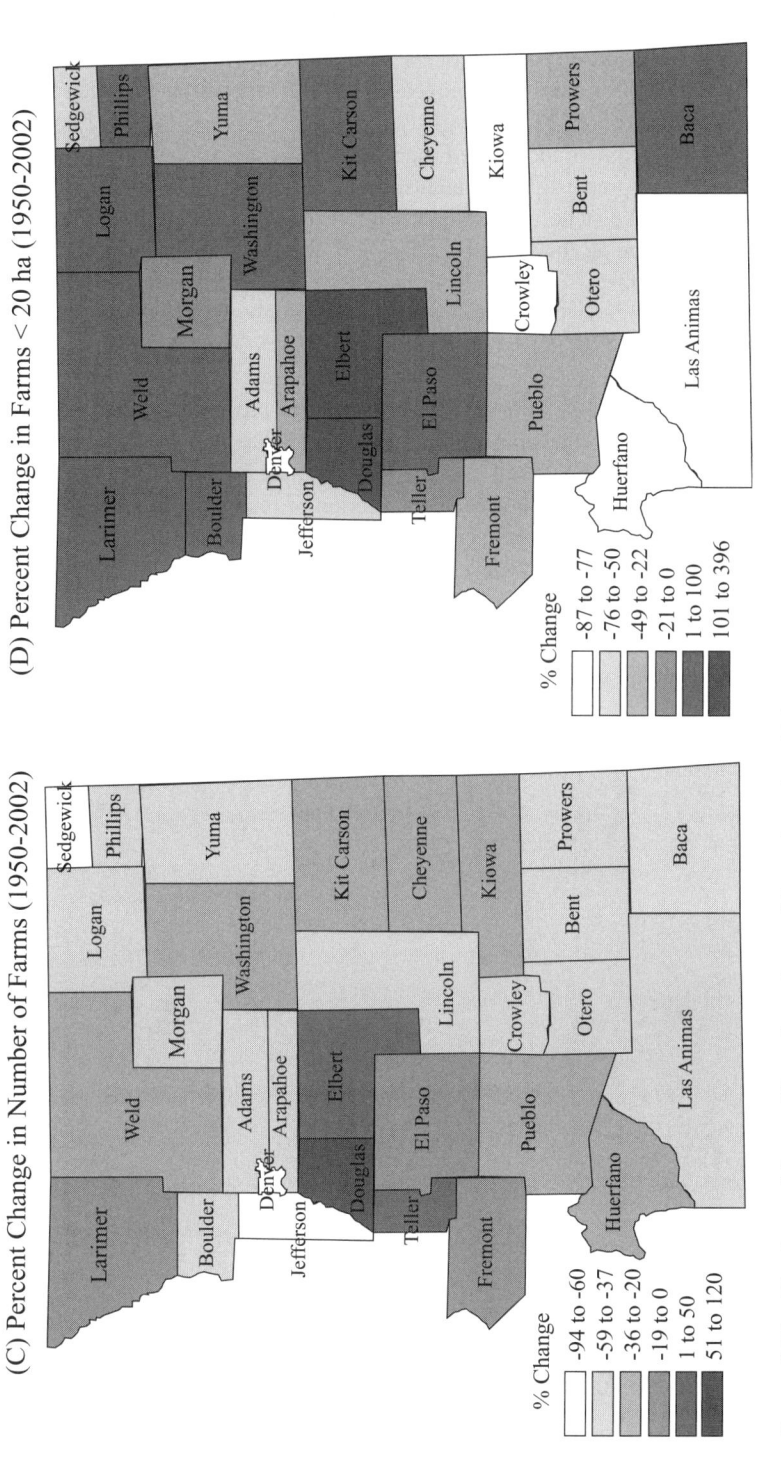

Figure 19.1 Numbers of farms (C), and numbers of farms less than 20 ha (D) since 1950 for the available census data for eastern Colorado (U.S. Census). The numbers of farms have increased because the average farm size is decreasing in the areas closest to urban growth along the Front Range of the Rockies.

Colorado Department of Local Affairs [2004]). During the same period, eastern Colorado lost 0.7 million ha, or 7%, of its agricultural land to other uses, with significant losses (10%) occurring in rangeland and other pasturelands (data from U.S. Census of Agriculture [2004]).

Much of the development along the Front Range has occurred as rural residential or exurban development that fragments larger agricultural lands into smaller parcels, particularly the richest bottomlands formerly devoted to crops. In counties along the Interstate Highway 25 corridor, both the total number of farms and the number of small farms (<20 ha) have increased dramatically (27% and 52%, respectively) during the past 20 years, whereas the numbers of farms and small farms have declined by 4% and 6% in rural counties (U.S. Census of Agriculture, 2004) (Fig. 19.1C, D). In addition to the loss of habitat, urban and exurban development leads to an increase in the number of nonnative predators such as domestic cats and dogs, and in the number of exotic and invasive plant species (Fig. 19.2) (Hanson et al., 2005; Maestas et al., 2003). Aquatic habitats also are threatened by groundwater depletion and the construction of reservoirs that accompany development, altering natural flows and turbidity, and fragmenting continuous river networks. Oil, gas, and coal development, and especially wind turbine farms (Drewitt and Langston, 2006), are emerging as possible threats to the native shortgrass steppe and wildlife in the region.

Figure 19.2 A homestead from the 1800s with current exurban growth in the background. (Photo by Justin Derner.)

On the substantial area of northern Colorado (≈80,000 ha) in public lands such as the Pawnee National Grassland (PNG), the primary land use is livestock production by local cattleman's associations. However, the PNG is currently experiencing dramatic increases in recreational use (S. Curry, personal communication, June 2006). Visitors from nearby urban areas pursue activities such as bird watching, hunting or recreational shooting, biking, hiking, and riding horses or motorized vehicles at the PNG.

Land-Use Intensification Reduces Native Species and Increases Invasive Species

Although plant and animal communities of the shortgrass steppe contain relatively few specialized or endemic species, human activities and changes in land use have led to significant declines of several native species. Thirty-eight of the 62 species of conservation concern in Colorado occur in the shortgrass steppe (Table 19.1), including three federally protected species (least tern, piping plover, black-footed ferret) and three species (black-tailed prairie dog, mountain plover, swift fox) recently petitioned for listing under the Endangered Species Act. Half of the species (18) are found in and/or near aquatic or wetland habitats, including eight species of fish and five amphibians. A number of grassland songbirds found in taller grass and shrub habitats (cassin's sparrow, brewer's sparrow, lark sparrow, grasshopper sparrow, lark bunting, western meadowlark) have suffered alarming population declines regionwide during the past three decades (2.2±1.9% decline per year [data from Rocky Mountain Bird Observatory; Colorado Division of Wildlife, 2003a]), whereas other species at risk, such as mountain plovers and long-billed curlews, are positively associated with heavily grazed grasslands.

Current wildlife conservation efforts (e.g., Colorado Division of Wildlife, 2003b) have emphasized the important ecological role of prairie dogs as prey for raptors and mammalian carnivores, and as ecosystem engineers that keep grasslands short and create burrows used by a number of other species (Kotliar et al., 1999; Stapp, 1998). Five species in Table 19.1 are closely associated with prairie dog colonies, which have declined during the past century as a result of habitat modification, widespread eradication programs, and the introduction of plague (as discussed later) (Antolin et al., 2002). In 2000, the U.S. Fish and Wildlife Service (2000) ruled that the black-tailed prairie dog warranted threatened status, but postponed formal action, citing higher priority threats to other species.

Although the petition was ultimately denied in 2004, as part of the assessment required by the petition process, states with prairie dog populations developed a multistate conservation plan (Luce, 2003), which led to a more complete inventory of the status of the black-tailed prairie dog throughout its range. An aerial survey conducted in the eastern plains of Colorado in 2001 found at least 255,000 ha of prairie dog colonies (White et al., 2005), which was 14 times more than the estimate of 17,806 ha with colonies used in the listing petition. This more accurate estimate exceeded the state's conservation target set out in the multistate plan (Luce, 2003).

Table 19.1 Species of Conservation Concern in the Shortgrass Steppe on the Eastern Plains of Colorado

Group	Conservation Status[a]	Species of Concern on Plains, %
Fishes[w]		35 (8/23)
Plains minnow	SE	
Suckermouth minnow	SE	
Brassy minnow	ST	
Arkansas darter	ST	
Plains orangethroat darter	SC	
Iowa darter	SC	
Stonecat	SC	
Flathead chub	SC	
Amphibians		
Northern cricket frog[w]	SC	86 (6/7)
Great Plains narrowmouth toad[w]	SC	
Northern leopard frog[w]	SC	
Wood frog[w]	SC	
Plains leopard frog[w]	SC	
Couch's spadefoot toad	SC	
Reptiles		
Yellow mud turtle[w]	SC	80 (8/10)
Triploid checkered whiptail	SC	
Texas horned lizard	SC	
Roundtail horned lizard	SC	
Common kingsnake	SC	
Texas blind snake	SC	
Massasauga	SC	
Common garter snake	SC	
Birds		
Least Tern[w]	FE,SE	53 (10/19)
Piping Plover[w]	FT,ST	
Western Snowy Plover[w]	SC	
Mountain Plover[p]	SC	
Long-billed Curlew[w,p]	SC	
Plains Sharp-tailed Grouse	SE	
Columbian Sharp-tailed Grouse	SC	
Lesser Prairie-Chicken	ST	
Burrowing Owl[p]	ST	
Ferruginous Hawk[p]	SC	
Mammals		
Black-footed ferret[p]	FE,SE	31 (4/13)
Black-tailed prairie dog[p]	SC	
Northern pocket gopher	SC	
Swift fox	SC	

[a]Conservation status: FE = federal endangered; FT = federal threatened; FC = federal candidate; SE = state endangered; ST = state threatened; Sc = state species of special concern.
Species denoted with a "w" are associated with aquatic systems, wetlands, riparian vegetation, or playa lakes and beaches. Associates of shortgrass steppe prairie dog colonies are denoted by a "p."
(From Colorado Division of Wildlife [2003a].)

Accurate assessments of the prairie dog abundance before European settlement do not exist. Vermiere et al. (2004) suggested that as the result of cattle grazing, by the early 20th century, prairie dog populations may have increased significantly over presettlement levels. Clearly, prairie dogs and associated wildlife on the shortgrass steppe represent an interesting case study with regard to species of concern, with little population data extending back farther than a hundred years, as well as changing scientific understanding and public perceptions.

Partly as a result of passage of the Environmental Protection Act, a ban on poisoning of prairie dogs on the PNG has been in effect since the 1970s. Shortly after the petitions for listing were filed in 1998, prairie dogs were protected from recreational shooting, and, at approximately the same time, outbreaks of plague became less extensive, possibly as a result of persistent drought. From 1998 to 2004, the total area and number of prairie dog towns increased exponentially on both the CPER (Derner et al., 2006) and the PNG (Fig. 19.3). Although after 2004, plague epizootics killed several of the largest towns, the total number of towns has continued to increase, as new towns become established and previously inactive ones are recolonized (Fig. 19.3). As of 2007, shooting of prairie dogs in Colorado was restricted only during the breeding season, from March 1 to June 14. How future changes in land use, recreational shooting, and plague will influence population fluctuations in prairie dogs and associated species remains to be seen.

Although intensified land use resulted in declines in native habitat and animal species, another important change in the shortgrass steppe has been the introduction of exotic plant species (Rickets et al., 1999). Disturbances associated with

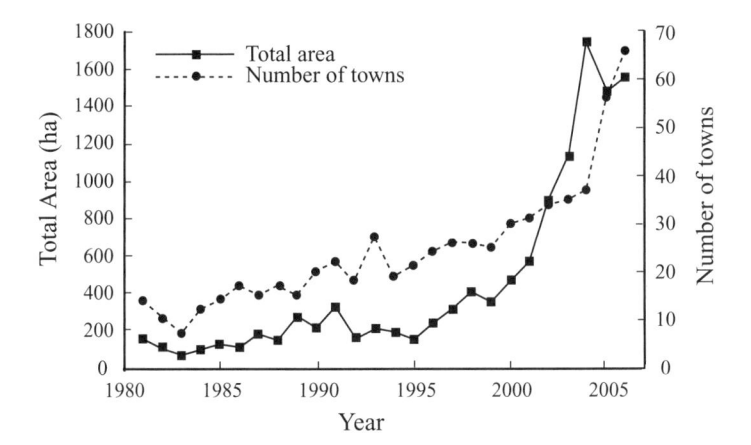

Figure 19.3 Extent of active black-tailed prairie dog towns on federally owned land on the CPER and PNG, Colorado, between 1981 and 2006. Despite outbreaks of plague that reduce total active area and lead to increases in the number of small towns (Stapp et al., 2004), prairie dogs have expanded dramatically, and currently cover between 1.5 to 2% of the 87,000-ha area. (Data from the PNG are courtesy of the U.S. Forest Service, with special thanks to Mark Ball and Elizabeth Humphrey.)

cropping (tillage, irrigation, fertilization) and residential development (tillage, irrigation, fertilization, but also pavement, road construction, and horticulture) increase the spread of plant propagules, both intentionally and unintentionally. For the shortgrass steppe, however, these disturbances function to increase the probability of survival and spread of these exotic plants, which are already part of the regional flora (Barkley et al., 1986). An important characteristic of the shortgrass steppe is that exotic plants fail to expand into uplands. The prevailing explanation for this is that native grasses—under the normal shortgrass steppe conditions of water and nutrient limitation and livestock grazing—have a competitive advantage over the exotics (Betz, 2001; Milchunas et al., 1992). However, roadsides, cropland without weed control, and abandoned croplands provide substantial opportunities for expansion of exotic plant species. Subdivision of private lands into smaller ranchettes or horse properties is very likely to increase the regional dominance of invasive and weedy species.

Land-Use Intensification Reduces Ecosystem Services

What about the effects of intensified human land use on ecosystem services such as carbon storage, the capture or production of greenhouse gases, and regional energy and water cycling? Clearly, both irrigated row–crop agriculture and urbanization increase net primary production (NPP) above that of the native shortgrass steppe. For instance, Kaye et al. (2005) estimated that within the developed area of Larimer County, Colorado, urban lawns accounted for up to 30% of the regional aboveground net primary production (ANPP) and 24% of regional soil respiration, although they occupy only 6.4 % of the area. Aboveground NPP, and hence food production, may be enhanced per unit area on irrigated cropland over native steppe, but energy-intensive additions of fertilizer, herbicides, and pesticides, coupled with transportation costs, are likely to reduce or potentially reverse net carbon capture and could lead to net carbon export at the regional scale. Carbon cycling is clearly accelerated by these land-use practices, but the regional effects on net ecosystem production or net carbon storage are unknown for the shortgrass steppe

Methane and nitrous oxide, two other important greenhouse gases with considerably stronger greenhouse effects than CO_2, are also naturally produced and consumed. Bacteria in native shortgrass steppe represent a considerable sink for methane (Mosier et al., chapter 14, this volume), but rates of methane uptake in irrigated crops and urban lawns are less than half that of the native steppe (Kaye et al., 2004; Mosier et al., chapter 14, this volume). In addition, urban lawns and irrigated crops produce 10 times more nitrous oxide than the native steppe, compounding their contribution to higher levels of greenhouse gases in the atmosphere.

Lastly, recent analyses demonstrate a significant effect of irrigated land and urban development on regional energy balance (Pielke et al., 1997), with feedbacks to precipitation. Recent atmosphere–biosphere simulation analyses (Stohlgren et al., 1998) suggest that irrigation in the Plains has resulted in decreases in July temperatures and increased stream flows in the adjacent Rocky Mountains. Despite these dynamics in which irrigation influences the Front Range, very dry years in the shortgrass steppe still occur.

Livestock Grazing Is a Management Tool in the Shortgrass Steppe

Much of the shortgrass steppe remains as native grassland because the climate is too dry to support extensive farming without irrigation. Livestock grazing continues to be the primary land use in the region and, when managed correctly, is viewed by many as compatible with multiple conservation goals (Knopf and Rupert, 1996; Maestas et al., 2003). Overgrazing may be a problem locally, especially in sensitive riparian areas (Fleischner, 1994), but the relatively low productivity of the shortgrass steppe makes widespread overgrazing economically unsustainable. In much of the western United States, one of the most economically and ecologically significant changes in rangeland ecosystems is the arrival and spread of invasive plants. Exotic plants can reduce forage quantity and quality, alter ecosystem function, and reduce biological diversity (DiTomaso, 2000; Masters and Sheley, 2001). As mentioned earlier, the native shortgrass steppe is highly resistant to exotic plant invasion, and cattle grazing intensity is inversely related to exotic species richness (Milchunas et al., 1990, 1992). Thus, domestic livestock may be an effective means of controlling the spread of some exotic plants.

It is likely that traditional ranching with uniform grazing has resulted in a less heterogeneous grassland than existed before European settlement, in part because of changes to natural disturbance regimes (including fire), and in part to changes in grazing patterns caused by the fencing in of open range and the switch from native to domesticated grazers. Recently, Fuhlendorf and Engle (2001) proposed that grazing management should "enhance heterogeneity instead of homogeneity to promote biological diversity and wildlife habitat on rangelands grazed by livestock." page 625. They advocated management approaches that simultaneously address the objectives of conservation biologists, ecologists, and rangeland managers. Contemporary approaches to grazing management incorporate the view that herbivores and the rangeland manager are both components of a complex ecosystem (Whalley, 2000) that requires multiple-use management to accommodate a diverse array of products and services.

Implicit is the notion that grazing management influences both the structure and function of rangeland ecosystems, and also influences ecosystem services. Rangeland managers have attempted to accommodate this by using cattle to manipulate species composition (Milchunas and Lauenroth, 1993), plant community structure (Milchunas and Lauenroth, 1989; Sala et al., 1986), and spatial heterogeneity of vegetation (Adler et al., 2001). Ecosystem services include maintenance or enhancement of biological diversity, carbon sequestration, increased water quality, and reduction of invasive weeds. Collectively, contemporary approaches to grazing management address the need for more sustainable rangelands while satisfying the public's recreational and aesthetic desires. As Briske (1993) stated "the sustainability of grazed systems is a more fundamental issue than grazing optimization." page 24.

However, current ecosystem-oriented approaches to grazing management alone are not able to achieve all conservation goals, being somewhat limited in the ability to create or modify habitats to the extent required for specific wildlife species of concern. In some cases, private land managers may be willing to modify management practices to produce species-specific habitat modifications, especially if financial support through direct payments or economic incentives is available. For example, seasonal or short-term intensive grazing may be used to create mosaics of habitat for species of grassland birds such as mountain plovers, which require extremely short vegetation for nesting and chick rearing (Knopf and Miller, 1994). In other cases, such as in riparian areas, the ecological costs of livestock grazing may prohibit its use as a management tool (Fleischner, 1994).

Conservation goals often vary between groups of users of the shortgrass steppe. At times, discord can arise between these groups in trying to develop a single management procedure to achieve diverse and sometimes divergent products and outcomes. One particular challenge is to devise approaches that meet the recreational needs of the public. On the shortgrass steppe, using grazing as a primary management tool has the advantage of being compatible with ecosystem and species-level approaches to conservation, with the socioeconomic status of rural communities, and with the ranching way of life (Knight et al., 2002).

Conservation Strategies

> *A prairie like that, one big enough to carry the eye clear to the sinking, rounding horizon, can be as lonely and grand and simple in its forms as the sea. It is as good a place as any for the wilderness experience to happen; the vanishing prairie is as worth preserving for the wilderness idea as the alpine forest.*
>
> —Wallace Stegner (1969, p. 152.)

Although rapid population growth along the eastern edge of the Rocky Mountains poses a great threat to the native shortgrass steppe, large tracts of it persist. Many of these tracts are on private lands, but many others are in public landholdings. To be effective, conservation approaches require the cooperation of many diverse entities, including private landholders, public agencies, land trusts, conservancies, and environmental groups. Increasing concerns about loss of rural cultures, populations, and agricultural/rural landscapes have led to conservation assessments that focus on ecosystem services (Theobald, 2003), and in some cases, economic incentives such as conservation easements (Kabii and Horwitz, 2006; Newburn et al., 2005). Conservation easements are an increasingly important land use that will shape the landscape into the future.

We expect that future conservation strategies with the greatest chances of success will do the following:

1. Advocate ecosystem-level conservation based on sound science, taking into account that agriculture-based economies provide the best hope for conserving large tracts of open space in the region (Knight et al., 2002). For the most part, this will consist of livestock production, because the shortgrass steppe is extremely slow to recover from tilling that disrupts the soil. In abandoned croplands, landowners should be encouraged to plant native species, including local genetic stocks, for revegetation and soil recovery efforts.

2. Promote grazing practices that restore structural and compositional heterogeneity to grasslands on a landscape scale, while minimizing detrimental effects on the most ecologically sensitive areas (e.g., riparian areas). Ideally, management at small scales (i.e., pastures, allotments) would collectively enhance heterogeneity at larger scales (i.e., watersheds, landscapes), such that the desired outcome would consist of a mosaic of natural habitats with differing serial stages of vegetation comprising an array of heights and types of plants that enhance ecosystem services (e.g., biodiversity) and amenities (recreation, wildlife viewing) for the public. Restoration of natural grazers such as bison and prairie dogs in large remaining tracts of grassland is one approach to meeting these goals that will enhance the conservation value of these lands, particularly with respect to restoring native dominant organisms.
3. Provide incentives to ranchers and farmers to maintain their land as functional open space, and target areas where large, contiguous areas of grasslands can be protected from development. Effective conservation of the shortgrass steppe will require protection of the most ecologically valuable and threatened tracts. Although some of the recent conservation efforts have focused on the important ecological role that prairie dogs play in the shortgrass steppe, we should recognize that some tracts of land may include areas not capable of supporting colonies of prairie dogs. Conservation easements between private landholders and local land trusts and national organizations allow for the purchase or protection of lands and development rights. Federal programs developed under the Farm Bill, such as the Farm and Ranch Lands Protection Program, Grasslands Reserve Program, provide funds and support for these agreements.

Urban open spaces, even areas that retain shortgrass plant communities or prairie dogs, may be population sinks or ecological traps (Battin, 2004; Kristan, 2003) for many species. Although these areas have recreational, educational, or aesthetic utility, they may be of minimal ecological value for plant and animal communities, or for other ecosystem services, because of their small size and isolation. However, they provide recreation for some, and are very important educational opportunities for urban residents to learn about the values and characteristics of grasslands.

Implications of Global Change

Jack A. Morgan, Daniel G. Milchunas, and William K. Lauenroth

Rates of Global Change

Today, scientists broadly agree that greenhouse gases like CO_2, nitrous oxide, and methane will continue to increase in the earth's atmosphere, although there is still debate about the rate at which these gas emissions are increasing, and the consequences for climate change around the world. Releases of significant quantities of CO_2, nitrous oxide, and methane into the atmosphere began during the Industrial Age, and continue today at increasingly higher rates (Parry et al., 2007). Anthropogenically driven global warming was detected during the 20th century, and is expected to increase significantly during the 21st century. Predicted changes in precipitation dynamics are complex and variable among world regions, but are expected to lead to more intense storms, which will increase surface runoff and erosion in some regions (Campbell et al., 1997). Nitrogen availability is increasing globally (Vitousek et al., 1997), particularly in grassland regions, as a result of high rates of fertilization coupled with urbanization (Burke et al., 2002). With few exceptions, the potential interactions of these and other less studied global change factors (ultraviolet radiation, ozone) have yet to be evaluated in multifactor experiments, nor have management practices been considered in the context of most global change experiments. So although global change is an agreed-upon certainty, its extent and impact are still unknown. The effects of global change are likely to be experienced in virtually all world ecosystems, including the shortgrass steppe, where we expect sensitivity to be high.

Water and Temperature Will Continue to Be Important

There are good reasons to suspect that the shortgrass steppe may be sensitive to several features of global change. Like most terrestrial ecosystems, spatial and temporal variations in temperature and precipitation affect the geographic extent of the shortgrass steppe (Lauenroth and Milchunas, 1991; Stephenson, 1990), the balance and productivity of species within its different regions (Alward et al., 1998; Coffin and Lauenroth, 1996; Epstein et al., 1997, 2002; VEMAP Members, 1995; Lauenroth et al., chapter 17, this volume), the structure of the plant community (Alward et al., 1998; Milchunas et al., 1989; Lauenroth, chapter 5, this volume), and the cycling of key nutrients (Burke et al., 1997a,b; Burke et al., chapter 13, this volume) and water availability (Lauenroth and Bradford, 2006; Lauenroth and Milchunas, 1991). Water is a key environmental driving variable in this semiarid grassland, with variations in annual amount and distribution causing large changes in ANPP (Lauenroth and Sala, 1992; Milchunas et al. 1994), and ultimately resulting in plant community shifts (Milchunas and Lauenroth, 1995). Thus, any fluctuations in water status, whether driven directly by altered precipitation patterns or indirectly through increased evapotranspiration under warmer temperatures, are certain to have impacts on the ecology, land use, and overall economics of this grazing land. Increases in temperature will alter the boundaries and species composition of the shortgrass steppe, and may increase productivity with extension of the growing season, provided temperature increases are not so severe that the desiccation response prevails.

Carbon and nitrogen cycling, and carbon storage are also sensitive to changes in precipitation and temperature (Burke et al., 1997a; Epstein et al., 2002; Lauenroth et al., 2004). Simulation results (Parton et al., chapter 15, this volume) suggest that total ecosystem carbon will decrease under all scenarios that include an increase in temperature, because of the sensitivity of decomposition to increased temperature. Increases in precipitation alone would likely result in increases in all fluxes, including NPP, respiration, and greenhouse gas fluxes to the atmosphere. In these simulations, the only climatic change that results in increased carbon storage is addition of nitrogen; all other changes result in either no change or losses of total carbon and nitrogen. Current long-term studies on the CPER support these simulations with respect to warming, wetting, and drying.

There are strong interrelations between climate and land-use management in the shortgrass steppe (Lauenroth et al., chapter 1, this volume). After grazing, the second most important land management type on the shortgrass steppe is dryland winter wheat (Lauenroth et al., 1999, 2000), managed with a fallow rotation. The western edge of the shortgrass steppe lies at the margin of viability for dryland wheat, and any decreases in precipitation or increases in evapotranspiration rates (both are likely with increasing temperature) may reduce the feasibility of growing wheat. Although wheat is the dominant crop by area, irrigated crops are by far the largest source of revenue; these crops are dependent upon a reliable source of surface or groundwater. Decreases in precipitation, combined with competition from urban subdivisions for water, could decrease the amount of irrigated cropland on the shortgrass steppe.

Responses to Carbon Dioxide Are Important as Well

Increases in atmospheric CO_2 are likely to have a greater impact on the shortgrass steppe than on many other ecosystems, largely because of water relations and the sensitivity of this region to water. Although increases in atmospheric CO_2 have been studied mostly in terms of the effects of enhanced photosynthesis on ecosystem functioning, the indirect effects on soil–plant water relations can be important as well. Increasing ambient CO_2 levels induces stomatal closure in most herbaceous species (Kimball and Idso, 1983; Morrison and Gifford, 1983; Wand et al., 1999), resulting in altered seasonal water dynamics and increased plant water use efficiency (Morgan et al., 2004b). This water relations response to elevated CO_2 may account in large part for the strong and consistent production increases to elevated CO_2 in the shortgrass steppe (LeCain et al., 2003; Milchunas et al., 2004; Morgan et al., 2001, 2004b; Nelson et al., 2004). Water-driven responses appear to be the dominant mechanism behind CO_2-induced production increases in semiarid systems and during dry spells in more mesic, native (or seminative) grasslands (Morgan et al., 2004b; Niklaus et al., 1998; Owensby et al., 1996b). Collectively, these findings suggest that global change will primarily affect systems like the shortgrass steppe through the indirect effects on water relations and interactions with temperature.

Nutrient Cycles Will Be Altered by Increased Carbon Dioxide, but Slowly

One of the more consistent responses to elevated CO_2, especially in systems exhibiting relatively large increases in NPP, is a decline in shoot nitrogen concentration (King et al., 2004; Milchunas et al., 2005; Morgan et al., 2004a). Although lower tissue nitrogen concentration may be partially attributed to the higher nutrient use efficiency that plants experience at elevated CO_2 (Drake et al., 1997), the accumulation of nitrogen in organic compounds at elevated CO_2, and the inability of the soil to release nitrogen quickly enough to meet the increased growth are more likely to be important on the shortgrass steppe (Luo et al., 2004; Reich et al. 2006a,b; Zak et al., 2000). Low-productivity ecosystems such as the shortgrass steppe are characterized by slow decomposition and nutrient mineralization, and low rates of nutrient supply (Burke et al., 1997a; Epstein et al., 2002; Wardle et al., 2004), and may be particularly vulnerable to the feedback of CO_2 on soil nitrogen supply. This low level of soil biological activity may account for the lack of significant effects of elevated CO_2 on fluxes of CO_2, methane, NO_x, and nitrous oxide during the course of a 5-year CO_2 enrichment experiment at the CPER (Mosier et al., 2002), despite significant effects on plant production (Milchunas et al., 2004; Morgan et al., 2001, 2004a). Slow nutrient cycling in such systems may also increase the time required for responses to perturbations such as increased CO_2 or altered temperature. This means that results from relatively short-term studies (<10 years) will not necessarily suffice for predicting the effects of incremental global change during the next hundred years (Morgan, 2002; Zak et al., 2000).

Species Responses to Carbon Dioxide Matter, But Are Difficult to Predict

Results from experiments conducted in the shortgrass steppe provide some insight into how individual species and functional groups will respond to increasing atmospheric CO_2 (Morgan et al., 2004b, 2007; Nowak et al., 2004). Increases in productivity and cover have tended to be more pronounced for C_3 than C_4 grasses (Morgan et al. 2004a; 2007), as predicted by differences in photosynthetic pathways (Polley, 1997). However, the response of C_3 grasses is not via photosynthesis, but by greater seedling recruitment of the C_3 perennial grass *Stipa comata*. Furthermore, photosynthesis, water relations, and productivity of dominant shortgrass steppe grass species, C_3 and C_4 alike, can respond to CO_2 (Hunt et al., 1998; LeCain et al., 2003; Morgan et al., 1994). Species shifts in the shortgrass steppe caused by increasing CO_2 will ultimately depend on complex interactions among the plant community, the soils, and the climate (Morgan et al., 2004a), and will not be easily predicted by differences in a single attribute like the photosynthetic pathway. Incorporating multiple plant mechanisms into models that accurately predict species responses to CO_2 and other environmental variables will be a challenge.

An example of surprising responses is the 40-fold increase in aboveground biomass of the C_3 subshrub *Artemisia frigida* during 5 years of CO_2 enrichment, which involved the interaction of several plant mechanisms and a drought (Morgan et al., 2007). Basic information on how critical plant species acquire and use water to produce biomass, participate in soil–plant nutrient cycling (King et al., 2004; Mosier et al., 2003), and recruit new individuals (e.g., Peters et al., chapters 6 and 7, this volume) must be integrated into models to predict confidently how productivity and plant species composition in the shortgrass steppe will shift in the face of increasing atmospheric CO_2.

Ultraviolet Radiation Is an Important Current Control and Its Importance May Increase in the Future

Another aspect of climate change is the potential for alterations in levels of ultraviolet radiation reaching the earth's surface, which may increase because of depletion of ozone, or may decrease if cloudiness or particulates in the atmosphere increase. Ultraviolet radiation affects decomposition rates in the shortgrass steppe and similar dry areas of the world (Austin and Vivanco, 2006; Parton et al., 2007b). A short-term study of ultraviolet effects on plant production and decomposition at the shortgrass steppe suggests complex interactions with elevated atmospheric CO_2. In contrast to increases in production with elevated CO_2, ultraviolet radiation decreased primary production (Milchunas et al., 2004). These interactions among various aspects of climate change and different species responses to CO_2, temperature, and ultraviolet radiation highlight the challenges associated with predicting the effects of global change.

Future Directions in Climate Change Research and Implications for Management

Our experience in the shortgrass steppe indicates that global change is likely to have important effects primarily through its impact on water relations, but that feedbacks involving nutrient cycles may be important as well. Primary production is responsive to CO_2, and nitrogen-releasing reactions appear unable—in the short term—to meet the higher nitrogen demand completely when CO_2 increases. Thus, the carbon-to-nitrogen ratio of the system may increase, with repercussions for plant nutrient status, soil biology, and whole-system nitrogen cycling (Luo et al., 2004; Reich et al., 2006b; Zak et al., 2000). An important economic consequence may be a reduction in forage quality below levels for sustainable livestock production (Milchunas et al., 2005). This response has been observed in other grasslands (Körner, 2000; Owensby et al., 1996a), and seems to be one of the more predictable responses in grasslands that experience significant production responses to CO_2. Changes in species composition such as those that occurred in the CO_2 enhancement experiment (Morgan et al., 2004a) may contribute to a decline in forage quality, although our ability to extrapolate this response to other regions of the shortgrass steppe is uncertain because of the variability of species composition in plant communities (Lauenroth and Milchunas, 1991).

As we design future global change experiments, considerable care should be taken in selecting which factors to impose in multifactor field experiments, and which ones we must be content to model. Considerations for selection criteria include current knowledge gaps, cost, ability to impose realistic treatments, ability to extrapolate and model results, and potential uses of research for management and policy.

Ecological Consequences of a Changing Disease Environment

Michael F. Antolin

Changing land use on the grassland, combined with increases in CO_2, climate change, and invasions by exotics, will also create new opportunities for pathogens to spread, with as-yet unknown ecological consequences. It is increasingly apparent that local prevalence and outbreaks of infectious diseases can be influenced by large-scale climatic patterns like those driven by El Niño Southern Ocillation (ENSO), the periodic surface warming of the southern Pacific (Anderson et al., 2004; Harvell et al., 2002; Pascual and Dobson, 2005; Stapp et al., 2004). In most cases, the mechanisms associated with ENSO-driven disease cycles are unknown, but it is hypothesized that changes in temperature and precipitation trigger cascades that increase or decrease primary productivity. In turn, changing resources alter host density and contact rates between hosts and/or the insect vectors that transmit the pathogens (Dobson and Foufopoulos, 2001; Gratz, 1999; Jones et al., 1998; Parmenter et al., 1999). Perturbations within ecological communities lead to changing epidemiological conditions and rates of disease transmission, with either amplification of pathogens to epidemic levels, or declines back to low-level persistence. Within these broad patterns, however, the mechanisms that permit both disease outbreaks and low-level persistence remain to be explored (Harvell et al., 2002).

The changing disease environment on the shortgrass steppe may have three direct ecological consequences: (1) change in fauna and flora mortality caused by exotic pathogens, and resulting changes in community structure; (2) altered frequency of disease outbreaks triggered by climate cycling between droughts and cooler periods with more abundant rainfall; and (3) greater exposure of humans and their domestic animals to pathogens when they move to exurban homes surrounded by the grasslands. A fourth possibility—that native pathogens of both plants and animals increase in frequency—will be considered at the end of this section.

We can illustrate the effects of a changed disease environment by the best-studied disease system on the shortgrass steppe: the black-tailed prairie dogs and sylvatic plague (the name given to the disease when it cycles in wild populations). The case of the black-tailed prairie dog illustrates how single species may be affected, but also shows that in many instances the changes could be communitywide, especially if the pathogen infects keystone species like prairie dogs (Antolin et al., 2002; Kotliar et al., 1999; Stapp, 1998). Besides, we may expect a continued stream of new introductions of both plant and animal pathogens, much as we see an onslaught of introduced weedy plants and animals (Anderson et al., 2004; Dobson and Foufopoulos, 2001). Recently introduced animal pathogens include West Nile virus, which found a good home in the western United States, where an effective mosquito vector (*Culex taraslis*) already resides, and the monkeypox virus, which was introduced into the United States via the exotic pet trade from Africa, but apparently failed to establish. Furthermore, climate change may play a role in the severity of pathogen introductions. For instance, plant fungal and bacterial pathogens show greater rates of introduction and emergence with climate change (especially increased moisture), whereas plant viruses seem to be unaffected by climate (Anderson et al., 2004).

Plague, caused by the bacterium *Yersinia pestis*, is primarily a disease of rodents spread by fleas, although it is best known for the high mortality it caused historically during human epidemics (Poland and Dennis, 1998). The eastern extent of plague in wild rodents is near the 97th meridian in southern Texas, extending northward to the 102nd meridian in North Dakota (Antolin et al., 2002; Barnes, 1993). In the absence of significant climate change, plague is not expected to spread farther eastward, because previous introductions into ports along the Atlantic and Gulf coasts of the United States failed to establish sylvatic plague in the eastern part of the United States. Worldwide plague foci persist in semiarid ecosystems with populations of burrowing rodents, including the western United States (Poland and Dennis, 1998).

The first large-scale die-offs of prairie dogs in northeastern Colorado were reported in 1948 (Ecke and Johnson, 1952), and plague has recurred here since that time (Centers for Disease Control and Prevention, unpublished records). Outbreaks of plague decimate local populations of prairie dogs, with mortality usually reaching 100% (Barnes, 1993). Long-term monitoring (1981 to present; Fig. 19.3) on the PNG revealed that prairie dogs persist as a metapopulation, with local extinctions followed by recolonization 2 to 4 years later (Roach et al., 2001; Stapp et al., 2004, chapter 8, this volume). By changing the population dynamics of prairie dogs, plague has had devastating effects on the black-footed ferret, possibly the most endangered mammal in North America. Abundance of three bird species (burrowing owl, ferruginous hawk, and mountain plover) also tends to fluctuate along with prairie dog populations (Antolin et al., 2002).

The effects of climate on incidence of plague were first noted in northern New Mexico and Arizona, where rainfall, summer temperature, and habitat characteristics (e.g., piñon pine and juniper) all influence transmission of plague from rodents to fleas to humans (Enscore et al., 2002; Parmenter et al., 1999). Our own studies (Stapp et al., 2004) suggest a similar link between prairie dog die-offs and cooler/moister summers on the shortgrass steppe during ENSO years. It remains to be seen whether the climate affects the pathogen directly, alters the reproduction of flea vectors, or changes dynamics of the entire rodent community (Gage and Kosoy, 2005).

How will the disease environment change in the future? We have already considered some effects of exotic pathogens like plague, but this question brings up the fourth disease possibility: strong emergence of native pathogens that currently exist at low levels. Disease emergence may have two sources: direct environmental changes by humans, and emergence as a consequence of climate change. A clear example of human impacts is the emergence of the bacterial pathogen that causes rabbit fever (tularemia) in prairie dogs that were captured and held in captivity for the exotic pet trade (Avashia et al., 2004). Similarly, domestic animals can transmit disease to wild populations: The last black-footed ferrets in Wyoming were taken into captivity to avoid an outbreak of canine distemper virus. With larger populations of both humans and their domesticated plants and animals, the risk of pathogen transmission from wild to domesticated populations is bound to increase, and vice versa (Weiss and McMichael, 2004). The effects of climate change are more complex, because some conditions favor pathogen spread while others slow it (Harvell et al., 2002). For instance, drought could reduce disease risk in plant species, because pathogenic fungi and bacteria require relatively moist conditions for transmission. On the other hand, drought could increase spread of animal pathogens if animals aggregate around scarce resources (e.g., water) during droughts. Great consideration also must be given to changes in geographic distribution in relation to climate change. Pathogens that are common in warmer southern regions of North America may spread northward as temperatures increase, especially if their most common hosts also shift their ranges northward.

In sum, changes in the disease environment on the shortgrass steppe parallel closely the changes expected after the introduction of exotic animals and plants, the effects of human disturbances on plant and animal communities, and the influence that climate change will have on the range and abundance of potential host species. The greatest ecological effects will be seen in alteration of keystone species like prairie dogs, or dominant species like blue grama and buffalograss. Regardless, the possibilities warrant continued vigilance and surveillance.

Conclusion

Ingrid C. Burke and William K. Lauenroth

Long-term research in the shortgrass steppe has taught us one very important lesson: The steppe changes slowly in response to environmental changes that fall within its evolutionary and developmental history, a history that has included drought and herbivory. Because of the character of these two forces, much of the richness of the shortgrass steppe exists belowground. The most rapid changes in the absence of humans occurred during and after the Pleistocene, and were associated with arid times that produced high rates of erosion and deposition (Kelly et al., chapter 3, this volume). However, even these changes took place over thousands of years, and we now think that soil instability occurred because *B. gracilis* was not present. The dominant organisms and structure of the native steppe have been relatively constant since then, even in the face of extended drought.

Now and into the foreseeable future, the shortgrass steppe is facing unprecedented changes wrought by humans: land-use change, global change, and changes in diseases. How will the shortgrass steppe respond? Except in the most dramatic of changes (exurbanization, cultivation management, combined with drought and wind), we are confident of three things: (1) the native portions of the shortgrass steppe will respond slowly; (2) many of the changes will be hidden from view, occurring below the soil surface, where by far the largest proportion of biological activity occurs; and (3) long-term sustainability will be determined by the landscape to regional-scale context of the changes. Continued research over long periods, including research that applies the newest techniques, that increases our understanding belowground, and that allows us to scale to the regional view, will give us the largest likelihood of predicting the future.

The sun burns the snow high on the mountains
It runs and it grows as it falls
Silt and soil
Down it boils
Down to the valleys,
The gold river rolls to the plains

The rangeland lies high, up from the river
The coolies are dry where the shortgrass grows
Fields of hay, cottonwood shade
Green patch of home through the high dusty land
The river flows

Early evening light, boys practice roping
The day fades away, the night rolls on.....
Lives of pride
Men who ride
They keep the old skills that came up the trail from Mexico

The long river winds through green years and dry years
Brand 'em in the spring, ship 'em in the fall
The new colt foaled, the mare grows old
Cycle of changes in this changeless land where the shortgrass grows
In this changeless land where the shortgrass grows
    Ian and Sylvia Tyson, "The Short Grass"

References

Adler, P. B., D. A. Raff, and W. K. Lauenroth. 2001. The effect of grazing on the spatial heterogeneity of vegetation. *Oecologia* **128**:465–479.

Alward, R. D., J. K. Detling, and D. G. Milchunas. 1998. Grassland vegetation changes and nocturnal global warming. *Science* **283**:229–231.

Anderson, P. K., A. A. Cunningham, and N. G. Patel. 2004. Emerging infectious diseases of plants: Pathogen pollution, climate change and agrotechnology drivers. *Trends in Ecology and Evolution* **19**:535–544.

Antolin, M. F., P. Gober, B. Luce, D. E. Biggins, W. E. Van Pelt, D. B. Seery, M. Lockhart, and M. Ball. 2002. The influence of sylvatic plague on North American wildlife at the landscape level, with special emphasis on black-footed ferret and prairie dog conservation. *Transactions North American Wildlife and Natural Resources Conference* **67**:104–127.

Austin, A. T., and L. Vivanco. 2006. Plant litter decomposition in a semi-arid ecosystem controlled by photodegradation. *Nature* **442**:555–558.

Avashia, S. B., J. M. Petersen, C. Lindley, M. E. Schriefer, K. L. Gage, M. Cetron, A. Thomas, T. A. Demarcus, D. K. Kim, J. Buck, J. A. Monteneiri, J. L. Lowell, M. F. Antolin, M. Y. Kosoy, L. G. Carter, M. C. Chu, K. Hendricks, T. David. D. L. Dennis, and J. L. Kool. 2004. First reported prairie dog-to-human tularemia transmission, Texas, 2002. *Emerging Infectious Diseases* **10**:483–486.

Barkley, T. M., R. E. Brookse, K. Schofield, R. L. McGregor, and 11 members of the Great Plains Flora Association. 1986. *Flora of the Great Plains.* University Press of Kansas, University of Kansas, Lawrence, Kans.

Barnes, A. M. 1993. A review of plague and its relevance to prairie dog populations and the black-footed ferret, pp. 28–37. In: J. L. Oldemeyer, D. E. Biggins, and B. J. Miller (eds.), *Proceedings of the symposium on the management of prairie dog complexes for the reintroduction of the black-footed ferret.* U.S. Department of the Interior biological report no. 13.U. S. Fish and Wildlife Service, Denver.

Battin, J. 2004. When good animals love bad habitats: Ecological traps and conservation of animal populations. *Conservation Biology* **18**:1482–1491.

Betz, D. 2001. *Dynamics of exotic species in the Pawnee National Grasslands, CO, USA.* Masters thesis. Colorado State University, Fort Collins, Colo.

Briske, D. D. 1993. Grazing optimization: A plea for a balanced perspective. *Ecological Applications* **3**:24–26.

Brown, D. G., K. M. Johnson, T. R. Loveland, and D. M. Theobald. 2005. Rural land-use trends in the conterminous United States, 1950–2000. *Ecological Applications* **15**:1851–1863.

Burke, I. C., W. K. Lauenroth, G. Cunfer, J. E. Barrett, A. R. Mosier, and P. Lowe. 2002. Nitrogen in the central grasslands region of the U.S. *BioScience* **52**(9):813–823.

Burke, I. C., W. K. Lauenroth, and D. G. Milchunas. 1997a. Biogeochemistry of managed grasslands in Central North America, pp. 85–102. In: E. A. Paul, E. T. Elliott, K. Paustian, and C. V. Cole (eds.), *Soil organic matter in temperate agroecosystems.* CRC Press, Boca Raton, Fla.

Burke, I. C., W. K. Lauenroth, and W. J. Parton. 1997b. Regional and temporal variability in aboveground net primary productivity and net N mineralization in grasslands. *Ecology* **78**(5):1330–1340.

Campbell, B. D., D. M. Stafford Smith, and G. M. McKeon. 1997. Elevated CO2 and water supply interactions in grasslands: A pastures and rangelands management perspective. *Global Change Biology* **3**:177–187.

Coffin, D. P., and W. K. Lauenroth. 1996. Transient responses of North-American grasslands to changes in climate. *Climatic Change* **34**:269–278.

Colorado Department of Local Affairs. 2004. *Economic and demographic information system.* Online. Available at www.dola.state.co.us/is/cedishom.htm.

Colorado Division of Wildlife. 2003a. *Colorado listing of endangered, threatened and wildlife species of special concern.* Online. Available at http://wildlife.state.co.us/species_cons/list.asp.

Colorado Division of Wildlife. 2003b. *Conservation plan for grassland species in Colorado. Final plan 26 Nov 2003.* Online. Available at http://wildlife.state.co.us/species_cons/Grasslands_Species_Conservation/conservationplan.asp.

Derner, J. S., J. K. Detling, and M. F. Antolin. 2006. Are livestock weight gains affected by black-tail prairie dogs? *Frontiers in Ecology and the Environment* **4**:459–464.

DiTomaso, J. M. 2000. Invasive weeds in rangelands: Species, impacts, and management. *Weed Science* **48**:255–265.

Dobson, A., and J. Foufopoulos. 2001. Emerging infectious pathogens of wildlife. *Philosophical Transactions of the Royal Society London* **B56**:1001–1012.

Drake B. G., M. A. Gonzàlez-Meler, and S. P. Long. 1997. More efficient plants: A consequence of rising atmospheric CO_2. *Annual Review of Plant Physiology and Plant Molecular Biology* **48**:609–639.

Drewitt, A. L., and R. H. W. Langston. 2006. Assessing the impacts of wind farms on birds. *Ibis* **148**:29–42.

Ecke, D. H., and C. W. Johnson. 1952. *Plague in Colorado and Texas. Part I. Plague in Colorado.* Public health monograph no. 6. U.S. Government Printing Office, Washington, D.C.

Enscore, R. E., B. J. Biggerstaff, T. L. Brown, R. F. Fulgham, P. J. Reynolds, D. M. Engenthaller, C. E. Levy, R. R. Parmenter, J. A. Montenieri, J. E. Cheek, R. K. Grinnell, P. J. Ettestad, and K. L. Gage. 2002. Modeling relationships between climate and the frequency of human plague cases in the southwestern United States, 1960-1997. *American Journal of Tropical Medicine and Hygiene* **66**:186–196.

Epstein, H. E., I. C. Burke, and W. K. Lauenroth. 2002. Regional patterns of decomposition and primary production rates in the U.S. Great Plains. *Ecology* **83**:320–327.

Epstein, H. E., W. K. Lauenroth, I. C. Burke, and D. P. Coffin. 1997. Productivity patterns of C_3 and C_4 functional types in the U.S. Great Plains. *Ecology* **78**:722–731.

Fleischner, T. L. 1994. Ecological costs of livestock grazing in western North America. *Conservation Biology* **8**:629–644.

Fuhlendorf, S. D., and D. M. Engle. 2001. Restoring heterogeneity on rangelands: Ecosystem management based on evolutionary grazing patterns. *BioScience* **51**:625–632.

Gage, K. L., and M. Y. Kosoy. 2005. Natural history of plague: Perspectives from more than a century of research. *Annual Review of Entomology* **50**:505–528.

Gratz, N. G. 1999. Emerging and resurging vector-borne diseases. *Annual Review of Entomology* **44**:51–75.

Gunderson, L., C. Folke, and M. Janssen. 2005. Integrating ecology and society to navigate turbulence. *Ecology and Society* **10**:39.

Hansen, A. J., R. L. Knight, J. M. Marzluff, S. Powell, K. Brown, P. H. Gude, and A. Jones. 2005. Effects of exurban development on biodiversity: Patterns mechanisms, and research needs. *Ecological Applications* **15**(6):1893–1905.

Harvell, C. D., C. E. Mitchell, J. R. Ward, S. Altizer, A. P. Dobson, R. S. Ostfeld, and M. D. Samuel. 2002. Climate warming and disease risks for terrestrial and marine biota. *Science* **296**:2158–2162.

Hautaniemi, L. S., and M. P. Gutmann. 2005. Isolated elderly in the U.S. Great Plains, the roles of environment and demography in creating a vulnerable population. *Annales de Demographie Historique* **2**:81–108.

Hunt, H. W., E. T. Elliott, J. K. Detling, J. A. Morgan, and D.- X. Chen. 1998. Responses of a C_3 and C_4 perennial grass to elevated CO_2 and climate change. *Global Change Biology* **2**:35–47.

Jones, C. G., R. S. Ostfeld, M. P. Richard, E. M. Schauber, and J. O Wolff. 1998. Chain reactions linking acorns to gypsy moths and Lyme disease risk. *Science* **279**: 10232–6.

Kabii, T., and P. Horwitz. 2006. A review of landholder motivations and determinants for participating in conservation convenanting programmes. *Environmental Conservation* **33**:11–20.

Kaye, J. P., I. C. Burke, A. R. Mosier, and J. P. Guerschman. 2004. Methane and nitrous oxide fluxes from urban soils to the atmosphere. *Ecological Applications* **14**:975–981.

Kaye, J. P., R. L. McCulley, and I. C. Burke. 2005. Carbon fluxes, nitrogen cycling, and soil microbial communities in adjacent urban, native and agricultural ecosystems. *Global Change Biology* **11**:575–587.

Kimbal, B. A., and S. B. Idso. 1983. Increasing atmospheric CO_2: Effects on crop yield, water use and climate. *Agricultural Water Management* **7**:55–72.

King, J. Y., A. R. Mosier, J. A. Morgan, D. R. LeCain, D. G. Milchunas, and W. J. Parton. 2004. Plant nitrogen dynamics in shortgrass steppe under elevated atmospheric CO_2. *Ecosystems* **7**:147–160.

Knight, R. L., W. C. Gilgert, and E. Marston (eds.). 2002. *Ranching west of the 100th meridian. Culture, ecology and economics.* Island Press, Washington, D.C.

Knopf, F. L., and B. J. Miller. 1994. *Charadrius montanus:* Montane, grassland or bare-ground plover. *Auk* **111**:504–506.

Knopf, F. L., and J. R. Rupert. 1996. Reproduction and movements of Mountain Plovers breeding in Colorado. *Wilson Bulletin* **108**:28–35.

Körner, C. 2000. Biosphere responses to CO_2 enrichment. *Ecological Applications* **10**:1590–1619.

Kotliar, N. B., B. W. Baker, A. D. Whicker, and G. Plumb. 1999. A critical review of assumptions about the prairie dog as keystone species. *Environmental Management* **24**:177–192.

Kristan, W. B., III. 2003. The role of habitat selection in population dynamics: Source–sink systems and ecological traps. *Oikos* **103**:457–468.

Lauenroth, W. K., and J. B. Bradford. 2006. Ecohydrology and the partitioning AET between transpiration and evaporation in a semiarid steppe. *Ecosystems* **9**:756–767.

Lauenroth, W. K., I. C. Burke, and M. P. Gutmann. 1999. The structure and function of ecosystems in the central North American grassland region. *Great Plains Research* **9**:223–259.

Lauenroth, W. K., I. C. Burke, and J. M. Paruelo. 2000. Patterns of production and precipitation-use efficiency of winter wheat and native grasslands in the central Great Plains of the United States. *Ecosystems* **3**:344–351.

Lauenroth, W. K., H. E. Epstein, J. M. Paruelo, I. C. Burke, M. R. Aguiar, and O. E. Sala. 2004. Potential effects of climate change on the temperate zones of North and South America. *Revista Chilena de Historia Natural* **77**(3):439–453.

Lauenroth, W. K., and D. G. Milchunas. 1991. Short-grass steppe, pp. 183–226. In: R. T. Coupland (ed.), *Ecosystems of the world 8A: Natural grasslands.* Elsevier, Amsterdam.

Lauenroth, W. K., and O. E. Sala. 1992. Long-term forage production of North American shortgrass steppe. *Ecological Applications* **2**:397–403.

LeCain, D. R., J. A. Morgan, A. R. Mosier, and J. A. Nelson. 2003. Soil and plant water relations, not photosynthetic pathway, primarily influence photosynthetic responses in a semi-arid ecosystem under elevated CO_2. *Annals of Botany* **92**:41–52.

Luce, R. J. 2003. *A multi-state conservation plan for the black-tailed prairie dog,* Cynomys ludovicianus, *in the United States: An addendum to the Black-tailed Prairie Dog Conservation Assessment and Strategy, 3 Nov 1999.*

Luo, Y., B. Su, W. S. Currie, J. S. Dukes, A. Finzi, U. Hartwig, B. Hungate, R. E. McMurtrie, R. Oren, W. J. Parton, D. E. Pataki, M. R. Shaw, D. R. Zak, and C. B. Field. 2004. Progressive nitrogen limitation of ecosystem responses to rising atmospheric carbon dioxide. *BioScience* **54**:731–739.

Maestas, J. D., R. L. Knight, and W. C. Gilgert. 2003. Biodiversity across a rural land-use gradient. *Conservation Biology* **17**:1425–1434.

Masters, R. A., and R. L. Sheley. 2001. Principles and practices for managing rangeland invasive plants. *Journal of Range Management* **54**:502–517.

Michener, J. A. 1974. *Centennial.* Fawcett Publishing, Greenwich, Conn.

Milchunas, D. G., J. R. Forwood, and W. K. Lauenroth. 1994. Forage production across fifty years of grazing intensity treatments in shortgrass steppe. *Journal of Range Management* **47**:133–139.

Milchunas, D. G., J. Y. King, A. R. Mosier, J. C. Moore, J. A. Morgan, M. H. Quirk, and J. R. Slusser. 2004. UV radiation effects on plant growth and forage quality in a shortgrass steppe ecosystem. *Phytochemical Phytobiology* **79**:404–410.

Milchunas, D. G., and W. .K. Lauenroth. 1989. Three-dimensional distribution of plant biomass in relation to grazing and topography in the shortgrass steppe. *Oikos* **55**:82–86.

Milchunas, D. G., and W. K. Lauenroth. 1993. Quantitative effects of grazing on vegetation and soils over a global range of environments. *Ecological Monographs* **63**:327–366.

Milchunas, D. G., and W. K. Lauenroth. 1995. Inertia in plant community structure: State changes after cessation of nutrient enrichment stress. *Ecological Applications* **5**:452–458.

Milchunas, D. G., W. K. Lauenroth, and P. L. Chapman. 1992. Plant competition, abiotic, and long- and short-term effects of large herbivores on demography of opportunistic species in a semiarid grassland. *Oecologia* **92**:520–531.

Milchunas, D. G., W. K. Lauenroth, P. L. Chapman, and M. K. Kazempour. 1989. Effects of grazing, topography, and precipitation on the structure of a semiarid grassland. *Vegetatio* **80**:11–23.

Milchunas, D. G., W. K. Lauenroth, P. L. Chapman, and M. K. Kazempour. 1990. Community attributes along a perturbation gradient in a shortgrass steppe. *Journal of Vegetation Science* **1**:375–384.

Milchunas, D. G., A. R. Mosier, J. A. Morgan, D. R. LeCain, J. Y King, and J. A. Nelson. 2005. Elevated CO_2 and defoliation effects on a shortgrass steppe: Forage quality versus quantity for ruminants. *Agriculture, Ecosystems and Environment* **111**:166–184.

Millennium Ecosystem Assessment Series. 2003. *Ecosystems and human well-being: Current state and trends. Findings of the Condition and Trends Working Group.* Island Press, Washington, D.C.

Morgan, J. A. 2002. Looking beneath the surface. *Science* **298**:1903–1904.

Morgan, J. A., H. Hunt, C. A. Monz, and D. R. LeCain. 1994. Consequences of growth at two carbon dioxide concentrations and two temperatures for leaf gas exchange in *Pascopyrum smithii* (C_3) and *Bouteloua gracilis* (C_4). *Plant, Cell and Environment* **17**:1023–1033.

Morgan, J. A., D. R. LeCain, A. R. Mosier, and D. G. Milchunas. 2001. Elevated CO_2 enhances water relations and productivity and affects gas exchange in C_3 and C_4 grasses of the Colorado shortgrass steppe. *Global Change Biology* **7**:451–466.

Morgan, J. A., D. G. Milchunas, D. R. LeCain, M. S. West, and A. R. Mosier. 2007. Carbon dioxide enrichment alters plant community structure and accelerates shrub growth in the shortgrass steppe. *Proceedings of the National. Academy of Sciences USA* **104**(37):14724–14729.

Morgan, J. A., A. R. Mosier, D. G. Milchunas, D. R. LeCain, J. A. Nelson, and W. J. Parton. 2004a. CO_2 enhances productivity, alters species composition, and reduces forage digestibility of shortgrass steppe vegetation. *Ecological Applications* **14**:208–219.

Morgan, J. A., D. E. Pataki, C. Körner, H. Clark, S. J. Del Grosso, J. M. Grünzewig, A. K. Knapp, A. R. Mosier, P. C. D. Newton, P. A. Niklaus, J. B. Nippert, R. S. Nowak, W. J. Parton, H. W. Polley, and M. R. Shaw. 2004b. Water relations in grassland and desert ecosystems to elevated atmospheric CO_2. *Oecologia* **140**:11–25.

Morrison, J. I. L., and R. M. Gifford. 1983. Stomatal sensitivity to carbon dioxide and humidity. A comparison of two C_3 and two C_4 grass species. *Plant Physiology* **71**:789–796.

Mosier, A. R., J. A. Morgan, J. Y. King, D. LeCain, and D. G. Milchunas. 2002. Soil–atmosphere exchange of CH_4, CO_2, NO_x, and N_2O in the Colorado shortgrass steppe under elevated CO2. *Plant and Soil* **240**:201–211.

Mosier, A. R., E. Pendall, and J. A. Morgan. 2003. Soil–atmosphere exchange of CH_4, CO_2, NO_x, and N_2O in the Colorado shortgrass steppe following five years of elevated CO_2 and N fertilization. *Atmospheric Chemistry and Physics Discussions* **3**:2691–2706.

Nelson, J. A., J. A. Morgan, D. R. LeCain, A. R. Mosier, D. G. Milchunas, and W. J. Parton. 2004. Elevated CO_2 increases soil moisture and enhances plant water relations in a long-term field study in the semi-arid shortgrass steppe of Northern Colorado. *Plant and Soil* **259**:169–179.

Newburn, D., S. Reed, P. Berck, and A. Merenlender. 2005. Economics and land-use change in prioritizing private land conservation. *Conservation Biology* **19**:1411–1420.

Niklaus, P. A., D. Spinnler, and C. Körner. 1998. Soil moisture dynamics of calcareous grassland under elevated CO_2. *Oecologia* **117**:201–208.

Nowak, R. S., D. S. Ellsworth, and S. D. Smith. 2004. Functional responses of plants to elevated CO_2: Do photosynthetic and productivity data from FACE experiments support early predictions? *New Phytologist* **162**(2):253–280.

Owensby, C. E., R. C. Cochran, and L. M. Auen. 1996a. Effects of elevated carbon dioxide on forage quality for ruminants, pp. 363–371. In: C. Körner and F. Bazzaz (eds.), *Carbon dioxide, populations and communities*. Physiological ecology series. Academic Press, San Diego.

Owensby, C. E., J. M. Ham, A. K. Knapp, C. W. Rice, P. I. Coyne, and L. M. Auen. 1996b. Ecosystem level responses of tallgrass prairie to elevated CO_2, pp. 147–162. In: G. W. Koch and H. A. Mooney (eds.), *Carbon dioxide and terrestrial ecosystems*. Academic Press, San Diego, Calif.

Parmenter, R. R., E. P. Yadav, C. A. Parmenter, P. Ettestad, and K. L. Gage. 1999. Incidence of plague associated with increased winter–spring precipitation in New Mexico. *American Journal of Tropical Medicine and Hygiene* **61**:814–821.

Parry, M., O. Canziani, J. Palutikof, P. van der Liden and C. Hanson (eds.) 2007. *Climate change 2007: Imapcts, adaptation and vulnerability*. Cambridge University Press, Cambridge.

Parton, W. J., M. P. Gutmann, and D. S. Ojima. 2007a. Long term trends in population, farm income, and crop production in the Great Plains. *BioScience* **57**:738–747.

Parton, W. J., M. P. Gutmann, and W. R. Travis. 2003. Sustainability and historical land use change in the Great Plains: The case of Eastern Colorado. *Great Plains Research* **13**:97–125.

Parton, W. J., W. L. Silver, I. C. Burke, L. Grassens, M. E. Harmon, W. S. Currie, J. Y. King, E. Carol Adair, L. A. Brandt, S. C. Hart, and B. Fasth. 2007b. Global-scale similarities in nitrogen release patterns during long-term decomposition. *Science* **315**:361–364.

Pascual, M., and A. Dobson. 2005. Seasonal patterns of infectious diseases. *PLoS Med* **2**(1):e5.

Pielke, R. A., Sr., T. J. Lee, J. H. Copeland, J. L. Eastman, C. L. Ziegler, and C. A. Finley. 1997. Use of USGS-provided data to improve weather and climate simulations. *Ecological Applications* **7**:3–21.

Poland, J. D., and D. T. Dennis. 1998. Plague, pp. 545–558. In: A. S. Evans and P. S. Brachman (eds.), *Bacterial infections in humans, epidemiology and eontrol*. 3rd ed. Plenum Publishing, New York.

Polley, H. W. 1997. Implications of rising atmospheric carbon dioxide concentration for rangelands. *Journal of Range Management* **50**:561–577.

Reich, P. B., S. E. Hobbie, T. Lee, D. S. Ellsworth, J. B. West, D. Tilman, J. M. H. Knops, S. Naeem, and J. Trost. 2006a. Nitrogen limitation constrains sustainability of ecosystem response to CO_2. *Nature* **440**:922–924.

Reich, P. B., B. A. Hungate, and Y. Luo. 2006b. Carbon–nitrogen interactions in terrestrial ecosystems in response to rising atmospheric carbon dioxide. *Annual Review Ecological Systems* **37**:611–636.

Rickets, T. H., E. Dinerstein, D. M. Olson, C. J. Loucks, W. Eichbaum, D. DellaSala, K. Kavanagh, P. Hedao, P. T. Hurley, K. M. Carney, R. Abell, and S. Walters. 1999. *Terrestrial ecoregions of North America: A conservation assessment*. Island Press, Washington, D.C.

Roach, J. L., P. Stapp, B. VanHorne, and M. F. Antolin. 2001. Genetic structure of a metapopulation of black-tailed prairie dogs. *Journal of Mammalogy* **82**:946–959.

Sala, O. E., M. Osterheld, R. L. C. Leon, and A. Soriano. 1986. Grazing effects upon plant community structure in subhumid grasslands of Argentina. *Vegetatio* **67**:27–32.

Samson, F., and F. L. Knopf. 1994. Prairie conservation in North America. *BioScience* **44**:418–421.

Stapp, P. 1998. A reevaluation of the role of prairie dogs in Great Plains grasslands. *Conservation Biology* **12**:1253–1259.

Stapp, P., M. F. Antolin, and M. Ball. 2004. Patterns of extinction in prairie-dog metapopulations: Plague outbreaks follow El Niño events. *Frontiers in Ecology and the Environment* **2**:235–240.

Stegner, A. E. 1969. *The sound of mountain water*. Doubleday, New York.

Stephenson, N. L. 1990. Climatic control of vegetation distribution: The role of the water balance. *American Naturalist* **135**:649–679.

Stohlgren, T. J., T. N. Chase, R. A. Pielke, T. G. F. Kittell, and J. S. Baron. 1998. Evidence that local land use practices influence regional climate, vegetation, and streamflow patterns in adjacent natural areas. *Global Change Biology* **4**(50):495–504.

The Nature Conservancy. 2003. The Nature Conservancy's GIS website. Online. Available at http://gis.tnc.org/data/MapbookWebsite.

Theobald, D. M. 2003. Targeting conservation action through assessment of protection and exurban threats. *Conservation Biology* **17**:1624–1637.

Theobald, D. M. 2004. Placing exurban land use change in a human modification framework. *Frontiers in Ecology and Environment* **2**(3):139–144.

Theobald, D. M. 2005. Landscape patterns of exurban growth in the USA from 1980 to 2020. *Ecology and Society* **10**:32.

U.S. Census Bureau. 2000. United States Census 2000. Online. Available at www.census.gov/main/www/cen2000.html.

U.S. Census of Agriculture. 2004. United States Department of Agriculture National Agricultural Statistics Service, 2002 Census of Agriculture. Online. Available at www.nass.usda.gov/census/census02/volume1/co/index2.htm.

U.S. Fish and Wildlife Service. 2000. *12-Month finding for a petition to list the black-tailed prairie dog as threatened*. Federal Register 65 FR 5476 5488, 4 February 2000.

VEMAP Members. 1995. Vegetation/ecosystem modeling and analysis project: Comparing biogeography and biogeochemistry models in continental-scale study of terrestrial ecosystem responses to climate change and CO_2 doubling. *Global Biogeochemical Cycling* **9**:407–437.

Vermeire, L. T., R. K. Heitschmidt, P. S. Johnson, and B. K. Sowell. 2004. The prairie dog story: Do we have it right? *BioScience* **54**:689–695.

Vitousek, P. M., J. D. Aber, R. W. Howarth, G. E. Likens, P. A. Matson, D. W. Schindler, W. H. Schlesinger, and D. G. Tilman. 1997. Technical report: human alteration of the global nitrogen cycle: sources and consequences. *Ecological Applications* **7**:737–750.

Wand, S. J. E., G. F. Midgley, M. H. Jones, and P. S. Curtis. 1999. Responses of wild C_4 and C_3 grass (Poaceae) species to elevated atmospheric CO_2 concentrations: A meta-analytic test of current theories and perceptions. *Global Change Biology* **5**:723–741.

Wardle, D. A., R. D. Bardgett, J. N. Klironomos, H. Setälä, W. H. van der Putten, and D. H. Wall. 2004. Ecological linkages between aboveground and belowground biota. *Science* **304**:1629–1633.

Weiss, R. A., and A. J. McMichael. 2004. Social and environmental risk factors in the emergence of infectious diseases. *Nature Medicine* **10**:S70–S76.

Whalley, R. D. B. 2000. Grasslands, grazing animals and people: How do they all fit together? *Tropical Grasslands* **34**:192–198.

White, G. C., J. R. Dennis, and F. M. Pusateri. 2005. Area of black-tailed prairie dog colonies in eastern Colorado. *Wildlife Society Bulletin* **33**(1):265–272.

Zak, D. R., K. S. Pregitzer, J. S. King, and W. E. Holmes. 2000. Elevated atmospheric CO2, fine roots and the response of soil microorganisms: A review and hypothesis. *New Phytologist* **147**:201–222.

Index

Abandoned agricultural fields, 107–111
Aboveground biomass, 424
Aboveground net primary production
  biomass components, 296t
  carbon dioxide effects, 294–295
  Central North American Grassland region, 271–274
  for crops, 279
  definition of, 270
  description of, 270
  distributions, 275f–276f
  grazing effects on, 395, 424, 429–430, 436
  interannual fluctuations in, 290f
  landscape effects, 283, 289–290
  land use effects on, 278–281
  leaf area index, 271–272
  mean annual precipitation and, 279
  nitrogen effects, 290–294
  normalized difference vegetation index, 271, 271f
  precipitation effects, 274–277, 288–289, 291f, 298, 300, 321, 431
  regional-scale spatial patterns and controls, 274–281
  resource availability effects on, 293
  satellite-based remote sensing for assessment of, 281
  site-scale spatial patterns and control, 281–283
  soil texture effects, 277–278, 281–283
  spatial patterns, 298–300
  temporal patterns, 284–298, 300–301
  in urban ecosystems, 332–333
  water-holding capacity, 277–278
  winter wheat, 278–280
Agriculture
  aboveground net primary production affected by, 278
  drought effects, 63
  Dust Bowl effects, 63
  history of, 58
  irrigated, 58, 59f
  mammals affected by, 168–170
  nonirrigated, 59
Agropyron smithii, 285, 397, 410–411
Air masses, 15
Alfalfa, 168
Altithermal, 6–7, 56
Ammonia, 318, 329, 343, 356, 365, 379
Ant(s)
  colony distribution, 226
  harvester, 217, 220, 224, 229–230, 232, 234, 237, 470
  wood, 224
Ant nests, 93–96, 236
Aquatic habitats, 489
Arctic air, 16
Argillic horizons, 48–49
Arkansas, 31
Armadillos, 136
Artemisia filifolia, 71
Artemisia frigida, 498
Arthropods, 232–233, 468–472
Artificial disturbances, 99–101
Atmospheric carbon dioxide, 66, 342, 377, 497
Atriplex canescens, 397, 453–454, 460
Autumn, 17

Badgers
  burrows created by, 96
  description of, 136
  illustration of, 138f
  population density of, 147t
Bankard soil, 38
Beetles, 217, 219, 224, 234
Belowground biomass, 295, 296t, 424
Belowground food webs
  agricultural practices effect on, 261–262
  carbon dioxide levels and, 262–264
  climate changes and, 262–263
  components of, 249
  connectedness of, 250, 251f
  description of, 248–249
  energy flow description of, 251–254

511

Belowground food webs (*continued*)
   feeding rates, 252–253
   functional, 252t, 254–257
   grazing effects on, 260
   interaction patterns, 257–258
   mineralization rate, 253
   resource utilization in, 264
   rhizosphere-based model, 249–250
   summary of, 264–266
Belowground net primary production, 295–298, 296t, 301
Biogeochemical processes
   carbon, 307–310
   climatic controls, 313–315
   fauna, 325–326
   future environmental changes effect on, 333–335
   irrigated agriculture effects on, 330
   nitrogen, 310–312
   overview of, 306–307
   physiographic control, 316–321
   plant litter, 321
   vegetation effects, 321–326
Biomass
   aboveground net primary production and, 296t
   belowground, 295, 296t, 424
   grazing and, 423–427
   horizontal placement of, 321
   precipitation effects on, 424
Biota
   case study of, 43
   fauna, 325–326
   paleoclimate and, 43–46
   soil formation affected by, 42–47
Birds. *See also specific species*
   at Central Plains Experimental Range, 187, 196, 198f
   climate effects, 193–195
   community attributes, 183–184
   courtship displays by, 188
   description of, 181–183
   diet of, 187
   ecosystem role of, 199–200
   future needs of, 200–201
   grassland, 183, 195
   grazing effects, 190–192, 466–468, 474–476
   habitats for, 183, 188
   insect regulation by, 200
   long-term trends, 195–199
   microclimate effects, 193–195
   migratory nature of, 187
   nesting by, 188
   population density of, 183
   population trends, 195
   prairie dogs and, 159–160
   range maps for, 185–186
   raptors, 183
   reproductive failure, 188
   species of, 183–188, 202–208
   types of, 202–208
   vegetation effects, 188–190, 466
   weather effects, 193–195
Bison, 55–56, 58, 156, 389–391, 447, 462–463, 477
Black-footed ferrets, 165, 501
Black-tailed jackrabbits, 136, 463
Blizzards, 16
Blue grama. *See Bouteloua gracilis*
Boom-and-bust economy, 65
Boomers, 59
Bouteloua gracilis
   badger burrows effect on, 96
   basal cover, 233
   cattle fecal pats effect on, 90–93
   at Central Plains Experimental Range, 77
   characteristics of, 71
   description of, 8, 70–72, 79, 133, 460
   drought resistance of, 111
   fire effects on, 168
   gaps in resource space caused by death of, 87, 120
   grazing effects on, 405, 406t–407t, 408, 415, 438
   growth of, 285
   history of, 104
   nitrogen effects on, 293
   post-disturbance recovery of, 86
   prairie dogs' effect on, 106
   recovery of, 109–110, 113, 123
   rooting by, 285
   seed dispersion, 110, 127, 129
   skunk burrows effect on, 96
Brewer's Sparrows, 191
Bromus tectorum, 422
Buchloë dactyloides
   in abandoned agricultural fields, 108
   at Central Plains Experimental Range, 77
   characteristics of, 71
   description of, 8, 70–72, 133–134, 460
   drought resistance of, 111
   fire effects on, 168
   prairie dogs' effect on, 106

Index

recovery of, 108
seed dispersion, 110
Buffalograss. *See Buchloë dactyloides*
Buffalo jumps, 56
Burrows
  badgers, 96
  mammal construction of, 156–157
  pocket gophers, 164
  prairie dogs, 158–159
  skunk, 96

Cacti, 58
Cactus, 416, 417t
Calcic horizons, 48
Cambic horizons, 48
Carabid beetles, 219, 231
Carbon
  description of, 307–310
  nitrogen and, ratios between, 324
  redistribution of, 322
  ultraviolet radiation effects on, 315
Carbon cycling, 314, 496
Carbon dioxide
  aboveground net primary production affected by, 294–295
  atmospheric, 342, 377, 497
  description of, 66
  elevated levels of, 262–264
  nutrient cycles affected by, 497–498
  residual effects of, 363–365
  soil water and, 286
  species responses to, 498–499
  trace gas exchange affected by, 360–363
  vulnerability assessment modeling, 377–378
Carbon flow, 251, 257
Carbon isotope labeling, 297
Cassin's Sparrows, 191, 467
Caternary patterns, 317, 319
Cattle
  Agropyron smithii consumption by, 397
  diet selectivity by, 397
  Dust Bowl effects on, 64
  fecal pats, 90–93, 161
  foraging patterns, 397–399
  history of, 59, 60f
  management tool use of, 493–494
  present-day uses, 65
Cattle grazing. *See also* Grazing
  birds and, 190–192
  bison grazing vs., 462–463
  controversies associated with, 438

  description of, 8, 62, 71, 134
  erosional processes affected by, 326–327
  grassland fauna affected by, 192
  mammals affected by, 166–168
  nitrogen affected by, 327–329, 365
  prevalence of, 447
  rodent populations affected by, 167
  simulations of, 400f–401f
  soil compaction caused by, 168
  soil organic matter affected by, 329
  statistics regarding, 447
  vegetation affected by, 191
Central grassland region
  aboveground net primary production, 271–274
  description of, 3, 4f
  history of, 55
  semiarid zone of, 7
  spatial gradients of, 313
Central Plains Experimental Range
  aboveground net primary production at, 292f
  bird species at, 187, 196, 198f
  Bouteloua gracilis at, 77
  Buchloë dactyloides at, 77
  community types on, 78
  description of, 9–11
  eolian deposits, 36
  graminoids in, 75
  grasshopper species at, 218
  insect species at, 218
  long-term grazing treatments, 460
  parent materials, 35–37
  plant biomass at, 75, 76f
  soil in, 38, 39f, 41–42, 281–282
  trace gas exchange at, 344
  vegetation in, 73–74
CENTURY ecosystem simulation model, 307, 374, 376, 378–382
C_4 grasses
  Bouteloua gracilis. See Bouteloua gracilis
  Buchloë dactyloides. See Buchloë dactyloides
  description of, 5–6, 8
  growth of, 285–286
  illustration of, 396f
  photosynthesis of, 286
  soil moisture and, 45
  trace gas exchange and, 356–357
  water-plus-nitrogen treatment effects, 105
Chestnut-collared Longspurs, 181, 182f, 191
Clay, 30

514   Index

Clementsian succession model, 85, 85f, 108
Climate
  air masses, 15
  birds and, 193–195
  case study of, 49–51
  changes in, 262–263
  description of, 7–8, 14
  drought, 26
  evaporation, 23–24
  geographic factors that control, 14–15
  hail, 25
  humidity, 15, 20, 22f
  land-use management and, 497
  mammal distribution affected by, 141–142
  plague incidence and, 501
  precipitation, 7, 17–18, 21–23
  reasons for studying, 51
  semiarid, 7, 48
  snowfall, 24, 25f
  soil formation affected by, 47–51
  summary of, 26–27
  temperature, 7, 18–20, 20f
  thunderstorms, 25
  tornadoes, 16, 25–26
  trace gas exchange affected by, 347–351, 367
  variability of, 194
  weather patterns, 16–18
  wind, 15, 25
Cold temperatures, 18–19
Colorado Piedmont
  description of, 31, 34
  playas in, 40
Conifer species, 7
Conservation, 494–495
Conservation easements, 495
Conservation Reservation Program, 170, 190
Convection, 15
Coronado, 57
Cotton rats, 136
Cottontail rabbits, 136, 149, 161, 463
Cover, 80–82, 143–144
Coyotes, 147t, 148, 161
CPER. *See* Central Plains Experimental Range
C_3 plants, 44–45
Crassulacean acid metabolism, 284–285
Critical grazing pressure, 448
Crop agriculture
  aboveground net primary production for, 278–279
  drought effects, 63

Dust Bowl effects, 63
  history of, 58
  irrigated, 58, 59f
  mammals affected by, 168–170
  nonirrigated, 59
  types of, 300
Cyclogenesis, 16

Darkling beetles, 219
DAYCENT model, 374–375, 378–382, 384
Decomposition, 315, 499
Deer mice, 145, 466
Denitrification, 349, 366, 380
Desert cottontails, 136, 149, 161, 463
Dew point temperature, 20, 22f
Disease environment, 500–502
Disturbances
  abandoned agricultural fields, 107–111
  animal-induced, 237
  ant nest sites, 93–96
  artificial, 99–101
  badger burrows, 96
  Bouteloua gracilis recovery after, 86
  cattle fecal pats, 90–93
  characteristics of, 84, 88–90
  crop-related, 492
  description of, 112
  drought, 111–112
  energy flow alterations secondary to, 265
  factors that affect, 112
  fertilization, 104–105
  gap dynamics model of, 86–87
  grazing, 259–260, 416, 419–420, 438
  insects as cause of, 237
  intermediate-size, 88, 101–104, 416
  large-scale, 88, 104–112
  levels of, 84
  long-term responses to, 154
  low-intensity, 112
  military vehicle tracking, 105–106
  old roads, 104
  patches caused by, 86–87, 237
  pocket gophers as cause of, 96–99, 165
  prairie dogs, 106–107
  recovery from, 88
  simulation model of, 87, 119
  skunk burrows, 96
  small-scale, 88–101
  soil organic matter pools affected by, 325
  white grubs, 101–104, 237
Drought, 26, 63, 111–112, 496
Dryland agriculture, 62

Dune formation, 6–7
Dung beetles, 234
Dust Bowl, 62–64, 170
Dust storms, 63
Dwarf shrubs, 75

Ecosystem(s)
  birds' role in, 199–200
  conservation efforts, 494
  insects and, 217, 232–239
  land-use intensification effects on, 492
  resource-limited, 314
Ecosystem modeling
  CENTURY model, 307, 374, 376, 378–382
  DAYCENT model, 374–375, 378–382, 384
  land–atmosphere interactions, 377
  overview of, 373–375
  PHOENIX model, 374
  regional impact assessment, 376–377
  ROOT model, 374
  site-specific impact assessments, 375–376
  vulnerability assessments, 377–378, 383–384
Ecosystem respiration, 363
Edgar soil, 36–37
Eleodes extricata, 219, 224, 226f
Eleodes obsoleta, 219
Elevation, 15
ELM Grassland model, 373–375
El Niño Southern Oscillation, 500
Energy flow food web, 251–254
Environmental Protection Act, 491
Eolian sand, 6, 34, 37, 72
Erosion, 326–327
Europe
  early settlements, 58–61
  explorers from, 57–58
Evaporation, 23–24, 285
Exotic plants, 493

Fecal pats, 90–93
Ferruginous hawks, 181, 182f, 198f
Fertilization, 104–105
Fire, 168
Food webs, belowground
  agricultural practices effect on, 261–262
  carbon dioxide levels and, 262–264
  climate changes and, 262–263
  components of, 249
  connectedness of, 250, 251f
  description of, 232, 248–249
  energy flow description of, 251–254
  feeding rates, 252–253
  functional, 252t, 254–257
  interaction patterns, 257–258
  mineralization rate, 253
  resource utilization in, 264
  rhizosphere-based model, 249–250
  summary of, 264–266
Forage production, 428
Forbs, 409–410
Forest soil, 43
Formica obscuripes, 224, 226
Fourwing saltbush, 134
Functional food web, 254–257

Gap dynamics
  description of, 84, 86–87, 112
  simulation model based on, 87, 119, 374
GEM model, 376
Geography
  climate affected by, 14–15
  features of, 7
Glacial/interglacial cycle, 6, 34
Global change
  future directions in research on, 499
  rates of, 495–496
Global warming, 66
Goodnight, Charles, 60
Gophers
  burrowing by, 164
  description of, 136
  disturbances caused by, 96–99, 165
  grasshopper mice and, 164
  mounds, 142–143, 163–165
  soil properties and, 142
Graminoids, 75, 94
Grasses
  Bouteloua gracilis. See *Bouteloua gracilis*
  Buchloë dactyloides. See *Buchloë dactyloides*
  in Great Plains, 6
Grasshopper mice
  description of, 136, 137f, 143–144, 149–152, 466
  pocket gophers and, 164
  prairie dogs and, 161
Grasshoppers, 218–219, 221–222, 227–228, 231–232, 234, 470
Grassland Biome, 139, 199
Grassland birds, 183, 195

Grazing. *See also* Cattle grazing
  abiotic conditions that affect, 429
  aboveground net primary production affected by, 395, 424, 429–430, 436
  Agropyron smithii affected by, 410–411
  annuals response to, 411
  ant nest sites affected by, 93–94
  Atriplex canescens, 453–454
  biodiversity affected by, 472
  biomass distribution, 423–427
  birds and, 190–192
  bison, 55–56, 58, 156, 389–391, 447, 462–463, 477
  Bouteloua gracilis affected by, 405, 406t–407t, 408, 415, 438
  cactus protection against, 416, 417t
  compensatory regrowth after, 431–433
  controversies associated with, 438
  description of, 8, 62
  as disturbance, 259–260, 416, 419–420, 438
  erosional processes affected by, 326–327
  evolutionary history of, 389–391, 418, 437
  excessive, 420
  forage production, 428, 449
  forage quality and, 433–436
  grasshopper, 222–223, 229, 234, 431
  grassland fauna affected by, 192
  heavy, 408–409
  historical uses, 107
  importance of, 155–156
  insects affected by, 240–241
  intensity of, 392–395, 433–436, 473
  mammals affected by, 166–168
  management of, 493–494
  nitrogen affected by, 327–329, 365, 434f–435f
  Opuntia polyacantha affected by, 415
  plant community effects
    population dynamics, 408
    productivity, 427–436
    species, 412, 413f–414f, 436–437
    structure, 404–427
  prairie dogs' effect on, 465
  prevalence of, 62
  profitability of, 455
  rabbit populations affected by, 167
  regrowth after, 431–433
  regulation of, 391
  research on, 447–454
  rodent populations affected by, 167, 465–466
  roots affected by, 425, 427
  seasonal variations, 402–404
  seed production effects, 430–431
  Shortgrass Steppe Long-Term Ecological Research contributions, 11
  short-stature vs. tall-stature grasslands, 418
  simulations of, 400f–401f
  soil compaction caused by, 168
  soil organic matter affected by, 329
  species endangerment secondary to, 476
  stocking rates, 448–450, 454–455
  summary of, 436–438
  supplemental forages, 452–454
  of swales, 402
  systems, 450–452
  topography effects, 411
  ungrazed vs. disturbed communities, 419–420, 437
  vegetation affected by, 191
Great Plains
  bird species in, 184–185
  grasses in, 6
  land-use distress signs, 66
Great Plains Framework for Agricultural Resource Management, 455
Greenhouse gases, 342, 495–496
Ground beetles, 219
Ground squirrels, 136, 137f
Grubs, white, 101–104, 229, 237

Hail, 25
Harvester ants, 217, 220, 224, 229–230, 232, 234, 237, 470
Hastings soil, 50
Hawks, 181, 182f
Herbivores. *See also* Bison; Cattle
  description of, 155–156, 227–229
  energy flow, 392–395
  foraging patterns, 397
  grazing intensities, 392–395
Heterotrophs, 241
High-pressure weather systems, 16
Hillslopes, 40
Holocene
  characteristics of, 6–7
  stream terraces, 34
  vegetation, 46
Homestead Act, 58
Horizonation, 37
Horned Larks, 159, 181, 466, 474
Horses, 56–57

Humidity, 15, 20, 22f
Huston hypothesis, 416

IBP Grassland Biome project, 10, 151–154, 249, 293, 373
Insects
  abundance of, 239–240
  beetles, 217, 219
  bird regulation of, 200
  broad-scale distribution patterns for, 222–227
  broad-scale processes, 235
  community interactions, 227–232
  daily movements by, 221–222
  description of, 215–216
  disturbances caused by, 237
  ecosystem processes and, 232–239
  energy flows, 232–233
  fine-scale processes, 233–235
  in food webs, 232
  grasshoppers, 218–219, 221–222, 227–228, 231–232, 234, 470
  grazing and, 222–223, 240–241
  harvester ants, 217, 220, 224, 229–230, 232, 234, 237, 470
  microhabitat patterns of, 221–222
  nutrient flows, 232–233
  plant–herbivore interactions, 227–229
  population density of, 218–219, 220f
  population dynamics and distribution, 217–221
  as predators, 230–231
  as prey, 231–232
  seed–granivore interactions, 229–230
  soil influences on, 223
  soil roles of, 235–237
  spatial distribution of, 221–226
  spatial scaling of, 239–240
  temporal dynamics of, 218–221
  termites, 217, 224
  types of, 216t
  vegetation effects on, 222, 225f
  weather effects on, 221
International Geosphere Biosphere Program, 344
Irrigated crops, 8
Irrigation
  energy costs and, 65
  history of, 58, 59f
  underground water for, 64

Jackrabbits, 136, 152, 463
Juniperus osteosperma, 71

Kangaroo rats, 147, 170
Kochia scoparia, 323, 420–421, 421f

Lagomorphs, 463–464
Land use
  aboveground net primary production affected by, 278–281
  cattle grazing, 8, 62, 134
  climate effects, 497
  description of, 5f
  Dust Bowl effects, 62–64
  early European settlements, 58–61
  in early twentieth century, 61–62
  ecosystem services affected by, 492
  European explorers, 57–58
  future of, 65–66, 485
  history of, 55–66, 391
  net primary production affected by, 278–281
  present day, 65
  trace gas exchange affected by, 357–359, 367
Land-use distress, 66
Land Utilization Project, 64
Lark Bunting, 181, 182f, 466
Latitude, 15
Leaf area index, 271–272
Lightning, 25
Livestock grazing. *See* Cattle grazing; Grazing
Livestock production. *See also* Cattle
  domestic, 58
  Dust Bowl effects on, 64
  prevalence of, 65
Low-pressure weather systems, 16

Macroarthropods, 468–470, 472
Mammals. *See also specific mammal*
  biogeographic patterns, 136
  burrow construction by, 156–157
  crop production effects on, 168–170
  in cultivated areas, 169
  description of, 132–133
  evolutionary patterns of, 136–137
  exploitation of, 165–166
  extinction of, 6
  factors that affect
    climate, 136–137
    food availability, 146–148
    soil, 142–143
    species interactions, 148–151
    vegetation structure, 143–146, 170
    weather, 136–137

518 Index

Mammals (*continued*)
  grazing effects, 166–168
  habitat for, 133–134
  herbivory by, 155–156
  human activities that affect, 165–170
  livestock grazing effects on, 166–168
  mound construction by, 156–157
  nitrogen introduced by, 157
  pesticide effects on, 169
  population monitoring programs, 140
  rabbits, 136
  research of, 139–141
  simulation models of, 156
  summary of, 170–171
  types of, 134–139, 135t
Mammoths, 55–56
McCown's Longspurs, 181, 474
Meadow voles, 136
Mean annual potential evapotranspiration, 274
Mean annual precipitation, 279
Mesoscale convective systems, 17
Methane
  atmospheric, 365
  carbon dioxide effects on, 360–361
  climatic variability in, 347
  description of, 342–343, 345
  land use effects on, 358, 359f
  nitrogen additions' effect on, 356
  oxidation of, 362
  plowing effects on emission of, 360
  seasonal variations in, 350–351
  sinks for, 492
  soil water and temperature effects on, 349
  uptake of, 359f
  variations in fluxes of, 366–367
Microarthropods, 471
Microtopography, 100–101, 322
Microwatersheds, 40
Military vehicle tracking, 105–106
Mounds
  pocket gophers, 142–143, 163–165
  prairie dogs, 159
Mountain Plovers, 188, 190, 467
Mule deer, 462
Multidecadal drought, 26

Native Americans, 56–57
Nematodes, 471–472, 476
Net primary production
  aboveground. *See* Aboveground net primary production

  belowground, 295–298, 296t, 301
  estimates of, 427
  precipitation effects on, 274–277, 288–289, 291f, 298, 321, 382, 430
  spatial patterns, 298–300
  temporal patterns, 284–298
Nitrate, 318, 379
Nitrogen
  aboveground net primary production affected by, 290–294
  annual losses of, 365–367
  carbon and, ratios between, 324
  carbon dioxide effects on, 497–498
  cattle grazing effects on, 327–329, 365
  cycling of, 323, 496
  distribution and fluxes of, 310–312
  grazing effects on, 327–329, 365, 434f–435f
  losses of, 365–367
  in lowland swales, 399
  nutrient cycling affected by alterations in availability of, 312
  topographic differences in deposition of, 318
Nitrogen mineralization, 111, 315
Nitrogen oxides, 345, 347
Nitrous oxide
  carbon dioxide effects on, 360–361, 363
  denitrification-related emissions, 380
  description of, 342–343
  fluxes in, 348
  land use effects on, 358, 359f
  nitrogen additions' effect on, 352–356
  plowing effects on emission of, 360
  soil water and temperature effects on, 349
  sources of, 343, 348–349
  variations in fluxes of, 366–367
Nonsaline lowlands, 78
Normalized difference vegetation index, 271, 271f, 281, 286–287
North America, 3, 4f

Oklahoma Panhandle, 61
Oligocene, 5
Olney soil, 50
Opuntia polyacantha, 415
Organic matter pools
  carbon, 307–310
  nitrogen, 310–312
Overgrazing, 493

Paleocene, 4–5
Paleoclimate, 43–46

Paleosols, 38, 46, 48–49
Particulate organic matter, 307
Pasimachus elongatus, 231
Passerines, 160
Patch dynamics, 86, 237
Pawnee National Grasslands, 9, 107, 391, 489
Permafrost, 265
Pesticides, 169
PHOENIX model, 374
Physiography, 32f–34f
Plague, 142, 163, 500–501
Plains Village, 56
Plants. *See also* Vegetation
  Bouteloua gracilis. See Bouteloua gracilis
  Buchloë dactyloides. See Buchloë dactyloides
  C_3, 44–45
  classification of, 285
  community structure, grazing effects on, 390, 404–427
  crassulacean acid metabolism, 284–285
  description of, 395–404
  ecophysiology of, 284–286
  litter, 321
  photosynthetic pathways, 324
Playa, 40
Pleistocene, 6, 502
Plovers, 188, 190, 467
Plowing, 360
PNG. *See* Pawnee National Grasslands
Pocket gophers
  burrowing by, 164
  description of, 136
  disturbances caused by, 96–99, 165
  grasshopper mice and, 164
  mounds, 142–143, 163–165
  soil properties and, 142
Pocket mice, 136, 147
Pogonomyrmex occidentalis, 93, 95
Population growth in Colorado, 485–488, 486f–487f
Prairie, 9
Prairie dogs
  animals affected by, 161–162
  avian species and, 159–160
  biology of, 162
  burrows constructed by, 158–159
  colonies of, 162
  description of, 136, 464–465
  disturbances caused by, 106–107, 136
  ecological impact of, 157–163

extirpation of, 158, 165–166
future of, 491
genetic relatedness among, 162
grasshopper mice and, 161
grazing effects, 465, 491
mortality causes, 142
mounds constructed by, 159
as pests, 158
plague susceptibility by, 163, 501
poisoning of, 491
populations of, 158
shooting of, 158, 166
site selection by, 163
soil types preferred by, 142
studies of, 158–159
vegetation affected by, 156
weather effects on distribution of, 141
Precipitation
  aboveground net primary production affected by, 274–277, 288–289, 291f, 321, 432
  description of, 7, 17–18
  ecophysiology of plants affected by, 285
  evaporation of, 23–24
  mean annual, 279
  net primary production affected by, 274–277, 288–289, 291f, 298, 321, 382, 430
  rates of, 21–23, 274, 390, 460
  regional patterns in, 313
  seasonality of, 285–286, 300
  spatial variability of, 313–314
  standardized index, 26
  temporal variability of, 314–315
Predation hypothesis, 416
Pronghorn, 136, 138f, 165, 169, 460, 462
Prosopis spp., 58

Quercus havardii, 71

Rabbits, 136, 148–149, 152, 167
Railroads, 59
Rainfall. *See* Precipitation
Rain shadow, 15
Range livestock industry, 59, 60f
Raptors, 183
Red-winged Blackbirds, 190
Relative humidity, 20
Remmit soil, 38
Research
  approaches and sites for, 9–11
  global change, 499

520 Index

Research (*continued*)
   grazing, 447–454
   Shortgrass Steppe Long-Term Ecological Research. *See* Shortgrass Steppe Long-Term Ecological Research
Resource islands, 321
Rhizosphere, 249–250
Riparian areas, 73, 133, 183
Roads, 104
Roadside vegetation, 145
Robber flies, 231–232
Rodents. *See also specific rodent*
   description of, 136, 140
   grazing effects, 167, 465–466
   livestock grazing effects on, 167
   population density of, 153
   temporal variation in, 151
   vegetation for, 145
Root biomass, 308
ROOT model, 374

Salsola iberica, 420–421, 421f
Saltbush, 134, 143–144
Sand
   Edgar, 36–37
   Eolian, 34, 37, 72
   in soil, 34
   Valent, 36–37
Schizachyrium scoparium, 71
Seasons. *See also specific season*
   patterns in, 16–18
   precipitation range based on, 22
Seeds
   Bouteloua gracilis, 110, 127, 129
   Buchloë dactyloides, 110
   dispersal of, 124–125
   grazing effects on production of, 430–431
   soil texture effects on, 430
Settlers, 60
Sheep, 463
Shortgrass prairie, 9
Shortgrass steppe
   definition of, 133
   earliest human contact, 55–57
   future of, 171, 484–506
   history of, 4–5
   land use in, 5f
   location of, 7, 133, 390
   size of, 7
Shortgrass Steppe Long-Term Ecological Research
   cover types at, 80–82

description of, 9–11, 140
ecological modeling, 373
IBP Grassland Biome studies integrated with, 151–155
site vegetation, 72–82
Shrews, 136
Simulations/simulation modeling
   CENTURY model, 307, 374, 376, 378–382
   DAYCENT model, 374–375, 378–382, 384
   description of, 119
   future of, 129
   grazing, 400f–401f
   of mammals, 156
   PHOENIX model, 374
   ROOT model, 374
   single-plot, 122–123
   STEPPE. *See* STEPPE
   summary of, 128–129
Skunk burrows, 96
SMART model, 455
Snowfall, 24, 25f
Soil
   after abandonment of fields, 110
   age of, 37
   Bankard, 38
   biota
      description of, 236, 249
      tillage effects on, 261–262
   calcium carbonate accumulation in, 30
   characteristics of, 30
   clay accumulation in, 30
   density of, grazing effects on, 326–327
   forest, 43
   formation of
      biota effects, 42–47
      climate effects on, 47–51
      parent material effects, 31–37
      time effects, 37–38, 39f
      topography effects, 39–42
      vegetation effects on, 43, 46t
   greenhouse gases in, 342
   Hastings, 50
   heterogeneity of, 30–31
   insects and, 223, 235–237
   mammal distribution affected by, 142–143
   moisture of, 21–22
   Olney, 50
   organic matter inputs, 43
   paleosols in, 38, 46, 48–49
   playa, 40, 41f

Remmit, 38
texture of
  aboveground net primary production affected by, 277–278, 281–283
  biogeochemical patterns controlled by, 316
  seed production affected by, 430
  topographic variation and, 320–321
  trace gas exchange affected by, 351–352, 354f
tillage of, 261
water movement in, 39–40
Soil–atmosphere exchange of trace gases
  carbon dioxide effects, 360–363
  climate effects, 347–351
  controls over, 347–365
  description of, 344
  landscape position effects, 351–352
  land-use changes effect on, 357–359
  plant species effect on, 356–357
  soil nitrogen additions, 352–356
  soil texture effects, 351–352, 354f
  weather effects, 347–351
Soil organic carbon, 261, 307–310
Soil organic matter
  cultivation and recovery effects on, 329–331
  disturbances' effect on, 325
  grazing effects on, 329
  processes that affect, 322
  topographic patterns of, 319
  urbanization effects, 331–333
Soil respiration, 333, 367
SOILWAT, 127–128
Soil water
  description of, 286, 313
  methane fluxes and, 349
  nitrous oxide fluxes and, 349
South Platte River, 34
Sparrows, 191
Species interactions, 148–151
Spring, 16–17, 25
SPUR model, 375–376
Squirrels, 136, 137f
Standardized precipitation index, 26
STEPPE
  description of, 120–121
  forest models vs., 120
  growth, 122
  mortality, 121
  recruitment, 121
  seed dispersal, 124–125

single-plot simulations, 122–123
soil texture, 125–126
SOILWAT simulations and, 127–128
spatially explicit simulations, 124–126
uses of, 119
Stocking rates, 448–450
Stream downcutting, 34
Stresses, 104–105
Succession
  Clementsian model of, 85, 85f, 108
  description of, 84
  gap dynamics model of, 86–87
  old field models of, 85–86, 109f
  stages of, 107–108
Summer, 15, 17
Swales, 290, 399, 402
Swift foxes, 136, 138f, 147t, 148, 165
Sylvatic plague, 500

Taylor Grazing Act, 391
Temperature
  description of, 7, 18–20, 20f
  ecophysiology of plants affected by, 285
  mammal distribution affected by, 141
  regional patterns in, 313
  spatial variability of, 313–314, 496
  temporal variability of, 314–315
  trace gas exchange and, 350
Tenebrionids, 219, 221, 234
Termites, 217, 224
Tertiary, 31
Texas Panhandle
  cattle in, 61
  description of, 60–61
  irrigation in, 64
  water tables in, 65
Thunderstorms, 25, 263
Tillage of soil, 261
Time, 37–38, 39f
Topography
  description of, 133–134
  elements of, 39
  grazing affected by, 411
  soil formation affected by, 39–42
Tornadoes, 16, 25–26
Trace gases. *See also specific gas*
  atmospheric levels of, 342–343
  at Central Plains Experimental Range, 344
  estimation of, at long-term sites, 343–347
  historical perspective of, 343–344
  nitrogen oxides, 345, 347

522 Index

Trace gases (*continued*)
  plowing related changes in, 360
  rationale for measuring, 344–345
  soil–atmosphere exchange of
    carbon dioxide effects, 360–363
    climate effects, 347–351
    controls over, 347–365
    description of, 344
    landscape position effects, 351–352
    land-use changes effect on, 357–359
    plant species effect on, 356–357
    soil nitrogen additions, 352–356
    soil texture effects, 351–352, 354f
    weather effects, 347–351
Tracking by military vehicles, 105–106
Triticum aestivum
  description of, 8
  Dust Bowl effects on, 63
  history of, 61
  in Oklahoma Panhandle, 61
Trophic structure
  belowground food web. *See* Belowground food web
  disturbances effect on, 258
  grazing effects on, 259–261
Tropical grasslands, 368
Tularemia, 501

Ultraviolet radiation, 315, 499
Underground water, 64
Urbanization, 331–333

Valent soil, 36–37
Vegetation. *See also* Plants
  basal cover of, 460
  biogeochemical processes affected by, 321–326
  birds and, 188–190, 466
  characteristics of, 8–9
  community types, 71–82
  distribution of, 33f
  grazing effects on, 191

Holocene, 46
insect distribution affected by, 222, 225f
mammal distribution affected by, 143–146, 170
overview of, 70–71
patchy areas of, 134
pocket gophers' effect on, 96–99
prairie dogs' effect on, 106
roadside, 145
soil formation affected by, 43, 46t
temperature effects on, 19
water erosion from rainfall prevented by, 22

Water, 496
Water-plus-nitrogen treatments, 105
Water tables, 65
Weather
  birds and, 193–195
  denitrification, 349
  insect populations affected by, 221
  mammal distribution affected by, 141–142
  patterns of, 16–18
  trace gas exchange affected by, 347–351, 367
West Nile virus, 500
Wheat-fallow agriculture, 330, 357
White grubs, 101–104, 229, 237
White-tailed jackrabbits, 136, 464
Wildlife conservation, 489
Wind
  description of, 15, 25
  playas created by, 40
Winter, 16, 18–19
Winter wheat, 168, 278–279
Wolf spiders, 230–231
Wood ants, 224
Woodrats, 136

Yersinia pestis, 501